# Lecture Notes in Artificial Intelligence          9011

Subseries of Lecture Notes in Computer Science

More information about this series at http://www.springer.com/series/1244

Ngoc Thanh Nguyen · Bogdan Trawiński
Raymond Kosala (Eds.)

# Intelligent Information and Database Systems

7th Asian Conference, ACIIDS 2015
Bali, Indonesia, March 23–25, 2015
Proceedings, Part I

 Springer

*Editors*
Ngoc Thanh Nguyen
Ton Duc Thang University
Ho Chi Minh city
Vietnam

and

Wroclaw University of Technology
Wroclaw
Poland

Bogdan Trawiński
Wroclaw University of Technology
Wroclaw
Poland

Raymond Kosala
Bina Nusantara University
Jakarta
Indonesia

ISSN 0302-9743
Lecture Notes in Artificial Intelligence
ISBN 978-3-319-15701-6
DOI 10.1007/978-3-319-15702-3

ISSN 1611-3349     (electronic)

ISBN 978-3-319-15702-3     (eBook)

Library of Congress Control Number: 2015932661

LNCS Sublibrary: SL7 – Artificial Intelligence

Springer Cham Heidelberg New York Dordrecht London

Printed on acid-free paper

Springer International Publishing AG Switzerland is part of Springer Science+Business Media
(www.springer.com)

# Preface

ACIIDS 2015 was the seventh event in the series of international scientific conferences for research and applications in the field of intelligent information and database systems. The aim of ACIIDS 2015 was to provide an internationally respected forum for scientific research in the technologies and applications of intelligent information and database systems. ACIIDS 2015 was co-organized by Bina Nusantara University, Indonesia and Wrocław University of Technology, Poland in cooperation with Ton Duc Thang University, Vietnam and Quang Binh University, Vietnam, and with IEEE Indonesia Section and IEEE SMC Technical Committee on Computational Collective Intelligence as patrons of the conference. It took place in Bali, Indonesia during March 23–25, 2015.

Conferences of series ACIIDS have been well established. The first two events, ACIIDS 2009 and ACIIDS 2010, took place in Dong Hoi City and Hue City in Vietnam, respectively. The third event, ACIIDS 2011, took place in Daegu, Korea, while the fourth event, ACIIDS 2012, took place in Kaohsiung, Taiwan. The fifth event, ACIIDS 2013, was held in Kuala Lumpur in Malaysia while the sixth event, ACIIDS 2014, was held in Bangkok in Thailand.

We received more than 300 papers from about 40 countries all over the world. Each paper was peer reviewed by at least two members of the International Program Committee and International Reviewer Board. Only 117 papers with the highest quality were selected for oral presentation and publication in the two volumes of the ACIIDS 2015 proceedings.

Papers included in the proceedings cover the following topics: semantic web, social networks and recommendation systems, text processing and information retrieval, intelligent database systems, intelligent information systems, decision support and control systems, machine learning and data mining, multiple model approach to machine learning, innovations in intelligent systems and applications, artificial intelligent techniques and their application in engineering and operational research, machine learning in biometrics and bioinformatics with applications, advanced data mining techniques and applications, collective intelligent systems for e-market trading, technology opportunity discovery and collaborative learning, intelligent information systems in security and defense, analysis of image, video and motion data in life sciences, augmented reality and 3D media, cloud-based solutions, Internet of things, big data, and cloud computing.

Accepted and presented papers highlight new trends and challenges of intelligent information and database systems. The presenters showed how new research could lead to new and innovative applications. We hope you will find these results useful and inspiring for your future research.

We would like to express our sincere thanks to the Honorary Chairs, Prof. Harjanto Prabowo (Rector of the Bina Nusantara University, Indonesia) and Prof. Tadeusz Więckowski (Rector of the Wrocław University of Technology, Poland) for their supports.

Our special thanks go to the Program Chairs, Special Session Chairs, Organizing Chairs, Publicity Chairs, and Local Organizing Committee for their work for the conference. We sincerely thank all members of the International Program Committee for their valuable efforts in the review process which helped us to guarantee the highest quality of the selected papers for the conference. We cordially thank the organizers and chairs of special sessions which essentially contributed to the success of the conference.

We also would like to express our thanks to the Keynote Speakers (Prof. Nikola Kasabov, Prof. Suphamit Chittayasothorn, Prof. Dosam Hwang, and Prof. Satryo Soemantri Brodjonegoro) for their interesting and informative talks of world-class standard.

We cordially thank our main sponsors, Bina Nusantara University (Indonesia), Wrocław University of Technology (Poland), Ton Duc Thang University (Vietnam) Quang Binh University (Vietnam), and patrons: IEEE Indonesia Section and IEEE SMC Technical Committee on Computational Collective Intelligence. Our special thanks are due also to Springer for publishing the proceedings, and to other sponsors for their kind supports.

We wish to thank the members of the Organizing Committee for their very substantial work and the members of the Local Organizing Committee for their excellent work.

We cordially thank all the authors for their valuable contributions and other participants of this conference. The conference would not have been possible without their supports.

Thanks are also due to many experts who contributed to making the event a success.


March 2015                                                          Ngoc Thanh Nguyen
                                                                    Bogdan Trawiński
                                                                    Raymond Kosala

# Organization

## Honorary Chairs

Harjanto Prabowo      Bina Nusantara University, Indonesia
Tadeusz Więckowski      Wrocław University of Technology, Poland

## General Chairs

Ngoc Thanh Nguyen      Wrocław University of Technology, Poland
Ford Lumban Gaol      Bina Nusantara University, Indonesia

## Program Chairs

Bogdan Trawiński      Wrocław University of Technology, Poland
Raymond Kosala      Bina Nusantara University, Indonesia
Tzung-Pei Hong      National University of Kaohsiung, Taiwan
Hamido Fujita      Iwate Prefectural University, Japan

## Organizing Chairs

Harisno      Bina Nusantara University, Indonesia
Suharjito      Bina Nusantara University, Indonesia
Marcin Maleszka      Wrocław University of Technology, Poland

## Special Session Chairs

Dariusz Barbucha      Gdynia Maritime University, Poland
John Batubara      Bina Nusantara University, Indonesia

## Publicity Chairs

Diana      Bina Nusantara University, Indonesia
Adrianna Kozierkiewicz-
   Hetmańska      Wrocław University of Technology, Poland

## Local Organizing Committee

Togar Napitupulu      Bina Nusantara University, Indonesia
Nilo Legowo      Bina Nusantara University, Indonesia
Benfano Soewito      Bina Nusantara University, Indonesia
Fergyanto      Bina Nusantara University, Indonesia
Zbigniew Telec      Wrocław University of Technology, Poland
Bernadetta Maleszka      Wrocław University of Technology, Poland
Marcin Pietranik      Wrocław University of Technology, Poland

## Steering Committee

| | |
|---|---|
| Ngoc Thanh Nguyen (Chair) | Wrocław University of Technology, Poland |
| Longbing Cao | University of Technology Sydney, Australia |
| Tu Bao Ho | Japan Advanced Institute of Science and Technology, Japan |
| Tzung-Pei Hong | National University of Kaohsiung, Taiwan |
| Lakhmi C. Jain | University of South Australia, Australia |
| Geun-Sik Jo | Inha University, Korea |
| Jason J. Jung | Yeungnam University, Korea |
| Hoai An Le-Thi | Paul Verlaine University – Metz, France |
| Toyoaki Nishida | Kyoto University, Japan |
| Leszek Rutkowski | Częstochowa University of Technology, Poland |
| Suphamit Chittayasothorn | King Mongkut's Institute of Technology Ladkrabang, Thailand |
| Ford Lumban Gaol | Bina Nusantara University, Indonesia |
| Ali Selamat | Universiti Teknologi Malaysia, Malyasia |

## Keynote Speakers

| | |
|---|---|
| Nikola Kasabov | Auckland University of Technology, New Zealand |
| Suphamit Chittayasothorn | King Mongkut's Institute of Technology Ladkrabang, Thailand |
| Dosam Hwang | Yeungnam University, Korea |
| Satryo Soemantri Brodjonegoro | Indonesian Academy of Sciences, Indonesia |

## Special Sessions Organizers

1. *Multiple Model Approach to Machine Learning (MMAML 2015)*

| | |
|---|---|
| Tomasz Kajdanowicz | Wrocław University of Technology, Poland |
| Edwin Lughofer | Johannes Kepler University Linz, Austria |
| Bogdan Trawiński | Wrocław University of Technology, Poland |

2. *Special Session on Innovations in Intelligent Systems and Applications (IISA 2015)*

| | |
|---|---|
| Shyi-Ming Chen | National Taiwan University of Science and Technology, Taiwan |

3. *Special Session on Innovations in Artificial Intelligent Techniques and Its Application in Engineering and Operational Research (AITEOR 2015)*

| | |
|---|---|
| Pandian Vasant | Universiti Teknologi PETRONAS, Malaysia |
| Vo Ngoc Dieu | HCMC University of Technology, Vietnam |
| Irraivan Elamvazuthi | Universiti Teknologi PETRONAS, Malaysia |

Mohammad Abdullah-Al-Wadud       King Saud University, Saudi Arabia
Ahamed Khan                      Universiti Selangor, Malaysia
Timothy Ganesan                  Universiti Teknologi PETRONAS, Malaysia
Perumal Nallagownden             Universiti Teknologi PETRONAS, Malaysia

4. *Special Session on Analysis of Image, Video and Motion Data in Life Sciences
   (IVMLS 2015)*

Kondrad Wojciechowski            Polish-Japanese Institute of Information
                                    Technology, Poland
Marek Kulbacki                   Polish-Japanese Institute of Information
                                    Technology, Poland
Jakub Segen                      Gest3D, USA
Andrzej Polański                 Silesian University of Technology, Poland

5. *Special Session on Machine Learning in Biometrics and Bioinformatics with
   Application (MLBBA 2015)*

Piotr Porwik                     University of Silesia, Poland
Marina L. Gavrilova              University of Calgary, Canada
Rafał Doroz                      University of Silesia, Poland
Krzysztof Wróbel                 University of Silesia, Poland

6. *Special Session on Collective Intelligent Systems for E-market Trading, Technology
   Opportunity Discovery and Collaborative Learning (CISETC 2015)*

Tzu-Fu Chiu                      Aletheia University, Taiwan
Chia-Ling Hsu                    Tamkang University, Taiwan
Feng-Sueng Yang                  Aletheia University, Taiwan

7. *Special Session on Intelligent Information Systems in Security & Defence
   (IISSD 2015)*

Andrzej Najgebauer               Military University of Technology, Poland
Dariusz Pierzchała               Military University of Technology, Poland
Ryszard Antkiewicz               Military University of Technology, Poland
Ewa Niewiadomska-
   Szynkiewicz                   Warsaw University of Technology, Poland
Zbigniew Tarapata                Military University of Technology, Poland
Richard Warner                   IIT/Chicago-Kent College of Law, USA
Adam Zagorecki                   Cranfield University, UK

8. *Special Session on Advanced Data Mining Techniques and Applications
   (ADMTA 2015)*

Bay Vo                           Ton Duc Thang University, Vietnam
Tzung-Pei Hong                   National University of Kaohsiung, Taiwan
Bac Le                           Ho Chi Minh City University of Science, Vietnam

9. *Special Session on Cloud-Based Solutions (CBS 2015)*

| | |
|---|---|
| Ondrej Krejcar | University of Hradec Králové, Czech Republic |
| Vladimir Sobeslav | University of Hradec Králové, Czech Republic |
| Peter Brida | University of Žilina, Slovakia |
| Kamil Kuca | University of Hradec Králové, Czech Republic |

10. *Special Session on Augmented Reality and 3D Media (AR3DM 2015)*

| | |
|---|---|
| Atanas Gotchev | Tampere University of Technology, Finland |
| Janusz Sobecki | Wrocław University of Technology, Poland |
| Zbigniew Wantuła | ADUMA S.A., Poland |

11. *Special Session on Internet of Things, Big Data, and Cloud Computing (IoT 2015)*

| | |
|---|---|
| Adam Grzech | Wrocław University of Technology, Poland |
| Andrzej Ruciński | University of New Hampshire, USA |

## International Program Committee

| | |
|---|---|
| Muhammad Abulaish | Jamia Millia Islamia, India |
| El-Houssaine Aghezzaf | Ghent University, Belgium |
| Haider M. AlSabbagh | University of Basra, Iraq |
| Toni Anwar | Universiti Teknologi Malaysia, Malaysia |
| Ahmad Taher Azar | Benha University, Egypt |
| Amelia Badica | University of Craiova, Romania |
| Costin Badica | University of Craiova, Romania |
| Emili Balaguer-<br>  Ballester | Bournemouth University, UK |
| Zbigniew Banaszak | Warsaw University of Technology, Poland |
| Dariusz Barbucha | Gdynia Maritime University, Poland |
| John Batubara | Bina Nusantara University, Indonesia |
| Ramazan Bayindir | Gazi University, Turkey |
| Maumita Bhattacharya | Charles Sturt University, Australia |
| Maria Bielikova | Slovak University of Technology in Bratislava, Slovakia |
| Veera Boonjing | King Mongkut's Institute of Technology Ladkrabang, Thailand |
| Mariusz Boryczka | University of Silesia, Poland |
| Urszula Boryczka | University of Silesia, Poland |
| Abdelhamid Bouchachia | Bournemouth University, UK |
| Stephane Bressan | National University of Singapore, Singapore |
| Peter Brida | University of Žilina, Slovakia |
| Piotr Bródka | Wrocław University of Technology, Poland |

Andrej Brodnik — University of Ljubljana, Slovenia
Grażyna Brzykcy — Poznań University of Technology, Poland
The Duy Bui — VNU University of Engineering and Technology, Vietnam
Robert Burduk — Wrocław University of Technology, Poland
David Camacho — Universidad Autónoma de Madrid, Spain
Frantisek Capkovic — Institute of Informatics, Slovak Academy of Sciences, Slovakia
Oscar Castillo — Tijuana Institute of Technology, Mexico
Dariusz Ceglarek — Poznań School of Banking, Poland
Stephan Chalup — University of Newcastle, Australia
Bao Rong Chang — National University of Kaohsiung, Taiwan
Somchai Chatvichienchai — University of Nagasaki, Japan
Rung-Ching Chen — Chaoyang University of Technology, Taiwan
Shyi-Ming Chen — National Taiwan University of Science and Technology, Taiwan
Suphamit Chittayasothorn — King Mongkut's Institute of Technology Ladkrabang, Thailand
Tzu-Fu Chiu — Aletheia University, Taiwan
Kazimierz Choroś — Wrocław University of Technology, Poland
Dorian Cojocaru — University of Craiova, Romania
Phan Cong-Vinh — NTT University, Vietnam
Jose Alfredo Ferreira Costa — Universidade Federal do Rio Grande do Norte, Brazil
Keeley Crockett — Manchester Metropolitan University, UK
Boguslaw Cyganek — AGH University of Science and Technology, Poland
Ireneusz Czarnowski — Gdynia Maritime University, Poland
Piotr Czekalski — Silesian University of Technology, Poland
Paul Davidsson — Malmö University, Sweden
Roberto De Virgilio — Universita' degli Studi Roma Tre, Italy
Tien V. Do — Budapest University of Technology and Economics, Hungary
Pietro Ducange — University of Pisa, Italy
El-Sayed M. El-Alfy — King Fahd University of Petroleum and Minerals, Saudi Arabia
Vadim Ermolayev — Zaporozhye National University, Ukraine
Rim Faiz — University of Carthage, Tunisia
Victor Felea — Alexandru Ioan Cuza University of Iasi, Romania
Thomas Fober — University of Marburg, Germany
Dariusz Frejlichowski — West Pomeranian University of Technology, Poland
Mohamed Gaber — Robert Gordon University, UK
Patrick Gallinari — LIP6 - University of Paris 6, France
Dariusz Gąsior — Wrocław University of Technology, Poland

Andrey Gavrilov              Novosibirsk State Technical University, Russia
Janusz Getta                 University of Wollongong, Australia
Dejan Gjorgjevikj            Saints Cyril and Methodius University of Skopje,
                               Macedonia
Daniela Godoy                ISISTAN Research Institute, Argentina
Gergö Gombos                 Eötvös Loránd University, Hungary
Fernando Gomide              State University of Campinas, Brazil
Vladimir I. Gorodetsky       St. Petersburg Institute for Informatics
                               and Automation, Russia
Janis Grundspenkis           Riga Technical University, Latvia
Adam Grzech                  Wrocław University of Technology, Poland
Slimane Hammoudi             ESEO Institute of Science and Technology, France
Habibollah Haron             Universiti Teknologi Malaysia, Malaysia
Tutut Herawan                University of Malaya, Malaysia
Francisco Herrera            University of Granada, Spain
Bogusława Hnatkowska         Wrocław University of Technology, Poland
Huu Hanh Hoang               Hue University, Vietnam
Natasa Hoic-Bozic            University of Rijeka, Croatia
Tzung-Pei Hong               National Univesity of Kaohsiung, Taiwan
Wei-Chiang Hong              Hangzhou Dianzi University, China
Mong-Fong Horng              National Kaohsiung University of Applied
                               Sciences, Taiwan
Sheng-Jun Huang              Nanjing University of Aeronautics
                               and Astronautics, China
Zbigniew Huzar               Wrocław University of Technology, Poland
Dosam Hwang                  Yeungnam University, Korea
Dmitry Ignatov               National Research University Higher School
                               of Economics, Russia
Lazaros Iliadis              Democritus University of Thrace, Greece
Hazra Imran                  Athabasca University, Canada
Mirjana Ivanovic             University of Novi Sad, Serbia
Konrad Jackowski             Wrocław University of Technology, Poland
Chuleerat Jaruskulchai       Kasetsart University, Thailand
Joanna Jędrzejowicz          University of Gdańsk, Poland
Piotr Jędrzejowicz           Gdynia Maritime University, Poland
Janusz Jeżewski              Institute of Medical Technology and Equipment
                               ITAM, Poland
Gordan Jezic                 University of Zagreb, Croatia
Geun-Sik Jo                  Inha University, Korea
Kang-Hyun Jo                 University of Ulsan, Korea
Jason J. Jung                Yeungnam University, Korea
Janusz Kacprzyk              Polish Academy of Sciences, Poland

| | |
|---|---|
| Rajesh Reghunadhan | Central University of Bihar, India |
| Przemysław Różewski | West Pomeranian University of Technology, Szczecin, Poland |
| Leszek Rutkowski | Częstochowa University of Technology, Poland |
| Henryk Rybiński | Warsaw University of Technology, Poland |
| Alexander Ryjov | Lomonosov Moscow State University, Russia |
| Virgilijus Sakalauskas | Vilnius University, Lithuania |
| Daniel Sanchez | University of Granada, Spain |
| Juergen Schmidhuber | Swiss AI Lab IDSIA, Switzerland |
| Bjorn Schuller | Technical University Munich, Germany |
| Jakub Segen | Gest3D, USA |
| Ali Selamat | Universiti Teknologi Malaysia, Malaysia |
| Alexei Sharpanskykh | Delft University of Technology, The Netherlands |
| Quan Z. Sheng | University of Adelaide, Australia |
| Andrzej Siemiński | Wrocław University of Technology, Poland |
| Dragan Simic | University of Novi Sad, Serbia |
| Gia Sirbiladze | Tbilisi State University, Georgia |
| Andrzej Skowron | University of Warsaw, Poland |
| Adam Słowik | Koszalin University of Technology, Poland |
| Janusz Sobecki | Wrocław University of Technology, Poland |
| Kulwadee Somboonviwat | King Mongkut's Institute of Technology Ladkrabang, Thailand |
| Zenon A. Sosnowski | Białystok University of Technology, Poland |
| Serge Stinckwich | University of Caen Lower Normandy, France |
| Stanimir Stoyanov | Plovdiv University "Paisii Hilendarski", Bulgaria |
| Jerzy Świątek | Wrocław University of Technology, Poland |
| Andrzej Świerniak | Silesian University of Technology, Poland |
| Edward Szczerbicki | University of Newcastle, Australia |
| Julian Szymański | Gdańsk University of Technology, Poland |
| Ryszard Tadeusiewicz | AGH University of Science and Technology, Poland |
| Yasufumi Takama | Tokyo Metropolitan University, Japan |
| Pham Dinh Tao | INSA-Rouen, France |
| Zbigniew Telec | Wrocław University of Technology, Poland |
| Krzysztof Tokarz | Silesian University of Technology, Poland |
| Behcet Ugur Toreyin | Çankaya University, Turkey |
| Bogdan Trawiński | Wrocław University of Technology, Poland |
| Krzysztof Trawiński | European Centre for Soft Computing, Spain |
| Maria Trocan | Institut Superieur d'Electronique de Paris, France |
| Hong-Linh Truong | Vienna University of Technology, Austria |
| Olgierd Unold | Wrocław University of Technology, Poland |
| Pandian Vasant | Universiti Teknologi PETRONAS, Malaysia |

| | |
|---|---|
| Joost Vennekens | Katholieke Universiteit Leuven, Belgium |
| Jorgen Villadsen | Technical University of Denmark, Denmark |
| Bay Vo | Ton Duc Thang University, Vietnam |
| Yongkun Wang | University of Tokyo, Japan |
| Izabela Wierzbowska | Gdynia Maritime University, Poland |
| Marek Wojciechowski | Poznań University of Technology, Poland |
| Dong-Min Woo | Myongji University, Korea |
| Michał Woźniak | Wrocław University of Technology, Poland |
| Marian Wysocki | Rzeszow University of Technology, Poland |
| Guandong Xu | University of Technology Sydney, Australia |
| Xin-She Yang | Middlesex University, UK |
| Zhenglu Yang | University of Tokyo, Japan |
| Lean Yu | Chinese Academy of Sciences, AMSS, China |
| Slawomir Zadrozny | Systems Research Institute, Polish Academy of Sciences, Poland |
| Drago Žagar | University of Osijek, Croatia |
| Danuta Zakrzewska | Lodz University of Technology, Poland |
| Faisal Zaman | Dublin City University, Ireland |
| Constantin-Bala Zamfirescu | Lucian Blaga University of Sibiu, Romania |
| Katerina Zdravkova | Ss. Cyril and Methodius University in Skopje, Macedonia |
| Aleksander Zgrzywa | Wrocław University of Technology, Poland |
| Jianwei Zhang | National University Corporation Tsukuba University of Technology, Japan |
| Min-Ling Zhang | Southeast University, China |
| Zhongwei Zhang | University of Southern Queensland, Australia |
| Zhi-Hua Zhou | Nanjing University, China |

## Program Committees of Special Sessions

*Multiple Model Approach to Machine Learning (MMAML 2015)*

| | |
|---|---|
| Emili Balaguer-Ballester | Bournemouth University, UK |
| Urszula Boryczka | University of Silesia, Poland |
| Abdelhamid Bouchachia | Bournemouth University, UK |
| Robert Burduk | Wrocław University of Technology, Poland |
| Oscar Castillo | Tijuana Institute of Technology, Mexico |
| Rung-Ching Chen | Chaoyang University of Technology, Taiwan |
| Suphamit Chittayasothorn | King Mongkut's Institute of Technology Ladkrabang, Thailand |
| José Alfredo F. Costa | Federal University of Rio Grande do Norte, Brazil |
| Bogusław Cyganek | AGH University of Science and Technology, Poland |
| Ireneusz Czarnowski | Gdynia Maritime University, Poland |

| | |
|---|---|
| Patrick Gallinari | Pierre et Marie Curie University, France |
| Fernando Gomide | State University of Campinas, Brazil |
| Francisco Herrera | University of Granada, Spain |
| Tzung-Pei Hong | National University of Kaohsiung, Taiwan |
| Konrad Jackowski | Wrocław University of Technology, Poland |
| Piotr Jędrzejowicz | Gdynia Maritime University, Poland |
| Tomasz Kajdanowicz | Wrocław University of Technology, Poland |
| Yong Seog Kim | Utah State University, USA |
| Bartosz Krawczyk | Wrocław University of Technology, Poland |
| Kun Chang Lee | Sungkyunkwan University, Korea |
| Edwin Lughofer | Johannes Kepler University Linz, Austria |
| Héctor Quintián | University of Salamanca, Spain |
| Andrzej Siemiński | Wrocław University of Technology, Poland |
| Dragan Simic | University of Novi Sad, Serbia |
| Adam Słowik | Koszalin University of Technology, Poland |
| Zbigniew Telec | Wrocław University of Technology, Poland |
| Bogdan Trawiński | Wrocław University of Technology, Poland |
| Krzysztof Trawiński | European Centre for Soft Computing, Spain |
| Olgierd Unold | Wrocław University of Technology, Poland |
| Pandian Vasant | Universiti Teknologi PETRONAS, Malaysia |
| Michał Woźniak | Wrocław University of Technology, Poland |
| Zhongwei Zhang | University of Southern Queensland, Australia |
| Zhi-Hua Zhou | Nanjing University, China |

*Special Session on Innovations in Intelligent Systems and Applications (IISA 2015)*

| | |
|---|---|
| I-Cheng Chang | National Dong Hwa University, Hualien, Taiwan |
| Shyi-Ming Chen | National Taiwan University of Science and Technology, Taipei, Taiwan |
| Po-Hung Chen | St. John's University, New Taipei City, Taiwan |
| Shou-Hsiung Cheng | Chienkuo Technology University, Changhua, Taiwan |
| Mong-Fong Horng | National Kaohsiung University of Applications, Taiwan |
| Wei-Lieh Hsu | Lunghwa University of Science and Technology, Taoyuan, Taiwan |
| Feng-Long Huang | National United University, Miaoli, Taiwan |
| Pingsheng Huang | Ming Chuan University, Taoyuan County, Taiwan |
| Bor-Jiunn Hwang | Ming Chuan University, Taoyuan County, Taiwan |
| Huey-Ming Lee | Chinese Culture University, Taipei, Taiwan |
| Li-Wei Lee | De Lin Institute of Technology, New Taipei City, Taiwan |
| Chung-Ming Ou | Kainan University, Taoyuan County, Taiwan |

| Jeng-Shyang Pan | Harbin Institute of Technology, China |
| Victor R.L. Shen | National Taipei University, New Taipei City, Taiwan |
| Chia-Rong Su | Chang Gung University, New Taipei City, Taiwan |
| An-Zen Shih | Jinwen University of Science and Technology, Taiwan |
| Chun-Ming Tsai | Taipei Municipal University of Education, Taiwan |
| Wen-Chung Tsai | Chaoyang University of Technology, Taichung, Taiwan |
| Cheng-Fa Tsai | National Pingtung University of Science and Technology, Taiwan |
| Cheng-Yi Wang | National Taiwan University of Science and Technology, Taipei, Taiwan |
| Chih-Hung Wu | National Taichung University of Education, Taiwan |

*Special Session on Innovations in Artificial Intelligent Techniques and Its Application in Engineering and Operational Research (AITEOR 2015)*

| Gerhard-Wilhelm Weber | Middle East Technical University, Turkey |
| Junzo Watada | Waseda University, Japan |
| Kwon-Hee Lee | Dong-A University, South Korea |
| Hindriyanto Dwi Purnomo | Satya Wacana Christian University, Indonesia |
| Charles Mbohwa | University of Johannesburg, South Africa |
| Gerardo Maximiliano Mendez | Instituto Tecnológico de Nuevo León, Mexico |
| Leopoldo Eduardo Cárdenas Barrón | Tecnológico de Monterry, Mexico |
| Petr Dostál | Brno University of Technology, Czech Republic |
| Erik Kropat | Universität der Bundeswehr München, Germany |
| Timothy Ganesan | Universiti Teknologi PETRONAS, Malaysia |
| Vedpal Singh | Universiti Teknologi PETRONAS, Malaysia |
| Gerrit Janssens | Hasselt University, Belgium |
| Suhail Qureshi | UET Lahore, Pakistan |
| Monica Chis | SIEMENS Program and System Engineering, Romania |
| Michael Mutingi | National University of Singapore, Singapore |
| Ugo Fiore | University of Naples Federico II, Italy |
| Utku Kose | Uşak University, Turkey |
| Nuno Pombo | University of Beira Interior, Portugal |
| Kusuma Soonpracha | Kasetsart University, Thailand |
| Armin Milani | Islamic Azad University, Iran |

Leo Mrsic                        University College of Law and Finance Effectus
                                     Zagreb, Crotia
Igor Litvinchev                  Nuevo Leon State University, Mexico
Goran Klepac                     Raiffeisen Bank, Croatia
Herman Mawengkang                University of Sumetera Utara, Indonesia

*Special Session on Analysis of Image, Video and Motion Data in Life Sciences (IVMLS 2015)*

Aldona Drabik                    Polish-Japanese Institute of Information
                                     Technology, Poland
Leszek Chmielewski               Warsaw University of Life Sciences, Poland
André Gagalowicz                 Inria, France
David Gibbon                     AT&T, USA
Celina Imielinska                Vesalius Technologies, USA
Ryszard Klempous                 Wrocław University of Technology, Poland
Ryszard Kozera                   Warsaw University of Life Sciences, Poland
Marek Kulbacki                   Polish-Japanese Institute of Information
                                     Technology, Poland
Aleksander Nawrat                Silesian University of Technology, Poland
Lyle Noakes                      The University of Western Australia, Australia
Jerzy Paweł Nowacki              Polish-Japanese Institute of Information
                                     Technology, Poland
Eric Petajan                     Directv, USA
Gopal Pingali                    IBM, USA
Andrzej Polański                 Polish-Japanese Institute of Information
                                     Technology, Poland
Andrzej Przybyszewski            University of Massachusetts, USA
Jerzy Rozenbilt                  University of Arizona, Tucson, USA
Jakub Segen                      Gest3D, USA
Aleksander Sieroń                Medical University of Silesia, Poland
Konrad Wojciechowski             Polish-Japanese Institute of Information
                                     Technology, Poland

*Special Session on Machine Learning in Biometrics and Bioinformatics with Application (MLBBA 2015)*

Marcin Adamski                   Białystok Technical University, Poland
Ryszard Choraś                   Uniwersytet Technologiczno-Przyrodniczy, Poland
Nabendu Chaki                    University of Calcutta, India
Rituparna Chaki                  University of Calcutta, India
Rafał Doroz                      University of Silesia, Poland
Marina Gavrilova                 University of Calgary, Canada
Phalguni Gupta                   Indian Institute of Technology Kanpur, India

| Anil K. Jain | Michigan State University, USA |
|---|---|
| Hemant B. Kekre | NMIMS University, India |
| Vic Lane | University of London, UK |
| Davide Maltoni | Università di Bologna, Italy |
| Ashu Marasinghe | Nagaoka University of Technology, Japan |
| Nobuyuki Nishiuchi | Tokyo Metropolitan University, Japan |
| Javier Ortega-García | Universidad Autónoma de Madrid, Spain |
| Giuseppe Pirlo | Università degli Studi di Bari, Italy |
| Piotr Porwik | University of Silesia, Poland |
| Arun Ross | Michigan State University, USA |
| Khalid Saeed | Białystok Technical University, Poland |
| Refik Samet | Ankara University, Turkey |
| Michał Woźniak | Wroclaw University of Technology, Poland |
| Krzysztof Wróbel | University of Silesia, Poland |

*Special Session on Collective Intelligent Systems for E-market Trading, Technology Opportunity Discovery and Collaborative Learning (CISETC 2015)*

| Ya-Fung Chang | Tamkang University, Taiwan |
|---|---|
| Peng-Wen Chen | Oriental Institute of Technology, Taiwan |
| Kuan-Shiu Chiu | Aletheia University, Taiwan |
| Tzu-Fu Chiu | Aletheia University, Taiwan |
| Chen-Huei Chou | College of Charleston, USA |
| Chia-Ling Hsu | Tamkang University, Taiwan |
| Fang-Cheng Hsu | Aletheia University, Taiwan |
| Kuo-Sui Lin | Aletheia University, Taiwan |
| Min-Huei Lin | Aletheia University, Taiwan |
| Yuh-Chang Lin | Aletheia University, Taiwan |
| Pen-Choug Sun | Aletheia University, Taiwan |
| Leuo-Hong Wang | Aletheia University, Taiwan |
| Ai-Ling Wang | Tamkang University, Taiwan |
| Henry Wang | Chinese Academy of Sciences, China |
| Feng-Sueng Yang | Aletheia University, Taiwan |
| Ming-Chien Yang | Aletheia University, Taiwan |

*Special Session on Intelligent Information Systems in Security & Defence (IISSD 2015)*

| Ryszard Antkiewicz | Military University of Technology, Poland |
|---|---|
| Leon Bobrowski | Białystok University of Technology, Poland |
| Urszula Boryczka | University of Silesia, Poland |
| Mariusz Chmielewski | Military University of Technology, Poland |
| Rafał Kasprzyk | Military University of Technology, Poland |
| Jacek Koronacki | Polish Academy of Sciences, Poland |

Leszek Kotulski                      AGH University of Science and Technology,
                                     Poland
Krzysztof Malinowski                 Warsaw University of Technology, Poland
Andrzej Najgebauer                   Military University of Technology, Poland
Ewa Niewiadomska-Szynkiewicz  Warsaw University of Technology, Poland
Dariusz Pierzchała                   Military University of Technology, Poland
Jarosław Rulka                       Military University of Technology, Poland
Zenon A. Sosnowski                   Białystok University of Technology, Poland
Zbigniew Tarapata                    Military University of Technology, Poland
Richard Warner                       IIT/Chicago-Kent College of Law, USA
Marek Zachara                        AGH University of Science and Technology,
                                     Poland
Adam Zagorecki                       Cranfield University, UK

*Special Session on Advanced Data Mining Techniques and Applications (ADMTA 2015)*

Bay Vo                               Ton Duc Thang University, Vietnam
Tzung-Pei Hong                       National University of Kaohsiung, Taiwan
Bac Le                               Ho Chi Minh City University of Science, Vietnam
Chun-Hao Chen                        Tamkang University, Taiwan
Chun-Wei Lin                         Harbin Institute of Technology Shenzhen Graduate
                                     School, China
Wen-Yang Lin                         National University of Kaohsiung, Taiwan
Guo-Cheng Lan                        Industrial Technology Research Institute, Taiwan
Yeong-Chyi Lee                       Cheng Shiu University, Taiwan
Le Hoang Son                         Ha Noi University of Science, Vietnam
Le Hoang Thai                        Ho Chi Minh City University of Science, Vietnam
Vo Thi Ngoc Chau                     Ho Chi Minh City University of Technology,
                                     Vietnam
Van Vo                               Ho Chi Minh University of Industry, Vietnam

*Special Session on Cloud-Based Solutions (CBS 2015)*

Ana Almeida                          Porto Superior Institute of Engineering, Portugal
Oliver Au                            The Open University of Hong Kong, Hong Kong
Zoltan Balogh                        Univerzita Konštantína Filozofa v Nitre, Slovakia
Jorge Bernardino                     Polytechnical Institute of Coimbra, Portugal
Peter Brida                          University of Žilina, Slovakia
Jozef Bucko                          Technical University of Košice, Slovakia
Tanos Costa                          Franca Military Institute of Engineering, Brazil
Bipin Desai                          Concordia University, Canada
Ivan Dolnak                          University of Žilina, Slovakia
Elsa Gomes                           Porto Superior Institute of Engineering, Portugal

| | |
|---|---|
| Alipio Jorge | University of Porto, Portugal |
| Ondrej Krejcar | CBAR, University of Hradec Králové, Czech Republic |
| Kamil Kuca | Biomedical Research Center, University Hospital of Hradec Králové, Czech Republic |
| Juraj Machaj | University of Žilina, Slovakia |
| Norbert Majer | The Research Institute of Posts and Telecommunications (VUS), Slovakia |
| Goreti Marreiros | Porto Superior Institute of Engineering, Portugal |
| Peter Mikulecký | University of Hradec Králové, Czech Republic |
| Marek Penhaker | VSB – Technical University of Ostrava, Czech Republic |
| Teodorico Ramalho | University of Lavras, Brazil |
| Maria Teresa Restivo | Universidade do Porto, Portugal |
| Tiia Ruutmann | Tallinn University of Technology, Lativia |
| Ali Selamat | Universiti Teknologi Malaysia, Malaysia |
| José Salmeron | Universidad Pablo de Olavide, Seville, Spain |
| Vladimir Sobeslav | University of Hradec Králové, Czech Republic |
| Vassilis Stylianakis | University of Patras, Greece |
| Jan Vascak | Technical University of Košice, Slovakia |

*Special Session on Augmented Reality and 3D Media (AR3DM 2015)*

| | |
|---|---|
| Jędrzej Anisiewicz | ADUMA, Poland |
| Robert Bregovic | Tampere University of Technology, Finland |
| Piotr Chynał | Wrocław University of Technology, Poland |
| Bogusław Cyganek | AGH University of Science and Technology, Poland |
| Irek Defee | Tampere University of Technology, Finland |
| Piotr Hrebeniuk | Cohesiva, Poland |
| Marcin Sikorski | Polish-Japanese Institute of Information Technology, Poland |
| Mårten Sjöström | Mid Sweden University, Sweden |
| Jarmo Viteli | University of Tampere, Finland |
| Jakub Wójnicki | ADUMA MOBILE, Poland |

*Special Session on Internet of Things, Big Data and Cloud Computing (IoT 2015)*

| | |
|---|---|
| Jorgi Mongal Batalla | National Institute of Telecommunications, Poland |
| Adam Grzech | Wrocław University of Technology, Poland |
| Jason Jeffords | DeepIS, USA |
| Krzysztof Juszczyszyn | Wrocław University of Technology, Poland |

# Contents – Part I

**Intelligent Database Systems**

**Intelligent Information Systems**

**Innovations in Intelligent Systems and Applications**

# Contents – Part II

## Internet of Things, Big Data and Cloud Computing

## Artificial Intelligent Techniques and Their Application in Engineering and Operational Research

# Semantic Web, Social Networks and Recommendation Systems

# Interactive Refinement of Linked Data: Toward a Crowdsourcing Approach

Boonsita Roengsamut[1](✉) and Kazuhiro Kuwabara[2]

[1] Graduate School of Information Science and Engineering, Ritsumeikan University,
1-1-1 Noji-Higashi, Kusatsu, Shiga 525-8577, Japan
mb369@hotmail.com
[2] College of Information Science and Engineering, Ritsumeikan University,
1-1-1 Noji-Higashi, Kusatsu, Shiga 525-8577, Japan

**Abstract.** This paper proposes an approach in which a system extracts information interactively from a user to refine linked data. A multilingual frequently-asked-questions (FAQs) database in the domain of rental apartments is used as a test-bed. This database includes a domain ontology represented using linked data. It contains the relationships between part of the floor plan of a rental apartment and FAQ entries, which are derived using the domain ontology. When a user finds an error in the relationship, an interactive process is initiated so that even a casual user who is not a domain expert can contribute to fixing the error. The proposed method also makes use of pictures so that the language barrier can be lowered. Since the proposed approach is targeted to a casual user, it can be used in crowdsourcing the refinement of linked data.

**Keywords:** Linked data · RDF · Crowdsourcing · Ontology

## 1 Introduction

Linked data is a key technology to realize the *Web of Data* − in other words, the Semantic Web − and many data are published as linked data [3]. Interlinking plays an important role in linked data applications such as interlinking music datasets [10] and media contents [7]. A method to add additional Resource Description Framework (RDF) links by using an identity resolution component has also been proposed [5].

New links may be created using inference based on existing data and domain ontologies, but they may contain errors if there is a problem in the domain ontology. Suppose that problems are found in linked data when the application is used. It is difficult to update the ontology. In this paper, we present an approach for a casual user, or a non-expert, to fix a problem interactively.

We use a multilingual rental apartment FAQ system [6] as a test-bed for our proposed approach. In this system, the ontology is constructed from the FAQ entries about problems that often occur while living in rental apartments. One of the usages of the ontology in this system is to classify the FAQ entries according to the corresponding parts of the apartment floor plan [11].

© Springer International Publishing Switzerland 2015
N.T. Nguyen et al. (Eds.): ACIIDS 2015, Part I, LNAI 9011, pp. 3–12, 2015.
DOI: 10.1007/978-3-319-15702-3_1

When a user finds that the FAQ entry is classified incorrectly, an interactive refinement process is initiated. This interactive process is designed to use inputs from casual users (non-experts). By aggregating the inputs from many casual users, we can expect to obtain reasonably correct results. In addition, in order to overcome the language barrier in a multilingual environment, we apply a similar approach to interlinking involving pictures.

The remainder of the paper is structured as follows. The next section discusses some related works from the viewpoint of ontology refinement. Section 3 presents the multilingual rental apartment FAQ system that we used as a testbed. Section 4 describes the proposed method of refining and updating data in ontology by casual users. Section 5 explains the picture function. Section 6 discusses how the system could be expanded from the crowdsourcing aspect. The final section concludes our paper along with directions for future work.

## 2   Related Works

There are many approaches for ontology refinement. One of those approaches is ontology learning. Ontologies can help humans and machines communicate more concisely. The Semantic Web relies heavily on formal ontologies to structure data for comprehensive machine understanding. Ontology learning can greatly help ontology engineers construct good ontologies. Maedche and Staab [8] presented a comprehensive ontology-learning framework that includes several steps of import, extraction, pruning, refinement, and evaluation. The authors of this paper consider ontology learning as semi-automatic machine learning with human intervention for constructing ontologies for the Semantic Web.

Ontology learning focuses on automatic or semi-automatic construction of ontologies that can overcome the difficulties of manual ontology construction. An approach for ontology learning from text, proposed in [15], was evaluated against the existing manually-created ontologies of the same domains. The results proved to be very promising.

Currently, new approaches mostly involve contributive systems or crowd-sourcing techniques because of their low set-up costs and efficiency in quality. A good example is the JeuxDeMots project [9], in which a lexical network is constructed with the help of a game with a purpose and thousands of players that are appealed to contribute on lexical and semantic relations between terms. A strategy based on an inference mechanism, which formulates new relations between terms from an already existing relationship in the network, has been proposed to consolidate the lexical network.

## 3   Rental Apartment FAQ System

We use a multilingual application [6], with FAQs and a domain ontology as an example application to demonstrate how we implement a multilingual ontology, automatically link linked data, and visually present the concepts of the ontology. The ontology we have developed involves rental apartments. This application aims to support international students when faced with the difficulty of explaining a problem, such as an air conditioner not working in their apartment, to their Japanese apartment superintendent due to language barriers. Instead of dealing with language difficulties and wasting time explaining the situation to the Japanese apartment superintendent, the user can overcome those obstacles by using this application.

The system infers the association between a question entry and part of the floor plan by using morphological analysis to extract the keywords from the question statement. The keyword extraction results are stored in the RDF database with a link to the question. The system then searches for a thesaurus entry with the same label as the keyword. However, most of the time, the extracted keywords might relate to multiple parts of the floor plan. For example, from the question statement *The water in the toilet keeps running*, the keywords *water* and *toilet* will be extracted. The word *toilet* is related to the *Bathroom* and so is *water*. However, there is also relevance between *water* and *Kitchen*. With accurate priority rules, the system will be able to determine that this question is associated with *Bathroom*, not with *Kitchen* [11].

Data are represented in RDF, which is written in the Turtle language. Apache Jena Fuseki is used to store the RDF data and execute a SPARQL query. While the previous implementation implemented the RDF data mainly in Japanese; here, the RDF data is translated into English to create a more international system. The user interface is constructed to test the English database.

## 4   Ontology Refinement Process

We have noticed problems with the system caused by missing or wrong links in ontology. Let us suppose, for example, a user enters the question sentence *Shower is broken*, and the system returns the most closely related part of the floor plan as *Kitchen*, which is wrong for the user. Right now, this problem has been solved by manually adding the correct keyword or updating the correct link to the ontology [11]. The downside of this approach is that it is very costly and requires knowledge from experts. Our proposed approach allows casual users to become involved in the revision process of the ontology development. We aim to enhance the system to be more cost effective, while maintaining a good quality of correctness of data.

### 4.1   Refinement Protocol

Missing or incorrect links in ontology can lead to misunderstandings and bad user experiences. Our proposed method of ontology refinement will be able to overcome this problem at a reasonable cost. In order to refine the ontology, interaction between the user and the computer is described in the Unified Modeling Language (UML) diagram [2] as shown in Figure 1. For example, the user types a sentence or a keyword in the search box. Keywords are extracted from the input

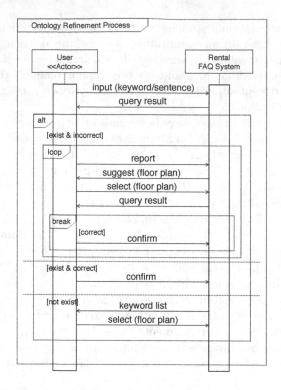

**Fig. 1.** Ontology refinement process diagram

and matched with words in the rental apartment ontology. If the extracted key-word has no match with any keywords in the ontology, a user will select the most closely related part of the floor plan from the possible list and assign it to the extracted keyword. The result will be kept in the temporary ontology initially. After the most frequently selected keyword is calculated, it will be added to the ontology. If the extracted keyword matches words in the ontology, the user will check whether the resulting link between the words is correct or not. If the link is correct, the user will be asked to confirm the answer. If the user confirms the answer, it will be added to the ontology.

## 4.2   Sample Scenario

There are three cases to consider here: (1) No keyword exists in the ontology; (2) Some keywords exist in the ontology, while others do not; and (3) The extracted keywords, which exist in the ontology, have the wrong link.

**Case 1: No keyword exists in the ontology.** If there is no existing keyword in the ontology, all keywords related to the part of the floor plan that the user has searched for will be listed and suggested to the user. The selected answer will be kept in a temporary ontology. If the majority percentage of the users agree on the same keyword, that keyword will be updated to the real ontology. For instance, the user inputs the sentence, *Toilet flush does not work.* This sentence is processed and the keywords, *Toilet* and *flush*, are extracted, ignoring words such as verbs, tentative verbs, or modal auxiliaries. These two keywords are combined to create one keyword, e.g., *Toilet flush*. *Toilet flush* is then searched for in the ontology to find a matching keyword, but it does not exist. The parts of the floor plan are listed so that the user can select the most suitable part, e.g., *Toilet*, that should be matched with the extracted keyword. The result is kept in the temporary ontology first and processed later to be updated to the real ontology.

**Case 2: Some keywords do not exist.** If some keywords exist and others do not, the parts of the floor plan relating to the existing keywords are displayed. When a user notices that the displayed results are an incorrect link or not the most relevant keyword, this fault will be reported. More related links are suggested to the user, and the reasons why the mismatched result was obtained are presented. The user selects the part of the floor plan that is most relevant to the keywords, a new temporary ontology is created, and the query result is recalculated and displayed to the user. If the user confirms that it is correct, the most frequently selected keyword is updated to the ontology. If the user thinks it is still incorrect, they have to report it again. For example, the user types a sentence, *The hot water in the sink is not hot.* The keywords extracted from this sentence are *hot water* and *sink*.

If one of these keywords exists in the ontology, the most closely related part of the floor plan e.g. *hot water* is shown to be related to *Bathroom*. If the user wants *hot water* to be related to *Kitchen*, an incorrect link is reported. More related links to that part of the floor plan are suggested to the user, and the system provides reasons to explain why this keyword has the wrong link or it does not exist in the ontology and suggests other related parts of the floor plan in a dropdown list so that the user can choose the most appropriate one from the list. That is, *hot water* has a wrong link in the sense that the most closely related part of the floor plan should be *Kitchen*. *Sink* does not exist in the ontology; thus, a list is presented to the user, who will select the most closely related part of the floor plan from the list presented.

After the part of the floor plan is assigned to each keyword, the total number of related floor plan parts, which users assign to each of the keywords are counted and kept in a temporary ontology, e.g., the keyword *hot water* has five votes for *Bathroom* and three votes for *Kitchen*. The user then confirms whether the recalculated query result is correct. If it is correct, the part of the floor plan with the highest number of user votes is updated to the real ontology. If it is incorrect, the user reports that the result is still incorrect, and the process to assign the most closely related part of the floor plan is initiated again.

**Case 3: Keyword has the wrong link.** If the extracted keywords are incorrectly linked to the part of the floor plan, the user reports that the provided part of the floor plan is wrong. Other related parts of the floor plan are suggested to the user with reasons, and the user selects the most relevant floor plan part to that keyword. The query result is recalculated and displayed to the user. The system asks the user to confirm whether it should be updated to the ontology or not. If the user confirms that it is correct, the result is processed to update the real ontology. If the user points out that it is still incorrect, the user must report the incorrect link once again. The user, for example, searches by typing *Electricity does not come out from the socket*. The keyword *socket* is extracted from the sentence and matched with words in the ontology. The part of the floor plan related to *socket* is *Kitchen*. The incorrect result is reported by the user because the user means a socket in the *Toilet*. After the user assigns *socket* to *Toilet*, in this case, the system processes the most relevant part of the floor plan and the result becomes *Living room* instead of *Toilet*. Because the keyword *socket* has received 10 votes for *Living room* and two votes for *Toilet*, thus, *socket* is most closely related to *Living room* in the ontology.

## 5 Using Pictures

We propose the idea of incorporating a picture function to the system because when things get broken, the user may not know what to call them. We want our user to get the best experience out of using our FAQ system. Therefore, when in doubt about the name of the broken device, the user can take a picture of the broken item and upload it to the system. This will lower the language barrier meaning the question statement can be in any language. Additionally, the question statement can be linked to the most closely related part of the floor plan without keyword extraction.

A photo will be posted on the system's board for another user's help. Other users will be able to add data, e.g., device names, to the posted photo. Data will be queried for duplications. If the input word exists in the ontology, a link between that photo and the input word will be created. Conversely, the word will be added to the ontology if it does not exist in it. Then, a link between the word, picture, and the relevant part of the floor plan will be created.

The picture function will close the gap between the machine and the human user and should overcome any language barriers. Users do not have to type a single word to interact with the FAQ system, thus providing the user with a better experience.

### 5.1 Protocol

As shown in Figure 2, the user uploads a photo of the device with which he/she is having trouble. That photo will be posted on the board for other users to name. The name that a user enters will be queried to establish whether it already exists in the ontology or not. If the name exists in the ontology, a link between the

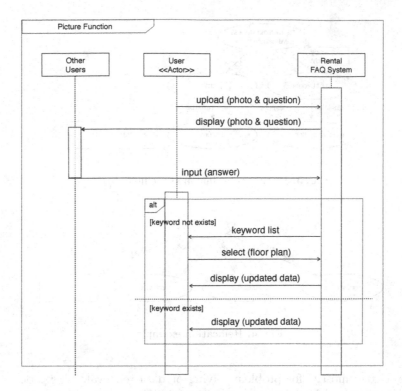

**Fig. 2.** Picture function

picture and the name will be created. If the name does not exist in the ontology, the name will be added to the ontology. The link between the name, picture, and the most relevant part of the floor plan will be created.

## 5.2 Sample Scenario

In the case that only one device exists in the picture when the picture is uploaded, the question statement, *Something is stuck in the ventilation fan*, is linked with it. The link between the question statement and `fp:Bathroom` is derived by the ontology as follows (Figure 3). `pict:1` represents the uploaded picture and `:depicts` indicates a relationship exists between `pict:1` and `word: ventilation_fan`. The `rdf:type` and `:uploaded` relationships are also appended to indicate that `pict:1` has a picture type and was uploaded on date `2015-01-01`. The `word:ventilation_fan` has a relationship shown by `:contains` with `fp: Bathroom`, meaning that the ventilation fan is in the bathroom. In this manner, the query result of the part of the floor plan of `pict:1` would be *Bathroom*.

## 6 Discussion

Crowdsourcing is the process of openly distributing/outsourcing a task usually performed by a specific employee or supplier to a large network of people to

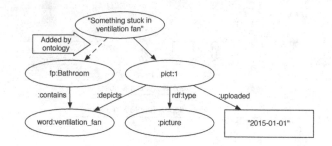

**Fig. 3.** Picture contains one specified object

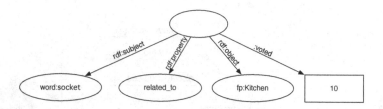

**Fig. 4.** Reification example

achieve a goal; mostly, for problem solving or data retrieval. This process can either be performed by a large group of individuals or in a collaborative peer-production fashion. The key to crowdsourcing is its openness and the use of a large network of contributors [1]. At the current stage, our proposed method of interactively refining the linked data is more of an interactive process of human computation. This can be treated as the first step towards the crowdsourcing concept.

Much research uses the concept of human computation in the Semantic Web [4]. There are several tools for interlinking RDF data using human computation [13]. For example, the problem of ontology alignment is translated into microtasks and executed via the labor market [12]. In order to use this interactive process in a crowdsourcing concept, we need to aggregate the results from human input and create a temporary ontology for which an appropriate ontology representation is mandatory. The temporary ontology representation issue is one of our concerns.

In order to store the number of votes on the relevant part of the floor plan for each keyword in a temporary ontology, the representation of the data is an important issue that must be addressed. There are three solutions that can fix this problem.

1. *Keep data in a table outside the RDF graph.*
   The data is kept in the table outside the RDF. Data that are stored in a table format can be easily represented and updated. It is also a compact and efficient way to store data. Unfortunately, in this case, it is difficult to

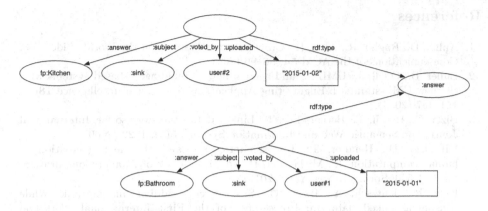

**Fig. 5.** User's answers in the RDF format

manage because we have to use SQL query language to query the data from a table while, at the same time, using SPARQL query to query the data from the RDF.

2. *Use a form of reification.*

   The RDF directly supports reification. In order to represent a single triple, we need three triples for the reification, and :voted link is added to store the number of votes as shown in Figure 4.

3. *Record users' answers in the RDF format.*

   In this method, the number of votes is not necessarily added to the RDF, as shown in Figure 5. An SQL query is needed to query and count the number of votes. The number of votes does not have to be updated in the RDF.

# 7    Conclusion and Future Work

The FAQ rental apartment system focuses on casual users. The system's ontology can be easily refined or updated through basic interaction between a human user and a computer. To ensure the best user experience with an effective ontology refining result, we propose a system that accepts not only texts input but also images. Users can add images to define any particular terminology. Our system is currently being implemented and the evaluation plan is being made to show that our system works seamlessly and effectively.

In addition, we plan to apply the crowdsourcing technique, in particular, gamification into our work. The goal is to ensure users enjoy using our system and, at the same time, help refine the system ontologies. It will be a fun-to-use application that can deliver useful and effective results. This falls into the concept of *games-with-a-purpose* that can update and refine ontologies, *e.g.,* OntoGame [14]. Since crowdsourcing is mainly used for large scale information and knowledge [16], we also plan to extend the proposed approach to handle the refinement of large-scale datasets.

# References

1. Anhai, D., Raghu, R., Alon, H.: Crowdsourcing systems on the World-Wide Web. Communications of the ACM **54**, 86–96 (2011)
2. Bauer, B., Odell, J.: UML 2.0 and agents: how to build agent-based systems with the new UML standard. Engineering Applications of Artificial Intelligence **18**(2), 141–157 (2005)
3. Bizer, C., Heath, T., Berners-Lee, T.: Linked data - the story so far. International Journal on Semantic Web and Information Systems **5**(3), 1–22 (2009)
4. DiFranzo, D., Hendler, J.: The semantic web and the next generation of human computation. In: Michelucci, P. (ed.) Handbook of Human Computation, pp. 523–530. Springer, New York (2013)
5. Isele, R., Jentzsch, A., Bizer, C.: Silk server - adding missing links while consuming linked data. In: Proceedings of the First International Workshop on Consuming Linked Data (COLD 2010) (2010). http://ceur-ws.org/Vol-665/IseleEtAl_COLD2010.pdf
6. Kinomura, S., Kuwabara, K.: Developing a multilingual application using linked data: a case study. In: Bădică, C., Nguyen, N.T., Brezovan, M. (eds.) ICCCI 2013. LNCS, vol. 8083, pp. 120–129. Springer, Heidelberg (2013)
7. Kobilarov, G., Scott, T., Raimond, Y., Oliver, S., Sizemore, C., Smethurst, M., Bizer, C., Lee, R.: Media meets semantic web – how the BBC uses DBpedia and linked data to make connections. In: Aroyo, L., et al. (eds.) ESWC 2009. LNCS, vol. 5554, pp. 723–737. Springer, Heidelberg (2009)
8. Maedche, A., Staab, S.: Ontology learning for the semantic web. IEEE Intelligent Systems **16**(2), 72–79 (2001)
9. Manel, Z., Mathieu, L., Alain, J.: Inference and reconciliation in a crowdsourced lexical-semantic network. Computación y Sistemas **17**(2), 147–159 (2013)
10. Raimond, Y., Sutton, C., Sandler, M.: Automatic interlinking of music datasets on the semantic web. In: Proceedings of the Linked Data on the Web Workshop (2008). http://CEUR-WS.org/Vol-369/paper18.pdf
11. Saito, Y., Roengsamut, B., Kuwabara, K.: Incremental refinement of linked data: ontology-based approach. In: Nguyen, N.T., Attachoo, B., Trawiński, B., Somboonviwat, K. (eds.) ACIIDS 2014, Part I. LNCS, vol. 8397, pp. 133–142. Springer, Heidelberg (2014)
12. Sarasua, C., Simperl, E., Noy, N.F.: CROWDMAP: crowdsourcing ontology alignment with microtasks. In: Cudré-Mauroux, P., et al. (eds.) ISWC 2012, Part I. LNCS, vol. 7649, pp. 525–541. Springer, Heidelberg (2012)
13. Simperl, E., Wölger, S., Thaler, S., Norton, B., Bürger, T.: Combining human and computation intelligence: the case of data interlinking tools. International Journal of Metadata, Semantics and Ontologies **7**(2), 77–92 (2012)
14. Siorpaes, K., Hepp, M.: Games with a purpose for the semantic web. IEEE Intelligent Systems **23**(3), 50–60 (2008)
15. Tegos, A., Karkaletsis, V., Potamianos, A.: Learning of semantic relations between ontology concepts using statistical techniques. In: Proceedings of the High Level Information Extraction Workshop (HLIE 2008), ECML-PKDD 2008 Conference (2008)
16. Xin, P., Muhammad, A., Christof, E.: Collaborative software development platforms for crowdsourcing. IEEE Software **31**(2), 30–36 (2014)

# Synthetic Evidential Study as Augmented Collective Thought Process -- Preliminary Report

Toyoaki Nishida[1(✉)], Masakazu Abe[1], Takashi Ookaki[1], Divesh Lala[1],
Sutasinee Thovuttikul[1], Hengjie Song[1], Yasser Mohammad[1], Christian Nitschke[1],
Yoshimasa Ohmoto[1], Atsushi Nakazawa[1], Takaaki Shochi[2,3], Jean-Luc Rouas[2],
Aurelie Bugeau[2], Fabien Lotte[2], Ming Zuheng[2], Geoffrey Letournel[2],
Marine Guerry[3], and Dominique Fourer[2]

[1] Graduate School of Informatics, Kyoto University, Sakyo-ku, Kyoto, Japan
{nishida,christian.nitschke,ohmoto,nakazawa.atsushi}
@i.kyoto-u.ac.jp, {abe,ookaki,lala,thovutti,song}
@ii.ist.i.kyoto-u.ac.jp, yasserfarouk@gmail.com
[2] LaBRI, Bordeaux, France
{takaaki.shochi,jean-luc.rouas,aurelie.bugeau,zming,
geoffrey.letournel,fourer}@labri.fr,
fabien.lotte@inria.fr
[3] CLLE-ERSS UMR5263 CNRS, Bordeaux, France
{takaaki.shochi,marine.guerry}@labri.fr

**Abstract.** Synthetic evidential study (SES) is a novel approach to understanding and augmenting collective thought process through substantiation by interactive media. It consists of a role-play game by participants, projecting the resulting play into a shared virtual space, critical discussions with mediated role-play, and componentization for reuse. We present the conceptual framework of SES, initial findings from a SES workshop, supporting technologies for SES, potential applications of SES, and future challenges.

**Keywords:** Group learning assistance · Intelligent virtual agents · Role play

## 1 Introduction

A collective thought process becomes more and more critical in the network age as a means for bringing together limited intelligence embodied by natural or artificial agents. A powerful methodology is needed to make collective thought processes effective. Methodologies such as brainstorming or mind map have been invented but they are mostly the third-person understanding and are limited in terms of actuality. Their output can only appeal to people through narratives or other pedagogical media. It is pretty hard for the ordinary audience to share the thought in terms of vivid and immersive understanding, or first-person understanding, of the output unless enough background knowledge is shared. The problem might be solved if it is presented as an interactive movie, but a huge cost would be required for that. Even with the existing state-of-the-art technology, however, it still appears beyond our scope to build a tool that allows for creating meaningful low-cost movies.

© Springer International Publishing Switzerland 2015
N.T. Nguyen et al. (Eds.): ACIIDS 2015, Part I, LNAI 9011, pp. 13–22, 2015.
DOI: 10.1007/978-3-319-15702-3_2

A less challenging, but still useful goal might be to build an intelligent tool that would allow people to progressively build a story base, which is a background setting consisting of pieces of story scenes, each of which consists of events played by one or more role actors with reference to the physical or abstract background. A story base may serve as a mother from which individual stories and games may be spawned. We assume that a story base will greatly help professional storytellers and game players produce high-quality content.

The long-term goal of this project is to establish a powerful method for allowing everybody to participate in a collective thought process for producing a story base for a given theme. There is a huge area of applications in education and entertainment. In addition, we believe that the project benefits science and technology. On the scientific hemisphere, it will open up a new methodology for investigating in-situ human behaviors. On the engineering hemisphere, it will significantly benefit not only content production but also product prototyping and evaluation.

In this paper, we portray a novel approach, called *synthetic evidential study* (SES), for understanding and augmenting collective thought process through substantiated thought by interactive media. The proposed approach draws on authors' previous work, including conversational informatics, human-computer interaction, computer vision, prosody analysis and neuro-cognitive science [1]. In what follows, we present the overview of SES, preliminary implementation of its components, and future perspectives.

## 2    Overview of SES

The present version of SES basically consists of four stages. The first stage is role-play by participants. Actors are invited to play a given role to demonstrate their first-person interpretation in a virtual space. A think aloud method is used so the audience can hear the background as well as the normal foreground speech. Each actor's behaviors are recorded using audio-visual means. The second stage is projecting role-play into a shared virtual space. The resulting theatrical play as an interpretation is recorded and reproduced for criticism by the actors themselves. The third stage is critical discussions with mediated role-play. It permits the participants or other audience to share the third-person interpretation played by the actors for criticism. The actors revise the virtual play until they are satisfied. The understanding of the given theme will be progressively deepened by repeatedly looking at embodied interpretation from the first- and third- person views. The final stage is componentization for reuse. The mediated play is decomposed into components and stored in the story base.

For illustration, suppose a handful people become interested in some scene of Romeo and Juliet by William Shakespeare. First, the participants will set up a SES workshop comprising the stages *a-c*, where the participants may either start from scratch or take up a previous piece of interpretation from the story base, criticize it, and produce their own, depending on their interest. Each one of them is asked to demonstrate her or his first-person interpretation for the scene, by following the events and expressing her or his thought as a behavior that she or he thinks the role would have acted for each event. Reproducing the behavior of a role, i.e. Romeo or

Juliet, in each given scene will allow the participant to feel the role's mental and emotional state, resulting in deeper and immersive understanding of the scene. On the second stage, they criticize with each other to improve the shared interpretation. The third person perspectives would allow the participants to gain the holistic understanding of the scene. Discussions permit the participants to know other possibilities of interpretation and their strength and weakness.

In order to maximally benefit from the above-mentioned aspects of SES, we need a powerful computational platform. It should be built on a distributed platform as participants would like to participate in from geographically distant points. We have found that the game engine, Unity 3D[1] in particular, best fits this purpose. It allows us to share a virtual space with complex objects and animated characters. Reproducing participants' theatrical role-play as Unity objects allows for implementing the stage *d* to progressively construct a story base for a community. It will enable its members to exploit the components of interpretation to build and share sophisticated knowledge about subjects of common interest.

The conceptual framework for the SES support technology consists of a shared virtual space technology for interfacing the users with the story world and the discussion space technology for supporting criticism and improving play as interpretation.

Virtual space technology plays a significant role to configure vast varieties of conversational environments. Consider the actors are asked to interpret the balcony scene of the Romeo and Juliet play. Although a physical setting for the balcony scene is critical for interpreting the details, it is too expensive and hence infeasible for ordinary cases unless the studio is available. Another difficulty arises when the number of participants is not enough. Whereas a virtual environment inhabited by non-playable characters solves difficulties in general, problems remain regarding how to generate qualified settings and non-playable characters for interpretation. Our technical contributions mostly address the second issue, while we utilize existing techniques for the first issue. Furthermore, the interface should be immersive and gesture-driven so participants can concentrate on the SES activities. Our technological supports for SES consist of the shared virtual space, virtual character realization, and discussion support.

We draw on technologies that have been developed in pursuit of conversational informatics. Our technology for SES not only supports and records conversations in a virtual immersive environment and projects the behaviors of the actors to those of synthetic characters but also analyzes interactions. It will help the participants deepen their interpretation even from the viewpoint of the second person or the interactant through interaction, which is only available by the virtual technology.

SES significantly extends the horizon of conventional pedagogy, as the participants of the SES workshops may be able to learn to view a phenomenon from multiple angles including not just active participants but also the perspective of other role players in real time if the agent technology is fully exploited. It should be extremely effective for social education, such as one for anti-bullying, as it allows the participants to "experiment" social affairs from different perspectives, which is almost impossible otherwise. From scientific points of view, the SES enables the understanding of human behaviors in a

---

[1] http://www.unity3d.com

vast variety of complex situations. As a result, one can design experiments far more realistic than conventional laboratory experiments. For example, the experimenters can slip into the SES session cues or distractors in a very natural fashion.

In what follows, we elaborate the SES workshop, the supporting technologies we have developed so far, and how SES is applied to conduct study human behaviors.

## 3      SES Workshop

The SES workshop is a joint activity open for every group of people to figure out a joint interpretation of a given theme by bringing together prior interpretations of participants. Repetition of acting together and discussion is a critical feature of a SES workshop[2]. In order to gain the practical features of the SES workshop, we conducted a preliminary workshop to gain initial insights about SES. We chose a story called Ushiwaka and Benkei[3] because it was very popular in Japan though the details are not well considered as it is rather a fiction told for children though it is partly based on historical fact. We had conducted one 1-hour session in which four participants repeated two discussion-play cycles. Due to the limitation of our measurement facility at that time, each one-role player acted for his part (Fig. 1a). The actor's motion was recorded by a Kinect[4] and projected as the behavior of a Unity agent. This is the topic of the subsequent discussions (Fig. 1b).

(a)   Play and record                              (b) Criticize and improve

**Fig. 1.** Snapshots from the preliminary SES workshop

According to an informal *a posteriori* interview, the participants were able to be well involved in the discourse and obtained a certain degree of immersive understanding. In fact, we observed that the participants got interested in the details of the story such as how to handle a long sword. It motivated the participants to collect more information from the net and reflected that on their role play acts. In addition, the SES

---

[2] In fact, a single player can conduct the SES workshop by leveraging the SES technology of virtual shared space and characters, if somebody prefers working alone.

[3] Ushiwaka and Benkei story goes like this. When Ushiwaka, a young successor to a noble Samurai family which once was influential but which was killed by the opponents, walked out of a temple in a mountain in the suburbs of Kyoto where he was confined, to wander around the city as daily practice, he met Benkei, a strong priest Samurai on the Gojo Bridge. Although Benkei tried to punish him as a result of having been provoked by a small kid Ushiwaka, he couldn't as Ushiwaka was so smart to avoid Benkei's attack. After a while, Benkei decided to become a life-long guard for Ushiwaka.

[4] http://www.xbox.com/Kinect

workshop seems to work well for integrating partial knowledge of participants by acting and discussion together.

## 4    Computational Platform for SES

The conversation augmentation technology [1] supports immersive interactions made available by a 360-degree display and surround speakers and audio-visual sensors for measuring the user's behaviors (Fig. 2). The "cell" can be connected with each other or with other kinds of interfaces such as a robot so that the users can participate in interactions in a shared space.

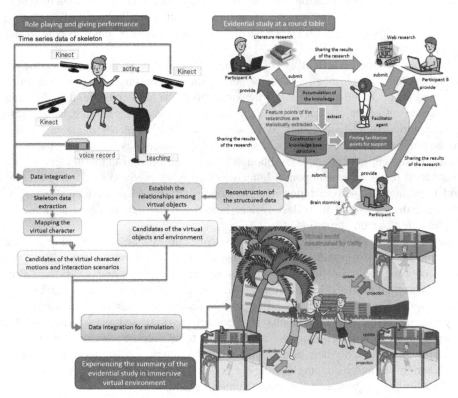

**Fig. 2.** The computational platform for SES

It allows to project the behaviors of a human to those of an animated character who habits in a shared virtual space. The computational platform is coupled with the Unity platform so the participants can work together in a distributed environment.

A virtual space builder called FCWorld [2] enables the immersive environment to be linked to external software like Google Street View to benefit from various content available on the net. FCWorld was used to implement virtual network meetings with the Google Street View background.

## 5    Virtual Character Realization

Animated characters are critical elements of SES. Roughly speaking, a virtual character realization consists of two phases: character appearance generation and behavior generation.

As known in nonverbal communication, not only their behaviors but also their appearance may significantly influence the nature of interactions. We want to make the appearance of characters as unique as possible so they maximally reflect the target interpretation. An avatar with the face of an existing human might be expressive but too specific. Designing a game character is too expensive for ordinary SES sessions. We believe the de-identification approach [3] is the most effective. The idea is to take an existing human face and remove identifiable features while maximally retaining the expressive aspects. Ideally, voice de-identification should be coupled with visual de-identification.

Behavior generation is another key technical topic in virtual character realization. On the one hand, sophisticated behavior generation is necessary to express subtleties of mental state. On the other hand, participants should be easy and natural so they can concentrate on essential issues. Although recent advanced technologies using inexpensive depth-color sensors such as Kinect exist, there is plenty of room for improvement, such as full body behavior generation that integrates facial expression and bodily movement generation [1]. Although technologies are available for generating point cloud image representation of humans, more work needs to be done to accommodate point cloud representation into the game platform. Alternative approach is to employ crowd sourcing to collect typical behaviors that can serve as a prototype adaptable for a given purpose [4].

Learning by imitation is a powerful framework. So far, we have develop a basic platform and a more powerful imitation engine which monitors the behavior of the target continuously, autonomously detects recurrent signals and infers causality among observed events [5]. Imitation can play several roles at different stages of the SES session. For example, direct pose and motion copying can be used during role play to generate the preliminary behaviors that will be discussed and improved by the group. Our pose copying system is based on a modular decomposition of the problem that simplifies the extension to non-humanoid characters. During motion update, a combination of motion copying and the correction by repetition technology we developed earlier [6] can be used to improve the motions smoothly.

## 6    Discussion Support

Another critical phase of the SES support technology is discussion support. Our discussion support consists of the round-table support and the full-body interface. The ultimate goal of the former is to build a chairperson agent who can support discussions by estimating distribution of opinions, engagement, and emphasizing points. We have implemented several prototypes [7-9]. These prototypes focused on interactive decision-making during which people dynamically and interactively change the focusing points. Our technologies allow to capture not only explicit social signals that clearly manifest on the surface but also tacit and ambiguous cues by integrating audio-visual and physiological sensing. As for the latter, we exploit Kinect technology to measure and criticize physical display of interpretation played together one or more local participants [1]. Since the system we developed is easy to setup measuring environment for human motion capture, we can conduct SES anywhere indoors.

## 7    First-Person View by Corneal Imaging

Capturing the first person view of the world provides a valuable means for estimating the mental status of a human either in a role-playing game or in discussion. In fact, we have found that first-person view may bring about quite different emotional state from the third-person view in human-robot interaction [10]. By exploiting the fact that the cornea of a human reflects the surrounding scene over a wide field of view, our corneal imaging technology allows for determining the point of gaze (PoG) and estimating the visual field from reflecting light at the corneal surface using a closed-form solution. Compared to the existing approaches, our method achieves equipment and calibration-free (PoG), depth-varying environment information and peripheral vision estimation. In particular, the first and the third are very important to be used as human interface devices and beyond the current human view understandings that uses only the 'point' of the gaze information [11].

## 8    Prosody Analysis

Prosody analysis is a key technology for estimating and distinguishing social affects, such as laughter and smile, in face-to-face communication. Production and interpretation of social affects is indispensable for obtaining in-depth understanding of the SES sessions. So far, our prosody analysis and corpus-building technologies have been applied to analysis of laughter/smile/sad speech and cross-cultural communication. Intended affective meanings are conveyed by various modalities such as body movements, gestures, facial expressions as well as vocal expressions. In particular, vocal expressiveness of these affects is intensively studied since the 90's. Recently, Riliard et al. envisage the prosodic variations, which are used to encode such social affects. They also try to identify the characteristics of these prosodic codes in competition with others (e.g. syntactical and lexical prosodic configuration) [14]. Such prosodic analysis focusing on social affective meaning may be implemented by cross-cultural communication processing. Thanks to this methodological approach some universal and culture-specific prosodic patterns were identified even for the same label of affect (e.g. surprise).

Combined with the SES technology, our prosody analysis techniques will be extended to multi-modal prosody analysis, powerful enough to investigate complex social-emotional communication. Coupled with brain activity visualization technique such as [13], it will allow us to investigate brain activities underlying the social affects.

## 9    SES-Based Human Behavioral Science

SES can serve as a novel platform for conducting human behavioral study as it allows the researchers to build a sophisticated experiment environment.

(1) Building multi-modal corpus for cross-cultural communication. Socio-emotional aspects are critical in understanding cross-cultural communication. Other cross cultural studies on affective speech are conducted in cross cultural paradigm among four languages: Japanese, American English, Brazilian Portuguese and French [14,15]. The results showed that subjects of different cultural origins shared about 60 % of the global representation of these expressions, that 8% are unique to modalities, while 3 % are unique to language background. The results indicate that even if specific cultural details punctually play an important role in the affective interpretations, it may also emphasize the fact that, a great deal of information is already shared amongst speakers of different linguistic backgrounds.

Therefore, the SES platform permits researchers to set up complex situations to quantitatively investigate socio-emotional aspects of cross-cultural communication in the shared virtual space so that culture-dependent cues can be observed and contrasted in multiple desired conditions, as a natural extension to our method of analyzing prosodic aspects.

(2) Dance lesson. Measure and criticize the detailed quantitative aspects of motion. Coupled with group annotation tools, we can analyze the detailed nonverbal behaviors of tutor-student interactions. The group annotation tools are useful for building consensus because they automatically extract and propose feature points for criticizing. The scheme can be used to analyze how people criticize the features of played actions [1].

(3) Laughter. In this study, we mostly depend on physiological sensors to make subtle distinctions among hidden laughter, enforced laughter, and genuine laughter. Especially hidden laughter is an important cue to get the true response. The SES framework most fits more comprehensive study on comedic plays that induce laughter on the audience [16].

(4) Virtual basketball. In this study, we shed light on how social signals from beginning players and non-playable characters in motion can be used to read intentions from friends and opponents. In the future, the study may be extended to highlight the group discussions for criticizing and improving performance [17].

(5) Cultural crowd. This study aims at gaining first-person understanding of crowd in a different culture. Currently, we are focusing on queuing behaviors in a different culture, trying to identify social signals and norm in a given culture, by conducting a contrastive study [18].

(6) Physiological study and evaluation of social interactions in the virtual space. Being able to assess how observers perceive, from an affective (e.g., emotions) and cognitive (e.g., attention, engagement) point of view, the play of the actors, could provide very interesting insights to refine and improve the play and/or the story. Similarly measuring such affective or cognitive states during collaborative tasks performed in the joint virtual space could be used to study, assess and then optimize the

collaborative work. Interestingly enough, we and others have shown that such affective and cognitive states could be measured and estimated in brain signals (electroencephalography) [19] or in other physiological signals (heat rate variability, galvanic skin response, etc.) [20]. The SES would thus provide a unique test bed to study physiological based assessment of both affective and cognitive experience and distance collaboration tasks.

## 10   Concluding Remarks

In this paper, we introduced synthetic evidential study (SES) as a novel approach for understanding and augmenting collective thought process. We presented the conceptual framework of SES, initial findings from a SES workshop, supporting technologies for SES, and potential applications of SES. We believe that SES has many applications ranging from science to engineering, such as content production, collaborative learning and complex human behavior analysis. Future challenges include, among others, virtual studio, layered analysis of human behavior, and multimodal prosody analysis.

The SES paradigm has opened up numerous new challenges as well as opportunities. Among others, we have recognized three challenges as the most useful to enhance the current framework of SES. The first is virtual studio that permits user to set up and modify a complex terrain and objects on the fly. The second is analysis of participants' action and separation of layers, as a participant' behavior refers to either the action of the role or meta-level action such as a comment to her or his action. The third is a methodology for guiding SES sessions to bring about intended effects on the participants, depending on the purpose. For example, if the purpose of a given SES activity is to help the participants design a new industrial product, it should be very helpful if the participants are encouraged to consider potential product usage scenarios in a comprehensive fashion.

**Acknowledgments.** This study has been carried out with financial support from the Center of Innovation Program from Japan Science and Technology Agency, JST and AFOSR/AOARD Grant No. FA2386-14-1-0005, JSPS KAKENHI Grant Number 24240023, the French State, managed by the French National Research Agency (ANR) in the frame of the "Investments for the future" Programme IdEx Bordeaux (ANR-10-IDEX-03-02), Cluster of excellence CPU. We are grateful for Peter Horsefield who helped us improve the presentation of this paper.

## References

1. Nishida, T., Nakazawa, A., Ohmoto, Y., Mohammad, Y.: Conversational Informatics–A Data-Intensive Approach with Emphasis on Nonverbal Communication. Springer (2014)
2. Lala, D., Nitschke, C., Nishida, T.: Enhancing communication through distributed mixed reality. In: Ślęzak, D., Schaefer, G., Vuong, S.T., Kim, Y.S. (eds.) AMT 2014. LNCS, vol. 8610, pp. 501–512. Springer, Heidelberg (2014)

3. Letournel, G., Bugeau, A., Ta, V.-T., Domenger, J.-P., Gallo, M.C.M.: Anonymisation fine de visages avec préservation des expressions faciales, Reconnaissance de Formes et Intelligence Artificielle (RFIA) 2014, Rouen, France (2014)

4. Han, X., Zhou, W., Jiang, X., Song, H., Zhong, M., Nishida, T.: Utilizing URLs position to estimate intrinsic Query-URL relevance. In: Proc. ICDM 2013, pp. 251–260 (2013)

5. Mohammad, Y., Nishida, T.: Robust learning from demonstrations using multidimensional SAX. Presented at 14th International Conference on Control, Automation and Systems (ICCAS 2014), Gyeonggi-do, Korea (2014)

6. Mohammad, Y., Nishida, T.: NaturalDraw: interactive perception based drawing for everyone. In: Proc. IUI 2007, pp. 251–260 (2007)

7. Ohmoto, Y., Miyake, T., Nishida, T.: Dynamic estimation of emphasizing points for user satisfaction evaluations. In: Proc. the 34th Annual Conference of the Cognitive Science Society, pp. 2115–2120 (2012)

8. Ohmoto, Y., Kataoka, M., Nishida, T.: Extended methods to dynamically estimate emphasizing points for group decision-making and their evaluation. Procedia-Social and Behavioral Sciences 97, 147–155 (2013)

9. Ohmoto, Y., Kataoka, M., Nishida, T.: The effect of convergent interaction using subjective opinions in the decision-making process. In: Proc. the 36th Annual Conference of the Cognitive Science Society, pp. 2711–2716 (2014)

10. Mohammad, Y., Nishida, T.: Why should we imitate robots? Effect of back imitation on judgment of imitative skill. Int. J. of Soc. Robotics (published online)

11. Nitschke, C., Nakazawa, A., Nishida, T.:I see what you see: point of gaze estimation from corneal images. In: Proc. 2nd IAPR Asian Conference on Pattern Recognition (ACPR), pp. 298–304 (2013)

12. Rilliard, A., De Moraes, J., Erickson, D., Shochi, T.: Social affect production and perception across languages and cultures - the role of prosody. Leitura 52 (forthcoming)

13. Frey, J., Gervais, R., Fleck, S., Lotte, F., Hachet, M., Teegi: Tangible EEG Interface, ACM User Interface Software and Technology (UIST) symposium (2014)

14. Rilliard, A., Erickson, D., De Moraes, J., Shochi, T.: Cross-cultural perception of some Japanese expressions of politeness and impoliteness. In: Linguistic Approaches to Emotions, in Context, pp. 251–276. John Benjamins, Amsterdam (2014)

15. Fourer, D., Shochi, T., Rouas, J.-L., Aucouturier, J.-J., Guerry, M.: Prosodic analysis of spoken Japanese attitudes. Proc. Speech Prosody 7, 149–153 (2014)

16. Tatsumi, S., Mohammad, Y., Ohmoto, Y., Nishida, T.: Detection of Hidden Laughter for Human-agent Interaction. Procedia Computer Science 35, 1053–1062 (2014)

17. Lala, D., Mohammad, Y., Nishida, T.: A joint activity theory analysis of body interactions in multiplayer virtual basketball. In: 28th British Human Computer Interaction Conference, Southport, UK (2014)

18. Thovuttikul, S., Lala, D., van Kleef, N., Ohmoto, Y., Nishida, T.: Comparing people's preference on culture-dependent queuing behaviors in a simulated crowd. In: Proc. ICCI*CC 2012, pp. 153–162 (2012)

19. Frey, J., Mühl, C., Lotte, F., Hachet, M.: Review of the use of electroencephalography as an evaluation method for human-computer interaction. In: International Conference on Physiological Computing Systems (PhyCS 2014), pp. 214–223 (2014)

20. Fairclough, S.: Fundamentals of physiological computing. Interacting with Computers 21, 133–145 (2009)

# Exploiting Ontological Reasoning in Argumentation Based Multi-agent Collaborative Classification

Zhiyong Hao[✉], Bin Liu, Junfeng Wu, and Jinhao Yao

National University of Defense Technology, Changsha 410073,
Hunan, People's Republic of China
haozhiyongphd@gmail.com

**Abstract.** Argumentation-based multi-agent collaborative classification is a promising paradigm for reaching agreements in distributed environments. In this paper, we advance the research by introducing a new domain ontology enriched inductive learning approach for collaborative classification, in which agents are able to constructing arguments taking into account their own domain knowledge. This paper focuses on classification rules inductive learning, and presents Arguing SATE-Prism, a domain ontology enriched approach for multi-agent collaborative classification based on argumentation. Domain ontology, in this context, is exploited for driving a paradigm shift from traditional data-centered hidden pattern mining to domain-driven actionable knowledge discovery. Preliminary experimental results show that higher classification accuracy can be achieved by exploiting ontological reasoning in argumentation based multi-agent collaborative classification. Our experiments also demonstrate that the proposed approach out-performs comparable classification paradigms in presence of instances with missing values, harnessing the advantages offered by ontological reasoning.

**Keywords:** Argumentation · Prism algorithm · Collaborative classification · Domain ontology

## 1 Introduction

Recently, argumentation has been recognized as a powerful technique used by intelligent agents for reaching agreements that harmonize the conflicts in distributed environments [6]. Furthermore, it provides a natural means of dealing with conflicts that greatly resemble the way in which humans come to a consensus [9]. Thus, Argumentation based technique are gaining increasingly interests in intelligent system research community. More recently, since the intensive research in data mining and knowledge discovery, different argumentation based system has also been proposed for multi-agent learning [8], multi-agent classification [12], and knowledge extraction [13].

Collaborative classification in distributed environments, on the other hand, is an important task for agents. The basic idea behind a collaborative classification problem is that classification can be conducted as a process whereby several agents can have different viewpoints about the category of a given case, according to their own

© Springer International Publishing Switzerland 2015
N.T. Nguyen et al. (Eds.): ACIIDS 2015, Part I, LNAI 9011, pp. 23–33, 2015.
DOI: 10.1007/978-3-319-15702-3_3

experience data collection. Our previous research [5] shows that Argumentation-based Multi-Agent Collaborative Classification (A-MACC) is an effective classification paradigm in this very kind of situation. However, certain classification problems with some samples size [15] and partially specified instances [14], such as space target classification, often generate classifiers that overfit training data. Here we explore the question that classifying agents may have access to some background domain knowledge, how can we exploit such knowledge to improve the classification performance? To do this, we propose the use of ontological reasoning in argumentation based multi-agent collaborative classification.

This paper presents and evaluates a domain ontology enriched approach for A-MACC paradigm using the 'separate and conquer' approach, called Arguing SATE-Prism. The main aim of Arguing SATE-Prism is to improve the classification performance in distributed environments, using ontological reasoning. It is noted that all the classifier agents can perform modular classification rules inductive learning with Prism algorithms [10], i.e. a modern representative of 'separate and conquer' approach for inducing classification rules. The key idea is that arguments can be generated using domain ontology enriched inductive learning algorithms dynamically. The proposed approach is evaluated empirically on a dataset from UCI Machine Learning Repository and a real dataset from space target classification domain.

The remainder of this paper is organized as follows. Section 2 briefly reviews the related background works and defines the basic terminology we use in this paper. In section 3 we present Arguing SATE-Prism, a domain ontology enriched approach. Section 4 evaluates Arguing SATE-Prism on related datasets. Section 5 contains some concluding remarks and future works.

## 2 Preliminaries

Classification is a common task in machine learning, which has been solved using various approaches [3]. Many classification algorithms have been proposed in the last decades, including decision trees, artificial neutral networks and support vector machines. This paper focus on modular classification rules inductive learning for the reason that it can be integrated with argumentation seamlessly. Generally, in a classification rules inductive learning' problem, a training examples set $E = \{e_i | e_i = (x_i, c_i), i = 1, \cdots m\}$, where $x_i$ is an element of feature space $X$, representing a training example $e_i$'s vector of attribute-value, $c_i$ is an element of concept space $C = \{c_1, \cdots, c_n\}$, representing training example $e_i$'s category. Thus, a training example $e_i = (v_{1,i}, \cdots, v_{m,i}, c_i)$, where $v_{j,i}$ is the value of $A_j (j = 1, \cdots l)$ for $x_i = (v_{1,i}, \cdots, v_{m,i})$. The hypotheses $H$ induced from the training examples are often represented in the form of a rules set. Thus, the classification process for an instance $x \in X$ is to find a rule $h \in H$, such that $h(x)=c$.

Generally, classification rule induction algorithms can be categorized into two different approaches, the 'divide and conquer' and the 'separate and conquer'. The former approach induces classification rules in the intermediate form of decision trees, whereas the latter one induces a set of modular rules directly. As already noted [5], the main drawback of the 'divide and conquer' approach is its tree structure. As a modern representative of the 'separate and conquer' approach, the original Prism algorithm was designed by Cendrowska [1] to induce directly a set of 'modular' rules that do not necessarily fit into a decision tree representation, thus avoiding the redundant terms of decision trees. Prism family of algorithms generally induce rule sets that tend to overfit less compared with decision trees algorithms, especially for noisy datasets [10]. The basic Prism algorithm generates the rules concluding each of the possible classes in turn, and each rule is generated term by term, which is in the form of attribute-value pairs. The attribute-value term added at each step is chosen to maximize the separation between the classes.

Following Fürnkranz et al's formulation [4], the features can be defined as the basic elements of rules, represented by attribute-value pairs. Thus, given a set of training examples, described by attribute values, a classification rules inductive learning system constructs a set of rules of the form:

$$\text{IF } \mathbf{f}_1 \wedge \mathbf{f}_2 \wedge \cdots \mathbf{f}_L \text{ THEN } Class = y_i$$

The *rule antecedent* is a logical conjunction of features, where a feature $\mathbf{f}_t$ ($t = 1, 2, \cdots L$) is a test that checks whether the instance to be classified has the specified property or not. The rule consequent is a class label, described by the values of classification attributes. In this attribute-value framework, a feature $\mathbf{f}_t$ typically has the form $A_j = v_{j,i}$ for nominal attributes, and $A_j > v$ or $A_j \leq v$ for numeric attributes, where $v$ is a threshold value that does not need to appear in the training examples. In this paper, we focus on nominal attributes for the reason that ontological reasoning can be easily exploited in such rules.

Given a classification rule $h$, its performance measure can be described by two metrics, *coverage* and *accuracy*. The *coverage* of a rule $h$ is:

$$coverage \ (h) = \frac{|\{e \in E | h \sqsubseteq e\}|}{|E|}$$

where $|E|$ is the total number of training examples. In this paper, $|\cdot|$ represents the total number of elements of a set. We write $h \sqsubseteq e$ when a training example $e$ is covered by a classification rule $h$.

Moreover, the *accuracy* of a rule $h$ is:

$$accuracy \ (h) \ = \frac{|\{e \in E | c(e) = cons(h) \wedge h \sqsubseteq e\}|}{|\{e \in E | h \sqsubseteq e\}|}$$

where, $c(e)$ is the example $e$'s category, and $cons(h)$ is the consequent of the rule $h$.

In the task of Multi-Agent Inductive Learning (MAIL) [8], a collection of agents with inductive learning capabilities try to learn the same hypotheses from different sets of training examples. One typical case of MAIL is distribute rule learning, where several agents firstly learn rules from different training datasets on their own, and then attempt to verify that the rules learned by each of them are in agreement with the data seen to all the agents. This paper focuses on the task of collaborative classification performed by a collection of agents with inductive learning capabilities, called Multi-Agent Collaboration Classification (MACC). The intuitive idea of MACC is to collaborative so that final classification results are in agreement by all the classifier agents. Sepecifically, MACC is defined as follows:

**Definition 1 （Multi-Agent Collaborative Classification, MACC）**
Given: a multi-agent system $\mathcal{A} = \{Ag_1, \cdots, Ag_n\}$, where each agent has an individual set of training examples, $E_1, \cdots, E_n$, an instance $x$ to be classified,
Find: for each agent $Ag_i$, a learned classification rule $h$, such that $h(x)=c$, and $h$ is consistent with all the sets of examples, $E_1, \cdots, E_n$.

We say that a classification rule $h$ is consistent with respect to a multi-agent system $\mathcal{A}$ when $h$ is consistent with all the sets of examples of the agents in $\mathcal{A}$. That is to say, it is a consistent classification result. Argumentation has been recognized as a powerful tool for conflicts resolution through multi-party arguments game. In the next section, we will present our proposed approach, called Arguing SATE-Prism.

# 3    Arguing SATE-Prism

This section presents Arguing SATE-Prism, a domain ontology enriched approach for multi-agent collaborative classification based on argumentation. The intuition behind Arguing SATE-Prism is to further improve the classification performance in A-MACC paradigm, by exploiting ontological reasoning. In particular, we propose applying Prism family of algorithms to induce 'modular' classification rules from the data on distributed sites. And the domain ontology enriched arguments used by each classifier agent are derived from each data sites. It is noted that the argumentation dialogues produced by Arguing Prism [5], can provide an efficient collaborative classification in multi-agent systems. In this section, we first present an overview of Arguing Prism, in subsection 3.1. Then, rules induction methods for generating arguments using domain ontology enriched Prism algorithms, which named SATE-Prism, are detailed in subsection 3.2.

## 3.1    Argumentation Model

Arena [13] is a dialectical analysis model for multiparty argument games to evaluate rules learned from different past instances. The model provides a novel way that can transform the multiparty arguments games into two-party argument games using ideas from the Arena Contest of Chinese Kungfu. As investigated by Yao et al [13], Arena has a capability in learning and performs well, and thus it provides a feasible way to evaluate the rules from different classifier agents for classification.

The Arena model is used here to allow any number of classifier agents to engage in a dialogue process, the aim of which is to classify a new instance collaboratively. Each classifier agent formulate arguments for one advocated classification or against the classification advocated by other agents, using 'modular' classification rules induction algorithms described in section 2.

As already stated, each participant agent has its own local repository of data in forms of data instances. These agents produce reasons for or against certain classifications by inducing rules from their own datasets using Prism family of algorithms. The antecedent of every classification rule represents a set of reasons for believing the consequent. The classification rules induction provides several different types of speech acts, which can be employed to perform the argumentation dialogues.

The Arena model used for Arguing Prism allows a number of classifier agents to argue about the classification of a new instance. Each classifier agent argues for a particular classification or against other classifications. Arguments for or against a particular classification are made with reference to an agent's own rule set induced from individual repository, in forms of data instances. Each instance consists of a set of attribute-value pairs and a single class-value pair indicating a particular classification.

Generally, an argumentation framework is composed by a finite set of arguments and an attack relation among the arguments. Let us first define the classification arguments considered by our approach:

**Definition 2 (Classification Argument)**
A Classification Argument= <*CLR, L_confidence*> is a 2-tuples, where *CLR* is a classification rule induced from the repository of an individual agent, *L_confidence* is the confidence of a rule argument following the Laplace probability estimation procedure [8]. *L_confidence* is mainly used to prevent estimations too close to 0 or 1 when very few instances are covered by inductive learning algorithm.    It is noted that the *L_confidence* should be higher than a given threshold for a legal classification argument. Further details for the arguments are described in subsection 3.2.

For the purpose of exploiting domain ontology, we devised three kinds of speech acts (also called moves) for each participant agent as follows:

(1) Proposing a classification claim. There is only one kind of proposing moves, which allows a new classification argument with its *L_confidence* higher than a given threshold to be proposed.

(2) Attacking a classification claim. Moves intended to show that a classification argument proposed by some other agent should not be considered decisive with respect to the current instance to be classified. There are three kinds of attacking moves in our system.

(3) Refining a classification claim. Moves that enable a classification rule to be refined to meet an attack. There is only one refining moves is implemented in Arguing SATE-Prism. This move allows the addition of new premises to a previously proposed classification argument so as to increase the *L_confidence* of the rule.

The arguments exchanged via the moves described above are stored in a central data structure, called *dialectical analysis tree* [13], which is maintained by the referee of Arena model. Having introduced the moves in the Arena model, the realization of these moves is detailed through the argumentation dialogue protocol in Arena. Assuming that we have a new instance to be classified, and a number of classifier agents participating in the Arena model, the argumentation dialogue protocol operates as follows:

Before the start of the dialogue, the referee of Arena model randomly selects one participate agent as a *master* to begin.

(1) At the first round, the *master* proposes a new classification argument, such that its *L_confidence* is higher than a given threshold. The referee establishes a new *dialectical analysis tree*, whose root represents the *master*'s proposing move. If the *master* fails to play an opening move, then the referee selects anther one to commence the dialogue. If all the classifier agents fail to propose an opening move, then the dialogue terminates with failure.

(2) In the second round, the other participate agents attempt to defend or attack the proposing argument, using any kinds of move described previously. If all the agents fail to play a move, the dialogue terminates, and the instance is classified according to the class prompted by the *master*. Otherwise, the *dialectical analysis tree* is updated with submitted moves.

(3) The argumentation process continues until the *master* is defeated, then another round of argumentation begins, the protocol moves to (1).

(4) If two subsequent rounds passed without any new moves being submitted to the *dialectical analysis tree*, or if numerous rounds have passed without reaching an agreement, the referee terminates the dialogue.

More details of the realization of Arguing SATE-Prism proposed in this paper will no longer be demonstrated here owning to the limitation of space. Once an argumentation dialogue has terminated, the status of the corresponding *dialectical analysis tree* will indicate the 'winning' class and its corresponding classification rule.

## 3.2    Domain Ontology Enriched Prism for Generating Arguments

Generally, ontologies represent a shared, formal understanding of a domain theory, where the term 'shared' refers to an agreement within a community of experts over the description of their domain, and 'formal' indicates the representation of this agreement in some sort of computer-understandable format. Domain ontology consists of the background knowledge that a domain expert would deploy in reasoning about a specific situation [2]. For example, a space target domain ontology is a shared and formal understanding of domain knowledge, which can be used in reasoning about a specific target, e.g. its category.

The hierarchical relations in domain ontology between concepts, which are also known as class-subClass relations in ontology, can be used to make generalization over the values of attributes while learning classification rules for constructing arguments in Arguing SATE-Prism. We focus on domain ontologies defined over values of an attribute, namely Semantic Attribute-value Tree (SAT). In what follows, we define SAT, and introduce the notions of Laplace probability of SAT nodes.

### Definition 3 (Semantic Attribute-value Tree, SAT)
A semantic attribute-value tree (SAT) associate with an nominal attribute $A$, noted as SAT($A$), is a tree rooted at $A$. All the leaves of the tree correspond to the possible primitive values of the attribute $A$. The other nodes of the tree correspond to generalized values of the attribute $A$. The arrows between nodes are 'sub Class of' relationships between the corresponding attribute values.

**Definition 4 (Laplace probability of SAT nodes)**
A Laplace probability of each non-leave SAT($A$) node $N$

$$Lp(\text{SAT}(A)_N) = \frac{t_N^C + 1}{t_N + 2}$$

Where SAT($A$)$_N$ is the non-leave node $N$ of SAT($A$) , $t_N^C$ is the total number of training instances covered by node $N$ in SAT($A$) and the target class $C$, and $t_N$ is total number of training instances covered by node $N$ in SAT($A$). We add 1 to the numerator and 2 to the denominator following the Laplace correction, which basically avoiding extreme probability when counting from very few instances.

Having introduced the speech acts in subsection 3.1, the realization of arguments for these moves using ontology enriched Prism algorithms is described in this subsection. A classifier agent can generate classification arguments using any inductive learning algorithms capable of learning modular classification rules. For the purpose of simplicity, however, we use Prism family of algorithms mentioned in section 2 in this paper.

Since rules for generating arguments are learned through inductive learning techniques, their validity may not be ensure. Thus only the classification arguments satisfying some confidence threshold are accepted for the collaborative classification task. In this paper, we use Prism classification rule learning algorithm in our domain ontology enriched approach. The proposed algorithm is called as SATE-Prism, namely SAT Enriched Prism. The main idea of SATE-Prism is to replace the values of nominal attribute with more general concepts iteratively during the bottom-up inductive learning of Prism algorithm. Specifically, SATE-Prism has at least three advantages as follows. Firstly, both Prism inductive learning algorithms and SAT constructing methods follow the bottom-up process, thus combining SAT with Prism is more naturally, comparing with decision trees [14]. Secondly, combining SAT with Prism will increase the accuracy for partially specified data. Thirdly, combining SAT with Prism will obtain a good robustness in face of instances with missing values.

---

**Step 1**: Calculate the probability $Lp$ of each target class for each leave node of SAT($A$), where the probability $Lp$ follows definition 4.

**Step 2**: Polymerization according to the established SATs, for each non-leave node with $m$ direct child node, its probability is calculated follows

$$Lp = \frac{m + \sum_i^m t_N^C(i)}{2m + \sum_i^m t_N(i)}$$

where $i$ refer to the $i$th child node.

**Step 3**: Repeat step 1 to 2 until all the node of SAT($A$) have been counted, except for the root node.

---

Fig. 1. Computing the counts based on the given SAT($A$)

SATE-Prism works bottom-up, starting at an arbitrary leave node of each SAT for nominal attribute values, and constructs modular classification rules for each mutually exclusive target classes using the most generalized attribute values. Generally, the SATE-Prism algorithm consists of the following two steps:

(1) Computing the counts for each non-root node of SAT on attribute $A$ (Figure 1)

(2) For each target class $Ci$ in turn, starting with the complete training set each time, and building modular rule classifiers based the counts obtained in (1) (Figure 2)

The method for computing the counts based on the given SAT, described in Figure 1, starting with calculating the Laplace probability for each leave node. Then following a bottom-up process, we can obtain all the Laplace probability of non-root node of a SAT.

---

**Step 1**: Calculate the probability of the target class $Ci$ for each attribute-value pair based on the counts of SAT.

**Step 2**: Select the pair with the largest probability and create a subset of the training set comprising all the examples with the selected pair.

**Step 3**: Repeat 1 and 2 for the subset until it only contains examples of class $Ci$. The induced rule is the conjunction of all the attribute-value pairs selected.

**Step 4**: Remove all the examples covered by the rule from the training set.

**Step 5:** Repeat step 1 to 4 until all the examples of class $Ci$ have been removed.

---

**Fig. 2.** Generating classification rules based the Counts of SAT

For each target class, the method described in Figure 2 also follows a bottom-up process, thus it is easy to combine these two procedures together, and develop the SATE-Prism algorithm for generating arguments in multi-agent dialogues game as described in subsection 3.1.

## 4     Evaluation

This section presents a preliminary empirical evaluation of the proposed approach. We demonstrate that it is possible to use domain ontology to improve A-MACC's classification performance. To do this, we consider two experimental scenarios. Firstly, Arguing SATE-Prism's classification performance is evaluated in terms of classification accuracy. Secondly, SATE-Prism's performance is also evaluated in terms of its tolerance to missing attribute values.

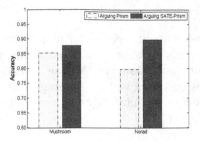

**Fig. 3.** Classification accuracy

In our experiments, we firstly use the datasets from the UCI Machine Learning Repository. The experiments reported in this paper were conducted on the commonly used datasets in ontology guided machine learning, the Mushroom Toxicology Dataset. This dataset is used for mushroom toxicology classification problems with 22 nominal attributes, and 17 of them are able to construct SATs following Taylor et al [11]. Secondly, we consider the real dataset from space target classification problem, which is the Norad catalog dataset (which comes from the North American Aerospace Defense Command).

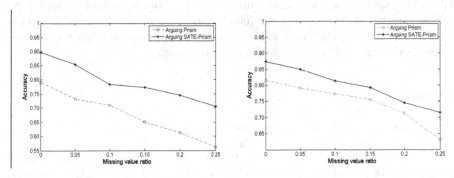

**Fig. 4.** Tolerance to missing values (the left side represents the Mushroom dataset and the right side represents the Norad dataset)

For each dataset, the classification accuracy of Arguing SATE-Prism is estimated using Ten-fold Cross-Validation (TCV), and we report the average results for different approaches. The results of our first experiments are presented in Figure 3. It is shown that SATE-Prism can increase classification accuracy on both two datasets, especially the real space target dataset Norad.

In order to explore the performance of Arguing SATE-Prism on datasets with missing attribute values, the two datasets with a pre-specified percentage (0%, 5%, 10%, 15%, 20%, 25%) of missing values were generated. For each dataset, the experiments also compare Arguing SATE-Prism algorithm with Arguing Prism. Arguing Prism adopted the estimating missing values with most frequently appeared ones, which is commonly used procedure when dealing with missing values [7]. While Arguing

SATE-Prism takes the domain ontology for benefits in learning classification rules for generating arguments.

The results of our second group of experiments are presented in Figure 4. In general, it can be observed that for both datasets the classification accuracy decreases with an increasing missing value ratio. However, Arguing SATE-Prism shows a higher tolerance to missing values compared with its Arguing Prism. This tolerance to missing values may due to generalization advantages offered by the domain ontology.

## 5    Conclusion

This work has presented Arguing SATE-Prism, a domain ontology enriched approach for multi-agent collaborative classification based on argumentation. The key idea is that domain ontology can be exploited for driving a paradigm shift from traditional data-centered hidden pattern mining to domain-driven actionable knowledge discovery. In our approach, ontological reasoning is used to aiding agents for generating arguments, and further to improve the classification performance in A-MACC paradigm. Arguing SATE-Prism has been evaluated empirically by preliminary experiments. The results showed that not only higher classification accuracy can be achieved by exploiting ontological reasoning in argumentation based multi-agent collaborative classification, but also robust classifier is able to be learned from datasets with missing attribute values. In future works, we would like to explore new frameworks for multiple classifiers which are not only learned from their own repository, but also learned from argumentation, i.e. online learning.

## References

1. Cendrowska, J.: PRISM: An algorithm for inducing modular rules. International Journal of Man-Machine Studies **27**(4), 349–370 (1987)
2. Emele, C.D., Norman, T.J., Şensoy, M., Parsons, S.: Exploiting domain knowledge in making delegation decisions. In: Cao, L., Bazzan, A.L.C., Symeonidis, A.L., Gorodetsky, V.I., Weiss, G., Yu, P.S. (eds.) ADMI 2011. LNCS, vol. 7103, pp. 117–131. Springer, Heidelberg (2012)
3. Fisch, D., et al.: So near and yet so far: New insight into properties of some well-known classifier paradigms. Information Sciences **180**(18), 3381–3401 (2010)
4. Fürnkranz, J., Gamberger, D., Lavrac, N.: Foundations of rule learning. Springer (2012)
5. Hao, Z., et al.: Arguing Prism: An Argumentation Based Approach for Collaborative Classification in Distributed Environments. Database and Expert Systems Applications. Springer International Publishing (2014)
6. Heras, S., Botti, V., Julián, V.: An ontological-based knowledge-representation formalism for case-based argumentation. Information Systems Frontiers, 1–20 (2014)
7. Li, H., et al.: An interval set model for learning rules from incomplete information table. International Journal of Approximate Reasoning **53**(1), 24–37 (2012)
8. Ontanón, S., Plaza, E.: Multiagent inductive learning: an argumentation-based approach. In: Proceedings of the 27th International Conference on Machine Learning (ICML-10) (2010)

9. Parsons, S., Sierra, C., Jennings, N.: Agents that reason and negotiate by arguing. Journal of Logic and computation **8**(3), 261–292 (1998)
10. Stahl, F., Bramer, M.: Jmax-pruning: A facility for the information theoretic pruning of modular classification rules. Knowledge-Based Systems **29**, 12–19 (2012)
11. Taylor, M.G., Stoffel, K., Hendler, J.A.: Ontology-based Induction of High Level Classification Rules. In: DMKD (1997)
12. Wardeh, M., Coenen, F., Capon, T.B.: PISA: A framework for multiagent classification using argumentation. Data & Knowledge Engineering **75**, 34–57 (2012)
13. Yao, L., et al.: Evaluating the valuable rules from different experience using multiparty argument games. In: Proceedings of the 2012 IEEE/WIC/ACM International Joint Conferences on Web Intelligence and Intelligent Agent Technology, vol. 02. IEEE Computer Society (2012)
14. Zhang, J., Honavar, V.: Learning decision tree classifiers from attribute value taxonomies and partially specified data. In: Proceedings of the 20th International Conference on Machine Learning (ICML-03) (2003)
15. Zhu, F.-Y., Qin, S.-Y.: Small-shaped space target recognition based on wavelet decomposition and support vector machine. International Conference on Wavelet Analysis and Pattern Recognition, vol. 3. IEEE (2007)

# Architecture of Desktop Presentation Tool for E-Learning Support and Problem of Visual Data Transfer over Computer Network

Michal Kökörčený and Agáta Bodnárová[✉]

Faculty of Informatics and Management, University of Hradec Králové,
Hradec Králové, Czech Republic
michal@74.cz, agata.bodnarova@uhk.cz

**Abstract.** The aim of the Desktop-projector project is to create a software tool (application) for more effective teaching in computer classrooms, especially in case where a classic projector cannot be used. The Desktop-projector is typically used for showing presentations or other activities carried out on the teacher's desktop on other desktops. The teacher's desktop is broadcasted to other computers on a local area network. The article describes software architecture of the Desktop-projector solution and discusses problems with visual data transfer over computer network, such as UDP datagram loss and sufficient performance (speed) of the transfer. After considering the results from this article we have obtained the optimal configuration values required for proper setting of the application's parameters, ensuring good enough quality of the Desktop-projector solution.

**Keywords.** Desktop · Projector · UDP · Data loss · Performance · Socket · Buffer

## 1    Introduction

The aim of the Desktop-projector project is to create software tool (application) for more effective teaching in computer classrooms, especially in cases where there is not a classroom or portable data projector, or where it is needed to transfer images from different computers that cannot be operatively connected to the data projector. Furthermore, the experiment described in the paper [8] shows that the desktop broadcasting software was better if there was a large amount of data on screen and, on the other hand, the classic data projector was better if there was a small amount of data on screen. In this experiment cognitive effects of a classic data projector were compared with an application for desktop broadcasting in a computer laboratory [8]. The quantitative results of cognitive effects on users shows that, in certain cases, desktop broadcasting software provides more effective teaching and learning compared with other tools [8].

This article presents a desktop-sharing tool, the Desktop-projector, which was created for the needs of our educational organization. The desktop-projector is a program for broadcast (transfer) of the desktop (screen) of one computer to other

© Springer International Publishing Switzerland 2015
N.T. Nguyen et al. (Eds.): ACIIDS 2015, Part I, LNAI 9011, pp. 34–45, 2015.
DOI: 10.1007/978-3-319-15702-3_4

computers on the local area network (LAN). Desktop Projector is typically used for showing presentations or activities carried out with one computer to other computers (such as showing the teacher's computer desktop to students). This article describes the basic principles of the solution, the problems encountered during the research, and their possible solutions.

## 2 Comparing with Other Architectures

Before the development of the Desktop-projector we tried to find and test other, already existing software products. Several solutions are available, both commercial/proprietal and free, typically open source products. In general, these applications are based on a few architecture types and principles. Nevertheless each architecture has significant disadvantages and limitations, which led to research and development of our own software solution. These architectures are hereinafter described and compared with our requirements.

- Architecture 1: One receiver and many transmitters

The first architecture uses one computer in the role of receiver and many computers in the role of transmitter – but at the same time the only one transmitter can be active. Therefore it is always a one-to-one connection. This is suitable in some cases, where you have a classic data projector connected to one computer in a classroom – then you can project on a screen or on any other computer over a local area network and this data projector.

This solution of desktop broadcasting system does not meet our requirements – contrary to this architecture, we require many receivers and one transmitter, and data transfer one-to-many. Furthermore, in case of one-to-one data transfer it is possible to use a lossless TCP connection. Thus the entire solution is relatively simple; in this case we no need to solve problems described in this article, such as datagram loss, optimal size of screen block, throughput performance etc.

Conclusion: This type of architecture is not appropriate and applicable to our needs.

- Architecture 2: Periodic capture and transfer of whole screen

Another architecture used in the existing solutions is based on the very simple principle – a screenshot is taken periodically (in a defined time period), then the whole picture (whole screen) is compressed and periodically transmitted over a local area network. The disadvantage of this solution is obvious – heavy CPU load, heavy network load, low frame rate (FPS – frames per second) and low throughput performance. This architecture has relatively good results with low screen resolutions, e.g. 640x480 or 800x600 pixels, where the size of each frame (in bytes) is rather small. Nevertheless this is very inefficient solution for high screen resolutions, e.g. 1680x1050 or 1920x1080 pixels or higher, because of the data size rises exponentially according to the screen resolution. Therefore most of these applications have limitation for maximum supported screen resolution, typically 800x600 pixels etc.

Conclusion: This is a simple architecture, which has poor results for high screen resolution. Therefore it is not suitable for our needs – nowadays high screen resolution is standard – this architecture is very inefficient.

- Architecture 3: VNC based solutions

There are several desktop broadcasting systems based on VNC (Virtual Network Computing) software tools. This architecture uses special techniques to capture changed parts of screen (not whole screen). For these purposes operating system hooks are used which catch relevant operating system messages related to screen updates (not every messages). Then, the coordinates and size of changed part are getting from the message and consequently captured from the screen. This is a very efficient solution, because only changed parts are detected and don't need to be captured periodically – these are captured only when a message is received, i.e. when screen is updated.

On the other hand, there are certain disadvantages to such architecture. The main problem is how to catch all relevant messages so that no screen update has been omitted (i.e. has not been detected). VNC tools relies on undocumented messages, there is not ensured compatibility with various operating system versions etc. Additionally, on Windows platform these messages are related to the standard GDI interface – using this technique we cannot capture various graphics effects, such as anti-aliasing. This can be accomplished using DirectX only, by means of front buffer capture. The next problem of the VNC approach is that these applications are primarily designed for one-to-one communication, typically via TCP connection. In case of many receivers, multiple simultaneous TCP connections exist at the same time. It produces heavy network load and low throughput performance.

Conclusion: This is very efficient solution, which has several disadvantages and low performance in one-to-many communication. Existing VNC tools have not solved problems described in this article. VNC tools were created for different purposes.

- Other solutions

The only similar software that has the required features and functionality is the TightProjector. This application is described and available at [12]. Nevertheless detailed comparison with our application Desktop-projector was not possible. TightProjector software is based on other principles; it contains a completely different configuration in relation to the problems covered in this article. TightProjector is a commercial application, source code is not available and we cannot compare its architecture with our solution. Additionally, the application TightProjector was not functional in the test environment in which our application, the Desktop-projector, was tested and on which performance tests were carried out as it is described in the following chapters.

Conclusion: TightProjector [12] is the only one desktop broadcasting system which has the required features and functionality. Nevertheless direct comparison is not possible due to slightly different principles and configuration, unavailability of source code and problems with its functionality.

# 3    Problem Definition

The basic requirement is the ability of projection of one computer's desktop (screen) to desktops (screens) of other computers. Visual information is transmitted via LAN. One computer can be in the role of the visual information transmitter. The role of receivers is not limited; it is available for several PCs in the same LAN. Usually the number of simultaneous receivers is under forty. The real-time transmission of continuous video signal (streaming) is not required. The solution is based on the need of capture and transmission of the transmitter's desktop to other computers desktops on the local network – this difference significantly affects the concept of the solution and the resulting software architecture.

The aim is to achieve as fast transmission as possible of visual information from the transmitting computer and then display on the receiving computers screens. The entire transfer of visual information, respectively the transfer of the changes in this visual information, is not necessarily in real-time mode (it would be very CPU consuming and bandwidth consuming of the transmission channel), but be sufficient to meet the basic requirements for application carried out during the presentation of the activities on the computer. Sufficient speed can be regarded as our empirical experience showed, several frames per second (e.g., 3 to 5 frames per second).

During the work on this project we were faced with several complications. The most serious problem is the occasional UDP datagrams loss during the transmission of visual information. During the transfer of the image of the computer's desktop screen some blocks are  not processed and remain in the old state – without redrawing. From the user perspective it is very unsuitable. The number of lost datagrams in this case is a clear metric for quality evaluation of the solution. The aim is to identify the cause of the UDP datagram loss – which is quite difficult – and then make adjustments in the application to eliminate the problem or at least reduce it. Generally, the datagram loss problem may be caused both in the transmitter or receiver site and at different levels: at the network level, application layer level or at the level of operating system.

The goal of our work and this article is to introduce the software architecture of Desktop-projector solution to obtain the optimal configuration values required for proper setting of the application's parameters, ensuring good quality of the solution. The main criteria are: minimum datagram loss (minimum of lost screen blocks), maximum data throughput (speed of visual data transfer) and minimum CPU consumption (primarily on the transmitter side).

# 4    Description of Current Solution

The basic principles of the Desktop-projector application, its problems, and possible solutions were described in our paper [4]. In this article we will focus on introduction of software architecture and presentation of results of extended and deeper performance testing. The current version of Desktop-projector is designed as a standalone application on a Win32 platform in Delphi XE Professional. However the choice of the technology does not significantly affect the problem of implementation of this project, which are described in this article. The overall architecture of the Desktop-projector solution is shown in Fig. 1.

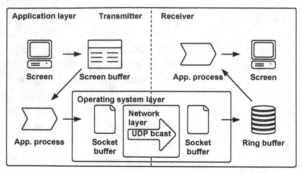

**Fig. 1.** The overall architecture of the solution

To increase the efficiency of the application process of the broadcast transmission, the main process has been divided into three separated threads.

The first thread (grabbing thread) provides full screen capture and detection of changed blocks. The entire screen was divided into equal blocks – actually tested block sizes from 12x12 pixels to 48x48 pixels. The grabbing process is implemented as follows: get full-screen image, divide the screen into blocks, and detect changed blocks. The results of the impact of different block size screen on the transmission quality are described in the next chapter. Visual data are stored in two screen buffers – the first buffer contains the last screenshot, the second buffer contains last but one screenshot. The results of comparison of these buffers – changed screen blocks – are stored in a state buffer. Only changed blocks, not the entire screen, are transferred. This solution is effective in case of bandwidth consumption on the local network. The disadvantage of the solution is higher CPU consumption of the transmitter due to the need of continuously changes detection.

The second thread (transmitting thread) provides visual data compression and sending via the UDP protocol. This process is implemented as follows: get information about changed screen blocks from the state buffer, get changed blocks from the first screen buffer, compress visual data (currently RLE algorithm is used, we plan to use ZLIB compression), and transmit changed blocks using UDP (User Datagram Protocol) datagrams. There are processed only these screen blocks, which are in the state buffer marked as changed. Both operations – screen capture and detection of changed blocks; and visual data compression and sending – are time consuming, so the division into separated threads allows much better usage of resources. The operations do not block each other waiting for their completion.

An important part of the architecture is the continual refresh process, which is implemented by a separated thread. This process ensures refreshing (re-sending) one entire row of screen blocks in a defined time period. Thus the entire screen is fully refreshed (row by row) within approximately 3 seconds. This solution is necessary because of UDP broadcasting was chosen as a data transfer mechanism. It is a one way data transport – there is no backward communication from receivers to transmitter. In case the receiver is started later, during the transmission, this receiver do not have full image information and cannot repaint the whole screen. Therefore the continual refresh process (thread) provides full repaint within relatively short time period, without the necessity of backward communication. Another important function of the continual refresh process (thread) is elimination of lost datagrams, where some blocks are not processed and remain in the old state (without redraw). In this way lost screen blocks are re-drawn in a short time period. Occasional UDP datagrams loss during the transmission of visual information is the most serious problem of the solution.

The process of receiving is relatively trivial, there are several consecutive steps: receiving of UDP datagram, storing and retrieving from ring buffer, decompression of screen block and block displaying (drawing).

In comparison with the complexity of the transmitter and receiver module – receiver has very simple architecture, the goal is to ensure minimum datagram loss (minimum of lost screen blocks), there are high performance requirements – on the other hand, transmitter has more complex architecture, the goal is to ensure maximum data throughput (speed of visual data transfer) and minimum CPU consumption. Identification of the cause(s) of UDP datagram loss is highly problematic, even after several months of development we failed to find a final and reliable solution. To solve the previously defined problem we have performed several tests described in the next chapter.

## 5     Performance Testing

This article follows the results published in our paper [4]. Based on these published results we have performed extended and deeper performance testing and optimization of parameters of the Desktop-projector application – the results are described hereinafter. The goal of the extended performance testing is to obtain the optimal configuration values required for proper setting of the application's parameters, especially to determine the optimal screen block size regard to the UDP datagrams loss and performance (speed) of visual data transmission between the transmitter and receiver.

For these measurements 100 Mbps Ethernet network with a simple L2 switch was used. The computers in the following configuration were used: CPU 2.4 GHz, 4 GB RAM, screen resolution of 1680x1050 pixels and OS Windows XP / Vista. The tests were used for different screen block sizes: 12x12, 16x16, 24x24, 32x32, 40x40 and 48x48 pixels. The measurements were performed on application level of Desktop-projector receivers (not transmitter).

The first of the tests was focused on the UDP datagram loss detection dependence on the screen block size. The results of the experiment are shown in Fig. 2. As opposed to conditions and results published in our paper [4] we have performed this experiment in two alternatives: standard UDP transfer, where larger datagrams are fragmented into several Ethernet frames according to the Maximum Transmission Unit (MTU); and mode, where UDP datagrams are not fragmented – whole UDP datagram is transferred as one Ethernet frame, which can be greater than MTU [5]. This mode is turned on by setting the DF (Do not Fragment) bit [11] on network socket. See Winsock API for more details.

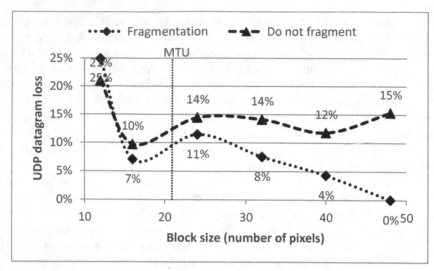

**Fig. 2.** UDP datagram loss dependence on the screen block size

As already mentioned, in contrast to the results listed in articles [2] and [6], the UDP datagrams loss rate is reduced by increasing the size of IP datagrams. An interesting jump was seen between the opposite trend of loss between the values of 16x16 pixels (7 % / 10 %) and 24x24 pixels (11 % / 14 %). This behavior was confirmed by repeated measurement. The size of a datagram in case of 16x16 pixel large block is 768 B, in case of 24x24 pixel block is 1728 B (without compression). It is because the 24x24 pixel large blocks datagram oversize the MTU. The Maximum Transmission Unit (MTU) is the maximum length of data that can be transmitted by a protocol in one instance. For example, the MTU of Ethernet (by default 1500 B) is the largest number of bytes that can be carried by an Ethernet frame (excluding the header and trailer) out [10]. The 24x24 pixel large block's datagram is divided into two packets, causing a greater loss of packets at the network layer. However, by expanding the screen blocks, there is a gradual downtrend in the rate of loss – it confirms the hypothesis that the loss of datagrams occurs at the application layer rather than at the network layer.

A very important comparison is between fragment and do not fragment transfer mode (DF bit). For 12x12 and 16x16 pixel large blocks, where size of a datagram is lesser than MTU, are results similar – there is no significant difference, whole datagram is always transferred within one Ethernet frame. Nevertheless, for 24x24 pixels and larger blocks, where a datagram is greater than MTU, it produces noticeable different results. As it is shown in Fig. 2, do not fragment transport mode (DF bif) gives worse loss rate than standard mode – the greater is the screen block the higher (worse) is the loss rate. It is because the larger Ethernet frames take more time to transfer and thus there is a higher probability of conflict with other frames on the network layer, which causes the loss of UDP datagram. Therefore, from the point of view of measurement of datagram loss, it is preferred to use standard transport mode instead of do not fragment mode (DF bit).

The second test was focused on the dependence of the speed of visual data transmission on the screens block size. Results of the experiments are shown in Fig. 3. According to the assumptions found in articles [2] and [6], the data throughput between the transmitter and receiver increases with the increasing size of transmitted datagrams.

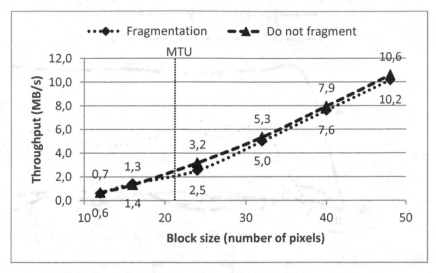

**Fig. 3.** Data transfer performance dependence on the screen block size

Based on the results of performance tests it can be considered more appropriate to use larger values of screen block size. On the other hand, larger screens blocks load more traffic, because in small change (such as a change of one pixel) more data is required to transfer. Therefore it is necessary to choose a value that would ensure a sufficiently small datagram loss and sufficiently large data throughput. Based on the results of this test and results published in our paper [4], the optimal value appears to 40x40 pixels.

Comparison of fragment and do not fragment transfer (DF bit) modes shows, that there is no significant difference between these transport modes. Respectively, do not fragment transfer mode (DF bit) produces slightly better performance (data throughput) than standard mode for 24x24 pixels and larger blocks; while for 12x12 and 16x16 pixel large blocks, where size of a datagram is lesser than MTU, are results the same. Considering of results of this test and previous test and chosen 40x40 pixel block size leads to the conclusion that standard transfer mode is preferred instead of using do not fragment mode (DF bit). Do not fragment mode produces only slightly better performance (data throughput) but significantly worse UDP datagram loss.

Next performed tests investigated the influence of sender's socket buffer size and receiver's socket buffer size on UDP datagram loss and data throughput. For these tests the standard transfer mode was used (instead of do not fragment mode).

In our paper [4] we stated that the size of the sender's socket buffer has minimum influence on the results. Contrary to this statement, the extended performance testing proved that the best results (minimum UDP datagram loss) are achieved when sender's socket buffer size is set to value 0 kB. In our prior work presented in the paper [4] we tested sender's socket buffer size in the range 16 kB – 256 kB, which is the cause of wrong assumptions. In these tests we have used the range 0 kB – 256 kB. Results of the experiment are shown in Fig. 4 and Fig. 5.

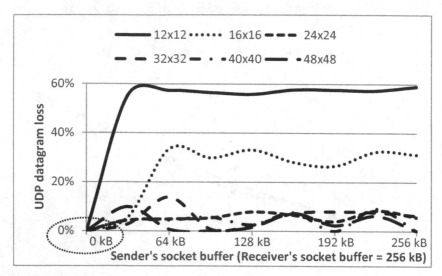

**Fig. 4.** UDP datagram loss dependence on the sender's socket buffer size

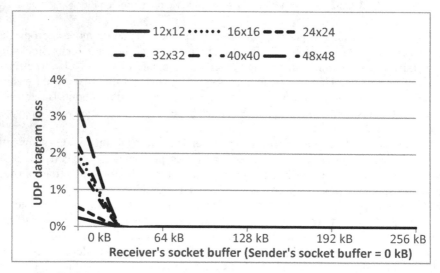

**Fig. 5.** UDP datagram loss dependence on the receiver's socket buffer size

The zero data loss has always been reached when sender's socket buffer size was equal to 0 kB and the receiver's socket buffer size was greater than 0 kB. When sender's socket buffer size is equal to 0 kB, the transmitting thread is blocked (waiting) until all data are sent to the network; when sender's socket buffer size is greater than 0 kB, data are placed into memory and sent separately – the transmitting thread is not blocked (waiting) for the operation completion. Therefore 0 kB sender's socket buffer size causes small delay between two sent datagrams (because of transmitting thread blocking), which gives the receivers more time to incoming datagrams processing – it causes lower data loss on the application level of the receiver.

Fig. 6 shows UDP datagram loss dependency on receiver's socket buffer size when sender's socket buffer size was set to value 32 kB. In this case we obtained worse results and higher data loss than for value 0 kB (compare Fig. 6 vs. Fig. 5). You can see similar results for greater values of sender's socket buffer size (64 kB, 96 kB, 128 kB etc.).

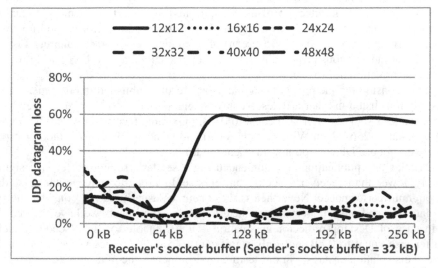

**Fig. 6.** UDP datagram loss dependence on the receiver's socket buffer size

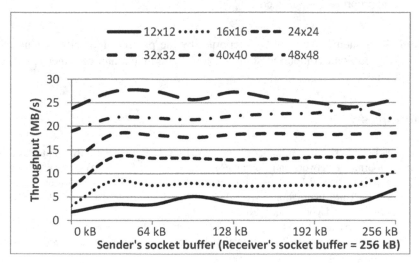

**Fig. 7.** Data transfer performance dependence on the sender's socket buffer size

On the other hand, data throughput performance, as it is shown in Fig. 7, does not significantly depend on the sender's socket buffer size. Respectively, data throughput is slightly lower (worse) when sender's socket buffer size is equal to 0 kB and higher (better) when sender's socket buffer size is greater than 0 kB. These results confirm the assumptions found in articles [2] [6] [3].

## 6      Conclusion

In this article we have introduced an e-learning tool – the Desktop-projector – and we have described the architecture of the solution and discussed problems of visual information transfer over the LAN using the UDP. Desktop-projector is used for transmission (transfer) of the desktop (screen) of one computer to other computers on the LAN. The most serious problem is the occasional UDP datagram loss during the fast visual information transmission.

After considering the results from this paper, results published in our paper [4] and assumptions listed in other articles, we have decided to use: 40x40 pixel screen block size, sender's socket buffer size equal to 0 kB, maximum receiver's socket buffer size (it is equal to 256 kB on Windows platform) and standard IP transfer mode (respectively do not use DF bit – do not fragment mode).

Articles and publications recommend using a sender's socket buffer size significantly lower than receiver's socket buffer size In order to achieve the best data throughput performance. Nevertheless these sources usually do not consider the rate of data loss and other important parameters. In our experiments we have showed that – in case of Desktop-projector application – it is suitable to use a sender's socket buffer size equal to 0 kB, not only significantly lower than receiver's socket buffer size. This solution ensures – in our testing environment – the minimum (or zero) UDP datagram loss and relatively high data throughput (not maximal throughput), which is the most appropriate for our needs.

**Acknowledgement.** The paper is supported by the project of specific science Smart networking & cloud computing solutions and Economical and managerial aspects in Biomedicine.

## References

1. Gamess, E., Surós, R.: An upper bound model for TCP and UDP throughput in IPv4 and IPv6. Journal of Network and Computer Applications Archive **31**(4), 585–602 (2008)
2. Gu, Y., Grossman, R.L.: Optimizing UDP-based protocol implementations. In: Proceedings of the Third International Workshop on Protocols for Fast Long-Distance Networks (PFLDnet 2005), Lyon, France (2005)
3. Gu, Y., Grossman, R.L.: Using UDP for Reliable Data Transfer over High Bandwidth-Delay Product Networks. The International Journal of Computer and Telecommunications Networking Archive **51**(7), 1777–1799 (2007)

4. Kokorcseny, M., Bodnarova, A.: Desktop presentation tool for e-learning support and the high-speed UDP datagram loss. In: Proceedings of the International Conference on E-Learning and E-Technologies in Education (ICEEE 2012), Lodz, Poland, pp. 21–25 (2012)
5. Lee, J.Y.: Optimum UDP packet sizes, in ad hoc networks. IEICE Trans. Commun., Japan (2002)
6. Sawashima, H., et al.: Characteristics of UDP packet loss: effect of TCP traffic. In: Proceedings of the INET 1997, Engineering 3-1 (1997)
7. Xylomenos, G.: TCP and UDP performance over a wireless LAN. In: Proceedings of the INFOCOM 1999 – Eighteenth Annual Joint Conference of the IEEE Computer and Communications Societies, New York, USA, (1999)
8. Yamanoue, T., et al.: Comparing a video projector and an inter-PC screen broadcasting system in a computer laboratory. In: Proceedings of the 38th Annual Fall Conference on SIGUCCS, Norfolk, Virginia, USA, pp. 229–234 (2010)
9. Internet Direct (Indy), TIdUDPServer.OnUDPRead.
http://www.indyproject.org/docsite/html/frames.html?frmname=topic&frmfile=index.html
10. MTU manipulation. http://packetlife.net/blog/2008/nov/5/mtu-manipulation/
11. Setsockopt function: Microsoft (2012).
http://msdn.microsoft.com/enus/library/windows/desktop/ms740476(v=vs.85).asp
12. TightProjector Software. http://www.tightvnc.com/projector/

# The Comparison of Creating Homogeneous and Heterogeneous Collaborative Learning Groups in Intelligent Tutoring Systems

Jarosław Bernacki and Adrianna Kozierkiewicz-Hetmańska[✉]

Department of Information Systems, Wroclaw University of Technology, Wroclaw, Poland
{jaroslaw.bernacki,adrianna.kozierkiewicz}@pwr.edu.pl

**Abstract.** Collaborative learning is a method of gaining knowledge in groups. This method is often used in Intelligent Tutoring Systems (ITS). ITS can adapt learning process to students' abilities, learning styles or preferences. Moreover, ITS allows to create collaborative learning groups of students. Such groups could be homogeneous or heterogeneous. It is often said that heterogeneity in groups improves learning effects [3,8]. In this paper an original algorithm for creating heterogeneous groups is proposed. Results of heterogeneous and homogeneous groups were compared and research has shown that students working in heterogeneous groups achieved better results than students in homogeneous groups. It points out that to suitable assign students to groups is a very important matter.

## 1 Introduction

*"Collaborative learning is a method of teaching and learning in which students team together to explore a significant question or create a meaningful project"* [9]. Many teachers and educationalists emphasize the benefits of group learning called also team learning . The main advantage of collaborative learning includes interpersonal development. Students learn to relate to their peers, to communicate with others. It is helpful for pupils with low social skills. Students are actively involved in learning process. Each learner has opportunity to give his response, exchange ideas and get feedback from another members of team. The team work is helpful to find creative solution which reflects a wide range of students' perspectives. Additionally, cooperative learning positively influences student's motivation and improves self-esteem [8]. Our previous research pointed out that learners working in groups (generated by the proposed algorithm) achieved better results than those, who were working alone [4].

The growth of popularity of the Internet and web-based systems caused the growth of popularity of collaborative learning not only in traditional in class learning but also in distance learning. The aforementioned benefits of collaborative learning are the reason for searching for the more effective strategy for collaboration learning. The collaborative learning requires creating proper working teams and recommending the proper learning materials and tasks. The first problem focuses on creating proper

© Springer International Publishing Switzerland 2015
N.T. Nguyen et al. (Eds.): ACIIDS 2015, Part I, LNAI 9011, pp. 46–55, 2015.
DOI: 10.1007/978-3-319-15702-3_5

working group. The questions which appear: how many members should a learning group contain? How different or similar its members should be? What kind of groups is better: heterogeneous or homogeneous groups? How to create learning groups? The problem of recommendation of learning material suitable for students based on their characteristics, needs are preferences was presented in work [14]. This paper focuses on a method which creates an effective collaborative learning group. In our previous work [4] we assumed that student in intelligent tutoring system is represented by a set of user's data which contains information about student's demographic data, abilities, personal character traits, interests and learning styles (according to Felder and Silverman model). Next, based on the learning styles, system decides whether student prefers working in groups or working alone. All students which are suitable for working in groups are matched to the appropriate group. In this work, we describe two methods: for creating homogeneous and heterogeneous groups. Our approach will be tested in specially implemented environment and finally, we will try to answer which strategy of creating learning groups is better.

The rest of the paper is organized as follows: Section 2 contains a short overview of methods and strategies for creating learning groups and obtained results. In Section 3 we propose our strategy for creating homogeneous and heterogeneous group. In Section 4 the results of comparison strategy for creating homogeneous and heterogeneous group are presented. Section 4 contains also statistical analysis of obtained results. Section 5 concludes this work.

## 2    Related Work

Cooperative learning is working together in a group in order to achieve particular goals [10]. In [8] it is showed that cooperative learning has many advantages. It teaches students that knowledge should be shared with other team-mates or it is important to discuss with each other when students present different opinions. The most effective work is when team members work well together in order to achieve their goals [15,18]. It is essential that they must attentive listen to each other and have clear roles and tasks. Cooperative learning can be based on students that are mixed-ability [8], however it could be difficult to provide a proper environment for each learner [18]. Very often it is assumed that a group should be created by students with similar abilities [4,7] learning styles [11] or personalities [14].

There are many approaches in setting the optimal number of group members. Usually, it is recommended that group should be as small as possible [10]. In [6] it is said, that group should consist of 4-5 people. In [1,14] there is proposed, that effective collaborative learning can be done in groups of two people (which is also called "peer learning"). Researches show, that peer learning is very effective method of gaining knowledge [4,16].

Researchers present many approaches for creating groups. In general, they are divided into homogeneous and heterogeneous groups.

A homogeneous group is a kind of group, where participants are selected in a way, that they differ from each other as little as possible. Generally the differences are demographic, such as: gender, age, nationality. Usually, if students create group by themselves, they form a homogeneous group [8].

According to [8], groups should be heterogeneous in view of e.g. nationality, ethnicity or gender. Heterogeneity can increase learning effects in collaborative learning [3]. High-ability students can explain material to low-ability mates, so they can remember it longer.

However, there are some disadvantages of forming heterogeneous groups, for example high-ability students can have difficulties in explaining the material, so low-ability learners will not want to listen them. Moreover, students explaining material do not bother, whether other mates understood the explanation. In effect, learners have negative attitude to each other [2,10]. Another problem is that when in a group there are high-ability, medium-ability and low-ability learners, they can interact as teacher-student excluding medium-ability mates [17].

In [3] grouping was realized with Ant Colony Optimization algorithm in order to maximize heterogeneity of members in group.

Approach in the work [15] is based on Felder / Silverman model of learning styles. Heterogeneity is created as connection of students with opposite learning styles, e.g. active and reflective.

In [2] there are created homogeneous and heterogeneous groups. Before grouping, each student was asked to answer questions in a prepared quiz. Then, students were divided into three populations: high-achieving students (those, who had the best results in the quiz), average-achieving students and low-achieving students. Homogeneous groups were created as follows: two high-achievers, two average-achievers and two low-achievers. Heterogeneous groups were generated with one high-achiever, one average-achiever, one low-achiever and one "random" student who could be from any group. In experiment, homogeneously generated groups gained better results.

## 3      Strategies for Creating Collaborative Learning Groups

Our main aim is to create effective homogeneous and heterogeneous learning groups. The general idea of the proposed algorithm is to check whether students are suitable for working in groups by analysing their learning styles (based on Felder and Silverman model [11]). Felder and Silverman proposed model where the learner's behaviour is considered in four dimensions: perception, reception, processing and understanding. Every dimension is bipolar:

- processing: *active* or *reflective*
- perception: *sensitive* or *intuitive*
- receiving: *visual* or *verbal*
- understanding: *sequential* or *global*

For creating collaborative learning groups we consider only processing dimension in order to determine if student is suitable for working in groups. Based on the Index of Learning Styles (ILS) [13], intelligent tutoring system decides that students is suitable (or not) for learning in groups. The active students are offered  cooperation with others students, whereas reflective students are recommended for working alone. Other dimensions of ILS are not useful in creating learning groups in our approach.

Personal traits can be determined with the use of a typology proposed by Carl Gustav Jung in 1920 [19]. Jung observed that people have certain preferences for "management" with their own energy, regeneration, methods of gathering information or making decisions. In this way, he described 4 dimensions, where each dimension consists of 2 bipolar values:

- Dimension E-I: **E**xtroversion / **I**ntroversion
- Dimension S-N: **S**ensing / i**N**tuition
- Dimension T-F: **T**hinking / **F**eeling
- Dimension J-P: **J**udging / **P**erceiving

Based on above dimensions, there are $2^4 = 16$ personality types. Myers-Briggs Type Indicator (MBTI) can be used for determining Jung's typology. This test is usually constructed as a questionnaire where user indicates 1 of 2 values in each of above dimensions. MBTI is also often used to determine which profession matches the given personality type [19].

If student registers to the intelligent tutoring system and fills some psychological questionnaires, then system creates a learner profile. In this paper we assumed that the learner's profile is represented as a tuple of values defined as follows:

$$t : A \rightarrow V ,$$

where:

$A$ - finite set of profile attributes, $V$ - attribute values, $V = \bigcup_{a \in A} V_a$ , $\forall_{a \in A} (t(a) \in V_a)$.

In our approach, we assumed that learner's profile contains student's learning style (only part concerns on processing dimension), student's behavior measured by the Myers-Briggs Type Indicator, some demographic data which allows to distinguish students in the intelligent tutoring system, the initial knowledge level assessed by student oneself and final knowledge level assessed by a proper test. The content of the learner's profile is presented in Table 1.

**Table 1.** The content of the learner profile

| Attribute name | Attribute domain |
| --- | --- |
| Login | sequence of symbols |
| Password | sequence of symbols |
| Processing | {active, reflective}, |
| General attitude (GA) | {extroversion, introversion} |
| Information receiving (IR) | {sensing, intuition} |
| Decision making (DM) | {thinking, feeling} |
| Relationship with outside world (R) | {judging, perceiving} |
| Knowledge level (KL) | {1,2,3} |
| Result of the final test | [0%,100%] |

The students profiles are the basis of creating heterogeneous and homogeneous groups described in the next Subsections. When students are matched into a proper group, they are offered the learning material. The problem of recommendation of learning material for students needs, preferences and learning styles were described in [14]. In Intelligent tutoring systems students have possibility to use a tool such as "chat", that make it possible to contact each other and allow to discuss about presented learning materials and tasks [4].

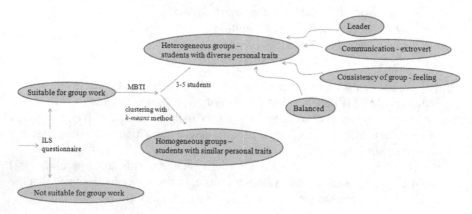

**Fig. 1.** The general idea of method for creating collaborative learning groups

## 3.1     Creating Homogeneous Groups by Using *k-means* Algorithm

Many authors underline that similar people learn in the similar way and should be working together [9,12,14,16]. Based on those assumption we propose strategy for creating homogeneous groups by using *k-means* algorithm [4].

Groups are created as clusters which consist of students with similar characteristics (the similar personality type and knowledge level). In the first step $k$ centroids are chosen in random way and then every object from database is assigned to the closest cluster by using a distance function. In our case, object (student) is described by learner's profile. The distance between learner's profiles could be calculated in the following way:

$$d(t_1,t_2) = \sum_{i=1}^{5} \delta_a \qquad (1)$$

where: $\delta_a = \begin{cases} 0 & if \ t_1(a)=t_2(a) \\ 1 & otherwise \end{cases}$ ; $a \in \{GA, IR, DM, R, KL\}$

The new centroids are recalculated as the mean location of the values that are assigned to this cluster. This steps are repeated until the positions of centroids are not changing. The final clusters contain students characterized by the similar personal traits and abilities. Students belonging to the same cluster, are paired/grouped in the random way and start to learn in group.

## 3.2    Creating Heterogeneous Groups

In [5] authors chose the critical factors for effective teams. Authors claim that a proper chosen leadership, communication, cohesion and the proper balanced heterogeneity of group, influences on a team productivity. Based on Jung's theory of psychological types (described in Section 3) some types of people are natural leaders [5]. Persons who check out as a leader are described by the following composition of personal character traits (where the first type of person has the strongest potential for leadership and the last type the weakest) [20]:

1) extroversion, sensing, thinking, judging
2) extroversion, intuition, thinking, judging
3) introversion, intuition, feeling, judging
4) introversion, sensing, feeling, judging
5) introversion, sensing, thinking, judging

We used this information for creating effective learning groups. If we assumed that in intelligent tutoring system $n$ students have been registered then the number of groups $k$ is computed as the integer part of division of number of students and 4: $k=[n/4]$. The first step of our procedure is to choose the $k$ leaders of $k$ groups among all students registered in intelligent tutoring systems. The leaders are chosen by comparing the user's profile with the list of traits described above. If in our database is no enough students suitable for leader's role the rest of leaders are chosen as the closest to any of five above types. In the second step, for each leader a student is assigned who is the furthest to him (by using distance function defined in Section 3.1). Next, to each group the subsequent student is assigned. Student is assigned to a concrete group from database, if satisfied is the following condition (student who is the furthest from all members of this group):

$$\sum_{i=1}^{s} d(t_{\max}, t_i) = \max_{j \in \{1,\dots,n\}} \sum_{i=1}^{s} d(t_j, t_i) \tag{2}$$

This step is repeated until all groups are not smaller than 4. In this way we obtain $k$ groups and each of them contains exactly 4 members. Each student which is not assigned yet is tested to which group is the furthest and next assigned to this group as the fifth member.

The main aim of proposed algorithm for creating heterogeneous group is to obtain well-balanced groups. We tested our algorithm in specially implemented environment. We assumed that only personal traits are considered. We chose randomly 400 learners' profiles. In our database were 210 people characterized as extrovert and 190 as introvert, 189 with attribute value equal sensing and 211 with attribute value equal intuition, 216 with attribute value equal thinking and 184 with feeling, 197 judging persons and 203 perceiving persons.

From created learners' profiles, 50 leaders were chosen and for each leader three members were joined according to described procedure. The dimensions of Jung's Theory are bipolar, so we check that in each created group the average number of person characterized as extroverts is equal 2.1, sensing persons is equal 1.89, thinking students is equal 2.16 and judging people is equal 1.97.

In the first step we tested if samples come from normal distribution. The results of Lilliefors test (for significance level equal $\alpha = 0.05$) allow to reject the hypothesis about the normality of the distribution of analyzed features. For the further analysis the Wilcoxon (signed-rank) test was used. The results of statistical analysis are presented in Table 2.

**Table 2.** The results of the Wilcoxon test

| Sample | Statistical test value | p-value |
|--------|------------------------|---------|
| Extrovert | 2.073897 | 0.038089 |
| Sensing | 1.512613 | 0.130378 |
| Thinking | 1.220215 | 0.222383 |
| Judging | 2.536132 | 0.011208 |

If we assume the significance level equals $\alpha = 0.01$ then we cannot reject hypothesis that median equal is 2.1, 1.89, 2.16 and 1.97 for sample extrovert, sensing, thinking and judging, respectively. It means, that each group is consists of members with well-balanced personal character traits. It confirmed, that our algorithm works properly.

## 4      Experimental Results

The main goal of our experiment is to check if heterogeneous groups achieve better results in learning than homogeneous groups. The experiment was conducted on a group of 82 people (students) who used specially implemented prototype of intelligent tutoring system. This prototype is a web application, written in PHP with MySQL database. Each student must register in the system and give some essential data, such as login, password and some demographical data. Then, students log into system and fill an ILS Questionnaire and MBTI test, which results are stored in our database. Next, system generates heterogeneous or homogeneous teams. Students are divided into heterogeneous and homogeneous populations in a random way. When students are grouped (according to procedures described in Section 3.1 and 3.2) , they may contact each other, using a tool named "chat". They all also have been presented with teaching material concerning computational complexity and list of tasks. Only leader of a group (selected in a way described in Subsection 3.2) can approve answers for questions in tests/tasks.

Students taking part in our experiment study computer science at postgraduate studies. They are characterized by diversity of learning styles and personal types.

Students learnt in generated groups of 3 or 4 people about basics of computational complexity and tried to solve a test, which included 5 questions with 4 possible answers (only 1 answer was correct). For each correct answer students got 1 point; otherwise - 0 points.

The comparison of the learning's efficiency in heterogeneous and homogeneous groups was based on the statistical analysis between results of a test in both populations.

In our experiment, students were divided into following populations: 42 students working in heterogeneous groups and 40 students working in homogeneous groups.

The statistical analysis used the following data (samples):

    (1)  The results obtained by heterogeneous groups
    (2)  The results obtained by homogeneous groups

Some of the results are presented in Figure below.

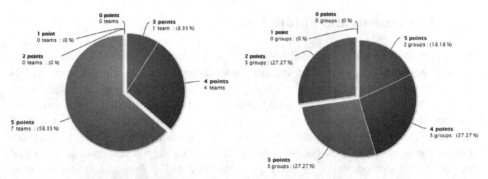

**Fig. 2.** The results of heterogeneous (on the left side) and homogeneous groups

In heterogeneous population, seven groups obtained maximum number of points in test. Four groups achieved 4 points and one group got 1 point. There were no results with 2, 1 or 0 points.

In homogeneous population, two groups achieved 5 points, three groups received 4 points. Also three groups obtained 3 and 2 points. No groups received 1 or 0 points.

The whole analysis was made at significance level $\alpha = 0.05$. Before selecting a proper test, we used Lilliefors test in order to examine, if mentioned samples come from normal distribution. Results of the tests are presented in Table 3.

**Table 3.** The results of the Lilliefors test

| Sample | Statistical test value | p-value |
|--------|------------------------|---------|
| (1) | 0.289848 | 0.00231 |
| (2) | 0.207879 | 0.203311 |

The analysis showed that sample with results of heterogeneous groups do not come from a normal distribution (the null hypothesis was rejected). Therefore, to perform further analysis, a non-parametric tests were used.

In order to compare medians, above samples were tested with U Mann-Whitney test. The statistical test value was equal 40.5 and *p-value* = 0.034561, so it means that null hypothesis cannot be accepted - the analyzed samples differ significantly for assumed significance level.

**Table 4.** Sum of ranks and the mean of the ranks for two samples

| Sample | Sum of rank | Average rank group |
|--------|-------------|--------------------|
| (1) | 206.5 | 15.884615 |
| (2) | 118.5 | 9.875 |

The sum of rank and average rank groups is definitely higher in case of heterogeneous groups, which shows Table 4. This means that heterogeneous groups achieved clearly better results than homogeneous population, so our algorithm for creating heterogeneous groups works.

## 5     Conclusions and Future Works

In this work we compared methods of creating heterogeneous and homogeneous groups. An original algorithm for creating heterogeneous groups was proposed and its effectiveness was analyzed. For this purpose, the Jung's typology was used. If possible, each heterogeneous group consisted of: leader of the group, extrovert person ensuring the communication, a "feeling" person providing consistency of a group. Additionally, the main aim of proposed algorithm for creating heterogeneous group is to obtain well-balanced group. The experimental results confirmed that our proposed method works properly and generated well-balanced learning group in respect of personal character traits.

Base on statistical analysis, an experiment showed that students working in heterogeneous groups achieved better results in learning than students working in homogeneous groups. This means that construction of a group based on method proposed in this paper (described in Subsection 3.2) increases the efficiency of learning and it is essential for effective learning.

In future work, it is considered to propose a method for recommending teaching material for heterogeneous groups. In homogeneous groups it is relatively easy to propose material adapted to student's needs or preferences because students have similar learning styles or personal traits. However, teaching material for heterogeneous groups should be "flexible" to their needs. It is also planned to check, what is the optimal cardinality of groups - is it better, when group contains 4-5 students or when it is a group of two students. Moreover, social ties and comparison of interactions between students in different types of the groups during learning process will be considered.

**Acknowledgment.** This work was partially supported by the European Commission under the 7th Framework Programme, Coordination and Support Action, Grant Agreement Number 316097, ENGINE - European research centre of Network intelliGence for INnovation Enhancement (http://engine.pwr.wroc.pl/).

# References

1. Aleven, V., Belenky, D.M., Olsen, J.K., Ringenberg, M., Rummel, N., Sewall, J.: Authoring collaborative intelligent tutoring systems. In: Looi, C.-K., Walker, E. (eds.) Proceedings of the Workshops at the 16th International Conference on Artificial Intelligence in Education AIED 2013, Memphis, USA (2013)
2. Baer, J.: Grouping and Achievement in Cooperative Learning, Coll Teach 51 no.4 Fall 2003, WN: 0328803828010 (2003)
3. Graf, S., Bekele, R.: Forming heterogeneous groups for intelligent collaborative learning systems with ant colony optimization. In: Ikeda, M., Ashley, K.D., Chan, T.-W. (eds.) ITS 2006. LNCS, vol. 4053, pp. 217–226. Springer, Heidelberg (2006)
4. Bernacki, J., Kozierkiewicz-Hetmańska, A.: Creating collaborative learning groups in intelligent tutoring systems. In: Hwang, D., Jung, J.J., Nguyen, N.-T. (eds.) ICCCI 2014. LNCS (LNAI), vol. 8733, pp. 184–193. Springer, Heidelberg (2014)
5. Bradley, J.H., Hebert, F.J.: The effect of personality type on team performance. Journal of Management Development **16**(5), 337–353 (1997)
6. Clifford, M.: Facilitating Collaborative Learning: 20 Things You Need to Know From the Pros (2012). http://www.opencolleges.edu.au/informed/features/facilitating-collaborative-learning-20-things-you-need-to-know-from-the-pros/ (last access: April 18, 2014)
7. Collaborative Learning Heterogeneous Versus Homogeneous Grouping. http://www.ukessays.com/essays/education/collaborative-learning-heterogeneous-versus-homogeneous-grouping-education-essay.php (last access: October 20, 2014)
8. Cooper, J., Prescott, S., Cook, L., Smith, L., Mueck, R., and Cuseo, J.: Cooperative learning and college instruction: Effective use of student learning teams. California State University Foundation, Long Beach, CA. Homogeneous or heterogeneous groups? (1990)
9. Cooperative and Collaborative Learning. http://www.thirteen.org/edonline/concept2class/coopcollab/ (last access: October 20, 2014)
10. Das, J., Rai, N., Samsudin, S.: Cooperative learning: heterogeneous vs homogeneous grouping. In: Proc. of APERA Conference 2006, Hong Kong, November 28–30, 2006
11. Felder, R.M., Silverman, L.K.: Learning and teaching styles in engineering education. Engr. Education **78**(7), 674–681 (2002)
12. Garcia, E., Romero, C., Ventura, S.: Data mining in course managements systems: Moodle case study and tutorial. Elsevier Science **51**(1), 368–384 (2008)
13. ILS Questionnaire. http://www.engr.ncsu.edu/learningstyles/ilsweb.html (last access: March 31, 2014)
14. Kozierkiewicz-Hetmańska, A.: A method for scenario recommendation in intelligent e-learning systems. Cybernetics and Systems **42**(2), 82–99 (2011)
15. Martin, E., Paredes, P.: Using learning styles for dynamic group formation in adaptive collaborative hypermedia systems. In: ICWE Workshops 2004, pp. 188–198 (2004)
16. McLaren, B.M., Rummel, N., Tchounikine, P.: Computer Supported Collaborative Learning and Intelligent Tutoring Systems, pp. 447–463. Springer, Berlin (2010)
17. Santrock, John W.: Educational Psychology, 2nd edn. McGraw-Hill, New York (2004)
18. Slavin, R.E.: Cooperative learning. Review of Educational Research. **50**, 315–345 (1980)
19. The Myers Briggs Type Indicator. http://www.myersbriggs.org/my-mbti-personality-type/mbti-basics/ (last access: October 20, 2014)
20. Types of personality. http://zawodowe.info/charakterystyka-typologi-wg-mbti/ (last access: October 21, 2014) (in Polish)

# Analyzing Music Metadata on Artist Influence

Marek Kopel[(✉)]

Wroclaw University of Technology, Wybrzeze Wyspianskiego 27,
50-370 Wrocław, Poland
marek.kopel@pwr.edu.pl
http://www.ii.pwr.wroc.pl/kopel

**Abstract.** The paper focuses on analyzing how music artist influence one another. This analysis is a part of evaluation of the music metadata for being used as Semantic Web data source. The music dataset case study shall reveal problems to be solved before enabling the data to be usable for automatic inferencing by Web 3.0 user agents. The described part of the research is finding the authors and performers of the most covered works. The analysis is based on the musicbrainz dataset, mostly on relationship metadata stored in l_entity_entity tables. Results are presented and the main problems of the dataset and analysis approach are discussed.

**Keywords:** Influence · Music · Dataset · Musicbrainz · Semantic web

## 1 Introduction

Artists often speak about their influences being works of other artists. In case of music artists the influence may come from different aspects of the work. The composition side of a song may inspire artists e.g. by its arrangement, harmony, instrumentation, beat, dynamics or sound. Lyrics may bring inspiration with the story they tell and message they bring, One may also appreciate their form or figures of speech that lyricist used. The influence may also come from the artists themselves, apart from the music. Examples of this influence may be Beatlemania - fan frenzy towards The Beatles [1] and 'Freddie For A Day' - tribute to Freddie Mercury and fund raising for fighting AIDS [2].

This paper focuses on the measurable aspects of artists influencing one another. The 'measurable aspects' are standardized facts of using other artists work in own creations. Using means not only literally sampling other artists recordings, but also performing others' songs. recording covers and medleys, using citations, mixing and creating mash ups.

Other research in this field had been published only in recent years along with the access to datasets that would allow for the musical influence analysis. Two examples of those works are: a more general analysis of music genres influencing one another published in [3] and a more specific case study on musical influence within a single genre: synthpop and a Depeche Mode as a best representative of the genre - published in [4].

© Springer International Publishing Switzerland 2015
N.T. Nguyen et al. (Eds.): ACIIDS 2015, Part I, LNAI 9011, pp. 56–65, 2015.
DOI: 10.1007/978-3-319-15702-3_6

Even though artists may not know each other personally and may even not be familiar with others' work, they still create a social network. Just like employees of a corporation - who had never met - do. This musical social network based on information from online databases is also referred to as "social networks of artists" in [5]. Actually the whole music business ecosystem forms a social network that may give more information on artists' influence. Having included into the network people like producers, sound engineers, record label's management etc. may reveal other channels (network paths) the influence spreads. Also those people are responsible for the form of a final version of artists work (they may be artists themselves) and may contribute to creating artist clusters within the network. A common example is Motown Records which gathered most influential black artists of the '60s. Another idea concern including in the network people influencing artists personal life. With the technology of today's social media and information on fans interaction with artist's work, it may be even possible to measure the influence of artists on people as a human kind.

The rationale for this analysis of online music data is the evaluation of the possibility of exposing the data so they are machine readable. Since the relationships of influence are not as straightforward, as e.g. being an author or being a band member, simple converting it to RDF triples would not work. So first we try to make the relationships be human readable by applying it to the use case of finding most covered authors and performers.

## 2    Dataset

As mentioned earlier, in this study, the influence is measured using indexed facts from music industry. The two main facts proving that an artist is influenced by another are:

- recording own version of someone else's song - called covering,
- using part of another's artist recording in own recording - called sampling.

The biggest database of music recordings and its releases - musicbrainz [6] - indexes those facts, therefore it is the one we use for building our social network. Musicbrainz offers open content and that is why it is used by another open content dataset - Freebase [7]. Freebase is collaborative knowledge base which metadata feed The Knowledge Graph [8], which in turn is Googles take on Semantic Web [9].

In the experiments the database snapshot from 2014.05.14 is used locally. This PostgreSQL database weights 17GB and contains - among others - information on: over 13 millions of recordings of almost half a million works by over 0.8 million artists in over 1.3 million releases.

There exist services like DBTune or Data Incubator: MusicBrainz that would wrap musicbrainz data in RDF framework with SPARQL endpoints, making the data part of the Linked Open Data (LOD) project [10]. This allows querying the data in the way Semantic Web approach had envisioned, i.e. using SPARQL syntax. But there are 2 decisive factors in favour of using plain SQL. First: most of the relationship information used in this study is not accessible via SPARQL, since the relationships are not explicit within this SQL schema. It will is discussed in detail below. The second factor is that considering the size of the data - where each SQL subquery would run on average for 10 minutes - it is cheaper to use simple SQL without the overhead of transformation from relational database to hugely redundant RDF/XML serialization. In this aspect this is a preliminary research which shall lead to enabling inference using music semantic data within RDF framework.

The social network uses artists as nodes. The edges reflect the fact of covering, sampling, making a tribute to an artist by another artist. Those facts would give the edges a direction of a relationship. A relationship that would make an undirected edge would be the fact named in musicbrainz: collaboration, tribute, supporting, vocal supporting, member of, subgroup. From those explicitly declared relationships between artists (e.g. Sting - member of - The Police), we may derive more relationships. E.g. the influence of Lennon on McCartney and vice versa is not stated explicitly, but seems obvious, considering the facts that they co-authored most of Beatles' songs and played in one band for over 10 years. Another example of that 'hidden relationship' is the influence of band Smile on band Queen, coming from the fact, that half of the members of the latter (May and Taylor) were also members of the former. Some of the collaboration facts may also be extracted e.g. from the artist_credit table, which contains variations of artist names and pieces of text to join the artist names, e.g. "Queen & David Bowie".

## 3    What is a Cover?

For the first round of experiments the fact of recording a cover was used as a network edge. The fact (a link between recording and work) is stored in table l_recording_work in rows with a special link_type. The link_type used for specifying covers has name: "performance", description: "This is used to link works to their recordings." and link_phrase: "live medley:medley including a partial instrumental cover recording of". So this type of links is used for any performance of a work, which may seem to be a problems. The very first recording of a work is also treated here as a covers. Is it incorrect? Well, it is all a question of semantics. The intuitive definition of cover seems obvious, but with all the border cases it gets complicated.

Let us assume, that we explicitly declare: "if an artist is making the first recording of a work then all his later recordings should be excluded from the set of the work's covers". This still leaves in question all later recording by the original artist in collaboration with another artist (different artist_credit). But even is the following - seemingly obvious - case, the rule won't apply. 'Yesterday' - originally recorded by The Beatles - is later recorded by Paul McCartney - author and original performer (as Beatles member) of the song and released under his own name. Is the release really a cover? This is, probably, why musicbrainz puts all of those cases in one link_type named: "performance".

# 4    Finding Influential Artists: Authors vs. Performers

The default relationship - linking artist with work - stored in table l_artist_work refer to an author-artist with a role of 'composer', 'lyricist' or plainly 'writer'. The aggregated version (grouping all the roles by artist name) of top 50 authors of that data is shown on bar diagram in figure 1. It was obtained using the following query:

```
SELECT a.name, COUNT(DISTINCT ac.name) as c
FROM l_recording_work lrw
JOIN recording r ON r.id=lrw.entity0
JOIN l_artist_work law ON law.entity1=lrw.entity1
JOIN artist a ON a.id=law.entity0
JOIN artist_credit ac ON ac.id=r.artist_credit
JOIN link l ON l.id=lrw.link
JOIN link_type lt ON lt.id=l.link_type
WHERE lt.name='performance'
GROUP BY a.name
ORDER BY c desc;
```

This way the influential artists would be e.g. John Lennon and Paul McCartney, Mick Jagger and Keith Richards or Jimmy Page and Robert Plant. This is correct by when mentioning influences artist would rather names of the performers, in this case The Beatles, The Rolling Stones and Led Zeppelin, respectively, i.e. the bands, which those authors performed their works in. Here the connections are obvious, but many times the performer is not the authors of the performed works. When someone points to Elvis Presley as a source of her musical inspiration, she may not be aware of the fact that most of Elvis hit songs were written by Leiber/Stoller partnership. On the other hand, when listing Jerry Leiber and Mike Stoller as most influential artists, it may not be explicit, that huge part of that success is due to Presley's performances. Hence, Presley - "the King of Rock and Roll" - would not be on the list himself.

## 4.1    Discovering the First Performer of a Work

For this purpose another relationship between artist and work shall also be extracted. The relationship that gives an artist the "virtual ownership" of a work, so that the work is said to be "her song". Usually this relationship is equivalent to 'first performer of work'. One would hope to get list of all performances of a work by joining links from tables l_artist_recording and l_recording_work (since going straight to l_artist_work one shall get the authors) with link_type set to 'performer'. Unfortunately, the information in those tables is stored at more detailed level than we need for our purpose. E.g. in case of the original recording of 'Yesterday' we will get Paul McCartney with two link_types: 'vocal' and 'instrument (guitars)'. We will not find a record for The Beatles with link_type 'performer' as we intend.

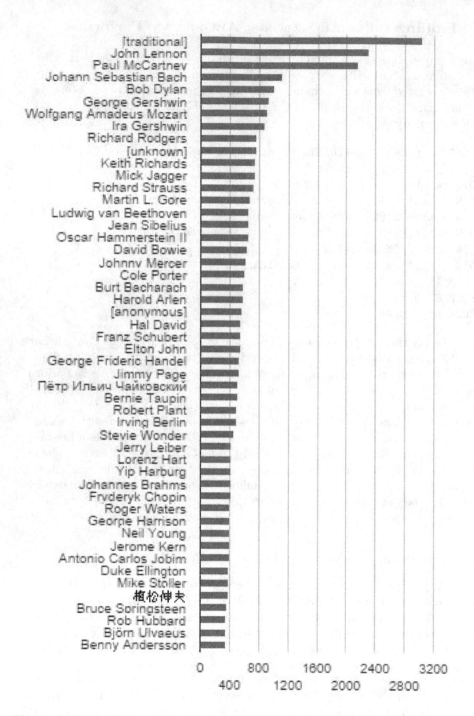

**Fig. 1.** Authors of the most covered works. The horizontal scale shows number of distinct artist_credits that covered (released a performance) of a work.

Another approach is needed for finding the 'first performer of work'. Examining the musicbrainz database schema, we may find that instead of using the relationships from table l_artist_recording, we may join the recording with the table artist_credit, which would give the desired performers of recordings. But for finding the first performer, which means: the first recording of a work, information on recording dates is also needed. This information is stored in the link record corresponding to a l_recording_work record. The link table holds the begin_date and end_date for each l_entity_entity relationship, in this case for the first and last day of making the recording. As one may imagine this type of information is pretty specific and usually inaccessible, this is why this crowd-sourced database lacks the information for many of the l_recording_work records.

But there is another information on date that concerns establishing the first performer of a work. It is the date of releasing a recording. In our approach it may be even more desired, since a recording could not make the performer influential until the song is released. Inspecting the EER diagram shows that there is no direct access to the release date from the recording entity. It actually requires a path of 4 joins: recording - track - medium - release - release_country. The release_country table holds the dates of a releasing a recording in different countries. In some cases the date may be found when replacing the last table in the path with table release_unknown_country. Comparing the earliest releases dates of recordings of each work with the recording dates from link entity it turns out that in  40% of works the former are earlier dates than the latter and they fill out the blank spots where latter dates were 'NULL'. The earlier date from the two for each work determines the first performer. The results are show in the table 1.

Now, having the first performer for each work, a final aggregation can be made to show the 'artist whose songs have been covered by the highest number of distinct artists'. That aggregation, presented on bar diagram in figure 2, gives the most influential music artists (in the meaning from the beginning of this section).

## 5   Discussion

The highest numbers for the most influential performers are a little skewed because of the aggregation method. Since aggregation by artist name summed up the numbers of artists covering 'their songs', the sumes may have an overhead for not considering the distinct covering artists. This means the more songs of a single artist have been covered, the more it is likely that the same artists were covering those songs and thus are counted again for each song. This only applies to a few the top first performers, as only they have more than a single work performed first. As shown in table 1, the most skew would be The Beatles score since in the top 100 of most covered works they have 14 songs. On the following positions: Frank Sinatra - 6 first recordings of the top 100 most covered songs, BBC Symphony Orchestra - 5, Charlie Parker - 4, Nat King Cole - 4, Boston Symphony Orchestra - 4, Django Reinhardt - 2 and London Philharmonic Orchestra - 3.

**Table 1.** Most covered works with mapped first performers based on the earliest dates (1st date) of the recording or release of the work. Last column states the number of covers of each work by distinct artists

| work | 1st performer | 1st date | covers nr |
|---|---|---|---|
| Star of the County Down | BBC Symphony Orchestra | 1945-01-02 | 227 |
| The Christmas Song | King Cole Trio | 1946-06-14 | 212 |
| Over the Rainbow | Judy Garland | 1938-10-07 | 205 |
| Yesterday | The Beatles | 1965-08-06 | 185 |
| Suite Nr. 3 D-Dur, BWV 1068: II. Air | Pau Casals | 1916-05-05 | 181 |
| Summertime | George Gershwin | 1935-10-14 | 173 |
| Eleanor Rigby | The Beatles | 1966-08-05 | 167 |
| Stardust | Hoagy Carmichael | 1927-10-31 | 157 |
| Moon River | Henry Mancini | 1945-01-02 | 154 |
| Night and Day | Django Reinhardt | 1938-01-31 | 142 |
| Fly Me to the Moon (In Other Words) | Nat King Cole | 1961-12-22 | 133 |
| Your Song | Three Dog Night | 1970-03-31 | 131 |
| Imagine | John Lennon | 1971-09-09 | 128 |
| Ain't No Sunshine | The Temptations | 1972-01-11 | 128 |
| The Look of Love | Burt Bacharach | 1967-01-29 | 122 |
| Come Together | The Beatles | 1969-09-26 | 119 |
| Silent Night | Frank Sinatra | 1945-08-27 | 111 |
| White Christmas | Frank Sinatra | 1944-11-14 | 109 |
| Sunny | Booker T. & The MG's | 1967-05-23 | 108 |
| Hey Jude | The Beatles | 1968-08-26 | 108 |
| Bridge Over Troubled Water | Simon & Garfunkel | 1969-11-11 | 101 |
| What a Wonderful World | Louis Armstrong | 1968-01-01 | 100 |
| Blackbird | The Beatles | 1968-11-22 | 99 |
| Someone to Watch Over Me | Frank Sinatra | 1944-10-11 | 98 |
| Michelle | The Beatles | 1965-12-03 | 96 |
| O Fortuna | The Philadelphia Orchestra | 1960-04-24 | 95 |
| West Side Story: Somewhere | Manny Albam | 1957-10-10 | 95 |
| Something | The Beatles | 1969-09-26 | 93 |
| And I Love Her | The Beatles | 1964-06-20 | 92 |
| The Man I Love | Django Reinhardt | 1939-08-25 | 91 |
| Light My Fire | The Doors | 1967-01-04 | 90 |
| The Surrey With the Fringe on Top | Alfred Drake | 1943-10-20 | 90 |
| (I Can't Get No) Satisfaction | The Rolling Stones | 1965-05-27 | 90 |
| Paint It Black | The Rolling Stones | 1966-05-07 | 90 |
| Can't Help Falling in Love | Elvis Presley | 1961-03-23 | 89 |
| Johnny B. Goode | Chuck Berry | 1957-12-29 | 87 |
| Let It Be | Aretha Franklin | 1970-01-15 | 87 |
| I Got Rhythm | Stphane Grappelli | 1935-10-13 | 86 |
| 's Wonderful | Ella Fitzgerald | 1958-08-10 | 86 |
| I'm in the Mood for Love | The Nat King Cole Trio | 1946-03-15 | 85 |
| Embraceable You | The Nat King Cole Trio | 1943-12-15 | 85 |
| Also sprach Zarathustra: I. Einleitung | Geoff Love and His Orchestra | 1972-05-30 | 85 |
| They Can't Take That Away From Me | Charlie Parker | 1950-07-05 | 84 |
| I Didn't Know What Time It Was | Charlie Parker | 1949-11-30 | 84 |
| Enjoy the Silence | Depeche Mode | 1990-02-05 | 82 |
| Jupiter, the Bringer of Jollity | BBC Symphony Orchestra | 1945-01-02 | 82 |
| Time After Time | Frank Sinatra | 1946-09-17 | 82 |
| My Funny Valentine | Frank Sinatra | 1953-11-05 | 81 |
| Time After Time | Cyndi Lauper | 1983-10-14 | 80 |
| Knockin' on Heaven's Door | Bob Dylan | 1973-07-13 | 79 |

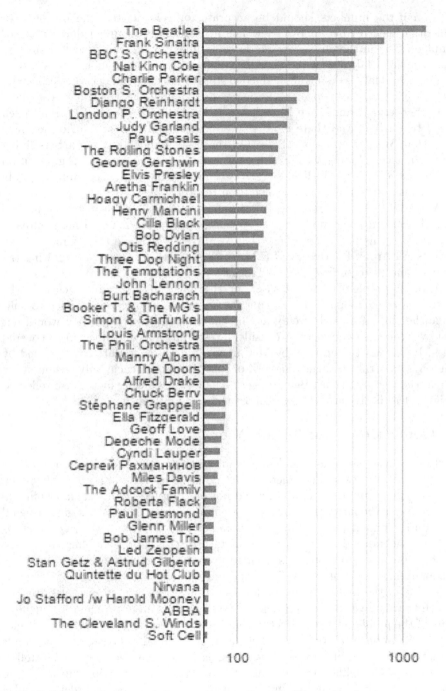

**Fig. 2.** Original performers of the most covered works. The vertical axis groups works (of different authors) by performers from table 1. The horizontal log scale is analogical to figure 1, showing the number of distinct artists covering each work.

Even if the numbers should be accurate for this matter, the list of most influential artist is biased by a few factors. The crowdsourced dataset is not complete. The authorship of works is sometimes not known, which results in bars in figure 1 labeled as: [traditional], [unknown] or [anonymous].

The record labels do not show the real state of music social network, especially today, when the most covering is being documented on YouTube. And most popular covering bands, like Pomplamoose or Walk Off The Earth aren't even signed to a label. And there is no way of accessing the cover information now other than Web scraping YouTube. On the other hand all of the artists living before the record industry took off are not included in the analysis. E.g. intuition tells that J.S.Bach would have been a very influential performer, unfortunately none of his performances could have been recorded.

Also 'the first performer' assumption fails to embrace the recursion in covering. Recursion mean making a cover based on a performance that is already a cover. E.g. this is a known fact, that Wet Wet Wet made their famous cover 'Love Is All Around' based on R.E.M. version of the song and not knowing the original, author recording by The Troggs.

Another problem of the dataset, or the music industry it reflects, is the author copyright versus the real contribution. It is a common practice to split the authorship of a work evenly among the band members, so they would not fight over the money. But let us consider "Yesterday", as one of the most covered songs. It was written entirely by McCartney, but the metadata say "penned by Lennon/McCartney", suggesting 50% of contribution for each, only because they had agreed to co-sign each Beatles song? The latter, skewing fact is also reflected in figure 1 as both authors have almost the same influence.

## 6    Conclusions and Future Work

Most of the analysis pitfalls and shortcomings come from the massiveness of the music social network and the metadata to describe it. This determines that each dataset aggregating the metadata could not be complete. Even the music industry that forms the network is so inconsistent, at so many levels, that gathering all the data for influence analysis is usually impossible. The low level, SQL analysis of the musicbrainz dataset is crucial for future research on enabling intelligent agents inferencing upon the music semantic data, following the trend of Web 3.0. The analysis revealed gaps (NULLs) and inconsistencies in the information storage. Some of the problems were addressed in the analysis, nevertheless results of each that large scale analysis should be treated as an approximation at best. Some of the problems come directly from the database design. Guideline "Prefer Specific Relationship Types" make it hard for a global analysis, as it demands extracting and deriving upper level relationships from the low level, explicit declarations. Guideline "Do not cluster" concerning the relationship results in a sparse network of links making it even harder to traverse in order to find a desired relationship path.

Some of the problems come directly from the database design. Guideline "Prefer Specific Relationship Types" make it hard for a global analysis, as it demands extracting and deriving upper level relationships from the low level, explicit declarations. Guideline "Do not cluster" concerning the relationship results in a sparse network of links making it even harder to traverse in order to find a desired relationship path.

Some of the problems may be avoided while doing a more specific analysis. Limiting the range of the data used in the analysis and focusing on lower level relationship is better supported by musicbrainz design. This kind of analysis is planned for future work in the field of musical influence. One type of such analysis is focusing on different periods of time, analyzing e.g. the most influential authors and performers of each decade and thus analyzing the influence dynamics.

**Acknowledgments.** This work was supported by the European Commission under the 7th Framework Programme, Coordination and Support Action, Grant Agreement Number 316097, ENGINE - European research centre of Network intelliGence for INnovation Enhancement (http://engine.pwr.wroc.pl/).

# References

1. Taylor, A.J.W.: Beatlemania—A Study in Adolescent Enthusiasm. Br. J. Soc. Clin. Psychol. **5**(2), 81–88 (1966)
2. What Is Freddie For A Day?. http://www.mercuryphoenixtrust.com/site/whatisffad. Accessed 20 Oct 2014
3. Bryan, N.J., Wang, G.: Musical influence network analysis and rank of sample-based music. In: ISMIR, pp. 329–334 (2011)
4. Collins, N.: Computational analysis of musical influence: a musicological case study using MIR tools. In: ISMIR, pp. 177–182 (2010)
5. Jacobson, K., Sandler, M., Fields, B.: Using audio analysis and network structure to identify communities in on-line social networks of artists. In: ISMIR, pp. 269–274 (2008)
6. Swartz, A.: Musicbrainz: A semantic web service. IEEE Intell. Syst. **17**(1), 76–77 (2002)
7. Bollacker, K., Evans, C., Paritosh, P., Sturge, T., Taylor, J.: Freebase: a collaboratively created graph database for structuring human knowledge. In: Proceedings of the 2008 ACM SIGMOD International Conference on Management of Data, New York, NY, USA, pp. 1247–1250 (2008)
8. Singhal, A.: Introducing the knowledge graph: things, not strings. http://googleblog.blogspot.com/2012/05/introducing-knowledge-graph-things-not.html. Google Blog May 2012. Accessed 20 Oct 2014
9. Shadbolt, N., Hall, W., Berners-Lee, T.: The Semantic Web Revisited. IEEE Intell. Syst. **21**(3), 96–101 (2006)
10. Bizer, C., Heath, T., Berners-Lee, T.: Linked data-the story so far. Int. J. Semantic Web Inf. Syst. **5**(3), 1–22 (2009)

# How to Measure the Information Diffusion Process in Large Social Networks?

Dariusz Król[(✉)]

Department of Information Systems, Wrocław University of Technology,
Wrocław, Poland
Dariusz.Krol@pwr.edu.pl
http://www.ii.pwr.edu.pl/~krol/eng_index.html

**Abstract.** This paper forces to think critically about the actual list of developed statistics to measure the information diffusion in complex networks, for example about universally popular structural-based performance metrics and especially lately about dynamic performance metrics. Therefore, somewhat similar to the latter yet an alternative set of measurements for diffusion process is proposed, i.e. the *scope*, *speed* and *failure* of spread. Particular attention is paid to analyse measures for generic diffusion-based algorithm using *pull* strategy and the *time–to–recovery* parameter in large social networks.

**Keywords:** Data propagation · Spreading phenomenon · Cascading behaviour · Measurement · Performance metric · Epidemic threshold

## 1 Introduction

*What are major scientific challenges for diffusion phenomenon?*

Based on practical experience there is a good reason to believe that some of the core concepts related to real world networks may be counter-intuitive to commonly held beliefs. Frequently, even small individual change or activity may contribute towards large-scale action including disastrous consequences. For example, in a formation of interconnected networks, when nodes in one network depend on nodes in another, not serious disturbance in one network can cascade through the entire system often to be big enough to cause a serious disturbance or even to collapse it [2].

Here, in our present study, we are interested in better understanding and harnessing such an intriguing diffusion phenomenon. This paper represents a continuation and an extension of our previous work on the same topic [10–12]. If the previous one concentrated on diffusion properties, algorithms and strategies now we analyse selected network statistics that can be specifically applied to any diffusion situation.

Addressing this phenomenon involves tackling three major scientific challenges. The first, absorbing technical knowledge, is the gathering of large-scale data on spreading problem. The second challenge is the application of formal

© Springer International Publishing Switzerland 2015
N.T. Nguyen et al. (Eds.): ACIIDS 2015, Part I, LNAI 9011, pp. 66–74, 2015.
DOI: 10.1007/978-3-319-15702-3_7

models to make it possible to quantify and qualify the effect of infection mechanism. The third challenge concerns the development of well-defined measures to properly determine the network and process properties, and this is consistent with our main objective of the paper.

There are numerous existing statistics based on network analysis advanced to exhibit a substantial list of diffusion properties. For example, [1] presents an overview of 30 network topology metrics including second moments and their pairwise correlations. Most of these come from social science and the network theory and seems to represent their suitability for diffusion studies. However, none of these metrics has been thorough evaluated for large-scale graph generated by diffusion-based algorithm using *pull* strategy and *time–to–recovery* parameter. In this paper we go over these characteristics, examine their feasibility, advantages, and disadvantages, and make some experimental research for generating diffusion graphs in large scale. The work first objective is to provide a comprehensive and accessible review of the main measurements which then can be used to estimating important properties of diffusion mechanism. The second objective is to provide some understanding on how we can effectively measure the information diffusion process in large networks.

This paper is organized as follows. Section 2 reports the actual list of developed diffusion measures for social informatics. Then in Section 3, we use Matlab simulation to study the diffusion mechanism and to verify our measurement setup. Finally, Section 4 concludes the paper.

## 2    Diffusion Measures for Social Informatics

*How quickly, widely and robustly does social diffusion spread?*

Formally, in network theory we distinguish between two different classes of measures: the simple version computed for each selected component (nodes, links), and the aggregated version computed for a whole structure. Thus, a measure to quantify the importance of a single component $i$ during diffusion $\wp(\mathsf{G}, G)$ with action $\mathsf{G}$ over the network described by a directed connected graph $G$ is a mapping $\sigma : N \cup L \to \Re$ which generally assigns a non-negative real number to each component $i \in N \cup L$, where a higher value of $\sigma$ contractually indicates a greater importance of a component[1]. Consequently, an aggregated measure to quantify importance of a whole process $\wp(\mathsf{G}, G)$ is a mapping $\phi : triplets \to \Re$ which usually assigns a non-negative real number to each generated set of *triplets*.

Whereas there are dozens of various network measures, including three key concepts - the average path length, clustering coefficient, and degree distribution, it is practically impossible to find a set of diffusion measures that are suitable for

---

[1] A diffusion $\wp(\mathsf{G}, G)$ with action $\mathsf{G}$ over the network is described by a directed connected graph $G = (N, L)$ where $N$ is the node set and $L$ is the set of directed links containing connected pairs of nodes $(i, j)$ unfolds in discrete time-steps $t \geq 0$ and is defined as an ordered sequence of triplets $(i, j, t)$. Each triplet corresponds to a single interaction event at a time-step $t$ between a pair $(i, j)$, where $i, j = 1..N$. For more details refer to [10].

all principles of social informatics. However, some may be more useful than others in particular contexts. On the other hand, selection of appropriate measures can enhance the efficiency of diffusion control and management. A marketable measure for the process gives us a number to sense how large it is (*scope*), how rapid it grows or decays (*speed*), and finally how robust it is (*failure*).

For instance, in compartmental models such as Susceptible → Infective → Recovered (SIR) model and Susceptible → Infective → Susceptible (SIS) model, the effective spreading rate $\tau$ and the effective reproductive rate $r$ are commonly used [4,8]. In particular, the concept of a threshold [7] for diffusion takeoff is directly related to this reproduction number $r$. If $r > 1$, the process starts and eventually settles down to an endemic equilibrium. If $r < 1$, the process likely dies out. In this case, the final number of infected individuals will be approximately $\frac{n}{1-r}$, where $n$ is the number of initial seeds. However, these measures are not explicitly related to the topological characteristics of the social networks and so are not quite adequate for assessing the process of diffusion. Therefore, it becomes necessary to propose new measures that timely integrate all corresponding characteristics *scope*, *speed* and *failure* (see Section 3).

Centrality measures were among the first proposed measures that focus on identifying the activity prominence of the individual in a social network. The most common are degree, closeness, betweenness and eigenvector centrality [13]. Degree centrality of a node is defined as a number of its immediate neighbors. Closeness centrality of a node describes the efficiency of action diffusion from one node to all others. Betweenness centrality of a node is defined as the percentage of shortest paths across all possible pairs of nodes that pass through given node. Eigenvector centrality presents how differences in degree can propagate through a network. More recently, to the category of identifying the most influential spreaders has joined k-shell index [9]. However, due to computational complexity and partly monotonicity, centrality measures are not well suited for social diffusions with thousands or millions of triplets. Only eigenvector centrality and the k-shell index seem to be a computationally efficient tools for capturing a large diffusion process and remains the object of our future research.

Since plain diffusion moves according to the availability of active connections, the process captures the interplay between temporal and topological constraints. Frequently, in this process links are not available at any moment and thus both the activity and the topology affects the dynamics, cf. `diffusion loop` with `threshold condition` as expounded in [10]. In every social diffusion, investigation of spreading actions, ordering them in time and analysing the *scope–speed–failure* relationship is very important. Hence in Table 1, the notions of diffusion distance, centrality, efficiency and robustness are introduced. Each measure is restricted to a given time-step interval which may be specified by parameters $[t_1, t_2]$.

The measure $\sigma_1$ allows identifying the diffusion path length, what is crucial for many standard centrality indices like betweenness and flow centrality, which in turn are based on counting the number of shortest paths.

**Table 1.** Diffusion measures for a given step-time interval

| Measure | Meaning | Definition |
|---|---|---|
| $\sigma_1$:diffusion distance $d_{ij}^{\wp(G)}(t_1,t_2)$ | The number of steps it takes for action to spread from node $i$ to node $j$. | |
| $\sigma_2$:diffusion centrality $c_k^{\wp(G)}(t_1,t_2)$ | The number of paths a diffusion process can pass through a given node. | $\displaystyle\sum_i\sum_j \frac{\|(i,k,j)\|}{\|(i,j)\|}, i\neq j$ |
| $\phi_1$:diffusion efficiency $E^{\wp(G)}(t_1,t_2)$ | The efficiency of diffusion process from one node to all others. | $\displaystyle\frac{1}{N^{\wp(G)}(N^{\wp(G)}-1)}\sum_{i\neq j\in 1..N}\frac{1}{d_{ij}^{\wp(G)}(t_1,t_2)}$ |
| $\phi_2$:diffusion robustness $R_{t_d}^{\wp(G)}(t_1,t_2)$ | The ability of diffusion process to maintain its activity after damage at a given time-step $t_d$. | $1-\dfrac{\lvert E^{\wp(G)}(t_1,t_d)-E^{\wp(G)}(t_d,t_2)\rvert}{E^{\wp(G)}(t_1,t_2)}$ |

To find diffusion source and measure the importance (influence [6]), defined as the number of ways a diffusion process can spread from a given node(s), we can utilize the second measure $\sigma_2$, where $\|(i,k,j)\|$ denotes the number of diffusion paths from node $i$ to node $j$ that pass through intermediary node $k$, and $\|(i,j)\|$ denotes the number of diffusion paths from node $i$ to node $j$, respectively. The diffusion centrality takes all diffusion paths, not only the shortest like in betweenness centrality, into account. In our case, it can be regarded as how much a node can facilitate communication to other nodes in the system.

Last two measures $\phi_1$ and $\phi_2$ address the problem of efficiency and robustness. The efficiency in the diffusion between individuals $i$ and $j$ is inversely proportional to the distance between them. It can be thought as a measure of how efficiently action is exchanged over a network, given that the nodes are communicating with all other nodes concurrently. Consecutively, the diffusion robustness can be defined as the drop in performance when the selected components are attacked and blocked. After such damage at a step $t_d$ some diffusion paths will not exist any more, thus, we expect that the diffusion efficiency will change the value evidently.

Although we have in mind the maximum simplicity, calculating diffusion measures requires considering all paths in a diffusion graph at every step, which is still computationally intractable and needs more investigation. Therefore, next we test *scope*, *speed* and *failure* which may be equivalents to distance, centrality, efficiency and robustness.

# 3    Experimental Results

*How does the diffusion scope, speed and failure can be practically applied?*

Since random graphs are widely used to compare with real networks, in the experiments, we have used the Erdős-Rényi model [5], the first ever proposed algorithm for the formation of random graphs[2]. All topological and statistical operations were implemented using Matlab software[3] (version 8.3.0.532). The size of the networks is 100K nodes and single value of the average node degree around 3 have been used in the experiments. We consider deterministic, explicitly given, thresholds[4]. Each node initially chooses a fixed threshold which represents the minimum fraction of active neighbours necessary for its activation. This is to account for our lack of knowledge of the exact threshold value of each individual. To avoid bias in random selection, we repeated the simulations 100 times in each experiment and calculated the average.

As pointed out in the introduction, we execute the spreading algorithm with the *pull* strategy on random graph with two essential parameters. We are based on the diffusion algorithm proposed in [12], where a susceptible nodes become infected if at least one of their neighbours is infected and aggregated influence is larger than the diffusion threshold[5]. The other applies to the parameter *time-to-recovery*, which is the number of simulation steps in which a given node may be infected and in consequence may spread the infection to others.

In terms of *scope*, *speed* and *failure* the simulation results are clear and cogent. It is important to note the difference in scales of the left and right-hand panels in Figs 1 and 2 after all. For example, Fig. 1 depicts the results of 100 simulation runs of the *pull* strategy on Erdős-Rényi graph with $\theta = 0.1, 0.3, 0.7, 0.9$ and seeds randomly generated and expressed in 1000 basis points (bps) of all nodes[6]. The graph shows the fraction of infected nodes (on the Y axis) that the process spreads to as a function of the simulation step (discrete time on the X axis). We use *scope* as a standardized number of successfully infected nodes, whereas *failures* illustrates a number of unsuccessfully infected nodes. The *speed* line can be thought of as the rate at which the diffusion process covers a generic graph. For *scope*, *speed* and *failures* we plot three different outlines: the average case, and the 5th and 95th percentiles.

---

[2] Erdős-Rényi model selects with equal probability pairs of nodes from the graph set of nodes and connects them with a predefined probability.

[3] Matlab software is available at http://www.mathworks.com.

[4] The epidemic threshold defined in SIR/SIS models is inversely proportional to the largest eigenvalue of the adjacency matrix of the graph. The smaller the largest eigenvalue, the larger robustness of a network against the spread of data. [3].

[5] We recall that the aggregated influence depends on the strategy. For *push* the maximum influence between a given node and all infected neighbours needs to exceed $\theta$, while for *pull* the sum of influences from all infected neighbours need to exceed $\theta$ for the diffusion to take place.

[6] Generally, the relationship between percentage changes and basis points is that 1 percentage point change denotes 100 basis points, and 0.01 percentage points denotes 1 basis point.

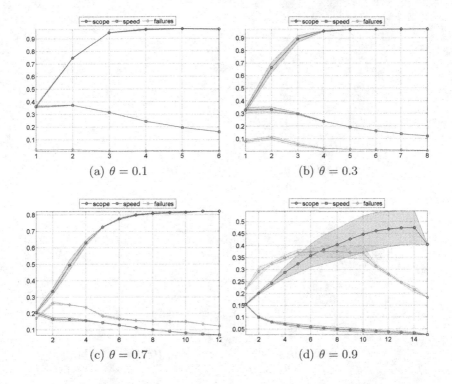

(a) $\theta = 0.1$

(b) $\theta = 0.3$

(c) $\theta = 0.7$

(d) $\theta = 0.9$

**Fig. 1.** The *scope–speed–failure* (blue–red–green) relationship w.r.t. the fraction of the affected vertices (Y axis) and a finite number of time-steps (X axis) for different values of $\theta$. The results are the averages of 100 network realizations for Erdős-Rényi model depicted in the inter–percentile range 5th–95th, with $ttr = 10$ in the *pull* strategy and randomly selected seeds expressed in 1000 bps

From panels 1(a) and 1(b) we infer that after 3 simulation steps the process spreads on average to 90% of available nodes. However, when threshold $\theta$ grows to 0.9 in panel 1(d), the mean outbreak affects only 45% of all nodes. Additionally, a significant increase in the inter–percentile range parallel to the extend to which time-step involves w.r.t. the growing threshold in the latter panel 1(d) is evident.

The diffusion dynamics differs from case to case in general and the *failure* measure in particular. The initial phase of spreading largely determines the overall diffusion *speed*, and when the *failure* grows fast, the *speed* actually began to decrease considerably (cf. Figs 1(a) and 1(d)).

Fig. 2 shows pairwise comparisons of the diffusion networks - 1000 vs 3000 bps - developed with the *pull* strategy, different values of *ttr* and fixed $\theta = 0.9$. It is expected that this time we get confirmation that *ttr* and the seedset are vital parameters, which influence the *scope* and *speed* of the process. In left panels 2(a), 2(c), 2(e) it may seem that the improvements of the scope is dependent only on *ttr*. However, right panels 2(a), 2(c), 2(e) represent a broader

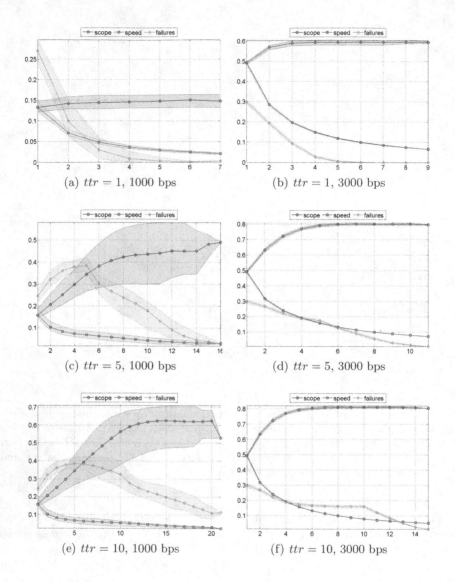

**Fig. 2.** *Scope* (in blue) as a function of time-steps compared with *speed* (in red) and *failure* (in green) for different values of *ttr* and multiple random seeds expressed in bps. All pairwise diffusion networks - 1000 vs 3000 bps - are developed with the *pull* strategy and $\theta = 0.9$. Data presented in the figure have been averaged running in 100 different network configurations

*scope*, whereas a *failure* is limited. As far as $ttr \geq 5$ is concerned time-steps can be shortened by 25% of the generic diffusion steps.

# 4   Summary

*What conclusion is to be drawn from this?*

This paper has surveyed measurements related to diffusion mechanisms developed in complex networks, particularly in large random graphs. We have touched upon a few themes of explaining how the information spreads within and across a variety of real world networks, briefly explaining the concept of measure, the motivation and the challenge, while referring to articles where more details can be found.

We are still far from a satisfactory adherence to reality and proper utilization of diffusion measures. The main limitation of our analysis is that, we have no opportunity to directly relate proposed metrics to the real world networks. Moreover, when the network size grows as large as in online social network, or a mobile phone network, the average clustering coefficient in a simple random graph (e.g. the Erdős-Rényi model) approaches zero and the geodesic distance between any two vertices approaches infinity. From this perspective, we need to create a random graph with clustering (RGC). Therefore, our new experiments will include another algorithm to generate RGC, for example, generalized version of Gleeson's algorithm [14]. By the similar reasoning, we should apply time-dependent threshold condition in order to detect tie creation and deletion in dynamic networks.

We hope that this paper provides an advanced point for researches to use the diffusion characteristics in more effective manner. Although we have noticed some improvement to diffusion statistics, especially on *scope–speed–failure* correlation, they remain based on *pull* strategy and the *time–to–recovery* parameter, whereas many real world systems use *push* strategy from one to another. Thus, we plan to investigate the diffusion characteristics for *push* and hybrid *push-pull* strategy, respectively.

**Acknowledgments.** This work was supported by the European Commission under the 7th Framework Programme, Coordination and Support Action, Grant Agreement Number 316097, ENGINE - European research centre of Network intelliGence for INnovation Enhancement (http://engine.pwr.wroc.pl/).

# References

1. Bounova, G., de Weck, O.: Overview of metrics and their correlation patterns for multiple-metric topology analysis on heterogeneous graph ensembles. Phys. Rev. E **85**, 016117 (2012)
2. Buldyrev, S.V., Parshani, R., Paul, G., Stanley, H.E., Havlin, S.: Catastrophic cascade of failures in interdependent networks. Nature **464**(7291), 1025–1028 (2010)
3. Chakrabarti, D., Wang, Y., Wang, C., Leskovec, J., Faloutsos, C.: Epidemic thresholds in real networks. ACM Trans. Inf. Syst. Secur. **10**(4), 1–26 (2008)
4. Dodds, P.S., Watts, D.J.: A generalized model of social and biological contagion. Journal of Theoretical Biology **232**(4), 587–604 (2005)
5. Erdös, P., Rényi, A.: On the evolution of random graphs. Publ. Math. Inst. Hung. Acad. Sci. **5**, 17–61 (1960)

6. Goyal, A., Bonchi, F., Lakshmanan, L.V.: Learning influence probabilities in social networks. In: Proceedings of the Third ACM International Conference on Web Search and Data Mining, WSDM 2010, pp. 241–250. ACM, New York (2010)

7. Granovetter, M.: Threshold models of collective behavior. American Journal of Sociology **83**(6), 1420–1443 (1978)

8. Jacquez, J.A., O'Neill, P.: Reproduction numbers and thresholds in stochastic epidemic models. i. homogeneous populations. Math. Biosciences **107**(2), 161–186 (1991)

9. Kitsak, M., Gallos, L.K., Havlin, S., Liljeros, F., Muchnik, L., Stanley, H.E., Makse, H.A.: Identification of influential spreaders in complex networks. Nat. Phys. **6**(11), 888–893 (2010)

10. Król, D.: On modelling social propagation phenomenon. In: Nguyen, N.T., Attachoo, B., Trawiński, B., Somboonviwat, K. (eds.) ACIIDS 2014, Part II. LNCS, vol. 8398, pp. 227–236. Springer, Heidelberg (2014)

11. Król, D.: Propagation phenomenon in complex networks: Theory and practice. New Generation Computing **32**(3–4), 187–192 (2014)

12. Król, D., Budka, M., Musiał, K.: Simulating the information diffusion process in complex networks using push and pull strategies. In: Proceedings of the European Network Intelligence Conference, ENIC 2014, pp. 1–8. IEEE, New York (2014)

13. Sun, J., Tang, J.: A survey of models and algorithms for social influence analysis. In: Aggarwal, C.C. (ed.) Social Network Data Analytics, pp. 177–214. Springer, US (2011)

14. Wang, C., Lizardo, O., Hachen, D.: Algorithms for generating large-scale clustered random graphs. Network Science **2**, 403–415 (2014)

# A Method for Improving the Quality of Collective Knowledge

Van Du Nguyen[✉] and Ngoc Thanh Nguyen

Department of Information Systems, Faculty of Computer Science
and Management, Wroclaw University of Technology, Poland
{van.du.nguyen,ngoc-thanh.nguyen}@pwr.edu.pl

**Abstract.** Collective knowledge is considered as a representative of a collective consisting of autonomous members. In the case if members' knowledge states have to reflect some real world knowledge for example weather forecasts then the quality of collective knowledge is an important issue. The quality is measured by the difference between the collective knowledge and the real world knowledge. In this work, a method for improving the quality of collective knowledge is proposed by taking into account the number of members in a collective. For this aim, we experiment with different number of collective members using multi-dimensional vector structure to determine how the number of collective members influences the quality of collective knowledge. According to our experiments, collectives with more members will give better solutions than collectives with fewer members.

**Keywords:** Collective knowledge · Knowledge integration · Consensus choice

## 1 Introduction

Nowadays, using many experts for solving problems in the real world is more and more popular because it is helpful in giving more proper solutions. However, each expert is a good specialist in one or two dimensions whereas problems in the real worlds are multi-dimensional, then the given solutions may conflict with each other. Collective knowledge is the knowledge that determined from a set of knowledge states given by collective members and is understood as the common knowledge of a collective consisting of autonomous members [8]. In Nguyen [6] based on consensus methods, a lot of algorithms for determining collective knowledge for different knowledge representations such as logical, relational structures and ontology have been worked out. Although, determining the knowledge of a collective is an important task, there exists another important issue with its quality. The quality is a quantity reflects how good a collective knowledge is and measured by taking into account the difference between the collective knowledge and the real knowledge state [6]. Also in that work, with some restrictions the author has proved that using the collective knowledge is safer than any knowledge state in a collective. Clearly, in the case if all of the members in a collective have the same distance to the real knowledge state then the collective knowledge is closest to the real knowledge state in comparison with all members in a collective.

© Springer International Publishing Switzerland 2015
N.T. Nguyen et al. (Eds.): ACIIDS 2015, Part I, LNAI 9011, pp. 75–84, 2015.
DOI: 10.1007/978-3-319-15702-3_8

This work is inspired by taking into account the number of members in a collective. For example, we consider the following collectives when experts are invited to give their opinions about a real world as follows: $X = \{yes, no\}$, $X' = \{yes, no, yes\}$, $X'' = \{yes, no, yes, yes\}$. With collective $X$, it's difficult to infer which one can be considered as the representative of the collective. However, for collective $X'$ member "*yes*" is more probable and this member is the most probable in the case of tive$X''$. From this example we can see that the number of members in a collective plays an important role not only in determining the representative of a collective but also in evaluating the quality of that representative. The problem we investigate in this work is formulated as follows: For a subject/matter in the real world (i.e. the weather state for a future day), a number of experts is requested to give their opinions. We assume that there exist a real knowledge state about that subject/matter exists and it is not known at the moment when experts are asked for giving their opinions. In addition, the experts' knowledge states (experts' opinions) in a collective are represented by a multi-dimensional vector. Each dimension represents a value of an attribute of a subject/matter in the real world. The quality is based on the distance from the real knowledge state to the collective knowledge. We experiment with different number of collective members. After that, by means of experiment analysis we aim to determine the influence of the number of collective members on the quality of collective knowledge. In other words, whether solving a subject/matter in the real world with more experts is better in getting more proper solution than that with fewer experts. That is the main problem what we investigate in this work. To our best knowledge, this aspect has not been investigated in the literature, although it has very important practical impact.

The remaining part of this paper is organized as follows. In Section 2 we present the analysis of related works. The basic elements of consensus choice, collective of knowledge states, quality of collective knowledge are presented in Section 3. In Section 4 we present experimental results and their evaluation to determine the influence of the number of members on the quality of collective knowledge. Finally, some conclusions and future works are pointed out in Section 5.

## 2    Related Works

In this section, we present some researches and approaches related to using consensus reaching processes to improve the agreement level among experts in solving group decision making (GDM); using many experts for solving multi-dimensional problems in the real world through statistical analysis; evaluating the quality of collective knowledge.

Firstly, In GDM problems, there exists a set of alternatives then experts are invited to give their preferences over alternatives. Consensus reaching processes play an important role in determining a solution that acceptable to all experts. They are applied to improve the agreement level among experts. There exist a lot of approaches to support consensus reaching process with difference type of preference structures such as: preference relations, preference orderings, utility vectors, etc [3] or numerical or linguistic information. In the case when the experts' opinions are not close enough to the consensus then experts can discuss and modify their opinions. There are two mechanisms supporting experts to discuss and modify their opinions such as feedback and no feedback. With the feedback mechanism, some advice is generated and provided experts on how to modify their opinions to make them closer to the consensus [3]. Inversely, instead of implementing feedback mechanism, other consensus models automatically update the opinions or important weights of those experts to make their opinions be close enough to others of the group [12].

In general, the aforementioned works aim at improving the agreement level in solving GDM problems. However, in our approach, there exists a real knowledge state about a subject/matter in the real world and each member in a collective reflects it to some degree. The real knowledge state is not known at the moment when experts are asked for giving their opinions. Then consensus methods are used to determine a representative for the set of opinions given by experts.

The next problem relates to using a group of people to solve some problems in the real world. In [9] the author has presented an interesting statistical analysis for the television game show, *Who Wants to Be a Millionaire,* only 65% of solutions given by specialists were on target whereas 91% of solutions given by audiences were on target. From this fact, the author has stated that *a group of people is more intelligent than single person* in solving some problems. This is because each member is a good specialist in one or two dimensions, whereas problems in the real world are multi-dimensional. Similarly, in [10] the author has presented several experiments to prove that the decision determined from a group of people in general is better than decision of single person. This also means *more experts are better than single expert.* These experiments can be considered as one of the most important motivations for researches related to using many experts to solve some common problems in the real world.

In [6, 7] the author has presented some interesting proofs related to the quality of collective knowledge as follows: the collective knowledge is not worse than the worst member in the collective; in the case of knowledge states reflect the real knowledge state to the same degree, then its knowledge should be better than all knowledge states. In [2] the authors have investigated analyzing the influence of the consistency degree of a collective on the quality of its knowledge using binary vector structure. In this approach, the quality of collective knowledge has based on the average of distances from the collective members to its knowledge. Through experiment analysis the authors have shown that the inconsistency degree of a collective has influence on the quality of its knowledge. That is, the larger the average of distances between collective members, the better the quality of its knowledge. In addition, in the case of the number of members in a collective is large enough, the more different the opinions of collective members, the better the credibility of collective knowledge.

In summary, the problem of the relationship between the number of collective members and the quality of its knowledge has not been investigated in the aforementioned works. Owning to this problem one can know how many members which should enough for a given problem.

## 3     Determining Knowledge of a Collective

### 3.1     Consensus Choice

A consensus choice has usually been understood as a general agreement in situations where parties have not agreed on some matter [1]. There are two well-known criteria used for processing inconsistency of knowledge from different sources such as: the minimal sum of distances $(O_1)$/squared distances $(O_2)$ between the collective knowledge and the collective members respectively. These criteria play an important role in determining the knowledge of a collective because of satisfying these postulates; it implies satisfying the majority of other postulates [5; 6]. Although criterion $O_1$ is more popular than $O_2$ for consensus choice and some optimal tasks, criterion $O_2$ is better than $O_1$ in some situations [6]. In this work, criterion $O_2$ is used to determine the knowledge of a collective.

### 3.2     Collective of Knowledge States

By U we denote a set of objects representing the potential elements of knowledge referring to a concrete real world. The elements of U can represent, for example, logic expressions, tuples, etc. Symbol $2^U$ denotes the powerset of U that is the set of all subsets of U. By $\Pi_k(U)$ we denote the set of all k-element subsets (with repetitions) of set U for $k \in N$ (N is the set of natural numbers), and let

$$\Pi(U) = \bigcup_{k=1}^{\infty} \Pi_k(U)$$

Thus $\Pi(U)$ is the set of all non-empty finite subsets with repetitions of set U. A set $X \in \Pi(U)$ can represent the knowledge of a collective where each element $x \in X$ represents knowledge of a collective member. Note that X is a multi-set. We also call X a collective knowledge profile. Set U can contain elements which are inconsistent with each other.

In this work, based on Euclidean space, a collective of knowledge states *(or collective for short)* is described as follows:

$$X = \left\{ x^{(i)} = \left( x_1^{(i)}, x_2^{(i)}, \dots, x_m^{(i)} \right) : i = 1, 2, \dots, n \right\}$$

where $x_k^{(i)} \in R, \ k = 1, 2, \dots m$

It is a multi-dimensional vector including a set of experts' opinions about a subject/matter in the real world.

### 3.3     Knowledge of a Collective

The knowledge of a collective is a knowledge state that considered as a representative for the set of knowledge states in a collective. The representative is not necessary one of those knowledge states. It's determined from the set of knowledge states and belongs to the set U. In this work, the knowledge of a collective is a knowledge state satisfying criterion $O_2$ as follows:

$$d^2(x, X) = \min_{y \in U} d^2(y, X)$$

where x is the knowledge of collective X and $d^2(x, X)$ is the sum of the squared distances from x to the members in X.

### 3.4     Quality of Collective Knowledge

In the case of existing a real knowledge state about a subject/matter in the real world, the quality of collective knowledge is measured by taking into account the difference from the knowledge of a collective to the real knowledge state [7]. Each member in a collective reflects the real knowledge state to some degree but the degree remains unknown because of incomplete and uncertain situations. The quality is defined as follows:

$$\hat{d}(x, X) = 1 - d(x, r^*)$$

where $x$ is collective knowledge of collective X, $r^*$ is the real knowledge state.

The quality of collective knowledge is based on its distance to the real knowledge state. The closest one to the real knowledge state is the best one.

## 4     Preliminary Experiments

### 4.1     The Proposed Method

As mentioned in the previous section, there are many approaches related to the problem improving the agreement level among experts. The experts could modify their opinions according to some advice generated from a moderator (feedback mechanisms) or the experts' opinions could be automatically updated to reach an optimal agreement level among experts. In this work, the number of members in a collective is taken into account for improving the quality of its knowledge. For this aim, based on Euclidean space, we experiment with different numbers of collective members. The set U is a set of points in the circle of radius 1.0, center (0, 0) (Fig. 1).

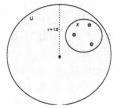

**Fig. 1.** Set U

Each point presents an expert's opinion about a real world. A collective is a set with repetitions of k members belonging to the set U. For example, from Fig. 1 we have a collective with 3 members($X = \{x^{(1)}, x^{(2)}, x^{(3)}\}$) as follows:

**Table 1.** A collective with 3 members

|          | $x_1$ | $x_2$ |
| -------- | ----- | ----- |
| $x^{(1)}$ | 0.24  | 0.49  |
| $x^{(2)}$ | 0.73  | 0.65  |
| $x^{(3)}$ | 0.76  | 0.21  |

The experts' opinions are randomly generated according to the following parameters:

**Table 2.** Parameters of collectives

| MIN_MEM | MAX_MEM | COLLECTIVE |
| ------- | ------- | ---------- |
| 3       | 50      | 10, 20, 50 |
| 3       | 100     | 10, 20, 50 |

The *MIN_MEM/MAX_MEM* value presents the minimal/maximal number of members in a collective. The *COLLECTIVE* value is the number of collectives used with each k-member collective. For example, we experiment with each k-member collective is repeated 10/20/50 times. This will be helpful in giving more objective results about the relationship between the number of collective members and the quality of its knowledge. Additionally, the quality of k-member collectives is the average of the qualities of collectives.

For a real world, the real knowledge state is not known at the moment when experts are requested for giving their opinions about a real world. Simulating a real knowledge state is thus an important issue. In this work, we assume that a real knowledge state is an element from the set U and the knowledge states in a collective must be close enough to that element. From this assumption, a real knowledge state $r^*$ is a random element which satisfies:

- o   $r^* \in U$,
- o   $\forall i = 1, 2, \dots, n : d(r^*, x^{(i)}) \leq 1.0$

Meaning a real knowledge state is a member belonging to the set U which the distance to the furthest member in a collective cannot greater than 1.0. The procedure starts by generating a collective with the maximum number, then the real knowledge state of each collective is also generated in this step *(set R)*. After that the number of members is lowered to the minimal number (each step by 1 member). The real knowledge state of each collective is also used to evaluate the quality of collective knowledge when the number of members is lowered. The procedure is repeated for the rest number of collectives. Concretely, the procedure for determining qualities of collective knowledge is described as follows:

**Algorithm 1:** Determining qualities of collective knowledge of collectives

---

**Input:** min/max – minimal/maximal number of members
      k – number of collectives
**Output:** qualities of collective knowledge of collectives

    **Initial phase:**
        $X = \{X_1, X_2, ..., X_k\}$ – set of collectives
        $X_i = \{x^{(1)}, x^{(2)}, ..., x^{(max)}\}$ – members of collective $X_i$
        $R = \{r_1, r_2, ... \ r_k\}$ – real knowledge states of k collectives
        $Q = \varnothing$
**BEGIN**
**For** $n = \{max, max - 1, ..., min\}$ **do**
    SUM = 0.0
    **For** each $X_i$ from $X$ **do**
            Determine the knowledge of the collective $(x_i^*)$
            $SUM \mathrel{+}= d(x_i^*, r_i)$
            Remove a member from $X_i$
    $Q = Q \cup SUM/k$
**END**

---

## 4.2   Experimental Results and Evaluation

As mentioned in previous section, we experiment with parameters according to Table 2. However, we only present experimental results of collectives with parameters (3, 50, 20) and (3, 100, 20) because of the page limitation.

**Table 3.** Experiment results

| Members | Quality |
|---------|---------|
| 3 | 0.6981 |
| 4 | 0.6986 |
| ⋮ | ⋮ |
| 49 | 0.9159 |
| 50 | 0.9186 |

Table 3 presents qualities of collective knowledge of collectives from 3 to 50 members. Each quality value is the average of qualities of 20 collectives with the same k members. Generally, from this table we can see that the knowledge of collectives with more members will closer to the real knowledge state than that of collectives with fewer members. It means the higher the number of members in a collective, the better the quality of collective knowledge. However, in order to have a good understanding about the relationships between these values, we present an analysis of experiments to determine how the number of members in a collective influences the quality of its knowledge. In other words, whether solving a problem in the real world with more experts are better in getting more proper solution than that with fewer experts. First of all, we consider the following charts:

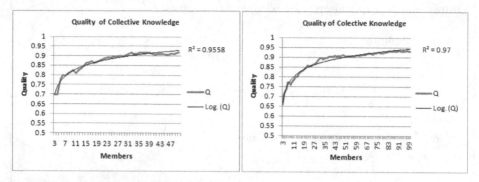

**Fig. 2.** Collective knowledge quality of collectives

According to Fig. 2 when the number of collective members is larger, the quality of collective knowledge better. Generally, collectives with more members will give a better solution to the real knowledge state than collectives with fewer members. However, we need to know how strong the relationship is? For this aim, a statistical test (correlation coefficient) is used to evaluate the strength of the relationship. The data from our experiments do not come from a normal distribution (according to the Shapiro-Wilk tests). Therefore, the Spearman correlation is used to measure the relationship between the number of members and the quality of collective knowledge.

**Table 4.** Correlation coefficient between the number of members and the quality

| Parameters | rho | p-value |
|---|---|---|
| (3, 50, 20) | 0.924 | < 2.2e-16 |
| (3, 100, 20) | 0.986 | < 2.2e-16 |

From Table 4, the relationship between the number of members in a collective and the quality of collective knowledge is very strong. In addition, the *p-values* are very small. Therefore, we can state that the correlation coefficient in this case is statistically significant and the relationship is reliable. Concretely, when the number of members in a collective increases; the difference between the collective knowledge and the real knowledge state decreases. Generally, the statement *the larger the number of collective members, the better the quality of its knowledge* is true. The better quality is the closer one to the real knowledge state. From this fact, inviting more experts to solve a problem in the real world is considered as an approach for improving the quality of collective knowledge. Of course, this statement should be formally proved and should be taken into account more specific conditions. In addition, the cost problem and the number of experts are also taken into account. This should be the subject of future work.

According to [6], the quality of collective knowledge is based on the average of distances from the collective knowldge to the members in a collective. Thus adding a member which is the knowledge of a collective leading to improve the quality of collective knowledge. In this work, however, the quality is measured by the difference between the real knowledge state and the collective knowledge. Then adding this kind of member does not cause the quality of collective knowledge to be changed. For more concrete, we consider the following theorem:

**Theorem 1.** *For given collective*

$$X = \left\{ x^{(i)} = \left( x_1^{(i)}, x_2^{(i)}, \dots, x_m^{(i)} \right) : i = 1, 2, \dots, n \right\}$$

*Let vector* $x = \{x_1, x_2, \dots, x_m\}$ *be its knowledge satisfying criterion* $O_2$.
*Let* $X' = X \,\dot\cup\, \{x\}$ *and* $x'$ *be knowledge of collective* $X'$; *then*

$$\hat{d}(x, X) = \hat{d}(x', X')$$

**Proof.** According to [6] $x$ is also the knowledge of collective $X'$. Thus $x$ and $x'$ are identical then we have:

$$\hat{d}(x, X) = d(x, r^*)$$

and

$$\hat{d}(x', X') = d(x', r^*) = d(x, r^*)$$

The theorem 1 is proved.                                                                 ♦

From this theorem, the quality of collective knowledge will not be better or worse if added member is the knowledge of a collective. This is only useful in improving the consistency degree of collectives. Therefore, in the case of existing a real knowledge state and the members in a collective reflect it to some degree, this kind of member is not taken into account in improving the quality of collective knowledge.

## 5      Conclusions

The paper has proposed a method for improving the quality of collective knowledge by taking into account the number of members in a collective. By means of experiment analysis using multi-dimensional vector structure, the authors have stated that the number of members in a collective has influence on the quality of its knowledge. Concretely, the knowledge of a collective with more members is better than that with fewer members. This approach could be considered as a method for improving the quality of collective knowledge. Besides, the quality is unchanged if added members are the knowledge of a collective. The future works should be investigating the problem with other structures such as: binary vector, relational structure, etc. In addition, we also aim to propose a mathematical model to prove the influence of the number of collective members on the quality of its knowledge. To the best of our knowledge, paraconsistent logics could be useful [4]. This approach can be used for evaluating the trust degrees of agent in a multi-agent system [11].

# References

1. Day, W.H.E.: The consensus methods as tools for data analysis. In: Bock, H.H. (ed), Proceedings of IFCS 1987 Classification and re-lated methods of data analysis, pp. 317–324. North-Holland (1987)
2. Gębala, M., Nguyen, V.D., Nguyen, N.T.: An analysis of influence of consistency degree on quality of collective knowledge using binary vector structure, New Trends in Computational Collective Intelligence, Studies in Computational Intelligence (eds. Camacho, D., Kim, S.-W., Trawiński, B., 2015), pp. 3–13. Springer International Publishing (2015)
3. Herrera-Viedma, E., Herrera, F., Chiclana, F.: A consensus model for multiperson decision making with different preference structures. IEEE Transactions on Systems, Man and Cybernetics, Part A: Systems and Humans 3(32), 394–402 (2002)
4. Nakamatsu, K., Abe, J.: The paraconsistent process order control method. Vietnam Journal of Computer Science 1(1), 29–37 (2014)
5. Nguyen, N.T.: Using consensus methods for solving conflicts of data in distributed systems. In: Jeffery, K., Hlaváč, V., Wiedermann, J. (eds.) SOFSEM 2000. LNCS, vol. 1963, pp. 411–419. Springer, Heidelberg (2000)
6. Nguyen, N.T.: Advanced methods for inconsistent knowledge management. Springer-Verlag, London (2008)
7. Nguyen, N.T.: Inconsistency of knowledge and collective intelligence. Cybernetics and Systems 39(6), 542–562 (2008)
8. Nguyen, N.T.: Processing inconsistency of knowledge in determining knowledge of collective. Cybernetics and Systems 40(8), 670–688 (2009)
9. Shermer, M.: The science of good and evil. Henry Holt, New York (2004)
10. Surowiecki, J.: The wisdom of crowds, Anchor (2005)
11. Sliwko, L.: Nguyen N.T.: Using Multi-agent Systems and Consensus Methods for Information Retrieval in Internet. International Journal of Intelligent Information and Database Systems 1(2), 181–198 (2007)
12. Wu, Z., Xu, J.: A consistency and consensus based decision support model for group decision making with multiplicative preference relations. Decision Support Systems 3(52), 757–767 (2012)

# Text Processing
# and Information Retrieval

# A Machine Translation System for Translating from the Polish Natural Language into the Sign Language

Wojciech Koziol[1], Hubert Wojtowicz[2]([✉]), Daniel Szymczyk[3],
Kazimierz Sikora[4], and Wiesław Wajs[5]

[1] Faculty of Mathematics and Nature, Chair of Computer Science,
University of Rzeszow, Rzeszow, Poland
[2] Faculty of Mathematics and Nature, Interdisciplinary Centre for Computational
Modelling, University of Rzeszow, Rzeszow, Poland
hubert.wojtowicz@gmail.com
[3] Institute of Physiotherapy, University of Rzeszow, Rzeszow, Poland
[4] Faculty of Polish Language, Chair of History of Language and Dialectology,
Jagiellonian University, Cracow, Poland
[5] Faculty of Electrical Engineering, Institute of Automatics,
AGH University of Science and Technology, Cracow, Poland

**Abstract.** A scheme for machine translation of texts between sign language and phonic language is proposed in the paper. The approach is based on the creation of substitution grammar of the sign language through the realization of semantic markers in the database of the Polish language and the application of graph structures implemented in the Prolog language. A modular architecture of the translation system implemented using this approach is described in the paper. Communication and interoperation between the system's modules is explained. Details of the principles of operation of the translation system are discussed. As a proof of concept an example of the translation process is presented.

## 1 Introduction

Automatic translation of texts of phonic natural languages into the sign language is a complex task both in terms of linguistic and computer science aspects. Its complexity cannot be reduced to the quantity of lexis. Disparities confirmed in lexicographical sources of a few thousands to tens of thousands of lexical units between phonic and sign languages, can not hide the fact, that the sign language is as efficient tool for communication as the phonic language. Therefore no substantial basis exists for identifying language based on signs with a limited communication code. This situation requires focusing attention on proper construction of the plane of reference and basics for the equivalence of translation. At the current stage of our research on supporting communication between hearing and deaf people, a unilateral translation of statements from the Polish natural language into the sign language has been implemented. It should be emphasized, that the attempts undertaken worldwide to automate the analysis

N.T. Nguyen et al. (Eds.): ACIIDS 2015, Part I, LNAI 9011, pp. 87–96, 2015.
DOI: 10.1007/978-3-319-15702-3_9

of visual information, expressed in the form of gestural code, so far have not yielded satisfactory results. Difficulties in automating the translation process stem from the specifics of visual communication and the perception of sign language gestures. The prospect of developing a functional tool for translation from the sign language to the phonic language slips the further away, the better the deficiencies of currently developed translators are recognized. The efficiency of these systems is limited to the recognition of rarely more than 100 signs and only on a pre-declared narrow repertoire of the simplest sentences. In a situation, in which the system effectively recognizes only individual gestures and their isolated sequences, the usefulness of such tools (also in the analytical procedure) becomes illusory. Many significant goals in the field of automatic translation as of yet remain unrealized. In this situation undertaking research on the processing and translation of sign language messages into the phonic natural language seems to be fully justified. The research undertaken by the authors has a very important social dimension. In accordance with the directives of European Union's social policy, the results of the conducted research can be used to counteract social exclusion and to help in overcoming the barriers, which isolate deaf people in society. This research can also help to improve the quality of life of the relatively large group of deaf and hearing impaired people (in Poland this group numbers about 50000 people).

## 2    A Translation System Architecture

A complete translation system architecture consists of two communication pathways. First of them concerns the route that the message sent by a hearing person must traverse in order to be read and understood by a deaf person. The second concerns the route which must be passed by a message sent by a deaf person to be read and understood by a hearing person, not speaking in the sign language. The term 'route' is used in a sense, which from the point of computer science view entails a set of procedures necessary in the process of translation, which are taking advantage of certain data sets. These procedures provide the conversion of the Polish natural language sentences into the equivalent sign language messages in the form of hand gestures with face mimicry, arranged in the appropriate order and comprehensible for the deaf people. Implementation of any of the communication pathways is a complex process. Even a brief reflection on the architecture of both pathways allows to notice, that in both cases it is necessary to include grammar and semantics knowledge-bases for the words of the Polish natural language. Particularly important is the role of semantic analysis for the second pathway, because the source information originating from a deaf person is often incomplete. Usually this information is abbreviated to the minimum and deprived of grammar properties. Additionally the information components may be presented in an inverted order. Lack of grammatical collocation makes the system dependent only on features of semantic collocation, which appropriately recognized allow to restore the full content of sentence in the Polish language on the basis of connotation scheme of the verb. For the second pathway realization of hand gesture recognition task is also necessary. It involves the selection of the appropriate gesture acquisition technology, accumulation of the gestures' patterns database and creation of an effective gesture recognition method capable of working in diverse environments. It should be noted that currently no complete solution exists for the recognition of sign language gestures, translation of gestures into text and synthesis of Polish natural language sentences on the basis of translated text components. The architecture of the translation system realized as practical part of this research entails one-way communication, i.e. process of entering the message generated by a hearing person into the system, message processing into the form of sign language gestures, and finally visualization of translated message in the form of sign language gestured using 3D technology - which means implementation of first communication pathway [1].

## 2.1 Modular Architecture of the System

In the early stages of the system implementation it's been assumed that due to a degree of complexity the system should have a modular architecture. This solution provides scalability, allows combining of various information technologies and also allows the use of appropriate programming paradigms. Splitting of the system into the modules also allows the deployment of individual modules on different machines connected through the network. In the Fig. 1 a modular architecture of the system realizing translation of the Polish natural language to the sign language is shown.

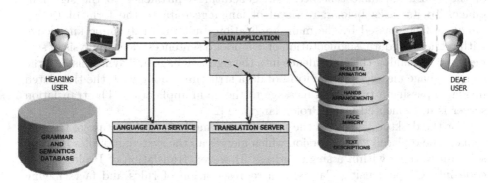

**Fig. 1.** Scheme of the translation system architecture

In the system following modules can be distinguished:

**Main module** - also described as main application - denoted in red colour. It constitutes the most important part of the system responsible for interaction with users. It gets the text in Polish language from the console and passes it to the language data service. It also retrieves the data from language data service and passes them to the translation server. It receives from the translation server the translated content in the form of data structures storing list of gestures to be signed along with hands arrangements and face mimicry of the 3D model.

Finally it visualizes the gestures in 3D environment. The main application thus serves the role of presentation layer of the system. It is implemented in the C# language.

**Language data service** - a process responsible for handling access to grammar and semantics databases. It obtains from the main application a text entered by a user and parses it into a list of words. It searches the grammar and semantics databases for the overloads of the particular word. Overloads mean the occurrence of the same word in different meanings, for different parts of speech and in form of conglomerates, i.e. multi-word phrases, which together constitute a separate semantically meaningful entity. The translation service retrieves a complete set of overloads for a particular word. One of the overloads, in a course of an analysis, will match the semantic meaning of the particular entity being a part of the segment or a sentence. The service also retrieves a set of morphological features for the particular word and attributes determining syntactic and semantic collocation of a given unit with other units in a sentence. In addition an attribute labeled *sign_name_blender* is collected as a unique test key representing name of the sign gesture in the database. The field storing the value of this key connects sign gestures in the main database with lexemes in the grammar and semantics database. The retrieved set of information is transformed into the form of appropriate facts of Prolog language, and then returned to the main application, which passes it to the translation server. The language data service is implemented using C# language.

**Translation server** - a process responsible for performing deep analysis of the Polish language sentences and generating equivalence to the sign language. To its tasks belong: retrieval of language data in the form of Prolog language facts passed by the main application; building of dynamic knowledge with retrieval of data; segmentation of compound sentences; deep analysis of segments; determining the equivalence into the sign language and synthesis into the sign language messages in the form of data structures along with the translated message, passing a translated message to the main application. The translation server is implemented using Prolog language [2].

For the desktop version of the application a modular architecture of the system can be replaced with a version which merges all the system modules into one solution working within a single process. However, translation of Prolog source code into C# and using classes as a representation of rules and facts having specific data structure is inconvenient. For the desktop version of the application there exists a possibility of disposing of a PostgreSQL database server along with its database management system. It simplifies software installation process on an end user machine. In order to bypass the need to install PostreSQL server in the desktop version of the application it is necessary to serialize the language database. However in this case the serialized files have to be loaded into RAM and de-serialized every time the program is started. This process takes significant amount of time, even in case of binary serialization. Additionally operational memory is burdened with large amounts of data, throughout the entire period of application operation. Application of SQLite database technology instead of

PostgreSQL is ruled out due to the fact, that SQLite has a much poorer range of data types - it lacks the array type used for representation of language data in the PostgreSQL database system.

## 2.2 Communication and Interoperation between the System's Modules

Modular solution to the translation system architecture brings many benefits like: flexibility, scalability, partitioning of the problem into smaller parts, possibility of distributing computations etc. Modular approach creates the necessity of providing the appropriate data link between the modules of the system. Due to the fact, that in the future it is planned to make the application available on the Internet, for the data link between modules TCP/IP network protocol was chosen. For the desktop version of the application, taking into account that all computations are carried out on a single machine, TCP/IP protocol can be substituted by streams, shared memory or other inter-CPU communication mechanism. In the Fig. 2 architecture of the system with specification of data links between modules is shown.

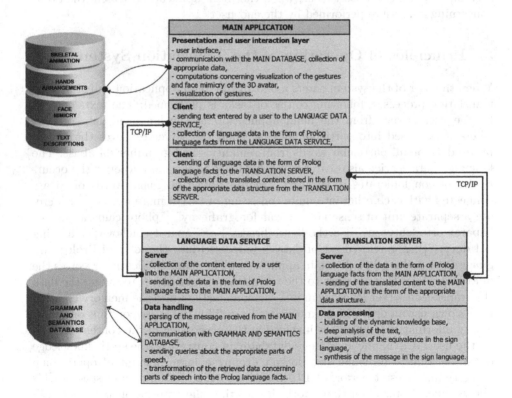

**Fig. 2.** Scheme of translation system architecture with specification of data links between modules

The main application contains two threads of TCP/IP client. In the first thread it is connected to the language data service, to which it passes text entered from the console by the end user. The service parses the received text, retrieves appropriate data from the language data base and sends it back to the main application in the form of Prolog language facts. The second client thread concerns the connection between main application and translation server. It passes the data obtained from language data service to the translation server. Then the translation server performs analysis and translation to the sign language, and afterwards it sends the resulting translation to the main application. On the basis of the obtained data the main application carries out visualization of the content in the sign language.

Language data service and translation server contain TCP/IP server threads. These threads monitor the appropriate ports for the data sent by the client, which means the main application. After the receipt of the data, it is processed and sent back to the client. Interoperation in the double client-server architecture continues until the moment of sending the finishing phrase 'quit' to the client. This phrase is the information for the processes of the language data service and to the translation server instructing them to finish their operations in the system. The 'quit' phrase is sent, when the main application detects the event concerning its closure performed by the end user.

## 3    Principles of Operation of the Translation System

When the user of the system enters a text in the main application and starts the translation process, a following course of tasks is performed: The texts entered by the user is sent from the main application to the language data service, where it is parsed into particular words. Then the server retrieves the appropriate data on all particular words from semantics and grammar database. The language data service also searches the database taking into account the occurrences of conglomerates in the text. (In the Polish language words often are connected with each other into units consisting of two or more words, which create a separate unit of sense i.e. "aparat fotograficzny" ["photo camera"] is not "aparat ortodontyczny" ["orthodontic braces"] or "aparat słuchowy" ["hearing aid device"]). A retrieved set of data is processed into the form of Prolog language facts and send to the main application. The main application receives the data and passes them further to the translation server. After receiving the data the translation server loads them in a dynamic fashion into the memory creating for its own use an operational knowledge base. This database contains only necessary informations, which are involved in the processing of the content entered by the user. A dynamic knowledge database, in contrast to the static knowledge database, due to its small size doesn't occupy a large amount of operational memory and doesn't perform CPU-intensive searches of large data spaces. This allows much faster algorithms work. Re-starting the translation process erases the dynamic knowledge database from the operational memory and re-creates it anew on the basis of the current text entered by the user. After the construction of the dynamic knowledge-base in the memory, algorithms carrying out segmentation of compound sentences found in the text are executed. These algorithms analyze relationships of hipotaxis and parataxis of compound sentences. The segmentation process involves the creation of an array of structures describing segments, which are formed for every possible overload of verb or participle. For every clause of the compound sentence many segments can be generated, depending on the number of overloads of particular verb/participle present in the dynamic knowledge-base. The structure describing the segment contains: default beginning and end range of the segment; segment identifier; number of overloads and position of verb/participle, which fulfills the role of predicate of the segment; identifier of the structure containing structures of required and matched elements, which were identified in the segment; and lastly structures created in the process of equivalence determination. The ranges determining breadths of segments are used in the further stages of the analysis during the search for particular phrases and elements inside the segment. After completion of the segmentation, follows the process of building a dynamic knowledge base in the form of structures describing particular segments of the sentences. These structures contain syntactic and semantic requirements, which define rules of syntactic and semantic collocations of verb predicates in particular segments with obligatory nominal phrases for the particular segment. The requirements concerning occurrences of certain words in the segment are also defined during this stage. For example for the third person in the imperative mode the predicate requires the occurrence of modal expression "niech" ["let"], and the predicate expressed in past tense requires the occurrence of auxiliary word created from the verb "być" ["to be"] in the specific form i.e. "będę, będziesz, będzie," [variations of the verb "to be" in Polish language]. In the next stage the algorithms searching for required elements are executed for all segments. The manner of conducting the search is based on requirements structures defined in the previous stage. The search returns obligatory nominal phrases for the segments taking into account positions of subject, direct object and indirect object. In the next phase comparison of structures defining requirements with structures obtained in the search (matching) phase is conducted. If for the structure describing the segment exists a discrepancy between required and found elements then the structure in this form is considered as incomplete and deviant. It is therefore removed. By the elimination of such structures only those overloads remain, which satisfy the conditions of the segment. The system accepts incomplete structures only in the case, when the match is missing only for a single connotation field, and in the semantic scheme there exists no expansion slot for the subordinate clause. An additional condition for the acceptance of such an incomplete structure is the occurrence after a given segment of the analyzed text of an appropriate expansion slot announcing the incorporation of the subordinate clause. Carrying out a process of incomplete segments elimination, doesn't guarantee that after its completion only one target variant of the segment remains for every clause of the compound sentence. It happens that for the verb having multiple meanings, connotative schemes are similar to each other or even identical. In this case, the system will generate a greater number of correct segments. The selection of the single most appropriate scheme will then be made on the basis of statistics. In practice, this procedure boils down to the selection of the first completed structure. This is done based on the fact that the creators of "Dictionary of Syntactic Generative Verbs of the Polish Language" [3], while arranging the sequence of schemes for verbs characterized by polysemy, described them according to the frequency of their occurrence in the text. At this stage, the number of segments stored in the segment table is equal to the actual number of segments, i.e. the number of clauses of the compound sentence. The selection of an appropriate connotation structure for every clause of the compound sentence is followed by a stage of the search of non-compulsory parts of the segment, i.e. adverbials and attributives (attributive designations) belonging to the nominal phrases' structures. Identification of obligatory elements ultimately determines the breadths of all segments and closes the process of text analysis. It must be added that all elements identified in the course of the analysis are stored in the structure of identified elements as overloads identifiers. Completion of the analysis allows for the execution of processes of equivalence determination and message synthesis into the sign language. Determination of equivalence involves creation of the structure, which is symmetric to the structure of identifiers, which contains the found words in the form of identifiers of their overloads. Every overload stored in the dynamic knowledge-base has an argument, in which a pointer is coded to the animation of the gesture stored in the main database. This argument's name is *sign_name_blender*. The system will therefore iterate through the entire structure of identifiers and will build a structure of equivalences, changing identifiers of overloads to the values of the *sign_name_blender* arguments of these overloads. In case when the value of the *sign_name_blender* argument is not set, the system will add animation, for which the word will be spelled using finger spelling system. Because the symmetry of the structure of equivalences retains the information about nominal phrases, the system is able to arrange all phrases identified inside the segments according to the order dictated by the syntax of the sign language. Recognition of the type of sentence and its internal structure allows also to specify, which words in the sentence should additionally carry an information about the face mimicry of the 3D model. As a result of the synthesis process, a list of sequentially aligned segments containing the information enabling generation of model's animations is created. Each of the segments contains the identifier of the sign language gesture, face mimicry identifier and text, uttered by the avatar in the course of signing the message. On the basis of data received, the main application retrieves from the main database appropriate binary objects, which describe spatial movements of the model for the sign language gestures and deformations of lips and face mimicry. In the later stage transients for gestures and face mimicry are computed. Finally in the end application a complex data structure is created, for controlling of the model, on the basis of which a 3D animation is assembled and then used for relaying the content of the sign language message.

## 4   The Results of Text Analysis Process

The functioning of the system is described using an example of analysis of the observation sentence. For technical reasons only few key frames of 3D animation, which form the sequence of sign gestures, are presented (Fig. 4). Below the analyzed sentence written in the Polish natural language and its English translation is presented, followed by the appropriate sequence of signs:

Boli cię głowa.
Your head hurts.
[ciebie] + [boleć] + [głowa]

The sequence shown in Fig. 3 corresponds to the translation of the text from the natural language into the appropriate form of sign language sentence. From the set of overloads, of the word "head", which are in the dynamic database, the system chooses the overload, which obligatorily puts as a subject a human body part, which means it contains [human_part] label.

```
overload([0,0], głowa, rzeczownik, [m,1_p,r_z], 118254,
         [[null,pause,[human_part,animal_part],[],[]]], [glowa]).
overload([0,1], głowa, rzeczownik, [m,1 p,r z], 118256,
         [[null,pause,[plus_human],[],[]]], [madry,czlowiek] ).
overload([0,2], głowa, rzeczownik, [m,1_p,r_z], 118258,
         [[null,pause,[amateur],[],[]]], [czlowiek] ).
overload([0,3], głowa, rzeczownik, [m,1_p,r_z], 118259,
         [[null,pause,[official_position],[],[]]], [szef] ).
overload([0,4], głowa, rzeczownik, [m,1_p,r_z], 118260,
         [[null,pause,[plant_part,artifact_part],[],[]]], [glowa] ).
```

**Fig. 3.** Selection of the most appropriate overload for the analyzed word "head"

In a similar manner the system defines and recognizes the position of a human object (having the qualities of consciousness) in the form of pronoun. In the processed text, both of these elements (parts of sentence) are distinguished by inflection features: Subject (S) = nominative case, Direct object (O1) = accusative. Propositional modality of an interrogative sentence is expressed by the assignment of an appropriate sign gesture to the avatar.

## 5   Summary

Analytical procedures described in the paper are currently tested on the level of compound sentence structures. Works are hindered by objective problems related with weak unification and standardization of lexical unit of the sign

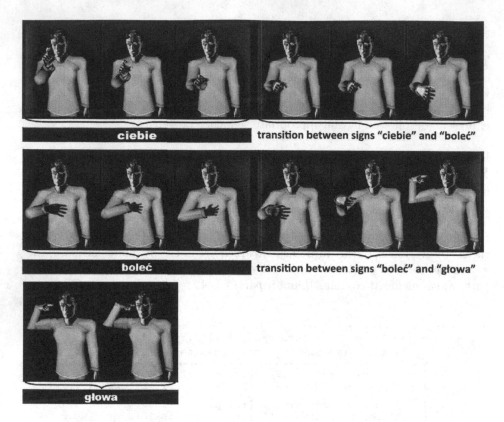

**Fig. 4.** Visualization of the translated sentence in the sign langauge

language (different conceptual variations and variability realization limiting the discernibility function of the spatial attributes). Overcoming this problem is in the conviction of authors a necessary condition in implementing the comprehensive translation based on gesture recognition from the sign language into the phonic language.

## References

1. Koziol, W., Wojtowicz, H., Sikora, K., Wajs, W.: A system for visualization of the polish sign language gestures. In: Proc. of the 16th Int. Conf. on Intelligent Multimedia Systems and Services (IMMS), vol. 254. IOS Press Frontiers of Artificial Intelligence (2013)
2. Koziol, W., Wojtowicz, H., Sikora, K., Wajs, W.: Analysis and synthesis of the system for processing of sign language gestures and translatation of mimic subcode in communication with deaf people. In: Graña, M., Toro, C., Howlett, R.J., Jain, L.C. (eds.) KES 2012. LNCS, vol. 7828, pp. 61–70. Springer, Heidelberg (2013)
3. Polanski, K.: Dictionary of Syntactic Generative Verbs of the Polish Language. Ossolineum, Wroclaw-Warszawa-Krakow-Gdansk (1980-1992)

# Graph-Based Semi-supervised Learning for Cross-Lingual Sentiment Classification

Mohammad Sadegh Hajmohammadi[1(✉)], Roliana Ibrahim[2], and Ali Selamat[2]

[1] Department of Computer Engineering, Sirjan Branch,
Islamic Azad University, Sirjan, Iran
hajmohammadi@iausirjan.ac.ir
[2] Software Engineering Research Group, Faculty of Computing,
Universiti Teknologi Malaysia, 81310 UTM Skudai, Johor, Malaysia
{roliana,aselamat}@utm.my

**Abstract.** Cross-lingual sentiment classification aims to use labelled sentiment data in one language for sentiment classification of text documents in another language. Most existing research works rely on automatic machine translation services to directly transfer information from one language to another. However, different term distribution between translated data and original data can lead to low performance in cross-lingual sentiment classification. Further, due to the existence of differing structures and writing styles between different languages, using only information of labelled data from a different language cannot show a good performance in this classification task. To overcome these problems, we propose a new model which uses sentiment information of unlabelled data as well as labelled data in a graph-based semi-supervised learning approach so as to incorporate intrinsic structure of unlabelled data from the target language into the learning process. The proposed model was applied to book review datasets in two different languages. Experiments have shown that our model can effectively improve the cross-lingual sentiment classification performance in comparison with some baseline methods.

**Keywords:** Cross-lingual · Sentiment classification · Graph-based · Semi-supervised learning

## 1 Introduction

Text sentiment classification refers to the task of determining the sentiment polarity (e.g. positive or negative) of a given text document [1]. Recently, sentiment classification has received considerable attention in the natural language processing research community due to its many useful applications such as opinion summarization [2] and online product review classification [3].

Up until now, different approaches have been employed in sentiment classification. These approaches can be divided into two main groups, namely; unsupervised and supervised methods. The unsupervised methods classify text documents based on the polarity of words and phrases contained in the text [4, 5]. This group of methods needs a sentiment lexicon to distinguish between the positive and negative terms. In contrast,

© Springer International Publishing Switzerland 2015
N.T. Nguyen et al. (Eds.): ACIIDS 2015, Part I, LNAI 9011, pp. 97–106, 2015.
DOI: 10.1007/978-3-319-15702-3_10

supervised methods train a sentiment classifier based on labelled corpus using machine learning classification algorithms [6, 7]. The performance of these methods intensively depends on the quantity and the quality of labelled corpus as the training set.

Based on these two groups of methods, sentiment lexicons and annotated sentiment corpora can be seen as the most important resources for sentiment classification. However, since most recent research studies in sentiment classification have been presented in the English language, there are not enough labelled corpus and sentiment lexicons in other languages [8]. Further, manual construction of reliable sentiment resources is a very difficult and time-consuming task. Therefore, the challenge is how to utilize labelled sentiment resources in one language for sentiment classification in another language. This subsequently leads to an interesting research area called cross-lingual sentiment classification (CLSC).

The most direct solution of this problem is the use of machine translation systems to directly project the information of data from one language into the other language [9-12]. The most existing research works develop a sentiment classifier based on the translated labelled data from the source language and use this classifier to determine the sentiment polarity of test data in the target language [13, 14]. Machine translation can be employed in the opposite direction by translating the test documents from the target language into the source language [15, 16]. In this situation, the sentiment classifier is trained based on the original labelled data in the source language and then applied to the translated test data. A few number of research works used both direction of translation to create two different views of the training and the test data to compensate some of the translation limitations [9, 10]. But because the training set and the test set are from two different languages with different intrinsic structures and writing styles and also originate from different cultures, these methods cannot reach the performance of monolingual sentiment classification methods in which the training and test samples are from the same language. Recently, some research works try to incorporate unlabelled document from the target language into the learning process of sentiment classification to fill the gaps between original and translated documents [9-11, 17, 18]. Although using unlabelled data from the target language can help to improve the classification performance, CLSC cannot reach the performance of mono-lingual sentiment classification because intrinsic structure of documents in the target language is fixed and different from the documents in the source language. Therefore, incorporating the intrinsic structure of documents in the target language is expected to result in better performance in CLSC. In fact, a good CLSC model should uses the information of the source language data while following the structure of the target language documents.

In this paper, a new model of CLSC is designed by taking into account the labelled documents in the source languages as well as the intrinsic structure of unlabelled documents in the target language. This model is based on the graph-based semi-supervised learning approach.

## 2     Related Works

Cross-lingual sentiment classification has been extensively studied in recent years. These research studies are based on the use of annotated data in the source language (always English) to compensate for the lack of labelled data in the target language. Most approaches focus on resource adaptation from one language to another language with few sentiment resources. For example, Mihalcea et al.[19] generated subjectivity analysis resources into a new language from English sentiment resources by using a bilingual dictionary. Wan [20] used unsupervised sentiment polarity classification in Chinese product reviews. He translated Chinese reviews into different English reviews using a variety of machine translation engines and then performed sentiment analysis for both Chinese and English reviews using a lexicon-based technique. Finally, he used ensemble methods to combine the results of analysis.

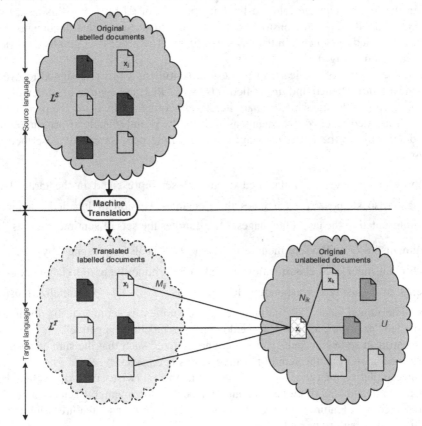

**Fig. 1.** Graph construction process in graph-based model

In another work, Wan [9] used the co-training method to overcome the problem of cross-lingual sentiment classification. In this paper, he exploited a bilingual co-training approach to leverage annotated English resources to sentiment classification in Chinese reviews. In this work, firstly, machine translation services were used to translate English labelled documents (training documents) into Chinese and similarly, Chinese unlabeled documents into English. The author used two different views (English and Chinese) in order to exploit the co-training approach into the classification problem. In an early work, Hajmohammadi et al. [11] tried to utilize multiple source languages in the process of CLSC. They showed that using more source languages can help to cover more information of sentiment terms in the classification process. To the best of our knowledge, graph-based method has not yet been investigated in the field of cross-lingual sentiment classification.

## 3     Proposed Model

This model is designed so as to incorporate the intrinsic structure of review documents in the target language into the learning process of CLSC. For this task, two different weighted graph are constructed based the translated labelled documents and original unlabelled documents in the target language. The process of graphs construction is illustrated in Fig. 1.

At the beginning of the learning process, a sentiment score is assigned to every document in both labelled and unlabelled sets. After that, the sentiment scores of review documents in the unlabelled set are iteratively computed by using the predefined labels of translated labelled documents as well as the pseudo-labels of original unlabelled documents in the target language. This learning process can be described as follows:

1. Suppose, $U$ denotes the unlabelled document set represented in the target language. Also suppose $L^T$ , denotes the translated version of labelled document set represented in the target languages. $Y_U$ denotes the sets of sentiment scores for documents in $U$ . The sentiment score set of $L^T$ is also represented by $Y_L$ .

2. Traditional supervised classification is used to determine the pseudo-labels of documents in $U$ using corresponding labelled sets, $L^T$. $Y_U$ is initialized using these determined labels. The initial label of a document is set to 1, if the document is labelled "positive", and to -1, if the document is labelled "negative".

3. The sentiment scores in each score set are normalized such that the sum of positive scores becomes 1 and the sum of negative scores becomes -1.

4. Cosine similarity measure is used to compute the pairwise similarity values between two documents (both labelled and unlabelled documents). Each document is represented by a feature vector, each entry of which contains a feature weight. TF-IDF is used as feature weights.

5. A graph is constructed based on the labelled and unlabelled documents represented in the target language. The nodes of this graph represent documents in $L^T$ and $U$ . The edges of this graph represent the content similarities between documents in $U$ and documents in $L^T$ . A similarity matrix, $M$ , is created from the documents in $L^T$ and $U$ and normalized such that the sum of each row becomes 1. The normalized matrix is sorted in descending order for every row in order to find the nearest neighbors of a document.

6. A matrix $\tilde{M}$ is used to denote the $k$-nearest neighbors of $U$ in the labelled set. Therefore, $Y_U$ , the sentiment scores of unlabelled documents , can be computed as follows:

$$Y_U^{(k)}(i) = \sum_{j \in \tilde{M}_i} (M_{ij} \times Y_L(j))$$  (1)

Where $Y_U^{(k)}(i)$ represents the sentiment score of $i$th document in $U$ at the $k$th iteration, and $Y_L(j)$ represent the sentiment score of $j$th document in $L^T$.

7. In the same way, a graph is constructed using only unlabelled documents. The nodes of this graph represent documents in $U$ and the edges denote the similarities between unlabelled documents. A similarity matrix, $N$ , is created from these similarity scores and also normalized such that the sum of each row becomes 1. The normalized matrix is sorted in descending order for every row in order to find the nearest neighbors of a document.

8. A matrix $\tilde{N}$ is used to denote the k-nearest neighbors of documents in $U$ . Therefore, $Y_U$ , the sentiment scores of unlabelled documents, can be computed as follows:

$$Y_U^{(k)}(i) = \sum_{j \in \tilde{N}_i} (N_{ij} \times Y_U^{(k-1)}(j))$$  (2)

9. In order to incorporate the sentiment scores of neighbors document in both labelled and unlabelled sets, the above iterative formulas are combined and two new iterative formulas are obtained to compute $Y_U$ as follows:

$$Y_U^{(k)} = \alpha M Y_L + \beta N Y_U^{(k-1)}$$  (3)

Where $\alpha$ and $\beta$ demonstrate the relative effect of labelled and pseudo-labelled data in final sentiment score computation and $\alpha+\beta=1$.

10. $Y_U$ is normalized at every iteration such that the sum of positive scores becomes 1 and the sum of negative scores becomes -1. This normalization process is needed for algorithm convergence. The iterative process is continued until convergence.

11. A sentiment label is assigned to each document in unlabelled pool according to calculated sentiment scores in $Y_U$ . If the sentiment score is in the range of 0 to +1, then the document is labelled as "positive". If this score is between -1 and 0, then the document is labelled as "negative".

The convergence of the algorithm occurs when the difference between the sentiment scores calculated at two consecutive steps of algorithm for all unlabelled examples falls below the certain threshold.

As described in this process, the sentiment score for each unlabelled document is calculated based on two different graphs. One graph is constructed to connect the unlabelled documents to the labelled documents and another graph is constructed to represent the inter-connection of unlabelled documents. Consequently, the sentiment score of an unlabelled document is computed by incorporating the similarities of that document to the labelled documents as well as its similarities to other pseudo-labelled (unlabelled) documents. This means that each unlabelled document receives a sentiment score from both labelled and unlabelled examples. Due to the existence of similar intrinsic structures among unlabelled documents, incorporating their sentiment scores is expected to improve the performance of CLSC in compare to other methods.

# 4     Evaluation

In this section, we evaluate our proposed approach in CLSC on two different languages in the book review domains and compare it with some baseline methods.

## 4.1     Datasets

Two different evaluation datasets have been used in this paper.

- English-Japanese dataset (En-Jp): This dataset contains Amazon book review documents in English and Japanese languages. This dataset was used by Prettenhofer and Stein [21] .
- English-Chinese dataset (En-Ch): This dataset was selected from Pan reviews dataset [18]. It contains book review documents in English and Chinese languages.

Table 1 shows the characteristics of these two datasets. All review documents in the source language (English) are translated into the target languages using the Google translate engine[1]. In the Japanese text document, we applied MeCab[2] segmenter software to segment the reviews; while Chinese documents were segmented by the Stanford Chinese word segmenter[3]. In the feature extraction step, unigram and bi-gram patterns were extracted as sentimental patterns. To reduce computational complexity, especially in density estimation, we performed feature selection using the information gain (IG) technique. We selected 5000 high score unigrams and bi-grams as final features. Each document is represented by a feature vector, each entry of which contains a feature weight. We used TF-IDF as feature weights.

---

[1] http://translate.google.com/

[2] http://mecab.googlecode.com/svn/trunk/mecab/

[3] http://nlp.standfor.edu/software/segmenter

**Table 1.** Charactristices of datasets used in the evaluation

| Dataset | Domain | Languages | | Total documents | Positive documents | Negative documents |
|---|---|---|---|---|---|---|
| En-Ch [18] | Book review | Source Language | English | 2000 | 1000 | 1000 |
| | | Target Language | Chinese | 4000 | 2000 | 2000 |
| En-Jp [21] | Book review | Source Language | English | 2000 | 1000 | 1000 |
| | | Target Language | Japanese | 4000 | 2000 | 2000 |

## 4.2    Baseline Methods

The following baseline methods are implemented in order to evaluate the effectiveness of proposed models.

— Co-training: This is the traditional co-training algorithm which was used in the study by [9, 22]. It uses labelled data from the source language and unlabelled data from the target language in two views.
— Structural Correspondence Learning model (SCL): This model was implemented as introduced in [13]. The Google Translate service was used to map words in the source vocabulary to the corresponding translation in the target vocabulary. Other parameters were set as used in [13].
— Transductive SVM in the source language (TSVM): This method uses the well-known transductive learning model based on support vector machine (SVM) for sentiment classification. In this model a transductive SVM is trained based on the translated labelled document and original unlabelled documents.

## 4.3    Results and Discussion

In this section, the proposed model is compared with three baseline methods. In the proposed algorithm, $\alpha$ and $\beta$ were set to 0.4 and 0.6 respectively, which indicates the contribution from unlabelled data is a little more important than that from labelled data. The threshold also was set to 0.1e-08 for convergence condition. The parameter $k$ was set to 30 in $k$-nearest neighbor matrix. Cosine measure was used to determine the content similarity between documents.

Table 2 and Table 3 show the numerical results for comparing the proposed model and the baseline methods. As we can see in these tables, the proposed model can show a good performance in compare to all of the baseline methods and obtained the best accuracy in all datasets.

**Table 2.** Performance comparison in English-Japanese (En-Ch) dataset (best results are reported in bold-face type)

| Methods | Accuracy | Positive | | | Negative | | |
|---|---|---|---|---|---|---|---|
| | | Pre | Rec | F1 | Pre | Rec | F1 |
| Proposed model | **73.81** | 79.27 | 64.30 | 71.00 | 70.10 | **83.27** | **76.12** |
| Co-Training | 73.32 | 77.17 | 66.75 | 71.59 | 70.38 | 79.90 | 74.84 |
| SCL | 70.58 | 70.89 | 69.24 | 70.06 | 70.28 | 71.90 | 71.08 |
| TSVM | 71.75 | 71.60 | **71.85** | **71.73** | **71.90** | 71.64 | 71.77 |

**Table 3.** Performance comparison in English-Japanese (En-Jp) dataset (best results are reported in bold-face type)

| Methods | Accuracy | Positive | | | Negative | | |
|---|---|---|---|---|---|---|---|
| | | Pre | Rec | F1 | Pre | Rec | F1 |
| Proposed model | **72.72** | **75.18** | 67.83 | **71.32** | 70.70 | 77.61 | 73.99 |
| Co-Training | 72.27 | 74.54 | 67.73 | 70.96 | 70.40 | 76.80 | 73.45 |
| SCL | 69.50 | 72.89 | 62.39 | 67.23 | 67.26 | 76.60 | 71.63 |
| TSVM | 69.02 | 69.00 | **69.07** | 69.03 | 69.04 | 68.97 | 69.00 |

Compared to the co-training and SCL models, proposed model shows better overall accuracy in all datasets. This is due to the taking into account the intrinsic structure of documents in the target language during the sentiment scores prediction process.

In compare to transductive SVM (TSVM), the proposed model shows better performance in almost all datasets. This means that, incorporation of document similarities has a beneficial effect in the sentiment score prediction process.

# 5     Conclusion

In this paper, we have proposed a new graph-based semi-supervised learning model to improve the performance of cross-lingual sentiment classification. In the proposed model, automatic machine translation was used to project the information of source language documents into the target languages. Two different graphs were constructed based on the similarity measure between the labelled and unlabelled document and among unlabelled documents. The sentiment score of each unlabelled document was then computed through propagation of sentiment scores of labelled and unlabelled documents. This model was applied to the cross-lingual sentiment classification dataset in two different languages and the performance of the proposed model was compared with some baseline methods. The experimental results show that this model can improve the performance of CLSC in compare to the baseline methods.

# References

1. Liu, B., Zhang, L.: A Survey of Opinion Mining and Sentiment Analysis. In: Aggarwal, C.C., Zhai, C. (eds.) Mining Text Data, pp. 415–463. Springer US(2012)
2. Ku, L.W., Liang, Y.T., Chen, H.H.: Opinion extraction, summarization and tracking in news and blog corpora. In: Proceedings of AAAI-2006 Spring Symposium on Computational Approaches to Analyzing Weblogs (2006)
3. Kang, H., Yoo, S.J., Han, D.: Senti-lexicon and improved Naïve Bayes algorithms for sentiment analysis of restaurant reviews. Expert Syst. Appl. **39**(5), 6000–6010 (2012)
4. Turney, P.D.: Thumbs up or thumbs down? semantic orientation applied to unsupervised classification of reviews. In: Proceedings of the 40th Annual Meeting on Association for Computational Linguistics, Philadelphia, Pennsylvania: Association for Computational Linguistics (2002)
5. Taboada, M., et al.: Lexicon-based methods for sentiment analysis. Comput. Linguist. **37**(2), 267–307 (2011)
6. Pang, B., Lee, L., Vaithyanathan, S.: Thumbs up? sentiment classification using machine learning techniques. In: Proceedings of the ACL-02 conference on Empirical methods in natural language processing, Association for Computational Linguistics (2002)
7. Moraes, R., Valiati, J.F., Neto, W.P.G.: Document-level sentiment classification: An empirical comparison between SVM and ANN. Expert Syst. Appl. **40**(2), 621–633 (2013)
8. Montoyo, A., Martínez-Barco, P., Balahur, A.: Subjectivity and sentiment analysis: An overview of the current state of the area and envisaged developments. Decis. Support Syst. **53**(4), 675–679 (2012)
9. Wan, X.: Bilingual co-training for sentiment classification of Chinese product reviews. Comput. Linguist. **37**(3), 587–616 (2011)
10. Hajmohammadi, M.S., Ibrahim, R., Selamat, A.: Bi-view semi-supervised active learning for cross-lingual sentiment classification. Inf. Process. Manage. **50**(5), 718–732 (2014a)
11. Hajmohammadi, M.S., Ibrahim, R., Selamat, A.: Cross-lingual sentiment classification using multiple source languages in multi-view semi-supervised learning. Eng. Appl. Artif. Intell. **36**, 195–203 (2014b)
12. Balahur, A., Turchi, M.: Comparative experiments using supervised learning and machine translation for multilingual sentiment analysis. Computer Speech & Language (2013)
13. Prettenhofer, P., Stein, B.: Cross-language text classification using structural correspondence learning. In: Proceedings of the 48th Annual Meeting of the Association for Computational Linguistics, pp. 1118–1127. Association for Computational Linguistics, Uppsala, Sweden (2010)
14. Perea-Ortega, J.M., et al.: Improving polarity classification of bilingual parallel corpora combining machine learning and semantic orientation approaches. J. Am. Soc. Inform. Sci. Technol. **64**(9), 1759–1962 (2013)
15. Banea, C., Mihalcea, R., Wiebe, J.: Multilingual subjectivity: are more languages better? In: Proceedings of the 23rd International Conference on Computational Linguistics, pp. 28–36. Association for Computational Linguistics: Beijing, China (2010)
16. Balahur, A., Turchi, M.: Comparative experiments using supervised learning and machine translation for multilingual sentiment analysis. Comput. Speech Lang. **28**(1), 56–75 (2014)
17. Hajmohammadi, M.S., Ibrahim, R., Selamat, A.: Density based active self-training for cross-lingual sentiment classification. In: Jeong, H.Y., Obaidat, M.S., Yen, N.Y., Park, J.J. (eds.) Advanced in Computer Science and Its Applications. LNEE, vol. 279, pp. 1053–1059. Springer, Heidelberg (2014c)

18. Pan, J., Xue, G.-R., Yu, Y., Wang, Y.: Cross-lingual sentiment classification via bi-view non-negative matrix tri-factorization. In: Huang, J.Z., Cao, L., Srivastava, J. (eds.) PAKDD 2011, Part I. LNCS(LNAI), vol. 6634, pp. 289–300. Springer, Heidelberg (2011)
19. Mihalcea, R., Banea, C., Wiebe, J.: Learning multilingual subjective language via cross-lingual projections. In: Proceedings of the 45th Annual Meeting of the Association of Computational Linguistics (2007)
20. Wan, X.: Using bilingual knowledge and ensemble techniques for unsupervised Chinese sentiment analysis. In: Proceedings of the Conference on Empirical Methods in Natural Language Processing, pp. 553–561. Association for Computational Linguistics, Honolulu (2008)
21. Prettenhofer, P., Stein, B.: Cross-Lingual Adaptation Using Structural Correspondence Learning. ACM Trans. Intell. Syst. Technol. 3(1), 1–22 (2011)
22. Wan, X.: Co-training for cross-lingual sentiment classification. In: Proceedings of the Joint Conference of the 47th Annual Meeting of the ACL and the 4th International Joint Conference on Natural Language Processing of the AFNLP, pp. 235–243. Association for Computational Linguistics: Suntec, Singapore (2009)

# An Adaptation Method for Hierarchical User Profile in Personalized Document Retrieval Systems

Bernadetta Maleszka[✉]

Wroclaw University of Technology, Wybrzeze Wyspianskiego 27,
50-370 Wroclaw, Poland
Bernadetta.Maleszka@pwr.edu.pl

**Abstract.** Nowadays personalization systems becomes more and more famous. Usually, such systems gather information about a user to recommend him better results: web pages, documents, etc. Important aspect of modeling user procedures is to keep the profile up-to-date according to changes of user interests. In this paper a hierarchical model of user profile is considered. Connections between terms in a hierarchy reflects generalization relation. A set of assumption for profile structure is proposed. For such an user profile a method for its adaptation is presented and a quality criterion for this method is proposed. Experimental evaluation has shown that quality criterion is satisfied.

**Keywords:** Hierarchical user profile · User preference · User profile adaptation · Evaluating retrieval systems

## 1 Introduction

Traditional information retrieval systems, such as those based on content-based and collaborative filtering, tend to use fairly simple user models: vector of relevant documents or set of interests [1]. On the other hand it was shown that term dependence is a natural consequence of language use. Its successful representation has been a long standing goal for information retrieval research [14]. A generalization relation between terms is considered in hierarchical user model.

Hierarchical model can be treated as generalization of vector model and was previously considered in literature e.g. [7], [17] and by author in works [9], [10] and [11]. In this paper a method for user profile adaptation method is presented and methodology of simulations evaluation is described.

The main assumptions for the personalization system are as follows: user has a preference that the system does not know. The system builds a user model which consists of demographic data and user profile (hierarchical structure of weighted terms). It is desirable that profile should be close to user preference. Both user preference and profile should have hierarchical structure to have ability to compare them. A profile for a new user is generated based on profiles of similar users (with similar interests and demographic data). When user interacts with

© Springer International Publishing Switzerland 2015
N.T. Nguyen et al. (Eds.): ACIIDS 2015, Part I, LNAI 9011, pp. 107–116, 2015.
DOI: 10.1007/978-3-319-15702-3_11

the system, his profile is adapted based on his activities (queries and set of relevant documents to those queries). A set of assumptions about the profile structure are presented and method for adapting the profile is described.

To check the quality of proposed system a methodology of experimental evaluations is proposed. The experiments are performed without using real users which would be time-consuming process. The main idea of experiments are as follows. In the system we have many users that are clustered into $k$ groups. Based on this partition we determine a set of significant demographic data using method described in our previous work [13]. In the next step a new users are registering to the system. A non-empty beginning profile is determined and recommended for each new user. Then all users are interacting (ask queries and point relevant documents). After one or two blocks of sessions it is checked if quality criterion is satisfied.

The rest of the paper is organized as follows. In Section 2 we present a short survey of adaptation methods for hierarchical profile structure. The model of documents set, user preference and profile are presented in Section 3. The algorithm for profile adaptation method is also described. Section 4 contains information about experimental methodology and obtained results. In the last Section 5 we gather the main conclusions and future works.

## 2   Related Works

User profile should contain information about user to be used to personalized search. Gauch et al. [4] differentiate keyword profiles and concept profiles. Concept profiles are similar to keyword profiles in that often they are represented as vectors of weighted features, but the features represent concepts rather than words or sets of words. Various mechanisms are applied to express how much the user is interested in each topic. The simplest technique is a numerical value, or weight, associated with each topic.

A specific kind of concept profiles are hierarchical profiles while flat set of concepts does not contain information about relation between concepts. In literature of user modeling, many aspects about relations between nodes – terms are considered. The most often one can find the following interpretation of links:

- links reflect generalization relation between terms [12];
- if the term is on lower depth in the tree, it is more important [12];
- user profile is often a part of greater hierarchical taxonomy e.g. genre taxomony for movies [15];
- each node contains cluster of terms; terms on lower depth correspond to long-term user interests, while terms in leaves are short-term user interests [6], [18];
- a hierarchy is built based on current user activities – current user hierarchy is generated after each session and then is merged with existing user profile [3].

An exemplary methodology for the construction of a concept hierarchy which is constructed based on a set of user specified documents on a topic of interest is presented by Nanas et al. [14]. The first step of construction is to identify the most appropriate terms for building the concept hierarchy (user profile). Then, term weighting is used to assess the specificity of terms to the documents' underlying topic, i.e. their ability to distinguish documents about that topic from the rest of the documents in the collection. In the next step the associations between terms are found based on co-occurrence of these terms in documents.

In our system we assume that user profile is a part of greater structure, relation between nodes is generalization relation and the depth of node in hierarchy reflects its importance in user interests.

Hierarchical user structure needs specific method to learn and adapt user profile based on user activities.

Li et al. [8] propose dynamic adaptation strategies for long-term and short-term user profile. The taxonomic hierarchy for their long-term model is a part of the Google Directory [5]. In the Google Directory, each web page is classified into a topic. In the ,,adding" operation, topics associated with the clicked pages are added into the user topic tree click by click. Also, each node in the user topic tree has a value of the number of times the node has been visited. A short-term model caches the most recently clicked pages with a fixed size that is determined by the ability of the search engine. The procedure of updating short-term user profile is compared to cache management problem: which pages should be deleted from cache to have a place for new pages. The idea of adaptation process is based on changing degree of importance in long-term model observing short-term profile.

Another idea of adaptation method is presented by Yang et al. [18]. They consider two aspects of adaptation: how often adapt user profile and how to do it. The frequency of adaptation is usually dependent on when a session has finished. Such an approach allows to perform statistically tests because session contains many user queries and relevant documents. User profile also consists of long-terms and short-terms. Hierarchical model is generated using agglomerate clustering in data mining and pruning method to delete the most specific terms.

In this paper a method for generating and adapting user profile is also based on long-term and temporary profile. The adapting method uses collaborative recommendation. We propose a novel approach to evaluate the system.

## 3    Model of Retrieval System

In this section we provide necessary definitions and assumption about our system.

Documents in this system is described by the set of weighted terms:

$$d_i = \{(t_j^i, w_j^i) : t_j^i \in T \wedge w_j^i \in [0.5, 1), j = 1, 2, \ldots, m_i\} \tag{1}$$

where $t_j^i$ is a term from terms set $T$, $w_j^i$ is weight of this term, and $m_i$ is a number of terms is document $d_i$. We assume that term should belong to this set when its weight $w_j^i \in [0.5, 1)$. Terms with smaller weights are omitted.

## 3.1   User Preference and User Profile

Both user preference and user profile has hierarchical structure described by the following definition.

**Definition 1.** $H_0$-*based hierarchical user profile*
  $H_0$-based hierarchical user profile is a pair:

$$t = (H_0, w_t),$$

*where:*

– $H_0$ *is a thesaurus.*
– $w_t : V_0 \rightarrow [0,1]$ *is a function that assigns weights to nodes of $H_0$ in profile t. The weight reflects the level of user interests in considered term.*

The space that contain all hierarchical user profiles based on $H_0$ is denoted by **H**.

In our system we assume that user preference is determined by set of relevant documents. Based on the set, mean values of weights are calculated for each term that occur in document description.

To compare two profiles or profile with preference we use the following distance measures:

$$d_1(t_1, t_2) = \frac{\sum_{v \in V} \alpha^{h(v)} |w_{t_1}(v) - w_{t_2}(v)|}{\sum_{v \in V} \alpha^{h(v)}} \tag{2}$$

and

$$d_2(t_1, t_2) = \sqrt{\frac{\sum_{v \in V} \alpha^{2 \cdot h(v)} (w_{t_1}(v) - w_{t_2}(v))^2}{\sum_{v \in V} \alpha^{2 \cdot h(v)}}} \tag{3}$$

where $t_1$ i $t_2$ are hierarchical profiles, $w_t(v)$ is a weight of node $v$ in profile $t$, $h(v)$ is a depth of node $v$, and $\alpha$ is in the interval of $(0,1)$.

The importance of weight is considered in dependence of node's depth – the deeper node in tree, the less important impact of its weight on distance between trees. This property was presented previously in [12].

Both measures are non-negative and symmetric and meet triangle inequality condition. Both distances are equal to 0 for the same trees.

## 3.2   Constraints for User Profile

In hierarchical structure of user preference and profile we assume that terms (nodes) are in generalization relation. We propose the following constraints that result from specification of information retrieval domain.

– **Key Constraint (K0).** User query consists of terms not belonging to the same path in thesaurus $H_0$. In single query user should not use terms connected by generalization relation.

– **Minimum Total Frequency (K1).**

$$\sum_{v \in V} w_t(v) \geq 1$$

If the user query contains exactly one term in each query, the sum of frequencies of all terms is equal to 1. In a single query, the user can use more than one term (usually user query contains from 2 to 4 terms [2]). When calculating frequencies of each term in this situation, the sum of those frequencies should be greater than 1.

– **Maximum Path Frequency (K2).**

$$\forall_{p \in P} \sum_{v \in p} w_t(v) \leq 1$$

where $P$ is the set of all paths in the thesaurus. This constraint is a consequence of K0. If the user can use only one term from a given path when formulating a query, the sum of frequencies of all terms on this path should not be greater that 1. On the other hand, if the sum of weights on the same path were greater than 1, then when generating queries to model the user it would be possible to select randomly the parent and child node. This would be contrary to K0 constraint.

The profile should satisfy all these conditions. If any of these conditions is not met, the following normalization procedure should be performed [11].

1. Check if K1 is met – *If K1 is not met, then automatically K2 is met. To satisfy K1, recalculating (multiplying by a constant $c_1$) all weights in the tree is required. It is possible to assume a constant $c_1$ such that the sum of all weights is equal to 1. After recalculating, K1 is satisfied. At the same time, K2 will still be met, as at most the sum of weights on a single path will be equal to 1.*
2. Check if K2 is met – *If K2 is not met, then automatically K1 is met. To satisfy K2, we select the path with the largest sum of weights and recalculate (divide by a constant $c_2$) all the weights in the tree, so that the elements in this path sum up to 1. After recalculating, K2 is met. At the same time K1 is met, as at worst it will be equal to 1 (the sum of elements on that single path).*

**Remark.** Constraint K1 refers to lower boundary of weights sum. Similarly, it is possible to give upper boundary, which can be treated as additional constraint: Maximum Total Frequency (K3)

$$\sum_{v \in V} w_t(v) \leq b$$

where $b$ is the maximum number of terms in a single user query. Assuming that in every query user asks the maximum allowed number of terms $b$, the sum of the terms frequency is equal to $b$. When real user interacts with the system, he formulates queries composed of a few terms. Usually these terms are taken from user preferences. The studies described in [2] showed that more than 85% user queries consist of a maximum of three terms. Due to this fact, we can assume that $b = 3$.

### 3.3 A Method for User Profile Adaptation

User profile can be obtain in the following ways:

1. beginning profile is empty – terms and its weights are determined based on first user session;
2. beginning profile is recommended for user based on similar users according to his demographic data. The profile is an integrated profile for the group.

In the first case, the weight of term is calculated as a mean values of terms weights in relevant documents. Additional assumption is as follows: if user is interested in the term, he asks about this term more frequent. We can also assume that weight of term is proportional to frequency of term occurrence in user queries.

If user profile does not satisfy key assumption, it is necessary to rescale weights according to normalization procedure. This procedure is presented in Section 3.2.

In the second case, beginning profile is a representative profile of group of users and it satisfy key assumption K0 – K3.

In this paper we differentiate the following profiles:

– temporary profile – a profile determined based on current user activities;
– recommended profile – a representative profile of user's group;

The main aim of adaptation procedure is to change value of term weight based on user current queries. Profile is changed after each block of session. A temporary profile is determined based on terms from current user queries. Weight of term is calculated as average value of weights of the term in relevant documents in this session (Algorithm 1). After fix number of sessions (block of session) the temporary profile is merged with current user profile. The algorithm guarantees that key assumptions are satisfied.

## 4    Experimental Evaluation

### 4.1    Quality Criterion for Adaptation Method

During registration process, a new user is asked to fill a questionnaire about his demographic data. Based on these data he is classified into the group with similar values of demographic data. The first profile is recommended to the user. This profile contains integrated knowledge about whole group of users (a representative of this group). To check the quality of such procedure we propose the following criterion.

**Algorithm 1.** A method for determine a temporary user profile.

**Input**: $Q$ – a set of term and relevant documents that contain those terms;
**Output**: a temporary user profile
**foreach** *term $kw \in Q$* **do**
  Calculate an average value of terms' weights $w(kw)$ in relevant documents
  on current session;
  If profile does not satisfy key assumptions K0 – K3, run normalization
  procedure for this profile.

**Definition 2.** *Quality criterion for adaptation method*
    *If the user was classified into group $G_1$, then a distance between his profile
and centroid profile for group $G_1$ after a few sessions is the smallest distance
among distances between his profile and centroids of all other groups after a few
sessions.*

$$UP(0) = UP_i^* \Rightarrow d(UP(1), UP_i^*(1)) = min_{j=1,2,...,K} d(UP(1), UP_j^*(1))$$

*where: UP(0) is a user profile is session $s = 0$, $UP_i^*(1)$ is centroid of i-th group
in session $s = 1$ and $K$ denotes number of groups.*

The quality criterion is an intuitive assumption to be satisfied: when a new
user is classified into group of users with similar interests, his activities should be
closed to activities of other users in this group. This criterion should be satisfied
only with a few first blocks of sessions, while preference of each user can change
and each group evolves. After many blocks of sessions the partition should be
recalculated. In another case, system could not evolve despite of the fact that
users as elements of this system evolve (theirs preferences are changing with
time).

## 4.2    Experiments Methodology

The aim of performed experimental researches is to check if quality criterion is
satisfied. Due to the fact that performing experiments with real users is time-and-
cost consuming we propose a method to simulate user preference and activities
in document retrieval system. A details of simulations were described in our
previous works [9], citeaciids13. Here we present only idea of the methodology.

1. Generate set of documents (set of terms, documents and determine which
   terms describe each document).
2. Generate group of users and determine users profiles (profiles should be
   initially adapted based on previous users queries).
3. Determine partition of users into $k$ groups based on user profiles using $k$-
   means algorithm.
4. Determine a subset of demographic data that are the most important in
   classification procedure (a method was described in [13]).

5. Do the following instruction $l$ times:
   (a) Generate a new user.
   (b) Based on his demographic data classify him to the right group and recommend him a non-empty user profile.
   (c) Determine user profile after a few blocks of sessions based on current user queries and recommended user profile.
   (d) Check if result profile satisfies adaptation quality criterion.

User profiles determined in step no. 2 should be initially adapted while they are basis for clustering process. They should be close to user preferences.

## 4.3   Results of Simulations

At the beginning we generate set of terms and set of documents. Each document is described by a few weighted terms. Next, a set of users is generated (demographic data and preference based on set of relevant documents) and user behaviour is simulated and user profile is built and adapted. As a result each user has profile that is close to his preference. Users are clustered into $k$ group based on theirs profiles using $k$-means algorithm. A minimal set of significant demographic data is determined for obtained partition – the algorithm was described in details in [13] and [9]. Here we present only brief idea of this algorithm.

The users are described by two independent set of attributes – by demographic data and usage data (profile). In the first step, the users are clustered based on usage data. The partition of users is used to determine the minimal set of attributes coming from the demographic data that the partition obtained using selected demographic data is as close as possible to partition obtained in the first step.

Using proposed method it is enough to ask user only about actually important demographic information that user needs to introduce when starting interaction with the system. Each group of users is described by minimal set of attributes determined in the previous step. A new user should be classified into the group of users which are the most similar in terms of demographic information. The system does not need to wait until user has enough usage data to recommend him some propositions. In our approach a non-empty profile will be recommended for the new user. Based on the knowledge about other users in the same group the first profile will be determined as a centroid of this group. While user is interacting with the system and his usage data are gathered, his profile is adapted to his preference.

Experiments were performed for the following values of parameters:

- number of groups $k = 10$;
- number of user in basis group: 100;
- number of new users: 10;
- number of adaptations: 1, 2 and 3;
- number of significant attributes: 4.

We assume that 4 demographic data is significant feature – it means that all users are divided into 16 groups according to demographic data (single attribute can have binary value). Some of these groups can be empty,so we assume only 10 groups when we divide users according to theirs profiles.

The main objective of experiments is to check if the user was classified into proper group – it means that after a few blocks of sessions, distance between user profile and centroids of all groups is the smallest for the group that user belongs to. The experiments were performed for 1, 2 and 3 blocks of sessions. In single simulation a number of user's group was compared with number of centroid of the cluster that the distance between centroid and profile was the smallest one. Average obtained results are presented in Table 1. The quality criterion for adaptation method is satisfied in 55% − 65% cases for one or two blocks of sessions. This value decreases for more blocks of sessions. It can be a reason to make a new partition for all groups of users. Quality criterion for adaptation

**Table 1.** Simulation results: average value of measure for quality criterion for subsequent blocks of sessions.

| No. of adaptation | 1 | 2 | 3 |
|---|---|---|---|
| Average value | 55% | 65% | 40% |

method should not be satisfied for more numbers of blocks (considering the beginning partition). Otherwise, an evolution of whole system is not possible – it means that if user was classified at the beginning, it is impossible to change the group (even if his profile is closer to centroid of another group).

## 5    Summary and Future Works

A hierarchical model of user profile is considered in this paper. Links between nodes in a hierarchy reflects generalization relation. A set of assumption for profile structure is proposed. A key assumption is that user query would not contain terms correlated by generalization relation. It implies the rest of assumption which are formally described. For such an user profile a method for its adaptation is presented and a quality criterion for this method is proposed. Experimental evaluation has shown how quality criterion is satisfied.

In the future works we plan to enrich user profile with context information to avoid disambiguation in user queries.

**Acknowledgments.** This research was partially supported by Polish Ministry of Science and Higher Education.

## References

1. Adomavicius, G., Mobasher, B., Ricci, F., Tuzhilin, A.: Context-Aware Recommender Systems, pp. 67–80. Association for the Advancement of Artificial Intelligence (2011)

2. Clarke, C.L.A., Cormack, G.V., Tudhope, E.A.: Relevance ranking for one to three term queries. Information Processing & Management **36**(2), 291–311 (2000)
3. Yang, F.-Q., Sun, T.-L., Sun, J.-G.: Learning hierarchical user interest models from Web pages. Wuhan University Journal of Natural Sciences **11**(1), 6–10 (2006)
4. Gauch, S., Speretta, M., Chandramouli, A., Micarelli, A.: User profiles for personalized information access. In: Brusilovsky, P., Kobsa, A., Nejdl, W. (eds.) Adaptive Web 2007. LNCS, vol. 4321, pp. 54–89. Springer, Heidelberg (2007)
5. Google Directory. http://directory.google.com
6. Kim, H.R., Chan, P.K.: Learning implicit user interest hierarchy for context in personalization. In: Proceedings of the 8th International Conference on Intelligent User Interfaces, pp. 101–108. ACM (2003)
7. Li, S., Wu, G., Hy, X.: Hierarchical user interest modeling for Chinese Web pages. In: Proceedings of ICIMCS 2011, pp. 164–169 (2011)
8. Li, L., Yang, Z., Wang, B., Kitsuregawa, M.: Dynamic adaptation strategies for long-term and short-term user profile to personalize search. In: Dong, G., Lin, X., Wang, W., Yang, Y., Yu, J.X. (eds.) APWeb/WAIM 2007. LNCS, vol. 4505, pp. 228–240. Springer, Heidelberg (2007)
9. Maleszka, B.: Methods for User Personalization in Document Retrieval Systems Using Collective Knowledge. PhD thesis, Wroclaw University of Technology (2014)
10. Maleszka, M., Mianowska, B., Nguyen, N.-T.: A heuristic method for collaborative recommendation using hierarchical user profiles. In: Nguyen, N.-T., Hoang, K., Jędrzejowicz, P. (eds.) ICCCI 2012, Part I. LNCS (LNAI), vol. 7653, pp. 11–20. Springer, Heidelberg (2012)
11. Maleszka, M., Mianowska, B., Nguyen, N.T.: A method for collaborative recommendation using knowledge integration tools and hierarchical structure of user profiles. Knowledge-Based Systems **47**, 1–13 (2013)
12. Manouvrier, M., Rukoz, M., Jomier, G.: A generalized metric distance between hierarchically partitioned images. In: Proceedings of the 6th International Workshop on Multimedia Data Mining: Mining Integrated Media and Complex Data, MDM 2005, pp. 33–41. ACM, New York (2005)
13. Mianowska, B., Nguyen, N.T.: A method for collaborative recommendation in document retrieval systems. In: Selamat, A., Nguyen, N.T., Haron, H. (eds.) ACIIDS 2013, Part II. LNCS (LNAI), vol. 7803, pp. 168–177. Springer, Heidelberg (2013)
14. Nanas, N., Uren, V., Roeck, A.: Building and applying a concept hierarchy representation of a user profile. In: Proceedings of SIGIR. ACM (2003)
15. Pogacnik, M., Tasic, J., Meza, M., Kosir, A.: Personal Content Recommender Based on a Hierarchical User Model for the Selection of TV Programmes. User Modeling and User-Adapted Interaction **15**, 425–457 (2005)
16. Sieg, A., Mobasher, B., Lytinen, S., Burke, R.: Concept based query enhancement in the ARCH search agent. In: Proceedings of the 4th International Conference on Internet Computing, IC 2003 (2003)
17. Wang, J., Li, Z., Yao, J., Sun, Z.Q., Li, M., Ma, W.-Y.: Adaptive user profile model and collaborative filtering for personalized news. In: Zhou, X., Li, J., Shen, H.T., Kitsuregawa, M., Zhang, Y. (eds.) APWeb 2006. LNCS, vol. 3841, pp. 474–485. Springer, Heidelberg (2006)
18. Yang, F., Sun, T., Sun, J.: Learning Hierarchical User Interest Models from Web Pages. WUJNS Wuhan University Journal of Natural Sciences **11**(1), 6–10 (2006)

# Distributed Web Service Retrieval Method

Adam Czyszczoń$^{(\boxtimes)}$ and Aleksander Zgrzywa$^{(\boxtimes)}$

Institute of Informatics, Wrocław University of Technology,
Wybrzeże Wyspiańskiego 27, 50-370 Wrocław, Poland
{adam.czyszczon,aleksander.zgrzywa}@pwr.edu.pl
http://www.ii.pwr.edu.pl

**Abstract.** This research addresses the problem of Web Service Retrieval by presenting a method for Web scale distributed crawling and indexing of Web Services. This paper includes the retrieval of both SOAP and RESTful Web Services and consists of overall architecture of the retrieval system, architecture of the crawler and distributed indexing approach that adapts Google's MapReduce algorithm. Moreover, the research takes into account new and still unsolved problem of RESTful Web Service identification and data extraction from HTML API documentation pages. The research includes implementation of the approach that allowed to conduct preliminary experiments.

**Keywords:** Distributed retrieval method · Web service retrieval · Web service · SOAP · RESTful · Indexing · Crawling · MapReduce

## 1  Introduction

The task of Web Service Retrieval (WSR) is to find relevant services in large data collection satisfying a query that describes the information need of a user. There is a huge number of Web Services distributed over the Internet. The number of Web Services available on the Web represents large data collection that the index construction of such a data cannot be performed efficiently as a single process or in not scalable manner. In order to index Web Services efficiently we introduce the distributed indexing approach which allows to split the indexing process into parallel tasks distributed across nodes in a computer cluster, and to partition the index into separate parts. To accomplish such a task our method is an application of the MapReduce approach – introduced by Google programming model for processing large datasets [1].

On the other hand, there are various steps that need to be applied in order to collect the data about Web Services so they can be later indexed. In recent years RESTful Web Services have become popular because of their flexibility and ease of implementation. Therefore, many service providers have resigned from SOAP Web Services in the favor of lightweight RESTful solutions. However, descriptions of the services of this class are most frequently given in HTML documents in the form of API (Application Programming Interface) documentation. Therefore finding data about RESTful Web Services is new and still

© Springer International Publishing Switzerland 2015
N.T. Nguyen et al. (Eds.): ACIIDS 2015, Part I, LNAI 9011, pp. 117–126, 2015.
DOI: 10.1007/978-3-319-15702-3_12

unsolved problem that is beyond the scope of the current approaches on SOAP Web Service retrieval that rely on formal WSDL (Web Service Description Language) service descriptions [2]. To solve this problem, we propose in this paper a scalable Web Crawler with intelligent tools for automatic and effective RESTful Web Service identification and information extraction. This approach is founded on an algorithm that uses binary classification of link structure patterns.

## 2    Related Work

This paper is a continuation of our research on Web Service Retrieval methods and provides an outline for a general framework of such a retrieval system. This research is a consolidated combination of methods regarding Web Service distributed indexing [3] and crawling [2,4]. These methods are adapted and refined in order to meet new criteria that result from the need of cooperation as a uniform system.

The first research on Web Service retrieval that included the concepts of web crawling ([5,6]) and inverted index structure ([7,8]) presented these topics in general. They also considered Web Service to be represented as single set of terms ("bag-of-words") and therefore index at the service level, while indexing at operational level allowed to achieve higher effectiveness [9]. Moreover, current research on WSR did not include distributed and scalable indexing for the whole Web Service Retrieval problem.

The MapReduce method was firstly used in 2003 to simplify construction of the inverted index for handling searches at Google [10]. However, the basic concept of distributed indexing that utilize this approach was presented in [11]. Proposed solution represented the key-value pairs of the mapper output as term-document pairs and to divide them into segments, where every segment was assigned to one reducer. After collecting all documents for a term by reducers, the resulting inverted index structure that was split into term partitions depend on the term's first character.

## 3    Overall Architecture

In this section we illustrate the overall architecture of a Distributed Web Service Retrieval system. The architecture of the proposed system is divided into three areas – crawling (steps 1-5), indexing (steps 6-8) and searching (steps 9-14). In the first step, the services are downloaded by many simultaneous *Web Service Crawler* processes (1). The crawler traverses the Web looking for web pages containing Web Services. Pages that include some services are prioritized according to certain rules described in our other paper [4]. The unique URLs of identified Web Services are stored as a list in the *Web Serive URLs* dataset (2). Remaining processes of Web Service crawlers get service URLs from the top of the list and download their content (3). Services that are not responding are tagged for further inspection and put on the bottom of the list. If a service is still not responding after several download attempts it becomes blacklisted for certain period of time $b$. After blacklist removal period is over, the service URL is removed from the list. The $b$ time must be long enough to ensure that potentially broken URLs do not go back on the list too often.

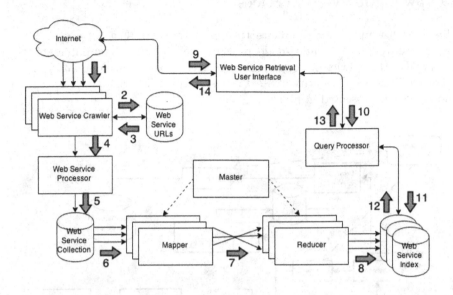

**Fig. 1.** Overall architecture of Web Service Retrieval system

In step (4) the service data extracted by the crawler nodes is passed to the *Web Service Processor*. This processor acts as a crawler's module. In case of SOAP services it parses subsequent sections of WSDL files. In case of RESTful services it uses intelligent data analysis tool to get proper sections of HTML API documentation pages. Collected in textual form and organized data about services is stored in a data structure that forms the *Web Service Collection* (5).

Steps 6-8 describe the MapReduce approach for distributed Web Service indexing. The input data from the collection is split into fixed-size chunks and distributed across the *Mapper* nodes (6). Those nodes execute the map function responsible for term weighting and modeling in the Vector Space Model (VSM). Data from mappers is passed to the *Reduce* nodes (7) that group terms of individual services. Reducers execute a vector merge formula on term-service pairs, producing the final inverted index divided into partitions (8). The number of reduce nodes determines the number of partitions. The MapReduce process is coordinated by the *Master* node.

After the given *Web Service Collection* is indexed it is possible to search through the data. In step (9) a user types a query in the *Web Service Retrieval User Interface*. The query is also modeled in the VSM in the same manner as services (10) and sent to the index (11) in order to retrieve Web Services that contain given query (12). The *Query Processor* translates data returned from index so it can be displayed in the user interface (13) and returned to the user as the search results list (14).

## 4   Crawling for Web Services

In this section we present the architecture of distributed *Web Service Crawler*. Crawling web pages is a multi-node process executed on a computer cluster. Every node of the crawler performs the steps necessary to download and process a web page in order to identify and extract services. The principle of operation of the crawler is presented in figure 2.

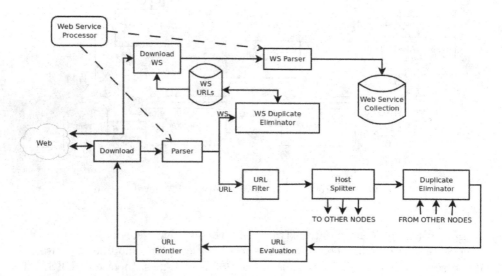

**Fig. 2.** Web Service Crawler architecture

The crawler gets an URL from the top of the list in the *URL Frontier*. Website located at this URL is downloaded and parsed. The *Parser* is responsible for processing the website and finding new URLs. Addresses are analyzed by the *Web Service Processor* module which divides them into two groups – the URLs that are links to Web Services and URLs that are links to web pages. Depending on the type, they are transmitted to different components.

In case of addresses that do not belong to Web Services, the URLs are filtered out using the *URL filter* in order to skip links that are forbidden to crawlers by Robots Exclusion Protocol. Addresses that can be accessed by crawlers are passed to the *Host Splitter*. This component sends the URLs directly to the *Duplicate Eliminator* of other nodes that are responsible for processing the URL of particular host. One crawler's node can process addresses of more than one host. Duplicate elimination allows to avoid downloading and processing the same websites many times by one node. Afterwards, the addresses are transmitted to the *URL Evaluation* component which analyzes the number of services found in the currently processed website and on that basis prioritizes its URLs. The higher the priority, the greater the chance that the address will be downloaded in subsequent stages of the crawling process. Prioritized URLs are added to the bottom of the *URL Frontier* list.

In case of URLs that belong to Web Services, the URLs are passed to the *WS Duplicate Eliminator* component which checks whether the address is unique. The component uses the *WS URLs* set that stores unique links to the services found on the currently viewed websites, by all active crawler's nodes. Links are added to the bottom of the list. At this point the execution process of a single node ends.

Another node of the crawler – dedicated to Web Services processing – gets the first WS URL from the list in *WS URLs*, downloads it using *Download WS* and parses using *WS Parser* component with the *Web Service Processor* module. The goal of this component is to extract information about service and store it as a *Web Service Collection*.

### 4.1   Web Service Processor

This module is responsible for two tasks depending if it is executed by a node responsible for website crawling or Web Service crawling. In the first case it checks if the given URL links to a WSDL document or if it is identified as a RESTful Web Service. The identification is conveyed using the approach described later in this section.

In the second case the module is responsible for proper parsing and service data extraction. In the situation of SOAP services the processor parses subsequent sections of WSDL files to obtain service name, description, version and components (see next section). In the case of RESTful services the process is as follows: the first step of RESTful Web Service identification is to determine whether a website actually contains an API documentation. If it does, such a website is parsed in order to collect URLs that belong to identified service and to collect information about the service.

Checking if a website is an API documentation is conveyed using the method presented in [12] that allows to calculate score of a page based on term frequency of certain keywords. Identification of RESTful Web Service is conveyed using binary classification of service's URI structure. The URI is encoded into pattern and send as a feature vector to the supervised learning algorithms that use ANN (Artificial Neural Network). Detecting information about service like for example root URI, sample URIs and resources description can be performed using a combination of regular expressions and method described in [12]. The details on the approach are described in [2].

## 5   Distributed Indexing

Before we introduce the distributed indexing approach we firstly need to briefly describe how Web Services are modelled in the Vector Space.

## 5.1  Modeling Services in the VSM

The file structure that allows to store such a data is called inverted index. It is a term-document matrix composed of dictionary of terms where for every term there is a list that records which documents the term occurs in [11]. In VSM the documents are represented as a vectors and every element in the document vector is calculated as weight that reflects the importance of a particular term that occurs in this document. To compute the weights the TF-IDF (Term Frequency-Inverse Document Frequency) scheme is used.

Therefore, we can define the VSM inverted index for WSR as the term-service matrix where the columns represent service vectors and rows represent term vectors. Afterwards, based on our previous work [13] we define Web Service to be composed of quadruple of elements where the first three represent parameters which correspond to service name, description and version, and the fourth represents the bag-of-words of all service components.

Having defined the inverted index for WSR and Web Service definition, we model Web Services in vector space in the following manner: each service parameter is represented as a vector composed of weights where each weight corresponds to a term from the bag-of-words of particular parameter. Weights are calculated using the TF-IDF for every service parameter separately. In result the VSM index is 4 times bigger than the standard one. To reduce it to its original size, we merge the weights using the MWV method presented in [13] and computed as the average weight of all service parameters.

## 5.2  The MapReduce Approach

The MapReduce model specifies the computation problem in terms of mapping and reducing function, and the underlying runtime system automatically parallelizes the computation across clusters of machines, handles machine failures, and schedules inter-machine communication to make efficient use of the network and disks [1]. The input data is split into fixed-size chunks and distributed across nodes executing mapping tasks. The map function processes a key/value pair to generate a set of intermediate key/value pairs and a reduce function merges all intermediate values associated with the same intermediate key [10]. Reduce tasks are distributed by partitioning the intermediate key space into parts using a partitioning function. Programming using the map end reduce functions makes computation problem linearly scalable [10].

In figure 3 we present the distributed indexing approach with MapReduce. The input *Web Service Collection* is split into $n$ parts called splits. The $m$ mapper nodes read the input splits where for every service's parameter it extracts its terms and calculate their weights using the TF-IDF scheme. The map function creates a vector of weights of every parameter with the number of elements equal to the number of unique terms in the service dictionary. Afterwards, for each term in service dictionary the intermediate key/value pairs are transmitted to the partitioner. The keys are composed of *term-serviceID* pairs and the values are represented by service parameters vectors. The intermediate files are stored

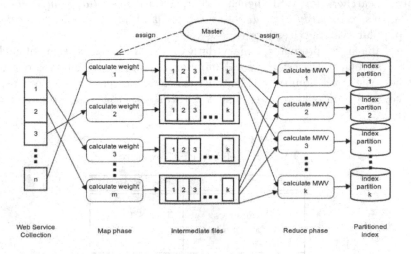

**Fig. 3.** Distributed Web Service indexing with MapReduce

on local mapper's disk. The MapReduce framework sorts the mapper output by the key pairs.

In the next step, the partitioner divides the intermediate keys into $k$ term partitions according to the first term's character and distributes them over $k$ reduce nodes. Afterwards, the reduce nodes iterate over their corresponding intermediate key/value pairs. For every term in given *serviceID* the reducer function reads the value represented by the parameters vectors and calculates the MWV value. This value is the final service vector. Because the reducers work only on terms they are assigned to. In result the reducing function can run in parallel and the distributed index is partitioned into $k$ partitions. For example if $k = 2$ the index is distributed over 2 nodes and divided into $a$-$m$ and $n$-$z$ partitions.

## 6    Evaluation

The evaluation of our research is divided into two parts. The first one concerns the methods that allow to collect Web Services using the proposed crawler architesture and the second one focuses on the proposed distributed indexing approach.

### 6.1    Preliminary Web Service Crawler Experiment

In this section we present crawling results of our implementation of *Web Service Crawler*. The implementation yet did not include the intelligent *Web Service*

*Processor* module that identifies RESTful services. Therefore, the introduced in this paper results do not contain any services of this class. The crawl was performed for the following public service repositories: *xmethods.net* (used by many researchers for WSR benchmarks), *service-repository.com*, *webservicex. net*, *venus.eas.asu.edu*, *visualwebservice.com* and *programmableweb.com* (free and popular Web Services directory).

In Table 1 we present summary data of the crawling experiment including some crawling statistics: the total number of pages crawled in order to find services, the total number of identified services, percentage between available services and identified services, and the total data that was processed during crawling.

Table 1. Web Service Crawler results

|  | Crawling results |
|---|---|
| Services | 662 |
| Parameters | 886 |
| Components | 18353 |
| Total elements | 19239 |
| Crawled pages: | 406277 |
| Services identified: | 3187 |
| Services avaliable: | 21% |
| Data processed (MB): | 6363 |

In order to create *Web Service Test Collection*, the above crawling data needs to be manually evaluated in terms of analysis for individual service/website elements and relevant services to certain keywords/phrases. This evaluation and resulting test collections were performed in our other paper [2].

## 6.2   Preliminary Distributed Indexing Experiment

The aim of this experiment was to conduct initial test that will indicate that the proposed approach can be linearly scalable. The experiment was performed on the data collected by the crawler. The MapReduce environment was pseudo-distributed because it was executed on a single machine with dual-core CPU where each core acted as one node in a pseudo-cluster. Therefore, only two tasks could be executed in parallel and only two index partitions could be produced effectively.

In figure 4 we present the average indexing time for different numbers of processing nodes in the pseudo-cluster that were executed as reduce tasks. The processes *no_MapReduce* and *MapReduce_1r* produced 1 index partition, and the *MapReduce_2r* produced 2 partitions.

The results showed that the one node MapReduce approach was 27% faster than a single process without MapReduce, and the two nodes MapReduce was 4% faster than one node MapReduce. Indexing time difference between one and two

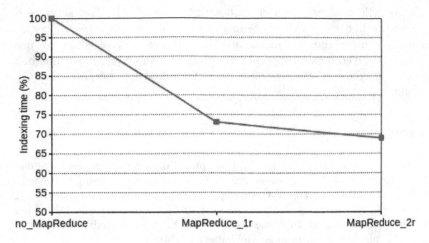

**Fig. 4.** Average indexing time for single process (*no_MapReduce*), 1 reduce task (*MapReduce_1r*) and 2 reduce tasks (*MapReduce_2r*)

nodes is not big, however for large datasets this may bring significant improvement. The datapoints on figure 4 indicate that proposed distributed indexing with MapReduce may be linearly scalable, however, to prove it a bigger dataset and computer cluster is required. Presented results also showed that the retrieval time using index divided into two partitions was 31% shorter than retrieval with one index partition. This fact also indicates the need for a larger number of partitions and cluster nodes.

## 7   Conclusions and Future Work

In this paper we presented a method of distributed Web Service Retrieval that includes Web scale distributed crawling and indexing of SOAP and RESTful Web Services. The research consists of overall architecture of a distributed Web Service Retrieval system that comprises the architecture of distributed Web Service crawler and distributed Web Service indexing approach. Additionally, the presented study takes into account the problem of RESTful Web Service identification and data extraction.

The research results indicate the direction for further research and experiments on the proposed methods of Web Service Retrieval. The preliminary experimental results of Web Service crawling provided data to develop Web Service test collection and to carry out the experiment on our distributed indexing approach. The preliminary experimental results of distributed indexing with MapReduce showed it is superior to the single process and show promising results for its scalable application. It also allowed to partition the inverted index which results in great retrieval time improvement.

In our future work we plan to implement intelligent tools for crawler that allow to identify and extract RESTful Web Services. Secondly, we will use the crawler to collect bigger dataset of both SOAP and RESTful services so we can prove the linear scalability of our distributed indexing approaches. On the other hand, other variants of distributed indexing with MapReduce are possible that can be derived depending on the balance of computation load put between mappers and reducers. All of these variants should be compared in terms of performance.

# References

1. Dean, J., Ghemawat, S.: Mapreduce: simplified data processing on large clusters. In: Proceedings of the 6th Conference on Symposium on Opearting Systems Design & Implementation, OSDI 2004, vol. 6, p. 10. USENIX Association, Berkeley (2004)
2. Czyszczoń, A., Zgrzywa, A.: Automatic RESTful web service identification and information extraction. In: Kwiecień, A., Gaj, P., Stera, P. (eds.) CN 2014. CCIS, vol. 431, pp. 318–327. Springer, Heidelberg (2014)
3. Czyszczoń, A., Zgrzywa, A.: The MapReduce approach to web service retrieval. In: Bădică, C., Nguyen, N.T., Brezovan, M. (eds.) ICCCI 2013. LNCS, vol. 8083, pp. 517–526. Springer, Heidelberg (2013)
4. Czyszczoń, A., Zgrzywa, A.: Propozycja skalowalnego pajaka usług interne-towych. In: Górski, J., Orłowski, C., (eds.) Integracja Systemów Informatycznych: Nowewyzwania, pp. 151–157. Pomorskie Wydawnictwo Naukowo-Techniczne PWNT (2011) (in Polish)
5. Dong, X., Halevy, A., Madhavan, J., Nemes, E., Zhang, J.: Similarity search for web services. In: Proceedings of the Thirtieth International Conference on Very Large Data Bases, VLDB 2004, vol. 30, pp. 372–383. VLDB Endowment (2004)
6. Platzer, C., Dustdar, S.: A vector space search engine for web services. In: Proceedings of the 3rd European IEEE Conference on Web Services (ECOWS 2005), pp. 14–16. IEEE Computer Society Press (2005)
7. Aiello, M., Platzer, C., Rosenberg, F., Tran, H., Vasko, M., Dustdar, S.: Web service indexing for efficient retrieval and composition. In: CEC/EEE, p. 63 (2006)
8. Wu, C., Chang, E.: Searching services "on the web": A public web services discovery approach. In: Eighth International Conference on Signal Image Technology and Internet Based Systems, pp. 321–328 (2007)
9. Peng, D.: Automatic conceptual indexing of web services and its application to service retrieval. In: Jin, H., Rana, O.F., Pan, Y., Prasanna, V.K. (eds.) ICA3PP 2007. LNCS, vol. 4494, pp. 290–301. Springer, Heidelberg (2007)
10. Dean, J., Ghemawat, S.: MapReduce: a flexible data processing tool. Commun. ACM **53**(1), 72–77 (2010)
11. Manning, C.D., Raghavan, P., Schütze, H.: Introduction to Information Retrieval. Cambridge University Press, New York (2008)
12. Steinmetz, N., Lausen, H., Brunner, M.: Web service search on large scale. In: Baresi, L., Chi, C.-H., Suzuki, J. (eds.) ICSOC-ServiceWave 2009. LNCS, vol. 5900, pp. 437–444. Springer, Heidelberg (2009)
13. Czyszczoń, A., Zgrzywa, A.: The concept of parametric index for ranked web service retrieval. In: Zgrzywa, A., Choroś, K., Siemiński, A. (eds.) Multimedia and Internet Systems: Theory and Practice. AISC, vol. 183, pp. 229–238. Springer, Heidelberg (2013)

# Discovering Co-author Relationship in Bibliographic Data Using Similarity Measures and Random Walk Model

Ngoc Tu Luong[1], Tuong Tri Nguyen[1(✉)], Jason J. Jung[2],
and Dosam Hwang[1]

[1] Department of Computer Engineering, Yeungnam University,
Gyeongsan 712-749, Korea
{luongngoctu,tuongtringuyen,dosamhwang}@gmail.com
[2] School of Computer Engineering, Chung-Ang University, Seoul 156-756, Korea
j2jung@gmail.com

**Abstract.** Discovering the research communities to bring techniques to the world is an interesting topic. In this paper we use the DBLP data to investigate the co-author relationship in a real bibliographic network and predict the interactions between co-authors. We analysis the research trend of authors and conferences based on extracted keywords from paper titles. We can understand research fields and change of research trend to find appropriate co-authors and conferences to submit our work. We also find potential co-authors for an existing author in DBLP data by using a variety of similarity measures and a random walk model. It can be useful for building a recommendation system.

**Keywords:** DBLP · Co-author · Similarity · Random walk · Bibliographic

## 1 Introduction

The science research communities are big and grow incessantly. There are many scientist and conferences in the world. The problem is bringing together scientists sharing similar interests, and bringing scientists to related conferences. We are interested in finding potential co-authors for researchers by discovering computer science communities in an author-conference social network, or finding appropriate co-author or conference by analyzing the research trend of authors or conferences.

We present experiments on DBLP [1]. There is a social network implicit in the DBLP database which includes information about authors, their papers and the conferences they published in [15]. We can extract keywords from the paper titles, co-author list, joined conference list and make some statistic for analyzing and predicting. The main contribution of this paper is to analyze the research trend of authors and conferences and to find future co-author relationships between existing authors in a bibliographic network by using variety of

© Springer International Publishing Switzerland 2015
N.T. Nguyen et al. (Eds.): ACIIDS 2015, Part I, LNAI 9011, pp. 127–136, 2015.
DOI: 10.1007/978-3-319-15702-3_13

similarity measures and random walk algorithm. From the results of each measures, we can compare the performance of measures.

The paper is organized as follows. In Sect. 2, some related works on co-author prediction are reviewed. Sect. 3 introduces different similarity measures that used to find similar authors. Additionally, an academic random walk model is described. Sect. 4 shows the experimental results on DBLP data. Finally, Sect. 5 draws a conclusion and future work of this study.

## 2    Related Works

The co-author relationship has been studied in many researches. In [14], the authors solve the problem of co-author relationship prediction in heterogeneous bibliographic network by using meta path-based relationship prediction model. In [7], the authors also focus on junior researchers, who do not have much publication information, and comb local and global network feature to improve the prediction performance. Otherwise in [5], the authors want to find experts, who have expertise on a specific academic topic, by using three model: a novel language weighted model, a topic-based model, and a hybrid model, which combines two earlier models. Other researches are based on random walk model to explore the co-author relationship. In [15], the authors discover the research communities by generating bipartite and tripartite graph models and using random walk model with restart to compute the relevance score between authors. In [11], the authors propose a model based on random walk with restart algorithm using three academic metrics to improve the quality and accuracy of collaboration recommendation.

In this paper, we use similarity measures to find similar authors for an existing author, based on his publication information: published papers, conferences which that author submitted papers, keywords in the tittles of his papers. We also use a random walk method with 3 academic metrics, proposed from [11], to make comparison of results.

## 3    Backgrounds

In this section we introduce some similarity measures that used to find similar authors based on keywords, papers and conferences. Besides, we show an academic random walk model for usage.

### 3.1    Similarity Measures

This section presents some known similarity measures and one proposal measure for calculating the similarity between two authors. Table 1 lists some notations which are used in similarity measures:

**Table 1.** Notation used in similarity measures

| Notation | Meaning |
|---|---|
| $P_A$ | the set of papers written by author A |
| $\|P_A\|$ | the number of papers written by author A |
| $P_B$ | the set of papers written by author B |
| $\|P_B\|$ | the number of papers written by author B |
| $\|P_A \bigcap P_B\|$ | the number of common papers written by 2 authors A and B |
| $\|P_A \bigcup P_B\|$ | the number of unique papers written by 2 authors A and B |
| $C_A$ | the set of conferences that author A submitted papers |
| $\|C_A\|$ | the number of conferences that author A submitted papers |
| $C_B$ | the set of conferences that author B submitted papers |
| $\|C_B\|$ | the number of conferences that author B submitted papers |
| $\|C_A \bigcap C_B\|$ | the number of common conferences that 2 authors A and B submitted papers |
| $\|C_A \bigcup C_B\|$ | the number of unique conferences that 2 authors A and B submitted papers |
| $K_A$ | the set of keywords appear in author A's papers |
| $\|K_A\|$ | the number of keywords appear in author A's papers |
| $K_B$ | the set of keywords appear in author B's papers |
| $\|K_B\|$ | the number of keywords appear in author B's papers |
| $\|K_A \bigcap K_B\|$ | the number of common keywords appear in papers of 2 authors A and B |
| $\|K_A \bigcup K_B\|$ | the number of unique keywords appear in papers of 2 authors A and B |

These information were extracted from DBLP dataset. And here are measures we used to calculate similarity between two authors A and B:

– Jaccard index (or Jaccard similarity coefficient) evaluates the similarity of two sets by the ratio of the size of the intersection of two sets to the size of their union [13]:

$$sim_{p\_jac}(A,B) = \frac{|P_A \bigcap P_B|}{|P_A \bigcup P_B|} \tag{1}$$

$$sim_{c\_jac}(A,B) = \frac{|C_A \bigcap C_B|}{|C_A \bigcup C_B|} \tag{2}$$

$$sim_{k\_jac}(A,B) = \frac{|K_A \bigcap K_B|}{|K_A \bigcup K_B|} \tag{3}$$

– Soergel similarity [4] evaluates the similarity of two sets by the ratio of the size of the intersection of two sets to the maximum number of papers/conferences/keywords of two authors:

$$sim_{p\_soer}(A,B) = \frac{|P_A \bigcap P_B|}{max(|P_A|,|P_B|)} \tag{4}$$

$$sim_{c\_soer}(A,B) = \frac{|C_A \bigcap C_B|}{max(|C_A|,|C_B|)} \tag{5}$$

$$sim_{k\_soer}(A, B) = \frac{|K_A \cap K_B|}{max(|K_A|, |K_B|)} \tag{6}$$

– Lorentzian similarity [6] evaluates the similarity of two authors by the logarithm of the number of intersection of papers/conferences/keywords of two auhors:

$$sim_{p\_lor}(A, B) = \ln\left(1 + \left|P_A \cap P_B\right|\right) \tag{7}$$

$$sim_{c\_lor}(A, B) = \ln\left(1 + \left|C_A \cap C_B\right|\right) \tag{8}$$

$$sim_{k\_lor}(A, B) = \ln\left(1 + \left|K_A \cap K_B\right|\right) \tag{9}$$

– Hamming distance between two sets is defined by the number of components in which they differ [13]:

$$sim_H(A, B) = \frac{1}{1 + dist_H(A, B)}$$
$$dist_H(A, B) = diff_C(A, B) + diff_K(A, B) + diff_P(A, B)$$
$$diff_C(A, B) = |C_A| + |C_B| - 2\left|C_A \cap C_B\right| \tag{10}$$
$$diff_K(A, B) = |K_A| + |K_B| - 2\left|K_A \cap K_B\right|$$
$$diff_P(A, B) = |P_A| + |P_B| - 2\left|P_A \cap P_B\right|$$

### 3.2  Academic Metrics for Random Walk Model

This section introduces Academic RWR [11]. This method is based on random walk with restart algorithm on the author-author graph. A random walk is the process by which randomly moving objects wander away from where they started [2]. The edge links in the graph are computed from three metrics: co-author order, latest collaboration time point and frequency of collaboration [11], to improve the accuracy.

– Co-author order [11]: is the order which the names of authors appeared in a paper. Consider two nodes $p_i$, $p_j$ in a co-author list. Measure of co-author order DCL (distance in coauthor list) is calculated by:

$$DCL(p_i, p_j) = \begin{cases} \dfrac{1}{i} + \dfrac{1}{j} & \text{if } j \leq 3 \\ \dfrac{1}{i} + \dfrac{2}{j} & \text{if } j > 3, i \leq 3 \\ \dfrac{2}{i} + \dfrac{2}{j} & \text{if } i > 3 \end{cases} \tag{11}$$

– Latest collaboration time point [11]: an author may have trend to collaborate with recent co-authors than with authors he co-authored long time ago.

Measure is calculated by using $LIM_t(p_i, p_j)$ (Link Importance):

$$LIM_t(p_i, p_j) = DCL(p_i, p_j) * k(t)$$
$$k(t) = \frac{t_i - t_0}{t_c - t_0} \qquad (12)$$

where k(t) is a function over time [11], $t_i$ is the latest time when two authors collaborated, $t_c$ is the current time, $t_0$ is the time when two authors first collaborated.

– Times of collaboration [11]: two authors who have collaborated many times in past may have high chance to work together again. The impact of different times of coauthoring is measured by:

$$LIM_{[t_1,t_2]}(p_i, p_j) = \sum_{t=t_1}^{t_2} LIM_t(p_i, p_j) = \sum_{t=t_1}^{t_2} DCL(p_i, p_j) * k(t) \qquad (13)$$

## 4  Experimental Results

### 4.1  Data Collection

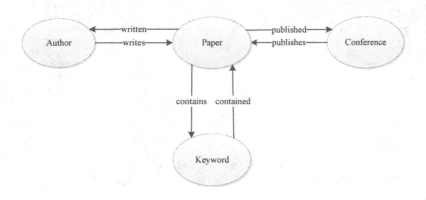

**Fig. 1.** DBLP bibliographic network schema

We use DBLP [1,9,10] data as an example of bibliographic network. The DBLP bibliographic network schema [14] is shown in the Fig. 1. The network contains 4 types of objects, namely Author, Conference, Paper, and Keyword (extracted from paper title). Links exist between authors and papers by the relations writes and written by, between papers and keywords by contains and contained by, between conferences and papers by publishes and published by [14].

The DBLP dataset is downloaded from the website http://www.informatik. uni-trier.de/~ley/db/. The data is stored in an XML file, contains information about conferences, journals, authors, and papers. DBLP indexes more than 18000 journal volumes, about 20000 conferences or workshops, more than 15000 monographs, over 2.3 million publications, published by more than 1.2 million authors [1]. We downloaded the file in September 2013 and used only publications for conferences. Any publication after that date and journal publications are not included in our experiment.

## 4.2 Data Extracting

We extracted the keywords from paper titles in DBLP data. In this experiment, we focused only on conference proceedings. Then we counted frequency of every keyword through each year. We manually selected some keywords which is topic-related words, e.g., data, database, Relational, etc. For example, some keywords extracted from author Jason J. Jung's paper titles are shown in Fig. 2.

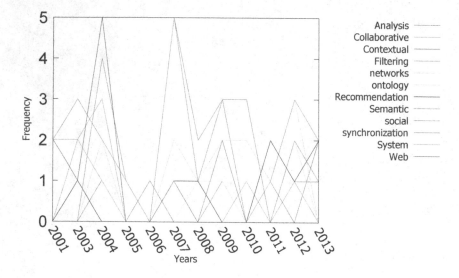

**Fig. 2.** Some keywords extracted from paper titles of author Jason J. Jung

The extracted keywords show the fields that author or conference interested in, and let us know the research trend of author or conference. Based on this, we can choose an appropriate conference to submit paper, or choose an appropriate author to cooperate. We can also predict or recommend future co-author who has same keywords for an existing author.

## 4.3    Results and Discussion

We choose author Philip S. Yu, a well-known researcher in data mining field, has published more than 400 conference papers (in our collected data), as source author. He had 245 coauthors before 2012. Then we continue finding co-authors of those 245 authors. There are total 4113 authors in the graph. The data was extracted from DBLP in the field of data mining involving 39 conferences and 38K authors. We selected data about papers, conferences, keywords in titles before year 2012. With each author as target author in that set, we use different measures to compute the similarity between source author and target author. Then we compare the result with the true co-authors list after year 2012 of author Philip S. Yu.

In this study, the Academic RWR [11] was implemented by building a 4113x4113 matrix, called S. Each element of S contained probability for each node $P_i$ skipping to next node $P_j$. The value of each elements is calculated by [11]:

$$S_{i,j} = \frac{W_{i,j}}{\sum_{P_k \in N(P_i)} W_{i,k}} \qquad (14)$$

where $N(P_i)$ is the set of neighbors of $P_i$ [11]. Initialize the rank score vector MR and the restart probability vector q as $(0,\ldots,1,\ldots,0)$, in which target node $P_i$ is set as 1 while others are set as 0 [11]. Initialize MR vector, the rank score of a node is calculated by [11]:

$$MR(p_i) = \frac{1-\alpha}{N} + \alpha \sum_{p_j \in M(p_i)} \frac{MR(p_j)}{L(p_j)} \qquad (15)$$

$M(p_i)$ is the set of nodes related to node $p_i$, $L(p_j)$ is the number of all the neighbors of node $p_j$, $\alpha$ denotes the probability of the walker continuing walking to the next node [11]. Iterate with some step, the iterative process is defined as [11]:

$$MR^{(t+1)} = \alpha SMR^{(t)} + (1-\alpha)q \qquad (16)$$

where $MR^{(t)}$ is the rank score vector at step t [11]. The rank score vector MR contains the score for each node [11]. Finally, we get nodes in the TOP N of the list MR to recommend to target node [11].

The precision and recall of measures are presented in table 2, 3, 4, 5. Besides that, Fig. 3 shows comparison between methods. We can see that the similarity measures computed on the number of common papers between two authors have better results than other measures. The result of ACRec method [11] is only better than the results of similarity measures which calculating on information about conferences or keywords. The Hamming distance computed by formula 10, which combined 3 information: papers, conferences, keywords, also have better result than ACRec method. Additionally, computing by using similarity measures is faster than ACRec. So, using similarity measures with information about papers of two authors will return better performance and results, especially on big data.

**Table 2.** Evaluation results by using paper information of authors

| Notation | Meaning | Precision | Recall | F_measure |
|---|---|---|---|---|
| *simp_jac* | Jaccard Similarity (1) | 0.3333 | 0.4814 | 0.3939 |
| *simp_soer* | Soergel Similarity (4) | 0.3333 | 0.4814 | 0.3939 |
| *simp_lor* | Lorentzian Similarity (7) | 0.3333 | 0.4814 | 0.3939 |

**Table 3.** Evaluation results by using conference information of authors

| Notation | Meaning | Precision | Recall | F_measure |
|---|---|---|---|---|
| *simc_jac* | Jaccard Similarity (2) | 0.0925 | 0.8148 | 0.1662 |
| *simc_soer* | Soergel Similarity(5) | 0.1111 | 0.8148 | 0.1955 |
| *simc_lor* | Lorentzian Similarity(8) | 0.1296 | 0.8148 | 0.2236 |

**Table 4.** Evaluation results by using keyword information of authors

| Notation | Meaning | Precision | Recall | F_measure |
|---|---|---|---|---|
| *simk_jac* | Jaccard Similarity(3) | 0.1481 | 0.8148 | 0.2507 |
| *simk_soer* | Soergel Similarity(6) | 0.1296 | 0.8148 | 0.2236 |
| *simk_lor* | Lorentzian Similarity(9) | 0.1296 | 0.8148 | 0.2236 |

**Table 5.** Evaluation results from Hamming distance and Academic Random Walk method

| Notation | Meaning | Precision | Recall | F_measure |
|---|---|---|---|---|
| *sim_ham* | Hamming distance (10) | 0.2592 | 0.8148 | 0.3933 |
| *acrec* | Academic RWR method [11] | 0.1666 | 0.7962 | 0.2756 |

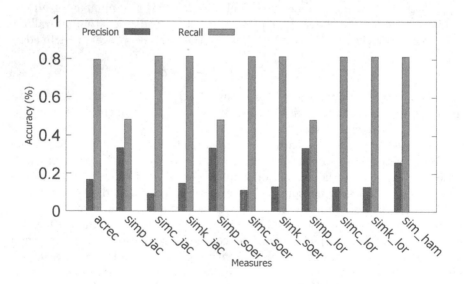

**Fig. 3.** Precision and recall of measures

# 5    Conclusions and Future Works

In this paper, we discover the bibliographic network through DBLP database. We used the keywords from paper titles to understand the research trend of conferences or authors and find similar conferences or authors on the network. We brought together different similarity measures, compared with a proposed method [11] based on Random Walk with Restart algorithm, used to find similar co-authors for an existing author and evaluate them on DBLP data. From the experiment results, the similarity measures from formula (1), (4) or (7) are the best choices. Since the experiment DBLP data only contains the titles of paper, it is hard to extract the correct topic of the paper. Some titles do not describe exactly or sometimes are not related to the content of the papers. Having more information about papers, e.g., abstracts, we can understand more correctly content of papers.

As future work, following the ranking algorithm [12] and the proposed model [8], we are planning to rank authors based on keywords, conferences in order to recommend to authors to take part in a conference that will be organized, or not.

**Acknowledgments.** This work was supported by the BK21+ program of the National Research Foun-dation (NRF) of Korea.

# References

1. Dblp (digital bibliography & library project) bibliography database. http://www. informatik.uni-trier.de/~ley/db/
2. Random walks. http://www.mit.edu/kardar/teaching/projects/chemotaxis(AndreaSchmidt)/random.htm
3. Sørensendice coefficient. http://en.wikipedia.org/wiki/S%C3%B8rensen%E2%80%93Dice_coefficient
4. Cha, S.H.: Comprehensive survey on distance/similarity measures between probability density functions. International Journal of Mathematical Models and Methods in Applied Sciences **1**(4), 300–307 (2007)
5. Deng, H., King, I., Lyu, M.R.: Formal models for expert finding on DBLP bibliography data. In: Proceedings of the 8th IEEE International Conference on Data Mining (ICDM 2008), Pisa, Italy, December 15–19, 2008, pp. 163–172 (2008)
6. Deza, E., Deza, M.: Dictionary of Distances. North-Holland (2006)
7. Han, S., He, D., Brusilovsky, P., Yue, Z.: Coauthor prediction for junior researchers. In: Greenberg, A.M., Kennedy, W.G., Bos, N.D. (eds.) SBP 2013. LNCS, vol. 7812, pp. 274–283. Springer, Heidelberg (2013)
8. Jung, J.J.: Ubiquitous conference management system for mobile recommendation services based on mobilizing social networks: A case study of u-conference. Expert Systems with Applications **38**(10), 12786–12790 (2011)
9. Ley, M.: The DBLP computer science bibliography: evolution, research issues, perspectives. In: Laender, A.H.F., Oliveira, A.L. (eds.) SPIRE 2002. LNCS, vol. 2476, p. 1. Springer, Heidelberg (2002)
10. Ley, M.: Dblp - some lessons learned. PVLDB **2**(2), 1493–1500 (2009)

11. Li, J., Xia, F., Wang, W., Chen, Z., Asabere, N.Y., Jiang, H.: Acrec: a co-authorship based random walk model for academic collaboration recommendation. In: 23rd International World Wide Web Conference, WWW 2014, Companion Volume, Seoul, Republic of Korea, April 7–11, 2014, pp. 1209–1214 (2014)
12. Pham, X.H., Nguyen, T.T., Jung, J.J., Hwang, D.: Extending HITS algorithm for ranking locations by using geotagged resources. In: Hwang, D., Jung, J.J., Nguyen, N.-T. (eds.) ICCCI 2014. LNCS, vol. 8733, pp. 332–341. Springer, Heidelberg (2014)
13. Rajaraman, A., Ullman, J.D.: Mining of Massive Datasets. Cambridge University Press, New York (2011)
14. Sun, Y., Barber, R., Gupta, M., Aggarwal, C.C., Han, J.: Co-author relationship prediction in heterogeneous bibliographic networks. In: International Conference on Advances in Social Networks Analysis and Mining, ASONAM 2011, Kaohsiung, Taiwan, 25–27 July 2011, pp. 121–128 (2011)
15. Zaïane, O.R., Chen, J., Goebel, R.: Mining research communities in bibliographical data. In: Zhang, H., Spiliopoulou, M., Mobasher, B., Giles, C.L., McCallum, A., Nasraoui, O., Srivastava, J., Yen, J. (eds.) WebKDD 2007. LNCS, vol. 5439, pp. 59–76. Springer, Heidelberg (2009)

# Intelligent Database Systems

# On Transformation of Query Scheduling Strategies in Distributed and Heterogeneous Database Systems

Janusz R. Getta$^{(\boxtimes)}$ and Handoko

School of Computer Science and Software Engineering,
University of Wollongong, Wollongong, Australia
{jrg,h629}@uow.edu.au

**Abstract.** This work considers a problem of optimal query processing in heterogeneous and distributed database systems. A global query submitted at a local site is decomposed into a number of queries processed at the remote sites. The partial results returned by the queries are integrated at a local site. The paper addresses a problem of an optimal scheduling of queries that minimizes time spend on data integration of the partial results into the final answer. A global data model defined in this work provides a unified view of the heterogeneous data structures located at the remote sites and a system of operations is defined to express the complex data integration procedures. This work shows that the transformations of an entirely simultaneous query processing strategies into a hybrid (simultaneous/sequential) strategy may in some cases lead to significantly faster data integration. We show how to detect such cases, what conditions must be satisfied to transform the schedules, and how to transform the schedules into the more efficient ones.

**Keywords:** Distributed heterogenous database systems · Data integration · Optimization of query processing

## 1 Introduction

Efficient data processing in the distributed and heterogeneous database systems is a critical factor for the successful implementations of global information systems. Performance of distributed applications strongly depends on the efficient algorithms that organize data processing in the distributed and heterogenous database systems. For instance, in the *MapReduce* programming model a user application that accesses data distributed over a number of remote sites simultaneously submits all of its sub-tasks to the remotes sites and later on integrates the partial results at a central site [4]. The simultaneous processing of sub-tasks makes *MapReduce* an efficient strategy when the amounts of processing and the amounts of data transmitted from the remote sites are more or less the same for all its sub-tasks. Unfortunately, a simultaneous processing strategy does not provide the best performance when one of the sub-tasks returns significantly

© Springer International Publishing Switzerland 2015
N.T. Nguyen et al. (Eds.): ACIIDS 2015, Part I, LNAI 9011, pp. 139–148, 2015.
DOI: 10.1007/978-3-319-15702-3_14

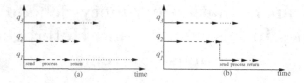

**Fig. 1.** Simultaneous (a) versus sequential (b) processing of tasks $q_1$ and $q_2$

larger amounts of data and/or when transmission speed is significantly lower than for the other sub-tasks. For example, consider the time diagrams in Fig. 1 when a global query has been decomposed into the queries $q_1$, $q_2$, and $q_3$ simultaneously processed at the remote sites. In the first case (a) data transmission of the results of $q_1$ dominates the total processing time. However, if some of data obtained from the processing of $q_2$ can be used to modify $q_1$ into $q_1'$ such that more processing can be done at a remote site then transmission of the results of $q_1'$ may take less time despite that processing of $q_1'$ follows processing of $q_2$, see case (b).

A partial order in which the individual queries are processed at the remote sites is called as a *query scheduling strategy*. In an entirely *sequential strategy* processing of a query at a remote site precedes processing of another query and the result of the first query can be used to modify the succeeding queries. In an entirely *simultaneous strategy* all queries are simultaneously submitted and processed at the remote sites. The efficiency of both strategies depends on the computational complexities of the individual tasks, computational power at a central and at the remote sites, amount of data transmitted over the networks, and data transmission speed of the networks used. Intuitively, a simultaneous strategy seems to be more efficient when the majority of query processing can be done at the remote sites and the amounts of data transmitted to a central site are small. A sequential strategy is more efficient when one or more tasks transmit the large amounts of data to a central site and it is possible to use the results of the other tasks to reduce the amounts of data to be transmitted later on. As usual the best solution is a hybrid one when some of the tasks are processed sequentially while the others simultaneously. Additional factors that significantly complicate data processing in distributed systems are a high level of autonomy and heterogeneity of the remote sites. The administrators of remote sites are usually very strict about performance and security of the managed systems and because of that they restrict external access to a read only mode without the rights to create and use the local data containers. It simply means that a central site cannot send a container with data to a remote site such that the container can be used for data processing there. Heterogeneity of remote sites means that organization of data, software and hardware used each at each site are different, which further limits any possible cooperation.

In this work we consider an environment of a heterogeneous and distributed database system where a user application issued at a *central site* accesses data at the remote sites. Then, it brings the partial results from the remote sites to a central site to integrate it into the final outcomes. In this work, we do not impose any restrictions on the compatibility of structures and contents of data containers at a central and the remote sites and we do not impose any assumptions about any level of "cooperation" between the sites. The only assumption is that the remote sites "display" a unified view of their data containers to the external user applications and are able to process the queries over the unified view. A global query issued at a central site is transformed into a set of queries $q_1, \ldots, q_n$ such that that each one of the queries accesses data from only one remote site. An expression $e(q_1, \ldots, q_n)$ integrates the partial results returned from the remote sites. Objective of this work is to find the formal backgrounds for the algorithms that schedule processing of the queries $q_1, \ldots, q_n$ at the remote sites and minimize the total processing time of $e(q_1, \ldots, q_n)$. In particular, we attempt to answers the questions when a sequential strategy is possible, when it is more efficient than a simultaneous strategy, what transformations of the queries must be applied to find a sequential strategy, what hybrid (simultaneous/sequential) strategies are possible for a given global query, and how to evaluate hybrid strategies.

The paper is organized in the following way. The next section overviews the previous works related to data processing in distributed systems. Section 3 present a model of query processing in a distributed system. and section 4 defines a global data model. The transformations of data processing strategies are presented in section 5 and evaluation of the strategies is explained in section 6. Section 7 concludes the paper.

## 2   Previous Work

Optimization of data processing in distributed systems has its roots in optimization of query processing in multidatabase and federated database systems [12]. One of the recent solutions to speed up distributed query processing in distributed systems considers the contents of cache in the remote systems and prediction of cache contents [11]. Wireless networks and mobile devices triggered research in mobile data services and in particular in location-dependent queries that amalgamate the features of both distributed and mobile systems [7]. An adaptive distributed query processing architecture is introduced at [14] where fluctuations in selectivity of operations, transmission speeds, and workloads of remote systems affect an order of distributed query processing.

In [15] the query sampling methods is used to estimated the query processing costs at the local systems. Query scheduling strategy in a grid-enabled distributed database proposed in [3] takes under the consideration so called "site reputation" for ranking response time of the remote systems. A new approach to estimation of workload completion time based on sampling the query interactions has been proposed in [1] and in [2]. Query monitoring can be used to collect information about expected database load, resource allocation, and expected size of the results [10].

The reviews of research on query scheduling and data integration are included in [8], [16]. The implementations of experimental data integration systems based on application of ontologies and data sharing are described in [13] and [6].

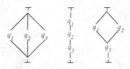

**Fig. 2.** The sample partial orders of processing the subqueries $q_1, q_2, q_3$

## 3   Query Processing in Distributed Systems

We consider a distributed and heterogeneous data base system where the data containers in the various formats like for example, relational, XML, object-relational, key-value, etc., are distributed over a number of highly autonomous remote sites. Each site "publishes" to all other sites a global view of data located at a site. A user application originated at a central site accesses data at the remote sites through a global query like $q(s_1:d_i, \ldots, s_n:d_n)$ where $d_i$ is a data container located at a remote site $s_i$. A global query is decomposed into $k$ queries $q_1, \ldots, q_k$ such that each query is processed at only one remote site and it is transformed into a *data integration expression* $e(q_1, \ldots, q_k)$. Let $Q = \{\top, \bot, q_1, \ldots, q_k\}$ be a set where $\top$ is a *start of processing* symbol and $\bot$ is an *end of processing* symbol. We define a partial order $P \subseteq Q \times Q$ such that $\langle q_i, q_j \rangle \in P$ if a query $q_i$ is processed before a query $q_j$. Then, $< Q, P >$ is a lattice where $sup(P) = \top$ and $inf(P) = \bot$ that represents a partial order of processing the queries $q_1, \ldots, q_k$ at the remote sites.

For instance, the lattices given in a Fig. 2 represent an entirely simultaneous strategy, entirely sequential strategy, and hybrid strategy where the queries $q_1$ and $q_2$ are processed simultaneously before $q_3$.

## 4   Global Data Model

A *global data model* provides a unified view of data stored at the remote sites. It amalgamates the contradictory requirements of generality with the very precise specifications of the basic operations.

A *data object* is defined as a pair $\langle id, t \rangle$ where $id$ is a unique object identifier at a given remote site and $t$ is a *description* of the object. A *description* is defined as a mapping $t : S \rightarrow dom(A)$ where $S$ is a set of *access paths* to the values of attributes, e.g. *address.street.house.flat*. Let $p.a$ be a path to a value of an attribute $a$. Then, a mapping $t$ satisfies a condition $t(p.a) \in dom(a)$ where $dom(a)$ denotes a domain of attribute $a$. A set of all access paths $S$ in a description of an object is called as a *schema of an object* and $dom(A) = \bigcup_{a \in A} dom(a)$. A *data container* $d$ is a set of data objects. A *schema of a data container* is a union of all schemas of all objects included in the container.

A set of operations on data containers includes the unary operations of *selection* and app *extraction*, and binary operations of *union, composition, semi-* and *anti-composition,* and *substitution.*

Let $d$ be a data container. An *access term* is defined as a triple $d.p.a.$ A *selection condition* $\phi$ is defined as a well-formed formula of prepositional calculus built from the access terms, relational operators, Boolean operators (*and*, *or*, *not*), constants, and brackets. Additionally, all access terms in a *selection condition* must start from a name of the same data container.

Let $d_i$ and $d_j$ be the data containers. A unary *selection* operation $\sigma$ on an argument $d_i$ is defined as $\sigma_\phi(d_i) = \{\langle id, t \rangle : \exists \langle id_i, t \rangle \in d_i \ and \ eval(\phi, \langle id_i, t \rangle)\}$ where a function *eval* evaluates a selection condition $\phi$ against the contents of a data object $\langle id_i, t \rangle$ into *true* or *false*. Note, that in a result of *selection* each object in a result of selection obtains a new identifier.

Let $S_i$ be a schema of a data container $d_i$ and let $S \subseteq S_i$. A unary *projection* operation $\pi$ of a data container $d_i$ on a schema $S$ is defined as $\pi_S(d_i) = \{\langle id, t \rangle: \exists \langle id_i, t_i \rangle \in d_i \ t = t_i[S]\}$ where $t_i[S]$ means restriction of a description $t_i$ to the access paths in $S$.

A binary *union* operation $\cup$ on the arguments $d_i$ and $d_j$ is defined as $d_i \cup d_j = \{\langle id, t \rangle : \exists \langle id_i, t \rangle \in d_i \ or \ \langle id_i, t \rangle \in d_j\}$.

A *constructor* operation $\theta$ is defined as $\theta : d_i \times d_j \rightarrow d_{ij}$ such that $\theta(\langle id_i, t_i \rangle, \langle id_j, t_j \rangle) = \langle id_{ij}, t_{ij} \rangle$ where $id_{ij}$ is an identifier of a new object and $t_{ij} = f(t_i, t_j)$ where $f$ is an expression that combines the descriptions $t_i$ and $t_j$ into a description $t_{ij}$ of a new object.

A *matching condition* $\psi$ is defined as a well-formed formula of prepositional calculus built from the access terms, relational operators, Boolean operators (*and*, *or*, *not*), constants, and brackets. Additionally, a *matching condition* consists only of the comparisons between the access terms that related to the different data containers.

A binary *composition* operation $\otimes_{\psi\theta}$ on the arguments $d_i$ and $d_j$ is defined as $d_i \otimes_{\psi\theta} d_j = \{\langle id_{ij}, t_{ij} \rangle : \exists \langle id_i, t_i \rangle \in d_i \ and \ \exists \langle id_j, t_j \rangle \in d_j \ eval(\psi, \langle id_i, t_i \rangle, \langle id_j, t_j \rangle)$ and $\langle id_{ij}, t_{ij} \rangle = \theta(\langle id_i, t_i \rangle, \langle id_j, t_j \rangle)\}$.

We say that operation $d_i \otimes_{\psi\theta} d_j$ is *semi-reversible* if $\pi_{S_i}(d_i \otimes_{\psi\theta} d_j) \subseteq d_i$ and $\pi_{S_j}(d_i \otimes_{\psi\theta} d_j) \subseteq d_j$. In the rest of this paper we consider only the composition operations which are semi-reversible on the schemas of its both arguments.

A *semi-composition* operation $\oplus_\psi$ on the arguments $d_i$ and $d_j$ is defined as $d_i \oplus_\psi d_j = \{\langle id, t_i \rangle : \exists \langle id_i, t_i \rangle \in d_i \ and \ \exists \langle id_j, t_j \rangle \in d_j \ eval(\psi, \langle id_i, t_i \rangle, \langle id_j, t_j \rangle)\}$.

An *anti-composition* operation $\ominus_\psi$ on the arguments $d_i$ and $d_j$ is defined as $d_i \ominus_\psi d_j = \{\langle id, t_i \rangle : \exists \langle id_i, t_i \rangle \in d_i \ and \ \forall \langle id_j, t_j \rangle \in d_j \ not \ eval(\psi, \langle id_i, t_i \rangle, \langle id_j, t_j \rangle)\}$.

Consider a data container $d_j = \{\langle id_1, t_1 \rangle, \ldots, \langle id_k, t_k \rangle\}$ and a matching condition $\psi(d_i.p_1.a_1, \ldots, d_i.p_m.a_m, d_j.p_1.b_1, \ldots, d_j.p_n.b_n)$. A *substitution* operator is denoted by $\psi \leftarrow d_j$ and it is defined as

$\psi(d_i.p_1.a_1, \ldots, d_i.p_m.a_m, d_j.p_1.b_1, \ldots, d_j.p_n.b_n) \leftarrow d_j =$
$\psi(d_i.p_1.a_1, \ldots, d_i.p_m.a_m, t_1(p_1.b_1), \ldots, t_1(p_n.b_n)) \ or \ \ldots or$
$\psi(d_i.p_1.a_1, \ldots, d_i.p_m.a_m, t_k(p_1.b_1), \ldots, t_k(p_n.b_n))$.

A *substitution* operator replaces all instances of access terms $d_j.s_1.b_1, \ldots, d_j.s_n.b_n$ in a matching condition $\psi$ with the values of all attributes taken from all objects in a data container $d_j$ and creates disjunction of all terms after the replacements. For example if $d_j = \{\langle id_1, t_1 \rangle, \langle id_2, t_2 \rangle\}$ and $t_1(s_1.name) = \textsf{James}$

and $t_2(s_1.name) = $ Mary then application of substitution operator $(d_i.s_1.name = d_j.s_1.name) \leftarrow d_j$ returns a matching formula $d_i.s_1.name = $ James $or$ $d_i.s_1.name = $ Mary.

The following equations hold for any data containers $d_i$, $d_j$ and any matching condition $\psi$.

$$d_i \oplus_\psi d_j = \sigma_{\psi \leftarrow d_j}(d_i) \tag{1}$$

$$d_i \ominus_\psi d_j = \sigma_{not(\psi \leftarrow d_j)}(d_i) \tag{2}$$

The equations listed above mean that the operations of *semi-* and *anti-composition* can always be replaced with a *filter* operation on the first argument while a *replacement* operation is applied to the second argument in a matching formula. If an expression $d_i \oplus_\psi d_j$ must be computed at a remote site that contains only a data container $d_i$ and a data container $d_j$ cannot be sent to the remote site then the computations of an expression with $\sigma_{\psi \leftarrow d_j}(d_i)$ replaces the computations of *anti-composition* at a remote site.

## 5    Transformations

We start from the simple transformations of simultaneous query scheduling strategies where two queries are simultaneously processed at the remote sites and their results are integrated with one of the arguments of *composition* operation. Next, we consider the complex transformations of the strategies where many queries are processed simultaneously and their results are integrated by an expression over many *composition* operations.

### 5.1    Simple Transformations

We consider simple a data integration expression $q_i \otimes_{\psi\theta} q_j$ where $q_i$ and $q_j$ are the queries to be processed at two different remote sites. If we expect that transmission of the results of $q_j$ will be significantly longer than transmission of the results of $q_i$ then it is worth to change a simultaneous schedule into sequential where $q_i$ is processed first and a part of it denoted by $x$ will be involved in processing of $q_j$. To find $x$ we rewrite a data integration expression into $q_i \otimes_{\psi\theta} (q_j \oplus_\psi x)$ where $x$ is an unknown data container that must be sent to a remote site where $q_j$ supposed to be processed.

We expect that a subexpression $(q_j \oplus_\psi x)$, when computed at a remote site, returns the results much smaller than the results of $q_i$. On the other hand, the results of the data integration expression must not change. Hence, to find $x$ we solve an equation

$$q_i \otimes_{\psi\theta} q_j = q_i \otimes_{\psi\theta} (q_j \oplus_\psi x) \tag{3}$$

There exists many solutions of an equation (3) above, e.g. $x$ equal to the results of $q_j$ or any superset of the results of $q_j$ satisfies the equation. We look for the smallest solution of the equation because we would like to minimize the amount of transmission to a remote site. An equation (3) can be transformed into an equivalent fixpoint equation and its fixpoint solution can be found using Kleene fix-point theorem [5].

$$x = x \cup \pi_{s_\psi}(q_i \otimes_{\psi\theta} q_j - q_i \otimes_{\psi\theta} (q_j \oplus_\psi x)) \cup (q_i \otimes_{\psi\theta} (q_j \oplus_\psi x) - q_i \otimes_{\psi\theta} q_j) \quad (4)$$

A projection $\pi_{s_\psi}$ on a schema $s_\psi$ of a matching condition $\psi$ is necessary because a schema of a data container $x$ does not need to include more attributes than it is used in a matching condition $\psi$. The solution of an equation (4) is obtained through the iterations starting from an empty data container $x_{(1)} = \emptyset$ and union of the results from each iteration. The smallest solution of the equation is equal to $x_{min} = \pi_{s_\psi}(q_i)$ which is consistent with our expectations. Then, the right hand side of equation (3) can be transformed into $q_i \otimes_{\psi\theta} (q_j \oplus_\psi \pi_{s_\psi}(q_i))$ and finally after application of equation (1) we obtain the final data integration expression $q_i \otimes_{\psi\theta}(\sigma_{\psi\leftarrow\pi_{s_\psi}(q_i)}(q_j))$. The expression is transformed into the following sequence of computations: $r_1 := q_i; r_2 := \sigma_{\psi\leftarrow\pi_{s_\psi}(r_1)}(q_j); result := r_1 \otimes_{\psi\theta} r_2$.

If in the computations of data integration expression with *semi-composition* $q_i \oplus_\psi q_j$ the results of $q_i$ are large and the results of $q_j$ are small then it is possible to compute $r_1 := \pi_{s_\psi}(q_j)$ first and then apply an equation (1) to replace the composition with $result := \sigma_{\psi\leftarrow r_1}(q_i)$ computed at a remote site. In the opposite case we obtain $r_1 := \pi_{s_\psi}(q_i); r_2 := \sigma_{\psi\leftarrow r_1}(\pi_{s_\psi}(q_j)); result := \sigma_{\psi\leftarrow r_2}(q_i)$.

If in the computations of *anti-composition* $q_i \ominus q_j$ the results of $q_i$ are large and the results of $q_j$ are small then it is possible to compute $r_1 := \pi_{s_\psi}(q_j)$ first and then apply an equation (2) to replace the composition with $result := \sigma_{not(\psi\leftarrow r_1)}(d_i)$ computed at a remote site. In the opposite case we obtain a plan: $r_1 := \pi_{s_\psi}(q_i); r_2 := \sigma_{\psi\leftarrow r_1}(\pi_{s_\psi}(q_j)); result := \sigma_{not(\psi\leftarrow r_2)}(q_i)$.

## 5.2   Complex Transformations

The simple transformations of two-argument data integration expressions described in the previous section can be systematically applied to find the complex transformations of n-argument data integration expressions. Consider a data integration expression $e(q_1, \ldots, q_n) = f(q_1, \ldots, q_k) \, \alpha_{\psi\theta} \, g(q_{k+1}, \ldots, q_n)$ where $\alpha \in \{\otimes, \oplus, \ominus\}$. Let $q_f = f(q_1, \ldots, q_k), q_g = g(q_{k+1}, \ldots, q_n)$.

Then, it is possible to determine whether a transformation from a simultaneous schedule to a sequential schedule is possible for $q_f$ and $g_g$ and if it is so, it is possible to find such transformation through systematic decomposition of the data integration expression into subexpression and find the transformations at each level of decomposition.

Without a significant loss of generality we consider an operation $\alpha_{\psi\theta}$ to be a composition $q_g \otimes_{\psi\theta} q_f$ and we transform a simultaneous schedule of processing $q_g$ and $q_f$ into a sequential one accordingly to the rules described earlier into $q_j := \pi_{S_\psi}(q_g)$ and $q_f := \sigma_{\psi\leftarrow q_j}(q_f)$. It means that in a sequential schedule an entire expression $g(q_{k+1}, \ldots, q_n)$ must be computed before a modified expression $\sigma_{\psi\leftarrow q_j}(f(q_1, \ldots, q_k))$.

The further transformations depend on a distributivity of $f(q_1, \ldots, q_k)$ over an operation of selection and distributivity of $g(q_{k+1}, \ldots, q_n)$ over an operation of projection. If it is possible to transform $\pi_{S_\psi}(g(q_{k+1}, \ldots, q_n))$ into $q'_g \, \alpha_{\psi'\theta'} \, q''_g$ such that $q'_g = \pi_{S_\psi}(g'(q_{k+1}, \ldots, q_m))$ and $q''_g = \pi_{S_\psi}(g''(q_{m+1}, \ldots, q_n))$ then its is possible to transform again the computations of $q'_g$ and $q''_g$ from the simultaneous into the sequential ones.

In the same way if it is possible to transform $\sigma_{\psi \leftarrow q_q}(f(q_1, \ldots, q_k))$ into $q'_f \alpha_{\psi'\theta'} q''_f$ such that $q'_f = \sigma_{\psi \leftarrow q_q}(f'(q_1, \ldots, q_i))$ and $q''_f = \sigma_{\psi \leftarrow q_q}(f''(q_{i+1}, \ldots, q_k))$ then its is possible to transform again the computations of $q'_g$ and $q''_g$ from the simultaneous into the sequential ones.

A process described above is recursively applied to to each subexpression of data integration expression until the operations of selection and projection are directly applied to the arguments.

As a simple example consider a data integration expression $(q_1 \otimes_{\psi_1 \theta} q_2) \oplus_{\psi_2} q_3$ where it is expected that the queries $q_2$ and $q_3$ return much smaller results than a query $q_1$. A simple transformation can be applied to change the processing of $q_1$ and $q_2$ from a simultaneous to a sequential one. It leads to a transformation of $q_1$ into $\sigma_{\psi_1 \leftarrow \pi_{S_{\psi_1}}(q_2)}(q_1)$. The second transformation can be obtained from the processing $q_3$ before $q_1$. An initial transformation $\sigma_{\psi_2 \leftarrow \pi_{S_{\psi_2}}(q_3)}(q_1 \otimes_{\psi_1 \theta} q_2)$ applies to a result of composition of $q_1$ and $q_2$. Assuming, that in this case selection is distributive over composition we obtain the following second transformation $\sigma_{\psi_2 \leftarrow \pi_{S_{\psi_2}}(q_3)}(q_1)$. The outcomes of both transformation can be merged into a single expression $\sigma_{\psi_2 \leftarrow \pi_{S_{\psi_2}}(q_3) \text{ and } \psi_2 \leftarrow \pi_{S_{\psi_2}}(q_2)}(q_1)$. It leads to a query scheduling strategy where $q_2$ and $q_3$ are simultaneously processed before $q_1$.

## 6    Evaluation of Query Scheduling Strategies

If in the example above distributivity of selection over composition in $\sigma_{\psi_2 \leftarrow \pi_{S_{\psi_2}}(q_3)}$ $(q_1 \otimes_{\psi_1 \theta} q_2)$ applies to both arguments of composition then is also possible to transform $q_2$ to $\sigma_{\psi_2 \leftarrow \pi_{S_{\psi_2}}(q_3)}(q_2)$. It means that it is possible to get more than one data integration plan where $q_3$ is processed before $q_1$ and $q_3$ is processed before $q_2$. To find an optimal processing plan we need information about the amounts of data to be transmitted over a network, transmission speed, amounts of time needed to process the queries at the remote sites and a cost function to calculate the total costs for each data integration plan represented by a lattice of queries. The total costs of processing a query $q_i$ can be estimated as $t_i = t_{s_i} + t_{p_i} + t_{r_i}$ where $t_{s_i}$ is time needed to send a query $q_i$ to a remote site, $t_{p_i}$ is time needed to process the query there, and $t_{r_i}$ is time needed to transmit the results to a central site. Each one of the parameters depend on the information listed above. When the queries $q_1, \ldots, q_n$ are processed simultaneously then their total processing time is equal to $max(t_1, \ldots, t_n)$. If the queries are processed sequentially and the results of a query $q_i$ are used to transform a query $q_{i+1}$ the the total processing time is equal to $t_1 + t'_2 + \ldots + t'_n$ where $t'_i$ are the processing times of transformed queries. If a data integration strategy is represented by a lattice $< Q, P >$ then a cost formula is derived in the following way. Let $p_i$ be a path from $\top$ to $\bot$ symbol in a lattice and passing through the nodes labeled with $q_{i_1}, \ldots q_{i_k}$. Then the costs of processing along a path $p_i$ is equal to $t_{i_1} + \ldots + t_{i_k}$. The costs of processing along all paths $p_1, \ldots p_n$ from $\top$ to $\bot$ symbol in a lattice is equal to $max(t_{p_1}, \ldots, t_{p_n})$. Then, an equality $max(a + b, a + c) = a + max(b, c)$ can be use to simplify a cost formula. For example, a cost formula derived for a data integration schedule in Fig. (2, case 3) $max(t_1 + t_3, t_2 + t_3)$ can be simplified to $t_3 + max(t_1, t_2)$.

# 7   Summary and Conclusions

This work is based on an observation that a significant difference between the amounts of data transmitted from the remote sites to a central site may have a negative impact on an overall time of data integration at a central site when a simultaneous query scheduling strategy is applied. Then, a transformation of a simultaneous strategy into a sequential or hybrid one speeds up data integration at a central site. This work shows when the transformations of query scheduling strategies are possible, how to perform it, when the transformations are beneficial, and how to evaluate the results.

Another interesting outcome of this work is a technique that embeds data into the queries through application of substitution operation. Substitution operation eliminates to some extent a problem of high level of autonomy of the remote sites that usually stop external users from transmitting data into the site and processing it there. A substitution operation allows for a safe processing of data obtained from another remote sites.

A system of operations proposed in this work allows for processing of any data containers as long as the access paths to the values of data items are provided and implemented by the owners of data. An interesting property of the system of operations is that it reduces to a standard relational algebra when the data containers include only homogeneous tuples or it reduces to XML algebra when the data containers include only XML documents, etc. For any structure of data objects included in data containers a system of operations needs the operations that select the objects that satisfy a given condition, operation that project the objects on a given sub-schema, union operation, operation that compares all pairs of objects and constructs new object from each pair, operation that compares objects from two containers and picks from one container the objects that match/do not match objects in the other container.

The transformation of query scheduling strategies mainly depend on the algebraic properties of a data integration expression and on the properties of composition operation. A composition operation must be semi-reversible such that it is possible to restore the subsets of the arguments from a result of an operation. A fixpoint equation (4) which is the basis for finding simple transformations is solvable when a function on its right hand side is monotonic. The complex transformations are possible when the subexpressions of data integration expression are distributive over the operations of selection and projection.

# References

1. Ahmad, M., Aboulnaga, A., Babu, S.: Query interactions in database workloads. In: Proceedings of the Second International Workshop on Testing Database Systems, pp. 1–6 (2009)
2. Ahmad, M., Duan, S., Aboulnaga, A., Babu, S.: Predicting completion times of batch query workloads using interaction-aware models and simulation. In: Proceedings of the 14th International Conference on Extending Database Technology, pp. 449–460 (2011)
3. Costa, R.L.-C., Furtado, P.: Runtime estimations, reputation and elections for top performing distributed query scheduling. In: Proceedings of the 2009 9th IEEE/ACM International Symposium on Cluster Computing and the Grid, pp. 28–35 (2009)
4. Dean, J., Ghemawat, S.: MapReduce: simplified data processing on large clusters. In: Proceedings of the 6th Symposium on Operating Systems Design and Implementation (2004)
5. Granas, A., Dugundji, J.: Fixed Point Theory. Springer-Verlag (2003)
6. Ives, Z.G., Green, T.J., Karvounarakis, G., Taylor, N.E., Tannen, V., Talukdar, P.P., Jacob, M., Pereira F.: The ORCHESTRA Collaborative Data Sharing System. SIGMOD Record (2008)
7. Ilarri, S., Mena, E., Illarramendi, A.: Location-dependent query processing: Where we are and where we are heading. ACM Computing Surveys 42(3), 1–73 (2010)
8. Lenzerini, M.: Data Integration: A Theoretical Perspective (2002)
9. Liu L., Pu, C.: A dynamic query scheduling framework for distributed and evolving information systems. In: Proceedings of the 17th International Conference on Distributed Computing Systems (1997)
10. Mishra, C., Koudas, N.: The design of a query monitoring system. ACM Transactions on Database Systems 34(1), 1–51 (2009)
11. Nam, B., Shin, M., Andrade, H., Sussman, A.: Multiple query scheduling for distributed semantic caches. Journal of Parallel and Distributed Computing 70(5), 598–611 (2010)
12. Ozcan, F., Nural, S., Koksal, P., Evrendilek, C., Dogac, A.: Dynamic Query Optimization in Multidatabases. Bulletin of the Technical Committee on Data Engineering 20(3), 38–45 (2011)
13. Thain, D., Tannenbaum, T., Livny, M.: Distributed computing in practice: the Condor experience: Research Articles. Concurrency Computing: Practice and Experience. 17(2–4), 323–356 (2005)
14. Zhou, Y., Ooi, B.C., Tan, K.-L., Tok, W.H.: An adaptable distributed query processing architecture. Data and Knowledge Engineering 53(3), 283–309 (2005)
15. Zhu, Q., Larson, P.A.: Solving Local Cost Estimation Problem for Global Query Optimization in Multidatabase Systems. Distributed and Parallel Databases 6(4), 373–420 (1998)
16. Ziegler, P.: Three Decades of Data Integration - All problems Solved? In: 18th IFIP World Computer Congress, vol. 12 (2004)

# An Approach of Transforming Ontologies into Relational Databases

Loan T.T. Ho$^{(\boxtimes)}$, Chi P.T. Tran, and Quang Hoang

Hue University of Sciences, Ho Chi Minh City, VietNam
{thuyloan13488,phuongchi0910,hquang10}@gmail.com

**Abstract.** The main purpose of the Semantic Web expresses the information as intelligent forms, enables better computers and people to work in cooperation. Ontologies, as meaning providers, play a key role in this effort. Currently, the OWL Web Ontology Language is used as a description ontology language that is the main of technique for storing ontologies. On the other hand, traditional relational database systems are often used as best mechanisms for storing, querying, manipulating the information, and have some benefits such as transaction management, security. For this reason, there is a need for storing ontologies represented by OWL into relational databases is proposed. Therefore, principles of mapping OWL concepts to relational database schemas are presented with an implemented tool, which is the purpose of this paper.

**Keywords:** Ontology · Relational database · OWL · Mapping · Transformation

## 1 Introduction

Nowadays, ontology is the platform of the Semantic web and has become more popular as the model of the information in specified domain. Ontologies play an important role by providing the means to align the knowledge performs. The language used to describe the ontology which is the Web Ontology Language OWL designed by W3C. OWL can be seen as an representative schema language that can be used to provide flexible access to data. Unfortunately, interpreting of schema statements in the OWL is different from explaining of similar statements in a relational database setting. This can lead to problems in data-centric applications, where OWLs interpretation of statements intended as constraints may be confusing and/or inappropriate [1].

In addition, for sharing OWL, we often publish OWL files on the web which can be used anywhere. This can lead to problems, how must OWL files be stored in databases and efficient processing this information by user applications. For this purpose to deal with storing ontology, Relational Database (RDB) is a good candidate that have proven capabilities to handle with large amounts of data, although ontologies can be stored in various databases such as relational, object or object-relational [2]. In particular, the advantages of relational database management systems are mature, performing, robust, reliable and available.

© Springer International Publishing Switzerland 2015
N.T. Nguyen et al. (Eds.): ACIIDS 2015, Part I, LNAI 9011, pp. 149–158, 2015.
DOI: 10.1007/978-3-319-15702-3_15

There are some approaches of transforming ontologies into relational database that were presented in the articles [3–8]. The authors only coped with main OWL structures, and some components of OWL can not be considered in those papers. Therefore, those approaches are mainly forthright and still incomplete, or obtained relational structures are not applicable for real information systems. In this paper, we propose another approach based on above, which can be used for automatically transforming from ontologies represented by OWL into RDB schemas. The major part of concepts are mapped into relational tables, relations and attributes, other semantics of constraints and properties are stored like metadata in special tables. Using this hybrid approach, it is probable to obtain appropriate relational structures and preserve semantic information ontology.

The rest of the paper is organized as follows. Section 2 gives a brief some concepts and the detail rules of transforming from ontology constructs into relational database schemas were described. In section 3, the implementation of the proposed approach is presented. Section 4 considers the related work. Finally, we conclude overall the paper in section 5.

## 2   OWL Concepts and Their Mapping to RDB Concepts

### 2.1   OWL Class

A class, which is a basic concept of the ontology, defines a group of individuals belong together. So, the transformation of classes is the most important step to support to convert other concepts of OWL later on. A named class is mapped into one database table. As the whole ontology has an unique name of the class, and even the name of instances is unique in the class, so this table is named with the name of the class and has a primary key that is created automatically by adding *ID* as a suffix at the end, such as a *Person* class is mapped to a *Person* table that has a *PersonID* primary key.

In addition, there are class hierarchies in OWL ontology. The fundamental taxonomic construct for classes is *rdfs:subClassOf*. So that, *rdfs:subClassOf* syntax is mapped to an inheritance relationship as Fig. 1. It means that the created table corresponds to the subclass. This table gets its primary key as a foreign key that relates to its superclass table. Besides, as a class also relates other classes, so one table is created for every class in the ontology with one-to-one relations between classes and their subclasses.

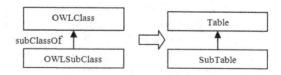

**Fig. 1.** Mapping OWL subclass construct to RDB

## 2.2   OWL Properties

In OWL, two types of properties are distinguished: Datatype properties specified a relation between instances of a class and data type values. Object properties specified a relation between instances of two classes. Furthermore, the authors in [9] said that properties can be single-valued or multivalued. If the cardinality of the property has a (maximum) value of 1, or the property is (inverse) functional, then the property is single-valued. In any other case, the property is multivalued.

**Object Property.** When transforming from OWL ontology into RDB schema, the object property is mapped to a foreign key or an intermediate table. Depending on the property is single-valued or multivalued, a relationship among the tables corresponded the classes can be one-to-many or many-to-many. In a case of many-to-many relation, an intermediate table must be created.

**Datatype Property.** In a class, for each datatype property is single-valued, it is mapped to a column in a table. This table corresponds to the class specified in the domain of the property. And the name of the column is same as the name of the datatype property. The type of the column that is specified in the range of the property is converted from XSD to SQL [9].

However, a datatype property is also multivalued, and SQL did not support multivalued columns. To deal with this problem, this property is mapped to a table. The name of the table is the datatype propertys name suffixed with *Value*. Its primary key is a combination of a corresponding column and a foreign key related to the table that corresponds to the class specified in the domain of the property. An example about a *hobby* property (i.e., a person has many hobbies). The property is mapped to a *hobbyValue* table. This table gets its primary key as a set of a *hobby* column with a varchar type and a *PersonID* foreign key that referenced to a *Person* table.

If a datatype property has a value restriction, then it is mapped to a corresponding column with a CHECK constraint. Such as a *typeOfProject* datatype property is restricted to have the same value for all instances of a *SoftwareProject* class.

```
<owl:DatatypeProperty rdf:ID="typeOfProject">
   <rdfs:domain rdf:resource="#SoftwareProject"/>
   <rdfs:range rdf:resource="&xsd:string"/>
</owl:DatatypeProperty> <owl:Class rdf:ID="SoftwareProject">
   <rdfs:subClassOf>
     <owl:Restriction>
       <owl:onProperty rdf:resource="#typeOfProject"/>
       <owl:hasValue rdf:resource=Software/>
     </owl:Restriction>
   </rdfs:subClassOf>
</owl:Class> CREATE TABLE SoftwareProject
 (typeOfProject VARCHAR,
  typeOfProject CHECK (typeOfProjecy = Software));
```

In addition to the RDF datatypes, OWL provides one added construct for defining a range of data values, namely an enumerated datatype. In the case, an enumerated datatype is mapped to a column with a CHECK constraint. An example about a *sex* property in a *Person* class with a list of values *Male* and *Female*.

```
<owl:DatatypeProperty rdf:ID=sex>
 <rdfs:domain rdf:resource=#Person/>
 <rdfs:range>
  <owl:DataRange>
  <owl:oneOf>
   <rdf:List>
    <rdf:first rdf:datatype=&xsd;string>Male
   </rdf:first>
   <rdf:rest>
     <rdf:List>
        <rdf:first rdf:datatype=&xsd;string>Female
     </rdf:first>
        <rdf:rest rdf:resource=&rdf;nil/>
     </rdf:List>
   </rdf:rest>
   </rdf:List>
  </owl:oneOf>
 </owl:DataRange>
</rdfs:range> </owl:DatatypeProperty> CREATE TABLE Person (sex
VARCHAR, sex CHECK IN (Male, Female));
```

## 2.3   Property Characteristics

In order to provide the mechanism for enhanced reasoning about a property, OWL used property characteristics.

**Symmetric Property.** Properties may be stated to be symmetric. If a property is symmetric and single  valued, then it is transformed to reflexive relation in RDB, i.e., the property is transformed to a foreign key. This key references to the same table that corresponds to both its domain and range class. For example, an *isSpouseOf* symmetric property (i.e., if one person is a spouse of another person, and vice versa) is mapped to an *isSpouseOf* column in a *Person* table, and this column is also a foreign key in the same table. The transformation of this property is shown as Fig. 2.

If a symmetric property is multivalued, then it is mapped to a table. This table gets its primary key as a set of foreign keys that reference to the table corresponded to both domain and range class of the property. For instance, a *hasFriend* symmetric property (i.e., if one person is a friend of many person, and vice versa) is mapped to a *Student-hasFriend* table. The transformation of this property is shown as Fig. 3.

**Fig. 2.** An example of the symmetric property that is single-valued

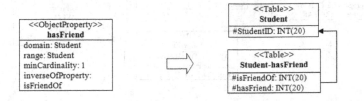

**Fig. 3.** An example of the symmetric property that is multivalued

**Transitive Property.** Properties may be stated to be transitive. The transformation of the transitive property is the same as the symmetric property. Such as a *hasAncestor* property is transitive property (i.e., if one person is an ancestor of another person, then the second person is an ancestor of the third person). The transformation of this property is shown as Fig. 4.

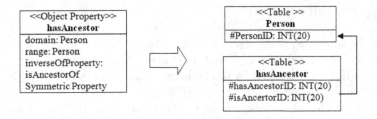

**Fig. 4.** An example of the transitive property that is multivalued

## 2.4   Property Restrictions

In OWL, we use a property restriction to constrain the range of a property in particular contexts following a variety of ways. Property restrictions can be applied both to data type properties and object properties. The context of an owl:Restriction can only be used for the various form such as the owl:allValues From, owl:someValuesFrom, owl:cardinality [10]. When transforming from OWL ontology into RDB, with aim of preserving all semantic information of ontology constraints, this information is saved in special tables like metadata tables. There are there metadata tables for each type of *AllValuesFrom, SomeValues-From, HasValue* restrictions and cardinality restrictions. The properties table stores the semantic of the properties in ontology. The above metadata tables are represented as follows:

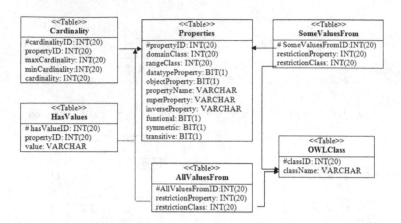

**Fig. 5.** Storing of OWL property restrictions in RDB

## 2.5   Individuals

In OWL, individuals are instances of classes. So that, after classes, properties, restriction constraints are mapped, we inserted all instances of classes into rows in the corresponding table. That is the last step of transforming from domain ontology into relational database.

## 3   Experiments

Based on the mapping rules, a transformation tool with java on the basis of Jena API is implemented. OWL is modeled by Protege 3.4, then we have carried out the tool. A graphical interface of the transformation tool is presented in Fig. 6. The interface shows the hierarchy of ontology classes as a tree construct, and buttons for transforming from ontology into relational database. And then, a generated text file as the output is described by SQL language. The connection between transformation tool and database sever using JDBC driver. Data are stored in My SQL 5.6 after that. We test by the UniversityOntology.owl example.

## 4   Related Work

In recent years, there are several approaches for addressing the issue of mapping from ontology representation in OWL to relational database [3–8]. Firstly, Kajal et al. [3] proposed a set of techniques to provide a mapping of an OWL ontology to a relational schema. The OWL2DB transformation algorithm consists of 8 steps and interprets OWL ontology like a graph. A shortcoming of that approach is only saving class, instances in RDB with other semantics represented in OWL. In 2006, based on the idea of the algorithm in [3], Lina et al. [4] developed this algorithm at a higher level, namely the transformation of constraints

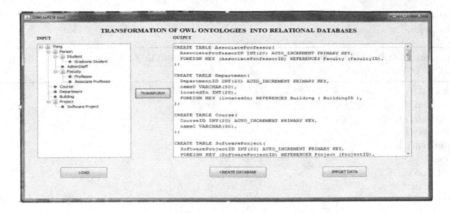

**Fig. 6.** The graphical interface of the transformation tool

in owl: Restriction syntax are solved. That is what the previous algorithm did not do. However, these authors said that the proposed algorithm was capable of transferring all OWL Lite and part of OWL DL syntax. Beside the direct mapping proposal, Jutas et al. [5] and Yanhui et al. [6] recommended a mediate approach for knowledge represented by ontology automatic transformation into conceptual data model, and then converted into relational database schema. The disadvantage of using ontology for conceptual data modelling is only main concepts that are transformed. In literature [7], the authors developed an ontology management system for storing data likes database management system. A single table in system called Fact Table is used for storing facts of ontology and a set of triggers which are fired when a fact is inserted, updated, or deleted from this table. However, the state of the art system is not still completed, and this transformation method does not preserve the real relational structure. On the other hand, a mechanism for generating the relational schema from a set of integrated XML files, which includes defining a set of mapping rules from the OWL ontology to the relational format, are presented by Brum et al. [8]. In Fig. 7, we summarize the above transformation proposals and tried to assemble the similar approaches in a group.

In addition, we analyzed and compared some of transformation proposals as shown in Table 1. The comparison shows that the above approaches is incomplete for mapping of OWL to relational database, just solving a part of OWL constructs. In some cases, the transformation method can be semi-automatic, e.g. they can require much user interaction. And finally, some of proposals are not implemented. Nevertheless, we realize that transformation of ontology to relational database is based on a set of rules called mapping rules, which can be extensible, because many other constructs in OWL ontology are not still considered and the direct mapping can be applicable, automatically implemented, had correctness. Therefore, this a reason why we choose a novel approach to transformation of ontologies to relational databases, which is the main contribution of this paper.

**Table 1.** Comparison of the proposals for transforming of OWL ontology to relational database

| Transforming of RDFS/ OWL construct | Kajal [3] | Lina [4] | Jutas [5] | Yanhui [6] | Juhnyoung [7] | Brum [8] |
|---|---|---|---|---|---|---|
| Class | + | + | + | + | + | + |
| subClassOf | + | + | - | + | + | - |
| oneOf, dataRange | - | - | + | + | - | - |
| disjoinWith | - | - | - | - | - | - |
| equivalentClass | - | - | - | - | - | - |
| unionOf | - | - | - | - | - | - |
| complemetnOf | - | - | - | - | - | - |
| intersectionOf | - | - | - | - | - | - |
| Inviduals | + | + | + | + | - | + |
| Datatype property | + | + | + | + | - | + |
| Object property | + | + | + | + | - | + |
| subPropertyOf | - | + | - | + | - | + |
| rdfs:domain, rdfs:range | - | + | + | + | - | + |
| Funtional property | - | + | - | - | - | - |
| Transitive property | - | - | - | - | - | - |
| Symmetric property | - | - | - | - | - | - |
| inverseOf | - | - | - | - | - | - |
| allValuesFrom | - | + | - | - | - | - |
| someValuesFrom | - | + | - | - | - | - |
| hasValue | - | + | - | - | - | - |
| cardinality | - | + | - | + | - | + |
| maxCardinality | - | + | - | + | - | + |
| minCardinality | - | + | - | + | - | + |
| xsd:datatype | + | + | + | + | - | + |
| Excess | 6/20 | 12/20 | 7/20 | 9/20 | 2/20 | 7/20 |
| Implementation | No | Yes | Yes | Yes | Yes | Yes |
| Degree of automation | automatic | semi-automatic | semi-automatic | semi-automatic | semi-automatic | automatic |
| Extensibility | Yes | Yes | No | No | Yes | No |

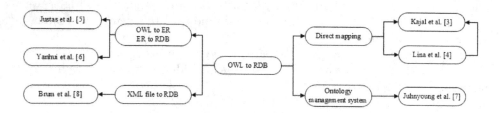

**Fig. 7.** The approaches of OWL ontology to relational database transformation

## 5   Conclusions

In this paper, we have an analysis and evaluation on some of approaches for transformation of ontology characterized as OWL into relational database.

Basing on the comparison among them, we have remarked that these proposals is incomplete. From the result of this, the combination of mapping rules is proposed in this paper. Currently, our approach is capable to transform the most of OWL DL concepts. We consider that ontology classes should be mapped to relational tables, properties are mapped to relations or attributes, instances corresponds to rows in the table. Semantic information about property restrictions and properties as symmetric, transitive, inverse functional are stored in tables like metadata. Using both direct mapping and metadata, we achieve applicable relational structure with preserving data and restrict losing the ontological constructs. A prototype tool of mapping rules which have OWL documents like the inputs and text files like the outputs, was implemented as an independent software.

However, some constructs are lost when transforming. For example, our method does not save the ontological semantics as complement class, intersection class, enumerated class. The constructs as subproperties, equivement property that are not corresponsive in RDB should be considered. In further, we intend to set up the transformation tool as a plug-in for a popular ontology editor Protege. We also extend this approach to forthcoming OWL2 that extends with current OWL.

# References

1. Ian, H., Boris, M., Ulrike, S.: Bridging the gap between OWL and relational database. Journal of Web Semantics: Science, Services and Agents on the World Wide Web 7(2), 74–89 (2009)
2. Lina, A., Christine, P., Stefano, S.: Reasoning with large ontologies stored in relational databases the OntoMinD approach. Data and Knowledge Engineering 69(11), 1158–1180 (2010)
3. Gali, A., Chen, C.X., Claypool, K.T., Uceda-Sosa, R.: From ontology to relational databases. In: Wang, S., Tanaka, K., Zhou, S., Ling, T.-W., Guan, J., Yang, D., Grandi, F., Mangina, E.E., Song, I.-Y., Mayr, H.C. (eds.) ER Workshops 2004. LNCS, vol. 3289, pp. 278–289. Springer, Heidelberg (2004)
4. Lina, N., Ernestas, V.: Transforming ontology representation form OWL to relational database. Information Technology And Control Kaunas Technology 35(3), 333–343 (2006)
5. Justas, T., Olegas, V.: A graph oriented model for ontology transfoemation into conceptual data model. Information Technology and Control 36(1A), 126–132 (2007)
6. Yahui, L., Chong, X.: An ontology-based approach to build conceptual data model. In: 9th International Conference on, Sichuan (2012)

7. Juhnyoung, L., Richard, G.: Ontology management for large-scale enterprise systems. Electronic Commerce Research and Applications **5**, 2–15 (2006)
8. Brum, S.D., de Campos, A., Piveta, E.K.: Mapping OWL ontologies to relational schemas. In: IEEE International Conference, Las Vegas (2011)
9. Irina, A., Nahum, K., Ahto, K.: Storing OWL ontologies in SQL relational database. Interational Journal of Electrical, Computer and Systems Engineering **1**(4), 242–257 (2007)
10. OWL Web Ontology Language Reference. http://www.w3.org/TR/owl-features/

# On Query Containment Problem
# for Conjunctive Queries Under Bag-Set
# Semantics

Victor Felea[1]($\boxtimes$) and Violeta Felea[2]

[1] Computer Science Department, Alexandru Ioan Cuza University of Iasi,
Iasi, Romania
felea@infoiasi.ro
[2] FEMTO-ST, Franche-Comté University, Besançon, France
violeta.felea@femto-st.fr

**Abstract.** The query containment is one of the fundamental problems
in the context of query processing and optimization. Concerning the
answers of queries for databases, three semantics were intensively analyzed. In the first one, called *set*, both databases and answers of queries
for databases are considered sets. In the second one, called *bag-set*,
databases are sets, but the answers are bags (multi-sets). In the third
one, called *bag*, both databases and answers are bags.

In this paper, we study the query containment problem for conjunctive queries (*CQs*) under *bag-set* semantics. It is not known whether this
problem is decidable. We give a characterization of this problem.

**Keywords:** Conjunctive queries · Query containment · Bag-set semantics

## 1 Introduction and Summary of Results

The problems of query containment, equivalence and minimization are fundamental problems in database processing and optimization.

The containment problem of queries is the following: given two queries $Q^1$
and $Q^2$, is it true that the answer of $Q^1$ for $D$ (denoted $Q^1(D)$) is included in
the answer of $Q^2$ for $D$ (denoted $Q^2(D)$), for every database $D$? This problem
was investigated for a large class of queries.

Concerning the answer of a query $Q$ for a database $D$, we distinguish several
semantics. In the first one, called set semantics, both the database $D$ and the
answer $Q(D)$ are considered sets (without duplicates). In the second one, called
bag-set semantics, the database $D$ is considered as set, but the answer $Q(D)$ as a
bag (multi-set). In the third one, called bag semantics, both the database $D$ and
the answer $Q(D)$ are bags. In recent years, a new semantics, called combined
semantics has been taken into consideration.

The query containment problem was studied for the class of conjunctive
queries($CQs$) in many works. Chandra and Merlin in [3] have shown that the

© Springer International Publishing Switzerland 2015
N.T. Nguyen et al. (Eds.): ACIIDS 2015, Part I, LNAI 9011, pp. 159–169, 2015.
DOI: 10.1007/978-3-319-15702-3_16

problems of containment, minimization and equivalence of $CQs$ queries under set semantics are *NP-complete*. Klug [8] studied conjunctive queries with comparison predicates under set semantics, and showed that this problem is in $\Pi_2^p$, and van der Meyden [9] proved that this problem is $\Pi_2^p$- complete.

Chaudhuri and Vardi [4] showed that under bag-set and bag semantics, the query-containment problem for $CQs$ -queries is $\Pi_2^p$ - hard. It is not known whether the query containment problem for $CQs$ queries is decidable under bag-set semantics and bag semantics. Brisaboa and Hernandez in [2] prove that the containment problems for $CQs$ queries under bag semantics can be tested on a finite set of canonical databases. They give a procedure that decides on the bag containment problem of $CQs$ queries in a large number of cases. In his thesis [5], Damigos investigates techniques for query optimization, using a set of views, focusing on the query containment, the query rewriting and the view selection.

A sub-class of $CQs$, which consists of $CQs$ queries with inequalities ($\neq$), denoted $CQIs$, was studied. Thus, Van der Meyden in [9] points out that the containment problem for $CQIs$ under set semantics is $\Pi_2^P$ -*complete*. In [7] the authors identify some cases when query containment for $CQIs$ can be characterized by to existence of homomorphisms. In [6], the authors study the query containment problem for $CQIs$ queries under bag-set and bag semantics, and they show that this problem is undecidable, using the *Hilbert's Tenth Problem* and some of its variants. In fact, this problem is undecidable even if the queries use only a single relation of arity 2 and the number of inequalities in the queries is at most some fixed constant.

In this paper, we study the query containment for the class $CQs$ under bag-set semantics, giving a characterization of this problem using equivalence classes defined on the set of variables from the first query, databases defined on these classes, injections on the classes and mapping generated by injections and the characteristics of the second query.

Let us consider an example.

*Example 1.* Assume the following database schema about companies:
$CS(cid, cname, ctry)$ - represents companies, $cid$ is a company identifier, primary key, $cname$ is the company name, $ctry$ is the country where $cname$ company is registered.
$CNTRS(cid1, cid2, year)$ - represents concluded contracts between the companies $cid1$ (as beneficiary) and $cid2$ (as supplier) in year $year$.

Let us take the following query:
To a company identified by $cid$, and a year $year$, find the number of contracts concluded by $cid$ company as beneficiary in year $year$. We can express this query as the following:
$Q : h(cid, year) : -\exists(cname, cid2, cname2, ctry, ctry2)CS(cid, cname, ctry) \wedge CS(cid2, cname2, ctry2) \wedge CNTRS(cid, cid2, year)$.

Replacing attributes by variables, we get the query:
$Q : h(x_1, x_2) : -\exists(y_1, y_2, y_3, y_4, y_5)\{CS(x_1, y_1, y_4) \wedge CS(y_2, y_3, y_5) \wedge CNTRS(x_1, y_2, x_2)\}$.

If we consider the constant $c_2$ instead of the variable $x_2$, we obtain:

$Q_{c_2} : h_1(x_1) : -\exists(y_1, y_2, y_3, y_4, y_5)\{CS(x_1, y_1, y_4) \wedge CS(y_2, y_3, y_5)\wedge$
$CNTRS(x_1, y_2, c_2)\}.$

If $d_2$ is another constant for $x_2$, we are interested in the following problem: for all companies $x_1$, the number of contracts having the company $x_1$ as beneficiary, in year $d_2$ is at least equal to the number of contracts in year $c_2$? This problem is equivalent to the containment problem: $Q_{c_2} \subseteq Q_{d_2}$ under $bag - set$ semantics.

## 2  Preliminaries

We consider finite relations defined on $dom$, that is an infinite denumerable set totally ordered, like the set of rational numbers. A relation $r$ defined on $dom$ is a set whenever its tuples are distinct. On the contrary, when each tuple of $r$ occurs one or more times, the relation is called a bag. We represent a bag as the set $\{(t_1 : m_1), \ldots, (t_n : m_n)\}$, where $m_i$ is the multiplicity of $t_i$, s.t. $m_i > 0$, and $t_1, \ldots, t_n$ are distinct.

**Definition 1.** *A conjunctive query (CQ) denoted Q has the following form:*

$$Q : h(\overline{x}) : -(\exists\overline{y})\{f(\overline{x}, \overline{y})\} \tag{1}$$

*where $\overline{x}$ is the vector of variables called free, $\overline{y}$ is the vector of variables called bound, that are existentially quantified, $f(\overline{x}, \overline{y})$ is a conjunction of atoms having variables from $\overline{x} \cup \overline{y}$, $h(\overline{x})$ is called the head of the query $Q$, and the right part is called its body.*

We only consider safe queries, where each variable $x_i$ from $\overline{x}$ must appear in an atom from the conjunction $f(\overline{x}, \overline{y})$.

An assignment mapping (or, simply, an assignment) is a mapping $\tau$ from the set of free and bound variables to the underlying domain $dom$. The assignment $\tau$ is extended to the vectors of variables: if $\overline{z} = (z_1, \ldots, z_n)$, then we define $\tau\overline{z} = (\tau(z_1), \ldots, \tau(z_n))$.

We use assignments to define the semantics of queries. Let us denote by $Ans_{bs}(Q, D)$, the answer of the query $Q$, for the database $D$, under the bag-set semantics. In the bag-set semantics, the relations are sets, but to determine the query answer we need to find all assignments that satisfy the body of query.

Let $\tau_1$ and $\tau_2$ be two assignments, s. t. the definition domain $D_1$ for $\tau_1$ is included in the definition domain $D_2$ of $\tau_2$. The notation $\tau_1 < \tau_2$ means the projection of $\tau_2$ on $D_1$ coincides to $\tau_1$. If $D$ is a database on $dom$, $\tau$ an assignment and $f(\overline{x}, \overline{y}) = R_1(\overline{w}_1) \wedge \ldots \wedge R_q(\overline{w}_q)$, then we denote by $D \models f[\tau]$ ([1]) the case when $\tau$ satisfies all atoms from $f(\overline{x}, \overline{y})$, that is $\tau R_i(\overline{w}_i) = R_i(\tau\overline{w}_i) \in D$, for each $1 \leq i \leq q$. To define formally the answer of a query under the bag-set semantics, let us denote by $Mapp(Q, D, \tau_0)$ the following set of assignments corresponding to a query $Q$, a database $D$ and an assignment $\tau_0$ defined on $\overline{x}$:

$$Mapp(Q, D, \tau_0) = \{\tau_1 | \tau_0 < \tau_1, D \models f[\tau_1], \tau_1 \text{ is defined on } \overline{x} \cup \overline{y}_1\}, \tag{2}$$

where $\tau_0 < \tau_1$ denotes that the projection of $\tau_1$ on $\overline{x}$ coincides to $\tau_0$. The answer of a query $Q$ for a database $D$ under the bag-set semantics is the following:

$$Ans_{bs}(Q, D) = \{(h(\tau_0 \overline{x}) : m) | m = |Mapp(Q, D, \tau_0)| \text{ and } m > 0\}, \qquad (3)$$

where $|S|$ denotes the cardinality of the set $S$.

Let $\tau_0$ be an assignment for $\overline{x}$. There is a partition on $\overline{x}$, denoted $Part(\tau_0)$ and defined as follows: $x_i Part(\tau_0) x_j$ iff $\tau_0(x_i) = \tau_0(x_j)$, where $x_i, x_j \in \overline{x}$. To a partition $\pi_0$ defined on $\overline{x}$, we associate a mapping denoted $\zeta_{\pi_0}$, defined from $\overline{x}$ into $Class_{\pi_0}$, the set of classes defined by the partition $\pi_0$ on $\overline{x}$ (a class is represented by one of its variable): $\zeta_{\pi_0}(x_i) = \widetilde{x}_{i\pi_0}$, where $\widetilde{x}_{i\pi_0}$ denotes the class defined by $\pi_0$, that contains the element $x_i$. This mapping is extended to vectors of variables, as for assignments. If a class from $Class_{\pi_0}$ consists of the variables $x_1, \ldots x_s$, then it is also denoted by $\widetilde{x_1 \ldots x_s}_{\pi_0}$. Moreover, associated to the assignment $\tau_0$, we consider an injection from $Class_{\pi_0}$ into $dom$, denoted $\sigma_{\tau_0}$ and defined as follows: $\sigma_{\tau_0}(\widetilde{x}_{i\pi_0}) = \tau_0(x_i)$, for each $\widetilde{x}_{i\pi_0}$ from $Class_{\pi_0}$. We have the equality: $\tau_0 = \sigma_{\tau_0} \circ \zeta_{\pi_0}$, where $\circ$ denotes the mapping composition. We have similar considerations for assignments from $\overline{x} \cup \overline{y}$ into $dom$.

Let us denote by $Rel(Q)$ the set of all relational symbols from $Q$. If $T$ is a database defined on $Class_\pi$ and has relational symbols from the set $\mathcal{R}$, then we say $T$ is a database on $(Class_\pi, \mathcal{R})$.

Let $\pi_0$ be a partition defined on $\overline{x}$. Now, we define the answer of a query $Q$ for a database $D$ and for a partition $\pi_0$, under the bag-set semantics. This answer will be denoted by $Ans_{bs}(Q, D, \pi_0)$, and it is as follows:

$$Ans_{bs}(Q, D, \pi_0) = \{(h(\tau_0 \overline{x}) : m) | m = |Mapp(Q, D, \tau_0)|, m > 0, Part(\tau_0) = \pi_0\} \qquad (4)$$

In the following, we need some operations like intersection, disjunctive union and inclusion defined on bags.

**Definition 2.** *Let $M_1$ and $M_2$ be two bags. Their intersection, disjunctive union, inclusion denoted $M_1 \cap M_2$, $M_1 \cup M_2$, $M_1 \subseteq_{bs} M_2$, respectively, are defined as follows:*
*(i) $M_1 \cap M_2 = \{(v : m) | \exists (v : n_1) \in M_1, \exists (v : n_2) \in M_2, \text{ s.t. } m = min\{n_1, n_2\}\}$*
*(ii) $M_1 \cup M_2 = \{(v : m) | (v : m) \in M_1, \text{ or } (v : m) \in M_2\}$, where $M_1 \cap M_2 = \emptyset$.*
*(iii) $M_1 \subseteq_{bs} M_2$ if for each $(v : n_1) \in M_1$, there exists $(v : n_2) \in M_2$ s.t. $n_1 \leq n_2$.*

Concerning these answers corresponding to partitions defined on $\overline{x}$, we have the following result:

**Proposition 1.** *Let $\pi_0^1$ and $\pi_0^2$ be two distinct partitions defined on $\overline{x}$ and $Part(\overline{x})$ the set of all partitions defined on $\overline{x}$. We have:*
*(i) $Ans_{bs}(Q, D, \pi_0^1) \cap Ans_{bs}(Q, D, \pi_0^2) = \emptyset$,*
*(ii) $Ans_{bs}(Q, D) = \cup_{\pi_0 \in Part(\overline{x})} Ans_{bs}(Q, D, \pi_0)$*

In the following definition, we specify formally, the notion of the containment of queries under bag-set semantics.

**Definition 3.** *Let $Q^1$ and $Q^2$ be two queries having the same head:*
$Q^i : h(\overline{x}) : -(\exists \overline{y}_i)[f_i(\overline{x}, \overline{y}_i)]$, $i = 1, 2$.
*We say that $Q^1$ is contained in $Q^2$, using the bag-set semantics, denoted $Q^1 \subseteq_{bs} Q^2$ if $Ans_{bs}(Q^1, D) \subseteq_{bs} Ans_{bs}(Q^2, D)$, for each database $D$ defined on* dom.

Using Proposition 1, we get the following remark:

*Remark 1.* We have: $Q^1 \subseteq_{bs} Q^2$ iff $Ans_{bs}(Q^1, D, \pi_0) \subseteq_{bs} Ans_{bs}(Q^2, D, \pi_0)$, for each $\pi_0$ defined on $\overline{x}$ and each database $D$ defined on dom.

Considering the statement from Remark 1, in the following, we focus on the containment query problem for two queries and a fixed partition $\pi_0$.

## 3    Parameters Corresponding to a Query and a Database

Firstly, we need to define a precedence relation between partitions. Let $\pi_0$ be a partition defined on $\overline{x}$ and $\pi_1$ a partition defined on $\overline{x} \cup \overline{y}_1$.

**Definition 4.** *We say that $\pi_0$ precedes $\pi_1$, denoted $\pi_0 \prec \pi_1$, if the following assertions yield:*
*(i) $x_i \pi_0 x_j$ implies $x_i \pi_1 x_j$, for each $x_i, x_j \in \overline{x}$ and*
*(ii) $t \pi_1 t'$ and $t, t' \in \overline{x}$ implies $t \pi_0 t'$.*

*Remark 2.* Let $\tau_0$ and $\tau_1$ be two assignments s.t. $\tau_0 < \tau_1$. Then $Part(\tau_0) \prec Part(\tau_1)$.

Let $\tau_0$ be an assignment defined on $\overline{x}$ and $\tau_i$ defined on $\overline{x} \cup \overline{y}_1$ s.t. $\tau_0 < \tau_i$. By Remark 2, we get $\pi_0 \prec \pi_i$. Let $\pi_0 = Part(\tau_0)$ and $\pi_i = Part(\tau_i)$. Let $Q^1$ be a query having the form as in Definition 3. Let $D$ be a database defined on dom. Let $M(Q^1, D, \tau_0)$ be the set of all $\tau$ defined on $\overline{x} \cup \overline{y}_1$ s. t. $\tau_0 < \tau$ and $\tau$ satisfies the body of $Q^1$, i.e.

$$M(Q^1, D, \tau_0) = \{\tau | \tau_0 < \tau, D \models f_1[\tau]\} = \{\tau_1, \ldots, \tau_m\} \qquad (5)$$

Associated to $\tau_i$, we define a sub-database of $D$, denoted $D_{\tau_i}$, as follows:

$$D_{\tau_i} = \{R(\overline{w}) | R(\overline{w}) \in D, \text{ and if } \overline{w} = (z_1, \ldots, z_k), \text{ then all } z_j \in range(\tau_i)\} \quad (6)$$

From the statement $D \models f_1[\tau_i]$, we obtain $D_{\tau_i} \models f_1[\tau_i]$. To $\tau_i$, we consider $\pi_i = Part(\tau_i)$, $\zeta_{\pi_i}$, $\sigma_i$ defined as in Section 2. We have: $\tau_i = \sigma_i \circ \zeta_{\pi_i}$. The injection $\sigma_i$ is defined on $Class_{\pi_i}$ into dom. Hence, $\sigma_i$ is bijective from $Class_{\pi_i}$ into $range(\sigma_i)$. Let $T_i = \sigma_i^{-1}(D_{\tau_i})$. From $D_{\tau_i} \models f_1[\tau_i]$, we obtain $T_i \models f_1[\zeta_{\pi_i}]$. The statement $T_i \models f_1[\zeta_{\pi_i}]$ means $f_1[\zeta_{\pi_i}] \subseteq T_i$. We have $\sigma_0 < \sigma_i$, $1 \leq i \leq m$.

By Remark 2, we get $\pi_0 \prec \pi_i$. Since the partitions from $\pi_1, \ldots, \pi_m$ can be duplicated, let $\pi_1, \ldots, \pi_h$ be the distinct elements from them. For each $\pi_s$, let $\sigma_s^1, \ldots, \sigma_s^{n_s}$ be the injections $\sigma_i$ that correspond to the partition $\pi_s$, i.e. $\sigma_s^j : Class_{\pi_s} \mapsto dom$, $1 \leq j \leq n_s$, $1 \leq s \leq h$ ($Part(\tau_s^j) = \pi_s$, $\tau_s^j = \sigma_s^j \circ \zeta_{\pi_s}$).

We have $m = \Sigma_{s=1}^h n_s$. Since $m > 0$, we have: $h > 0$ and there exists $s$, $1 \leq s \leq h$ s.t. $n_s > 0$. Let $T_s^j = (\sigma_s^j)^{-1}(D_{\tau_s^j})$, which is a database defined on $(Class_{\pi_s}, Rel(Q^1))$. When the integer $s$ is fixed, the databases $T_s^j$ can coincide for different values of $j$. Let us point out the distinct elements from them. Let $p_s$ be the number of these databases. Renumbering these databases, let $T_s^1, \ldots, T_s^{p_s}$ be the distinct databases. We obtain a surjective mapping $\eta_s$ from $\{1, 2, \ldots, n_s\}$ into $\{1, 2, \ldots, p_s\}$. Let $D'(\tau_0)$ be the union of all databases $D_{\tau_i}$. Using $\sigma_s^j$ and $T_s^j$, we express $D'(\tau_0)$ as follows:

$$D'(\tau_0) = \cup_{s=1}^h \cup_{j=1}^{n_s} \sigma_s^j(T_s^{\eta_s(j)}) \tag{7}$$

*Example 2.* Let the query $Q^1$ having the following structure:
$Q^1 : h(x_1 x_2) : -(\exists y_1)\{R(x_1, y_1) \wedge R(y_1, x_2)\}$, let $\tau_0(x_1, x_2) = (1, 2)$.
Let $D = \{R(1, 3), R(3, 2), R(1, 4), R(4, 2), R(1, 1), R(1, 2), R(1, 5)\}$. Let $\tau_1, \tau_2, \tau_3$ be the following assignments: $\tau_1(x_1, x_2, y_1) = (1, 2, 3)$, $\tau_2(x_1, x_2, y_1) = (1, 2, 4)$, $\tau_3(x_1, x_2, y_1) = (1, 2, 1)$. These are all the mappings that satisfy $D \models f_1[\tau_i]$ and $\tau_0 < \tau_i$. The partitions $\pi_0$, $\pi_i$ are the following: $\pi_0 = \{\widetilde{x_1}, \widetilde{x_2}\}$, $\pi_1 = \{\widetilde{x_1}, \widetilde{x_2}, \widetilde{y_1}\}$, $\pi_2 = \pi_1$, $\pi_3 = \{\widetilde{x_1 y_1}, \widetilde{x_2}\}$. The databases $D_{\tau_i}$ are as follows: $D_{\tau_1} = \{R(1, 3), R(3, 2), R(1, 1), R(1, 2)\}$, $D_{\tau_2} = \{R(1, 4), R(4, 2), R(1, 1), R(1, 2)\}$, $D_{\tau_3} = \{R(1, 1), R(1, 2)\}$. The injections $\sigma_i$ are as follows:
$\sigma_0(\widetilde{x_1}_{\pi_0}, \widetilde{x_2}_{\pi_0}) = (1, 2)$, $\sigma_1(\widetilde{x_1}_{\pi_1}, \widetilde{x_2}_{\pi_1}, \widetilde{y_1}_{\pi_1}) = (1, 2, 3)$,
$\sigma_2(\widetilde{x_1}_{\pi_1}, \widetilde{x_2}_{\pi_1}, \widetilde{y_1}_{\pi_1}) = (1, 2, 4)$, $\sigma_3(\widetilde{x_1 y_1}_{\pi_3}, \widetilde{x_2}_{\pi_3}) = (1, 2)$.
The databases $T_i$ are the following:
$T_1 = \{R(\widetilde{x_1}_{\pi_1}, \widetilde{y_1}_{\pi_1}), R(\widetilde{y_1}_{\pi_1}, \widetilde{x_2}_{\pi_1}), R(\widetilde{x_1}_{\pi_1}, \widetilde{x_1}_{\pi_1}), R(\widetilde{x_1}_{\pi_1}, \widetilde{x_2}_{\pi_1})\}$,
$T_2 = T_1$, $T_3 = \{R(\widetilde{x_1 y_1}_{\pi_3}, \widetilde{x_1 y_1}_{\pi_3}), R(\widetilde{x_1 y_1}_{\pi_3}, \widetilde{x_2}_{\pi_3})\}$
We have $h = 2$, $n_1 = 2$, $n_2 = 1$, $p_1 = 1$, $p_2 = 1$, $\sigma_1^1 = \sigma_1$ $\sigma_1^2 = \sigma_2$, $\sigma_2^1 = \sigma_3$, $T_1^1 = T_1^2 = T_1$, $T_2^1 = T_3$, $\eta_1(1, 2) = (1, 1)$, $\eta_2(1) = (1)$, $D'(\tau_0) = D - \{R(1, 5)\}$.

In the next sections, we use these parameters, but we prefer a partition $\pi_0$ and an injection $\sigma_0$, instead of an assignment $\tau_0$ (if $\pi_0$ and $\sigma_0$ are given, then $\tau_0$ is computed as: $\tau_0 = \sigma_0 \circ \zeta_{\pi_0}$; if $\tau_0$ is given, then $\pi_0 = Part(\tau_0)$ and $\sigma_0$ is defined by: $\sigma_0(\widetilde{t}_{\pi_0}) = \tau_0(t)$).

## 4     Generation of Mappings for $Q^2$ and Some Parameters

In Section 3, we have defined a number of parameters associated to a query $Q^1$, a database $D$, a partition $\pi_0$ on $\overline{x}$ and an injection $\sigma_0$ from $Class_{\pi_0}$ into $dom$. Let us denote by $PAR$ this vector of parameters, excepting $\sigma_0, \pi_0$:

$$PAR = (h, \pi_1, \ldots, \pi_h, p_1, \ldots, p_h, n_1, \ldots, n_h, \eta_1, \ldots, \eta_h,$$

$$T_s^j, \sigma_s^j, 1 \leq j \leq n_s, 1 \leq s \leq h). \tag{8}$$

Using the parameters from $PAR$ and the second query $Q^2$, we generate assignments defined on $\overline{x} \cup \overline{y_2}$. The operator for generation will be denoted by $GEN(Q^2, \sigma_0, \pi_0, PAR)$, where $\sigma_0 < \sigma_s^j$, $1 \leq j \leq n_s, 1 \leq s \leq h$, $\pi_0 \prec \pi_i$, $1 \leq i \leq h$. Let $Q^2$ having the form as in Definition 3, i.e.

$$Q^2 : h(\overline{x}) : -(\exists \overline{y}_2)\{f_2(\overline{x}, \overline{y}_2)\}, \text{ where } f_2(\overline{x}, \overline{y}_2) = S_1(\overline{z}_1) \wedge \ldots \wedge S_q(\overline{z}_q).$$

To define $GEN$ operator, we need to specify sets of tuples constructed using atoms $S_l(\overline{z}_l), 1 \le l \le q$, $\sigma_0, \pi_0$, and some of parameters specified in $PAR$.

$$M_l(Q^2, \sigma_0, \pi_0, PAR) = \{\psi^r_{s_l, j_l} | 1 \le s_l \le h, 1 \le j_l \le n_{s_l}, \psi^r_{s_l, j_l} : \overline{z}_l \mapsto Class_{\pi_{s_l}},$$

$$\psi^r_{s_l, j_l} S_l(\overline{z}_l) \in T^{\eta_{s_l}(j_l)}_{s_l}, \psi^r_{s_l, j_l}(t) = \tilde{t}_{\pi_0}, \forall t \in \overline{z}_l \cap \overline{x}\}, \text{ for each } 1 \le l \le q, \qquad (9)$$

where the index $r$ takes values from 1 to the number of mappings from $\overline{z}_l$ into $Class_{\pi_{s_l}}$ that satisfy the conditions from (9), the integer $l$ being fixed.

*Remark 3.* (i) Let $\psi$ be a containing mapping from $\overline{x} \cup \overline{y}_2$ into $\overline{x} \cup \overline{y}_1$ ([3]), $\pi_0$ a partition on $\overline{x}$, $\pi_1$ a partition on $\overline{x} \cup \overline{y}_1$ s.t. $\pi_0 \prec \pi_1$, $T_{\pi_1}$ a database on $Class_{\pi_1}$ s.t. $f_1[\zeta_{\pi_0}] \subseteq T_{\pi_1}$. Then we have $M_l(Q^2, \sigma_0, \pi_0, PAR) \ne \emptyset$, for each $1 \le l \le q$, and for each value of parameters from $PAR$ given in (8).
(ii) A necessary condition for the containment problem $Q^1 \subseteq_{bs} Q^2$ is $M_l(Q^2, \sigma_0, \pi_0, PAR) \ne \emptyset$, for each $1 \le l \le q$, and for each value of parameters from $PAR$.
(iii) The sets $M_l(Q^2, \sigma_0, \pi_0, PAR)$ depend only on $Q^1$ and $Q^2$.

**Proof.** (i) A containing mapping $\psi$ satisfies: $f_2[\psi] \subseteq f_1$ and $\psi(t) = t$, for each $t \in \overline{x}$. We define $\psi_l$ from $\overline{z}_l$ into $Class_{\pi_1}$ as follows: $\psi_l(\overline{z}_l) = (\zeta_{\pi_1} \circ \psi)|_{\overline{z}_l}$ (the projection of $(\zeta_{\pi_1} \circ \psi)$ on $\overline{z}_l$). We obtain: $\psi_l S_l(\overline{z}_l) = f_2[(\zeta_{\pi_1} \circ \psi)|_{\overline{z}_l}] \subseteq f_1[\zeta_{\pi_1}] \subseteq T_{\pi_1}$, and $\psi_l(t) = (\zeta_{\pi_1} \circ \psi)(t) = \zeta_{\pi_1}(t) = \tilde{t}_{\pi_1}$, for each $t \in \overline{x}$.
(ii) If we have $Q^1 \subseteq_{bs} Q^2$, then we also have $Q^1 \subseteq Q^2$ (the containment relation of type set). The last statement is equivalent to the existence of a containment mapping from $Q^2$ into $Q^1$ ([3]). $\qquad \square$

Let $\psi_1$ and $\psi_2$ be two mappings s.t. $\psi_1 : \overline{z}_i \mapsto Class_{\pi_\alpha}$, $\psi_2 : \overline{z}_j \mapsto Class_{\pi_\beta}$ and $i \ne j$. Let $\sigma_1, \sigma_2$ be two injections s.t. $\sigma_1 : Class_{\pi_\alpha} \mapsto dom$, $\sigma_2 : Class_{\pi_\beta} \mapsto dom$. Let $t$ be a variable from $\overline{y}_2 \cap \overline{z}_i \cap \overline{z}_j$. Generally, the value of $\sigma_1 \circ \psi_1$ for $t$ is different from $\sigma_2 \circ \psi_2$ for $t$, i.e. $(\sigma_1 \circ \psi_1)(t) \ne (\sigma_2 \circ \psi_2)(t)$. We are interested in the contrary case because we need to obtain assignments from $\overline{x} \cup \overline{y}_2$ into $dom$. For this, we consider $q$-tuples denoted $\tau = (\sigma^{j_1}_{s_1} \circ \psi^{r_1}_{s_1, j_1}, \ldots, \sigma^{j_q}_{s_q} \circ \psi^{r_q}_{s_q, j_q})$, where $\psi^{r_l}_{s_l, j_l} \in M_l(Q^2, \sigma_0, \pi_0, PAR)$, $1 \le l \le q$. From the definition of $\psi^{r_l}_{s_l, j_l}$ and its properties given in the definition of $M_l(Q^2, \sigma_0, \pi_0, PAR)$, we obtain the composition $\sigma^{j_l}_{s_l} \circ \psi^{r_l}_{s_l, j_l}$ is defined on $\overline{z}_l$ and has the values in $dom$. The following definition specifies the class of $q$-tuples necessary to generate assignments.

**Definition 5.** *Let $\tau$ be a $q$-tuple s.t. $\tau = (\sigma^{j_1}_{s_1} \circ \psi^{r_1}_{s_1, j_1}, \ldots, \sigma^{j_q}_{s_q} \circ \psi^{r_q}_{s_q, j_q})$, where $\psi^{r_l}_{s_l, j_l}$ belongs to $M_l(Q^2, \sigma_0, \pi_0, PAR)$, for each $1 \le l \le q$. The tuple $\tau$ is said to be compatible to $f_2(\overline{x}, \overline{y}_2)$ if for each $1 \le l < i \le q$ and for each variable $t \in (\overline{z}_l \cap \overline{z}_i)$, we have: $(\sigma^{j_l}_{s_l} \circ \psi^{r_l}_{s_l, j_l})(t) = (\sigma^{j_i}_{s_i} \circ \psi^{r_i}_{s_i, j_i})(t).$*

*Remark 4.* Let $\tau$ be a $q$-tuple as in Definition 5, compatible to $f_2(\overline{x}, \overline{y}_2)$. The tuple $\tau$ produces a mapping from $\overline{x} \cup \overline{y}_2$ into $dom$ s.t. $\tau S_l(\overline{z}_l) = (\sigma^{j_l}_{s_l} \circ \psi^{r_l}_{s_l, j_l}) S_l(\overline{z}_l)$, $1 \le l \le q$. We denote this mapping also by $\tau$.

Now, we define the set of mappings denoted $GEN(Q^2, \sigma_0, \pi_0, PAR)$ generated using the query $Q^2$ and the parameters $\sigma_0, \pi_0$ and those from $PAR$.

$$GEN(Q^2, \sigma_0, \pi_0, PAR) = \{\tau = (\sigma_{s_1}^{j_1} \circ \psi_{s_1, j_1}^{r_1}, \ldots, \sigma_{s_q}^{j_q} \circ \psi_{s_q, j_q}^{r_q}) | \psi_{s_l, j_l}^{r_l} \in$$
$$M_l(Q^2, \sigma_0, \pi_0, PAR), 1 \leq l \leq q, \tau \text{ is compatible to } f_2(\overline{x}, \overline{y}_2)\}. \quad (10)$$

*Remark 5.* Let $\tau$ be a mapping from $GEN(Q^2, \sigma_0, \pi_0, PAR)$ and $\tau_0 = \sigma_0 \circ \zeta_{\pi_0}$. The projection of $\tau$ on $\overline{x}$ coincides to $\tau_0$ ($\tau_0 = \tau|_{\overline{x}}$).

**Proof.** Let $t$ be a variable from $\overline{x}$. From the safety of $Q^2$, there exists $1 \leq l \leq q$ s.t. $t \in \overline{z}_l$. We have: $\tau(t) = (\sigma_{s_l}^{j_l} \circ \psi_{s_l, j_l}^{r_l})(t) = \sigma_{s_l}^{j_l}(\widetilde{t}_{\pi_0}) = \sigma_0(\widetilde{t}_{\pi_0}) = (\sigma_0 \circ \zeta_{\pi_0})(t) = \tau_0(t)$. The statement $\sigma_{s_l}^{j_l}(\widetilde{t}_{\pi_0}) = \sigma_0(\widetilde{t}_{\pi_0})$ is true because $\sigma_0 < \sigma_s^j$. $\qquad\square$

*Example 3.* Let $Q^1, D, \sigma_0, \pi_0, PAR$ be from Example 2. Let $Q^2$ be as follows:
$$Q^2 : h(x_1, x_2) : -(\exists y_2, y_3)\{R(x_1, y_2) \wedge R(y_3, x_2) \wedge R(y_2, y_3)\}.$$
To compute $M_1(Q^2, \sigma_0, \pi_0, PAR)$ using (9), we need to consider the following mappings: $\psi_{1,1}^1(x_1, y_2) = (\widetilde{x_1}_{\pi_1}, \widetilde{y_1}_{\pi_1})$, $\psi_{1,1}^2(x_1, y_2) = (\widetilde{x_1}_{\pi_1}, \widetilde{x_1}_{\pi_1})$,
$\psi_{1,1}^3(x_1, y_2) = (\widetilde{x_1}_{\pi_1}, \widetilde{x_2}_{\pi_1})$, $\psi_{2,1}^1(x_1, y_2) = (\widetilde{x_1 y_1}_{\pi_3}, \widetilde{x_1 y_1}_{\pi_3})$,
$\psi_{2,1}^2(x_1, y_2) = (\widetilde{x_1 y_1}_{\pi_3}, \widetilde{x_2}_{\pi_3})$. Moreover, we have $\psi_{1,2}^r = \psi_{1,1}^r, 1 \leq r \leq 3$. The set $M_1(Q^2, \sigma_0, \pi_0, PAR)$ consists of the following: $\psi_{1,1}^r, \psi_{1,2}^r, 1 \leq r \leq 3$, and $\psi_{2,1}^r, 1 \leq r \leq 2$. For $l = 2$ and $l = 3$, the sets $M_l(Q^2, \sigma_0, \pi_0, PAR)$ have similar forms. The set $GEN(Q^2, \sigma_0, \pi_0, PAR)$ consists of the following tuples (corresponding to the variables $(x_1, x_2, y_2, y_3)$): $(1, 2, 1, 3)$, $(1, 2, 1, 1)$, $(1, 2, 1, 4)$.

# 5    Containment Query Problem and GEN Operator

In this section, we establish a connection between the $GEN$ operator and the answer of the query $Q^2$ for a database computed using the parameters from $PAR$, $\pi_0$ and $\sigma_0$. Using the parameters $\sigma_s^j$, $\eta_s$, $T_s^i$, $\pi_0$ and $\sigma_0$, we compute the database $D'(\tau_0)$ as in (7) ($\tau_0 = \sigma_0 \circ \zeta_{\pi_0}$ and $\sigma_0 < \sigma_s^j$). The following Lemma establishes a connection between an element of $Ans_{bs}(Q^2, D'(\tau_0), \pi_0)$ and the number of mappings obtained by $GEN$ operator. More precisely, next lemma asserts that the multiplicity of the atom $h(\tau_0\overline{x})$ in $Ans_{bs}(Q^2, D'(\tau_0), \pi_0)$ is equal to $m$ iff the integer $m$ is the cardinality of the set obtained by applying the $GEN$ operator to the query $Q^2$ with the parameters $\sigma_0, \pi_0$ and those from $PAR$.

**Lemma 1.** *We have* $(h(\tau_0\overline{x}) : m) \in Ans_{bs}(Q^2, D'(\tau_0), \pi_0)$ *iff* $m = |GEN(Q^2, \sigma_0, \pi_0, PAR)|$, *where* $D'(\tau_0)$ *is specified in* (7) *and* $\tau_0 = \sigma_0 \circ \zeta_{\pi_0}$.

**Proof.** By the definition of the set $Ans_{bs}(Q^2, D'(\tau_0), \pi_0)$, we have $m = |Mapp(Q^2, D'(\tau_0), \tau_0)|$ and $m > 0$, where the set $Mapp(Q^2, D'(\tau_0), \tau_0)$ is computed by (2). So, the statement of Lemma is equivalent to the following:

$$\tau_0 < \tau_2, \tau_2 S_l(\overline{z}_l) \in D'(\tau_0), 1 \leq l \leq q, \text{ iff } \tau_2 \in GEN(Q^2, \sigma_0, \pi_0, PAR). \quad (11)$$

*only if* Assume that the assignment $\tau_2$ from $\overline{x} \cup \overline{y}_2$ into *dom* satisfies : $\tau_0 < \tau_2$, $\tau_2 S_l(\overline{z}_l) \in D'(\tau_0)$, for each $1 \leq l \leq q$. Using the expression of $D'(\tau_0)$ from (7), we obtain:

$$(\exists \text{ integers } s_l, j_l, 1 \leq s_l \leq h, 1 \leq j_l \leq n_{s_l}), \text{ s.t. } \tau_2 S_l(\overline{z}_l) \in \sigma_{s_l}^{j_l}(T_{s_l}^{\eta_{s_l}(j_l)}) \quad (12)$$

Let us denote by $\tau_2|\overline{z}_l$ the projection of $\tau_2$ on $\overline{z}_l$. From the statement (12), we get $range(\tau_2|\overline{z}_l) \subseteq range(\sigma_{s_l}^{j_l})$, which implies:

$$((\sigma_{s_l}^{j_l})^{-1} \circ \tau_2) S_l(\overline{z}_l) \in T_{s_l}^{\eta_{s_l}(j_l)}, 1 \leq l \leq q \quad (13)$$

Let us define the mappings $\psi_{s_l, j_l}$ from $\overline{z}_l$ into $Class_{\pi_{s_l}}$ as follows: $\psi_{s_l, j_l} = (\sigma_{s_l}^{j_l})^{-1} \circ \tau_2, 1 \leq l \leq q$. These mappings satisfy the following:
$\psi_{s_l, j_l}(t) = ((\sigma_{s_l}^{j_l})^{-1} \circ \tau_2)(t) = (\sigma_{s_l}^{j_l})^{-1}(\tau_2(t)) = (\sigma_{s_l}^{j_l})^{-1}(\tau_0(t)) = \sigma_0^{-1}(\tau_0(t)) = \zeta_{\pi_0}(t) = \tilde{t}_{\pi_0}$, for each $1 \leq l \leq q$ and $t \in \overline{z}_l \cap \overline{x}$. These relations and those from (13) imply $\psi_{s_l, j_l} \in M_l(Q^2, \sigma_0, \pi_0, PAR)$, for each $1 \leq l \leq q$. Since $\tau_2$ is compatible to $f_2$ by the hypothesis, it results that $\tau_2 \in GEN(Q^2, \sigma_0, \pi_0, PAR)$.

The *if* part is similar.                                                          □

Now, we can give a characterization for the problem: $Ans_{bs}(Q^1, D, \pi_0) \subseteq_{bs} Ans_{bs}(Q^2, D, \pi_0)$, for all databases $D$ on *dom*.

**Theorem 1.** *Let $Q^1$ and $Q^2$ be two queries as in Definition 3, an $\pi_0$ be a fixed partition on $\overline{x}$. We have:*

$$Ans_{bs}(Q^1, D, \pi_0) \subseteq_{bs} Ans_{bs}(Q^2, D, \pi_0), \text{ for each database } D \text{ iff}$$

$$|GEN(Q^2, \sigma_0, \pi_0, PAR)| \geq \Sigma_{s=1}^h n_s,$$

*for each $\sigma_0$, injection from $Class_{\pi_0}$ into dom and for all values of parameters from PAR, the vector of parameters specified in (8).*

**Proof.** *only if* Assume that $Ans_{bs}(Q^1, D, \pi_0) \subseteq_{bs} Ans_{bs}(Q^2, D, \pi_0)$, for each database $D$ on *dom*. To show $|GEN(Q^2, \sigma_0, \pi_0, PAR)| \geq m$, where $m = \Sigma_{s=1}^h n_s$, for each $\sigma_0$ and values of parameters from $PAR$, let us consider arbitrary, but fixed values, for these parameters. Let $\tau_0 = \sigma_0 \circ \zeta_{\pi_0}$. Let $D'(\tau_0)$ computed as in (7). By the hypothesis, we have: $Ans_{bs}(Q^1, D'(\tau_0), \pi_0) \subseteq_{bs} Ans_{bs}(Q^2, D'(\tau_0), \pi_0)$. Since $(h(\tau_0 \overline{x}) : m) \in Ans_{bs}(Q^1, D'(\tau_0), \pi_0)$, there exists the pair $(h(\tau_0 \overline{x}) : m_2)$ from $Ans_{bs}(Q^2, D'(\tau_0), \pi_0)$, s.t. $m_2 \geq m$. Using Lemma 1, we obtain: $|GEN(Q^2, \sigma_0, \pi_0, PAR)| = m_2 \geq m$.

*if* Assume that $|GEN(Q^2, \sigma_0, \pi_0, PAR)| \geq m$, where $m = \Sigma_{s=1}^h n_s$, for each $\sigma_0$ and for all values of parameters from $PAR$. We want to show $Ans_{bs}(Q^1, D, \pi_0) \subseteq_{bs} Ans_{bs}(Q^2, D, \pi_0)$, for each database $D$ defined on *dom*. Let $D$ be a given database and $(h(\tau_0 \overline{x}) : m) \in Ans_{bs}(Q^1, D, \pi_0)$. There exists distinct mappings $\tau_1, \ldots, \tau_m$ from $\overline{x} \cup \overline{y}_1$ into *dom* s.t. $D \models f_1[\tau_i]$ and $\tau_0 < \tau_i$. Let $D_{\tau_i}$ be the database constructed to $D$ and $\tau_i$, as in (6). Let $\pi_i = Part(\tau_i)$ and $\sigma_i$ a mapping from $Class_{\pi_i}$ into *dom*, defined as: $\sigma_i(\tilde{t}_{\pi_i}) = \tau_i(t)$, for each $t \in \overline{x} \cup \overline{y}_1$. The other values of the parameters from $PAR$ are defined as in Section 3. Let $D'(\tau_0)$ be the database defined in (7). We have: $(h(\tau_0 \overline{x}) : m) \in Ans_{bs}(Q^1, D'(\tau_0), \pi_0)$.

By the hypothesis, we have $m_2 = |GEN(Q^2, \sigma_0, \pi_0, PAR)| \geq m$. By Lemma 1, we get $(h(\tau_0 \overline{x}) : m_2) \in Ans_{bs}(Q^2, D'(\tau_0), \pi_0)$. Since $D'(\tau_0) \subseteq D$, we get: $Ans_{bs}(Q^2, D'(\tau_0), \pi_0) \subseteq_{bs} Ans_{bs}(Q^2, D, \pi_0)$. Thus, there exists $(h(\tau_0 \overline{x}) : m_3) \in Ans_{bs}(Q^2, D, \pi_0)$, with $m_2 \leq m_3$, i.e. $(h(\tau_0 \overline{x}) : m) \in Ans_{bs}(Q^2, D, \pi_0)$.     $\square$

The parameters $\pi_i$, $\sigma_i$ are not independent, because we have: $\sigma_0 < \sigma_s^j$, $1 \leq j \leq n_s, 1 \leq s \leq h$ and $\pi_0 \prec \pi_i, 1 \leq i \leq m$, where $m = \Sigma_{s=1}^h n_s$. On the other hand, the cardinalities of $\sigma_0$ and of $\sigma_s^j$ are infinite denumerable.

The next step is to give a statement equivalent to the right part of the assertion from Theorem 1, where the domain of injections $\sigma_0$ and $\sigma_s^j$ has a linear expression of integers $n_1, \ldots, n_h$ with constant coefficients as its cardinality.

Let $\varphi$ be an injection from $\cup_{s=1}^{s=h} \cup_{j=1}^{j=n_s} range(\sigma_s^j)$ into $dom$, and $PAR1$ the list of the parameters specified in (8), but containing $\varphi \circ \sigma_s^j$ instead of $\sigma_s^j$. Since the sets $M_l(Q^2, \sigma_0, \pi_0, PAR)$ do not depend on $\sigma_0, \sigma_s^j$, using the definition of $GEN$ operator, we obtain the following result:

**Lemma 2.** $|GEN(Q^2, \sigma_0, \pi_0, PAR)| = |GEN(Q^2, \varphi \circ \sigma_0, \pi_0, PAR1)|$

Let $m_s$ be cardinality of $Class_{\pi_s}$ and $n' = \Sigma_{s=1}^{s=h} m_s * n_s$. Using Lemma 2, we can consider the set $dom' = \{1, \ldots, n'\}$ as the range of the injections $\sigma_0$, and $\sigma_s^j$. Let $PAR2$ be the list of parameters given in (8), but the injections $\sigma_s^j$ have $dom'$ as their range. Using this fact, Theorem 1 and Remarks 1 and 3, we get:

**Theorem 2.** *Let $Q^1$ and $Q^2$ be two queries as in Definition 3. The following statements are equivalent:*
*(i) $Q^1 \subseteq_{bs} Q^2$,*
*(ii) $M_l(Q^2, \sigma_0, \pi_0, PAR) \neq \emptyset$, $1 \leq l \leq q$, $|GEN(Q^2, \sigma_0, \pi_0, PAR2)| \geq \Sigma_{s=1}^h n_s$, for each partition $\pi_0$ on $\overline{x}$, for each partition $\pi_s$ on $\overline{x} \cup \overline{y}_1$ s. t. $\pi_0 \prec \pi_s$ , for each natural value of $n_1, \ldots, n_h$, for each injection $\sigma_0$ from $Class_{\pi_0}$ into $dom'$, for each database $T_s^j$ on $Class_{\pi_s}$, for each injection $\sigma_s^j$ from $Class_{\pi_s}$ into $dom'$, $1 \leq j \leq n_s, 1 \leq s \leq h$, and for each surjective mapping $\eta_s$ from $\{1, \ldots, n_s\}$ into $\{1, \ldots, p_s\}$, $1 \leq s \leq h$. The parameters $h, \pi_1, \ldots, \pi_h, p_1, \ldots, p_h$ depend only on the query $Q^1$.*

## 6  Conclusion

We have given a characterization for the query containment problems using a family of injective mappings defined on the sets of equivalence classes on variables belonging to the first query. This characterization could be useful in the study of the decidability problem of query containment problem for $CQs$ queries.

## References

1. Abiteboul, S., Hull, R., Vianu, V.: Foundations of databases. Addison-Wesley Publishing Comp. (1995)
2. Brisaboa, N.R., Hernandez, H.J.: Testing bag-containment of conjunctive queries. Acta Informatica **34**, 557–578 (1997)

3. Chandra, A., Merlin, P.: Optimal implementation of conjunctive queries in relational data bases. In: ACM STOC (1977)
4. Chaudhuri, S., Vardi, M.: Optimization of real conjunctive queries. In: Proc. of the 12th ACM Symposium on the Principles of Database Systems, pp. 59–70 (1993)
5. Damigos, M.: Query Optimization under bag and bag-set semantics for multiple heterogeneous data sources, Ph.D. Thesis, NTUA, Athens (2011)
6. Jayram, T.S., Kolaitis, P.G., Vee, E.: The containment problem for real conjunctive queries with inequalities. In: PODS 2006, pp. 80–89 (2006)
7. Karvounarakis, G., Tannen, V.: Conjunctive queries and mappings with inequalities, Technical Report (2008)
8. Klug, A.: On conjunctive queries containing inequalities. JACM **35**(1), 146–160 (1988)
9. van der Meyden, R.: The complexity of querying infinite data about linearly ordered domains. Journal of Computer and System Sciences **54**(1), 113–135 (1997)

# A Belief-Based Bitemporal Database Model

Wajee Jiratanachit and Suphamit Chittayasothorn[✉]

Department of Computer Engineering, Faculty of Engineering,
King Mongkut's Institute of Technology Ladkrabang, Bangkok 10520, Thailand
wajee@sec.or.th, suphamit@kmitl.ac.th

**Abstract.** Current bitemporal data models use valid time and transaction time to indicate the time that a fact is valid in a domain, and the time that the fact is stored in a database respectively. In order to store historical data with different beliefs during the same valid time period, there need to be a time dimension which indicates the belief time; the time that the fact is believed to be true. This belief time is different from the conventional transaction time, which is the time that the fact is entered into the database. In this paper, a belief-based approach to bitemporal data model is presented together with its query and manipulation considerations.

**Keywords:** Temporal database · Bitemporal · Belief time · Valid time · Transaction time

## 1 Introduction

In conventional information systems, only facts which are currently true are stored in the database. In reality, facts change from time to time. Previously true facts are considered important and might be referred to by applications. There are research works which provide temporal aspects to databases [1-4]. They introduce two time dimensions i.e., valid time and transaction time. Valid time denotes the time when a fact is true in reality. Transaction time is the time when a database stored the fact. This is called a bitemporal data model. In reality, a fact may be entered with a delay to the database. A good example is the storing of historical data to the database. Two conflicting facts which are valid during the same period of time could be recorded in the same database because they were believed in different periods of time.

This paper proposes a belief time approach to the bitemporal data model. A belief-based bitemporal data model includes a valid and a belief time. A valid time is a time when a fact is believed to be true in a belief time period. With the belief time, historical data can be conveniently stored in the database.

## 2 Bitemporal Data Model

There are research works such as [1,4] that refer to bitemporal data model and its data manipulation operations. All of them use two time dimensions to represent the time

© Springer International Publishing Switzerland 2015
N.T. Nguyen et al. (Eds.): ACIIDS 2015, Part I, LNAI 9011, pp. 170–178, 2015.
DOI: 10.1007/978-3-319-15702-3_17

aspects: Valid time, the time when the fact is true in reality, and Transaction time, the time when the fact is recorded in the database. End valid time and end transaction time could be either a point of time or "forever" which states that they are still current. The start time is included but the end time is not included. This is called the close-open format.

We argue that only the valid time and the transaction time are not enough to solve these following problems.

1. Transaction time is supposed to be the time a fact is current in the database. Since it is the time that the information is entered, we argue that it is not suitable for such interpretation. It is possible that a fact is entered to the database earlier than the time when the fact is valid. Conversely, some facts may be entered to the database after they actually take place. Moreover, according to the interpretation, the data has to be entered to the database in the correct order as further described in (2).
2. For historical data, conflict information on the same topic may have the same or overlapping valid periods. If we want to enter these facts into the database now, the conventional valid time and the transaction time bitemporal database is not appropriate. The conflict information will not be allowed to be in the same database due to an integrity constraint called polyinstantiation constraint which is associated with the conventional bitemporal database.
3. In application development, choosing SQL operations (insert / update / delete) to perform data manipulation on bitemporal database is not an easy task. In order to insert a fact, its time may overlap with the existing ones. Many SQL operations may be required for a logical insert purpose. Before introducing a newly found fact into the database, the existing one has to be updated or deleted first. Such a system should be able to hide this complexity from the users.

Change of planned future facts is discussed in [5]. Changes made to facts which are valid in the past are previously considered not applicable. However, changes of the past which is the effects of an action according to the ramification problem is studied in [6]. Beliefs are studied by [7] in the multilevel secured database context without the consideration of validtime. In multilevel secured database, users are given different classification levels. [7] replaces each user with a time period. So, instead of having different users' information based on the same fact, different information based on different belief time of the same fact is recorded instead. These previous works refer to one time dimension only and are therefore not Bitemporal models.

## 3    The Proposed Belief-Based Bitemporal Model

From the problems of storing historical data as mentioned, there should be an aspect of time which enables a fact to be recorded based on a belief time period. We propose the "belief time" to indicate the time that fact is believed to be true. The fact with a valid time and a belief time is called "belief-based bitemporal fact". Two versions of the same fact, with the same or overlapping valid time periods, but contain different values because they are believed during different periods of time, could be stored. In conventional temporal database, they are regarded as a conflict and are not allowed to be in the database at the same time.

There are two dimensions of time in our Belief-based bitemporal data model, which are:

1. Valid time, a period of time when a fact is valid in a domain
2. Belief time, a period of time when a fact is believed to be true

Both valid and belief time are shown using the interval of time (start and end). The temporal labels for the Belief-based bitemporal data model are:

1. Start Valid Time (SV)
2. End Valid Time (EV), not included this point of time
3. Start Belief Time (SB)
4. End Belief Time (EB), not included this point of time

EV and EB could be point of time or "forever" which is shown as "9999-12-31"; the maximum value of the date data type.

**Table 1.** Example of storing a shape of the earth in the different believes

| planet_name | has_shape | SV | EV | SB | EB |
|---|---|---|---|---|---|
| Earth | Round | BC | '9999-12-31' | '1492-01-01' | '9999-12-31' |
| Earth | Flat | BC | '9999-12-31' | '1400-01-01' | '1492-01-01' |

From the above table, there was a belief from 1400 until 1491 that the earth's shape was flat. But since 1492, it was found that the earth is round. Using a belief time, it is possible to store a fact in the same valid time with different belief time for the different believes. The order of the data entries is not significant.

**Table 2.** Example of the distance of a planet to the sun

| plaet_name | has_distance_from_the_sun | SV | EV | SB | EB |
|---|---|---|---|---|---|
| Earth | 149.6 million km | BC | '9999-12-31' | '1996-10-12' | '9999-12-31' |
| Earth | 93 million km | BC | '9999-12-31' | '1672-01-01' | '1996-10-12' |

Let us demonstrate more examples of the application of the belief time concept. From Table 2, there was a belief since 1672 that the distance between the earth and the sun was 93 million kilometers. Later there is a new measurement technique which is called "parallax measurement". Using the new technique, since October 12, 1996 the astronomers found the new distance should be 149.6 million kilometers. This fact is believed since then until now, indicated by '9999-12-31'. The information may be entered long after the beliefs (evidences gathered and acknowledged). So, the transaction time is not relevant here. Moreover, both the valid time and belief time can be modified, in the case that new evidences are found. The valid time of a fact could be the current time or the past, and it could also be time in the future such as future planning. But the belief time could only start from the current time or the past.

# 4     Temporal Queries

Temporal operators [2] could be used in conditions of the query. In the case of queries on valid time, PV is the period of validity of the fact and PQ is the period of query (PQ). In the case of belief time, the period of belief (PB) is used instead of PV.

## 4.1     Unitemporal Query

A unitemporal query has a condition on only one type of time; valid or belief time. In relational database terminology, the query could be a selection, projection, or a join.

### Projections and Selections
A unitemporal selection specifies a valid time or a belief time which can be a point of time or a query time (QT) and period of time or a period of query (PQ). The time could be in the past, at the current, or in the future. Result will returns records that match conditions and the specified time.

Table 3. Example of suspected persons' locations

| suspected _person | stays_in | SV | EV | SB | EB |
|---|---|---|---|---|---|
| Mr.A | Town1 | '1997-01-01' | '2002-07-01' | '2005-01-01' | '9999-12-31' |
| Mrs.B | Town1 | '1990-01-01' | '9999-12-31' | '2005-01-01' | '9999-12-31' |
| Mr.C | Town1 | '2002-01-01' | '9999-12-31' | '2010-01-01' | '9999-12-31' |
| Mr.A | Town2 | '2002-07-01' | '9999-12-31' | '2011-01-01' | '9999-12-31' |

## 4.2     Bitemporal Queries

These queries are more complex to query because they refer to two time dimensions; valid and belief time. The query considers both valid and belief time which could be done using following scenarios.

### Projections and Selections
The query using a valid and a belief time could be done by drawing an area of both query times on the area of validity. If the query area is on the validity area of a record,

the result will return the record. Based on Table 3, consider the query: in the year 2010, who is believed to stay in Town1 during year 2000. The results are Mr. A and Mrs. B. The selection area is displayed as shown in the figure 1 below.

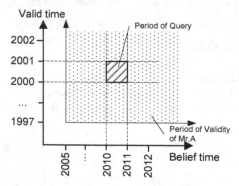

**Fig. 1.** A Bitemporal selection

There are four possible cases of bitemporal queries: 1) PV contains both valid time and belief time of PQ.  2) Valid time of PV overlaps valid time of PQ, and belief time of PV contains the one of PQ. 3) Belief time of PV overlaps belief time of PQ, and valid time of PV contains the one of PQ. 4) Valid times and belief times of both PV and PQ overlap.

**Bitemporal Joins**

Two tuples could be joined by merging two areas of two validity tuples first. Then, project by query area on the merged area. If the query area is included in merged area, the result will be returned. Consider the query: who is believed in year 2010 that stayed in the same place as Mrs. B in year 2000. The query is as in figure 2 below.

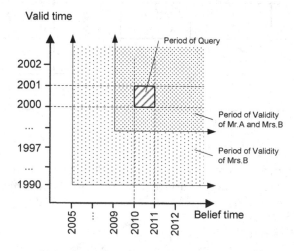

**Fig. 2.** Bitemporal join of Mr. A and Mrs. B with a query area

The result is Mr. A stayed in the same place as Mrs. B in the query time.

# 5    Data Manipulation

A valid time modification has been proposed in some researches by having only current modification or a modification from now or in the future only. But in the reality, there might be a new belief which is different from the one that is recorded in the past. For example, a detective found Mr. A was in Town1 during the whole year 2000, but later found that Mr. A only stay in Town1 since July, 2000. So the previous valid time might be modified.

The belief-based bitemporal data presentation allows the modification on fact, valid time, and/or belief time. Since some application recorded historical data, so the fact could be modified if the new evidence has been found:

— If a recorded fact is found that the belief time is shorter than the one that is currently recorded, an end belief time needs to be updated.
— If the evidence showed that it is valid in different period from the one that is recorded, the valid time must be updated.

In practice, it is difficult to let a user identify which database operation should be used. For example, if it was originally found that Mr. A was in Town1 between 2000 and 2012, but it was later found that Mr. A had moved to Town2 between 2012 and 2013. In this case, the application should perform updating the existing record and inserting the new one. The users should simply assert the fact. We therefore introduce new commands; ASSERT and REMOVE, as logical operations to free users from complicated underlying database operations.

## 5.1    Assertion

Since a belief could be in the past, present or future, user should be able to use such a command to assert the newly found fact without taking care of the existing ones. ASSERT command is introduced as an insert with confirmation command to let a management system performs update and/or delete the existing records and insert the new one. There are 16 possible scenarios of bitemporal assertion as shown in Appendix A.

According to Table 4, it was found and recorded that in 2010 Mr. A stayed in Town1 and he stayed there again in 2012 until July 2012. Then he moved to Town2 since July 2012 to the end of the year.

**Table 4.** Example of suspected persons' locations

| suspected _person | stays_in | SV | EV | SB | EB |
|---|---|---|---|---|---|
| Mr.A | Town1 | '2012-01-01' | '2012-07-01' | '2013-03-01' | '9999-12-31' |
| Mr.A | Town2 | '2012-07-01' | '2013-01-01' | '2013-03-01' | '9999-12-31' |
| Mr.A | Town1 | '2010-01-01' | '2011-01-01' | '2013-03-01' | '9999-12-31' |

If it was found in July 2013 that Mr. A actually was in Town3 in 2012, the bitemporal assertion in the asserting process has to be performed according to cases in Appendix A. The result is shown in Table 5.

**Table 5.** Example of a result after assertion

| suspected _person | stays_in | SV | EV | SB | EB |
|---|---|---|---|---|---|
| Mr.A | Bangkok | '2012-01-01' | '2012-07-01' | '2013-03-01' | **'2013-07-01'** |
| Mr.A | London | '2012-07-01' | '2013-01-01' | '2013-03-01' | **'2013-07-01'** |
| **Mr.A** | **Sydney** | **'2012-01-01'** | **'2013-01-01'** | **'2013-07-01'** | **'9999-12-31'** |
| Mr.A | Bangkok | '2010-01-01' | '2011-01-01' | '2013-03-01' | '9999-12-31' |

### 5.2    Removal

If it is found later that the evidence that identified Mr. A in Town1 in 2010 is false, the fact of Mr. A in that period should be deleted. Many records might be updated in this case. So a REMOVE command is introduced to let the system taking care of the 16 cases physical operations that are needed to be performed similar to the insertion operation described above.

## 6    Conclusion

This paper argues that the conventional bitemporal data model based on the valid time and transaction time as previously introduced is not adequate for representing historical data. A belief time purposed in this paper is the time that a fact is believed to be true during its validity. The belief time, together with the valid time, is suitable for representing facts which are changed according to different beliefs. Historical and archeological data with new evidences are suitable application domains for the belief-based bitemporal database systems.

## Appendix A. Sixteen Assertion cases

| | $SV_N$ $\leq SV_O$ | $EV_N$ $\geq EV_O$ | $SB_N$ $\leq SB_O$ | $EB_N$ $\geq EB_O$ | ☐ Existing fact ⌐⌐⌐ Asserted fact | Operations (N = new fact, O = existing fact) |
|---|---|---|---|---|---|---|
| 1 | Y | Y | Y | Y | Valid time / Belief time | Delete an existing fact, insert a new fact. |
| 2 | Y | Y | Y | N | | Update $SB_O = EB_N$, insert a new fact. |
| 3 | Y | Y | N | Y | | Update $EB_O = SB_N$, insert a new fact. |
| 4 | Y | Y | N | N | (1) (3) (2) | Update $EB_O = SB_N$, insert $<Val_O, SV_O, EV_O, EB_N, EB_O>$, insert a new fact. |
| 5 | Y | N | Y | Y | | Update $EV_O = SV_N$, insert a new fact. |
| 6 | N | Y | Y | Y | | Update $SV_O = EV_N$, insert a new fact. |
| 7 | N | N | Y | Y | (2) (3) (1) | Update $EV_O = SV_N$, insert $<Val_O, EV_N, EV_O, SB_O, EB_O>$, insert a new fact. |
| 8 | Y | N | Y | N | (2) (3) (1) | Update $SB_O = EB_N$, insert $<Val_O, EV_N, EV_O, SB_O, EB_N>$, insert a new fact. |
| 9 | Y | N | N | Y | (1) (2) (3) | Update $EB_O = SB_N$, insert $<Val_O, EV_N, EV_O, SB_N, EB_O>$, insert a new fact. |
| 10 | Y | N | N | N | (2) (1) (4) (3) | Update $SV_O = EV_N$, insert $<Val_O, SV_O, EV_N, SB_O, SB_N>$, insert $<Val_O, SV_O, EV_N, EB_N, EB_O>$, insert a new fact. |
| 11 | N | Y | Y | N | (3) (1) (2) | Update $SB_O = EB_N$, insert $<Val_O, SV_O, SV_N, SB_O, EB_N>$, insert a new fact. |

| 12 | N | Y | N | Y | | Update $EB_O = SB_N$, insert $<Val_O,SV_O,SV_N, SB_N, EB_O>$, insert a new fact. |
|----|---|---|---|---|---|---|
| 13 | N | Y | N | N | | Update $EV_O = SV_N$, insert $<Val_O,SV_N,EV_O, SB_O, EB_N>$, insert $<Val_O,SV_N,EV_O, SB_N, EB_O>$, insert a new fact. |
| 14 | N | N | Y | N | | Update $SB_O = EB_N$, insert $<Val_O,EV_N,EV_O, SB_O, EB_N>$, insert $<Val_O,SV_O, SV_N, SB_O, EB_N>$, insert a new fact. |
| 15 | N | N | N | Y | | Update $EB_O = SB_N$, insert $<Val_O,EV_N,EV_O, SB_N, EB_O>$, insert $<Val_O,SV_O, SV_N, SB_N, EB_O>$, insert a new fact. |
| 16 | N | N | N | N | | Update $EB_O = SB_N$, insert $<Val_O,SV_O,EV_O, SB_N, EB_O>$, insert $<Val_O,EV_N,EV_O, SB_N, EB_N>$, insert $<Val_O,SV_O, SV_N, SB_N, EB_N>$, insert a new fact. |

# References

1. Carvalho, A., Ribeiro, C., Sousa, A.A.: A spatio-temporal database system based on Time-DB and Oracle spatial. In: Tjoa, A.M., Xu, L., Chaudhry, S.S. (eds.) Research and Practical Issues of Enterprise Information Systems. IFIP, vol. 205, pp. 11–20. Springer, Boston (2006)
2. Date, C.J., Darwen, H., Lorentzos, N.A.: Temporal Data and The Relational Model. Morgan Kaufmann Publishers, California (2002)
3. Snodgrass, R.T.: Developing Time-Oriented Database Applications in SQL. Morgan Kaufmann Publishers, California (2000)
4. Jensen, C.S., Snodgrass, R.T.: Temporal Data Management. IEEE Transactions on Knowledge and Data Engineering 11, 36–44 (1999)
5. Sarda, N.L., Reddy, P.V.S.P.: Handling of Alternatives and Events in Temporal Databases. Knowledge and Information Systems 1, 337–368 (1999)
6. Papadakis, N., Antoniou, G., Plexousakis, D.: The ramification problem in temporal databases: Changing beliefs about the past. Data Knowledge Engineering 59, 397–424 (2006)
7. Gadia, S.K.: Applicability of temporal data models to query multilevel security databases: a case study. In: Etzion, O., Jajodia, S., Sripada, S. (eds.) Temporal Databases - Research and Practice. LNCS, vol. 1399, pp. 238–256. Springer, Heidelberg (1998)

# Architecture Dedicated to Data Integration

Jacek Dajda and Grzegorz Dobrowolski[✉]

AGH University of Science and Technology, al. Mickiewicza 30,
30-059 Krakow, Poland
{dajda,grzela}@agh.edu.pl

**Abstract.** The aim of the paper is to present a software architecture
dedicated to problem of heterogeneous data integration from a number of
rather big data sources. The potential capabilities of the architecture are:
distribution, decentralization, extensibility and support for code reuse.
By applying scalable Erlang technology and concept of plugin processes
the presented architecture seems to be interesting for wide range of application fields, in particular, threats detection in criminal analysis.

**Keywords:** Software system architecture · Data integration · Criminal
analysis

## 1 Introduction

Development of Internet and information technologies results in growing number
of information stored in various data stores, databases and other sources. These
sources, as developed for different needs and by different organizations, are organized in different ways. They are often called heterogenous which describe not
only the format and the technologies in which they are built but also the content
in the semantic sense. All of this makes a challenging problem for a data analyst
who intends to use information which such data carries in its whole.

Since business intelligence has been on a market it has operated on relational
database within given organization boundaries. What is more, this kind of analysis
is performed periodically, which means that there is a time for data preparation
and environment configuration. A typical example of supporting technology in
this case is ETL (Extract-Transform-Load) [3][2], but as mentioned, it assumes
that data preparation is carried out during data transformation and load processes
(so called materialized approach). What is more, ETL tools are not designed for
large distributed databases that dominate the current IT world.

Currently, as new parties (such as security forces) arrive into the ,,data-
world", a strong need arises for a new, more light-weight, but dynamic and
scalable approach that will handle recently identified BigData problems. The list
of adequate applications is numerous, e.g., criminal analysis aimed at detecting
potential threats or tracking criminals. The existing approaches [4] seem to be
limited to specific cases. Concepts based on services [5] or software agents [1]
occur to be interested, but they still remain in a research phase.

© Springer International Publishing Switzerland 2015
N.T. Nguyen et al. (Eds.): ACIIDS 2015, Part I, LNAI 9011, pp. 179–188, 2015.
DOI: 10.1007/978-3-319-15702-3_18

The paper proposes a concept of process-based architecture for integration that can handle distribution, decentralization and provides extensibility and support for code reuse. The concept is verified by Erlang-based implementation upon which a prototype integration system is assembled and executed.

## 2   Proposed Approach to Data Integration

In order to characterize a point of departure towards studies under suitable, effective and up-to-date approach to data integration, let us summarize general identification of it in the form of the following requirements.

1. Data sources (computerized) are of arbitrary types.
2. A human being is also a source.
3. The aim of the system is to put together (integrate) possibly all relevant information from the sources.
4. The process runs automatically but the human beings' intervention can occur to be indispensable.
5. The integration process has assumed duration.
6. The contents of the sources can change along the period.
7. For the integration purposes it can be necessary to process information somewhere in between the sources and computer representation of the aim.

Overall analysis of the above requirements leads to a few detailed observations with respect to a system which would fulfill them. The automatization turns our attention to data-driven systems. In the case of integration, some completeness conditions imposed on the concrete form of the aim could be defined. If these conditions fail, it triggers adequate actions of the system (dataflow). The automatization could be also extended to react to variability of the sources in time and other similar phenomena in the environment. Having found appropriate frequency, a timer could be enough to successfully monitor whatever we need.

The role of a human being in the system should be discussed deeper also. Primarily, he is a user. He defines the aim and is interested in obtained results to be as good as possible. It could be gained only the aim would be comprehensively and operationally done. But, it is almost impossible regarding we think of integration from the various sources. Then the user is also necessary in order to supervise or even control the process, mainly to accept relevance of some extracted information. In this way he becomes a source of not only the information but meta information also. Turning out perspective, we can see the system under consideration as supporting one dedicated to someone who can be called *analyst*.

The last requirement postulates processing of information. Despite a schema of it will not be to complicated due to integration purposes, we ought to assume that information streams from various sources will interweave and we face a workflow dictated by effects of interpretation of partial results and the analyst's decisions.

Creation of a framework for the workflow due to its dependency on the aim and characteristics of the sources is also the analyst's responsibility. The framework can be a priori described by scenarios that form a kind of functional requirements generated by the system field of application (domain). So, our system occurs to be scenario-driven.

The carried out short discussion shows what kind of systems is under consideration and, moreover, that searching for software architecture which could support such multi-conceptual and multi-functional system is the challenging task.

Among plenty of design decisions which ought to be undertaken to propose such architecture the most important are listed beneath.

1. The direct aim is represented by a data structure of the assumed semantics.
2. The integration process can be regarded as consecutive filling of the structure with information extracted from the sources. Partially filled the structure is used as a searching pattern each time (dataflow ones more).

A natural way of describing the aim semantics is a formal language of one of logics. Referring our discussion to very well known problems of knowledge sharing in the Web, it can be one of description logics languages (e.g., RDF or OWL). The chosen for integration sources can have such description mainly without special effort. In the case of relational data bases we have it already through their schemas and manuals; if needed, translation to one of the Web languages can be done. Analyzing the opposite profile of the sources – an unstructured text source – the situation is much harder but possible for resolving. Several ideas, even technologies can be adopted for the purpose. Of course, there is a problem of domain languages – in the last case some natural one. It is awaited that in majority of application cases ontological description of the system will be at most so complicated as terminologies what simplifies linguistic problems to understanding names. Because some processing modules are assumed to be present, they ought to have ontological description also. This way the whole system can acquire designation knowledge-driven.

Summarizing, searching for a software architecture that could effectively support the system of such reach characterization is worth for undertaking.

## 3  Overall Architecture for Integration

In order to realize the presented data integration approach, a prototype architecture is proposed that implements main assumptions and requirements described above.

The basic module of the architecture is a so-called *plugin* (Fig. 1 illustrates this concept). Although the name is taken from technological sphere, it has been chosen as a term for general architectural discussion for its expressiveness embracing the question of reconfiguration also. By applying the concept of the plugin, the architecture becomes extensible and can be adopted to specific needs

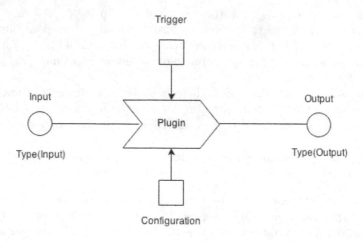

**Fig. 1.** A plugin

by assembling proper dataflows based on an available set of plugins. The plugin is an abstraction which realizes transformation of input data to a specific output.

A plugin has at least one input and one output that define its data processing ability. Plugins can be combined to form a flow (or flows) which is desired to accomplish the main goal of the system. It is obvious that the combination of plugins must take into account their compatibility. Then such combination represents a function which is superposition of the functions of the plugins used. Fig. 2 presents the general architecture of the system. So a plugin is described by an interface which assumes format of data which can arise on each input or output of it. One can say that the interface is defined by the types of data involved. Accordingly, semantics of data is bonded to the plugin function. If necessary, a plugin must produce a means for its internal configuration. It can be done with some simple script.

Because the system is built for data integration, the highest level force that triggers it (their plugins) is incompleteness of data obtained so far. Thereby the common mechanism of data-driven systems – incoming data triggers processing of them – must be extended. Our plugin is steered additionally by a special input which can be wired with a controlling module which introduces possibility to influence the data flow.

For the sake of clarity, plugins are grouped intro three categories: source (reading the content of Internet or other sources with some filtration and prepared input for processing plugins), final (closing the processing with the storing or presenting some information) and analytical plugins themselves which work according to arbitrary complicated algorithms.

When data emerges at an input of a plugin, it starts processing. This is the main rule for dataflow systems. Function of the described system is reacher than that. The controlling module provides a means for:

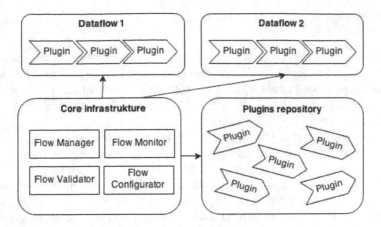

**Fig. 2.** General architecture

- making processing conditional on real time, e.g. when monitoring a changing source in some period of time, an input plugin, and consequently the flow, is triggered several times at given moments;
- allowing to re-start the system when new relevant information is obtained;
- allowing human intervention into the flow, e.g. to modify the integration goal or to check progress.

The above is obtained via the special input of a module. In the case of a processing plugin, one can understand that the special input is closed via the internal loop with the input.

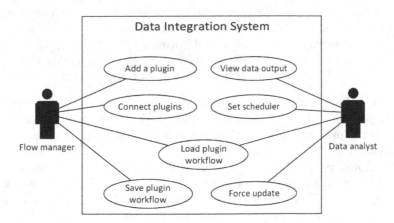

**Fig. 3.** Perspectives of system use

From the human being's point of view there are two following perspectives for the system:

- a flow manager,
- a data analyst.

The data analyst prepares a scenario for data integration. The scenario is done in more or less formal way and describes sources, applied processing and the goal of the proposed integration task. Not coming into details, the description gives the flow manager all necessary information for orchestrating the scenario in the system possibilities and conditions.

Basing on the scenario, the flow manager programs the appropriate dataflow by preparing necessary plugins and connecting them. Moreover he must arrange a special control structure according to the data analyst's indication.

In this way the data analyst becomes appropriately equipped or supported. Additionally, he can ask the flow manager during his work to re-program the whole structure by: changing it in the sense of connection, adding new plugins, removing another. Also the control structure can be a subject for modification and, if necessary, a new plugin constructed and integrated with the system. Each perspective is represented by several use cases which are shown in Fig. 3.

## 4    Notes on Implementation

To give some insights on how the proposed architecture can be realized, this section shortly overviews the most important solutions staying behind its prototype implementation.

The prototype is implemented in Erlang and Python languages. The core modules are realized in Erlang which provides scaling and parallel execution out-of-the box. This allows for embodiment of plugins into Erlang processes. Erlang offers poor support when analytical functions are programed (e.g., text mining) therefore the functions of plugins are implemented in Python and are called by the Erlang wrapper which forms a kind of the shell for a plugin.

The Plugin abstract class (see Fig. 4) provides the interface for: creating instances of plugin processes, changing their setting or connecting them. The following methods have to be implemented in order to create a new plugin:

init(Options: list) – is called for creation of the process (at its startup). It should perform all necessary initialization tasks, including obligatory creation of the process state and data structures, that would be later used as arguments for other functions.

verify(Options, State, Data) – checks whether the process data is up to date. The decision of how to perform this task is left for the plugin developer.

update(Options, State, Data, InputProcess, UpdatedInputProcess) – allows the process to update its data (and/or state).

return(Options, State, Data, Output, Type) – prepares the data that will be received as a result of processing.

The core infrastructure (see Fig. 2 ones more) is responsible for the plugins to be properly created and managed. Its main submodules are as follows.

```
<<behaviour>>
plugin

+init ( Options )
+verify ( Options, State, Data )
+update ( Options, State, Data, InputProcesses, UpdatedInputProcesses )
+return ( Options, State, Data, Output, Type )
+start_link ( Name, Module, Options, Persist )
+get_data ( Output, OutputType )
+update ( )
+update_with_ack ( )
+change_options ( NewOptions )
+change_persist_mode ( NewPersistMode )
+add_input_process ( Input, InputProcess )
+remove_input_process ( Input, InputProcess )
```

**Fig. 4.** Plugin abstract class

**Registry** is the Erlang process playing a central role in the system. The only proper way to instantiate a plugin is via the registry interface. For the same reasons (resolving identifiers, preserving correct system state and saving data) connecting/disconnecting processes and manual triggering of their updates is also a task performed by the registry.

**Repository** administers definitions and binary code of the plugins available in the system. During the start-up, it reads the configuration script (see next section for an example) and proliferates the configuration information.

**Scheduler** is a process for triggering the scheduled automatic restarts and managing their timing. A plugin instance could be, if awaited, enrolled for the restarting that occurs repeatedly after given amount of time.

**Persistence** is a module that allows the system to retain its state after the shutdown process. It provides a set of methods that enable saving the effects of occurring changes (creation of new plugin instance or connection, change of settings, etc.).

As mentioned, the flow can be executed either by manual restart or by scheduler triggering. These cases are almost identical and differ only in the source of the trigger. The execution mechanism is based on the update method in the Plugin class. Updating a plugin at the end of the flow resulst in consecutive updates of plugins that are inputs for current plugin. In this way, all the data stored in the processes (plugins) of flow are up-to-date and it this way the flow is executed.

## 5   INDECT MAS – Towards Working Example

As a next step INDECT-MAS a prototype system assembled for an exemplary but useful criminal analysis task has been built. On one hand it is a proof of

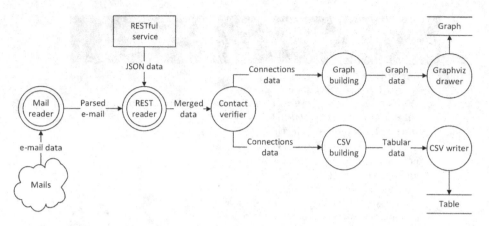

**Fig. 5.** Exemplary scenario

concept for the proposed architecture, on the other hand it answers to a scenario reproduced in Fig. 5.

The idea behind the scenario is to integrate two different sources of information and monitor them in order to produce up-to-date reports on organized crime gang structure and relationships. While investigating criminal activity, the police often tries to identify relationships between different people based on their phone calls, SMS and e-mails registered in different databases. The police data analyst wishes to have a possibly complete list of communicating suspects which can be easily done, e.g., in a form of graph.

The transcribed scenario forms the implemented flow. The data is taken from special REST service, while the connections are gathered from the e-mail correspondence. Results are awaited in a form of a graph (see exemplary results in Fig. 7) or .csv file. The flow consists of the following plugins:

**mail_reader** simple parsing of e-mail data (compliant with MIME standard) done in pure Erlang;

**rest_reader** obtaining names and addresses found in parsed e-mails from the RESTful service (programmed in Pyton);

**criminal_contact_verifier** analysis of e-mail and REST data in order to find connections;

**criminal_graph_preparation** reformatting data in order to apply a generic graph drawing plugin;

**graphviz_drawer** drawing the graph with use of GraphViz engine;

**criminal_csv_preparation** reformatting data for the CSV writing plugin;

**csv_writer** creation of CSV file containing all data found and integrated.

It should be emphasized that assembling a dataflow is a quite simple task. An extract from the configuration file for the presented scenario is shown in Fig. 6.

```
1  <plugins>
2      <source>
3          <plugin name="Mail Reader" module="mail_reader">
4              <option name="file" />
5
6              <output name="mails">
7                  <type name="mailType" />
8                  <type name="anyType" />
9              </output>
10         </plugin>
11     </source>
12     ...
13     <analytical>
14         <plugin name="REST Reader" module="rest_reader">
15             <option name="user" />
16             <option name="password" />
17             <option name="address" />
18             ...
19
20         </plugin>
21     ...
22     </analytical>
23     ...
24     <final>
25         <plugin name="Graphviz Drawer" module="graphviz_drawer">
26             ...
27         </plugin>
28     </final>
29     ...
30 </plugins>
```

**Fig. 6.** Sketch of the configuration file (dots mean lines omitted)

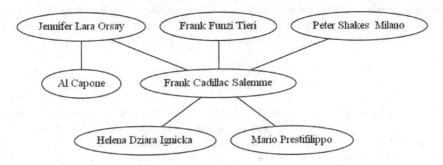

**Fig. 7.** Exemplary results in graphical form

Using the control structure, the prototype works, in fact, in the mode of monitoring of the e-mail source. After some period, the flow is restarted for a new portion of data and the new graph is produced with new suspects added.

# 6    Conclusions and Future Work

In the paper the software architecture dedicated to data integration i.e. acquiring and processing distributed, heterogeneous data, collected from the various sources also from the Internet has been proposed.

Systems built according to the architecture can be of the decision support kind and search results can be reacher as an effect of human and artificial intelligence synergy.

The potential capabilities of the architecture such as extensibility, support for code reuse and potential decentralization of processing based on plugins and Erlang processes makes it an interesting tool of relatively wide spectrum of application. There may be any possible data source connected to the system. The only condition is that the user is able to describe them ontologically – what gives a scheme of accessing the data – and to produce appropriate scanners that extracts the data. Therefore the data maybe imported from databases, spreadsheet files, blog servers, news servers, text files etc.

The presented idea of architecture aroused when building a prototype in projects dedicated to public security. In particular, the projects are aimed at supporting officers in gathering and analysing data coming from various sources of information. The prototype has been used to illustrate ideas presented in the paper.

The future work should include further evaluation of the concept by development of the prototype into a full-fledged distributed solution.

**Acknowledgments.** The research leading to these results has received funding from the research projects No. DOB-BIO6/08/129/2014 and DOB-BIO6/14/147/2014 funded by the Polish National Centre for Research and Development.

# References

1. Byrski, A., Kisiel-Dorohinicki, M., Dajda, J., Dobrowolski, G., Nawarecki, E.: Hierarchical multi-agent system for heterogeneous data integration. In: Bouvry, P., González-Vélez, H., Kołodziej, J. (eds.) Intelligent Decision Systems in Large-Scale Distributed Environments. SCI, vol. 362, pp. 165–186. Springer, Heidelberg (2011)
2. Casters, M., Bouman, R., Dongen, J.: Pentaho Kettle Solutions: Building Open Source ETL Solutions with Pentaho Data Integration. Wiley (2010)
3. Kimbal, R., Caserta, J.: The Data WarehouseETL Toolkit: Practical Techniques for Extracting, Cleaning, Conforming, and Delivering Data. Wiley (2004)
4. Schwinn, A., Schelp, J.: Data integration patterns. In: Abramowicz, W., Klein, G. (eds.) Business Information Systems, Preoceedings of BIS 2003, pp. 232–238 (2003)
5. Tatbul, N., Karpenko, O., Convey, C., Yan, J.: Data integration services. Brown University, Computer Science (2001)

# Intelligent Information Systems

# Global Logistics Tracking and Tracing
# in Fleet Management

Marcin Hajdul[1(✉)] and Arkadiusz Kawa[2]

[1] Institute of Logistics and Warehousing, Estkowskiego 6, 61-755 Poznań, Poland
marcin.hajdul@ilim.poznan.pl
[2] Poznań University of Economics, al. Niepodległości 10, 61-875 Poznań, Poland
arkadiusz.kawa@ue.poznan.pl

**Abstract.** Global real-time monitoring of means of transport and freight is necessary if we want to use the available resources in a sustainable way. Up-to-date information about the fleet and cargo can also improve the efficiency and effectiveness of the logistics processes organization in supply chains. Such monitoring also enhances transport security. Generally accessible solutions for global tracking, however, are expensive and quite unreliable. The purpose of this article is to present the concept of the T-Traco system, which is used for global tracking of vehicles and cargo. Its main advantage, in comparison to the existing solutions, is the use of the USSD communication channel, which is more reliable and much cheaper than GPRS and SMS.

**Keywords:** Global track & trace · Monitoring · Fleet management · GPS · USSD · Mobile application

## 1    Introduction

In logistics the flow of information is as important as the flow of things. Individual entities in the supply chain provide each other with information about the amount of the accumulated inventory, procurement, production and delivery time, etc. [1,2,3,4,12]. Enterprises have attempted to improve the information exchange process for many years. What does this mean in practice? It means complete elimination of paper documents and reduction of e-mail or phone communication. Increasingly, companies use information systems operating in the so-called cloud, which are available in an easy, cheap and safe way through a web browser from anywhere in the world. To improve the efficiency of the processes, a solution globally available via the Internet, enabling effective collaboration between existing businesses, applications and loads in the supply chain is needed [6,7, 11].

The access to the actual  information on the current state of a transport unit  in real -time is the key element of successful and efficient organization of logistic processes. At present there are solutions which use the GPS system (localization) and GPRS system (data transmission) which allow to track transport means and, from time to time, cargo units (mainly intermodal cargo units).

© Springer International Publishing Switzerland 2015
N.T. Nguyen et al. (Eds.): ACIIDS 2015, Part I, LNAI 9011, pp. 191–199, 2015.
DOI: 10.1007/978-3-319-15702-3_19

## 2     GPS and GPRS

GPS is a global system based on satellite radio signals. It allows to specify the exact position, speed and time anywhere on the globe. The system is also used in space navigation. It was constructed in such a way that the expectations of land and air forces and navies are met at any time, irrespective of the places on the globe, the weather, time of day or night. In turn, GPRS enables packet data transmission between the mobile device and the server. These data do not only concern the position, speed and time, but may also include other information about the vehicle. The advantage of GPRS is that the user pays for the actually sent or received number of bytes, not the time during which the connection is active.

The most obvious and common use of GPS is to supervise transport. In this kind of application, the moving transport object determines its position using an on-board GPS receiver. Data on the current position are transmitted via telecommunications links (usually via GPRS) to the traffic control center, for example to the parent company. Information transferred to the database can be supplemented by data on the parameters of the vehicle's movement. This requires equipping the vehicle with an on-board computer and its coupling with a GPS receiver. In addition, it is possible to transmit alarm signals or voice, depending on the communication capabilities of the terminal (phone) which the vehicle is equipped with [13].

Currently, most logistics companies take full advantage of the opportunities offered by GPS and GPRS systems. Senders, receivers and transport companies can observe the movement of their vehicles, and, thus the movement of freight, making vehicle management more efficient, and reaction to failure and other emergencies is faster. Providing current data on transport allows to continuously optimize transportation operations: routing, organization of turns, order of loading and unloading and so on. At any time, 24 hours a day, a dispatcher can obtain full information about the date, time, vehicle speed and the time and place of a single vehicle, a given group or all vehicles at the same time. In this way, s/he has full control over the vehicles that are up to several hundred kilometers away from the base. Constant monitoring of transport also allows to increase the scope of the control over a driver's work. Transport companies enrich their offer concerning information for the shipper about the state of the shipment and its current location [13].

Additional functions of the GPS-based systems are:

- Alarms and notifications
- Information on fuel consumption
- Monitoring temperature in the means of the transport
- Open door sensor
- Direct communication with drivers and operators
- Control of drivers' working time
- Planning and allocation of commisions
- Analysis of the driving style, proposing eco-driving
- Cooperation with other information systems (financial and accounting, fleet, SFA, CRM, etc.)
- Report on $CO_2$ emissions
- Records of journeys on toll roads
- Reporting bumps

These solutions, however, have one disadvantage: they become very expensive when used on a global scale. The costs of present solutions make them available only for medium and large local companies, in one country only, or optionally on the territory of EU. As a result, companies which use and sell logistic services are not able to fully use their logistic resources. Moreover, companies, according to the theory of systems, operate in a given environment and use and influencing its resources [8]. Therefore, if companies are not efficient, their processes become irrational, which in turn affects the environment in which they operate. As the result of this the traffic on roads increases alongside the emission of harmful substances or noise.

# 3    T-Traco

As the trend analysis conducted by the Institute of Logistics and Warehousing (Instytut Logistyki i Magazynowania - ILiM) indicates, companies are seeking solutions that will give them fast and cheap access to selected services. Communication is particularly important from the point of view of logistics. It is necessary, therefore, to use an innovative communication channel which will be cheaper and less unreliable. The solution to this problem is the T-Traco system, which is being developed by ILiM in collaboration with CallFreedom. T-Traco is a global system of intelligent cargo monitoring without roaming. This is possible through the use of the USSD (Unstructured Supplementary Service Data) communication channel [10].

The proprietary telecommunication system with own SIM card support platform and its own T-Traco SIM cards enables to use main cellular operators' connections all over the world. In Poland, the SIM cards use the infrastructure (masts) of all four operators: Orange, T-Mobile, Plus and Play. Also, the T-Traco SIM cards automatically change between the masts so that the best reach is guaranteed. This function works all over the world at almost 560 cellular operators on all continents.

T-Traco will ensure visibility of resources in real time, their location, speed, status, physical and chemical parameters and mutual communication with intelligent loads and transport service providers. T-Traco will use innovative services provided by Google Inc. that allow to visualize information sent by mobile devices mounted on vehicles / cargo units (so-called intelligent loads) on a map. In addition, the platform will combine information provided from intelligent cargo and Google servers concerning current traffic on roads and the current location of the cargo. On this basis, a user will know what time the load will reach the destination with accuracy of 1 minute. Additionally, intelligent load will also allow to identify the distance to the point of unloading with accuracy of 5 meters.

T-Traco will also offer a mobile version for truck drivers, train drivers or persons responsible for specific cargo transport. This will complement the whole system, in which two-way communication with a person who has the mobile version of the T-Traco application installed will be feasible. Thanks to that, it will be possible to transfer standardized status concerning both freight and transport globally. The user will be able to configure the status, for example: ready to be loaded, ready to be unloaded, loaded, unloaded, damaged during loading, damaged during unloading, delivery inconsistent with the order etc. What is more forwarder, through T-Traco mobile app, can send to the driver list of his new task with all necessary details.

**Fig. 1.** T-Traco and Google Maps

**Fig. 2.** Mobile version of the application T-Traco - used by the driver

The product operates in a cloud, therefore the costs of implementation are here excluded. Training is conducted as part of e-learning course, and is available for the users at any time, so there are no training costs. The user pays only a monthly fee which covers the costs of licensing and global communication. It is estimated that for 20 applications monthly individual cost will be 200% lower than the other competitive systems operating in Poland. At present there are no solutions in the world that would enable global tracking of intelligent cargo in real time with complete exclusion of roaming fees.

# 4     USSD vs. GPRS and/or SMS

What is mentioned, none of the present competitive systems in the world using the GPRS and/or SMS channels solve the problem of speed and efficiency of the messages exchanged between a device and a server in real time. For example, GPRS and/or SMS based channels are insecure with unpredictable delays in message delivery. Research described below support this statement.

The proposed solution with the USSD channel as a transport layer, together with the own telecommunications infrastructure, is a unique solution on a world scale. It also ensures more efficient, cheaper and substantially more stable operation as compared to the existing Track&Trace systems. All the available Track&Trace solutions that had been analyzed by the authors use the GPRS or SMS channels as a transmission layer for data exchange. Below are shown the advantages of USSD protocol vs. GPRS or SMS [5, 9]:

1.  USSD protocol operates as an active session, while SMS messaging operates in the transactional store-and-forward technology. Practically, it means that during active USSD session the server can communicate directly with a mobile device without any delays or not by means of agent services. In case of SMS communication, the message is first sent to the server (SMS Center) and then forwarded to the recipient. GPRS protocol operates on the same principle. The communication between a server and a mobile device is via http protocol with the use of agent/proxy server of a telephone operator (see fig. 3).

2.  Data delivery time via USSD protocol is much shorter than via SMS, as the USSD protocol is based on active session technology and not on the principle of agent servers, such as Proxy or SMS-C.

3.  The USSD protocol is available on every mobile device equipped with a GSM modem. The USSD does not require to use additional services on a mobile device, configuration of access points, SMS center addresses etc. USSD channel is made available every time a mobile device registers in the cellular network. This is particularly important in a roaming environment where problems with access to GPRS and SMS are frequent due to the fact that in roaming both SMS and GPRS are operated by agent companies.

4.  The USSD protocol based messages are always sent to HLR (Home Location Register) server of the home operator, even if the client is in roaming. This in turn means that the USSD protocol message exchange system will always operate in the same way, despite the location of a mobile device, or access to extra services (SMS or GPRS).

5.  The USSD protocol uses much lower resources of o device or of an operator's network as it "sewed in" on the control channel level of GSM protocol and does not require different channel allocation from the cellular network.

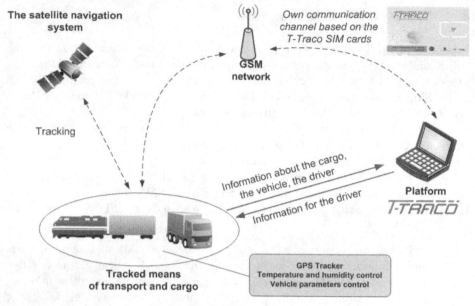

**Fig. 3.** T-Traco idea

## 5    Case Study

All the existing telecommunication solutions use the GPRS or SMS communication for the transmission of data to the web server. In GPRS http interface is used, where the information on the position of a tracked object is sent via GET or POST methods. In this case if the tracked object remains within roaming, GPRS services in roaming usually need additional operator's activation and the system is likely to lose its ability to transmit data to web server. Also, the fees for GPRS in roaming are much higher than local fees. When GPRS in not available in roaming many existing solutions attempt to send a SMS, which is much more expensive than the USSD, and still there is no guarantee that the message will reach its recipient. SMS may reach their recipients with a few weeks' delay because many SMS centers queue messages. Another important aspect is the load of a network. In GPRS communication each session requires a separate radio band (strip), while USSD protocol is always available when a mobile device logs on to GSM, because USSD remains in the control channel of GSM, so it is available always when a mobile device is within a cellular network. Due to the popularity of mobile internet services among end users and the development of mobile devices towards smartphones, tablets, etc. GPRS protocol is now very much loaded. This can be observed with "a naked eye" without conducting specialist research. Very often mobile internet fails to work on a popular smartphone where the traffic or crowds are big. With the USSD protocol the loss of data is not possible as this channel is never overloaded.

We conducted research on the reliability of GPRS and USSD protocols based communication. The research was made on selected routes in selected EU countries, as well as in United States and Japan. For example, on the 100 miles/162-km distance from Monterey/California/US to San Mateo/California/US two devices were compared (fig. 4).

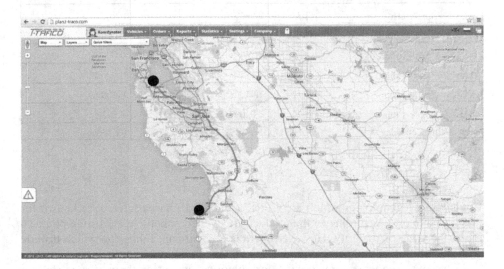

**Fig. 4.** Sample test route comparing GPRS and USSD data transmission

One device was sending information about vehicle position via GPRS, while the other device – via the USSD protocol with T-Traco SIM cards. The aim of this research was to track any delays in data transmission of the two alternative solutions. For the research purposes, mobile devices were delivered from the same producer (Sony Xperia S). On the 100miles/162-km distance each device was sending the total of 1188 batches with the information about geographical altitude and latitude, current speed and hour of measurement. The GPRS based device transmitted 24 batches with 1-minute delay, which means that as much as 86% of the information reached the server with delay (see tab.1). 816 batches, i.e. 68,7% of all data batches, were sent onto the server with a 10-30 minutes delay. When the USSD protocol was used, 99.2% of all data batches (1178 out of 594) reached the server without any delay. Only 10 batches were sent to the server with a one-minute delay, the reason for which was the change in the BTS transmitters of various operators (mobile devices with T-Traco cards automatically log on to this operator who has the strongest GSM signal in an area).

Table 1. T-Traco communication channel vs. traditional solutions

| Location | Location data sent by traditional GPRS channel | | Location data sent by T-Traco communication channel | |
|---|---|---|---|---|
| | pcs. | % | pcs. | % |
| Total number of sent locations | 1188 | 100% | 1188 | 100% |
| delay >= 30min <1h | 32 | 2,694% | 0 | 0,000% |
| delay >=10 min < 30min | 816 | 68,687% | 0 | 0,000% |
| delay >=5 min <10min | 92 | 7,744% | 0 | 0,000% |
| delay >= 3min <5min | 46 | 3,872% | 0 | 0,000% |
| delay >= 2min <3min | 12 | 1,010% | 0 | 0,000% |
| delay >=1min < 2 min | 24 | 2,020% | 10 | 0,842% |
| sent on time - no delay | 166 | 13,973% | 1178 | 99,158% |

# 6     Conclusions

Competitive technologies which are GPRS and/or SMS based have been known for many years now. According to the product life cycle competitive substitute products are at the stage of maturity. However, there is a significant limitation which thwarts global expansion of competitive products.

Nowadays companies are looking for solutions that offer them immediate and cheap access to selected services. Communication is particularly important from the point of view of logistics. T-Traco system fully responds to these needs.

T-Traco as a comprehensive product will be realized through the use of a globally innovative hardware system consisting of wireless mobile devices, desktop computer workstations (server) and means of transport and loading units connected with each other by a signal using USSD in GSM and UMTS networks. Thanks to this, direct communication between mobile devices will be possible. The primary advantage of using USSD operations is their global coverage and the lack of high roaming charges, which have so far prevented widespread use of intelligent load.

The most innovative function characteristics of T-Traco are as follows:

- Ability of global cargo and logistics tracking - flat worldwide rate, no roaming costs.
- Much better reliability than GPRS and/or SMS based solutions.
- Much higher exactness when tracking cargo units and vehicles – precision up to 5 m all over the world, which ensures security and certainty of a client's cargo location. The system sees even the slightest change in the tracked cargo or vehicle position and then identifies unauthorized change in the position. Thanks to this solution the client is sure that the dispatch of his cargo will be globally tracked, which ensures the overall security of a transaction.

- This solution allows automatic review of the position and the status of a cargo unit, the cargo and the vehicle. Therefore there will be no danger that the cargo is omitted by dispatchers or that it will be lost.

# References

1. Anholcer, M., Kawa, A.: Optimization of supply chain via reduction of complaints ratio. In: Jezic, G., Kusek, M., Nguyen, N.-T., Howlett, R.J., Jain, L.C. (eds.) KES-AMSTA 2012. LNCS, vol. 7327, pp. 622–628. Springer, Heidelberg (2012)
2. Balou, R.H.: The evolution and future of logistics and supply chain management. European Business Review **19**(4) (2007)
3. Chopra, S., Meindl, P.: Supply Chain Management: Startegy Planning and Operation. Prentice Hall, Upper Sadle River (2004)
4. Christopher, M.: Logistics and Supply Chain Management. Strategies for Reducing Cost and Improving Service. Financial Times Management, London (1998)
5. Dabas, A., Dabas, C.: Implementation of Real Time Tracking using Unstructured Supplementary Service Data. World Academy of Science, Engineering and Technology **30** (2009)
6. Golinska, P., Hajdul, M.: Multi-agent coordination mechanism of virtual supply chain. In: O'Shea, J., Nguyen, N.T., Crockett, K., Howlett, R.J., Jain, L.C. (eds.) KES-AMSTA 2011. LNCS, vol. 6682, pp. 620–629. Springer, Heidelberg (2011)
7. Golinska, P., Hajdul, M.: Virtual logistics clusters – IT support for integration. In: Pan, J.-S., Chen, S.-M., Nguyen, N.T. (eds.) ACIIDS 2012, Part I. LNCS, vol. 7196, pp. 449–458. Springer, Heidelberg (2012)
8. Hajdul, M.: Virtual collaboration in the supply chains – T-scale platform case study. In: Selamat, A., Nguyen, N.T., Haron, H. (eds.) ACIIDS 2013, Part II. LNCS, vol. 7803, pp. 449–457. Springer, Heidelberg (2013)
9. Herwono, I.: Performance Evaluation of GSM Signaling Protocols on USSD, Communication Networks. Aachen University of Technology, Aachen (2000)
10. http://www.t-traco.com
11. Kawa, A.: SMART logistics chain. In: Pan, J.-S., Chen, S.-M., Nguyen, N.T. (eds.) ACIIDS 2012, Part I. LNCS, vol. 7196, pp. 432–438. Springer, Heidelberg (2012)
12. Lambert, D.M., Knemeyer, A.M., Gardner, J.T.: Supply chain partnerships: model validation and implementation. Journal of Business Logistics **25**(2), 21–42 (2004)
13. Szymczak, M.: Satelitarna nawigacja pojazdów. System Navstar GPS, Eurologistics, April, 2001

# Human Activity Recognition Prediction for Crowd Disaster Mitigation

Fatai Idowu Sadiq[1], Ali Selamat[1,2(✉)], and Roliana Ibrahim[1]

[1] Faculty of Computing, Universiti Teknologi Malaysia, 81310 UTM Johor Bahru, Johor, Malaysia
sfatai2011@gmail.com, {aselamat,roliana}@utm.my
[2] UTM-IRDA Digital Media Center of Excelence,
Universiti Teknologi Malaysia, 81310 UTM Johor Bahru, Johor, Malaysia

**Abstract.** Context sensing and context acquisition have remained challenging issues in addressing the problems relating to Human Activity Recognition (HAR) for mitigation of crowd disasters. In this study, classification algorithms for higher accuracy of HAR which may be significantly low for effective stampede prediction in crowd disaster mitigation were investigated. The proposed HAR prediction model consists of mobile devices (mobile phone sensing) that can be used for monitoring a crowd scene in group movement: it employs tri-axial accelerometer sensors as well as other sensors like digital compass to capture relevant raw data from participants. In a previous study of stampede prediction, HAR accuracy of 92% was achieved by implementing J48, a Decision Tree, (DT) algorithm for context acquisition using a data mining tool. The implementation of the proposed model using K-Nearest Neighbour (KNN) algorithm with real time raw data collected with smartphones provided easily deployable context-awareness mobile Android Application Package (.apk) for effective crowd disaster mitigation and real time alert to avoid occurrence of stampede. The results gave 99.92% accuracy for activity recognition which outperforms the aforementioned study. Our results will forestall possible instances of false stampede alarm and reduce instances of unreported cases with higher accuracy if implemented in real life.

**Keywords:** Human Activity Recognition · Context awareness · Crowd disasters · Mitigation and Machine Learning

# 1    Introduction

Human activity is a common phenomenon in everyday life. These activities are found everywhere, i.e. ubiquitous either in offices or in our homes [1] . Nowadays, the presence of smartphones and their continuous growth has made recognition of human activities a possibility with the help of several inbuilt sensors that come with the smartphones[2, 3]. These sensors include accelerometer, gyroscope, digital compass etc; which are inbuilt sensors on the smartphones. It assigns real value estimates of acceleration along x, y, z coordinates from which velocity and displacement can be

© Springer International Publishing Switzerland 2015
N.T. Nguyen et al. (Eds.): ACIIDS 2015, Part I, LNAI 9011, pp. 200–210, 2015.
DOI: 10.1007/978-3-319-15702-3_20

measured [4]. It is often used as motion detector [5], and for body-posture sensing [6]. Previous studies have shown that HAR accuracy and mobile sensing towards context-awareness is a challenging problem in context-aware systems and applications [7-9].

However, human activity monitoring with the help of sensors on the smartphone is a new research area. Automatic recognition of user activities using different contextual data for enhancement of pervasive systems using context-awareness application is still in its infancy[2]. Meanwhile, context-aware computing has proved to be relevant in crowd disaster mitigation following a stampede occurrence that claimed lives of over hundred people in India [10]. Crowd disaster mitigation utilizes activity recognition to predict onset of a stampede in a crowd through the movement of participants and their behavioral patterns using HAR accuracy for crowd disaster mitigation[10].

## 2     Related Works for Crowd Monitoring

The origin of Crowd Monitoring Scenario (CMS) is traced back to 1995 when Close Circuit Televisions (CCTV) were used with pattern recognition for CMS [11]. This technique is conventionally manual, involves tedious efforts, and requires regular participation of security officers in physical appearance. As a result of the limitations Gomez, et.al [12], proposed Wireless Sensor Networks with wireless communication technique for better situational awareness of monitoring people in crowded areas. This study suffered from high rate of false alarm, and the use of low power sensors in its development due to low temperature and acoustic sensors used which makes it less reliable. The said limitations led to the introduction of an emerging technology with mobile phone sensing in [13]. Despite their efforts the new model still have shortcomings as a result; Ramesh et.al[10] proposed Context-Aware-Computing  together with Wireless Sensor Network (CAC-WSN) to address Human Activity Recognition (HAR) accuracy, poor real time information, high rate of false alarm for effective stampede prediction for mitigation of crowd disasters as identified  in Roggen et. al [13].  However, Ramesh et. al [10], was the first to introduce context-aware application using activity recognition to investigate HAR accuracy as a basis for effective stampede prediction for mitigation of crowd disasters by developing multi-context-fusion which suffers some drawbacks which include HAR accuracy which may be significantly low for stampede prediction is a motivation that led the study presented in this paper.

### 2.1     Problem Statement

Classification problem with HAR research has cut across many domains [14, 15]. Its introduction for handling the crowd disaster issue remains unresolved due to its wide spread across all nations over loss of life and properties. Likewise conventional manual crowd monitoring is a tedious effort that calls for dedication and participation of many security officers. Meanwhile, stampedes are the reasons for recent crowd disasters that have claimed lives of over hundreds of people in India. Previous research has recorded 92% activity recognition accuracy for onset stampede prediction [10]. The 92% accuracy may be significantly low for larger crowds with lots of activities which are inevitable in our environment. Hence, the need to investigate the possibility of achieving higher HAR accuracy in order to ascertain better reliability of real time message information alert for crowd disaster mitigation when the number of participants in the crowd increases is desirable. Figure 1 shows a typical crowd scene and indicates victims in a stampede from a crowd scenario.

**Fig. 1.** Typical crowd scene in an environment and stampede occurrence in a crowd [10]. http://www.recentnigerianjobs.com/2014/03/nigerian-immigration-service-exam.html[16].

## 3    Proposed Model for HAR Prediction

This section described the proposed model of the HAR prediction for mitigation of crowd disasters using mobile phone sensing with context-aware application deployedto each of the participant's smartphones as shown below. The application on the smartphones captures sensor data (i.e. activities such as jogging, standing, walking etc) using context by interacting with the people shown in the model and the application with the help of the context-aware applications through activity recognition by observing each of the participant's direction, location, behavior, movement pattern and their surrounding environment. The recognition is accomplished by exploiting the information retrieved from various sources such as environmental using the context sensing and acquisition in the mobile devices [17]. In order to capture human activities and store the sensor readings (data), the client is lunched and server configured. This model indicates server starting mode; thereafter the activity recognition commenced and synchronization takes place for sensor data readings on x-y-z-axes as shown in Figure 2, which produces the HAR dataset for classification in this paper with Weka data mining tool because; it is open source, flexible and permits us to use several classifiers in order to choose the one with the best accuracy performance.

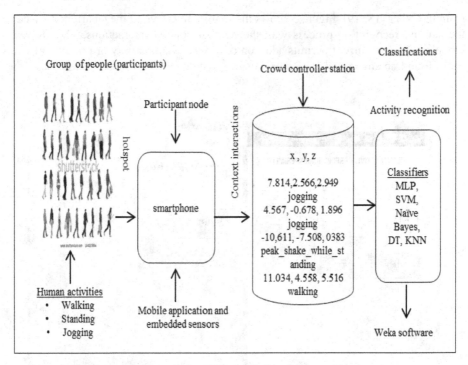

**Fig. 2.** Proposed model for HAR prediction using machine learning

## 3.1    Methodology

The method used in the baseline paper was for data collection and investigation into HAR prediction accuracy to predict stampede occurrence in crowded area. This study aims at exploring other methods capable of producing improve HAR accuracy against, the state-of-the-art 92% HAR accuracy which may be significantly low to cater for large crowd situation in Ramesh et al. [10]; this will facilitate effective stampede prediction for mitigation of crowd disasters in the environment. Figure 3 shows the mobile interface. The details of the methodology are as follows.

## 3.2    Pre-processing

The raw values from the smartphone's accelerometer were pre-processed before the feature extraction. For the pre-processing, we computed the mean for each of the axis x, y, z for accelerometer data captured to remove the random spikes and noise if any on the dataset [10]. The median value was obtained to treat the missing values found along the x, y, and z axes of the same dataset. Each of the three axes were analysed individually and the statistical metrics such as standard deviation and correlations were obtained to ascertain the stability of the dataset and to distinguish one activity from the other [18, 19]. Figure 3 shows the mobile interface of the smartphone used in the activity recognition process with the help of context interactions. The proposed model is flexible since it permits addition of any relevant activity or removal of irrelevant found, in any crowded area or scenario, which may be found necessary as times goes on.

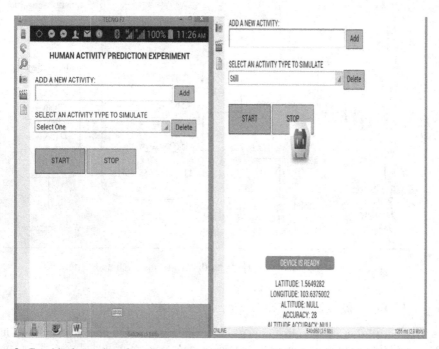

**Fig. 3.** Graphical user interface for sensor data capture of activity recognition for the propose model

## 3.3    Feature Extraction

Table 1 shows the extracted features. The activities such as jogging, walking, and standing were collected at the crowd controller indicated in Figure 2. Out of the 8 sensors programmed accelerometer sensors and digital compass values are found significant for our analysis and benchmarking with the HAR accuracy obtained in [10]. We only extracted accelerometer sensor x, y, z for analyzing each of the selected activities in our experiment. Activity recognition is regarded as a classification problem [8, 10, 20] hence the need for machine learning algorithms to help in decision making process, based on the collected information from the sensors data using context to make initial scientific hypothesis. In order to classify the human activities in a crowded area, as shown in Figure 2, five machine learning algorithms were investigated using our dataset to select the best performance classifier with higher accuracy.

Presently, there is no standard performance evaluation strategy for human activity recognition research: effort has been made for provision of public dataset [8, 21]. The dataset was randomly partitioned into two independent sets, with 70% assigned to training and the remaining 30% for testing [22]. The overall dataset, as the research progresses, may serve the purpose of public dataset in this domain as a contribution.

## 4    Experimental Setup

Researches in Activity Recognition (AR) have used 1 or more subjects [23] in conducting their experiment using acceleration data captured, evidence shown in [24]. In this paper a set of experiments were conducted to obtain the HAR dataset, 4 Subjects volunteers (Males) with ages between 21-30 years were picked for this task. Each of them was informed about the rules for choosing the activities while placing the smartphone such as (Samsung Galaxy X2, Samsung Galaxy Grand 2, and Gionee) on their hand and tied to their waist. The selected activities considered are *Standing, Walking, jogging, Still, Peak_shake_while_standing*. Each of the participants performed each of the activities specified for 10 minutes (600ms) using the same position which was chosen based on the evidence from previous studies[3, 25]. The data acquisition reading occurs every second. Example of raw data from the accelerometer in the sensor captured while recognising each activity in the experiment is as shown in Table 1. Figure 3 shows the mobile interface used for the activity recognitions of participants in the crowd. The correct classified activities are assigned (1) and incorrectly classified activities are assigned (-1); because we interested in the prediction of the class values in the training instances to obtain HAR accuracy in the simulation with weka; the outcome of the simulation form the basis in decision making to know if stampede will occur or not.

**Table 1.** HAR datasets collected with smartphone Sadiqsensor.apk

| Activities | x m/s$^2$ | y m/s$^2$ | z m/s$^2$ | Class values |
|---|---|---|---|---|
| Walking | -0.996 | 3.639 | 14.404 | 1 |
| walking | -10.4196 | 9.691729 | 10.30464 | -1 |
| Falling | -5.63116 | 3.44765 | -5.17148 | 1 |
| Falling | -3.94564 | -1.68552 | 8.810662 | 1 |

### 4.1    Experimental Results

The KNN (Ibl) based classification was carried out and a decision threshold were reached for human activity recognition to be determined. True Positive (TP rate) implies the rate of predicting correctly the overall possibilities and False Positive (FP rate) is the rate of predicting incorrectly to overall possibilities. Recall and precision indicates parameters that are of relevance. While the harmonic mean of precision and recall is the F-measure. Meanwhile the ROC area [10], presents the performance of a system as the threshold varied in Table 2. It shows the point where false negative, FN and false positive, FP meets; i.e. accuracy in this case is 99.92% (1).

**Table 2.** Simulation for HAR

| Metrics | Performance analysis of HAR accuracy prediction for activity recognition | | |
|---|---|---|---|
| | *Still* | *Walking* | *Running/jogging* |
| TP | 1 | 1 | 1 |
| FP | 0.001 | 0 | 0 |
| Precision | 0.998 | 1 | 1 |
| Recall | 1 | 1 | 1 |
| F-measure | 0.999 | 1 | 1 |
| ROC area | 1 | 1 | 1 |

Table 3 presents the confusion matrix of class activities' instances as part of data analysis in the simulation with WEKA for HAR prediction.

**Table 3.** Confussion Matrix for classified activities recognition

| a | b | c | d | e | Classified as | Difference |
|---|---|---|---|---|---|---|
| 775 | 0 | 0 | 0 | 0 | a=walking | 0 |
| 0 | 44 | 0 | 0 | 0 | b=jogging | 0 |
| 0 | 0 | 928 | 0 | 0 | c=Peakshake/wsta | 0 |
| 0 | 0 | 0 | 965 | 0 | d=Still | 0 |
| 0 | 0 | 0 | 2 | 886 | e=Standing | 2-incorrectly classify |

The prediction of HAR accuracy achieved using KNN from the simulation stood at 99.92% instead of 92% in [10] and 89% for EM using our real time HAR dataset with the proposed model. This is an improvement over an existing method in this domain. The confusion matrix for classified activities in the  simulation of the 5 activities considered for the subjects (participants) is shown in Table 3, which indicates that only 2 (0.0556 %)  of all the 3600 instances analysed, was incorrectly classified and regarded as FN. This is against 45 (7.554%) in [10]. The improved HAR accuracy obtained can form the basis of low probability for stampede occurrence as shown in Table 4. It explain comparison of our PM result using the same parameters with the method in the EM presented in [10]. Since the standard dataset is not common in this domain previous studies relied on raw dataset captured with smartphones. We therefore replicate the parameters used in previous studies (***) and use the same accelerometer sensors to provide the basis of comparisons. The standard datasets available is not suitable for crowd scenario, it is rather obtained with a hardware accelerometer sensors used for physical activities attached to the body, and not from a smartphones [21]. The circle shows the accuracy for the PM against the EM in this paper as shown in Figure 4.

**Table 4.** HAR accuracy prediction for the proposed approach PM and EM   (Benchmarking)

| Parameters description | Author's name | Existing Method (EM) | Proposed Method(PM) |
|---|---|---|---|
| | Ramesh, et'al 2013 | Our datasets | Our datasets |
| Mobile sensor | Accelerometer data sets (1) | Accelerometer data sets (1) | Accelerometer data sets (1) |
| No of instances | 302 *** | 3600 | 3600 |
| Training sets (TS) | 302 *** | 2520 | 2520 |
| Testing sets   (TE) | NSP | 1080 | - |
| Classifiers | DT (J48) | DT (J48) | KNN |
| TP (correctly classified instances) | 257 | 2251 | 2518 |
| FP (Incorrectly classified instances) | 45 | 269 | 2 |
| HAR accuracy | 92% | 89% | 99.92% |
| False Alarm | HIGH | HIGH | LOW |
| St-Prediction (STP) | LR | LR | HR |
| Crowd disaster * | Probability  is high | Probability is high | Probability is low |

Key:   *** = state-of-the-art-study; DT= decision tree algorithm; TS: (training set), TE :( Testing set); STP= Stampede prediction; LR =Less reliability; HR = High reliability. * = Decision.

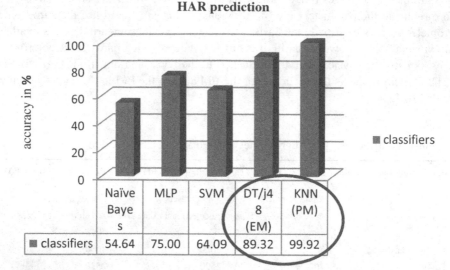

**Fig. 4.** HAR accuracy prediction performance evaluation for EM and PM visualisation

## 4.2    Conclusion and Future Work

The main contribution of this paper is the achievement of higher HAR accuracy prediction. The developed mobile application shows the reality for possible collections of large dataset that will be needed for the ongoing research. The improved 99.92% accuracy of HAR recorded in this paper as shown in Figure 4, will reduced the false negative alarm which can provide effective stampede prediction in crowd disaster area in the environment. This research has designed a mobile sensor interface that captured and carried out human activity recognitions prediction accuracy result that can be used to facilitate better stampede prediction for mitigation of crowd disaster in ongoing research.   In the future, we plan to extend this research by implementing KNN over graphical processing units (GPU) for stampede prediction in order to improve the performance of the system and enhance the algorithmic complexity of KNN for real time process of the entire system. To investigate stampede prediction algorithm and see how HAR can facilitate it. We hope to investigate the impact of other mobile sensors apart from the tri-axial accelerometer sensor on KNN and other classifiers with their influence on the activity recognition for mitigation of crowd disasters.

**Acknowledgement.** The Universiti Teknologi Malaysia (UTM) and Ministry of Education Malaysia under research Grant Vot 00M19 and Fundamental Research Funding from Ministry of Education Malaysia Vot 4F550 are hereby acknowledged for some of the facilities that were utilized during the course of this research work.

# References

1. Kose, M., Incel, O.D., Ersoy, C.: Online human activity recognition on smart phones. In: Workshop on Mobile Sensing: From Smartphones and Wearables to Big Data, pp. 11–15 (2012)
2. Wang, Y., Lin, J., Annavaram, M., Jacobson, Q.A., Hong, J., Krishnamachari, B., Sadeh, N.: A framework of energy efficient mobile sensing for automatic user state recognition, pp. 179–192
3. Kaghyan, S., Sarukhanyan, H.: Activity recognition using K-nearest neighbor algorithm on smartphone with Tri-axial accelerometer. In: International Journal of Informatics Models and Analysis (IJIMA), ITHEA International Scientific Society, Bulgaria, pp. 146–156 (2012)
4. Ravi, N., Dandekar, N., Mysore, P., Littman, M.L.: Activity recognition from accelerometer data. In: AAAI, **5**, pp. 1541–1546
5. DeVaul, R.W., Dunn, S.: Real-time motion classification for wearable computing applications. project paper (2001). http://wwwmedia.mit.edu/wearables/mithril/realtime.pdf
6. Foerster, F., Smeja, M., Fahrenberg, J.: Detection of posture and motion by accelerometry: a validation study in ambulatory monitoring. Computers in Human Behavior **15**(5), 571–583 (1999)
7. Riboni, D., Bettini, C.: COSAR: hybrid reasoning for context-aware activity recognition. Personal and Ubiquitous Computing **15**(3), 271–289 (2011)
8. Kaghyan, S., Sarukhanyan, H., Akopian, D.: Human movement activity classification approaches that use wearable sensors and mobile devices, pp. 86670O-86670O-12 (2013)
9. Chen, H.L.: An intelligent broker architecture for pervasive context-aware systems. University of Maryland, Baltimore County (2004)
10. Ramesh, M.V., Shanmughan, A., Prabha, R.: Context aware ad hoc network for mitigation of crowd disasters. Ad. Hoc. Networks **18**, 55–70 (2014)
11. Davies, A.C., Yin, J.H., Velastin, S.A.: Crowd monitoring using image processing. Electronics & Communication Engineering Journal **7**(1), 37–47 (1995)
12. Gomez, L., Laube, A., Ulmer, C.: Secure sensor networks for public safety command and control system, pp. 59–66
13. Roggen, D., Wirz, M., Tröster, G., Helbing, D.: Recognition of crowd behavior from mobile sensors with pattern analysis and graph clustering methods. arXiv preprint. arXiv:1109.1664 (2011)
14. Hildeman, A.: Classification of epileptic seizures using accelerometers (2011)
15. Wilde, A.G.: An overview of human activity detection technologies for pervasive systems
16. http://www.recentnigerianjobs.com/2014/03/nigerian-immigration-service-exam.html
17. Poppe, R.: Vision-based human motion analysis: An overview. Computer Vision and Image Understanding **108**(1), 4–18 (2007)
18. Figo, D., Diniz, P.C., Ferreira, D.R., Cardoso, J.M.: Preprocessing techniques for context recognition from accelerometer data. Personal and Ubiquitous Computing **14**(7), 645–662 (2010)
19. Anguita, D., Ghio, A., Oneto, L., Parra, X., Reyes-Ortiz, J.L.: Energy efficient smartphone-based activity recognition using fixed-point arithmetic. J. Univ. Comput. Sci. **19**, 1295–1314 (2013)
20. Silva, J.: Smartphone Based Human Activity Prediction (2012)
21. Xue, Y., Jin, L.: A naturalistic 3D acceleration-based activity dataset & benchmark evaluations, pp. 4081–4085

22. Anguita, D., Ghio, A., Oneto, L., Parra, X., Reyes-Ortiz, J.L.: A public domain dataset for human activity recognition using smartphones
23. Luštrek, M., Kaluža, B.: Fall detection and activity recognition with machine learning. Informatica **33**(2), 197–204 (2009)
24. Bao, Ling, Intille, Stephen S.: Activity Recognition from User-Annotated Acceleration Data. In: Ferscha, Alois, Mattern, Friedemann (eds.) PERVASIVE 2004. LNCS, vol. 3001, pp. 1–17. Springer, Heidelberg (2004)
25. Reiss, A., Hendeby, G., Stricker, D.: A competitive approach for human activity recognition on smartphones, pp. 455–460

# Frequencies Assignment in Cellular Networks

## Maximum Stable Approach

Ye Xu and Ibrahima Sakho[✉]

LITA, Université de Lorraine,
Ile du Saulcy, BP 794, 57045, Metz, France
ibrahima.sakho@univ-lorraine.fr

**Abstract.** With the limited number of communication frequencies and the increasing number of users, the problem of communication frequencies assignment without interference is more than ever at the heart of the development of cellular networks. This paper reports a heuristic assignment based on the scheduling of the cells and the scheduling of the maximum stables of the dependency graph of the cells network. The purpose of the maximum stables scheduling is to assign without interference a maximum number of frequencies to cells while the purpose of the cells scheduling is to minimize the number of the used frequencies. Therefore aim of the heuristic is to satisfy a maximum number of connection requests with a minimum number of frequencies. The heuristic is implemented and its performance evaluated for the well known network test Philadelphia-benchmark.

**Keywords:** Mobile phone networks · Cellular networks · Allocation of frequencies · Electromagnetic interference · Co-site constraint · Co-channel constraint · Adjacent constraint · Compatibility matrix · Philadelphia-benchmark · Clique · Stable

## 1 Introduction

Recent developments in the field of integration of electronic components and telecommunications technology led to the emergence of so compact and so light terminals that they are in turn become portable and mobile. The mobile phone is nowadays one of the most popular applications of the mobile devices technology. Unlike wired ones, mobile phones called mobile stations (MS) communicate using radio waves via base stations (BS) connected to the public telephone network through a switching center MS (MSC). Therefore, communications are a tributary of electromagnetic waves interference mainly when the frequencies distance constraints are not observed. Due to the limited number of frequencies allocated to mobile telephony, the geographical area covered by a network is divided into areas called cells so allowing the same frequency to be used without interference in different cells. But in view of the increasing number of MS and bandwidth needs of mobile phone services,

---

Co-author Ye Xu was in LITA from April to September 2014 for her Master 2 Thesis in Computer Science at the University of Lorraine.

N.T. Nguyen et al. (Eds.): ACIIDS 2015, Part I, LNAI 9011, pp. 211–220, 2015.
DOI: 10.1007/978-3-319-15702-3_21

it becomes increasingly difficult if not impossible to guarantee interference-free communications. It is therefore necessary to find an assignment strategy of available frequencies guaranteeing interference free communications between users. This paper presents an assignment method called Assignment by Maximum Stable (AMS) inspired of both a heuristic assignment described in [1] and the design methodologies of parallel algorithms whose a golden rule is to aim executing at each execution step of an application a maximum number of independent instructions. The principle of AMS is first to calculate all the maximum stables of the dependency graph of the considered cells network. The maximum stables are then scheduled in the descending order of their size in terms of the number of cells belonging to them. In the absence of maximum stables, cells are scheduled in the ascending order of their compatibility expressed by the distance that their frequencies must observe to avoid interferences. As stables are constituted of cells without constraints, a same frequency can then be allocated to all the cells of a same stable without interference satisfying so a maximum number of connection requests. On the other side, frequencies are assigned to cells according cells scheduling. This results in minimizing the number of used frequencies. The quality of the solutions generated by this process is very close to that of the best known algorithms but faster, thanks to assignment by maximum stables scheduled in the descending order of their size.

The rest of the paper is organized into four sections. Section 2 deals with the formulation of the frequencies assignment problem (FAP) in cellular phone networks. Section 3 presents the state-of-the-art on the solving methods reported in the literature and their analysis. Section 4 presents the alternative method that we propose called AMS. Section 5 presents the implementation and the performance analysis of AMS for several instances of the connection requests of Philadelphia-benchmark network. Section 6 concludes the paper.

## 2     Problem Formulation

Face to restricted available bandwidth for communication in mobile phones networks and the growing number of subscribers, spatial division multiplexing concept (SDM) has been introduced [2] to allow a same frequency to be reused. SDM principle consists to divide the service area, also called operation circle, into a number of cells, generally hexagonal as shown in Fig. 1, to allocate a same frequency to cells whose distances guarantee the absence of interference.

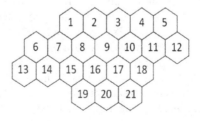

**Fig. 1.** Philadelphia-benchmark network

With SDM, a large geographic area can be covered without interference with a limited bandwidth. In the classical formulation of the FAP, a cellular network is considered as constituted of arbitrary n cells. Without loss of generality, it is assumed that frequencies are uniformly spaced in the radio frequencies spectrum and are mapped to positive numbers. The separation between frequencies takes into account the frequencies separation constraints of electromagnetic interference of radio waves. There are three main such constraints:

- Co-Site Interference Constraint (CSC): frequencies assigned to a same cell must be at a minimum distance,
- Adjacent Channel Interference Constraint (ACC): frequencies assigned to adjacent cells must comply with the frequencies minimum distance separation,
- Co-Channel Interference Constraint (CCC): co-assigned channels in different cells must comply with the minimum reuse distance.

These constraints are expressed in terms of the cells compatibility matrix, say C, each component of which, $c_{ij}$, indicates the distance constraint between any pair of frequencies assigned to the cells i and j. Table 1 illustrates such a matrix for the Philadelphia-benchmark network.

**Table 1.** An instance of the Philadelphia-benchmark network compatibility matrix

```
5 2 1 0 0 1 2 2 1 0 0 0 0 1 1 1 0 0 0 0 0
2 5 2 1 0 0 1 2 2 1 0 0 0 0 1 1 1 0 0 0 0
1 2 5 2 1 0 0 0 1 2 1 0 0 0 0 1 1 1 0 0 0
0 2 1 5 2 0 0 0 0 1 2 2 1 0 0 0 0 1 1 0 0
0 0 1 2 5 0 0 0 0 1 2 2 0 0 0 0 0 1 0 0 0
1 0 0 0 0 5 2 1 0 0 0 0 2 2 1 0 0 0 0 0 0
2 1 0 0 0 2 5 2 1 0 0 0 1 2 2 1 0 0 1 0 0
2 2 1 0 0 1 2 5 2 1 0 0 0 1 2 2 1 0 1 1 0
1 2 2 1 0 0 1 2 5 2 1 0 0 0 1 2 2 1 1 1 1
0 1 2 2 1 0 0 1 2 5 2 1 0 0 0 1 2 2 0 0 1
0 0 1 2 2 0 0 0 1 2 5 2 0 0 0 0 1 0 0 0 1
0 0 0 1 2 0 0 0 0 1 2 5 0 0 0 0 1 0 0 0 0
0 0 0 0 0 2 1 0 0 0 0 0 5 2 1 0 0 1 0 0 0
1 0 0 0 0 2 2 1 0 0 0 0 2 5 2 1 0 0 1 0 0
1 1 0 0 0 1 2 2 1 0 0 0 1 2 5 2 1 0 2 1 0
1 1 1 0 0 0 1 2 2 1 0 0 0 1 2 5 2 1 2 1 1
0 1 1 1 0 0 0 1 2 2 1 0 0 0 1 2 5 2 1 2 2
0 0 1 1 1 0 0 0 1 2 2 1 0 0 0 1 2 5 0 1 2
0 0 0 0 0 0 1 1 1 0 0 0 1 2 2 1 0 0 5 1 1
0 0 0 0 0 0 1 1 1 0 0 0 1 2 2 1 2 1 1 5 1
0 0 0 0 0 0 0 1 1 1 0 0 0 1 2 2 1 2 1 2 5
```

Thus, given a vector D that each component $d_i$ is the number of connection requests in cell i, a set of communication frequencies $f_k$ k = 1, 2, ..., m and a compatibility matrix C, the purpose of the FAP is to find a frequencies assignment that satisfies the above constraints using a minimum number of frequencies. More formally, let $\sum^b$ be the boolean summation, x be an nxm matrix that a component $x_{if} = 1$ (resp. 0) if frequency f is (resp. is not) allocated to cell i. The FAP comes down to find:

$$\underset{x\ \ f\ \ i}{Min} \sum \sum{}^b x_{if}$$

s.t.

$$\sum_f x_{if} = d_i, \ i = 1, 2, ..., n$$

$x_{if} + x_{ig} \leq 1$, if $|f - g| \leq c_{ii}$ $\forall f$, g allocated to i, for $i = 1, 2, \ldots, n$

$x_{if} + x_{jg} \leq 1$, if $|f - g| \leq c_{ij} \forall f$ (resp. g) allocated to i (resp. j) for i, j = 1, 2, \ldots, n

$x_{if} \in \{0, 1\}$.

## 3    Related Works

The literature reports several formulation models and solving methods for FAP. However they can be grouped in three classes of approaches: the graph-theory-based, the heuristic-based and the optimization-based approaches.

Graph-theory-based approaches know several variations according to the underlying graphs theory problems. In [3] FAP is proved identical to a graph coloring problem that is assigning colors to vertices of a graph such that two connected vertices do not share a same color. Indeed it suffices to identify each cell of the considered network to a node of the graph and each non null component of the compatibility matrix to an edge of the graph. The computational complexity of graph coloring is known to be NP-hard [4]. The minimum number of colors for coloring a graph is its chromatic number. The chromatic number problem of a graph is in connection with several equivalent other graph problems among which we can mainly cite: the maximum clique (resp. stable) problem that deals with the maximum number of vertices that are pair wise adjacent (resp. non-adjacent), the dominant set problem that addresses the determination of a largest subset of the vertices that are such that every vertex that does not belong to it has at least one common edge with one of its vertices [5]. FAP has been so also studied using each of these formulations [6 – 8]. Whatever is the approach, the problem is in general known to be NP-complete [5]. However, for particular graphs there are approaches with an algorithmic complexity of O(n log n) [9] where n is the number of vertices. Thus, regardless of the graph-theory-based approach FAP remains NP-complete. This is why cheaper approaches as heuristic-based ones have also been extensively explored.

Heuristics for FAP fall either in pure local search [10] or more efficient local search as simulated annealing [11] and tabu search [12]. Beyond local search, heuristics based on ant colonies paradigm have also been adapted for FAP [13]. It should be noticed that these heuristics can take advantage of their parallel nature to speed up their execution on appropriate parallel computers [14]. In the same vein, genetic algorithms paradigm has also been extensively used to solve FAP [15-17]. Unlike previous paradigm, genetic algorithms can lead to optimal or acceptable solution with appropriate individuals coding and crossover operator for both reducing the search space and making it closed mathematically speaking [18]. Other distributed heuristics approaches are evolutionary algorithms [19], multi-agent [20] and neural [21] models.

FAP was also discussed in terms of optimization in the strict sense of optimizing a criterion under some constraints [22]. The resulting models are generally non linear [23-24]. So, the main task of the approaches is the linearization of the models obtained. This is the case for example in [24]. The problem is formulated there for networks whose topology is a 2D grid in terms of linear integer programming with the goal of maximizing the number of simultaneous connections. Like the other approaches the exact resolution of such models is prohibitive because the models are deemed NP-complete [22].

This state-of-the-art clearly shows that FAP belongs to NP-complete problems class. Consequently, regarding the computing time that problems of this class may require to solve them, heuristics seem to be the most reasonable approaches to obtain in reasonable computing time approximate but certainly acceptable solutions.

In the sequel we propose a heuristic that we will refer to as assignment by maximum stables (AMS) based on the one in [1].

## 4    Assignment by Maximum Stables

This section is the main contribution of the paper. The proposed algorithm is based on the one described in [1] that builds, the one after the other, several solutions of a FAP then chooses the one using the smallest number of frequencies. Each solution is built according to a cells scheduling selected from the n! permutations of the n cells. For a given permutation $\pi$, frequencies are first assigned consistently with CSC to cell $\pi(1)$ connection requests then to cell $\pi(2)$ ones consistently with CSC and all other required constraints, then to cell $\pi(3)$ ones, etc. Consequently frequencies are partitioned in three classes: the used constituted of the assigned, those that can't be consistently and the non-assigned. This process is outlined below as Interference_Free_Frequencies_Assignment_1 algorithm (IFFA1).

```
Algorithm IFFA1
Input: n, D, C
Ouput: π* //the optimal π
Begin
   Generate a set Π of permutations of n elements
   For π ∈ Π Do
     For i from 1 to n
       While dᵢ > 0
          Assign next consistent frequency to cell π(i)
       Compute the current number of used frequencies
       If it is less than the previous one
          Set π* to π
   Return π*-based assignment
End.
```

The execution time of this algorithm is clearly prohibitive for large networks. Furthermore the quality of the solutions, in terms of number of used frequencies, clearly remains dependent on the considered permutations. However, the algorithm guarantees to assigned frequencies to be free of interference. In this paper we propose to overcome this prohibitive computing time while minimizing the number of frequencies used with the same quality in terms of total absence of conflicting frequencies. To do this we propose to make assignment according to a schedule that is the descending order of the size of the maximum stables of the dependency graph of the cells network. The assignment is really made as in IFFA1 but in assigning the next consistent frequency to all the cells belonging to a same stable. Actually at each step of the assignment a maximum number, all of the maximum stable, connection requests is satisfied while guaranteeing the absence of conflict between the assigned frequencies. In the absence of stables, cells are scheduled in the ascending order of their constraints expressed by the compatibility matrix and frequencies are then assigned as in IFFA1. Note that as benefits, interference relationships between certain cells tend to disappear. Thus cells dependency graph evolves and must be updated.

**Illustration:** We consider the FAP instance for Philadelphia-benchmark subnetwork constituted of the cells 4, 5, 6, 7, 8 and 9, and the respective connection requests 2, 3, 3, 3, 3, 2 under the constraints described by the compatibility matrix of Table 1. Fig. 2 shows the corresponding compatibility matrix and the induced dependency graph where only the dependencies between different nodes are represented.

|   | 4 | 5 | 6 | 7 | 8 | 9 |
|---|---|---|---|---|---|---|
| **4** | 5 | 2 | 0 | 0 | 0 | 1 |
| **5** | 2 | 5 | 0 | 0 | 0 | 0 |
| **6** | 0 | 0 | 5 | 2 | 1 | 0 |
| **7** | 0 | 0 | 2 | 5 | 2 | 1 |
| **8** | 0 | 0 | 1 | 2 | 5 | 2 |
| **9** | 1 | 0 | 0 | 1 | 2 | 5 |

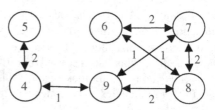

**Fig. 2.** The compatibility matrix extracted from Table 1 and the dependency graph of Philadelphia-benchmark subnetwork constituted of the cells 4, 5, 6, 7, 8 and 9

It admits {5, 6, 9}, {4, 6}, {4, 7}, {4, 8}, {5, 7} and {5, 8} as the only maximum stables. They are scheduled in the following order: {5, 6, 9}, {4, 7} and {8}. The purpose of the cells scheduling is to minimize the number of used frequencies, say $f_{max}$. This minimization depends on the cells scheduling. Indeed, let's consider for example the cells 6, 7 and 8. For the two scheduling (a) (6, 7, 8) and (b) (6, 8, 7), Table 2 shows the respective assignment tables, say $x^{(a)}$ and $x^{(b)}$ as denoted in Section 2: $x_{if} = 1$(resp. 0 not mentioned) if the frequency $f \in \{1, 2, 3, 4, 5\}$ is (resp. not) allocated to the cell $i \in \{6, 7, 8\}$. We obtain $f_{max}^{(a)} = 5$ and $f_{max}^{(b)} = 3$. The best assignment, $x^{(b)}$ is obtained with the one induced by the ascending order of the cells constraints.

**Table 2.** Frequencies assignment to cells 6, 7 and 8 according to the scheduling (6, 7, 8) and (6, 8, 7). $f_{max} = 5$ (resp. 3) for scheduling $x^{(a)}$ (resp. $x^{(b)}$)

| $x^{(a)}$ | 1 | 2 | 3 | 4 | 5 |
|---|---|---|---|---|---|
| **6** | 1 |   |   |   |   |
| **7** |   |   | 1 |   |   |
| **8** |   |   |   |   | 1 |

| $x^{(b)}$ | 1 | 2 | 3 | 4 | 5 |
|---|---|---|---|---|---|
| **6** | 1 |   |   |   |   |
| **7** |   |   | 1 |   |   |
| **8** |   | 1 |   |   |   |

Frequencies are essentially allocated by maximum stables then by cells. For this illustration, repetitively as many times as connection requests allow, compatible frequencies are allocated, in this order, to stables {5, 6, 9}, {4, 7}, {8}. Then similarly, compatible frequencies are allocated to the residual stables that is stables with residual connection requests namely $\{5, 6\}$, $\{7\}$ and $\{8\}$. The last residual connection requests come from cells 7 and 8. So to reduce $f_{max}$ it is necessary to schedule them according to their constraints with the last maximum stable that is $\{5, 6\}$. This leads to the scheduling (8, 7). The resulting assignment is given in Table 3. The repeated assignments are separated by the dotted line and the next repeated assignment by a double line.

**Table 3.** Best frequencies assignment for Philadelphia-benchmark subnet constituted of cells 4, 5, 6, 7, 8, 9 under the constraints of Table 1

| x | 1 | 2 | 3 | 4 | 5 | 6 | 7 | 8 | 9 | 10 | 11 | 12 | 13 | 14 | 15 | 16 |
|---|---|---|---|---|---|---|---|---|---|----|----|----|----|----|----|----|
| 4 |   | 1 |   |   |   |   |   |   | 1 |    |    |    |    |    |    |    |
| 5 | 1 |   |   |   |   |   | 1 |   |   |    |    |    | 1  |    |    |    |
| 6 | 1 |   |   |   |   |   | 1 |   |   |    |    |    | 1  |    |    |    |
| 7 |   |   | 1 |   |   |   |   |   | 1 |    |    |    |    |    |    | 1  |
| 8 |   |   |   | 1 |   |   |   |   |   |    | 1  |    |    | 1  |    |    |
| 9 | 1 |   |   |   |   |   | 1 |   |   |    |    |    |    |    |    |    |

The corresponding algorithm is as follows:

```
Algorithm IFFA2
D: Connection requests vector
d: the smallest number of residual connection requests
fmax: maximum number of used frequencies

Begin
  fmax ← 0
  Generate maximum stables of the dependency graph of the
  cells network
  While D is non-empty
    If there are stables
      Schedule the residual stables in the descending
      order of their size
    Else
      Schedule the residual cells in the ascending order of
      their constraint
    Repeat d times
      Assign compatible frequencies in their appearance to
      the residual stable cells in their scheduling order
    Update fmax // the last assigned frequency number
    Update D
End.
```

# 5    Implementation and Performance Evaluation

The proposed implementation language is the C programming language under Microsoft Visual Studio 2010 development environment. The evaluation machine is an Acer Aspire with Intel core I3 processor running at 2.4 GHz with 2 GB of RAM. For the simulation, we consider the Philadelphia-benchmark network with CSC = 5, ACC = 1 then 2 for the following two connection requests.

- $D^{(1)}$: (8, 25, 8, 8, 8, 15, 18, 52, 77, 28, 13, 15, 31, 15, 36, 57, 28, 8, 10, 13, 8)
- $D^{(2)}$: (5, 5, 5, 8, 12, 25, 30, 25, 30, 40, 40, 45, 20, 30, 25, 15, 15, 30, 20, 20, 25)

Table 4 compares its performance in terms of $f_{max}$ and execution time with the AFF1 one whose computations have been performed on a DEC ALPHA station.

**Table 4.** Simulation results in terms of the number of used frequencies and execution time

| No. | D | ACC | CSC | $f_{max}$ | | Exec. Time | |
| --- | --- | --- | --- | --- | --- | --- | --- |
| | | | | IFFA1 | IFFA2 | IFFA1 | IFFA2 |
| 1 | $D^{(1)}$ | 1 | 5 | 381 | 416 | 7.5 | 0.137 |
| 2 | $D^{(1)}$ | 2 | 5 | 463 | 547 | 9.5 | 0.171 |
| 3 | $D^{(2)}$ | 1 | 5 | 221 | 251 | 6.9 | 0.191 |
| 4 | $D^{(2)}$ | 2 | 5 | 273 | 376 | 7.7 | 0.164 |
| 5 | $D^{(1)}$ | 1 | 4 | 305 | 368 | 7.3 | 0.187 |
| 6 | $D^{(1)}$ | 2 | 4 | 465 | 539 | 8.4 | 0.134 |
| 7 | $D^{(2)}$ | 1 | 4 | 197 | 232 | 6.8 | 0.179 |
| 8 | $D^{(2)}$ | 2 | 4 | 278 | 362 | 7.5 | 0.152 |

IFFA1 $f_{max}$ are the best [1]. They outperform IFFA2 ones with a mean value of 63. But for each FAP instance, it has required, for the mentioned execution time, to create 1000 solutions while IFFA2 $f_{max}$ is obtained from one solution for the mentioned execution time. This latter characteristic is very important mainly for a dynamic context like mobile phone. IFFA2 $f_{max}$ can be improved by using more tuned stables scheduling. As proved by undergoing simulations this can be done in taking into account the residual maximum stables constraints too. Thus stables will be scheduled on both their size and their cells compatibility constraints for a better solution in one shot search in very short time as required by the mobile phone applications.

# 6    Conclusion

This paper addressed the FAP in cellular networks. From the related works section, FAP clearly belongs to NP-complete problems class. By nature the search for optimal solutions is prohibitive and therefore in practice the recourse to approximate solving methods and heuristics as genetic algorithms, ant colonies allow obtaining in reasonable computing time acceptable solutions mainly in dynamical context as mobile phones applications. This paper have proposed an assignment method called Assignment by Maximum Stable inspired of both an assignment heuristic described in [1] and the parallel algorithm design methodologies. The principle of AMS is first to calculate all the maximum stables of the dependency graph of the considered cells network induced by its cells compatibility matrix. Then the maximum stables are scheduled in the descending order of their size. In the absence of maximum stables, cells are scheduled in the ascending order of their compatibility. As stables are constituted of cells without constraints, a same frequency can then be allocated to all the cells of a same stable without interference satisfying a maximum number of connection requests. On the other side, assigning frequencies according cells scheduling induces minimizing the number of used frequencies. The quality of the solutions generated by this process is near to that of the best known algorithms but faster, thanks to the assignment by maximum stables scheduled in the descending order of their size, obtained in only one shot search. The paper also gives, for future works, some hints on the way that scheduling can be done to improve the quality of its solutions in terms of the number of the used frequencies.

# References

1. Chakraborty, G.: An Efficient Heuristic Algorithm for Channel Assignment Problem in Cellular Radio Networks. IEEE Trans. Veh. Technol. **50** (2001)
2. Aardal, K.I., et al.: Models and Solution Techniques for Frequency Assignment Problems. ZIB-Report 01–40 (2001)
3. Yu, F., Bar-Noy, A., Basu, P., Ramanathan, R.: Algorithms for channel assignment in mobile wireless networks using temporal coloring. In: Proceedings of the 16th ACM International Conference on Modeling, Analysis & Simulation of Wireless and Mobile Systems, pp. 49–58 (2013)
4. Martín, H.J.A.: Solving Hard Computational Problems Efficiently: Asymptotic Parametric Complexity 3-Coloring Algorithm. PLoS ONE **8**(1), e53437 (2013). doi:10.1371/journal.pone.0053437
5. Gavril, F.: Algorithms for Minimum Coloring, Maximum Clique, Minimum Covering by Cliques, and Maximum Independent Set of a Chordal Graph. doi:10.1137/0201013
6. Clark, B.N., Colbourn, C.J., Johnson, D.S.: Unit Disk Graphs. Ann. Discret. Math. **48**, 165–177 (1991)
7. Hurley, S., Smith, D.H., Thiel, S.U.: FAsoft: A system for discrete channel frequency assignment. Radio Science **32**(5), 1921–1939 (1998)
8. Balasundaram, B., Butenko, S.: Graph domination, coloring and cliques in telecommunications. In: Handbook of Optimization in Telecommunication (2006)

9.  Alzoubi, K.M., Wan, P.-J., Frieder, O.: Weakly-connected dominating sets and sparse spanners in wireless ad hoc networks. In: Proceedings of ICDCS, pp. 96–104 (2003)
10. Luna, F., et al.: Optimization algorithms for large-scale real-world instances of the frequency assignment problem. Soft. Comput. **15**, 975–990 (2011)
11. Björklund, P., Värbrand, P., Yuan, D.: Optimized planning of frequency hopping in cellular networks. J. Computers and Operations Research **32**(1), 169–186 (2005)
12. Zhang, Y.-B., Zhao, Y.-C., Xiong, H.: A tabu search algorithm for frequency assignment problem in wireless communication networks. In: Proceedings of Wicom 2009, pp. 2848–2851 (2009)
13. Parsapoor, M., Bilstrup, U.: Ant colony optimization for channel assignment problem in a clustered mobile ad hoc network. In: Tan, Y., Shi, Y., Mo, H. (eds.) ICSI 2013, Part I. LNCS, vol. 7928, pp. 314–322. Springer, Heidelberg (2013)
14. Alba, E.: Parallel Metaheuristics: A New Class of Algorithms. John Wiley & Sons, Inc. (2005)
15. Vidyarthi, G., Ngom, A., Stojmenovic, I.: A hybrid channel assignment approach using an efficient evolutionary strategy in wireless mobile networks. IEEE Trans. Veh. Technol. **54**, 1887–1895 (2005)
16. Colombo, G.: A genetic algorithm for frequency assignment with problem decomposition. J. of Mobile Network Design and Innovative Archive **1**(2), 102–112 (2006)
17. Luna, F., Nebro, A.J., Alba, E.: Durillo, J-J.: Large-Scale Real-World Telecommunication Problems Using a Grid-Based Genetic Algorithm. Engineering Optimization **40**(11), 1067–1084 (2008)
18. Ngo, C.Y., Li, V.O.K.: Fixed channel assignment in cellular radio networks using a modified genetic algorithm. IEEE Trans. Veh. Technol. **47**, 163–172 (1998)
19. Maximiano, M.-D., Vega-Rodriguez, M.A., Gomez Pilido, J.A., Sánchez-Pérez, J.M.: A hybrid differential evolution algorithm to solve a real-world frequency assignment problem. In: Proc. of International Multiconference on Computer Science and Information Technology, Wisia, pp. 201–205 (2008)
20. Elhachmi, J., Guenoun, Z.: Distributed Frequency Assignment Using Hierarchical Cooperative Multi-Agent System. Int. J. Communications, Network and System Sciences **4**, 727–734 (2011)
21. Funabiki, N., Okutani, N., Nis, S.: A Three-stage Heuristic Combined Neural Network Algorithm for Channel Assignment in Cellular Mobile Systems. IEEE Trans. Veh. Technol. **9**(2), 397–403 (2000)
22. Nemhauser, G., Wolsey, L.: Integer and Combinatorial Optimization. J. Wiley & Sons, Inc. (2014)
23. Aardal, K., van Hoesel, S.P.M., Koster, A.M.C.A., Mannino, C., Sassano, A.: Models and solution techniques for frequency assignment problems. Quarterly Journal of the Belgian, French and Italian Operations Research Societies **1**(4), 261–317 (2003)
24. Das, A.K., et al.: Optimization models for fixed channel assignment in wireless mesh networks with multiple radios. In: Proceedings of IEE SECON, pp. 463–474 (2005)

# Monitoring Lane Formation of Pedestrians: Emergence and Entropy

Jan Procházka and Kamila Olševičová[✉]

University of Hradec Králové, Rokitanského 62, 500 03 Hradec Králové,
Czech Republic
{jan.prochazka,kamila.olsevicova}@uhk.cz

**Abstract.** This paper deals with self-organization phenomenon in qualitative microscopic pedestrian simulation. The agent-based pedestrian model in NetLogo is presented. Within the model, the lane formation is identified as the emerging pattern growing from counter flows of individuals. Information entropy is applied in analytical component of the model with the aim to measure the level of self-organization. Experimental results are provided.

**Keywords:** Pedestrian simulation · Self-organization · Information entropy · Lane formation · Multi-agent systems · NetLogo

## 1 Introduction

Pedestrian and crowd motion models help us to identify and analyze spatial walking patterns under normal or competitive situations and to reproduce empirically observed crowd features. Moreover, similarly to models of fish schools or ant colonies, pedestrian simulations can serve as an experimental area for studying self-organization phenomena within computational methods. Being inspired by [3, 4, 6, 7, 9, 11] our idea was to apply information entropy to measure the level of self-organization. To achieve this we built a pedestrian model with analytical component for entropy calculation.

According the level of abstraction, there are three classes of pedestrian and crowd models: *microscopic models* which describe each pedestrian as a unique entity, *macroscopic models* which aggregate pedestrian dynamics by flows or densities and *mesoscopic models* which operate with velocity distributions. Theoretically, pedestrian and crowd simulations can build on *agent-based approach* where individuals are seen as autonomous, rational, adaptive entities. In *physics-based models* such as social-force models or fluid dynamics models, pedestrians are seen as particles which are under the pressure of external attractive and repulsive forces. In *cellular automata models*, main concept is the walkable space which is represented by the lattice; its each cell is occupied by nobody or one pedestrian and in each time step pedestrians move to unoccupied neighboring cells. In *queuing models* the space is reduced into the network of nodes and links, with pedestrians moving around it, but without details of dynamics inside nodes. For further details on pedestrian modelling principles, see e.g. [1, 2, 5, 8].

© Springer International Publishing Switzerland 2015
N.T. Nguyen et al. (Eds.): ACIIDS 2015, Part I, LNAI 9011, pp. 221–228, 2015.
DOI: 10.1007/978-3-319-15702-3_22

For our purpose, which is to measure the level of self-organization, agent-based microscopic simulations are relevant. From wide range of possible movement scenarios (for overview see e.g. [2]), *counter flow* was chosen to be simulated and one emerging pattern (*lane formation*) was observed. In following sections, our model is described, the encoding of lane formation is explained, application of entropy calculation is suggested and experimental results are provided.

## 2     Model Description

The model was implemented in NetLogo 5.1.0 [12], see Fig.1. Pedestrians are understood as autonomous, strictly decentralized, locally operating agents situated in the grid of 40x40 cells. The initial population of 200 agents is divided into two counter flows: 60 agents move from left to right (heading = 90), 140 agents move from right to left (i.e. heading = 180). The lane formation of minority (60 agents) is observed. Behavior of agents is defined by three rules corresponding to the principles of Reynold's Boid model [10]:

— Rule 1: to avoid direct collisions with other agents,
— Rule 2: to move to less occupied neighbourhood,
— Rule 3: to keep original heading.

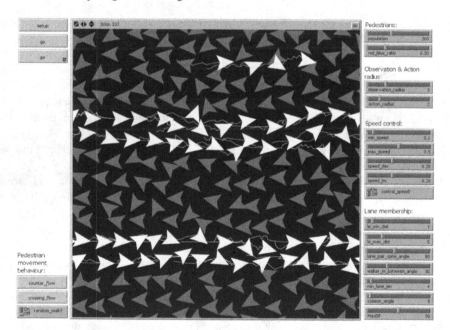

**Fig. 1.** Model interface with visualization of pedestrian lanes

Agent's observation radius $r_o$ is 3 cells, action radius $r_a$ is 2 cells around. Agent's minimum speed is 0.1 cell per time step, maximum speed is 0.5 cell per time step.

The speed increase (0.2 cell per time step) is applied in case the current speed was not maximal and the cell ahead was not occupied. The speed decrease (0.2 cell per time step) is applied in case the cell ahead was occupied by other agent or obstacle.

The lane formation phenomenon is characterized by several attributes. Formal specification of the lane membership is important for further analysis of this emerging structure growing from the counter flow. The lane is formed by individual agents who follow each other in queue that can be clearly distinguished. For two agents in the lane, minimum distance is 1 cell, maximum distance is 5 cells. The lane element is defined by a triplet of agents with the same heading, who meet following constraints:

— *closeness* – distance from the middle agent to outer agents of the triplet is within the defined interval $<lane\_radius_{min}, lane\_radius_{max}>$,
— *straightness* – angle between two links from the middle agent to the outer agents of the triplet is within the defined interval $<lane\_angle_{min}, \pi>$,
— *clearness* – there are no more agents between the middle and both outers agent of the triplet.

See Fig. 2 for example of how the lane membership of the agent *P1* is recognized.

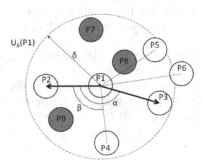

**Fig. 2.** Example of lane triplet defined by agents *P1*, *P2*, *P3*

The grid of cells is represented by the matrix. Each agent updates those matrix elements that correspond to the agent's current position and observation radius. Initially, all values in the matrix are set to 0. Then values are updated in each time step of simulation. Each agent updates values of cells inside its observation radius $r_o$. The value $V_{m,n}$ for the cell $[m, n]$ is

$$V_{m,n} = \sum_{\{P_i:|C_{m,n}P_i|\leq r_o\}} [(r_o - |P_i|)]$$

where the sum goes over all agents *Pi* in maximal distance of observation radius $r_o$ from the cell $[m, n]$. The matrix captures presence of agents within the grid and the mutual influence of agents: bigger value indicates higher number of agents nearby the cell (Fig. 3).

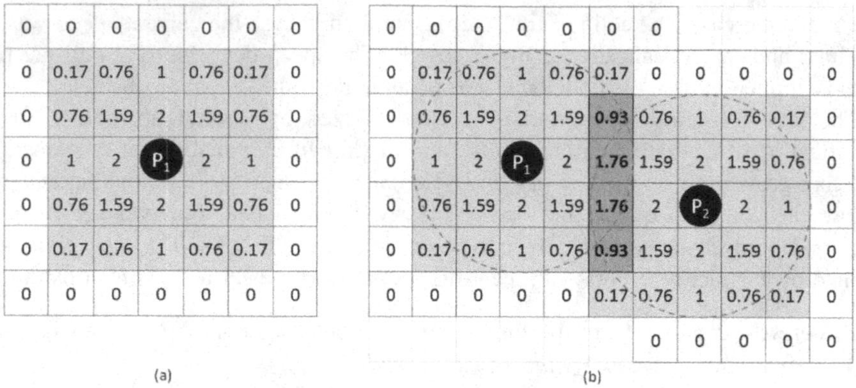

**Matrix (a):**

| 0 | 0 | 0 | 0 | 0 | 0 | 0 |
|---|---|---|---|---|---|---|
| 0 | 0.17 | 0.76 | 1 | 0.76 | 0.17 | 0 |
| 0 | 0.76 | 1.59 | 2 | 1.59 | 0.76 | 0 |
| 0 | 1 | 2 | P1 | 2 | 1 | 0 |
| 0 | 0.76 | 1.59 | 2 | 1.59 | 0.76 | 0 |
| 0 | 0.17 | 0.76 | 1 | 0.76 | 0.17 | 0 |
| 0 | 0 | 0 | 0 | 0 | 0 | 0 |

(a)

**Matrix (b):**

| 0 | 0 | 0 | 0 | 0 | 0 | 0 | | | | |
|---|---|---|---|---|---|---|---|---|---|---|
| 0 | 0.17 | 0.76 | 1 | 0.76 | 0.17 | 0 | 0 | 0 | 0 | 0 |
| 0 | 0.76 | 1.59 | 2 | 1.59 | 0.93 | 0.76 | 1 | 0.76 | 0.17 | 0 |
| 0 | 1 | 2 | P1 | 2 | 1.76 | 1.59 | 2 | 1.59 | 0.76 | 0 |
| 0 | 0.76 | 1.59 | 2 | 1.59 | 1.76 | 2 | P2 | 2 | 1 | 0 |
| 0 | 0.17 | 0.76 | 1 | 0.76 | 0.93 | 1.59 | 2 | 1.59 | 0.76 | 0 |
| 0 | 0 | 0 | 0 | 0 | 0.17 | 0.76 | 1 | 0.76 | 0.17 | 0 |
| | | | | | | 0 | 0 | 0 | 0 | 0 |

(b)

**Fig. 3.** Part of the matrix for agent P1 with $r_o = 3$ (left) and composition of values for proximate agents $P_1$ and $P_2$ (right)

To be able to choose the best next move, agents create their private updates of the matrix. Only those cells which are in action radius $r_a$ are updated. The action radius represents agent's decision area and it is smaller than observation radius $r_o$. The updating formula for matrix values in agent's action radius is

$$V_{m,n}^{Updated} = V_{m,n} - \log_{10}\left(MaxDif - |Heading_{desired} - Heading_{V_{m,n}}|\right)$$

where *MaxDif* is 90 degree, $Heading_{desired}$ is the target heading of the agent (90 or 180 for movement from left to right, or from right to left respectively) and $Heading_{V_{m,n}}$ is agent's current heading towards the cell *[m, n]*. Once individual update is calculated, agent chooses the best move to the less occupied cell, i.e. the cell with the minimum value.

## 3    Encoding Information on Lane Formation

The model is used for experimenting with the lane formation rising from the counter flow. To be able to quantify this emerging pattern using Shannon entropy, the information measure has to be specified. This measure has to be strongly related to the observed pattern, i.e. to distinguish between different configurations of agents in the grid with different numbers of lane-members. The measure is based on Hartley's information [4, 11] which applies five rules:

— Rule 1 – Messages are strings of characters from a fixed alphabet.
— Rule 2 – The amount of information contained in a message is a function of the total number of possible messages.
— Rule 3 – Alphabet with $s$ symbols produces $s^l$ possible messages of length $l$.
— Rule 4 – The amount of information contained in two messages is the sum of the information contained in both individual messages.

— Rule 5 – The amount of information in $l$ messages of the length one equals to the amount of information included in one message of length $l$.

These rules are applicable on our multi-agent pedestrian model. Then positions of agents in the grid of cells are fully expressed by one particular message.

For our purpose, the alphabet is the set of integers $\{1, 2, \dots, n\}$. Each value stands for the corresponding number of lane-members in particular segment of the grid. See Fig. 4 for example of a message for grid which is divided into 16 squares.

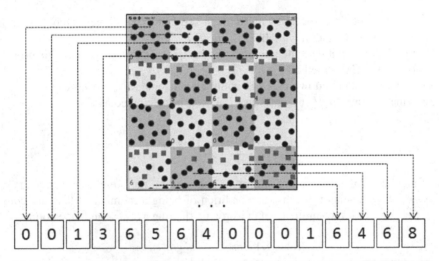

**Fig. 4.** Encoding of grid state: lane members are indicated by squares, other agents are indicated by circles, string values indicate numbers of lane members

## 4     Entropy of Pedestrian System

Experiment with information entropy measure consisted of two levels:

— Level 1 – preparation of   probabilities of grid states,
— Level 2 – simulation run and information entropy measurement.

Monte Carlo method was used at level 1 for calculation of probabilities of grid states, i.e. probabilities of appearance of each alphabet character (the integer number) within the encoded state message. Agents moved randomly throughout the grid, number of agents in 100 segments of 4x4 cells were monitored periodically. The probabilities were calculated as relative frequencies of numbers of agents and numbers of lane members within the grid at specific time points. Results for 100 segments of 4x4 cells areas are summarized in Tab. 1: Columns show frequencies of states occurrences: *Agents* column – all when counting agents without restrictions (all pedestrians are counted in the given grid segment), *Lane members* – when counting only pedestrians having the lane membership property within the given grid segment. The probabilities columns are relative frequencies calculated from state frequency and number of all observations. Probability $p_A$ relates to the segment state given by all pedestrians count, probability $p_{LM}$ relates to segment state given by only pedestrians having lane-member property.

During level 2 the simulation is run. Agents' behaviour swaps periodically between random walk and organized counter flow. This swapping enables us to observe differences in information entropy measurement for the two movement behaviours. Calculation of the information entropy of the agent system is based on the following assumptions:

— Count of agents in respective environment segments (with lane membership property) is a discrete random variable $X$ from $\{0,1,...,n\}$.
— The probability that the outcome of $X$ will be $x_i$ is $p_X(xi)$. We evaluate this probability by Monte-Carlo method.
— Information contained in a message about the outcome $X$ is $- \log p_X(x_i)$
— Information entropy $H_X$ of a message then can be calculated as:

$$H_X = - \sum_{i=1}^{n} p_X(x_i) \log p_X(x_i)$$

The calculation of information entropy is done during the simulation run every $10^{th}$ time step. Only agents fulfilling the condition of being a member of lane are considered. See Fig. 5 for the summary of entropy, its floating average and lane counts.

**Table 1.** Probabilities of grid states

| Number of agents in 4x4 cells area | Frequency | | Probabilities | |
|---|---|---|---|---|
| | All agents | Lane members | $p_A$ | $p_{LM}$ |
| 0 | 1093910 | 1674956 | 0.5469093 | 0.8374081 |
| 1 | 663932 | 254040 | 0.3319383 | 0.1270094 |
| 2 | 197750 | 59908 | 0.0988667 | 0.0299515 |
| 3 | 38383 | 9880 | 0.0191899 | 0.0049396 |
| 4 | 5511 | 1245 | 0.0027553 | 0.0006224 |
| 5 | 620 | 132 | 0.0003100 | 0.0000660 |
| 6 | 51 | 5 | 0.0000255 | 0.0000025 |
| 7 | 8 | 1 | 0.0000040 | 0.0000005 |
| 8 | 1 | - | 0.0000005 | - |
| 9 | 1 | - | 0.0000005 | - |
| 10 | - | - | - | - |
| Sum: | 2000167 | 2000167 | 1 | 1 |

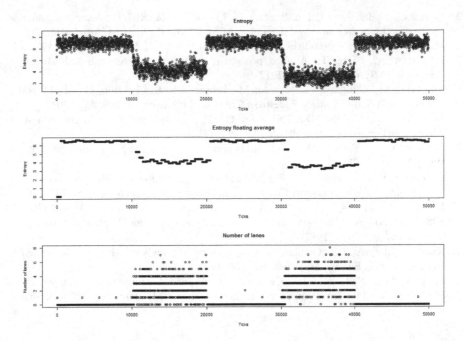

**Fig. 5.** Entropy, entropy floating average and lanes counts: random walk and organized counter flow swap after each 10.000 ticks elapsed

## 5    Conclusion

The level of self-organization of multi-agent system can be assessed using information entropy. This approach was demonstrated on counter flow pedestrian model in which our measure clearly identified the emergence of lanes. We intend to apply our method to recognize other types of crowd behaviour in pedestrian models, and more broadly to use it for assessment of self-organization in different kinds multi-agent systems.

**Acknowledgement.** This paper was supported by the University of Hradec Kralove, Faculty of Informatics and Management.

## References

1. Dai, J., Li, X., Liu, L.: Simulation of pedestrian counter flow through bottlenecks by using an agent-based model. Physica A: Statistical Mechanics and Its Applications **392**(9), 2202–2211 (2013)
2. Duives, D.C., Daamen, W., Hoogendoorn, S.P.: State-of-the-art crowd motion simulation models. Transportation Research Part C: Emerging Technologies **37**, 193–209 (2013)

3. Guy, S.J., van den Berg, J., Liu, W., Lau, R., Lin, M.C., Manocha, D.: A statistical similarity measure for aggregate crowd dynamics. ACM Trans. Graph. **31**(6), Article 190 (2012)
4. Hartley, R.V.L.: Transmission of Information. Bell Syst. Tech. J. **7**(3), 535–561 (1928)
5. Johansson, A., Kretz, T.: Applied pedestrian modelling. In: Agent-Based Models of Geographical Systems. Springer (2012)
6. Moussaïd, M., Guillot, E.G., Moreau, M., Fehrenbach, J., Chabiron, O., et al.: Traffic Instabilities in Self-Organized Pedestrian Crowds. PLoS. Comput. Biol. **8**(3) (2012)
7. Moussaïd, M., Helbing, D., Theraulaz, G.: How simple rules determine pedestrian behavior and crowd disasters. PNAS **108**(7) (2011)
8. O'Sullivan, D., Perry, G.L.W.: Spatial Simulation: Exploring Pattern and Process. Wiley-Blackwell (2013)
9. Parunak, H.V.D., Brueckner, S.: Entropy and self-organization in multi-agent systems. In: Proceedings of the International Conference on Autonomous Agents, pp. 124–130 (2001)
10. Reynolds, C.: Flocks, herds and schools: a distributed behavioral model. In: SIGGRAPH 1987: Proceedings of the 14th Annual Conference on Computer Graphics and Interactive Techniques, pp. 25–34 (1987)
11. Rioul, O., Magossi, J.C.: On Shannon's formula and Hartley's rule: beyond the mathematical coincidence. Entropy **16**, 4892–4910 (2014)
12. Wilensky, U.: NetLogo 5.1.0. Center for Connected Learning and Computer-Based Modeling. Northwestern University, Evanston, IL (2014). http://ccl.northwestern.edu/netlogo/

# Comparison of Algorithms for Multi-agent Pathfinding in Crowded Environment

Mariusz Hudziak, Iwona Pozniak-Koszalka,
Leszek Koszalka$^{(\boxtimes)}$, and Andrzej Kasprzak

Department of Systems and Computer Networks,
Wrocław University of Technology,
Wrocław, Poland
mariuszhudziak@gmail.com, {iwona.pozniak-koszalka,
leszek.koszalka,andrzej.kasprzak}@pwr.edu.pl

**Abstract.** The objective of this paper is to recommend the algorithms for planning the best paths for the simultaneously moving agents operated in the large crowded environment. The proposed approach consists of two parts. Firstly, a navigation mesh for passable regions in rectangular 2D environment is created using Quad-trees algorithm. A graph is created by connecting centers of regions, where weights in the graph are Euclidean distances between centers. In the second part, a path is found for each agent currently present in environment using Dijkstra or A* algorithm. To plan good path in each passable region actual density value is stored. Density information is further mapped on graph edges along with distance value. Agents reevaluate their paths accordingly to re-planning strategy. Three strategies are considered: periodical re-planning, periodical with initial re-planning and event-driven re-planning proposed by the authors. The six combinations of algorithms have been implemented and tested. Simulation experiments made using the created experimentation system showed that the approach with the own event-driven re-planning seems to be promising.

**Keywords:** Pathfinding · Multi-agent · Algorithm · Experimentation system · Simulation

## 1    Introduction

In virtual environment agents need to plan a good path from start point to the destination (goal) point. Usually, finding the shortest path is a simple and not time consuming task. However, in the crowded environment where many agents are moving - the traffic jams may occur, when same routes are taken by many agents. As it can be seen in Figure 1, in crowded environment such a solution is unfavorable. In this example agents take only the shortest route and congestion occurs. When making simulation experiments or playing games, with many agents present at work, while real-time solutions are needed, then the time is a very important index of performance. Also, when the planned path should avoid static obstacles and crowded regions of the environment, the solution to the pathfinding problem is not a simple task. In such

© Springer International Publishing Switzerland 2015
N.T. Nguyen et al. (Eds.): ACIIDS 2015, Part I, LNAI 9011, pp. 229–238, 2015.
DOI: 10.1007/978-3-319-15702-3_23

cases a more efficient agent's movement is desired. A sample solution with the diversified paths created by agents is illustrated in Figure 2 – it leads to more efficient agent's movement.

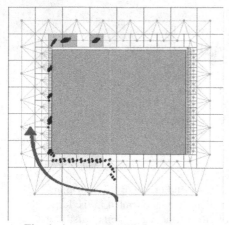

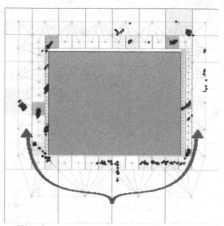

Fig. 1. Agents move along the shortest path causing congestion.

Fig. 2. Agents move on density aware path leading to lower congestion.

The aim of this paper is to solve the problem of finding a path for many agents in the crowded environment with obstacles - focusing on avoiding congested regions. We are looking not for the shortest path but the fastest path. The proposed solution is based on selecting the algorithms and combining them into whole solution (the complex algorithm). The two notions are important: the navigation mesh and the density. The navigation mesh is created from a given environment. The main idea is to represent passable regions of the environment by simpler representation, which can be further used for graph generation. The density of the crowd is measured as the number of the agents per square meter [1]. Higher the crowd density slower agents move through environment [2]. The agents are planning their paths avoiding crowded regions. It is done by applying the density information, which is computed as the overall size of agents in a given area divided by the total area size. When new agent enters the region, then the density value is increased. When agent leaves region, then the density is decreased. The actual density information in each region is stored, and next is mapped on the edge of the graph which goes into region – that way the weights within the graph are changing in time. New crowded regions could have emerged. The re-planning strategies are used to adapt to the changing environment. The overall solution leads to better utilization of different routes and faster agents speed.

The considered problem is represented in literature. Navigation mesh can be created in many different ways. The first approach consists in using the Quad-trees algorithm [3]. An approach called Probabilistic Roadmap Method (PRM) [4] uses some probabilistic distribution to set vertices on passable area and then connects them to form a graph. In this paper the first approach is applied, because PRM does not divide passable area into regions, which are needed to store density information. In the second part of the proposed solution, a pathfinding algorithm is used. The most known pathfinding algorithms are: Dijkstra algorithm described in [4] and A* approach presented in [5]. Moreover, different modifications of these algorithms exist, e.g., HPA* algorithm [6]. We implement the classic versions of these algorithms. First of all, we focus on the information about the density of changing environment. Interesting approach with the usage of density information in crowded environment is presented in [1]. It is based on generating a navigation mesh with *medial axis* and *Explicit Corridor Map* data structure [7], what allows for steering agents through environment using periodical re-planning strategies. In this paper, the two known re-planning strategies are considered, and a new strategy called event-driven re-planning is proposed. To evaluate the considered combinations of algorithms, the experimentation system consisted of three modules: Drawing, Navigation Mesh generation, and Multi-agent pathfinding simulation was designed. The first module is responsible for the creation of an environment with obstacles. In the second module, the navigation mesh and graph are generated. The third module allows observing the real-time simulation of moving agents and collecting data needed to calculate the introduced indices of performance.

The rest of the paper is organized as follows. In Section 2, the considered problem of pathfinding is formulated. Section 3 is devoted to presentation of the algorithms for solving the problem, in particular event-driven re-planning strategy. Section 4 contains the comparative analysis of quality of the considered algorithms made on the basis of simulation experiments. The conclusion and future plans appear in Section 5.

## 2    Problem Statement

The problem of multi-agent pathfinding is formulated as follows:

**Given**: 2D rectangular area (environment) transformed to a graph $G$ with vertices $V$ and edges $E$. Start and goal points: $v\_\{0\}$ and $v\_\{n\}$. The number of $N$ agents denoted by $a = 1, 2, ..., N$. Maximal agent speed $MAS$ and density $DEN$ in a given region at some point of time.

**To find**: The path $P$ from $v\_\{0\}$ to $v\_\{n\}$ in $G$ defined as a set of vertices being a subset of $V$. The expression (1) defines a length of path $P$ for agent $a$:

$$Pdist_a = \sum dist\left(v_i, v_{i+1}\right) \tag{1}$$

where $dist(v_i, v_{i-1})$ is the Euclidean distance between two vertices; $v_{num}$ is a number of vertices in a path. The formula (2) defines average speed of agent on a whole route.

$$avrspeed_a = \frac{\sum curspeed\left(a, t\right)}{g - s} \tag{2}$$

where $curspeed(a, t)$ is the current speed of the agent $a$ defined by the formula (3); $g$-$s$ is the number of moves made by agent $a$ through environment.

$$curspeed(a, t) = (1 - DEN) * MAS \tag{3}$$

where the density $DEN$ is the value dependent on the total number of agents in a current region in which agent $a$ is present at the time $t$, and the size of a region; $MAS$ is the maximal speed of the agent $a$. Remark: Higher the density $DEN$ in region on which agent is currently present at a given time, lower the agent speed.

**Such that**: The average agents travelling time defined by (4) is minimal:

$$TrvT = \frac{\sum \dfrac{Pdist_a}{avrspeed_a}}{N}$$

(4)

where $Pdist_a$ is the path length of $a$-th agent.

**Subject to the constraints**. The set of the static (not changing position in time) rectangular obstacles has to be taken into consideration.

# 3    Algorithms

**Navigation Meshes and Graphs Generation.** The considered input area (environment) is a rectangle with a set of impassable rectangular obstacles. This area is split into smaller passable regions using a modified Quad-trees algorithm. The modification leads to the creation of more paths, but gives more possibilities to find more diverse paths by pathfinding algorithms. After splitting the centers of regions are connected using the nearest k-neighbors strategy described in [8]. If connection leads through obstacle it is discarded. In such a way graph is created with vertices as centers and edges as connections. The weights on the edges are taken as Euclidean distances between centers. The implemented Quad-trees algorithm in pseudo-code is as follows:

```
function QUADTREE(Region R, Obstacles O, Out Out)
    if R.size ≤ MinWidthHeight then
        R.passable = False
        Out.insert(R)
    end if
    Regions ← SPLIT(R)
    for all reg in Regions do
        Split = False
        for all ob in O do
            if reg ∈ O then
                Split = True
            end if
        end for
        if Split = True then QUADTREE(reg, O, Out)
        else
            reg.passable = True
            Out.insert(reg)
        end if
    end for
end function
```

In the beginning the input area is treated as the first parameter $R$. Set of obstacles $O$ is provided as the second parameter. As an output of $QUADTREE$ function, arrays $Out$ of passable regions are returned. The $SPLIT$ function (as an output) provides four regions of a half width and a half height of an input region.

**Pathfinding.** In the next step, pathfinding algorithms are used to find a path for agents. Two algorithms can be used: Dijkstra algorithm for finding the shortest path in graph without negative weights, and a version of A* algorithm with Euclidean distance to the goal. Dijkstra or A* seeks for the shortest path from start to goal points without using density information. Paths are further stored and used by every new agent appearing in the region. These paths are reevaluated accordingly to re-planning strategy.

**Path Re-planning.** The shortest route found by pathfinding algorithms and taken by agents could lead to congestion in regions. Because agents decrease their speed when entering the congested region, these regions should be avoided. It can be done with one of the following re-planning strategies:

- Periodical re-planning,
- Periodical with the initial re-planning,
- Event-driven re-planning (proposed by the authors).

**Periodical Re-planning.** When re-planning strategy emits a signal that an agent should replan its path, then the used (A* or Dijkstra) algorithm is finding a new path with a new start point (a current position of an agent). Moreover, the current densities in regions are used to steer agents through environment. However, periodical re-planning strategy can replan paths after some time interval. The PERIODICAL function, for re-planning a single agent path depending on algorithm iteration, is presented in pseudo-code below:

```
function PERIODICAL(Iteration I, Agents A)
    for all agent in A do
        if agent.ID = I mod A.Size then
            UPDATEPATH(agent)
        end if
    end for
end function
```

**Periodical with Initial Re-planning**. This strategy is a little modification of the previous strategy. Each agent is re-planning its path on the start point. Such approach gives more diverse paths of agents from the beginning, because of density information. The previous assumptions about input parameters of the function also apply to PERIODICALINITIAL function. Pseudo-code of this strategy is presented below.

```
function PERIODICALINITIAL(Iteration I, Agents A)
    for all agent in A do
        if agent.ID = I mod A.Size then
            UPDATEPATH(agent)
        else if agent.IsNewAgent = True then
            UPDATEPATH(agent)
        end if
    end for
end function
```

The two above strategies are relatively simple and they do not cause remarkable time consuming. However, these strategies signal re-planning of agents paths rather randomly. The main stress in these strategies has to be put on the number of agents which replan their paths in each run of re-planning. The aim is to balance computational time of overall algorithm which highly depends on pathfinding algorithms and quality of the founded paths.

**Event-Driven Strategy**. This strategy is based on the specific events that may occur in the environment. The implementation of the created approach is presented in pseudo-code. The function EVENTDRIVEN implements the proposed idea. Firstly, if a region in an environment has the density higher than a given threshold $Th$, then it is marked as the dense. Secondly, if a path of a given agent leads through the dense region and agent is closer than $MinDist$ to the region center, then its path is updated.

```
function EVENTDRIVEN(Regions R, Agents A)
    for all r in R do
        if r.Density ≥ Th then
            for all a in A do
                isOnPath = ISREGIONONPATH(a, r)
                if isOnPath = True then
                    dist = DISTANCETOREGION(a, r)
                    if dist ≤ MinDist then
                        UPDATEPATH(a)
                    end if
                end if
            end for
        end if
    end for
end function
```

The distance to the dense region is limited because there is no need to replan path of the distant agents. Region could no longer be dense when agent approaches it after several iterations. Function IsRegionOnPath($a$, $r$) checks whether a path of a given current agent $a$ leads through a given region $r$. Function DistanceToRegion($a$, $r$) computes the distance from a position of a given agent $a$ to the center of the region $r$.

## 4    Investigation

**Experiments Design**. In experiments the number of 12 000 agents was put in the environment. Three agents, one for each pair of start and end points, were considered in any iteration. The following combinations of the algorithms were tested:

- D - Dijkstra algorithm without path re-planning strategy as benchmark comparison (reference),
- D/Per - Dijkstra algorithm with periodical path re-planning strategy,
- D/PerInit - Dijkstra algorithm with periodical initial path re-planning strategy,
- D/Event - Dijkstra algorithm with event-driven re-planning strategy,
- A* - A* algorithm without path re-planning strategy as benchmark comparison (reference),
- A*/Per - A* algorithm with periodical path re-planning strategy,
- A*/PerInit - A* algorithm with periodical initial path re-planning strategy,
- A*/Event - A* algorithm with event-driven re-planning strategy.

Remark: In every combination Quad-trees algorithm was applied. In each case, the following indices of performance were taken into consideration:

- avCT – the average computation time of single algorithm iteration,

- pthR – the amount of runs of pathfinding algorithms (Dijkstra or A*),
- maxA – the maximal simultaneous agents number,
- AvrA –the average simultaneous agents number (for more than 1500 agents),
- TrvTMs – the average travelling time of agents (in [$ms$]) for three pairs of start and end points.

In order to properly simulate the behavior of agents in the environment, the important assumption was made - the density in a region should not lower the speed of agents to zero, because such a situation could lead to stalling and some agents could never reach their goals. Therefore, the overcrowded places were treated as the obstacles that should be avoided. The created big environment was the rectangle of the size 1920x1080 pixels with rectangular obstacles (see Figure 3).

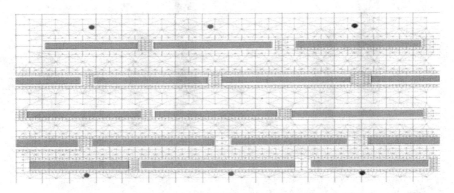

**Fig. 3.** Big environment of the size 1920x1080 pixels

In the figure, the dots represent start and goal points which were used as the pairs for each group of agents.

**Comparison of Algorithms**. In Table 1 the obtained results are presented.

**Table 1.** Indices of performance for the considered algorithms.

| Algorithm | avCT | TrvTMs | AvrA | pthR | maxA |
|-----------|------|--------|------|------|------|
| D | 0.570 | 6417.01 | 8102 | 3 | 11949 |
| D/Per | 28.813 | 4141.28 | 6804 | 60147 | 10700 |
| D/PerInit | 18.999 | 2985.22 | 5455 | 29571 | 8520 |
| D/Event | 37.674 | 2942.56 | 5355 | 55737 | 8527 |
| A* | 0.607 | 6417.01 | 8102 | 3 | 11949 |
| A*/Per | 26.340 | 4190.72 | 6743 | 68256 | 10632 |
| A*/PerInit | 16.019 | 2828.88 | 4957 | 30953 | 7920 |
| A*/Event | 35.499 | 3347.69 | 5690 | 58231 | 9063 |

The average travelling time was the best in case of using Dijkstra algorithm with the event re-planning. However, for A* algorithm slightly better was periodical re-planning. Explanation for such a situation could be that A* is a heuristic algorithm. Event re-planning strategy informs agents when they are close to the dense region. In some cases it could lead to strange behavior of the agent like going back using the same route - such a situation was sometimes observed during simulation. Taking into consideration the computational time it can be seen that for both pathfinding algorithms the periodical initial re-planning performed as the best. In A* case about 16 [*ms*] for iteration was needed. If 100 [*ms*] was available for the frame of 10 fps, then 84 [*ms*] was left for other tasks for CPU like graphics rendering or AI computation. It can be seen that combining Dijkstra with periodical initial re-planning gives ability to steer about 5000 agents simultaneously in real-time.

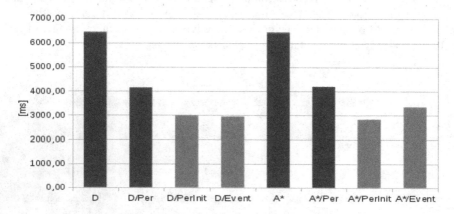

**Fig. 4.** Travelling time of agents in big environment.

It may be observed in Figure 4 that from the travelling time of agent's point of view (the most important index of performance) the best were event-driven re-planning and periodical with initial re-planning.

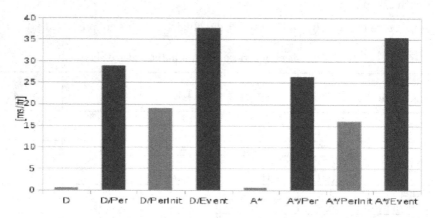

**Fig. 5.** Computational time of agents in big environment.

From the other side, when observing in Figure 5 the results obtained for computation of single iteration for algorithms, we see that event-driven re-planning needs more computational time than other re-planning algorithms. Thus, the event-driven re-planning is efficient but time consuming. When time of travelling and computational time have the same importance for the user then periodical with initial re-planning may be considered as the best re-planning techniques.

## 5    Conclusion and Future Plans

In this paper, the proposed solution to the multi-agent pathfinding problem was presented. It consists of a few steps: creating navigation mesh, generating graph, finding a path and finally re-planning agent's path. Agents use the density information to replan their paths. The overall solution gives good results in environment where many agents are moving simultaneously. Moreover, the real-time performance can be achieved. Re-planning strategies were combined with two pathfinding algorithms: A* and Dijkstra. Combination with A* algorithm gives slightly better computational time and comparable travelling time of agents.

The best results were obtained by periodical with initial re-planning strategy. Re-planning strategy can be also tuned to lower computation time or decrease travelling time of agents. It can be done by changing number of pathfinding algorithm runs. Higher its number lower could be travelling time, lower its number lower the computation time of the algorithm. Such a tuning could be useful in games with high fps number. For 25 fps the time of 40 [ms] is an upper limit for pathfinding, rendering graphics and other tasks, so resources for pathfinding task are limited. By worsening periodical with initial re-planning strategy 32 [ms] are left for other tasks. Also usage of other CPU cores could give good speed up of the overall algorithm.

Event-driven re-planning strategy designed by the authors gives very good travelling time of agents. Main drawback is computational time which is greater than for the known re-planning algorithms. However, in authors view event-driven re-planning can give possibility to further improvement.

The future research in this area should concentrate on two aspects: (i) It was observed that agents tend to create groups. When such a group is created congestion likely occurs. To avoid this situation the algorithm could recognize such a group and replan agent's path with some randomness. Such an action may give lower number of groups and speed up travelling time of the agents.

As the solution a combination of event-driven re-planning and periodical with initial re-planning can be thought; (ii) One way to increase performance of multi-agent pathfinding algorithm is to use higher number of pathfinding algorithm runs. To lower computation overhead of A*, its modification, i.e., HPA* algorithm can be used. It was proven to be up to 10 times faster than ordinary A* [6]. The combination of periodical with initial re-planning with HPA* could give very efficient overall solution, either.

Moreover, the experimentation system could be extended and improved by preparing and implementing new modules of the system to ensure designing multistage experiments [9] and processing output data [10] in the automatic way.

**Acknowledgement.** This work was supported by the statutory funds of the Department of Systems and Computer Networks, Faculty of Electronics, Wroclaw University of Technology, Wroclaw, Poland.

# References

1. Van Toll, W.G., Cook, A.F., Geraerts, R.: Real-time density-based crowd simulation. Computer Animation and Virtual Worlds Archive **23**(1), 59–69 (2012)
2. Daamen, W.: Modelling passenger flows in public transport facilities. PhD thesis T2004/6, Delft University of Technology (2004)
3. Hale, D.H.: A growth-based approach to the automatic generation of navigation meshes. Technical report, The University of North Carolina at Charlotte (2011)
4. Kavraki, L.E., et al.: Probabilistic roadmaps for path planning in high-dimensional configuration spaces. IEEE Transactions on Robotics and Automation **12**(4), 566–580 (1996)
5. Russell, S., Norvig, P.: Artificial Intelligence: A Modern Approach, 2nd edn., pp. 97–104. Prentice Hall (2003). ISBN 978-0137903955
6. Botea, A., Muller, M., Schaeffer, J.: Near optimal hierarchical path-finding. Report, Department of Computer Science, University Alberta (2004)
7. Geraerts, R.: Planning short paths with clearance using explicit corridors. In: Proc. to IEEE ICRA International Conference on Robotics and Automation (2010)
8. Geraerts, R.: Sampling-based motion planning: analysis and path quality. PhD thesis, The Utrecht University, The Netherlands (2006)
9. Kaminski, R.T., Koszalka, L., Pozniak-Koszalka, I., Kasprzak, A.: Evaluation and comparison of task allocation algorithms for mesh networks. In: Proc. 9th International Conference on Networks, pp. 104–108. IEEE Computer Society Press (2010)
10. Kmiecik, W., Wojcikowski, M., Koszalka, L., Kasprzak, A.: Task allocation in mesh connected processors with local search meta-heuristic algorithms. In: Nguyen, N.T., Le, M.T., Świątek, J. (eds.) Intelligent Information and Database Systems. LNCS, vol. 5991, pp. 215–224. Springer, Heidelberg (2010)

# A Sensor-Based Light Signal Controller

Abdul Mateen[1]([✉]), Adia Khalid[1], and Faheem Arif[2]

[1] Computer Science Department,
Federal Urdu University of Arts, Science and Technology, Islamabad, Pakistan
abdulmateen@fuuastisb.edu.pk, adia.khalid@gmail.com
[2] Computer Science Department, Military College of Signals,
National University of Sciences and Technology, Islamabad, Pakistan
fahim@mcs.edu.pk

**Abstract.** Traffic management especially in big cities is a salient problem worldwide. Developed countries had already given priority to solving it by applying modern traffic control and management approaches. Nevertheless, it is still a daunting challenge especially in third world countries. This problem is deemed a main contributor to wasting time, energy, money and other problems. This preliminary investigation presents a solution for managing vehicles at intersections controlled by light signals. Our solution tries to optimize the timing of each light signal in given that the volume of accumulated vehicles is estimated using environmental sensors such as a surveillance camera. The solution also takes care of the exceptions such as having an emergency or a rescue vehicle passing by. The system architecture is formalized using UML diagrams that show its main components and the interactions among them. Finally it is validated through simulation by taking various scenarios. The simulation results illustrate the efficiency of the proposed architecture.

**Keywords:** Traffic · Management · Autonomous · Controller

## 1 Introduction and Background

Number of vehicles that passes by a traffic signal at some specific time is known as traffic. The main elements of this include roads, vehicles and traffic signals. Traffic congestion may occur due to some interrelated real world problems such as more traffic flow than the available capacity, wrong or invalid lane turns, sudden and unauthorized pedestrian crossing, speed lane violations, police check posts, operation of the emergency services, etc. The traffic congestion problem can also be attributed to not optimizing the ON and OFF intervals of light signals.

The history of the traffic signal dates back to December, 1868 where the vehicle traffic is controlled at the British Parliament, London by J. P. Knight. The traffic signal at that time consists of two signals, i.e. red and green. The traffic lights switch from Red to Green light through a lever and vice versa. However, later on in 1912 modern traffic signals with red and green electric traffic signals [1] were introduced by Lester Wire who was an American. Later on, in 1914 an American traffic signal company put the traffic signals at the Ohio State that was consist of red and green lights with a buzzer. The 4-way traffic signal was introduced by the Willium Pots in

© Springer International Publishing Switzerland 2015
N.T. Nguyen et al. (Eds.): ACIIDS 2015, Part I, LNAI 9011, pp. 239–249, 2015.
DOI: 10.1007/978-3-319-15702-3_24

1920. First interconnected traffic signal was introduced in 1917 and installed at the Salt Lake City, USA with6 junctions and operated through human being (manually).

In general, there are three methods to control the flow of vehicles at signals: fixed, actuated and adaptive. In the fixed control method a static schedule that is derived from historical data is used. To overcome cons of this method, the actuated control is adopted where some vehicle detectors/ push buttons are used to convey information to the actuator module. After getting information about the traffic flow, the actuator takes action by allocating the least time interval needed to clear the stacked vehicles. Lastly, experts also used an adaptive control mechanism which assigns time according to the current traffic flow while incessantly learning from the past patterns.

The paper is structured as follows. Section 2 describes the state of the art related work germane to vehicle traffic control and management. Section 3 outlines the algorithm used to allot time for each signal to clear its traffic. Section 4 describes the system components and structure. Section 5 explains the system architecture through UML diagrams. Section 6 shows the comparative results of our architecture over the previous architectures. Finally, Section 7 concludes and sketches future directions.

## 2    Related Work

A lot of research related with the autonomic computing, agent oriented methodologies, vehicle traffic control and management is available in the form of research papers, white paper, technical reports and case studies. Some promising of these discussed here.

The research introduced a method [2] to controls traffic on roads through distributed way. Lee et al. [3] proposed an algorithm to get the parameters (for example vehicle quantity, speed etc) from the video image sequences. A model for an autonomous traffic control system is introduced by Alagar et al. [4]. A decentralized approach is adopted to manage the traffic at signals. The Vehicle Routing Problems (VRP) is elaborated by Thangiah et al. [5] and solution for the VRP is provided through agent oriented architecture. A simulation tool is used to optimize the timing of the traffic signal by Hewage et al. [6]. A multi-agent architecture is introduced [7] that provides traffic related information to drivers that assist them to make correct decisions during the driving.

Albagul et. al. [8] proposed an algorithm and simulation for the intelligent traffic signal which detects vehicles and provides the green signal time accordingly. The traffic congestion is removed by providing the appropriate time to signal through a mathematical function. The proposed algorithm for traffic control is simulated in MATLAB and countdown timer is also developed in Lab VIEW software.

Yi-Sheng Huang [9] provided a model for the traffic light control using the state chart diagram which provides the visual formalism of the system. Author performed the structure analysis of the state chart and used to show the 2, 6 and 8 phase traffic lights; and illustrated the concurrency, synchronization and causality of the system. The research also introduced a new methodology for modeling the system and named as concurrent state graph. The introduced methodology represents the concurrent states in complex state charts.

Shamshirband et. al. [10] introduced a multi-agent and Weighted Strategy Sharing (WSS) technique for cooperative learning in traffic management system. According to WSS, each agent measures the effectiveness of other agent in the system, assigns weight and learns from other accordingly. They also discussed the criteria by which expertness of the agents is calculated. According to author, the proposed technique is found to be effective after testing it for three traffic light system and compared it with non-cooperative agents. The said technique has some sever problems such as learning time is too high for large number of intersections and the other is very simple simulation that is not reflecting the real situation.

The major problem with human administrated traffic management system is poor response. On the other hand, automatic traffic management system allocation of constant time to each side creates the traffic congestion, waste of time and energy. Other problem with both systems is that they are oblivious to the special cases of rescue or emergency vehicles. In some countries, an emergency or rescue vehicles are assigned a dedicated lane, which is not feasible in third world countries.

We claim that our proposed solution herein provides an adaptive allocation solution of time and signal cycle time (i.e. total time for all four sides) while taking care of the exceptional situations of having emergency vehicles passing by.

## 3     Signal Allotted Time Algorithm

Traffic congestion is not an uncommon problem in big cities of modern as well as third world countries. Developed countries have already paid keen attention to traffic congestion problem and are solving it by adopting intelligent strategies. Pakistan is also facing the same situation and still implementing the static traffic management architecture which causes wastage of time and energy, insecure drive and delay in rescue services. Serious attention is required to resolve this issue by adopting intelligent and autonomous traffic management solution. The solution based case study is performed by taking the traffic data of Islamabad, capital of Pakistan.

The signal which is dependent only on its own traffic and is independent from the traffic flow of any other signal is called independent traffic signal. In case of independent traffic signal, following formulae are used to calculate the different parameters for the allocation of signal time:

AllotedTime = min(rows * time, threshold)

rows = round(vehicles / lanes)

time = avgSpeed / distance

distance = bufferDistance + avgVehicleLength

bufferDistance is the buffer distance between the vehicle and the signal, for the first row vehicles, and it is the safety distance that separates each car from another for the subsequent rows. This distance is typically two meters on average. avgVehicleLength is the average length of a typical vehicle, which is assumed to be four meters. Therefore, distance is the total distance needed for a vehicle to move from one row to another, which is six meters on average (bufferDistance + vehicleLength=6).

avgSpeed is the average speed of vehicles when they start to depart the signal area. It is assumed to be 20 km/hr. Accordingly, time is the time needed for a vehicle to advance to the next row; lanes is presumably three which represents the number of lanes each road has; vehicles is the number of vehicles as estimated by the monitoring camera. Based on these two parameters the number of rows accumulated, rows, at a given signal is computed.

The eventually allotted time, allotedTime, to a specific signal is the time needed to clear up all rows of vehicles, computed as rows*time, or the maximum prefixed time, threshold, which can be 15 seconds, for example.

For instance, suppose that we have four signals with vehicle distribution of 21, 54, 32, and 27, respectively. Each road consists of three lanes and the threshold time is preset to 30 seconds. time is 0.93 seconds (avgSpeed*distance = 20,000/3600 * 6). As a result:

Signal 1:        rows = round(21/3) = 7
                 allotedTime = min(7 * 0.93, 30) = 6.51 sec
Signal 2:        rows = round(54/3) = 18
                 allotedTime = min(18 * 0.93, 30) = 15 sec
Signal 3:        rows = round(32/3) = 11
                 allotedTime = min(11 * 0.93, 30) = 10.2 sec
Signal 4:        rows = round(27/3) = 9
                 allotedTime = min(9 * 0.93, 30) = 8.4 sec
                 Signal Time = Sum ((Signal Time)+Alert Time)
                 = (6.51 + 15 + 10.2 + 8.4) + 20
                 = 60.11 sec

where 20 sec is the total transition time for the alert/ yellow light

## 4    System Components and Structure

Our solution is based on autonomous agents who are capable of taking decisions and actions based on the traffic volume. The proposed system consists of three types of agents: environmental agent, observer agent and the knowledge-based agent. The *environmental agent* consists of a camera and a sound sensor, the *knowledge-base* agent is a data repository while the *observer agent* has the *Analyzer, Decision-Maker* and the *Learner* sub-agents.

Our system gathers information about the current traffic status via sensors, namely the camera and sonic sensor. This information is relayed to the Analyzer agent estimates the number of vehicles in the snapshot image. The vehicle count is passed to the Decision-Maker agent in order to allotting a suitable time interval.

If a rescue vehicle is approaching and is detected by the sonic sensor, the Decision-Maker agent retains the switch-to-green sequence, closes the other traffic signal after an alert time (e.g. 10 sec) and opens the traffic signal of the side from where emergency vehicle is coming. After the safe passage of rescue vehicle, the previous signal opening sequence is restored and an alert notification is sent to the next signals.

# 5    System Architecture

We use the UML diagrams to depict the system layout, the relationships among its components, and illustrate the functional and non-functional requirements. These UML diagrams are described as follows:

## 5.1    Use Case Diagram

Use case diagrams sketch the interactions between actors and system. In our architecture there are three actors: the camera, sound detector and actuator. Fig. 1 (a) shows the use-case diagram for the camera with three functions: set the camera, take the image of current traffic and send this information to the Actuator. Fig. 1 (b) shows the use-case diagram for the sound detector that has three functions: set the sound detector, recognize the sound of the emergency siren and send the alert information to the Actuator.

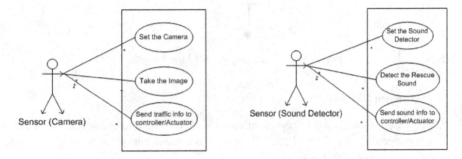

**Fig. 1.** The use-case diagram for the camera and sound detector

The use-case diagram for the Actuator is shown in Fig. 2. The Actuator starts its work upon receiving traffic info from the two sensors.

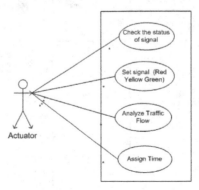

**Fig. 2.** Use case diagram for Actuator

## 5.2    State Chart Diagram

In UML, a state diagram represents the sequence of object states during their life time. State chart of the signal states which are wait, ready to go and go. As Fig. 3 shows, a light signal has three states: green (go) state, yellow (get ready), and red (stop). Green light remains either green or can be switched to yellow light; Yellow light can switch to red only while the red light remains red or switch to yellow.

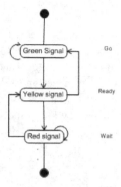

**Fig. 3.** State Chart diagram for Signal Lights

## 5.3    Sequence Diagram

Sequence diagrams are used to represent the system objects and the sequence of messages exchanged among them. Fig. 4 illustrates the most important messages passed between the agents of our system.

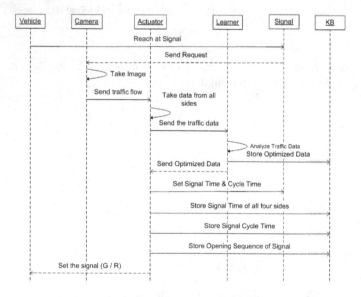

**Fig. 4.** Sequence Diagram

## 5.4    Collaboration Diagram

Collaboration diagrams as shown in Fig. 5 visually show the relationships between the agents.

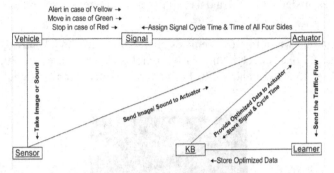

**Fig. 5.** Collaboration Diagram

# 6    Simulation Results

Simulation is a way to analyze and produce the solution of complex, risky and costly problems [11]. The traffic simulation will help us in analyzing, assessing and evaluating the traffic flow or congestion. Moreover it will also useful for determining the accuracy and effectiveness of the proposed traffic control architecture.

## 6.1    Traffic Simulation Entities

We are simulating only the vehicle traffic control architecture which is based on city traffic flow. In city traffic main features include vehicles, roads, intersection points (signals) and traffic lights.

## 6.2    Simulating the Environment

The simulation environment consists of vehicles (rescue and others), connected roads and traffic signals. Each road is divided into one or more lanes and operates in one direction. The environment that has been modeled is for independent traffic signal which consists of four roads connected in the form of signal.

## 6.3    Simulating the Vehicles

In simulation, vehicle is described with certain parameters such as size (length and width), location (x-axis, y-axis), color and speed. The x and y-position describes the location of vehicle and both are useful to move the vehicle in some specific direction. Number of vehicles and rescue vehicles on each signal are created randomly.

## 6.4    Independent Traffic Signal

The type of signal which is only depending on its own traffic flow is known as independent traffic signal. The parameters that were used to calculate are number of Lanes, vehicles at signal (1, 2, 3 and 4) and priority if there is some rescue vehicle.

**Scenario 1**
In the scenario (shown in Fig. 6), Vehicles at signal (1, 2, 3 and 4) having 3 lanes are 25, 20, 17 and 22 respectively. Number of vehicles in other cases are 39, 20, 17, 32; 20, 10, 7, 13; 5, 15, 11, 13 and 11, 8, 14, 9 while adopted opening signal sequence is 3, 2, 4 and 1.

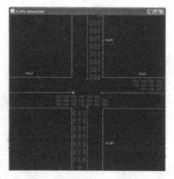

**Fig. 6.** Independent Traffic Signal Scenario 1

In Fig. 7, a few case of history based 100 sec cycle time and history based 70 sec cycle time approach shows minimum or equal average waiting time than our approach. However, the overall average waiting time of other approaches is greater than our proposed architecture. The range of average waiting time of our approach is 2.91 to 5.44, conventional approach with 4.74 to 15.5, history based 100 sec cycle time approach with 3.51 to 11.62 and history based 70 sec cycle time approach with 3.82 to 9.77. This represent that proposed architecture is more efficient.

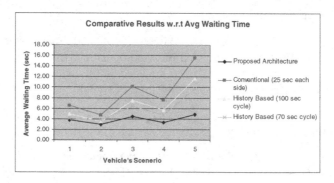

**Fig. 7.** Graph of Independent Traffic Signal Scenario 1

**Scenario 2**

In this scenario, Vehicles at signal (1, 2, 3 and 4) with three lanes are 25, 9, 19 and 13 respectively. In other four cases the number of vehicles as 16, 8, 13, 12; 19, 11, 16, 15; 10, 4, 8, 6; 21, 12, 19, 16 while the opening signal sequence is 2, 4, 3 and 1.

The Fig. 8 shows none of the approach has minimum or equal average waiting time than our proposed approach. The range of the average waiting time of our approach is 3.48 to 5.01, conventional approach with 7.55 to 17.92, history based 100 sec cycle time approach with 5.38 to 12.71 and history based 70 sec cycle time approach with. 4.79 to 11.42. This represents that our architecture depicted more efficient results than all other approaches.

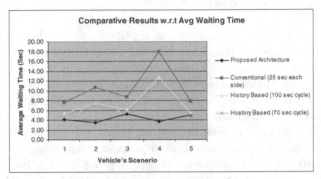

**Fig. 8.** Graph for Independent Traffic Signal Scenario 2

### 6.5    Independent Traffic Signal with Rescue Vehicle

Here, in this case a rescue vehicle appears from any side of the independent traffic signal. Priority of the signal from where rescue vehicle appears will have the highest priority even the number of vehicles on that side is greater than other signals.

Vehicles at signal (1, 2, 3 and 4) = 9, 7, 8, 6

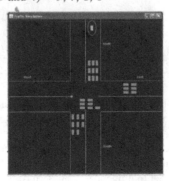

**Fig. 9.** Independent Traffic Signal with Rescue vehicle Scenario

Opening signal sequence (Before appearance of rescue vehicle) is 4, 2, 3 and 1. A rescue vehicle appears from the signal 4 as encircle in Fig. 9, when signal 2 was opened and after its appearance priority of the signal 4 increases. Due to this reason, signal 4 even that has larger number of vehicles opens before signal 3.

The opening signal sequence was 4, 2, 3 and 1. However, when signal 2 was opened the rescue vehicle appears from signal 4. After the appearance of rescue vehicle, priority of the side 4 will be increased. Due to this reason, signal 4 opens (even has larger number of vehicles than signal 3) before signal 3.

# 7    Conclusion and Future Work

In this article, we presented an autonomous traffic control solution for urban traffic signals. This solution is capable of handling the exceptional situation of controlling emergency vehicles. Among the advantages of this system are the adaptability as almost all the system parameters are gleaned from the environment via sensors.

The research is proposed to control the traffic signals intelligently and simulation design is based on different traffic signal's scenarios. The simulation is developed in JavaBeans IDE and in order to understand the results are presented in the form of graphs. The simulation results show that our proposed architecture saves much time and increase traffic flow. In summary our approach requires less human effort, dynamic decisions as per traffic flow, easy driving and reduction in accidents that ultimately saves precious lives, maximum traffic flow even on small roads, reduction in pollution and noise pollution, less fuel consumption, better social environment and cost reduction.

In future, we are working to integrate it with other systems such as traffic bulletin boards to share and update traffic congestion information with the public. These boards can be shown on main roads that will help drivers to change their direction/route if road is blocked or experiencing heavy traffic.

# References

1. Reference, N., Reference, H., Reference, R., Reference, B.: The Ins and Outs of the Joint Conference on Information Sciences. In: Proc. Of the Joint Conference on Information Sciences, pp. 200–204 (2003)
2. Bernon, C., Capera, D., Mano, J.-P.: Engineering Self-modeling Systems: Application to Biology. In: Artikis, Alexander, Picard, Gauthier, Vercouter, Laurent (eds.) ESAW 2008. LNCS, vol. 5485, pp. 248–263. Springer, Heidelberg (2009)
3. Tavladakis, K., Voulgaris, N.C.: Development of an autonomous adaptive traffic control system. The European Symposium on Intelligent Techniques, June 3-4, Greece (1999)
4. Stuart, R., Peter, N.: Artificial Intelligence. A Modern Approach, 2nd Edition (2008) ISBN No. 81-7758-367-0
5. Alagar, V.S., Muthiayen, D.: A rigorous approach to modeling autonomous traffic control systems. In: 6th International Syp. on Autonomous Decentralized Systems, 193–200, Italy (2003)

6. van Aart, C.: Organizational Principles for Multi-Agent Architectures. Series: WSSAT – Whitestein Series in Software Agent Technologies (2005) ISBN 3-7643-7213-2
7. Hewage, K.N., Ruwanpura, J.Y.: Optimization of Traffic Signal Light Timing Using Simulation. Winter Simulation Conference, pp. 1428–1433 (2004)
8. Casey, M.: MPEG-7 Sound Recognition Tools, Mitsubishi Electric Research Labs, Cambridge, MA, Unites States of America
9. Albagul, A., Hrairi, M., Wahyudi, Hidayathullah, M.F.: Design and Development of Sensor Based Traffic Light System. American Journal of Applied Sciences 3(3), 1745–1749 (2006)
10. Huang, Y.-S.: Design of Traffic Light Control Systems Using Statecharts. The Computer Journal 49(6) (2006)
11. Shamshirband, S.S., Shirgahi, H., Gholami, M., Kia, B.: Coordination between Traffic Signals Based on Cooperative. World Applied Sciences Journal 5(5), 525–530 (2008)

# Experimental Investigation of Impact of Migration Topologies on Performance of Cooperative Approach to the Vehicle Routing Problem

Dariusz Barbucha[✉]

Department of Information Systems, Gdynia Maritime University,
Morska 83, 81-225 Gdynia, Poland
d.barbucha@wpit.am.gdynia.pl

**Abstract.** A cooperative approach which integrates the asynchronous team paradigm with the island-based evolutionary algorithm concept is considered in the paper. Process of solving instances of the problem is carried-out by a set of software agents, each representing a heuristic algorithm, grouped in teams working on islands. All teams work in parallel and cooperate through periodic exchange of intermediary computation results. The process of forwarding results from one team to another (called migration) can be based on different topologies. The paper focuses on investigation of impact of the migration topologies on the performance of the proposed approach while solving instances of the Vehicle Routing Problem. Several migration models have been considered and experimentally compared in the paper.

**Keywords:** Agent-based optimization · Cooperative search · Asynchronous team · Island-based model · Multi-agent systems · Vehicle routing problem

## 1 Introduction

Among the approaches proposed for solving difficult optimization problems, special interest of researchers and practitioners is focused on hybridization of various methods which are able to produce a synergetic effect while solving instances of such problems. Different kinds of methods, forms of combining them into the effective problem-solving strategies, and technological advances, where jointly used, may offer effective tools for solving instances of the above problems.

Last years, one of the promising and intensively expanding directions of research, is the field of agent and multiple-agent systems [10]. A number of multiple-agent approaches integrated with some nature-inspired methods, proposed to solve different types of optimization problems grows systematically. One of them, where paradigms of the population-based methods, multiple agent systems and cooperative problem solving have been integrated, is the concept

© Springer International Publishing Switzerland 2015
N.T. Nguyen et al. (Eds.): ACIIDS 2015, Part I, LNAI 9011, pp. 250–259, 2015.
DOI: 10.1007/978-3-319-15702-3_25

of an asynchronous team (A-Team), originally introduced in [9]. In its fundamental form, the asynchronous team can be seen as a collection of *autonomous agents* that cooperate to solve a problem by dynamically evolving a population of solutions stored in the *common memories*.

The paper presents the Team of A-Teams (TA-Teams) approach, which main idea is to integrate the above mentioned team of asynchronous agent paradigm [9] with the island-based genetic algorithm concept [7]. In the island model, each island can exchange information with its neighbor island as defined in the graph of possible inter-island links commonly referred to as *migration topology*. As it has been shown in [3], employing the island model in parallel genetic algorithms can lead to increased algorithm performance what can be explained in terms of improved balance between exploitation and exploration of the solution space.

Technically the proposed approach is implemented in a multi-agent environment presented in [1] and next extended in [2]. Its main functionality focuses on organizing and conducting the process of search for the best solution using a set of agents representing a single-solution methods executed in parallel. During such execution, agents communicate asynchronously with each other but this communication is performed indirectly via the common, sharable memory (also called warehouse or pool of solutions). The novelty of the TA-Teams approach presented here is that agents are grouped in teams (single A-Teams) working on islands. Each team periodically communicates with other teams by sharing promising results with them according to predefined migration topology.

The paper focuses on different migration topologies and investigates the impact of them on performance of TA-Teams approach while solving instances of the Vehicle Routing Problem (VRP). The problem of impact of migration topologies on performance of island model has been also studied, for example, by Rucinski et al. [8]. Recently, Jędrzejowicz and Wierzbowska [6] studied different migration topologies used by TA-Teams approach to Euclidean Planar Traveling Salesman Problem (EPTSP). One of the goal of this paper is to extend their research by considering different migration topologies for another combinatorial optimization problem - VRP.

The rest of the paper includes the following sections. Section 2 provides main idea of the TA-Teams concept. Section 3 describes the computational experiment, including goal, description of the problem considered, main settings, and reports on the results. Finally, Section 4 concludes the paper and suggests directions for future research.

## 2    Team of A-Teams Concept

TA-Teams approach is based on the asynchronous teams (A-Teams) concept, originally introduced by Talukdar [9] as a result of integration of paradigms of the population-based methods and multiple agent systems. According to this concept, an asynchronous team can be seen as a collection of *autonomous agents* that cooperate to solve a problem by dynamically evolving a population of solutions stored in the *common memories*. Within an A-Team, each agent encapsulates a particular problem-solving method (exact or heuristic) allowing it to manipulate trial solutions accumulated in the memory. During the A-Team activity, memory is time varying: new members (solutions improved by agents) are continually added to the memory, while older members (worse solutions) are being erased, so the quality of the solutions gradually evolves over the time. The ground principle of asynchronous teams rests on combining these methods, which alone could be inept for the task, into effective problem-solving organizations, possibly creating a synergy effect. The observed combined effect of agents teamwork is often greater than the sum of their separate effects.

The idea of the TA-Teams presented here is to construct a number of single A-Teams and allow them to solve the same task in parallel by exploring different regions of the search space with the added process of communication between A-Teams. The architecture of the TA-Teams approach is presented in Fig. 1 and includes:

- the *set of single A-Teams* dedicated for solving instances of given optimization problem,
- the *communication protocol* assuring effective communication between the A-Teams during the whole process of search for the best solution.

## 2.1   Set of Single A-Teams

Single A-Team is a multi-agent architecture consisting of a sharable *memory*, which stores a population of individuals (solutions), a set of *agents* operating

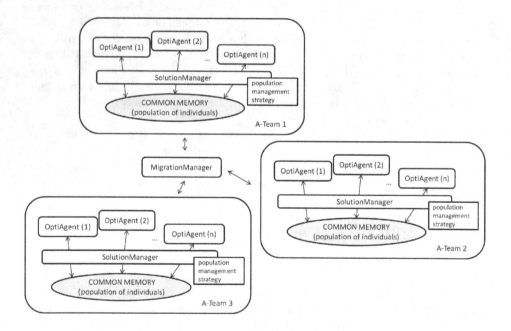

**Fig. 1.** Architecture of the Team of A-Teams for VRP

on individuals stored in the memory, and a *cooperation scheme*, which combines these agents into a single effective problem-solving strategy.

The role of the *memory* is to store the population of individuals and to accumulate results or trial solutions, processed by working agents. Each individual stored in the memory is represented in a form that reflects the characteristics of the problem being solved. The form should be also convenient to handle the calculations performed on it by search procedures. During the A-Team activity, memory dynamically changes: new members (solutions improved by agents) are continually added, while older members (worse solutions) are being erased, so the quality of the solutions gradually evolves over the time.

Two kinds of *agents* work within the A-Team: *OptiAgents* and *Solution-Manager*. *OptiAgents* encapsulate a particular single-solution problem-solving method, each, and operate on individuals stored in the memory during the process of search. It is expected that using different improvement algorithms executed by different agents increases chances for reaching the global optimum. *SolutionManager* agent acts as an intermediary between the memory and *OptiAgents*. It maintains the common memory and is responsible for managing the population of solutions. Moreover, it oversees the whole process of solving instances of given optimization problem.

Besides the set of effective methods, represented by *OptiAgents*, the ground principle of A-Team rests on combining these agents into a single effective problem-solving strategy. The *cooperation scheme* defined within A-Team is responsible for effective problem-solving organizations of the work of the agents while solving instances of given optimization problem, possibly creating a synergy effect.

The process of solving instances of given problem by such single A-Team is organized as a sequence of steps, including initialization and improvement phases. At first the initial population of solutions is generated and stored in the memory. Next, at the following computation stages, individuals forming the initial population are successively improved by autonomously working *OptiAgents*. The main steps of the single A-Team approach, repeated in loop until a stopping criterion is met, include:

1. Selecting a particular individual (solution) from the common memory by *SolutionManager* and sending it to autonomous, independently acting *OptiAgents*, which have already announced their readiness to act,
2. Improvement of solutions by these agents and sending them to back to the *SolutionManager*, and
3. Storing back (by *SolutionManager*) the potentially improved solution returned by *OptiAgents* in the common memory.

When the stopping criterion is met, the best solution in the population is taken as the final solution of the given problem instance.

The above process of searching for the best solution is performed in presence of *population management strategy*, which defines the set of rules determining how to choose solutions which are to be sent to *OptiAgents* for improvement and how to merge the improved solutions returned by *OptiAgents* with the whole population and when to stop the process of searching [1] .

## 2.2  Communication between Single A-Teams

As it was mentioned, TA-Teams approach allows for communication between single A-Teams. In the proposed approach it is supervised by a specialized agent called *MigrationManager* (see Fig. 1) and defined by a *migration strategy* including a number of parameters [6], [8]:

- The *number of A-Teams* (islands) existing in parallel,
- The *migration topology* means as an architecture in which an A-Team receives communication from another A-Team and sends communication to some other A-Team,
- The *migration size (rate)* telling how many individuals migrate from a source A-Team at a time,
- The *migration frequency* determining the length of time between migrations,
- The *migration policy* determining a rule telling how the received solution is incorporated into the common memory of the receiving A-Team.

The migration used in TA-Teams is asynchronous. With a given frequency *MigrationManager* sends messages to islands, pointing out to which islands current best solution should be send to. Each A-Team (that is an island), after receiving a message from *MigrationManager*, sends the current best solution to indicated island or islands.

Within TA-Teams each single A-Team works according to the rules defined in this A-Team. When one of the single A-Teams stops due to its population management strategy, the TA-Teams stops its computation, regardless of recent improvements in best solutions of the others A-Teams. The overall best result from common memories of all A-Teams in TA-Teams is taken as the final solution found for the task.

## 3  Computational Experiment

### 3.1  Goal and Main Settings

Computational experiment has been carried out in order to answer the following question: To what extent (if any) different migration topologies of exchanging information between single A-Teams influence computation results produced by the TA-Teams? The experiment has been performed on instances of the Vehicle Routing Problem, the network optimization problem, in which a set of given $N$ customers is to be served by the fleet of vehicles in order to minimize the service cost and satisfy several customer's and vehicle's constraints [5].

**Agents.** Five optimizing agents have been used by each single A-Team (see Table 1). They used dedicated methods operating on single ($R_i \in R$) or two ($R_i, R_j \in R$) randomly selected route(s) ($R$ - set of $m$ routes, $i \neq j$, and $i, j = 1, \ldots, m$).

**Table 1.** Agents and their characteristics

| Agent | Description |
|---|---|
| 3Opt | An implementation of the *3-opt* procedure operating on a single route $R_i$. Three randomly selected edges are removed and next remaining segments are reconnected in all possible ways until a new feasible solution (route) is obtained. |
| 2Lambda | A modified implementation of the dedicated *local search method* based on $\lambda$-*interchange local optimization* method. At most $\lambda$ customers are moved or exchanged between two selected routes $R_i$ and $R_j$. In the proposed implementation, it has been assumed that $\lambda = 2$. |
| 2LambdaC | Another implementation of the dedicated *local search method* which operates on two routes, and is based on exchanging or moving selected customers between these routes. Here, selection of customers to exchange or movement is taken in accordance to their distance to the centroid of their original route. First, a given number of customers from two selected routes $R_i$ and $R_j$ for which the distance between them and the centroid of their routes are the greatest are removed from their original routes. Next, they are moved to the opposite routes and inserted in them on positions, which give the smallest distance between newly inserted customers and the centroid of this route. |
| Cross1 | An agent which is implementation of the one-point crossover operator. Initially one point is randomly selected on each route $R_i$ and $R_j$, dividing these routes on two subroutes. Next, the first subroute of $R_i$ is connected with the second subroute of $R_j$, and the first subroute of $R_j$ is connected with the second subroute of $R_i$. |
| Cross2 | An implementation of the two-point crossover operator. Initially two points are selected randomly on each route $R_i$ and $R_j$, dividing these routes on three subroutes. Next, the middle parts (between crossing points) of each route are exchanged between considered routes. |

**Migration Settings.** In the reported experiment, the number of A-Teams has been set to 8 and the population size of each single A-Team has been set to 20, giving in total 160 individuals in the system.

The following migration topologies have been tested (see Fig. 2): *One Way Ring* - each A-Team receives communication from one adjacent A-Team and sends communication to another adjacent A-Team, the only uni-directional topology allowed, *Ring* - two-directional ring, *Broadcast* (or star topology), *Lattice* (of the size 2x4), *Torus*, and *Full* - fully connected topology with all possible communication paths between single A-Teams [6].

Additionally, a model called *Randomized* [6] has also been considered, with the overhead caused by the information flow between A-Teams reduced to minimum. The main idea behind this model is to exchange solutions between A-Teams whenever the current best solution in some A-Team's memory has not been changed by a fixed part of no improvement time gap. The source A-Team sends then appropriate message to the *MigrationManager* agent. Then *MigrationManager* chooses randomly other target A-Team and asks it for sending its

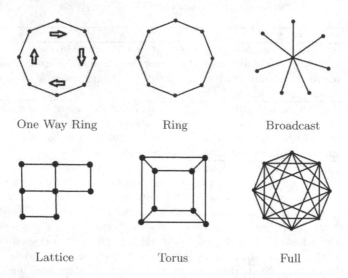

One Way Ring          Ring          Broadcast

Lattice          Torus          Full

**Fig. 2.** Migration topologies (each for 8 A-Teams) [6]

best solution to the source A-Team. The best solution taken from the source A-Team replaces the worst solution in the memory of the target A-Team.

The other migration settings are identical for all considered models:

- *Migration size = 1*: one current best solution is sent from the common memory of an A-Team to the common memory of another A-Team in one cycle,
- *Migration frequency = 0.5*: time between migrations between A-Teams,
- *Migration policy = best-worst*: the current best solution taken from the source population replaces the current worst solution in the target population.

The process of solving the instance by a single A-Team stops after a predefined amount of time (it has been decided to stop the process after 3 minutes).

**Other Settings.** The experiment involved 14 VRP instances of Christofides et al. [4], each of them containing 50-199 customers with capacity and, some of them, maximum length route restrictions. For each migration topology, each instance were repeatedly solved 10 times and the mean results from these runs were recorded.

Mean relative error - MRE (in %) from the optimal (or the best one) solution has been chosen as a measure of the quality of the results obtained by the proposed approach for different topologies.

All computations have been carried out on PC with Intel Core i5-2540M CPU 2.60 GHz and 8 GB RAM running under MS Windows 7 operating system.

## 3.2   Results

Results of the experiment are presented in Table 2 and Fig. 3. The first column of Table 2 includes names of the instances with the number of customers and type of the problem in brackets (C - only capacity constraints, CD - capacity and maximum length route restrictions). For each instance, the remaining columns of Table 2 contain results (MRE) produced by the proposed Team of A-Team approach for all considered migration topologies. Additionally, Fig. 3 presents MRE for each topology for two types of problems (C and CD), separately.

**Table 2.** Results (MRE from the best known solution) obtained by the Team of A-Teams approach to VRP for all tested instances and all migration topologies

| Instance | Migration topology | | | | | | |
|---|---|---|---|---|---|---|---|
| | One Way Ring | Ring | Broadcast | Lattice | Torus | Full | Randomized |
| vrpnc1 (50, C) | 0.00% | 0.00% | 0.00% | 0.00% | 0.00% | 0.00% | 0.00% |
| vrpnc2 (75, C) | 1.03% | 0.01% | 0.99% | 0.92% | 0.64% | 1.03% | 1.03% |
| vrpnc3 (100, C) | 0.79% | 0.43% | 0.43% | 0.48% | 1.17% | 0.98% | 0.73% |
| vrpnc4 (150, C) | 1.29% | 1.24% | 2.16% | 1.23% | 1.73% | 2.36% | 0.85% |
| vrpnc5 (199, C) | 4.25% | 4.77% | 3.71% | 4.06% | 3.71% | 4.64% | 4.01% |
| vrpnc6 (50, CD) | 0.00% | 0.00% | 0.00% | 0.00% | 0.00% | 0.00% | 0.00% |
| vrpnc7 (75, CD) | 1.68% | 0.35% | 1.52% | 1.35% | 1.43% | 0.35% | 0.53% |
| vrpnc8 (100, CD) | 0.00% | 0.00% | 0.00% | 0.00% | 0.00% | 0.11% | 0.19% |
| vrpnc9 (150, CD) | 3.46% | 4.03% | 2.80% | 3.31% | 4.11% | 3.07% | 3.14% |
| vrpnc10 (199, CD) | 4.90% | 4.57% | 4.92% | 4.34% | 4.09% | 4.83% | 4.66% |
| vrpnc11 (120, C) | 0.17% | 0.00% | 7.94% | 5.11% | 9.11% | 9.13% | 0.46% |
| vrpnc12 (100, C) | 0.64% | 0.00% | 0.97% | 0.95% | 0.00% | 0.00% | 0.64% |
| vrpnc13 (120, CD) | 2.03% | 1.95% | 2.11% | 2.43% | 1.96% | 2.36% | 2.34% |
| vrpnc14 (100, CD) | 0.00% | 0.00% | 0.00% | 0.00% | 0.00% | 0.00% | 0.00% |
| Average | 1.45% | 1.24% | 1.97% | 1.73% | 2.00% | 2.06% | 1.33% |

Analysis of the results allows for several observations. The first one is that the results produced by the proposed approach are satisfactory and close to the best known ones. Mean relative error depends on the instance solved. They do not exceed 5% for any instance but for most instances it is equal to 1-2%. Moreover, for most instances, MRE is relatively low regardless the migration topology applied. Best results have been observed for instances with smaller number of customers ($N \leq 150$).

The second observation refers to comparison of results produced by different TA-Teams with different migration topologies used for communication between single A-Teams. Taking into account value of MRE averaged over all instances, one can see that definitely best results have been obtained for topologies with smaller number of communication paths: *Ring*, *One Way Ring* and *Randomized* (MRE is equal to 1.24-1.45%). The results obtained for other topologies (*Broadcast*, *Torus* and *Full*) guarantee MREs equal to approx. 2%.

Through deeper analysis of the results, one can also conclude that the quality of the results depends on the group of the VRP problems to which the instances

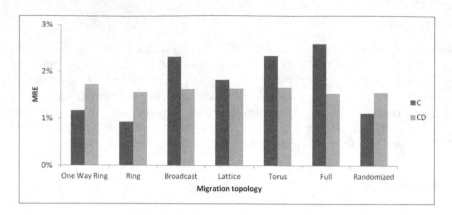

**Fig. 3.** Results (MRE from the best known solution) obtained by the Team of A-Teams approach to VRP for instances of two groups of problems and all migration topologies

belong to. By looking at Figure 3, it is easy to see that the above conclusion about outperformance of *Ring*, *One Way Ring* and *Randomized* topologies over the other ones holds for problems with only capacity constraints (C). On the other hand, for problems with capacity and maximum length route constraints (CD) the results are similar regardless the topology used for communication between single A-Teams.

And finally, an interesting conclusion can be drawn after analysis of the results presented here with the results presented in [6], where authors investigated an impact of migration topologies on performance of TA-Teams approach solving instances of EPTSP. According to [6], from all tested topologies by authors, the best results, in terms of MRE averaged over all runs and all instances, were obtained by the *Randomized* topology. The next best topologies were *One Way Ring* and *Ring*. On the other hand, the worst results have been obtained for *Broadcast* and *Full* topologies. Hence, one can conclude that the results obtained for VRP instances are similar to those obtained for EPTSP.

## 4   Conclusions

The paper focused on TA-Teams approach, integrating the asynchronous team paradigm with the island-based evolutionary algorithm concept. The process of solving the problem is carried-out by a set of agents, each representing a heuristic algorithm, operating on population of individuals (solutions) stored in the common sharable memory. Agents are grouped in teams, which periodically communicates with other by sharing promising results.

The main goal of the paper was to investigate experimentally of impact of different migration topologies of exchanging information between single A-Teams on performance of the approach. The experiment performed on instances of VRP confirmed that the choice of the migration topology may influence results obtained by the TA-Teams. What is also important to emphasize, the importance of proper choose of migration topology has been also confirmed by other authors [6] for EPTSP problem, which may suggest the general nature of observation.

Future research will focus on extending the proposed research by considering different migration parameters (for example migration frequency) as well as considering different problems to be solved.

# References

1. Barbucha, D., Czarnowski, I., Jędrzejowicz, P., Ratajczak-Ropel, E., Wierzbowska, I.: e-JABAT - an implementation of the web-based A-Team. In: Nguyen, N.T., Jain, L.C. (eds.) Intelligent Agents in the Evolution of Web and Applications. SCI, vol. 167, pp. 57–86. Springer, Heidelberg (2009)
2. Barbucha, D., Czarnowski, I., Jędrzejowicz, P., Ratajczak-Ropel, E., Wierzbowska, I.: Team of A-teams - a study of the cooperation between program agents solving difficult optimization problems. In: Czarnowski, I., Jędrzejowicz, P., Kacprzyk, J. (eds.) Agent-Based Optimization. SCI, vol. 456, pp. 123–141. Springer, Heidelberg (2013)
3. Cantú-Paz, E.: Efficient and Accurate Parallel Genetic Algorithms. Kluwer Academic Publishers, Norwell (2000)
4. Christofides, N., Mingozzi, A., Toth, P., Sandi, C. (eds.): Combinatorial optimization. John Wiley, Chichester (1979)
5. Golden, B.L., Raghavan, S., Wasil, E.A. (eds.): The Vehicle Routing Problem: Latest Advances and New Challenges. Operations Research Computer Science Interfaces Series, vol. 43. Springer, Heidelberg (2008)
6. Jędrzejowicz, P., Wierzbowska, I.: Impact of migration topologies on performance of teams of A-Teams. In: Grana, M., et al. (eds.) Advances in Knowledge-Based and Intelligent Information and Engineering Systems, pp. 1161–1170. IOS Press, Amsterdam (2012)
7. Martin, W.N., Lienig, J., Cohoon, J.P.: Island (migration) models: evolutionary algorithms based on punctuated equilibria. In: Bäck, T., et al. (eds.) Handbook of Evolutionary Computation, pp. C6.3:1–C6.3:16. Oxford University Press, New York (1997)
8. Rucinski, M., Izzo, D., Biscani, F.: On the impact of the migration topology on the Island Model. Parallel Computing 36, 555–571 (2010)
9. Talukdar, S., Baerentzen, L., Gove, A., de Souza, P.: Asynchronous Teams: Cooperation Schemes for Autonomous Agents. Journal of Heuristics 4(4), 295–321 (1998)
10. Wooldridge, M.: An Introduction to MultiAgent Systems. John Wiley & Sons, Chichester (2009)

# Decision Support and Control Systems

# Modelling a Robotic Cell and Analysis Its Throughput by Petri Nets

František Čapkovič$^{(\boxtimes)}$

Institute of Informatics, Slovak Academy of Sciences, Bratislava, Slovakia
Frantisek.Capkovic@savba.sk

**Abstract.** The place/transition Petri nets (P/T PN) and timed Petri nets (TPN) are used here for modelling the robotic working cell where the robot attends three machine tools processing in sequence the part in order to obtain the final shape of the product. The cell contains the input and output conveyers. The input raw material has to pass the way from the input conveyer consecutively through all machines to the output conveyer. The P/T PN model of the cell is constructed. Then, by means of supervision, the possible sequences of operations are ensured in order to avoid any ambiguity. Finally, the throughput of the robotic working cell is found by means of timed Petri nets (TPN) for all simple schedules. By means of comparison particular schedules as to the throughput the best throughput is found.

**Keywords:** Analysis · Modelling · Petri nets · Simulation · Timed petri nets

## 1 Introduction

Robotic cells are frequently used in flexible manufacturing systems (FMS). Excepting the robots the cell contain different kinds of machine tools, transport belts feeding a raw material or/and semi-product into the cell and carrying finished products away, etc. The robot performs complicated sequences of different operations. The cells, where the robot attends several machine tools processing a part in sequence (to obtain the final shape of the product), are frequently used. Usually, finding the maximal throughput of the cells is the subject of interest. Timed PN (TPN) are often used for solving such problems. Theory for sequencing of parts and robot moves introduced in [12] was evolved later in [3,5]. The theory was applied in [14, 15] using TPN and later in many other papers - e.g. in [4] for two machines, in [1] for three machines, in [13] for single-arm multi-cluster tools, etc. In this paper, in contrast of those approaches, another aspect of the problem is examined. It has rather an experimentally-computational character. Namely, the focus is on the characteristics of the PN model and corresponding simplicity at finding the throughput of particular simple schedules by means of simulation based on suitable TPN models. In such schedules exactly one part enters and one part leaves the cell in each working cycle. The main motivation is to find the PN model with the reachability

© Springer International Publishing Switzerland 2015
N.T. Nguyen et al. (Eds.): ACIIDS 2015, Part I, LNAI 9011, pp. 263–272, 2015.
DOI: 10.1007/978-3-319-15702-3_26

graph (RG) in the form of the single loop (corresponding to the working cycle) for each schedule of the cell. The RG branching is fully eliminated. To achieve such RG, at first the place/transition Petri nets (P/T PN) model of the cell is drawn based on the analogy with the structure of the real cell. The PN places represent the particular devices while the directed arcs with transitions symbolize the particular motions of the robot (they differ from schedule to schedule). The RG of an initial model of a schedule proposed empirically usually forks. The model properties can be improved by means of a supervisor and the desired RG without forks is obtained. The supervisor is synthesized in analytical terms. After assigning operation times to P/T PN transitions the expected TPN model arises. It serves for simulation purposes. Such a simple model accelerates the simulation process. The results for the particular schedules achieved by simulation can be compared and evaluated.

Let us introduce some preliminaries concerning PN and their supervision. P/T PN [7,8] are used here in the process of modelling the robotic cell. As to the structure they are bipartite directed graphs $< P, T, F, G >$ where $P$ is the set of places $p_i$, $i = 1, \ldots, n$; $T$ is the set of transitions $t_j$, $j = 1, \ldots, m$; $P \cap T = \emptyset$; $F$ is the set of directed arcs from places to transitions; $G$ is the set of directed arcs from transitions to places; $F \cap G = \emptyset$. P/T PN have their marking evolution (*dynamics*) formally expressed as $< X, U, \delta, \mathbf{x}_0 >$ where $X$ is the set of state vectors (marking the places); $U$ is the set of vectors of discrete events (state vectors of transitions); $X \cap U = \emptyset$; $\delta : X \times U \to X$ is the transition function; $\mathbf{x}_0$ is the initial state vector. The system form of the transition function $\delta$ is: $\mathbf{x}_{k+1} = \mathbf{x}_k + \mathbf{B}.\mathbf{u}_k$, where $\mathbf{B} = \mathbf{G}^T - \mathbf{F}$ and $\mathbf{F}.\mathbf{u}_k \leq \mathbf{x}_k$ for $k = 0, \ldots, N$. Here, $k$ is the discrete step of the dynamics development; $\mathbf{x}_k = (\sigma_{p_1}^k, \ldots, \sigma_{p_n}^k)^T$ is the $n$-dimensional state vector; $\sigma_{p_i}^k \in \{0, 1, \ldots, c_{p_i}\}$, $i = 1, \ldots, n$, express the states of atomic activities by 0 (passivity) or by $0 < \sigma_{p_i} \leq c_{p_i}$ (activity); $c_{p_i}$ is the capacity of $p_i$ as to the number of tokens; $\mathbf{u}_k = (\gamma_{t_1}^k, \ldots, \gamma_{t_m}^k)^T$ is the $m$-dimensional control vector; its components $\gamma_{t_j}^k \in \{0, 1\}$, $j = 1, \ldots, m$, represent occurring of elementary discrete events (e.g. starting or ending the activities, failures, etc.) by 1 (presence of the discrete event) or by 0 (absence of the event); $\mathbf{B}$, $\mathbf{F}$, $\mathbf{G}$ are matrices of integers; $\mathbf{F}$ is the incidence matrix of the directed arcs from the places to the transitions and $\mathbf{G}$ is the incidence matrix of the directed arcs from the transitions to the places; $(.)^T$ symbolizes the transposition.

The supervisor can improve the properties of the P/T PN model. For the supervisor synthesis the following general linear constraints [2,6] are utilized

$$\mathbf{L}_p.\mathbf{x} + \mathbf{L}_t.\mathbf{u} + \mathbf{L}_v.\mathbf{v} \leq \mathbf{b} \qquad (1)$$

where $\mathbf{L}_p$, $\mathbf{L}_t$, $\mathbf{L}_v$ are, respectively, the integer matrices with dimensionalities $(n_s \times n)$, $(n_s \times m)$, $(n_s \times m)$; $\mathbf{b}$ is $n_s$-dimensional integer vector; $\mathbf{v} = \mathbf{u}_0 + \mathbf{u}_1 + \ldots + \mathbf{u}_k$ is the Parikh's vector of PN. It contains information about how many times the particular transitions are fired during the development of the system (describing the P/T PN model) from $\mathbf{x}_0$ to $\mathbf{x}_k$. When $\mathbf{b} - \mathbf{L}_p.\mathbf{x} \geq \mathbf{0}$ is valid [6], the supervisor with the following structure and the initial state $^s\mathbf{x}_0$

$$\mathbf{F}_s = \max(\mathbf{0}, \mathbf{L}_p.\mathbf{B}, +\mathbf{L}_v, \mathbf{L}_t); \quad {}^s\mathbf{x}_0 = \mathbf{b} - \mathbf{L}_p.\mathbf{x}_0 - \mathbf{L}_v.\mathbf{v}_0 \tag{2}$$

$$\mathbf{G}_s^T = \max(\mathbf{0}, \mathbf{L}_t - \max(\mathbf{0}, \mathbf{L}_p.\mathbf{B} + \mathbf{L}_v)) - \min(\mathbf{0}, \mathbf{L}_p.\mathbf{B} + \mathbf{L}_v) \tag{3}$$

guarantees that constraints (1) are verified for the states of the system resulting from the initial state $\mathbf{x}_0$. Here, the max(.), min(.) are, respectively, the maximum and minimum operators for matrices. They are applied on the matrices being their operands element by element (i.e. entry by entry).

Because P/T PN are not able to yield information about time relations in the robotic cells, TPN [9,14,15] will be utilized too. Namely, they yield the time behaviour of marking and consequently make possible the performance evaluation. In general, time can be assigned to the transitions, places, arcs and tokens of P/T PN in order to obtain TPN. Here, TPN assigning time exclusively to P/T PN transitions (delays in deterministic case or a kind of probability distributions of timing in non-deterministic case) will be used because they play the most important role at the throughput of the robotic cells.

Finally, simulation results obtained by the HYPENS tool [10,11] in MATLAB will be displayed.

## 2    Problem Formulation

Consider the robotic cell given in Fig. 1. It contains three machine tools $M1$, $M2$, $M3$ and two transport belts $IN$ and $OUT$. The raw material picked up from $IN$ has to go in sequence through the three machines (where different operations are performed) before putting the finalized part on $OUT$. The problem of finding the throughput seems to be very simple because seemingly it is sufficient when the ROBOT (below $\Re$) realizes the 5 steps illustrated in Fig. 2 as the schedule a). However, because the operations performed by $\Re$ and by the machines take a time, it is necessary to find the best sequence (as to the shortest global processing time of the whole working cycle). Therefore, also the working cycles of other possible simple schedules have to be found and examined. As it was proved in

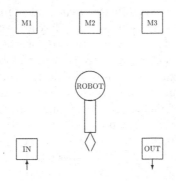

**Fig. 1.** The schematic view on the robotic cell

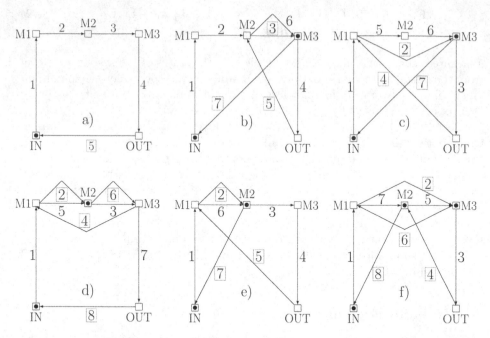

**Fig. 2.** The different simple schedules: a), b) and c) (in the upper row) and d), e) and f) (in the lower row). The squares represent the devices attended by ℜ. The big filled circles inside the squares represent the active devices at the beginning of the working cycle. The framed numbers placed at the directed arcs represent the movements of ℜ without any part while the non-framed ones represent the movements of ℜ with a part in its gripper. While ℜ is active permanently at any schedule and it already starts by gripping a part from IN, the machines active at the beginning of the working cycle at the particular schedules are different.

[12], in case of $n$ machines $M1, \ldots, Mn$, there exist $n!$ different simple schedules. In the simple schedule exactly one part enters and one part leaves the working cell in each working cycle. Other possibilities are not taken into account. Thus, in our case, there exist 6 different simple schedules. They are schematically displayed in Fig. 2. Each of the schedules has a different global processing time. Of course, the best schedule will be the schedule with the shortest global processing time. The throughput of such a configuration will be the best. From the point of view of the whole FMS (containing many working cells of different kinds) it is very important because its productivity depends on the throughput of the particular robotic cells. More details about the particular schedules are synoptically introduced in Tab. 1 where the dashed arrows mean the movements of the free ℜ (i.e. with the empty gripper) while the full (unbroken) arrows mean the movements of busy ℜ (with a part placed in its gripper). Now, let us model the particular schedules and analyse them according to the intention described above in the section 1.

**Table 1.** The scheme of the particular simple schedules

| Schedule | Causality of Operations and Activities in the Robotic Cell |
|---|---|
| a) | $IN \xrightarrow{1} M1 \xrightarrow{2} M2 \xrightarrow{3} M3 \xrightarrow{4} OUT \dashrightarrow^{5} IN$ |
| b) | $IN \xrightarrow{1} M1 \xrightarrow{2} M2 \dashrightarrow^{3} M3 \xrightarrow{4} OUT \dashrightarrow^{5} M2 \xrightarrow{6} M3 \dashrightarrow^{7} IN$ |
| c) | $IN \xrightarrow{1} M1 \dashrightarrow^{2} M3 \xrightarrow{3} OUT \dashrightarrow^{4} M1 \xrightarrow{5} M2 \xrightarrow{6} M3 \dashrightarrow^{7} IN$ |
| d) | $IN \xrightarrow{1} M1 \dashrightarrow^{2} M2 \xrightarrow{3} M3 \dashrightarrow^{4} M1 \xrightarrow{5} M2 \dashrightarrow^{6} M3 \xrightarrow{7} OUT \dashrightarrow^{8} IN$ |
| e) | $IN \xrightarrow{1} M1 \dashrightarrow^{2} M2 \xrightarrow{3} M3 \xrightarrow{4} OUT \dashrightarrow^{5} M1 \xrightarrow{6} M2 \dashrightarrow^{7} IN$ |
| f) | $IN \xrightarrow{1} M1 \dashrightarrow^{2} M3 \xrightarrow{3} OUT \dashrightarrow^{4} M2 \xrightarrow{5} M3 \dashrightarrow^{6} M1 \xrightarrow{7} M2 \dashrightarrow^{8} IN$ |

# 3   P/T PN Based Modelling of Schedules

The P/T PN places will represent the particular devices. The directed arcs among the places (containing the P/T PN transitions) will represent the particular movements of $\Re$. As it is clear from Fig. 2, the simplest structure of the P/T PN model has the schedule a). The model is given in Fig. 3 in the upper row left. There are practically no movements of $\Re$ without a part. Solely the movement of $\Re$ from $OUT$ to $IN$ (closing the working cycle) is performed without any part in the $\Re$ gripper. However, on the other hand, in this case $\Re$ has to wait at each machine for finishing the machining process. Hence, the time when $\Re$ is idle may be long. It is adverse for the throughput of the robotic cell. The time can be utilized more effectively. Namely, $\Re$ can perform other operations instead of waiting. The cases b) - f) point out the ways how to do this. However, at these models the structures of the schedules are more complicated (*trade-off*) than that in the schedule a). In the case a) the RG of P/T PN has directly the form of the working cycle without any branching (i.e. the model evolution is unambiguous). It corresponds to the first row of Tab. 1. However, in the other cases b) - f), when they are non-supervised, the branching of RG is very intensive and the evolutions of their models are ambiguous. Namely, there exist conflict situations because the PN places have several output transitions. The reason of using the supervision is to achieve the simple RGs also for the schedules b) - f). After synthesizing the suitable supervisors for their PN models the unambiguous forms (i.e. simple cycles) of their RGs are really achieved. They have the form identical with the further rows of Tab. 1. Even, the simple supervisors introduced below are sufficient for this. In Fig. 3 the supervisors are represented by the places $p_6$ - $p_{11}$ (in the cases b), c), e)) and $p_6$ - $p_{12}$ (in the cases d), f)). They were found in analytical terms by means of the PN-based supervision method (1) - (3) introduced in the section 1. Simply said, the mutual priorities $\pi(t_j)$ among the transitions $t_j$, $j = 1, \ldots, 7$ (in the cases b), c), e)) and $t_j$, $j = 1, \ldots, 8$ (in the cases d), f)) were predefined. Namely, $\pi(t_1) > \pi(t_2); \pi(t_2) > \pi(t_3); \pi(t_3) > \pi(t_4); \pi(t_4) > \pi(t_5); \pi(t_5) > \pi(t_6); \pi(t_6) > \pi(t_7); (\pi(t_7 > \pi(t_8))$. It means that the relations among the entries of the Parikh's vector $\mathbf{v}$ are the following: $v(t_1) > v(t_2); v(t_2) > v(t_3); v(t_3) > v(t_4); v(t_4) > v(t_5); v(t_5) > v(t_6); v(t_6) > v(t_7); (v(t_7) > v(t_8))$. The conditions

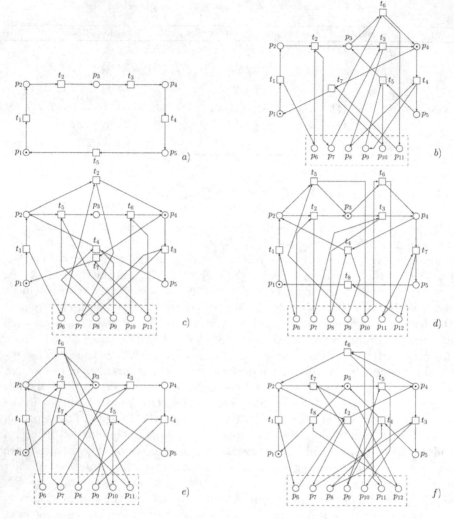

**Fig. 3.** The P/T PN models of the schedules: a) and b) (in the upper row); c) and d) (in the middle row); e) and f) (in the lower row). The dashed rectangles inclose the corresponding supervisors. More precisely, the supervisors are created by the inclosed places together with the directed arcs connecting them with the non-supervised models.

in the matrix form can be written as $\mathbf{L}_v.\mathbf{v} < \mathbf{0}$ where for 7 transitions (in case of the schedules b), c) and e))

$$\mathbf{L}_v = \begin{pmatrix} -1 & 1 & 0 & 0 & 0 & 0 & 0 \\ 0 & -1 & 1 & 0 & 0 & 0 & 0 \\ 0 & 0 & -1 & 1 & 0 & 0 & 0 \\ 0 & 0 & 0 & -1 & 1 & 0 & 0 \\ 0 & 0 & 0 & 0 & -1 & 1 & 0 \\ 0 & 0 & 0 & 0 & 0 & -1 & 1 \end{pmatrix}$$

For 8 transitions (in case of the schedules d) and f)) the matrix has one additional row and one additional column. The matrix $\mathbf{L}_v$ was utilized at the supervisor synthesis by means of the procedure (1) - (3), where $\mathbf{L}_p = \mathbf{0}$, $\mathbf{L}_t = \mathbf{0}$.

## 4   TPN Model, Simulation and Performance Evaluation

As it was premised, time relations depend on the machining times of the particular machines, on the time of moving $\Re$ between the devices, on its idle times, etc. These times will be assigned as the delays of the corresponding transitions of the P/T PN. Then, the TPN model will be at disposal. In the particular schedules the transition times are concerning the following activities of $\Re$:

a) $\tau(t_1)$ = picking up from $IN$ + moving to $M1$ + loading $M1$; $\tau(t_2)$ = waiting at $M1$ + unloading $M1$ + moving to $M2$ + loading $M2$; $\tau(t_3)$ = waiting at $M2$ + unloading $M2$ + moving to $M3$ + loading $M3$; $\tau(t_4)$ = waiting at $M3$ + unloading $M3$ + moving to $OUT$ + putting on $OUT$; $\tau(t_5)$ = moving from $OUT$ to $IN$.

b) $\tau(t_1)$ = picking up from $IN$ + moving to $M1$ + loading $M1$; $\tau(t_2)$ = waiting at $M1$ + unloading $M1$ + moving to $M2$ + loading $M2$; $\tau(t_3)$ = moving to $M3$; $\tau(t_4)$ = waiting at $M3$ + unloading $M3$ + moving to $OUT$ + putting on $OUT$; $\tau(t_5)$ = moving to $M2$; $\tau(t_6)$ = waiting at $M2$ + unloading $M2$ + moving to $M3$ + loading $M3$; $\tau(t_7)$ = moving from $M3$ to $IN$.

c) $\tau(t_1)$ = picking up from $IN$ + moving to $M1$ + loading $M1$; $\tau(t_2)$ = moving to $M3$; $\tau(t_3)$ = waiting at $M3$ + unloading from $M3$ + moving to $OUT$ + putting on $OUT$; $\tau(t_4)$ = moving to $M1$; $\tau(t_5)$ = waiting at $M1$ + unloading $M1$ + moving with the part to $M2$ + loading the part to $M2$; $\tau(t_6)$ = waiting at $M2$ + moving to $M3$ + loading into $M3$; $\tau(t_7)$ = moving empty to $IN$.

d) $\tau(t_1)$ = picking up from $IN$ + moving to $M1$ + loading the part to $M1$; $\tau(t_2)$ = moving to $M2$; $\tau(t_3)$ = waiting at $M2$ + unloading $M2$ + moving to $M3$ + loading into $M3$; $\tau(t_4)$ = moving to $M1$; $\tau(t_5)$ = waiting at $M1$ + unloading $M1$ + moving $M2$ + loading into $M2$; $\tau(t_6)$ = moving to $M3$; $\tau(t_7)$ = waiting at $M3$ + unloading $M3$ + moving to $OUT$ + putting on $OUT$; $\tau(t_8)$ = moving to $IN$.

e) $\tau(t_1)$ = picking up from $IN$ + moving to $M1$ + loading $M1$; $\tau(t_2)$ = moving to $M2$; $\tau(t_3)$ = waiting at $M2$ + unloading $M2$ + moving to $M3$ + loading $M3$; $\tau(t_4)$ = waiting at $M3$ + unloading $M3$ + moving to $OUT$ + putting on $OUT$; $\tau(t_5)$ = moving to $M1$; $\tau(t_6)$ = waiting at $M1$ + unloading $M1$ + moving to $M2$ + loading $M2$; $\tau(t_7)$ = moving from $M2$ to $IN$.

f) $\tau(t_1)$ = picking up from $IN$ + moving to $M1$ + loading $M1$; $\tau(t_2)$ = moving to $M3$; $\tau(t_3)$ = waiting at $M3$ + unloading $M3$ + moving to $OUT$ + putting on $OUT$; $\tau(t_4)$ = moving to $M2$; $\tau(t_5)$ = waiting at $M2$ + unloading $M2$ + moving to $M3$ + loading into $M3$; $\tau(t_6)$ = moving to $M1$; $\tau(t_7)$ = waiting at $M1$ + unloading $M1$ + moving to $M2$ + loading into $M2$; $\tau(t_8)$ = moving from $M2$ to $IN$.

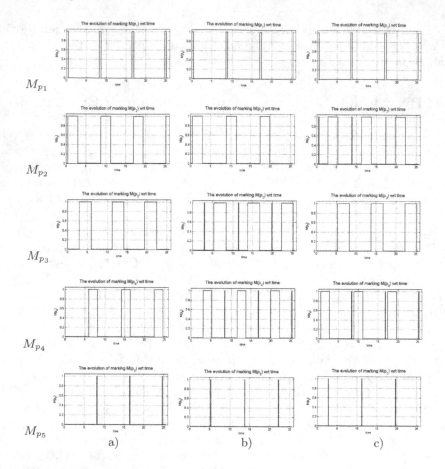

**Fig. 4.** TPN markings $M_{p_i} \in \{0, 1\}$, i.e. $\sigma_{p_i}$, $i = 1, \ldots, 5$, wrt. time for the schedules a) (left column), b) (middle column), c) (right column) on the time interval $< 0, 25 >$

## 4.1   Simulation Results

Consider the particular times in a time unit as follows: $\tau_{IN}$ = picking up from $IN = 0.15$; $\tau_{OUT}$ = putting on $OUT = 0.15$; $\tau_{mov}$ = each movement of free $\Re$ = 0.1; $\tau_{upMi}$ = uploading each machine = 0.2; $\tau_{outMi}$ = unloading each machine = 0.2; $\tau_{M1}$ = machining in $M1 = 2.0$; $\tau_{M2}$ = machining in $M2 = 2.5$; $\tau_{M3}$ = machining in $M3 = 1.75$. The results (TPN marking evolution) achieved for the time parameters are displayed in Fig. 4 and Fig. 5. In all schedules the marking of $p_5$, $M_{p_5}$ (lower pictures), expresses the number of the final products and the time in which they were put on $OUT$. Comparing these markings on the time interval $< 0, 25 >$ shows that the best throughput yields the schedule c). Other schedules succeed in the following order: f), b), e), a), d). Of course, when different time parameters are given, the throughput may be completely different.

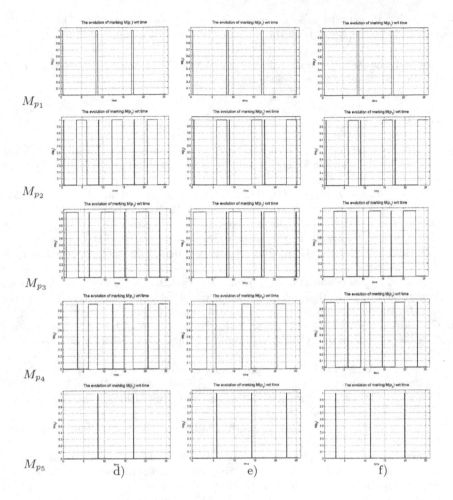

$M_{p_1}$

$M_{p_2}$

$M_{p_3}$

$M_{p_4}$

$M_{p_5}$

d)    e)    f)

**Fig. 5.** TPN markings $M_{p_i} \in \{0,1\}$, i.e. $\sigma_{p_i}$, $i = 1,\ldots,5$, wrt. time for the schedules d) (left column), e) (middle column), f) (right column) on the time interval $< 0,25 >$

## 5    Conclusion

The problem of the throughput of the concrete robotic cell was analysed and investigated by means of PN. P/T PN were used for the modelling the particular simple schedules and for improvement of their properties by means of supervision. TPN models, arising from such models by assigning the operation times to P/T PN transitions, were used in the simulation process. The simulation results yield the behaviour of the models in time. Comparing the results of the particular schedules the best schedule (as to the throughput) was found. Because the suitable PN models can be handled in analytical terms the proposed approach is simple and sufficiently general. Thus, it may be used also for other kinds of robotic cells. In general, also other kinds of rules can be used in (1) than the priorities of the steps used for the specific robotic cell examined here.

**Acknowledgments.** The author thanks the Slovak Grant Agency for Science VEGA for the support under grant # 2/0039/13.

# References

1. Abadi, I.N.K., Gholami, S.: Robot Movements in a Cyclic Multiple-Part Type Three-Machine Flexible Robotic Cell Problem. Transactions E: Industrial Engineering **16**, 36–54 (2009)
2. Čapkovič, F.: Cooperation of agents in manufacturing systems. In: Jedrzejowicz, P., Nguyen, N.T., Howlet, R.J., Jain, L.C. (eds.) KES-AMSTA 2010, Part I. LNCS (LNAI), vol. 6070, pp. 193–202. Springer, Heidelberg (2010)
3. Dawande, M., Geismar, H.N., Sethi, S.P., Sriskandarajah, C.: Sequencing and Scheduling in Robotic Cells: Recent Developments. Journal of Scheduling **8**, 387–426 (2005)
4. Fathian, M., Kamalabadi, I.N., Heydari, M., Farughi, H.: A Petri Net Model for Part Sequencing and Robot Moves Sequence in a 2-Machine Robotic Cell. Journal of Software Engineering and Applications **4**, 603–608 (2011)
5. Gultekin, H., Akturk, M.S., Karasan, O.E.: Scheduling in a Three-Machine Robotic Flexible Manufacturing Cell. Computers and Operation Research **34**, 2463–2477 (2007)
6. Iordache, M.V.: Methods for the Supervisory Control of Concurrent Systems Based on Petri Nets Abstraction. Ph.D. Thesis, University of Notre Dame, USA (2003)
7. Murata, T.: Petri Nets: Properties, Analysis and Applications. Proceedings of the IEEE **77**, 541–580 (1989)
8. Peterson, J.L.: Petri Nets Theory and the Modelling of Systems. Prentice-Hall Inc., Englewood Cliffs (1981)
9. Popova-Zeugmann, L.: Time Petri Nets: Theory, Tools and Applications. Part 1. http://www2.informatik.hu-berlin.de/~popova/1-part-short.pdf, Part 2. http://www2.informatik.hu-berlin.de/~popova/2-part-short.pdf
10. Sessego, F., Giua, A., Seatzu, C.: HYPENS: A matlab tool for timed discrete, continuous and hybrid Petri nets. In: van Hee, K.M., Valk, R. (eds.) PETRI NETS 2008. LNCS, vol. 5062, pp. 419–428. Springer, Heidelberg (2008)
11. Sessego, F., Giua, A., Seatzu, C.: HYPENS Manual. http://www.diee.unica.it/automatica/hypens/Manual_HYPENS.pdf
12. Sethi, S.P., Sriskandarajah, C., Sorger, G., Blazewicz, J., Kubiak, W.: Sequencing of parts and robot moves in a robotic cell. International Journal of Flexible Manufacturing Systems **4**, 331–358 (1992)
13. Zhu, Q.H., Qiao, Y.: Scheduling Single-Arm Multi-Cluster Tools with Lower Bound Cycle Time via Petri Nets. International Journal of Intelligent Control and Systems **17**, 113–123 (2012)
14. Zuberek, W.M.: Optimal schedules of manufacturing cells - modeling and analysis using timed Petri nets. In: IEEE International Symposium of Industrial Electronics -ISIE 1996, pp. 1055–1060. IEEE Press, New York (1996)
15. Zuberek, W.M., Kubiak, W.: Timed Petri Nets in Modelling and Analysis of Simple Schedules for Manufacturing Cells. International Journal of Computers and Mathematics with Applications **37**, 191–206 (1999)

# Algorithm to Plan Athlete's Prolonged Training Based on Model of Physiological Response

Krzysztof Brzostowski[(✉)], Jarosław Drapała,
Grzegorz Dziedzic, and Jerzy Świątek

Wrocław University of Technology, Wyb. Wyspiańskiego 27,
50-370 Wrocław, Poland
{krzysztof.brzostowski,jaroslaw.drapala,jerzy.swiatek}@pwr.edu.pl
http://www.pwr.edu.pl

**Abstract.** This paper proposes an algorithm to generate a long-term training for athletes. After introduction and a short review on methods of modelling the physiological response, the problem of planning prolonged training is formulated as optimization problem. In order solve this problem dynamical programming and model of physiological response was proposed. This model allows us to analyse the athlete's physiological response for different training loads. Based on this analysis and apply dynamical programming we proposed algorithm to design a plan of prolonged training with various training loads. In order to verify the proposed approach some simulation experiments were performed. Obtained results for our approach were compared with results obtained with use of algorithm generates a training plan without knowledge on the type of athlete's physiological response.

**Keywords:** e-Health · Mathematical modelling · Optimization · Dynamic programming

## 1 Introduction

Availability of mobile computation units such as smarthphones and various wearable sensors such as accelerometers, gyroscopes, heart rate monitors or GPS opens completely new areas for modern applications in sport [13], [14]. Currently on the market many different applications such as Endomondo, RunKeeper, Runtastic or SportsTracker are available.

Typical features of these applications are tracking sport session with use of GPS or heart rate monitor, generating statistics and calculating, e.g. burnt calories. Some of them have built-in pre-defined prolonged training plan. Usually the number of such predefined plans are limited and they are not adjusted to the user skills and abilities. Moreover, these applications are not able to analyse user's progress and cannot adapt prolonged training to these changes. Lack of this feature is significant flaw of such solutions because it is not possible to generate training plan in long-term horizon.

© Springer International Publishing Switzerland 2015
N.T. Nguyen et al. (Eds.): ACIIDS 2015, Part I, LNAI 9011, pp. 273–283, 2015.
DOI: 10.1007/978-3-319-15702-3_27

It must be emphasised that such applications cannot replace trainer for professionals but it can be useful to enhance cooperation between trainer and trainee. Moreover, these solutions for recreational athletes is desirable alternative for personal trainer.

There is literature in this field that describe methods to plan sport training [11], [21]. Some of such methods can be applied in modern mobile systems in order to support trainer of professional ones or recreational athletes.

It is possible to develop applications with more advanced functionalities. It can be done by combing mathematical modelling with data acquired from sensors such as GPS or heart rate. One of the example is injury or over-training prevention [16], [19], [8].

In this paper we consider the problem of generating training plan in long-term horizon. To this end well-known mathematical models are applied. Contribution of this work is development of a model combining two submodels: one for long-term effects of training and the second one for short-term simulation of athletes performing exercise. This constitutes a virtual athlete and may be treated as a testbed for long-term training routines. Proposed approach was verified by simulation studies.

## 2    Modelling of Physiological Response: A Review

Relationship between training and physical response for exercise can be modelled with use of general and dedicated models. As an example for the first case, in [18], [15], [17], the authors considered the application of Neural Networks for training plan generation. In these papers it was shown that – using Neural Networks – it is possible to predict output (e.g. physiological response of athlete in our case) based on input (e.g. training load). Unfortunately, it is not possible to find out what is the nature of relationship between them. Moreover it is not possible to establish relation between Neural Network's weights with athlete's vital signs.

In the second case (dedicated models), the final model is the results of analysis of underlying process and relationships between input(s) and output(s). Widely used model, in the field of modelling athlete's physiological response, is model proposed by Banister [5], [1], [2], [20]. The model is based on observation that athlete's performance is combination of fitness and fatigue responses to training load in a long-term horizon.

There are many applications of this model in sport science. In [4] authors reported application of Banister's model to predict weight lifters performance. The papers [9] and [7] present applications of this model to support the training of triathlon athletes and runners. This model can be used not only for athletes but – as it was reported in [3] – to plan a rehabilitation program for patients with coronary artery disease.

## 2.1   Banister's Model

The general idea behind Banitser's model is that athlete's performance is the results of fitness and fatigue effects as a response to physiological training. Fitness has a positive whilst fatigue has a negative impact on athlete's performance.

The relationship between training load and physical response is as follows [5]:

$$p(t) = p^* + w(t) * g(t) \qquad (1)$$

where: $p(t)$ is the physical response of athlete to exercise, $p^*$ stands for the initial physical state of the athlete, $w(t)$ is training load and $g(t)$ describes the response of the athlete's body to the training load.
Function $g(t)$ has the form:

$$g(t) = k_1 e^{-t/\tau_1} - k_2 e^{-t/\tau_2} \qquad (2)$$

where: $k_1$, $k_2$ are factors describing body response to fitness and fatigue respectively, $\tau_1$, $\tau_2$ are factors related to the time course of decay between training sessions for fitness and fatigue respectively.

The training load has cumulative nature. Taking it into account, the equation (1) can be rewritten in the form:

$$p(t) = p^* + \int_0^t w(t - t') \cdot g(t')dt \qquad (3)$$

where:

$$w(t) * g(t) = \int_0^t w(t - t') \cdot g(t')dt \qquad (4)$$

is the convolution.
Equation (3) can be represented in the discretized form:

$$\hat{p}_n = p^* + k_1 \sum_{i=1}^{n-1} w_i e^{-(n-i)/\tau_1} - k_2 \sum_{i=1}^{n-1} w_i e^{-(n-i)/\tau_2}. \qquad (5)$$

where: $n$ is the time interval (e.g. day). Parameters $k_1$, $\tau_1$ are connected with fitness and $k_2$, $\tau_2$ are related to fatigue. Because fitness has positive and fatigue has negative impact on athlete's physiological state it is clear that large ratios: $\frac{k_1}{k_2}$ and $\frac{\tau_1}{\tau_2}$ characterize athlete's with a better shape.

It is worth emphasising that proposed model is linear which means that non-linear elements of relationship describing training load and physiological response are not taken into account. Moreover parameters $k_1$, $\tau_1$ and $k_2$, $\tau_2$ are not only related to the athlete's but they are also connected with sport discipline [5]. For example the most important element that influences on marathon runner's performance is endurance. On the other hand, for the weight lifter important is strength. It means that mentioned parameters are connected with not only athlete but with sport discipline as well.

In [12] the authors indicate some limitations of Banister's model. One of them is the poor quality of performance prediction. This is due to ignoring time-invariant effects related to gain term of fatigue and fitness. Moreover, it is not easy to link model's parameters with vital signs of athlete.

Therefore, it is necessary to evaluate periodically the physiological response of the athletes and his/her current performance and update the model and the training plan.

## 3    Algorithm to Generate Long-Term Training Plan

All results in the work concern generating the runner's long-term training plans.

### 3.1    Training Load

In order to determine athlete's response to exercise (5) it is necessary to quantify training load. In [1] the authors proposed following formula to determine training load:

$$w(t) = D \cdot \Delta\text{HR} \cdot \alpha \exp\left(\beta\Delta\text{HR}\right) \tag{6}$$

where:

$$\Delta\text{HR} = \frac{\text{HR}_{ex} - \text{HR}_{rest}}{\text{HR}_{max} - \text{HR}_{rest}} \tag{7}$$

Factors $\alpha$ and $\beta$ have different values for male and female: $\alpha = 0.64$ and $\beta = 1.92$ and $\alpha = 0.86$ and $\beta = 1.67$ respectively [2]. These values define lactate profiles of male and female. The term $D$ is the time of exercise. $\text{HR}_{rest}$ is the resting heart rate and $\text{HR}_{max}$ is the maximum heart rate differ from sex and age. $\text{HR}_{rest}$ is measured when an athlete is in a neutrally temperate environment, relaxed and well-rested. Whereas $\text{HR}_{max}$ is determined by use of following formula [10]:

$$\text{HR}_{max} = 192 - 0.007 \cdot age^2 \tag{8}$$

where $age$ stands for athlete's age.

The next term in the model (7) is $\text{HR}_{ex}$. It is the heart rate during exercise. In practical application this is measured with the use of heart rate monitor (see for example [22]). In our case heart rate response $\text{HR}_{ex}$ is determined with use of heart rate response model [6].

### 3.2    Heart Rate Response Model

The mathematical model describing the response of the cardiovascular system to physical exercise has following form:

$$
\begin{aligned}
\dot{x}_1(t) &= -a_1 x_1(t) + a_2 x_2(t) + a_2 u^2(t) \\
\dot{x}_2(t) &= -a_3 x_2(t) + \phi(z(t)) \\
z(t) &= x_1(t) \\
\phi(z(t)) &:= \frac{a_4 x_1(t)}{1 + e^{-(x_1(t) - a_5)}}
\end{aligned}
\tag{9}
$$

where: $x_1$ is heart rate change from the rest $HR_{rest}$, $u$ denotes speed of the exerciser, $x_2$ may be considered as fatigue, caused by such factors as: – vasodilation in the active muscles leading to low arterial blood pressure, – accumulations of metabolic byproducts (e.g. lactic acid), – sweeting and hyperventilation. Fatigue cannot be directly measured, it may only be worked out on the basis of the HR. Parameters $a_1, \ldots, a_5$ take non-negative values. The values of these parameters are obtained by the estimation procedure, with use of data measured from few experiments. Such procedure was described in [6]. Each user may be characterized by different values of the model parameters. Moreover, each user has its own speed limit $u^{max}$ and fatigue limits $x_2^{max}$.

As a result of training, parameters of the HR model vary. We propose the following function relating the training load and variations of model parameters:

$$a_i(t+1) = a_i(t) + \frac{b(i) \cdot a_i(0)}{c} \cdot \frac{trimp(t)}{trimp^*} \cdot \frac{a_i(0)}{a_i(t)}, \tag{10}$$

where

$$b(i) = \{ \begin{matrix} 1, & i = 1 \ or \ i = 3 \\ -1, & i = 2, \ i = 4 \ or \ i = 5 \end{matrix}, \tag{11}$$

for $i = 1, 2, 3, 4, 5$, where $t$ is time measured in successive days of training, $a_1(t), a_2(t), a_3(t), a_4(t), a_5(t)$ are HR model parameters at successive days of training and $a_I(0)$ stands for initial values. $trimp$ denotes training load per day. Parameter $trimp^*$ is the upper limit of $trimp$ and its value may be fixed be an experiment. Parameter $c$ controls speed of changes of values of model parameters. We set $trimp^* = 100$ and $c = 1000$ as typical values for simulation study.

Putting it all together (equations 5, 6, 9, 10 and 11) we can illustrate our proposal in the Fig. 1.

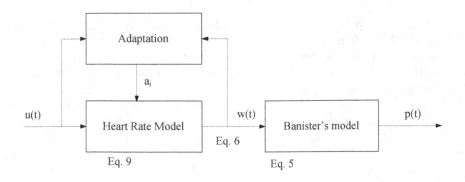

**Fig. 1.** System to generate long-term training plan

## 3.3    Problem Formulation

Based on recommendations on planning physiological training for different running types below some basic rules for our methods have been presented. The most important are:

- it is necessary to indicate start and end time of prolonged training;
- training plan is divided into two periods: active and regeneration phases;
- by period we mean one week.

The problem of generating long-term training for runners can be formulated as a dynamic programming problem. The formulation is as follows:
**For given**:

- model:
$$p_{n+1} = f(p_n, w_n) \quad \text{with initial conditions:} \quad p_0$$
  where: $w_n$ is decision (work load) in $n$th step, $p_n$ runner's performance in $n$th step (see 5);
- performance index:
$$Q_N(\bar{w}_N) = p_N$$

**find**:
$$\bar{w}_N^* = (\bar{w}_0^*, \bar{w}_1^*, \bar{w}_2^*, ..., \bar{w}_{N-1}^*)$$

such as:
$$(\bar{w}_0^*, \bar{w}_1^*, \bar{w}_2^*, ..., \bar{w}_{N-1}^*) = \min_{w_n \in \mathscr{R}} Q_N(\bar{w}_N) \qquad (12)$$

The dynamic programming routine solving the problem above is given below in the box on the next page.

## 4    Simulation Studies

In order to verify proposed algorithm simulation studies have been performed. To this end the model's parameters (5) are fixed as follows (see Tab. 1). In this table values of the parameters are calculated from experimental data and they are taken from the work [6]. Whilst the values of the parameters $\tau_1$, $\tau_2$, $k_1$, $k_2$ are selected based on guideline in the work [20].

In (Fig. 2) example solution for following set of parameters (see Tab. 1) are presented. In this case prolonged training was designed for 100 days.

It is worth to pay attention to the last 10 days of the generated plan. Because defined aim of the physiological training is reaching a maximum of performance at the end of 100 days preparation periods. In the theory of training the effect is well known and it is recommended to take it into account planning exercise routine. In our example proposed algorithm determines the plan in which the last few days have low intensity. These phases helps to reach the super compensation effect which allows to maximize performance at the end of preparation periods.

---

**Algorithm 1.** Algorithm for generating prolonged training.

---

1: **function** GENERATETRAININGLOAD($start, n, p^*$)
2:     ▷ $start$ - the first day for training generation
3:     ▷ $n$ - the number of day after which performance of athlete is evaluated
4:     ▷ $p^*$ - performance of athlete at the day $start$
5:     ▷ $tload[]$ - training load for days from $start$ to $n$
6:     ▷ $bestResult$ - the best result achieved by an athlete
7:     ▷ itwT - training period for intensive weeks, default value is 100 minutes
8:     ▷ ritwT - training period for restful weeks , default value is 0,5 * itwT
9:     $i \leftarrow 0$
10:    $tload \leftarrow$ []
11:    $bestResult \leftarrow 0$
12:    $curResult \leftarrow 0$                                  ▷ current result achieved by an athlete
13:    $bestTLoad \leftarrow 0$                                  ▷ optimal training load
14:    $t \leftarrow$ itwT or ritwT                                  ▷ depending on current period
15:    $s \leftarrow MaximumPosssibleSpeed(t)$
16:    **if** $t == 0$ **then**
17:        $ds \leftarrow [0]$
18:    **else**
19:        $ds \leftarrow [0,...,s]$                                  ▷ set of speeds for testing
20:    **for each** $s1$ in ds **do**
21:        $ods.InsertToTail(TrainingLoad(s1,t))$
22:    $i \leftarrow 0$
23:    **for each** o in ods **do**
24:        $curResult \leftarrow MeasureTrainingPerformance(start, p^*, tload, n)$
25:        **if** (start + 1 < n) **then**
26:            $tload \leftarrow GenerateTrainingLoad(start+1, n, p^*)$
27:        **if** curResult > bestResult **then**
28:            $bestResult \leftarrow curResult$
29:            $bestTLoad \leftarrow o$
30:            $bestTtime \leftarrow t$
31:            $bestTspeed \leftarrow ds[i]$
32:        $tload.popLast()$
33:        $i \leftarrow i + 1$
34:    ▷ the best proposal of training load, time and speed for a day
35:    $ot.InsertToHead(bestTLoad)$
36:    $ot.InsertToHead(bestTtime)$
37:    $ot.InsertToHead(bestTspeed)$
38:    **return** $ot, bestResult$

---

**Table 1.** Values of parameters for model (5)

| Parameter | Value |
|---|---|
| $a_1$ | $1,84$ |
| $a_2$ | $26,04$ |
| $a_3$ | $6,36 \times 10^{-2}$ |
| $a_4$ | $3,21 \times 10^{-3}$ |
| $a_5$ | $8,32$ |
| $\tau_1$ | $60$ |
| $\tau_2$ | $6$ |
| $k_1$ | $1,9 \times 10^{-3}$ |
| $k_2$ | $7,3 \times 10^{-3}$ |
| $age$ | $24\ [age]$ |
| $\mathrm{HR_{rest}}$ | $74\ [\mathrm{bps}]$ |

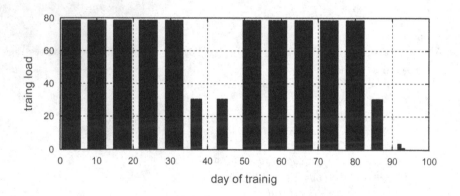

**Fig. 2.** Generated prolonged training plan

In the second experiments we want to compare results obtained with use of proposed algorithm (Fig. 3) and results acquired for algorithm where training loads were constants (Fig. 4). The rest rules of plan generation were the same as listed in (3).

In (Tab. 2) results of other simulation with different training loads were compared as well. It can be seen that application models (5), (9) and dynamical programming allow us to obtain better results than determined with use of constant training load.

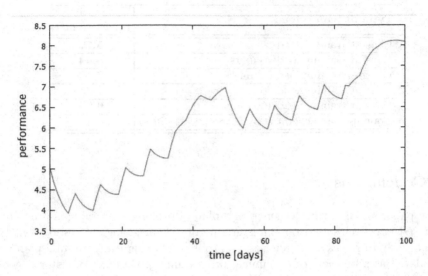

**Fig. 3.** Results of simulation for set of parameters (Tab. 1) and proposed algorithm to generate prolonged training

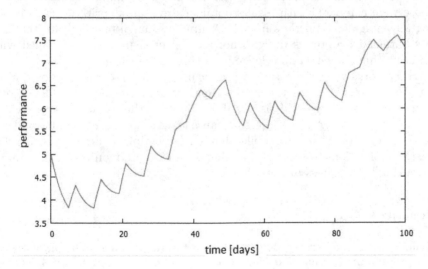

**Fig. 4.** Results of simulation for set of parameters (Tab. 1) and constant training load (80 units)

**Table 2.** Comparison results obtained for different constant training loads with results determined with use of proposed algorithm

| Type of traing load | Final performance |
|---|---|
| Constat training load 20 units | 5,72 |
| Constat training load 40 units | 6,44 |
| Constat training load 60 units | 7,17 |
| Constat training load 80 units | 7,35 |
| Constat training load 100 units | 6,02 |
| Training load generated by proposed algorithm | 8,09 |

# 5   Conclusions

In the paper an algorithm to generate a plan of prolonged training is presented. Proposed approach based on Banister's model of physiological response of the human body to physical exercises. The problem of generating training plan was formulated as a problem of dynamical programming. Because Banister's model is specific not only for concrete athlete but for the specific sport discipline we are focusing our attention on runners.

Results obtained for proposed algorithm based on model of physiological response were compared with results obtained from an algorithm which is not taking advantage of Banister's model. Computer simulations indicate that the results obtained for proposed algorithm are better than results obtained with use of algorithms in which knowledge from this model was not used.

Further works will be focused on practical application of the proposed algorithm. To this end some methods to estimate training load and athlete's performance must be designed. It is possible only with the use of sensors such as heart rate monitor or accelerometers. Because these sensors are available and can be easily integrated with e.g. mobile phone it is possible to design and develop pervasive applications to support planning exercise plan, monitoring of training and tracking the progress of the athlete.

# References

1. Banister, E.W., Calvert, T.W.: Planning for future performance: implications for long term training. Canadian Journal of Applied Sport Sciences **5**, 170–176 (1980)
2. Borresen, J., Lambert, M.I.: The quantification of training load, the training response and the effect on performance. Sports Medicine **39**(9), 779–795 (2009)
3. le Bris, S., Ledermann, B., Topin, N., Messner-Pellenc, P., Le Gallais, D.: A systems model of training for patients in phase 2 cardiac rehabilitation. International Journal of Cardiology **109**(2), 257–263 (2006)

4. Busso, T., Häkkinen, K., Pakarinen, A., Carasso, C., Lacour, J.R., Komi, P.V., Kauhanen, H.: A systems model of training responses and its relationship to hormonal responses in elite weight-lifters. European Journal of Applied Physiology and Occupational Physiology **61**, 48–54 (1990)
5. Calvert, T.W., Banister, E.W., Savage, M.V., Bach, T.: A systems model of the effects of training on physical performance. IEEE Transactions on Systems, Man and Cybernetics **2**, 94–102 (1976)
6. Cheng, T.M., Savkin, A.V., Celler, B.G., Su, S.W., Wang, L.: Nonlinear modeling and control of human heart rate response during exercise with various work load intensities. IEEE Transactions on Biomedical Engineering **55**(11), 2499–2508 (2008)
7. Millet, G.P., Candau, R.B., Barbier, B., Busso, T., Rouillon, J.D., Chatard, J.C.: Modelling the transfers of training effects on performance in elite triathletes. **23**, 55–63 (2002)
8. Morgan, W.P., Brown, D.R., Raglin, J.S., O'connor, P.J., Ellickson, K.A.: Psychological monitoring of overtraining and staleness. British Journal of Sports Medicine **21**, 107–114 (1987)
9. Morton, R.H., Fitz-Clarke, J.R., Banister, E.W.: Modeling human performance in running. Journal of Applied Physiology **69**, 1171–1177 (1990)
10. Gellish, R.L., Goslin, B.R., Olson, R.E., McDonald, A.U.D.R.Y., Russi, G.D., Moudgil, V.K.: Longitudinal modeling of the relationship between age and maximal heart rate. Medicine and Science in Sports and Exercise **39**(5), 822–829 (2007)
11. Goater, J., Melvin, D.: The Art of Running Faster. Human Kinetics (2012)
12. Hellard, P., et al.: Assessing the limitations of the Banister model in monitoring training. Journal of Sports Sciences **24**, 509–520 (2006)
13. Kirwan, M., Duncan, M.J., Vandelanotte, C., Mummery, W.K.: Using smartphone technology to monitor physical activity in the 10,000 Steps Program: a matched case-control trial. Journal of Medical Internet Research **14** (2012)
14. Lim, J.-E., Choi, O.-H., Na, H.-S., Baik, D.-K.: A context-aware fitness guide system for exercise optimization in U-health. IEEE Transactions on Information Technology in Biomedicine **13**, 370–379 (2009)
15. Linder, R., Mohamed, E.I., De Lorenzo, A., Pöppl, S.J.: The capabilities of artificial neural networks in body composition research. Acta Diabetologica **40**, s9–s14 (2003)
16. Nguyen, T.N., Su, S., Celler, B., Nguyen, H.: Advanced portable remote monitoring system for the regulation of treadmill running exercises. Artificial Intelligence in Medicine (2014)
17. Pfeiffer, M., Hohmann, A.: Applications of neural networks in training science. Human Movement Science **31**(2), 344–359 (2012)
18. Silva, A.J., et al.: The use of neural network technology to model swimming performance. Journal of Sports Science & Medicine **6**, 117–125 (2007)
19. Su, S.W., Huang, S., Wang, L., Celler, B.G., Savkin, A.V., Guo, Y., Cheng, T.M.: Optimizing heart rate regulation for safe exercise. Annals of Biomedical Engineering **38**, 758–768 (2010)
20. Taha, T., Thomas, S.G.: Systems modelling of the relationship between training and performance. Sports Medicine **33**(14), 1061–1073 (2003)
21. Zatsiorsky, V.M., Kraemer, W.J.: Science and practice of strength training. Human Kinetics (1995)
22. www.polar.com

# On a Simple Game Theoretical Equivalence of Voting Majority Games with Vetoes of First and Second Degrees

Jacek Mercik[1(✉)] and David Ramsey[2]

[1] Wroclaw School of Banking, Wroclaw, Poland
jacek.mercik@wsb.wroclaw.pl
[2] Wroclaw University of Technology, Wroclaw, Poland
david.ramsey@pwr.edu.pl

**Abstract.** Introducing a veto into the process of group decision making (voting, aggregating preferences) drastically changes the position of decision makers and, consequently, it changes their power index. In this paper we derive the Shapley-Shubik and Penrose-Banzhaf indices for a class of voting games with vetoes. We also present a way of constructing a simple voting game which is equivalent to a game with vetoes of first degree. This simplifies the calculation of power indices by allowing us to use standard algorithms which are available online.

**Keywords:** A priori power index · Veto · Majority game · Equivalence

## 1 Introduction

The analysis of vetoes in committee decision making has a wide spectrum of aspects. This article has three related goals. First, we want to present a way of calculating the two main power indices, namely the Shapley-Shubik and Penrose-Banzhaf indices, directly by analysing the number of winning and non-winning coalitions. In order to do this, we carry out an analysis of the power to create or to destroy a winning coalition when both conditional vetoes and unconditional vetoes are introduced. Second, in the case of unconditional vetoes, we derive an equivalent simple voting game where players do not have vetoes. We can also indirectly evaluate how a veto increases the power of a given player by considering two voting games in which the players have the same weights, but the player of interest only has power of veto in one of these games.

Almost all a priori power indices are defined as the ratio between the number of particular winning coalitions (with restrictions following from the assumptions made, mostly regarding the relations between the players) to the number of all possible coalitions (sometimes majority coalitions only). Therefore, knowing the number of winning coalitions allows us to evaluate a given power index and, what is probably more useful, enables us to estimate the likelihood of adopting new acts. Consequently, in practice, the number of winning coalitions including a particular decision-maker will be a measure of the power of that decision-maker acting under a given voting procedure.

© Springer International Publishing Switzerland 2015
N.T. Nguyen et al. (Eds.): ACIIDS 2015, Part I, LNAI 9011, pp. 284–294, 2015.
DOI: 10.1007/978-3-319-15702-3_28

The article is set up as follows. The next section outlines the concept of a simple voting game with veto (conditional and unconditional) and the calculation of various power indices. We also consider the concept of equivalency between voting games. Section 3 presents a way of calculating a power index for a simple voting game and how to take vetoes into account in the calculations. Section 4 presents the calculation of power indices for games in which voters have equal voting rights, but some have a veto of first degree (voting yes/no). This section describes a procedure to define an equivalent voting game without vetoes in which players have different weights. Section 5 presents the calculation of power indices in games where voters have equal voting rights, but some have a veto of second degree (voting yes/no). Finally, there are some conclusions and suggestions for future research.

## 2    Preliminaries

Let $N$ be a finite set of committee members, $q$ be a quota and $w_j$ be the voting weight of member $j$, where $j \in N$. In this paper, we consider a special class of cooperative games called weighted majority games (later called simple games or simple voting games). A weighted majority game G is defined by a quota $q$ and a sequence of nonnegative numbers $w_i$, $i \in n$, where we may think of $w_i$ as the number of votes, or weight, of player $i$ and $q$ as the threshold or quota needed for a coalition to win. We assume that q and wj are nonnegative integers. A subset of players is called a coalition.

A game on N is given by a map v : 2N→ R with v($\emptyset$) = 0. The space of all games on N is denoted by G. The domain $SG \subset G$ of simple games on N consists of all $v \in G$ such that: (i) $v(S) \in \{0,1\}$ for all $S \in 2^N$ ; (ii) $v(N) = 1$; (iii) v is monotonic, i.e. if $S \subset T$ then $v(S) \leq v(T)$ . A coalition S is said to be winning in $v \in SG$ if $v(S) = 1$ and losing otherwise. Therefore, passing a bill, for example, is equivalent to forming a winning coalition consisting of voters. A simple game $(N,v)$ is said to be proper, if and only if the following is satisfied: for all T ⊂ N, if $v(T) = 1$ then v $(N \backslash T) = 0$ .

We analyse only simple and proper games where players may vote either yes-no or yes-no-abstain, respectively.

If a given committee member can transform any winning coalition into a non-winning one by using a veto, then that veto is said to be of first degree. If the veto of a given committee member turns some, but not all, winning coalitions not including that member into non-winning coalitions, then that veto is defined to be of second degree (Mercik, 2011). This type of veto is illustrated by Example 1.

**Example 1.** Suppose (Mercik, 2014) that we wish to send a signal from A to B using the connections numbered from 1 to 5. Each of these connections can be either functioning or non-functioning. In order to transmit the signal, there needs to be a sequence of functioning connections starting at A and finishing at B. Hence, for a signal to be transmitted: (i) connection 3 must be functional, (ii) at least one of connections 1 and 2 must be functional, (iii) at least one of connections 4 and 5 must be functional. It follows that at least 3 of the five connections must be functional to transmit the signal, but not all sets of connections satisfying this condition lead to the transmission of the signal. The sets of functional connections which lead to the signal being transmitted are as follows: {1,2,3,4,5}, {1,2,3,4}, {1,2,3,5}, {1,3,4,5}, {2,3,4,5}, {1,3,4}, {1,3,5}, {2,3,4}, {2,3,5}.

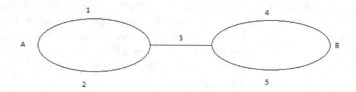

**Fig. 1.** Transfer of a signal between points A and B

Consider a voting game based on this scenario in which the connections represent voters. Each voter has a single vote and the event "a vote is passed" corresponds to the signal being transferred in the situation described above. Let the quota be 3 and assume that the players have the following veto powers:

a) player 3 can successfully veto any coalition,
b) player 1 (or player 2) can veto the coalition {3,4,5} (but e.g. player 1 cannot veto the coalition {2,3,4}, neither can player 2 veto the coalition {1,3,4}),
c) player 4 (or player 5) can veto the coalition {1,2,3} (but e.g. player 4 cannot veto the coalition {2,3,5}, neither can player 5 veto the coalition {1,3,4}).

Hence, player 3 has a veto of first degree and the remaining players have a veto of second degree. It can be seen that if we assume that the players outside of a coalition use their veto (rather than abstaining), then the set of winning coalitions coincides exactly with the sets of functional connections which lead to the signal being transmitted in the scenario above. As one may see above, the calculation of power indices for structured games (i.e. games where not all coalitions are allowed) seems to be relatively complicated, but even unstructured games (i.e. all coalitions are allowed) may be significantly complicated when conditional or unconditional vetoes are introduced. Therefore, if it is possible to replace a game with vetoes by a simple (but equivalent in the sense of power) majority game, then such a transformation of the problem is worth doing.

Formally speaking, we define a simple game to be equivalent to another simple game if: (i) the sets of players are the same, (ii) the set of winning coalitions is the same in both games.

Two games are said to be power equivalent with respect to a given index if: (i) the sets of players are the same, (ii) the powers of the players, as measured by the given index, in each game are the same.

If the above condition is satisfied for any power index, then we simply say that two games are power equivalent.

From the definition of any power index, it follows that if two games are equivalent, then they are power equivalent. However, it will be shown below that power equivalence is a weaker concept, i.e. when two games are power equivalent, they are not necessarily equivalent.

The following is a very trivial example. Consider a voting game with 3 players in which each player has weight 1 and none of the players have veto rights. Consider another game of those 3 players in which each player has weight 2 and none of the players have veto rights. If both of these games are majority games, by symmetry, the value of any power index for each player is equal to 1/3 in both games. These two games are power equivalent. Moreover, in both of these games any coalition containing at least two players is winning. Hence, these games are equivalent.

It should be noted that many simple voting games can be equivalent, even if we restrict the weights of players to be integers, consider majority voting and fix the weight of a given player. For example, consider the voting game with 3 players, where players 1 and 2 have weight $k$ and player 1 have weight 1. Since the sum of weights is $2k+1$, under majority voting the quota is $k+1$. It can be seen that any coalition containing at least two players is a winning coalition. Hence, any such game is equivalent to the two described above. Note that from this example it can be seen the relative weights of players do not always reflect the power they have.

Finally, consider the voting game with three players in which unanimity must be reached in order to pass a motion. By symmetry, the value of any power index for each of these players is equal to 1/3. Hence, this game is power equivalent to the games described above. However, in this game the only winning coalition consists of all of the players. Hence, this game is not equivalent to the games described above.

Using such an approach, we may indirectly calculate the power player $i$ gains by possessing a veto by comparing his/her power index in two games where the players have the same voting weights, but player $i$ has the right to veto in just one of the games.

# 3 Power Indices

A power index is a mapping $\varphi: SG \to R^n$. For each $i \in N$ and $v \in SG$, the $i^{th}$ coordinate of $\varphi(v) \in R^n, \varphi(v)(i)$, is interpreted as the voting power of player $i$ in the game $v$. In the literature, there are two dominant power indices: the Shapley-Shubik (Shapley, Shubik (1954)) power index and the Banzhaf power index (Penrose (1946), Banzhaf (1965)). Both are based on the concept of the Shapley value.

Now, let us have a look at the consequence of introducing a veto. First, consider a voting game with 3 players in which each player has weight 1 and none of the players have veto rights. By symmetry, the value of the power index for each player is equal to 1/3. Now suppose that player 1 has a veto of first degree. It follows that $v(\{1\}) = v(\{2\}) = v(\{3\}) = v(\{2,3\})=0$, $v(\{1,2\})=v(\{1,3\})=v(\{1,2,3\})=1$.

Considering the permutations of the players and which player is pivotal in each case (i.e. the player who changes a losing coalition into a winning coalition), we have

| Permutation | Pivotal Player |
|:---:|:---:|
| 1,2,3 | 2 |
| 1,3,2 | 3 |
| 2,1,3 | 1 |
| 2,3,1 | 1 |
| 3,1,2 | 1 |
| 3,2,1 | 1 |

It follows that the Shapley-Shubik power index for player 1 is 2/3 and for the other two players is 1/6.

Now consider the Banzhaf power index. The three winning coalitions are {1,2}, {1,3}, {1,2,3}. In the first two coalitions, both players involved are swing voters (when a swing voter leaves a coalition, it becomes losing). Only Player 1 is a swing voter in the third coalition. Hence, there are in total 5 swing votes, of which Player 1 has 3 and the other players have one each. It follows that the Banzhaf power index for player 1 is 3/5 and for the other two players is 1/5.

## 4    Games in Which Voters Have Equal Voting Rights, but Some Have a Veto of First Degree (voting yes/no)

Consider a game in which there are $n$ voters, each of whom have weight 1. Let the quota $q$ satisfy $n/2 < q < n$ (so at least a normal majority is needed). It is assumed that $q < n$, since when $q = n$ essentially all the players have a veto of first degree. Suppose that $k$ of these voters (labelled $1,2,\ldots,k$) have vetoes of first degree. First note that when $k = 0$ or $k = n$, the Shapley-Shubik power index (and Banzhaf power index) of each player is equal to $1/n$, due to symmetry. However, the ability of these committees to pass motions can be radically different. When $k = n$, then unanimity must be achieved to pass a motion. When $k = 0$ and $q = \lfloor 1 + \frac{n}{2} \rfloor$, then the voting system is standard majority voting. Here, $\lfloor x \rfloor$ represents the integer part of $x$. For $0 < k < n$, in order for a motion to be passed the following conditions must be satisfied:

a)   each of the $k$ players with veto rights must vote for the motion.
b)   whenever $k < q$, at least $q - k$ of the players without a veto must vote for the motion.

It can be seen that when $k \geq q$, then any swing or pivot voter must have veto power of first degree. It follows by symmetry that in this case the power index (Shapley-Shubik or Banzhaf) for the veto players is equal to $1/k$ and for the remaining players is equal to 0. Now suppose that $k < q$. We wish to calculate the power indices for the two types of player. We first consider the Shapley-Shubik power index for both types of player. Note that there are $n!$ permutations of the players. We now count the number of permutations in which a given non-veto player is a pivot voter. For player $i$, $i > k$, to be a pivot voter, the following condition must be satisfied:

all the veto players and $q-k-1$ non-veto players of the remaining $n-k-1$ non-veto players must appear before player $i$.

Note that there are $\binom{n-k-1}{q-k-1} = \frac{(n-k-1)!}{(q-k-1)!\,(n-q)!}$ ways of choosing the players who appear before player $i$ and $(q-1)!$ ways of ordering these players. The players that come after player $i$ must be the $(n-q)$ non-veto players who did not appear before player $i$. There are $(n-q)!$ orderings of these players. It follows that the power of a non-veto player is given by

$$\varphi_i(v) = \frac{(n-k-1)!(q-1)!(n-q)!}{(q-k-1)!\,(n-q)!n!} = \frac{(n-k-1)!(q-1)!}{(q-k-1)!n!}, \quad i > k. \tag{1}$$

Using the symmetry of the game and the fact that the power indices must sum to one, the Shapley-Shubik power index of a veto player is given by

$$\varphi_i(v) = \frac{1}{k}\left[1 - (n-k)\frac{(n-k-1)!(q-1)!}{(q-k-1)!n!}\right] = \frac{1}{k}\left[1 - \frac{(n-k)!(q-1)!}{(q-k-1)!n!}\right], \quad i \leq k.$$

For example, in the game with 4 players where two players have a veto of first degree and a majority is needed to pass a motion, i.e. $n = 4$, $k = 2$, $q = 3$, then $\varphi_1(v) = \varphi_2(v) = \frac{5}{12}$, $\varphi_3(v) = \varphi_4(v) = \frac{1}{12}$.

We now calculate the Banzhaf power index for both types of player in such games. From the two conditions for a motion to be passed, the number of winning coalitions is given by

$$\#(W) = \sum_{j=q-k}^{n-k} \binom{n-k}{j}.$$

In each of these coalitions, all the $k$ veto players are swing voters. A non-veto player in a winning coalition is a swing voter if and only if there are $(q-k)$ non-veto players in the coalition. There are $\binom{n-k}{q-k}$ such coalitions and $(q-k)$ such swing votes in each coalition. It follows that the total number of swing votes is

$$\#(S) = (q-k)\binom{n-k}{q-k} + k\sum_{j=q-k}^{n-k}\binom{n-k}{j}.$$

It follows that the Banzhaf index of a veto player is given by

$$\varphi_i(v) = \frac{\sum_{j=q-k}^{n-k}\binom{n-k}{j}}{(q-k)\binom{n-k}{q-k} + k\sum_{j=q-k}^{n-k}\binom{n-k}{j}}, \quad i \leq k. \tag{2}$$

As before, the Banzhaf indices of a non-veto player can be calculated using the symmetry of the game. It follows that

$$\varphi_i(v) = \frac{1}{n-k}[1 - k\varphi_1(v)], \quad i > k.$$

For example, in the game with 4 players where two players have a veto and a majority is needed to pass a motion, i.e. $n = 4$, $k = 2$, $q = 3$, then $\varphi_1^B(v) = \varphi_2^B(v) = \frac{3}{8}$, $\varphi_3^B(v) = \varphi_4^B(v) = \frac{1}{8}$.

We now show that such a game is equivalent to a given voting game with no powers of veto, but the veto players are each given a larger weight and the quota is adjusted accordingly. Consider the $n$ player game where $k$ players have veto of first degree and the quota is $q$. We now define a game without veto power, but with weighted votes, that is equivalent. Let $w_i$ be the weight of player $i$ and $\bar{w}$ be the sum of these weights. Assume that the minimum weight of the votes against a motion $m$ necessary to stop it being passed is the same in both games, i.e $m = n - q + 1$, and non-veto players are given a weight of 1. Hence, we define the quota to be $\bar{w} - n + q$. Note that by giving each of the veto players a weight of $m$ in the new game, then each of them essentially remains a veto player, since the quota cannot be attained if any veto player votes against the motion. Also, from the definition of the weights, it is simple to see that if at least $q-k$ non-veto players vote for a motion, in addition to the veto players, then the motion will be passed. It follows that the two games are equivalent, i.e. have the same set of winning coalitions. It should be noted that any higher weights given to the veto players would satisfy this equivalence relationship (as long as the quota is changed appropriately – in fact the veto players could all be given different weights ≥ m and the resulting game would still be equivalent).

**Example 2.** The United Nations Security Council: 5 permanent members, 10 non-permanent members, total 15 members. In this example n = 15, k = 5 and q = 9. First we calculate the Shapley-Shubik power index directly. There are 15! sequential coalitions, which means about 1.3 trillion permutations. A non-permanent member is pivotal only if it is the 9th player in the coalition, preceded by all five of the permanent members and three non-permanent members. From the arguments made above in the derivation of equation (1), this happens in $\frac{9!8!}{3!}$ , i.e. about 2.44 billion, sequential coalitions. Each non-permanent member has a Shapley-Shubik power index of 2.44 billion/1.3 trillion or 0.001865. Dividing the rest of the power equally among the 5 permanent members, we get a Shapley-Shubik power index for them of 0.196270.

Now, we construct an equivalent simple game without veto players: $m = n - q + 1 = 15 - 9 + 1 = 7$. Hence, the initial representation of the voting game played by the Security Council of the UN is [9; 1*, 1*, 1*, 1*, 1*, 1, 1, 1, 1, 1, 1, 1, 1, 1, 1], where veto players are marked by stars. This is equivalent to the simple game [39; 7, 7, 7, 7, 7, 1, 1, 1, 1, 1, 1, 1, 1, 1, 1]. One may check (for example: http://homepages.warwick.ac.uk/~ecaae/ssdirect.html) that the Shapley-Shubik power index values of this game are exactly the same as in the initial game with veto players.

Moreover, we can calculate the "net value" of the veto: 0.19627-1/15 ≈ 0.129603 (i.e. the Shapley-Shubik index "with veto" minus the index "without veto"). Generally, one may expect that a veto should increase the power of a player, i.e. the "net value of the veto" should be greater than 0.

Using equation (2), the Penrose-Banzhaf index of a veto member is approximately 0.139350. It follows that the Penrose-Banzhaf index of a veto member is 0.1 [1-5×0.13935]≈0.030325. Hence, it can be seen that a non-veto player has more power to break up a winning coalition than to form a winning coalition.

## 5   Games in Which Voters Have Equal Voting Rights, but Some Have a Veto of Second Degree (voting yes/no)

Consider a voting game composed of $n$ players with weight 1 where $k$ players have a veto of second degree. The quota is $q$ if no veto is used and $q'$ when at least one veto is used, where $q' > q$. We first derive the Shapley-Shubik power index of a non-veto player in this game using the approach based on sequences of voters as above. When a non-veto player is in position $q$ in this sequence, he/she is a pivot player if and only if all the veto players   and $(q-k-1)$ of the other non-veto players appear beforehand. This corresponds precisely to this non-veto player being a pivot player in the game where the veto players have a veto of degree one. When $k \geq q'$, then any player is a pivot player if and only if he/she appears in position $q'$. This is due to the fact that not all of the veto players can appear in the first $q'$-1 positions. Hence, the motion cannot be passed by a coalition composed of the players appearing before position $q'$, but any coalition of $q'$ members can pass a motion. It follows that when $k \geq q'$, the Shapley-Shubik power index of each player is $1/n$.

Finally when $k < q'$, a non-veto player is a pivot player when he/she appears in position $q'$ when $j$, $j < k$, veto players and $q'$-$j$-1 of the other non-veto players appear beforehand. It follows from these conditions that $m \leq j \leq k$-1, where $m=\max\{0,q'+k-n\}$. The number of ways of choosing these players is $\binom{k}{j}\binom{n-k-1}{q'-j-1}$. There are $(q'$-1)! ways of ordering these players and $(n-q')!$ ways of ordering the players that come after the pivot player. When $k < q$, summing the number of these sequences and dividing by the total number of possible sequences, $n!$, we obtain

$$\varphi_i(v) = \frac{1}{n!}\left[\frac{(n-k-1)!(q-1)!}{(q-k-1)!} + \sum_{j=m}^{k-1}\frac{(q'-1)!k!(n-k-1)!(n-q')!}{j!(k-j)!(q'-j-1)!(n-k-q'+j)!}\right], \quad i = k +$$

$1, k+2, \dots, n$ 

$$\tag{3}$$

Note that if $k \geq q$, then the first expression in this formula disappears. Using the symmetry of the game it follows that

$$\varphi_i(v) = \frac{1}{k}[1 - (n-k)\varphi_{k+1}(v)], \quad i = 1,2,\dots,k.$$

We now derive the Penrose-Banzhaf index. First, we consider which winning coalitions contain swing voters. Assume that $k < q$. Such coalitions can be split into the following four categories:

1) $W1$ : The set of coalitions with $k$ veto players and $q'$-$k$ non-veto players. Only veto players are swing voters. Hence, there are $k$ swing votes in each of these $\#(W_1) = \binom{n-k}{q'-k}$ coalitions

2) $W_2$ : The set of coalitions with $j$ veto players and $q'$-$j$ non-veto players, where $j<k$. Any member is a swing voter. Hence, there are $q'$ swing votes in each of these $\#(W_2) = \sum_{j=m}^{k-1}\binom{n-k}{q'-j}\binom{k}{j}$ coalitions, where $m=\max\{0, q'+k-n\}$.

3) $W_3$ : The set of coalitions with between $q+1$ and $q'$-1 members including all the veto players. Only veto players are swing voters. Hence, there are $k$ swing votes in each of these $\#(W_3) = \sum_{j=q+1}^{q'-1}\binom{n-k}{j-k}$ coalitions.

4)   $W_4$: The set of coalitions with $q$ players including all the veto players. All members are swing voters. Hence, there are $q$ swing votes in each of these $\#(W_4) = \binom{n-k}{q-k}$ coalitions.

It follows that the total number of swing votes is given by

$$\#(S) = k[\#(W_1) + \#(W_3)] + q'\#(W_2) + q\#(W_4).$$

A veto player has a swing vote in each of the coalitions in W1, W3 and W4.  By symmetry, such a player is in a proportion $j/k$ of coalitions with $j$ veto players. Hence, the number of swing votes of a veto player in coalitions from the set $W_2$ is given by

$$\#(V) = \sum_{j=m}^{k-1} \binom{n-k}{q'-j}\binom{k}{j}\frac{j}{k} = \sum_{j=m}^{k-1} \binom{n-k}{q'-j}\binom{k-1}{j-1}.$$

It follows that the Penrose-Banzhaf index of a veto player is given by

$$\varphi_i(v) = \frac{\#(W_1)+\#(W_3)+\#(V)+\#(W_4)}{k[\#(W_1)+\#(W_3)]+q\prime\#(W_2)+q\#(W_4)}. \tag{4}$$

The following example considers a similar game to the UN security council game considered above. We calculate power indices for this game and it is shown that there is a problem of deriving an equivalent weighted voting game.

**Example 3.** Suppose that in the UN security council voting game a veto may be over-ridden when at least 12 voters vote for a motion. Hence, we have $n = 15$, $k = 5$, $q = 9$, $q' = 12$. First we calculate the Shapley-Shubik power index for the non-veto players. Note that for a non-veto player to be a pivot player when he/she appears in position 12, then between 2 and 4 veto players must have appeared beforehand. From Equation (3), it follows that for $i = 6, 7,\ldots, 15$

$$\varphi_i(v) = \frac{1}{15!}\left[\frac{9!\,8!}{3!} + 11!\,5!\,9!\,3! \sum_{j=2}^{4} \frac{1}{j!\,(5-j)!\,(11-j)!\,(j-2)!}\right] \approx 0.053147.$$

It follows that the power of the veto players is given by

$$\varphi_i(v) = \frac{1}{5}[1 - 10\varphi_6(v)] \approx 0.093706.$$

It can be seen that the power of the veto players relative to the power of the non-veto players is much greater in the game where the vetoes are of first degree.

To calculate the Penrose-Banzhaf index, we first calculate the number of coalitions in which there are swing voters. Using the approach described above

$\#(W_1) = \binom{10}{7} = 120; \quad \#(W_2) = \sum_{j=2}^{4} \binom{10}{12-j}\binom{5}{j} = 335; \quad \#(W_3) = \sum_{j=10}^{11} \binom{10}{j-5} =$
$714; \quad \#(W_1) = \binom{10}{4} = 210. \quad \#(V) = \sum_{j=2}^{4} \binom{10}{12-j}\binom{4}{j-1} = 244.$

The total number of swing votes is given by

$$k[\#(W_1) + \#(W_3)] + q'\#(W_2) + q\#(W_4) = 10\,080.$$

The number of votes of a single veto player is given by

$$\#(W_1) + \#(W_3) + \#(W_4) + \#(V) = 1\,288.$$

It follows that the Penrose-Banzhaf index for a veto player is

$$\varphi_i(v) = \frac{1\,288}{10\,080} \approx 0.127778, \quad i = 1,2,3,4,5.$$

The index for a non-veto player is $\varphi_i(v) = 0.1[1 - 5\varphi_i(v)] \approx 0.036111$.

It may be surprising to note that these power indices are very similar to the ones derived in the original game. In comparison to vetoes of first degree, when vetoes of second degree are available, non-veto players have clearly more power in constructing a winning coalition. However, their power to disrupt a winning coalition is only marginally affected.

Now we try to define a weighted voting game which is equivalent to this game. Due to the symmetry of the game, we may assume that the weight of the veto players is $w$, where $w > 1$, and the weight of the non-veto players is 1. From the definition of the game considered, any coalition with 5 veto players and 3 non-veto players is a losing coalition and any coalition with 2 veto players and 10 non-veto players is a winning coalition. It follows that if this game is equivalent to a weighted voting game without vetoes, then $2w + 10 > 5w + 3$ (the sum of the weights in the winning coalition must be greating than the sum of the weights in the losing coalition). Hence, $w < 7/3$. On the other hand, a coalition with 5 veto members and 4 non-veto members is a winning coalition, whereas a coalition with 4 veto members and 7 non-veto members is a losing coalition. Arguing as above, $5w + 4 > 4w + 7$. Hence, $w > 3$. This gives a contradiction and it is clear that transforming such games to weighted voting games is problematic.

## 6    Conclusions

In this paper we have considered the effect of vetoes of first and second degree vetoes on voting games. The Shapley-Shubik and Penrose-Banzhaf indices for a class of voting games in which all players have the same weight, but some have power of veto, were derived. In the case of games with vetoes of first degree, we have presented a procedure for determining an equivalent weighted voting game. This procedure is useful, since algorithms are available to calculate power indices for such weighted voting games. However, the question of equivalence of weighted voting games with voting games with vetoes of second degree is a much more complex question and will be the subject of future research.

# References

1. Banzhaf III, J.F.: Weighted voting doesn't work: a mathematical analysis. Rutgers Law Review **19**, 317–343 (1965)
2. Mercik, J.: On a priori evaluation of power of veto. In: Herrera-Viedma, E., García-Lapresta, J.L., Kacprzyk, J., Fedrizzi, M., Nurmi, H., Zadrożny, S. (eds.) Consensual Processes. STUDFUZZ, vol. 267, pp. 145–156. Springer, Heidelberg (2011)
3. Mercik, J.W.: Classification of committees with vetoes and conditions for the stability of power indices. Neurocomputing Part C **149**, 1143–1148 (2015)
4. Penrose, L.S.: The Elementary Statistics of Majority Voting. Journal of the Royal Statistical Society **109**, 53–57 (1946)
5. Shapley, L.S., Shubik, M.: A method of evaluating the distribution of power in a committee system. American Political Science Review **48**(3), 787–792 (1954)

# Positivity and Stability of Time-Varying Discrete-Time Linear Systems

Tadeusz Kaczorek$^{(\boxtimes)}$

Faculty of Electrical Engineering, Bialystok University of Technology,
Wiejska 45D, 15-351 Bialystok, Poland
kaczorek@isep.pw.edu.pl

**Abstract.** The positivity and asymptotic stability of the time-varying discrete-time linear systems are addressed. Necessary and sufficient conditions for the positivity of the systems and sufficient conditions for asymptotic stability of the positive systems are established. The proposed stability tests are based on the norms of the system matrix. The effectiveness of the test are demonstrated on an example.

**Keywords:** Time-varying · Positive · Discrete-time · Positive stability

## 1 Introduction

A dynamical system is called positive if its trajectory starting from any nonnegative initial condition state remains forever in the positive orthant for all nonnegative inputs. An overview of state of the art in positive system theory is given in the monographs [7, 10] and in the papers [11-14]. Models having positive behavior can be found in engineering, economics, social sciences, biology and medicine, etc.

The Lyapunov, Perron and Bohl exponents and stability of time-varying discrete-time linear systems have been investigated in [1-9]. The positive standard and descriptor systems, and their stability have been analyzed in [13-16]. The positive linear systems with different fractional orders have been addressed in [14, 17] and the singular discrete-time linear systems in [15].

In this paper the positivity and asymptotic stability of the time-varying discrete-time linear systems will be investigated.

The paper is organized as follows. In section 2 the solution of the state-equation describing the time-varying discrete-time linear system is derived and necessary and sufficient conditions for the positivity of the systems are established. The asymptotic stability of the positive systems is addressed in section 3, where conditions for the stability are proposed. Concluding remarks are given in section 4.

The following notation will be used: $\Re$ - the set of real numbers, $\Re^{n \times m}$ - the set of $n \times m$ real matrices, $\Re_+^{n \times m}$ - the set of $n \times m$ matrices with nonnegative entries and $\Re_+^n = \Re_+^{n \times 1}$, $I_n$ - the $n \times n$ identity matrix.

© Springer International Publishing Switzerland 2015
N.T. Nguyen et al. (Eds.): ACIIDS 2015, Part I, LNAI 9011, pp. 295–303, 2015.
DOI: 10.1007/978-3-319-15702-3_29

## 2     Positive Time-Varying Discrete-Time Linear Systems

Consider the time-varying discrete-time linear system

$$x_{i+1} = A(i)x_i + B(i)u_i, \ i \in Z_+ = \{0,1,...\} \tag{2.1a}$$

$$y_i = C(i)x_i + D(i)u_i \tag{2.1b}$$

where $x_i \in \mathfrak{R}^n$, $u_i \in \mathfrak{R}^m$, $y_i \in \mathfrak{R}^p$ are the state, input and output vectors and $A(i) \in \mathfrak{R}^{n\times n}$, $B(i) \in \mathfrak{R}^{n\times m}$, $C(i) \in \mathfrak{R}^{p\times n}$, $D(i) \in \mathfrak{R}^{p\times m}$ are matrices with entries depending on $i \in Z_+$.

**Theorem 2.1.** The solution of equation (2.1a) for known initial condition $x_0 \in \mathfrak{R}^n$ and input $u_i \in \mathfrak{R}^m$, $i \in Z_+$ is given by

$$x_i = \Phi(i,0)x_0 + \sum_{j=0}^{i-1}\Phi(i,j+1)B(j)u_j, \ i \in Z_+ \tag{2.2a}$$

and

$$y_i = C(i)\Phi(i,0)x_0 + \sum_{j=0}^{i-1}C(i)\Phi(i,j+1)B(j)u_j + D(i)u_i, \ i \in Z_+ \tag{2.2b}$$

where

$$\Phi(k,j) = \begin{cases} I_n & \text{for } k = j \geq 0 \\ A(k-1)A(k-2)...A(j) & \text{for } k > j \geq 0 \end{cases}. \tag{2.2c}$$

**Proof.** Using (2.2a) and (2.1a) we obtain

$$A(i)x_i + B(i)u_i = A(i)\left[\Phi(i,0)x_0 + \sum_{j=0}^{i-1}\Phi(i,j+1)B(j)u_j\right] + B(i)u_i$$

$$= A(i)\Phi(i,0)x_0 + \sum_{j=0}^{i-1}A(i)\Phi(i,j+1)B(j)u_j + B(i)u_i \tag{2.3}$$

$$= \Phi(i+1,0)x_0 + \sum_{j=0}^{i}\Phi(i+1,j+1)B(j)u_j = x_{i+1}.$$

Substitution of (2.2a) into (2.1b) yields (2.2b). □

From (2.1a) and (2.2a) for $B(i)u_i = 0$, $i \in Z_+$ it follows that

$$x_i = \Phi(i)x_0, \ i \in Z_+ \tag{2.4a}$$

where

$$\Phi(i) = \Phi(i,0) = A(i-1)A(i-2)...A(0) \tag{2.4b}$$

is the solution of the equation

$$x_{i+1} = A(i)x_i, \ i \in Z_+. \tag{2.5}$$

From (2.4b) we have

$$\Phi(i+1) = A(i)\Phi(i), \ i \in Z_+. \tag{2.6}$$

**Definition 2.1.** The time-varying discrete-time linear system (2.1) is called the (internally) positive if and only if $x_i \in \Re_+^n$ and $y_i \in \Re_+^p$, $i \in Z_+$ for any initial conditions $x_0 \in \Re_+^n$ and all inputs $u_i \in \Re_+^m$, $i \in Z_+$.

**Theorem 2.1.** The time-varying discrete-time linear system (2.1) is positive if and only if

$$A(i) \in \Re_+^{n \times n}, \ B(i) \in \Re_+^{n \times m}, \ C(i) \in \Re_+^{p \times n}, \ D(i) \in \Re_+^{p \times m}, i \in Z_+. \tag{2.7}$$

**Proof.** Sufficiency. If the conditions (2.7) are satisfied then from (2.2) it follows that $x_i \in \Re_+^n$ and $y_i \in \Re_+^p$ for any $x_0 \in \Re_+^n$ and all inputs $u_i \in \Re_+^m$, $i \in Z_+$.

Necessity. Substituting $u_i = 0$, $i \in Z_+$ into (2.1a) and (2.1b) we obtain

$$x_{i+1} = A(i)x_i \in \Re_+^n \text{ and } y_i = C(i)x_i \in \Re_+^p, i \in Z_+ \tag{2.8}$$

for any $x_i \in \Re_+^n$ only if $A(i) \in \Re_+^{n \times n}$ and $C(i) \in \Re_+^{p \times n}$ for $i \in Z_+$.

Similarly, assuming $x_i = 0$ from (2.1) we obtain

$$x_{i+1} = B(i)u_i \in \Re_+^n \text{ and } y_i = D(i)u_i \in \Re_+^p, i \in Z_+ \tag{2.9}$$

for any $u_i \in \Re_+^m$ only if $B(i) \in \Re_+^{n \times m}$ and $D(i) \in \Re_+^{p \times m}$ for $i \in Z_+$. □

# 3    Stability of the Positive Systems

Consider the positive system (2.5) with the matrix (2.4b) satisfying the equation (2.6).

**Definition 3.1.** The positive system (2.5) is called asymptotically stable if the norm $\|x_i\|$ of the state vector $x_i \in \mathfrak{R}^n_+$, $i \in Z_+$ satisfies the condition

$$\lim_{i \to \infty} \|x_i\| = 0 \text{ for any finite } x_0 \in \mathfrak{R}^n_+ . \tag{3.1}$$

**Theorem 3.1.** The positive system (2.5) is asymptotically stable if the norm $\|A(i)\|$ of the matrix $A(i)$, $i \in Z_+$ satisfies the condition

$$\|A\| < 1 \text{ for } i \in Z_+ \tag{3.2a}$$

where

$$\|A\| \geq \max_{0 \leq i \leq \infty} \|A(i)\| \text{ for } i \in Z_+ . \tag{3.2b}$$

**Proof.** From (2.4) we have

$$\|x_i\| = \|A(i-1)A(i-2)...A(0)x_0\| \leq \|A(i-1)\| \|A(i-2)\| .. \|A(0)\| \|x_0\| \leq \|A\|^i \|x_0\| . \tag{3.3}$$

If the condition (3.2a) is satisfied then from (3.3) we have

$$\lim_{i \to \infty} \|x_i\| = 0 \text{ for any finite } x_0 \in \mathfrak{R}^n_+ \tag{3.4}$$

and by Definition 3.1 the positive system is asymptotically stable. □

**Theorem 3.2.** The positive system (2.5) is asymptotically stable if its system matrix $A(i) = [a_{jk}(i)] \in \mathfrak{R}^{n \times n}_+$ satisfies the condition

$$\max_{0 \leq j \leq n} \sum_{k=1}^{n} a_{jk}(i) < 1 \text{ for } i \in Z_+ \tag{3.5a}$$

or

$$\max_{0 \leq k \leq n} \sum_{j=1}^{n} a_{jk}(i) < 1 \text{ for } i \in Z_+ . \tag{3.5b}$$

**Proof.** Let $x_i = [x_{i1} \quad \cdots \quad x_{in}]^T$ and $\|x_i\| = \max_{0 \le j \le n}|x_{ij}|$ and $\|A(i)\| = \max_{0 \le j \le n} \sum_{k=1}^{n}|a_{jk}(i)|$.

Then it is easy to check [18] that the conditions of the norms of the vector $x_i$ and matrix $A(i)$ are satisfied. For the positive system (2.5) the condition (3.5a) or (3.5b) implies that $\|A(i)\| < 1$ for $i \in Z_+$ and by Theorem 3.1 the system is asymptotically stable. $\square$

**Theorem 3.3.** The positive system (2.5) is asymptotically stable if its system matrix

$$A(i) = \begin{bmatrix} 0 & 1 & 0 & \cdots & 0 \\ 0 & 0 & 1 & \cdots & 0 \\ \vdots & \vdots & \vdots & \ddots & \vdots \\ 0 & 0 & 0 & \cdots & 1 \\ a_0(i) & a_1(i) & a_2(i) & \cdots & a_{n-1}(i) \end{bmatrix} \in \mathfrak{R}_+^{n \times n} \tag{3.6}$$

satisfy the condition

$$\sum_{k=0}^{n-1} a_k(i) < 1 \text{ for } i \in Z_+. \tag{3.7}$$

**Proof.** By Lemma A.1 the condition (3.7) implies that the sum of entries of all rows of the matrix $A^n(i) \in \mathfrak{R}_+^{n \times n}$, $i \in Z_+$ is less one. In this case the condition (3.5a) is satisfied and by Theorem 3.2 the positive system is asymptotically stable since $i = kn \to \infty$ for $k \to \infty$. $\square$

**Example 3.1.** Consider the time-varying discrete-time linear system (2.5) with the matrix

$$A(i) = \begin{bmatrix} 0.1e^{-i} & 0 & 0.2 + 0.1\sin i \\ 0.1 + 0.1\cos i & 0.2 & 0.3 + 0.1e^{-2i}\cos i \\ 0.1 & 0.2 + 0.1e^{-i} & 1 + 0.4\cos 2i \end{bmatrix}, i \in Z_+. \tag{3.8}$$

Note that the matrix (3.8) has nonnegative entries for $i \in Z_+$. Therefore, by Theorem 2.2 the system (2.5) with (3.8) is positive one.

It is easy to check that the matrix (3.8) satisfies the condition (3.5a) since for $i = 0$ we have

$$A(0) = \begin{bmatrix} 0.1 & 0 & 0.2 \\ 0.2 & 0.2 & 0.4 \\ 0.1 & 0.3 & 1.4 \end{bmatrix} \in \Re_{+}^{n \times n}. \tag{3.9}$$

By Theorem 3.2 the positive system (2.5) with (3.8) is asymptotically stable. Note that the matrix (3.9) does not satisfy the condition (3.5b).

## 4     Concluding Remarks

The positivity and asymptotic stability of the time-varying discrete-time linear systems have been addressed. Necessary and sufficient conditions for the positivity of the systems have been established (Theorem 2.2). Sufficient conditions for asymptotic stability of the positive systems have been proposed (Theorem 3.1, 3.2, 3.3). It has been shown that the positive system (2.5) is asymptotically stable if the condition (3.5a) (or (3.5b)) is met. The effectiveness of the conditions has been demonstrated on example. The considerations can be easily extended to time-varying discrete-time linear systems with delays. An open problem is an extension of the conditions for fractional time-varying discrete-time linear systems.

**Acknowledgment.** This work was supported by National Science Centre in Poland.

## Appendix

**Lemma A.1.** If the matrix

$$A = \begin{bmatrix} 0 & 1 & 0 & \cdots & 0 \\ 0 & 0 & 1 & \cdots & 0 \\ \vdots & \vdots & \vdots & \ddots & \vdots \\ 0 & 0 & 0 & \cdots & 1 \\ a_0 & a_1 & a_2 & \cdots & a_{n-1} \end{bmatrix} \in \Re_{+}^{n \times n} \tag{A.1}$$

satisfies the condition

$$a_0 + a_1 + \ldots + a_{n-1} < 1 \tag{A.2}$$

then the matrix

$$A^n = \begin{bmatrix} a_{11} & \cdots & a_{1n} \\ \vdots & \cdots & \vdots \\ a_{n1} & \cdots & a_{nn} \end{bmatrix} \in \Re_{+}^{n \times n} \tag{A.3}$$

and

$$\sum_{j=1}^{n} a_{kj} > \sum_{j=1}^{n} a_{lj} < 1 \text{ for } k < l, \ k,l = 1,2,...,n. \tag{A.4}$$

**Proof.** Taking into account that for the matrix (A.1)

$$A^2 = \begin{bmatrix} 0 & 1 & 0 & ... & 0 \\ 0 & 0 & 1 & ... & 0 \\ \vdots & \vdots & \vdots & ... & \vdots \\ a_0 & a_1 & a_2 & ... & a_{n-1} \\ a_0 a_{n-1} & a_0 + a_1 a_{n-1} & a_1 + a_2 a_{n-1} & ... & a_{n-2} + a_{n-1}^2 \end{bmatrix} \in \Re_+^{n \times n} \tag{A.5}$$

we shall show that

$$a_0 + a_1 + ... + a_{n-1} > a_0 a_{n-1} + a_0 + a_1 a_{n-1} + ... + a_{n-2} + a_{n-1}^2. \tag{A.6}$$

Note that the inequality (A.6) is equivalent to

$$a_{n-1} > a_{n-1}(a_0 + a_1 + ... + a_{n-1}) \tag{A.7}$$

and this is true since (A.2) holds.

This inequality can be easily extended for the matrix $A^k$ for $k = 3,4,...,n$. This completes the proof. $\square$

For example for the matrix

$$A = \begin{bmatrix} 0 & 1 & 0 \\ 0 & 0 & 1 \\ 0.2 & 0.3 & 0.2 \end{bmatrix} \tag{A.8}$$

satisfying the condition (A.2) (since $0.2 + 0.3 + 0.2 = 0.7 < 1$) we have

$$A^3 = \begin{bmatrix} 0.2 & 0.3 & 0.2 \\ 0.04 & 0.206 & 0.304 \\ 0.0680 & 0.1312 & 0.2608 \end{bmatrix} \tag{A.9}$$

and

$$\begin{aligned} 0.2 + 0.3 + 0.2 = 0.7 &> 0.04 + 0.206 + 0.304 \\ = 0.55 &> 0.068 + 0.1312 + 0.2608 = 0.46 \end{aligned} \tag{A.10}$$

Lemma A.1 can be easily extended to the matrix (A.1) with entries $a_0(i)$, $a_1(i),...,a_{n-1}(i)$ depending on $i \in Z_+$.

# References

1. Czornik, A., Newrat, A., Niezabitowski, M., Szyda, A.: On the Lyapunov and Bohl exponent of time-varying discrete linear systems. In: 20th Mediterranean Conf. on Control and Automation (MED), Barcelona, pp. 194–197 (2012)
2. Czornik, A., Niezabitowski, M.: Lyapunov Exponents for Systems with Unbounded Coefficients. Dynamical Systems: an International Journal **28**(2), 140–153 (2013)
3. Czornik, A., Newrat, A., Niezabitowski, M.: On the Lyapunov exponents of a class of the secon order discrete time linear systems with bounded perturbations. Dynamical Systems: an International Journal **28**(4), 473–483 (2013)
4. Czornik, A., Niezabitowski, M.: On the stability of discrete time-varying linear systems. Nonlinear Analysis: Hybrid Systems. **9**, 27–41 (2013)
5. Czornik, A., Niezabitowski, M.: On the stability of Lyapunov exponents of discrete linear system. In: Proc. of European Control Conf., Zurich, pp. 2210-2213 (2013)
6. Czornik, A., Klamka, J., Niezabitowski, M.: On the set of Perron exponents of discrete linear systems. In: Proc. of World Congres of the 19th International Federation of Automatic Control, Kapsztad, pp. 11740–11742 (2014)
7. Czornik, A., Klamka, J., Niezabitowski, M.: About the number of the lower Bohl exponents of diagonal discrete linear time-varying systems. In: Proc. of the 11th IEEE International Conference on Control & Automation, Taichung, Taiwan, pp. 461–466, June 18–20 2014 r
8. Czornik, A.: The relations between the senior upper general exponent and the upper Bohl exponents. In: Proc. of the 19th International Conference on Methods and Models in Automation and Robotics, Międzyzdroje, Polska, pp. 897–902, September 02–05 2014 r
9. Niezabitowski, M.: About the properties of the upper Bohl exponents of diagonal discrete linear time-varying systems. In: Proc. of the 19th International Conference on Methods and Models in Automation and Robotics, Międzyzdroje, Polska, pp. 880–884, September 02–05 2014 r
10. Farina, L., Rinaldi, S.: Positive Linear Systems; Theory and Applications. Wiley, New York (2000)
11. Zhang, H., Xie, D., Zhang, H., Wang, G.: Stability analysis for discrete-time switched systems with unstable subsytems by a mode-dependent average dwell time approach. ISA Transactions **53**, 1081–1086 (2014)
12. Zhang, J., Han, Z., Wu, H., Hung, J.: Robust stabilization of discrete-time positive switched systems with uncertainties and average dwell time switching. Circuits Syst, Signal Process. **33**, 71–95 (2014)
13. Kaczorek, T.: Positive 1D and 2D systems. Springer, London (2001)
14. Kaczorek, T.: Positive linear systems consisting of n subsystems with different fractional orders. IEEE Trans. Circuits and Systems **58**(6), 1203–1210 (2011)
15. Kaczorek, T.: Positive descriptor discrete-time linear systems. Problems of Nonlinear Analysis in Engineering Systems **1**(7), 38–54 (1998)
16. Kaczorek, T.: Positive singular discrete time linear systems. Bull. Pol. Acad. Techn. Sci. **45**(4), 619–631 (1997)

17. Kaczorek, T.: Selected Problems of Fractional Systems Theory. Springer, Berlin (2012)
18. Kaczorek, T.: Vectors and Matrices in Automation and Electrotechics. WNT, Warszawa (1998)
19. Zhong, Q., Cheng, J., Zhong, S.: Finite-time $H_\infty$ control of a switched discrete-time system with average dwell time. Advances in Difference Equations, vol. 191 (2013)
20. Rami, M.A., Bokharaie, V.S., Mason, O., Wirth, F.R.: Extremel norms for positive linear inclusions. In: 20th International Symposium on Mathematical Theory of Networks and Systems, Melbourne (2012)

# Controllability of Discrete-Time Linear Switched Systems with Constrains on Switching Signal

Artur Babiarz, Adam Czornik,
Jerzy Klamka, and Michał Niezabitowski[✉]

Faculty of Automatic Control, Electronics and Computer Science,
Institute of Automatic Control, Silesian University of Technology,
Akademicka 16 Street, 44-101 Gliwice, Poland
{artur.babiarz,adam.czornik,jerzy.klamka,michal.niezabitowski}@polsl.pl
http://www.polsl.pl

**Abstract.** In this paper we consider the controllability problem for discrete-time linear switched systems. The problem consists of finding a control signal that steers any initial condition to a given final state regardless of the switching signal. In the paper a necessary and sufficient conditions for this type of controllability are presented. Moreover, we consider problems of controllability from zero initial condition and to zero final state.

**Keywords:** Controllability · Switched systems · Discrete-time · Hybrid systems

## 1 Introduction

Since the early work on state-space approaches to control systems analysis, it was recognized that certain nondegeneracy assumptions were useful, in particular in the context of optimal control. However, it was until Kalman's work [1] that the property of controllability was isolated as of interest in and of itself, as it characterizes the degrees of freedom available when attempting to control a system.

The study of controllability for linear systems has spanned a great number of research directions, and topics such as testing degrees of controllability, and their numerical analysis aspects, are still a subject of intensive research. This paper is devoted to controllability of discrete-time linear switched systems (see [2] for definition and motivation). Such a system can be seen as a collection of discrete stationary linear systems between which is followed by the switching signal. The

The research presented here were done by the authors as parts of the projects funded by the National Science Centre granted according to decisions DEC-2014/13/B/ST7/00755, DEC-2012/05/B/ST7/00065, DEC-2012/07/B/ST7/01404 and DEC-2012/07/N/ST7/03236, respectively. The calculations were performed with the use of IT infrastructure of GeCONiL Upper Silesian Centre for Computational Science and Engineering (NCBiR grant no POIG.02.03.01-24-099/13).

© Springer International Publishing Switzerland 2015
N.T. Nguyen et al. (Eds.): ACIIDS 2015, Part I, LNAI 9011, pp. 304–312, 2015.
DOI: 10.1007/978-3-319-15702-3_30

switching signal may model the phenomenon over which we have control (change of regulator parameters, gear ratio) or uncontrolled events (failures, changing of the operating point). Most of the works (see [3] - [19] and the references therein) on the controllability of hybrid systems is related to the first case and then, the controllability problem is formulated as a problem of finding control and switching signal which steers an initial condition to a given final state. In this case the switching signal plays a role of additional control. The work is connected to the second case and then, we are looking for a control, that regardless of the switching signal, steers an initial condition to a given final state. This situation is similar to a problem of controllability of jump linear systems ([20], [21]) but in our framework we do not have a probabilistic model of the switching signal. In addition, a new contribution of the paper is that we take into account the situation in which certain switching sequences are not possible. This situation often occurs in engineering practice.

## 2    Notation and Definitions

We consider a class of switched systems given by

$$x(k+1) = A(r(k))x(k) + B(r(k))u(k), \quad k \geq 0 \tag{1}$$

where $x(k) \in \mathbb{R}^n$ denotes the state vector, $r(k) \in \{1, ..., N\} =: S$ is the switching signal, $u(k) \in \mathbb{R}^m$ is the control input, $k = 0, 1, ...$ . Furthermore, for $r(k) = i$, $A_i := A(i)$ and $B_i := B(i)$ are constant matrices of appropriate sizes. Denote by $x(k, x_0, i_0, u)$ the solution of (1) under the control $u$ with initial condition $x_0$ at time $k = 0$ and switching signal satisfying $r(0) = i_0$. The control $u = (u(0), u(1), ...)$ is assumed to be such that $u(k)$ is of the form $f_k(r(0), r(1), ..., r(k))$. Each control of this form we will call as an admissible control. Let us introduce the following notation:

$$F(k, k) = I_{n \times n}$$

$$F(k, l, i_{k-1}, ..., i_l) = A(i_{k-1})A(i_{k-2})...A(i_l), \quad k > l \geq 0, \ i_{k-1}, ..., i_l \in S$$

$$F_r(k, l) = A(r(k-1))A(r(k-2))...A(r(l)), \ k > l \geq 0,$$

Using this notation we can write the solution of (1) in the following form

$$x(k, x_0, i_0, u) = F_r(k, 0)x_0 + \sum_{t=0}^{k-1} F_r(k, t+1)B(r(t))u(t), \quad k \geq 1 \tag{2}$$

or

$$x(k, x_0, i_0, u) = F(k, 0, r(k-1), ..., r(0))x_0+ \tag{3}$$

$$\sum_{t=0}^{k-1} F(k, t+1, r(k-1), ..., r(t+1))B(r(t))u(t), \quad k \geq 1.$$

We will also use the following notation

$$S_{i_0}^{(N)} = \{(i_0, ..., i_{N-1}) : i_0, ..., i_{N-1} \in S\}. \tag{4}$$

It will be convenient to have the elements of $S_{i_0}^{(N)}$ ordered in a sequence. In that purpose let order the elements of $S_{i_0}^{(N)}$ in lexicographical order i.e. they are ordered as follows

$$(i_0, 1, 1, ..., 1, 1), (i_0, 1, 1, ..., 1, 2), ..., (i_0, 1, 1, ..., 1, s),$$
$$(i_0, 1, 1, ..., 2, 1), (i_0, 1, 1, ..., 2, 2), ..., (i_0, 1, 1, ..., 2, s), ...$$
$$(i_0, s, s, ..., s, 1), (i_0, s, s, ..., s, 2), ..., (i_0, s, s, ..., s, s).$$

In many practical situations certain switches are impossible. It means that we have certain set $A$ of pairs $(i, j) \in S \times S$ such that it is impossible that $r(k) = i$, $r(k+1) = j$ for a $k = 0, 1, ...$. Withdraw from $S_{i_0}^{(N)}$ all the elements $(i_0, ..., i_{N-1})$ such that

$$(i_l, i_{l+1}) \in A \quad \text{for certain } l = 0, 1, ..., N - 1.$$

and denote by $\overline{S}_{i_0}^{(N)}$ the set obtained in this way. In this notation $\overline{S}_{i_0}^{(N)}$ is a sequence of all possible switching paths of the length $N$. By $\overline{\overline{s}}_{i_0}^{(N)}$ we will denote the number of elements of $\overline{S}_{i_0}^{(N)}$.

Fix a number $N > 0$ and a sequence $(i_0, i_1, ..., i_{N-1})$ of elements of $S$. Consider a matrix column blocks which are numbered successively by sequences: $i_0, \overline{S}_{i_0}^{(2)}, ..., \overline{S}_{i_0}^{(N)}$ and the block $(i_0, i_1, ..., i_k)$, $k = 0, 1, .., N - 1$ is given by

$$F(N, k + 1, i_{N-1}, ..., i_{k+1})B_{i_k}$$

and the others are equal to 0. Denote the matrix obtained in this way by $C(i_0, i_1, ..., i_{N-1})$ and by $G(i_0)$ - the matrix consisting of all $C(i_0, i_1, ..., i_{N-1})$ (as row blocks numbered by $\overline{S}_{i_0}^{(N)}$) for $(i_0, i_1, ..., i_{N-1}) \in \overline{S}_{i_0}^{(N)}$. Moreover, by $H(i_0) \in R^{n\overline{\overline{s}}_{i_0}^{(N)} \times m}$ let denote a matrix row blocks of which are numbered by the sequence $\overline{S}_{i_0}^{(N)}$, the block $(i_0, i_1, ..., i_{N-1})$ is given by $F(N, 0, i_{N-1}, ..., i_0)$. For example in the case when $S = \{1, 2\}$, $N = 3$, $A = \{(2, 1)\}$ and $i_0 = 1$, we have

$$G(1) = \begin{bmatrix} C(1, 1, 1) \\ C(1, 1, 2) \\ C(1, 2, 2) \end{bmatrix} =$$

$$\begin{array}{cccccc} (1) & (1,1) & (1,2) & (1,1,1) & (1,1,2) & (1,2,2) \\ [A_1^2 B_1 & A_1 B_1 & 0 & B_1 & 0 & 0] \\ [A_2 A_1 B_1 & A_2 B_1 & 0 & 0 & B_2 & 0] \\ [A_2^2 B_1 & 0 & A_2 B_2 & 0 & 0 & B_2] \end{array}$$

and

$$H(1) = \begin{bmatrix} A_1^3 \\ A_2 A_1^2 \\ A_2^2 A_1 \end{bmatrix}.$$

Moreover, let us denote by $f_1^{(k)}, ..., f_n^{(k)} \in \mathbb{R}^{nk}$ the vectors defined by

$$f_l^{(k)} = \left[ \begin{array}{c} e_l \\ e_l \\ \vdots \\ e_l \end{array} \right] \Big\} k \text{ times } e_l, \quad l = 1, ..., n$$

where $e_1, ..., e_n$ is the standard basis of $\mathbb{R}^n$.

We have the following definition:

**Definition 1.** *We say that system (1) is $i_0-$ controllable at time $N$ if, for all $x_0, x_1 \in \mathbb{R}^n$ there exists an admissible control $u$ such that*

$$x(N, x_0, i_0, u) = x_1. \tag{5}$$

*Analogically, we say that system (1) is $i_0-$ controllable at time $N$ to zero (from zero) if, for all $x_0 \in \mathbb{R}^n$ ($x_1 \in \mathbb{R}^n$) there exists a control $u$ such that*

$$x(N, x_0, i_0, u) = 0 \qquad (x(N, 0, i_0, u) = x_1). \tag{6}$$

*If the system (1) is $i_0-$ controllable at time $N$ ($i_0-$ controllable at time $N$ to zero, $i_0-$ controllable at time $N$ from zero) for all $i_0 \in S$ then we say that (1) is controllable at time $N$ (controllable at time $N$ to zero, controllable at time $N$ from zero).*

Observe that the controllability of each time-varying system corresponding to switching paths of the length $N$ is only the necessary, but not the sufficient condition for controllability at time $N$ of the system (1).

## 3   Main Results

The next theorem contains necessary and sufficient conditions for $i_0-$ controllability at time $N$ as well as $i_0-$ controllability at time $N$ from zero and to zero.

**Theorem 1.** *System (1) is $i_0-$ controllable at time $N$ from zero if and only if*

$$rank G(i_0) = rank \left[ G(i_0) \quad f_l^{\left( \overline{\overline{s}}_{i_0}^{(N)} \right)} \right], \text{ for all } l = 1, ..., n. \tag{7}$$

*System (1) is $i_0-$ controllable at time $N$ to zero if and only if*

$$Im H(i_0) \subset Im G(i_0), \tag{8}$$

*and it is $i_0-$ controllable at time $N$ if and only if*

$$rank G(i_0) = rank \left[ G(i_0) \quad f_l^{\left( \overline{\overline{s}}_{i_0}^{(N)} \right)} \right], \text{ for all } l = 1, ..., n, \tag{9}$$

*and*

$$Im H(i_0) \subset Im G(i_0). \tag{10}$$

Before entering the formal proof let us briefly discuss the main idea. Since the set $\overline{S}_{i_0}^{(N)}$ of all possible switching paths is finite, therefore the question about $i_0-$ controllability can be reformulated, similarly as for classical time-varying systems, as a question about existence of a solution of a finite set of linear equations. Nevertheless, now we must take into account the constrain that control $u(k)$ at time $k$ may depend only on the variables $r(0), ..., r(k)$ and must be independent of $r(k+1), ..., r(N)$. In the proof we obtain this by the proper definition of matrices $G(i_0)$ and $H(i_0)$.

*Proof.* Suppose that system (1) is $i_0-$ controllable at time $N$ from zero. Then for each $y \in \mathbb{R}^n$ there exists a control sequence $u(0), ..., u(N-1)$ such that

$$u(k) = g_k(i_0, r(1), ..., r(k)), \quad k = 0, ..., N-1,$$

and

$$x(N, 0, i_0, u) = y, \tag{11}$$

where $g_k$ is a function from $\overline{S}_{i_0}^{(k)}$ to $\mathbb{R}^m$ for $k = 0, ..., N-1$. It means that for any $(i_0, ..., i_{N-1}) \in \overline{S}_{i_0}^{(N)}$ the following

$$\sum_{t=0}^{N-1} F(N, t+1, i_{N-1}, ..., i_{t+1}) B(i_t) g_t(i_0, ..., i_t) = y$$

holds. This clearly forces that the system of equations

$$G(i_0)v = z,$$

where

$$z = \begin{bmatrix} t \\ \vdots \\ t \end{bmatrix} \left.\right\} \ \overline{\overline{S}}_{i_0}^{(N)} \ \text{times}$$

has a solution for each $t \in \mathbb{R}^n$. Since vectors $f_l^{\left(\overline{\overline{S}}_{i_0}^{(N)}\right)}$, $l = 1, ..., n$ form a basis of the space

$$\left\{ \begin{bmatrix} t \\ \vdots \\ t \end{bmatrix} \in \mathbb{R}^{n\overline{\overline{S}}_{i_0}^{(N)}} : t \in \mathbb{R}^n \right\},$$

Kronecker-Capelli Theorem (see e.g. [22]) implies that (7) holds.

Suppose now that (7) holds. Again by the Kronecker-Capelli Theorem the set of equations

$$G(i_0)v = z$$

where

$$z = \begin{bmatrix} y \\ \vdots \\ y \end{bmatrix} \left.\right\} \ \overline{\overline{x}} \ \text{times}$$

has a solution for each $y \in \mathbb{R}^n$. The above-mentioned fact implies that for each $y \in \mathbb{R}^n$ and each $(i_0, ..., i_{N-1}) \in \overline{S}_{i_0}^{(N)}$ there exists a sequence $g_k(i_0, ..., i_k)$, $k = 0, ..., N - 1$ such that

$$\sum_{t=0}^{N-1} F(N, t+1, i_{N-1}, ..., i_{t+1}) B(i_t) g_t(i_0, ..., i_t) = y.$$

If we define the control

$$u(k) = g_k(i_0, r(1)..., r(k))$$

then

$$x(N, 0, i_0, u) = y$$

and consequently system (1) is $i_0-$ controllable at time $N$ from zero.

Suppose that the system (1) is $i_0-$ controllable at time $N$ to zero. Then for each $y \in \mathbb{R}^n$ there exists a control sequence $u(0), ..., u(N-1)$ such that

$$u(k) = g_k(i_0, r(1), ..., r(k)), \quad k = 0, ..., N - 1,$$

and

$$x(N, y, i_0, u) = 0, \tag{12}$$

where $g_k$ is a function from $\overline{S}_{i_0}^{(k)}$ to $\mathbb{R}^m$, $k = 0, ..., N - 1$. From the last equation we get

$$\sum_{t=0}^{N-1} F(N, t+1, i_{N-1}, ..., i_{t+1}) B(i_t) g_t(i_0, ..., i_t) = -F(N, 0, i_{N-1}, ..., i_0) y.$$

and therefore

$$-H(i_0) y \in ImG(i_0)$$

what implies that (8) holds.

Assume now that the condition (8) holds. It means that for each $x_0 \in \mathbb{R}^n$ there exists $v$ such that

$$-H_X(i_0) x_0 = G_X(i_0) v.$$

This in turn implies that for each $(i_0, ..., i_{N-1}) \in \overline{S}_{i_0}^{(N)}$ there exists a sequence $g_k(i_0, ..., i_k)$, $k = 0, ..., N - 1$ such that

$$\sum_{t=0}^{N-1} F(N, t+1, i_{N-1}, ..., i_{t+1}) B(i_t) g_t(i_0, ..., i_t) =$$

$$-F(N, 0, i_{N-1}, ..., i_0) x_0.$$

If we define the control

$$u(k) = g_k(i_0, r(1)..., r(k))$$

then

$$x\left(N, x_0, i_0, u\right) = 0$$

and system (1) is $i_0-$ controllable at time $N$ to zero. In the same way we can show the part about $i_0-$ controllability at time $N$.

When we consider the time-varying system it is well known that the controllability from zero implies the controllability to zero [23] and that inverse implication is not true. The next example shows that for the switched system the controllability from zero does not imply the controllability to zero.

*Example 1.* Consider the system (1) with $S = \{1, 2\}$, $N = 2$, $A = \emptyset$.

$$A_1 = \begin{bmatrix} 1 & 2 \\ 3 & 1 \end{bmatrix}, \quad A_2 = \begin{bmatrix} -1 & 2 \\ 1 & -1 \end{bmatrix}, \quad B_1 = \begin{bmatrix} 0 \\ 2 \end{bmatrix}, \quad B_2 = \begin{bmatrix} 0 \\ 1 \end{bmatrix}$$

According to the notation we have

$$G(1) = \begin{bmatrix} C(1,1) \\ C(1,2) \end{bmatrix} = \begin{bmatrix} A_1 B_1 & B_1 & 0 \\ A_2 B_1 & 0 & B_2 \end{bmatrix} = \begin{bmatrix} 4 & 0 & 0 \\ 2 & 2 & 0 \\ 4 & 0 & 0 \\ -1 & 0 & 1 \end{bmatrix}$$

and

$$G(2) = \begin{bmatrix} C(2,1) \\ C(2,2) \end{bmatrix} = \begin{bmatrix} A_1 B_2 & B_1 & 0 \\ A_2 B_2 & 0 & B_2 \end{bmatrix} = \begin{bmatrix} 2 & 0 & 0 \\ 1 & 2 & 0 \\ 2 & 0 & 0 \\ -1 & 0 & 1 \end{bmatrix}.$$

Now it is clear that that the condition (8) is satisfied. The control which steers the zero initial condition to $\begin{bmatrix} x_1^{(0)} \\ x_2^{(0)} \end{bmatrix}$ at time $N = 2$ is given by

$$u\left(0\right) = \begin{cases} \frac{x_1^{(0)}}{4} & \text{if} \quad r(0) = 1 \\ \frac{x_1^{(0)}}{2} & \text{if} \quad r(0) = 2 \end{cases}$$

$$u\left(1\right) = \begin{cases} \frac{x_2^{(0)}}{2} - \frac{x_1^{(0)}}{4} & \text{if} \quad r(0) = 1, \ r(1) = 1 \\ x_2^{(0)} + \frac{x_1^{(0)}}{4} & \text{if} \quad r(0) = 1, \ r(1) = 2 \\ x_2^{(0)} + \frac{x_1^{(0)}}{2} & \text{if} \quad r(0) = 2, \ r(1) = 2 \\ \frac{x_2^{(0)}}{2} - \frac{x_1^{(0)}}{4} & \text{if} \quad r(0) = 2, \ r(1) = 1 \end{cases}.$$

From the other hand the system is not controllable to zero at time 2. In fact we have

$$H(1) = \begin{bmatrix} 7 & 4 \\ 6 & 7 \\ 5 & 0 \\ -2 & 1 \end{bmatrix}, \quad H(2) = \begin{bmatrix} 3 & -4 \\ -2 & 3 \\ 1 & 0 \\ -2 & 5 \end{bmatrix}$$

and

$$
\begin{bmatrix} 11 \\ 13 \\ 5 \\ 1 \end{bmatrix} \in H(1), \qquad \begin{bmatrix} -1 \\ 1 \\ 1 \\ 3 \end{bmatrix} \in H(2),
$$

but

$$
\begin{bmatrix} 11 \\ 13 \\ 5 \\ 1 \end{bmatrix} \notin G(1), \qquad \begin{bmatrix} -1 \\ 1 \\ 1 \\ 3 \end{bmatrix} \notin G(2).
$$

## 4   Conclusions

In the paper we presented the necessary and sufficient conditions for controllability (controllability to zero and controllability from zero) for linear discrete-time switched linear systems. These conditions are given in terms of relations consisting of ranks and images of matrices constructed on the base of the system coefficients. The proposed controllability concept is appropriate to the situation when the switching signal models unpredictable events, for example systems failures. Additionally, a new contribution of the paper is that we took into account the situation in which certain switching sequences are not possible. This situation often occurs in engineering practice.

**Acknowledgments.** The research presented here were done by the authors as parts of the projects funded by the National Science Centre granted according to decisions DEC-2014/13/B/ST7/00755, DEC-2012/05/B/ST7/00065, DEC-2012/07/B/ST7/01404 and DEC-2012/07/N/ST7/03236, respectively. The calculations were performed with the use of IT infrastructure of GeCONiI Upper Silesian Centre for Computational Science and Engineering (NCBiR grant no POIG.02.03.01-24-099/13).

## References

1. Kalman, R.E.: On the general theory of control systems. In: First IFAC Congress Automatic Control, Moscow, Butterworths, London, vol. 1, pp. 481–492 (1960)
2. Liberzon, D.: Switching in Systems and Control. Systems and Control: Foundations and Applications. Birkhauser, Boston (2003)
3. Klamka, J., Niezabitowski, M.: Controllability of switched linear dynamical systems. In: 18th International Conference on Methods and Models in Automation and Robotics, Miedzyzdroje, Poland, pp. 464–467 (2013)
4. Klamka, J., Czornik, A., Niezabitowski, M.: Stability and controllability of switched systems. Bulletin of the Polish Academy of Sciences - Technical Sciences **61**(3), 547–555 (2013)
5. Klamka, J., Niezabitowski, M.: Controllability of switched infinite-dimensional linear dynamical systems. In: 19th International Conference on Methods and Models in Automation and Robotics, Miedzyzdroje, Poland, pp. 171–175 (2014)
6. Krastanov, M.I., Veliov, V.M.: On the controllability of switching linear systems. Automatica **41**, 663–668 (2005)

7. Ge, S.S., Sun, Z., Lee, T.H.: Reachability and controllability of switched linear discrete-time system. IEEE Transactions on Automatic Control, AC **46**(9), 1437–1441 (2001)
8. Klamka, J., Niezabitowski, M.: Trajectory controllability of semilinear systems with multiple variable delays in control. In: Proceedings of the ICNPAA 2014 World Congress: 10th International Conference on Mathematical Problems in Engineering, Aerospace and Sciences, Narvik, Norway, July 15–18, 2014, vol. 1637, pp. 498–503 (2014)
9. Klamka, J., Ferenstein, E., Babiarz, A., Czornik, A., Niezabitowski, M.: Trajectory controllability of semilinear systems with delay in control and state. In: Proceedings of the 2nd International Conference on Control, Mechatronics and Automation, Dubai, United Arab Emirates, December 08–10, 2014
10. Chen, Q., Teng, Z., Hu, Z.: Bifurcation and control for a discrete-time prey-predator model with holling-IV functional response. International Journal of Applied Mathematics and Computer Science **23**(2), 247–261 (2013)
11. Zubowicz, T., Brdys, M.: Stability of softly switched multiregional dynamic output controllers with a static antiwindup filter: A discrete-time case. International Journal of Applied Mathematics and Computer Science **23**(1), 65–73 (2013)
12. Klamka, J., Czornik, A., Niezabitowski, M., Babiarz, A.: Controllability and minimum energy control of linear fractional discrete-time infinite-dimensional systems. In: 11th IEEE International Conference on Control & Automation, Taichung, Taiwan, pp. 1210–1214 (2014)
13. Sun, Z., Ge, S.S., Lee, T.H.: Controllability and reachability criteria for switched linear systems. Automatica **38**, 775–786 (2002)
14. Qiao, Y., Cheng, D.: On partitioned controllability of switched linear systems. Automatica **45**(1), 225–229 (2009)
15. Wang, Y., Qi, A.: A Lyapunov Characterization of Asymptotic Controllability for Nonlinear Switched Systems. Bulletin of the Korean Mathematical Society **51**(1), 1–11 (2014)
16. Lu, Q., Zuazua, E.: Robust null controllability for heat equations with unknown switching control mode. Discrete and Continuous Dynamical Systems **34**(10), 4183–4210 (2014)
17. Liu, X., Lin, H., Chen, B.: Structural controllability of switched linear systems. Automatica **49**(12), 3531–3537 (2013)
18. Xie, G., Wang, L.: Reachability realization and stabilizability of switched linear discrete-time systems. J. Math. Anal. Appl. **280**, 209–220 (2003)
19. Ji, Z., Lin, H., Lee, T.H.: A new perspective on criteria and algorithms for reachability of discrete-time switched linear systems. Automatica **45**, 1584–1587 (2009)
20. Czornik, A., Swierniak, A.: On controllability with respect to the expectation of discrete time jump linear systems. Journal of the Franklin Institute **338**, 443–453 (2001)
21. Czornik, A., Swierniak, A.: Controllability of Discrete Time Jump Linear Systems. Dynamics of Continuous, Discrete and Impulsive Systems, B: Applications and Algorithms **12**(2), 165–191 (2005)
22. Zhivetin, V.B.: Advanced Calculus, Lectures, vol. 1. Pensoft Publishers, Bulgaria (2007)
23. Klamka, J.: Controllability of Dynamical Systems. Kluwer Academic Publishers, Dordrecht (1991)

# Trajectory Controllability
# of Semilinear Systems with Delay

Jerzy Klamka, Adam Czornik, Michał Niezabitowski$^{(\boxtimes)}$, and Artur Babiarz

Faculty of Automatic Control, Electronics and Computer Science,
Institute of Automatic Control, Silesian University of Technology,
Akademicka 16 Street, 44-101 Gliwice, Poland
{jerzy.klamka,adam.czornik,michal.niezabitowski,
artur.babiarz}@polsl.pl
http://www.polsl.pl

**Abstract.** The finite-dimensional dynamical control system described by scalar semilinear ordinary differential state equation with delay is considered in this paper. The semilinear state equation contains both nonlinear perturbations and pure linear part. The concept of relative controllability on trajectory relative controllability for systems with point delay in control and in nonlinear term was extended. Finally, the remarks and comments on the relationships between different concepts of controllability were presented and the possible extensions proposed.

**Keywords:** Controllability · Trajectory relative controllability · Nonlinear finite - dimensional control systems · Semilinear control systems · Lipschitz condition · Delayed control systems

## 1 Introduction

One of the most important concept in mathematical control theory is controllability [1] - [7], which generally speaking means, that it is possible to steer dynamical control system from an arbitrary initial state to an arbitrary final state using the set of admissible controls. The research devoted the controllability was started in the 1960s by Kalman and refer to linear dynamical systems. Because the most of practical dynamical systems are nonlinear, that's why, in recent years various controllability problems for different types of nonlinear or semilinear dynamical systems have been considered in many publications and monographs. The study of controllability for linear systems has spanned a great number of research directions (e.g. fractional and/or positive [8] - [17]). Moreover, in the literature there are many different definitions of controllability for

The research presented here were done by the authors as parts of the projects funded by the National Science Centre granted according to decisions DEC-2012/05/B/ST7/00065, DEC-2012/07/B/ST7/01404, DEC-2012/07/N/ST7/03236 and DEC-2014/13/B/ST7/00755, respectively. The calculations were performed with the use of IT infrastructure of GeCONiL Upper Silesian Centre for Computational Science and Engineering (NCBiR grant no POIG.02.03.01-24-099/13).

© Springer International Publishing Switzerland 2015
N.T. Nguyen et al. (Eds.): ACIIDS 2015, Part I, LNAI 9011, pp. 313–323, 2015.
DOI: 10.1007/978-3-319-15702-3_31

different systems, namely approximate controllability [18] - [30], asymptotic controllability [31] - [34], complete controllability [35] - [44], constrained controllability [45] - [48], exact controllability [49] - [60], global controllability [61] - [66], null controllability [67] - [75], structural controllability [76] - [80] and trajectory-controllability [49], [81] - [83] (in short T-controllability).

The T-controllability of semilinear finite-dimensional first-order dynamical systems with delay in control and in nonlinear term is studied in this paper. Generally speaking, the T-controllability means, that it is possible to steer the dynamical control system from an arbitrary initial state to an arbitrary final state, along a prescribed trajectory, using the set of admissible controls. For simplicity of considerations, it is generally assumed that the values of admissible controls are unconstrained and that's why, T-controllability is a stronger notion than controllability. The motivation to study the T-controllability implies from the fact, that most of industrial processes are nonlinear or semilinear in nature.

The main goal of the paper is to extend the results given in papers [84], [85] to the semilinear dynamical system with delay in control and in nonlinear term.

## 2   System Description

First of all let us start from considering the semilinear nonstationary finite-dimensional control system described by the following scalar first-order ordinary differential equation with variable delay in control and state variables

$$\dot{x}(t) = a(t) x(t) + b(t) u(t - h(t)) + f(t, x(t), x(t - h(t))) \qquad (1)$$

defined for $t \in [t_0, T]$, $t > t_0$, where $h(t)$ is given differentiable variable delay. Because

$$t - h(t) > 0$$

is an increasing function of time $t$, then

$$\frac{d}{dt}(t - h(t)) > 0$$

and thus

$$\dot{h}(t) < 1,$$

$$\left(1 - \dot{h}(t)\right) > 0$$

and

$$h(T) \le T.$$

Moreover, $x(t) \in \mathbb{R}$, $u(t) \in \mathbb{R}$ are scalar state variable and scalar admissible unbounded control, respectively and $u \in L^2[t_0, T]$.

To simplify our considerations we assume zero initial conditions both for the state variable and control. Therefore, we have

$$x(t) = 0$$

and

$$u(t) = 0$$

for $t \in [t_0 - h(t_0), t_0]$. Moreover, we define the functions:

$$v(t) = t - h(t) \qquad (2)$$

for $t \in [t_0, T)$. Because $h(t)$ is differentiable, then the function

$$v : [t_0, T) \to \mathbb{R}$$

is also differentiable and strongly increasing ($\dot{v}(t) > 0$ in $[t_0, T]$).

Furthermore, let us assume that $a(t)$ and $b(t)$ are given scalar continuous functions defined on $t \in [t_0, T]$. As well, it is assumed that

$$f : ([t_0, T]) \times \mathbb{R} \times \mathbb{R} \to \mathbb{R} \qquad (3)$$

is given nonlinear function and measurable with respect to the first variable and continuous with respect to the second and third variable. Moreover, it should be notice that there is no constraints imposed on the state variable $x(t)$ and the admissible control values $u(t)$. To simplify our considerations we introduce an inverse function

$$r : [v(t_0), v(T)] \to [t_0, T], \qquad (4)$$

which is the inverse function to $v(t)$, hence

$$r(t) = t + h(t)$$

for $t \in [t_0, T)$.

Now, we will remind the well-known lemma [1] concerning relationships between the ordinary differential state equations with delay in control and the ordinary differential equations without delay in control.

**Lemma 1.** *Dynamical system (1) can be expressed as follows:*

$$\dot{x}(t) = a(t)\, x(t) + b_T(t)\, u(t) + f\left(t, x(t), x(t - h(t))\right) \qquad (5)$$

*for $t \in [t_0, T]$, where*

$$b_T(t) = F(t_0, r(t))\, b(r(t))\, \dot{r}(t) \qquad (6)$$

*for $t \in [t_0, v(T))$ and*

$$F(t_0, r(t)) = e^{\int_{t_0}^{r(t)} a(\tau) d\tau}. \qquad (7)$$

*Remark 1.* Let us observe, that function $b_T(t)$ strongly depends on time $T$.

Now, let us recall, for completeness of considerations, some fundamental definitions of controllability. It's well-known [1] that for the finite-dimensional first-order dynamical system (5) it is possible to define many different concepts of controllability.

As it was mentioned in the Introduction section, generally we may consider aproximate controllability and exact controllability for infinite-dimensional dynamical systems, or relative controllability and function controllability for dynamical systems with delays. In the next sections of the paper we will consider relative controllability, which is weaker concept than function controllability [7].

**Definition 1.** *Dynamical system (5) is said to be completely relatively controllable on $t \in [t_0, T]$ if for any $x_0$, $x_1 \in \mathbb{R}$ there exist a control $u \in L^2([t_0, T])$ such that the corresponding solution $x(\cdot)$ of (5) satisfies $x(T) = x_1$.*

*Remark 2.* It should be pointed out, that according to the Definition 1 there is no constraints imposed on trajectory between $t_0$ and $T$. Furthermore, let us observe that we do not know anything about trajectory along which the system moves.

From the real industrial processes point of view, it would be necessary to steer dynamical process from a given initial state to final state along the prescribed trajectory. Practically, it would be desirable to choose more suitable path which reduce cost of steering process. Hence, in the literature [85] appears the T-controllability notion.

Let $\tau$ be a given set of all differentiable functions $z(\cdot)$ defined on $t \in [t_0, T]$, which satisfy the initial and final conditions $z(t_0) = 0$, $z(T) = x_1$ and $z(\cdot)$ is a prescribed trajectory differentiable almost everywhere in $[t_0, T]$. Then for systems with delays in the state or in the control, we have the following definition of relative T-controllability.

**Definition 2.** *[85]. Dynamical system (5) is said to be relatively T-controllable if for any $z \in \tau$, there exists a control $u \in L^2([t_0, T])$ such that the corresponding solution $x(\cdot)$ of (5) satisfies $x(t) = z(t)$ in $[t_0, T]$.*

From Definitions 1 and 2 immediately follows the following corollary.

**Corollary 1.** *Suppose that system (5) is relatively T-controllable in time interval $[t_0, T]$. Then the dynamical system (5) is relatively completely controllable in time interval $[t_0, T]$.*

## 3 T-Controllability of Dynamical Systems with Constant Delay

In this section we will consider the special case of semilinear nonstationary finite-dimensional control system described by the following first-order differential equation with constant delay

$$\dot{x}(t) = a(t)x(t) + b(t)u(t - h) + f(t, x(t), x(t - h)) \tag{8}$$

for $t \in [0, T]$, $T > h$. Similarly to the previous section, also here we assume for simplicity of considerations the zero initial conditions both for the state variable and control, i.e.

$$x(0) = 0,$$

$$u(t) = 0$$

for $t \in [-h, 0)$, where

$$h > 0$$

is a constant delay.

Also, let us assume that the values of state $x(t)$ and control $u(t)$ belong to $\mathbb{R}$ and $a(t)$, $b(t)$ are continuous given functions defined on $t \in [0, T]$. Moreover, it is assumed that $f : ([0, T]) \times \mathbb{R} \times \mathbb{R} \to \mathbb{R}$ is a given function. Function $f$ is nonlinear and is measurable with respect to first variable and continuous with respect to second and third variables. Furthermore, it should be stressed that there is no constrians imposed on the control values $u(t)$. Using the method presented in the previous section we have

$$b_T(t) = F(0, t + h) b(t + h)$$

for $t \in [0, T - h)$ and

$$F(0, t + h) = e^{\int_0^{t+h} a(\tau) d\tau} \tag{9}$$

for $i = 0, 1, \ldots, M - 1$.

Furthermore, we can formulate the next corollary, which is sufficient condition for relative T-controllability of the considered system.

**Corollary 2.** *Suppose that:*

1. *the functions $a(t)$ and $b(t)$ are continuous on $t \in [t_0, T]$;*
2. *$b_T(t)$ do not vanish on $t \in [t_0, T]$;*
3. *Function $f(\cdot, \cdot)$ is Lipschitz continuous with respect to the second and third argument, i.e. there exists positive constants $\alpha_1$ and $\alpha_2$ such as*

$$|f(t, x, z) - f(t, y, z)| \leq \alpha_1 |x - y|$$

*and*

$$|f(t, x, z) - f(t, y, w)| \leq \alpha_2 |z - w|.$$

*Then, the first-order semilinear dynamical system with constant delay described by state equation (8) is relative T-controllable in a given time interval $[t_0, T]$ with respect to every differentiable trajectory $z(\cdot) \in \tau$.*

## 4    Example

Now, let us consider relative T-controllability of the simple illustrative example. Let the first-order finite-dimensional dynamical control system in control defined on a given time interval $[t_0, T]$ has the following form:

$$\dot{x}(t) = a(t) x(t) + b(t) u(t - h) + \sin x(t - h) \tag{10}$$

for $t \in [t_0, T]$, $T > h$, where

$$f(t, x(t), x(t - h)) = \sin x(t - h)$$

and functions $a(t)$ and $b(t)$ are continuous in $[t_0, T]$. Moreover, $f(x(t - h))$ satisfies Lipschitz condition. Hence, all assumptions of the Corollary 2 are satisfied and therefore, semilinear scalar dynamical system is relative T-controllable on $[t_0, T]$.

# 5  Conclusion

The sufficient conditions for relative T-controllability of semilinear differential equation with point delay in state variables and admissible controls have been formulated and proved in this paper. These conditions are the extension to the case of relative T-controllability of first-order dynamical control systems without delays [81]. Our results have been simplified to semilinear differential equation with constant point delay. Finally, it should be noticed, that the obtained results could be extended in many directions: control systems with point multiple delays both constant and variable; discrete-time [85]-[88]; switched systems [89], [90]; linear fractional discrete-time [91], [92] and infinite-dimensional systems [93].

**Acknowledgments.** The research presented here were done by the authors as parts of the projects funded by the National Science Centre granted according to decisions DEC-2012/05/B/ST7/00065, DEC-2012/07/B/ST7/01404, DEC-2012/07/N/ST7/03236 and DEC-2014/13/B/ST7/00755, respectively. The calculations were performed with the use of IT infrastructure of GeCONiI Upper Silesian Centre for Computational Science and Engineering (NCBiR grant no POIG.02.03.01-24-099/13).

# References

1. Klamka, J.: Controllability of Dynamical Systems. Kluwer Academic Publishers, Dordrecht (1991)
2. Ge, S.S., Sun, Z., Lee, T.H.: Reachability and controllability of switched linear discrete-time system. IEEE Transactions on Automatic Control **AC–46**(9), 1437–1441 (2001)
3. Qiao, Y., Cheng, D.: On partitioned controllability of switched linear systems. Automatica **45**(1), 225–229 (2009)
4. Sikora, B., Klamka, J.: On constrained stochastic controllability of dynamical systems with multiple delays in control. Bulletin of the Polish Academy of Sciences-Technical Sciences **60**(2), 301–305 (2012)
5. Klamka, J.: Controllability of dynamical systems. A survey, Bulletin of the Polish Academy of Sciences - Technical Sciences **61**(2), 335–342 (2013)
6. Klamka, J., Czornik, A., Niezabitowski, M.: Stability and controllability of switched systems. Bulletin of the Polish Academy of Sciences - Technical Sciences **61**(3), 547–555 (2013)
7. Klamka, J.: Relative and absolute controllability of discrete systems with delays in control. International Journal of Control **26**(1), 65–74 (1977)
8. Kaczorek, T.: Minimum energy control of fractional positive continuous-time linear systems. Bulletin of the Polish Academy of Sciences-Technical Sciences **61**(4), 803–807 (2013)
9. Balachandran, K., Kokila, J.: On the controllability of fractional dynamical systems. International Journal of Applied Mathematics and Computer Science **22**(3), 523–531 (2012)
10. Kaczorek, T.: Minimum energy control of positive continuous-time linear systems with bounded inputs. International Journal of Applied Mathematics and Computer Science **23**(4), 725–730 (2013)

11. Kaczorek, T.: Minimum energy control of fractional positive continuous-time linear systems. In: Proceedings of 18th International Conference on Methods and Models in Automation and Robotics, Miedzyzdroje, Poland, pp. 622–626 (2013)
12. Kaczorek, T.: An extension of Klamka's method of minimum energy control to fractional positive discrete-time linear systems with bounded inputs. Bulletin of the Polish Academy of Sciences-Technical Sciences 62(2), 227–231 (2014)
13. Kaczorek, T.: Minimum energy control of fractional positive continuous-time linear systems with bounded inputs. International Journal of Applied Mathematics and Computer Science 24(2), 335–340 (2014)
14. Kaczorek, T.: Minimum energy control of positive fractional descriptor continuous-time linear systems. Iet Control Theory and Applications 8(4), 219–225 (2014)
15. Kaczorek, T.: Necessary and sufficient conditions for the minimum energy control of positive discrete-time linear systems with bounded inputs. Bulletin of the Polish Academy of Sciences-Technical Sciences 62(1), 85–89 (2014)
16. Kaczorek, T.: A new formulation and solution of the minimum energy control problem of positive 2D continuous-discrete linear systems. In: Szewczyk, R., Zieliński, C., Kaliczyńska, M. (eds.) Recent Advances in Automation, Robotics and Measuring Techniques. AISC, vol. 267, pp. 103–114. Springer, Heidelberg (2014)
17. Kaczorek, T.: Minimum energy control of descriptor positive discrete-time linear systems. Compel - the International Journal for Computation and Mathematics in Electrical and Electronic Engineering 33(3), 976–988 (2014)
18. Klamka, J.: Approximate controllability of second order dynamical systems. Applied Mathematics and Computer Science 2(1), 135–146 (1992)
19. Liu, X., Liu, Z., Bin, M.: Approximate controllability of impulsive fractional neutral evolution equations with Riemann-Liouville fractional derivatives. Journal of Computational Analysis and Applications 17(3), 468–485 (2014)
20. Tao, Q., Gao, H., Zhang, B.: Approximate controllability of a parabolic integrodifferential equation. Mathematical Methods in the Applied Sciences 37(15), 2236–2244 (2014)
21. Debbouche, A., Torres, D.: Approximate controllability of fractional delay dynamic inclusions with nonlocal control conditions. Applied Mathematics and Computation 243, 161–175 (2014)
22. Mokkedem, F., Fu, X.: Approximate controllability of semi-linear neutral integrodifferential systems with finite delay. Applied Mathematics and Computation 242, 202–215 (2014)
23. Boyer, F., Olive, G.: Approximate controllability conditions for some linear 1d parabolic systems with space-dependent coefficients. Mathematical Control and Related Fields 4(3), 263–287 (2014)
24. Shen, L., Wu, Q.: Approximate controllability of nonlinear stochastic impulsive systems with control acting on the nonlinear terms. International Journal of Control 87(8), 1672–1680 (2014)
25. Boulite, S., Bouslous, H., El Azzouzi, M., Maniar, L.: Approximate positive controllability of positive boundary control systems. Positivity 18(2), 375–393 (2014)
26. Ji, S.: Approximate controllability of semilinear nonlocal fractional differential systems via an approximating method. Applied Mathematics and Computation 236, 43–53 (2014)
27. Fu, X., Lu, J., You, Y.: Approximate controllability of semilinear neutral evolution systems with delay. International Journal of Control 87(4), 665–681 (2014)
28. Guendouzi, T., Bousmaha, L.: Approximate Controllability of Fractional Neutral Stochastic Functional Integro-Differential Inclusions with Infinite Delay. Qualitative Theory of Dynamical Systems 13(1), 89–119 (2014)

29. Fujishiro, K., Yamamoto, M.: Approximate controllability for fractional diffusion equations by interior control. Applicable Analysis **93**(9), 1793–1810 (2014)
30. Sakthivel, R., Ganesh, R., Anthoni, S.: Approximate controllability of fractional nonlinear differential inclusions. Applied Mathematics and Computation **225**, 708–717 (2013)
31. Wang, Y., Qi, A.: A Lyapunov Characterization of Asymptotic Controllability for Nonlinear Switched Systems. Bulletin of the Korean Mathematical Society **51**(1), 1–11 (2014)
32. Motta, M., Rampazzo, F.: Asymptotic controllability and optimal control. Journal of Differential Equations **254**(7), 2744–2763 (2013)
33. Cai, X., Wang, X., Xu, X.: Asymptotic controllability of a class of discrete-time systems with disturbances. Journal of Systems Engineering and Electronics **20**(6), 1296–1300 (2009)
34. Tsinias, J.: Remarks on Asymptotic Controllability and Sampled-Data Feedback Stabilization for Autonomous Systems. IEEE Transactions on Automatic Control **55**(3), 721–726 (2010)
35. Davison, E.J., Kunze, E.: C.: Controllability of Integro-Differential Systems in Banach Space. SIAM Journal on Control and Optimization **8**(1), 489–497 (1970)
36. Wan, X., Sun, J.: Complete controllability for abstract measure differential systems. International Journal of Robust and Nonlinear Control **23**(7), 807–814 (2013)
37. Khartovskii, V.E., Pavlovskaya, A.T.: Complete controllability and controllability for linear autonomous systems of neutral type. Automation and Remote Control **74**(5), 769–784 (2013)
38. DAlessandro, D.: Equivalence between indirect controllability and complete controllability for quantum systems. Systems and Control Letters **62**(2), 188–193 (2013)
39. Khartovskii, V.E.: Complete controllability problem and its generalization for linear autonomous systems of neutral type. Journal of Computer and Systems Sciences International **51**(6), 755–769 (2012)
40. Wang, J., Zhou, Y.: Complete controllability of fractional evolution systems. Communications in Nonlinear Science and Numerical Simulation **17**(11), 4346–4355 (2012)
41. Semenov, Y.M.: On the complete controllability of linear nonautonomous systems. Differential Equations **48**(9), 1245–1257 (2012)
42. Yang, S., Shi, B., Zhang, Q.: Complete controllability of nonlinear stochastic impulsive functional systems. Applied Mathematics and Computation **218**(9), 5543–5551 (2012)
43. Sakthivel, R., Ren, Y.: Complete controllability of stochastic evolution equations with jumps. Reports on Mathematical Physics **68**(2), 163–174 (2011)
44. Shen, L., Shi, J., Sun, J.: Complete controllability of impulsive stochastic integro-differential systems. Automatica **46**(6), 1068–1073 (2010)
45. Berrahmoune, L.: A variational approach to constrained controllability for distributed system. Journal of Mathematical Analysis and Applications **416**(2), 805–823 (2014)
46. Balachandran, K., Kokila, J.: Constrained Controllability of Fractional Dynamical Systems. Numerical Functional Analysis and Optimization **34**(11), 1187–1205 (2013)
47. Karthikeyan, S., Balachandran, K.: Constrained Controllability of Nonlinear Stochastic Impulsive Systems. International Journal of Applied Mathematics and Computer Science **21**(2), 307–316 (2011)

48. El Hassan, Z., Fatima, G.: Regional Constrained Controllability Problem: Approaches and Simulations. International Journal of Control Automation and Systems **7**(2), 297–304 (2009)
49. George, R.K., Chalishajar, D.N., Nandakunaran, A.K.: Exact Controllability of Nonlinear Third Order Dispersion Equation. Journal of Mathematical Analysis and Applications **332**(2), 1028–1044 (2007)
50. Lu, X., Tu, Z., Lv, X.: On the exact controllability of hyperbolic magnetic Schrodinger equations. Nonlinear Analysis - Theory Methods and Applications **109**, 319–340 (2014)
51. Morancey, M.: Simultaneous local exact controllability of 1D bilinear Schrodinger equations. Annales de L Institut Henri Poincare-Analyse Non Lineaire **31**(3), 501–529 (2014)
52. Lu, Q.: Exact Controllability for Stochastic Transport Equations. SIAM Journal on Control and Optimization **52**(1), 397–419 (2014)
53. Lu, Q.: Exact controllability for stochastic Schrodinger equations. Journal of Differential Equations **255**(8), 2484–2504 (2013)
54. Zeng, Y., Xie, Z., Guo, F.: On exact controllability and complete stabilizability for linear systems. Applied Mathematics Letters **26**(7), 766–768 (2013)
55. Zheng, C., Zhou, Z.: Exact controllability for the fourth order Schrodinger equation. Chinese Annals of Mathematics Series B **33**(3), 395–404 (2012)
56. Leugering, G., Schmidt, E.J.P.G.: On Exact Controllability of Networks of Nonlinear Elastic Strings in 3-Dimensional Space. Chinese Annals of Mathematics Series B **33**(1), 33–60 (2012)
57. Duan, S., Hu, J., Li, Y.: Exact Controllability of Nonlinear Stochastic Impulsive Evolution Differential Inclusions with Infinite Delay in Hilbert Spaces. International Journal of Nonlinear Sciences and Numerical Simulation **12**(1–8), 23–33 (2011)
58. Li, T., Rao, B.: Strong (Weak) Exact Controllability and Strong (Weak) Exact Observability for Quasilinear Hyperbolic Systems. Chinese Annals of Mathematics Series B **31**(5), 723–742 (2010)
59. Sasu, A.L.: On exact controllability of variational discrete systems. Applied Mathematics Letters **23**(1), 101–104 (2010)
60. Ge, Z.Q., Zhu, G.T., Feng, D.X.: Exact controllability for singular distributed parameter system in Hilbert space. Science in China Series F-Information Sciences **52**(11), 2045–2052 (2009)
61. Zhirabok, A., Shumsky, A.: Approach to the analysis of observability and controllability in nonlinear systems via linear methods. International Journal of Applied Mathematics and Computer Science **22**(3), 507–522 (2012)
62. Chen, H., Sun, J.: A new approach for global controllability of higher order Boolean control network. Neural Networks **39**, 12–17 (2013)
63. De Leo, M., de la Vega, F., Constanza, S., Rial, D.: Global controllability of the 1d schrodinger-poisson equation. Revista de la Union Matematica Argentina **54**(1), 43–54 (2013)
64. Ibrahim, R.W.: Global controllability of a set of fractional differential equations. Miskolc Mathematical Notes **12**(1), 51–60 (2011)
65. Sun, Y.: Global controllability for a class of 4-dimensional affine nonlinear systems. In: Proceedings of the 30th Chinese Control Conference, Yantai, Peoples R. China, 22–24.07.2011, pp. 397–400 (2011)
66. Laurent, C.: Global controllability and stabilization for the nonlinear schrodinger equation on an interval. Esaim-Control Optimisation and Calculus of Variations **16**(2), 356–379 (2010)

67. Fernandez-Cara, E., Santos, M.C.: Numerical null controllability of the 1D linear Schrodinger equation. Systems and Control Letters **73**, 33–41 (2014)
68. Tao, Q., Gao, H., Yao, Z.: Null controllability of a pseudo-parabolic equation with moving control. Journal of Mathematical Analysis and Applications **418**(2), 998–1005 (2014)
69. Lu, Q., Zuazua, E.: Robust null controllability for heat equations with unknown switching control mode. Discrete and Continuous Dynamical Systems **34**(10), 4183–4210 (2014)
70. Shklyar, B.: Exact null-controllability of evolution equations by smooth scalar distributed controls and applications to controllability of interconnected systems. Applied Mathematics and Computation **238**, 444–459 (2014)
71. Fernandez-Cara, E., Limaco, J., de Menezes, S.B.: Null controllability for a parabolic-elliptic coupled system. Bulletin of the Brazilian Mathematical Society **44**(2), 285–308 (2013)
72. Mahmudov, N.I.: Exact null controllability of semilinear evolution systems. Journal of Global Optimization **56**(2), 317–326 (2013)
73. Louis-Rose, C.: Simultaneous null controllability with constraint on the control. Applied Mathematics and Computation **219**(11), 6372–6392 (2013)
74. Debbouche, A., Baleanu, D.: Exact Null Controllability for Fractional Nonlocal Integrodifferential Equations via Implicit Evolution System. Journal of Applied Mathematics, art. no. 931975 (2012)
75. Tenenbaum, G., Tucsnak, M.: On the Null-Controllability of Diffusion Equations. Esaim-Control Optimisation and Calculus of Variations **17**(4), 1088–1100 (2011)
76. Reissig, G., Hartung, C., Svaricek, F.: Strong Structural Controllability and Observability of Linear Time-Varying Systems. IEEE Transactions on Automatic Control **59**(11), 3087–3092 (2014)
77. Hartung, Ch., Reissig, G., Svaricek, F.: Necessary conditions for structural and strong structural controllability of linear time-varying systems. In: Proceedings of the European Control Conference (ECC), Zurich, Switzerland 17–19.07.2013, pp. 1335–1340 (2013)
78. Liu, X., Lin, H., Chen, B.: Structural controllability of switched linear systems. Automatica **49**(12), 3531–3537 (2013)
79. Hartung, Ch., Reissig, G., Svaricek, F.: Sufficient conditions for strong structural controllability of uncertain linear time-varying systems. In: Proceedings of the American Control Conference (ACC), Washington, DC, 17–19.06.2013, pp. 5875–5880 (2013)
80. Hartung, Ch., Reissig, G., Svaricek, F.: Characterization of strong structural controllability of uncertain linear time-varying discrete-time systems. In: Proceedings of the 51st IEEE Annual Conference on Decision and Control (CDC), HI, 10–13.12.2012, pp. 2189–2194 (2012)
81. Chalishajar, D.N., George, R.K., Nandakumaran, A.K., Acharya, F.S.: Trajectory controllability of nonlinear integro-differential system. Journal of the Franklin Institute **347**, 1065–1075 (2010)
82. Klamka, J., Niezabitowski, M.: Trajectory controllability of semilinear systems with multiple variable delays in control. In: Proceedings of the ICNPAA 2014 World Congress: 10th International Conference on Mathematical Problems in Engineering, Aerospace and Sciences, 15–18.07.2014, Narvik, Norway, 1637, pp. 498–503 (2014)

83. Klamka, J., Ferenstein, E., Babiarz, A., Czornik, A., Niezabitowski, M.: Trajectory controllability of semilinear systems with delay in control and state. In: Proceedings of the 2nd International Conference on Control, Mechatronics and Automation, Dubai, United Arab Emirates, 08–10.12.2014

84. Klamka, J.: Constrained controllability of semilinear systems with delayed controls. Bulletin of the Polish Academy of Sciences - Technical Sciences **56**(4), 333–337 (2008)

85. Kaczorek, T.: Computation of realizations of discrete-time cone-systems. Bulletin of the Polish Academy of Sciences - Technical Sciences **54**(3), 347–350 (2006)

86. Czornik, A., Swierniak, A.: On controllability with respect to the expectation of discrete time jump linear systems. Journal of the Franklin Institute **338**, 443–453 (2001)

87. Czornik, A., Swierniak, A.: On direct controllability of discrete time jump linear system. Journal of the Franklin Institute-Engineering and Applied Mathematics **341**(6), 491–503 (2004)

88. Czornik, A., Swierniak, A.: Controllability of Discrete Time Jump Linear Systems. Dynamics of Continuous, Siscrete and Impulsive Systems, B: Applications and Algorithms **12**(2), 165–191 (2005)

89. Czornik, A., Niezabitowski, M.: Controllability and stability of switched systems. In: 18th International Conference on Methods and Models in Automation and Robotics, Miedzyzdroje, Poland, pp. 16–21 (2013)

90. Klamka, J., Niezabitowski, M.: Controllability of switched linear dynamical systems. In: 18th International Conference on Methods and Models in Automation and Robotics, Miedzyzdroje, Poland, pp. 464–467 (2013)

91. Klamka, J., Czornik, A., Niezabitowski, M., Babiarz, A.: Controllability and minimum energy control of linear fractional discrete-time infinite-dimensional systems. In: 11th IEEE International Conference on Control & Automation, Taichung, Taiwan, pp. 1210–1214 (2014)

92. Kaczorek, T.: Fractional positive continuous-time systems and their reachability. International Journal of Applied Mathematics and Computer Science **18**(2), 223–228 (2008)

93. Klamka, J., Niezabitowski, M.: Controllability of switched infinite-dimensional linear dynamical systems. In: 19th International Conference on Methods and Models in Automation and Robotics, Miedzyzdroje, Poland, pp. 171–175 (2014)

# Machine Learning and Data Mining

# A New Pairing Support Vector Regression

Pei-Yi Hao[⊠]

Department of Information Management,
National Kaohsiung University of Applied Sciences, Kaohsiung, Taiwan
haupy@cc.kuas.edu.tw

**Abstract.** In this paper, the new pairing support vector regression (pair-SVR) algorithm is proposed to evaluate nonlinear regression models. In spirit of TSVR, the pair-SVR determines indirectly the regression function through a pair of nonparallel insensitive upper bound and lower bound functions solved by two smaller sized support vector machine (SVM)- type problems, which causes the pair-SVR not only have the faster learning speed than the classical SVR, but also be suitable for many cases, especially when the noise is heteroscedastic, that is, the noise strongly depends on the input value. Besides, the proposed approach improves the sparsity than that of TSVR by introducing an insensitive zone determined by a pair of nonparallel upper bound and lower bound function. Only points outside the insensitive zone are captured as SVs, and only those SVs determine the final regression model. In general, the number of SV is very few. This makes the prediction speed of pair-SVR is obviously faster than TSVR.

**Keywords:** Support vector machines · Support vector regression · Regression estimation · Noise heteroscedastic model

## 1 Introduction

Support Vector Machine (SVM) is a promising kernel-based learning algorithm for data classification and regression. It was first introduced by Vapnik et al. in 1995 as an approximate implementation of the structure risk minimization [1-2]. For achieving best generalization performance, SVM finds the best tradeoff between the model complexity and the learning ability according to the principle of the statistical learning theory. In many applications, SVM has been shown to outperform most traditional learning machines and has been introduced as powerful tools for solving classification and regression problems. Recently, Jayadeva et al. [3] proposed a twin support vector machine (TSVM) classifier for binary classification, motivated by GEPSVM [4]. TSVM determines two nonparallel hyperplanes by two smaller and related SVM-type problems, in which each hyperplane is closer to one of the two classes and is as far as possible from the other one. The strategy of solving two smaller QPPs rather than a single large QPP makes the learning speed of TSVM approximately 4 times faster than that of a classical SVM. TSVM has become one of the popular methods in machine learning because of its low computational complexity [5]. In the spirit of the

© Springer International Publishing Switzerland 2015
N.T. Nguyen et al. (Eds.): ACIIDS 2015, Part I, LNAI 9011, pp. 327–336, 2015.
DOI: 10.1007/978-3-319-15702-3_32

TSVM, Peng [6] has proposed a twin support vector regression (TSVR) for data regression. The TSVR indirectly optimizes the regression function through the ε-insensitive down- and up-bound functions. For this aim, a pair of smaller sized quadratic programming problems (QPPs) is solved instead of the large single QPP in the SVR. This strategy makes the TSVR work faster than the classical SVR. However, the major drawback of TSVR is its prediction speed is significantly slow due to the loss of sparsity. In TSVR, the number of basis function used for estimating the final regression function is equal to the number of training data points. This makes TSVR to be time-consuming in prediction for large-scale data set. In many real applications, the prediction speed is more important than training speed. Hence, it is necessary to improve the sparsity of TSVR, since a sparse regressor means a low economy for storage requirement and a high efficiency for the real time prediction.

In the spirit of TSVR, we propose a novel pairing support vector regression machine (pair-SVR), which determines two nonparallel bound function of regression model by solving two related SVM-type problems, each of which is smaller than in a conventional SVM. Similar to TSVR, our approach also aims at generating two non-parallel bound functions such that the insensitive zone enclosed by those two bound function includes as much training samples as possible. However, the formulation of proposed pair-SVR is totally different from that of TSVR and significantly improves the prediction speed and sparsity. The major advantage of the proposed pair-SVR is its efficiency for both training and prediction. We improve the sparsity of TSVR by introducing an insensitive zone determined by a pair of nonparallel upper bound and lower bound function. Only points outside the zone are captured as SVs, and only those SVs determine the final regression model. In general, the number of SV is very few. This makes the prediction speed of pair-SVR is obviously faster than TSVR. Besides, the strategy of solving two smaller QPPs rather than a single large QPP makes the learning speed of pair-SVR approximately 4 times faster than that of a classical SVR. Further, since the insensitive zone enclosed by a pair of nonparallel upper and lower bound function in the feature could has arbitrary shape in the original space, our approach are especially useful when the noise is heteroscedastic, that is, the noise strongly depends on the inputs.

# 2    Review of Support Vector Regression

## 2.1    Support Vector Regression (SVR)

Suppose we are given a training data set $\{(\mathbf{x}_1, y_1), ..., (\mathbf{x}_N, y_N)\} \subset \aleph \times R$, where $\aleph$ denotes the space of input patterns, for instance, $R^n$. In $\varepsilon$-SVM regression [2], the goal is to find a function f(x) that has at most $\varepsilon$ deviation from the actually obtained targets $y_i$ for all the training data, and at the same time, is as fat as possible. In other words, we do not care about error as long as they are less than $\varepsilon$, but will not accept any deviation larger than $\varepsilon$. An $\varepsilon$-insensitive loss function

$$|\xi|_\varepsilon := \begin{cases} 0 & if \ |\xi| \le \varepsilon \\ |\xi| - \varepsilon & otherwise \end{cases} \tag{1}$$

is used so that the error is penalized only if it is outside the $\varepsilon$-tube. Figure 1 depicts this situation graphically.

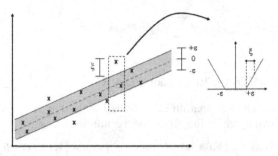

**Fig. 1.** The epsilon insensitive loss setting corresponding to a linear SV regression machine

To make the SVM regression nonlinear, this could be achieved by simply mapping the training patterns $x_i$ by $\Phi: \aleph \to F$ into some higher dimensional feature space $F$. A best fitting function $f(x) = w^t\Phi(x) + b$ is estimated in the feature space $F$. Flatness in the case means that one seeks small w. One way to ensure this is to minimize the Euclidean norm, i.e. $\|w\|^2$. Formally we can write this problem as a convex optimization problem by requiring:

$$\begin{aligned} \underset{w,b,\xi_i,\xi_i^*}{\text{minimize}} \quad & \frac{1}{2}\|w\|^2 + C\sum_{i=1}^{N}(\xi_i + \xi_i^*) \\ \text{subject to} \quad & y_i - (w^t\Phi(x_i) + b) \le \varepsilon - \xi_i \\ & (w^t\Phi(x_i) + b) - y_i \le \varepsilon - \xi_i^* \\ & \xi_i, \xi_i^* \ge 0 \qquad \forall i. \end{aligned} \tag{2}$$

The constant C>0 determines the trade off between the flatness of f(x) and the amount up to which deviations larger than $\varepsilon$ are tolerated. Using the Lagrangian theorem, we can formulate the dual problem as

$$\underset{\alpha_{1i},\alpha_{2i}}{\text{maximize}} \begin{cases} -\dfrac{1}{2}\displaystyle\sum_{i,j=1}^{N}(\alpha_{1i} - \alpha_{2i})(\alpha_{1j} - \alpha_{2j})\Phi(x_i)\cdot\Phi(x_j) \\ -\varepsilon\displaystyle\sum_{i=1}^{N}(\alpha_{1i} + \alpha_{2i}) + \sum_{i=1}^{\ell}y_i(\alpha_{1i} - \alpha_{2i}) \end{cases} \tag{3}$$

$$\text{subject to} \begin{cases} \displaystyle\sum_{i=1}^{N}(\alpha_{1i} - \alpha_{2i}) = 0 \\ \alpha_{1i}, \alpha_{2i} \in [0, C] \end{cases}$$

where $\alpha_{1i}, \alpha_{2i}$ are the nonnegative Lagrange multipliers. A key property of the SVM is that the only quantities that one needs to compute are scalar products, of the form $\Phi(\mathbf{x}) \cdot \Phi(\mathbf{y})$. It is therefore convenient to introduce the so-called kernel function $k$: $k(\mathbf{x}, \mathbf{y}) \equiv \Phi(\mathbf{x}) \cdot \Phi(\mathbf{y})$. The definition of kernel function k prevents the direct computation of inner-production in the high-dimensional feature space, which is very time-consuming, and makes the computation practical. Using this quantity, the solution of an SVM has the form:

$$f(\mathbf{x}) = \sum_{i=1}^{N} (\alpha_{1i} - \alpha_{2i}) k(\mathbf{x}_i, \mathbf{x}) + b. \tag{4}$$

## 2.2    Twin Support Vector Regression (TSVR)

The TSVR finds a pair of nonparallel functions around the data points. In general, it considers the following pair of functions for the nonlinear case:

$$f_1(\mathbf{x}) = \mathbf{w}_1^T \mathbf{K}(\mathbf{A}, \mathbf{x}) + b_1 \quad \text{and} \quad f_2(\mathbf{x}) = \mathbf{w}_2^T \mathbf{K}(\mathbf{A}, \mathbf{x}) + b_2$$

each one determines the $\varepsilon$-insensitive down- or up-bound function, respectively. The functions $f_1(\mathbf{x})$ and $f_2(\mathbf{x})$ are obtained by solving the following pair of QPPs:

$$\underset{\mathbf{w}_1, b_1}{\text{minimize}} \quad \frac{1}{2} \| \mathbf{Y} - \mathbf{e}\varepsilon_1 - (\mathbf{K}\mathbf{w}_1 + \mathbf{e}b_1) \|^2 + \frac{C_1}{N} \mathbf{e}^T \xi \tag{5}$$
$$\text{subject to} \quad \mathbf{Y} - (\mathbf{K}\mathbf{w}_1 + \mathbf{e}b_1) \geq \mathbf{e}\varepsilon_1 - \xi, \quad \xi \geq 0$$

$$\underset{\mathbf{w}_1, b_1}{\text{minimize}} \quad \frac{1}{2} \| \mathbf{Y} - \mathbf{e}\varepsilon_2 - (\mathbf{K}\mathbf{w}_2 + \mathbf{e}b_2) \|^2 + \frac{C_2}{N} \mathbf{e}^T \xi \tag{6}$$
$$\text{subject to} \quad \mathbf{Y} - (\mathbf{K}\mathbf{w}_2 + \mathbf{e}b_2) \geq \mathbf{e}\varepsilon_2 - \xi, \quad \xi \geq 0$$

where $\varepsilon_1$, $\varepsilon_2 \geq 0$ are the insensitive parameters. $C_1$, $C_2 \geq 0$ are the regularization parameters. $\mathbf{e}$ are vector of ones of $N$ dimensions. Y is the target vector $\mathbf{Y} = (y_1, .., y_N)^T$. $\xi$ is the slack vector $\xi = (\xi_1, ..., \xi_N)^T$. $\mathbf{K}(\mathbf{A}, \mathbf{x})$ is the column vector $(k(\mathbf{A}_1, \mathbf{x}), ..., k(\mathbf{A}_N, \mathbf{x}))^T$ where $\mathbf{A}_i$ are the ith training sample (row vector). $\mathbf{K}$ is the $N$ by $N$ kernel matrix such that $\mathbf{K}_{ij} = k(\mathbf{A}_i, \mathbf{A}_j)$. By considering the KKT conditions for the Lagrangian functions of (5) and (6), we obtain the dual QPPs, which are

$$\max \quad -\frac{1}{2}\alpha^T \mathbf{H}(\mathbf{H}^T\mathbf{H})^{-1}\mathbf{H}^T\alpha + \mathbf{f}^T \mathbf{H}(\mathbf{H}^T\mathbf{H})^{-1}\mathbf{H}^T\alpha - \mathbf{f}^T\alpha \tag{7}$$
$$\text{s.t.} \quad 0 \leq \alpha \leq \frac{C_1}{N}\mathbf{e}$$

and

$$\max \quad -\frac{1}{2}\beta^T \mathbf{H}(\mathbf{H}^T\mathbf{H})^{-1}\mathbf{H}^T\beta - \mathbf{h}^T \mathbf{H}(\mathbf{H}^T\mathbf{H})^{-1}\mathbf{H}^T\beta + \mathbf{h}^T\beta \tag{8}$$
$$\text{s.t.} \quad 0 \leq \beta \leq \frac{C_2}{N}\mathbf{e}$$

where $\mathbf{H} = [\mathbf{K}\ \mathbf{e}]$, $\mathbf{f} = \mathbf{Y} - \varepsilon_1\mathbf{e}$, and $\mathbf{h} = \mathbf{Y} + \varepsilon_2\mathbf{e}$.

After optimizing (7) and (9), we obtain the augmented vectors for $f_1(\mathbf{x})$ and $f_2(\mathbf{x})$, which are

$$\begin{bmatrix} \mathbf{w}_1 \\ b_1 \end{bmatrix} = (\mathbf{H}^T\mathbf{H})^{-1}\mathbf{H}^T(\mathbf{f}-\boldsymbol{\alpha}) \qquad \begin{bmatrix} \mathbf{w}_2 \\ b_2 \end{bmatrix} = (\mathbf{H}^T\mathbf{H})^{-1}\mathbf{H}^T(\mathbf{h}+\boldsymbol{\beta}) \qquad (9)$$

Then the estimated regressor is constructed by as follows:

$$f(\mathbf{x}) = \frac{1}{2}(f_1(\mathbf{x})+f_2(\mathbf{x})) = \frac{1}{2}(\mathbf{w}_1^T+\mathbf{w}_2^T)\mathbf{K}(\mathbf{A},\mathbf{x})+\frac{1}{2}(b_1+b_2) \qquad (10)$$

Noted, Eq. (9) shows the weight vector $\mathbf{w}_1$ and $\mathbf{w}_2$ of TSVR lacks the sparsity. The number of basis function used for estimating the final regression function is equal to the number of training data points. This makes the prediction speed of TSVR is slow, especially when the number of training data is huge. In many real applications, the prediction speed is more important than training speed. Hence, it is necessary to improve the sparsity of TSVR, since a sparse regressor means a low economy for storage requirement and a high efficiency for the real time prediction. In addition, the dual QPPs of TSVR (see(7) and (8)) have to compute the inversion of matrix $\mathbf{H}^T\mathbf{H}$ of size $(N+1)\times(N+1)$ with computational cost $O(N^3)$. This not only raises the computational cost of the TSVR, but also makes some successful learning algorithms of SVR, such as SMO, cannot be easily extended to the learning of TSVR provided the related matrices do not cached before learning.

## 3  Pairing Support Vector Regression Algorithm

In the spirit of TSVM, the proposed pair-SVR aims at generating a pair of nonparallel function $f_1(\mathbf{x})$ and $f_2(\mathbf{x})$ by solving two smaller SVM type QPPs, each of which determines the upper bound and lower bound of the insensitive zone, such that the insensitive zone encloses all training data points, and at the same time, is as small as possible. Following the concept of kernel-based learning, a non-linear function is learned by a linear learning machine in a kernel-introduced feature space while the capacity of the system is controlled by a parameter that does not depend on the dimensionality of the space. The basic idea is that a nonlinear regression function is achieved by simply mapping the input patterns $\mathbf{x}_i$ by $\Phi: R^n \rightarrow F$ into a high-dimensional feature space F. Hence, the proposed pair-SVR seeks to estimate the following two functions: $f_1(\mathbf{x}) = \langle \mathbf{w}_1 \cdot \Phi(\mathbf{x}) \rangle + b_1$ and $f_2(\mathbf{x}) = \langle \mathbf{w}_2 \cdot \Phi(\mathbf{x}) \rangle + b_2$. To estimate $f_1(\mathbf{x})$, the upper bound function of the insensitive zone, we force the upper bound function $f_1(\mathbf{x})$ to move downward by minimizing $\|\mathbf{w}_1\|^2$ and $b_1$ in the objective function, and at the same time, keep all training data $(\mathbf{x}_i, y_i)$ are below the upper bound function in the constraint. Hence, the problem of finding the $\mathbf{w}_1$ and $b_1$ is equivalent to the following optimization problem:

$$\underset{w_1,b_1,\xi_{1i}}{\text{minimize}} \frac{1}{2}\|\mathbf{w}_1\|^2 + b_1 + C_1 \sum_{i=1}^{N} \xi_{1i} \qquad (11)$$

subject to $\quad \langle \mathbf{w}_1 \cdot \Phi(\mathbf{x}_i) \rangle + b_1 \geq y_i - \xi_{1i}$ and $\quad \xi_{1i} \geq 0 \quad$ for i=1,...,N.

Minimization of $\|\mathbf{w}_1\|^2$ is equal to minimization of the model complexity of the regressor and would get a sparser regressoe. We can find the solution of this optimization problem in dual variables by finding the saddle point of the Lagrangian:

$$L = \frac{1}{2}\|\mathbf{w}_1\|^2 + b_1 + C_1\sum_{i=1}^{N}\xi_{1i} - \sum_{i=1}^{N}\alpha_{1i}\left[\langle\mathbf{w}_1\cdot\Phi(\mathbf{x}_i)\rangle + b_1 - y_i + \xi_{1i}\right] - \sum_{i=1}^{N}\beta_{1i}\xi_{1i} \quad (12)$$

where $\alpha_{1i}$ and $\beta_{1i}$ are the nonnegative Lagrange multipliers. Differentiating $L$ with respect to $\mathbf{w}_1$, $b_1$ and $\xi_{1i}$ and setting the result to zero, we obtain:

$$\partial L\big/\partial\mathbf{w}_1 = 0 \quad\Rightarrow\quad \mathbf{w}_1 = \sum_{i=1}^{N}\alpha_{1i}\Phi(\mathbf{x}_i), \quad (13)$$

$$\partial L\big/\partial b_1 = 0 \quad\Rightarrow\quad \sum_{i=1}^{N}\alpha_{1i} = 1, \quad (14)$$

$$\partial L\big/\partial\xi_{1i} = 0 \quad\Rightarrow\quad \alpha_i = C_1 - \beta_{1i} \text{ and } \alpha_{1i} \leq C_1, \quad (15)$$

Substituting Eqs. (13)-(15) into L, we obtain the following dual problem

$$\max\quad \frac{-1}{2}\sum_{i=1}^{N}\sum_{j=1}^{N}\alpha_{1i}\alpha_{1j}\langle\Phi(\mathbf{x}_i)\cdot\Phi(\mathbf{x}_j)\rangle + \sum_{i=1}^{N}\alpha_{1i}y_i \quad (16)$$

subject to

$$\sum_{i=1}^{N}\alpha_{1i} = 1, \qquad \alpha_{1i} \in [0, C_1].$$

Parameter $b_1$ can be determined from the Karush-Kuhn-Tucker (KKT) conditions:

$$\alpha_{1i}\left[\langle\mathbf{w}_1\cdot\Phi(\mathbf{x}_i)\rangle + b_1 - y_i + \xi_{1i}\right] = 0, \quad (17)$$

$$(C_1 - \alpha_{1i})\xi_{1i} = 0 \quad (18)$$

For some $\alpha_{1i} \in (0, C_1)$, we have $\xi_{1i} = 0$ and moreover the second factor in Eq. (17) has to vanish. Hence, $b_1$ can be computed as follows: $b_1 = y_i - \langle\mathbf{w}_1\cdot\Phi(\mathbf{x}_i)\rangle$ for some $\alpha_{1i} \in (0, C_1)$. Final, the upper bound function of the regressor is

$$f_1(\mathbf{x}) = \sum_{i=1}^{N}\alpha_{1i}k(\mathbf{x}, \mathbf{x}_i) + b_1. \quad (19)$$

To estimate $f_2(\mathbf{x}) = \langle\mathbf{w}_2\cdot\Phi(\mathbf{x})\rangle + b_2$, the lower bound function of the insensitive zone, intuitively, we should force the lower bound function $f_2(\mathbf{x})$ to move upward by maximizing $\|\mathbf{w}_2\|^2$ and $b_2$ in the objective function, and at the same time, keep all training data $(\mathbf{x}_i, y_i)$ are above the lower bound function in the constraint. However, maximizing $\|\mathbf{w}_2\|^2$ violates the principle of sparsity regressor. Hence, we use the following trick for estimating the lower bound function. First, we multiplies the desired target $y_i$ by -1 and estimates a mirroring function of $f_2(\mathbf{x})$. We force the mirroring function $\bar{f}_2(\mathbf{x}) = \langle \overline{\mathbf{w}}_2 \cdot \Phi(\mathbf{x}) \rangle + \bar{b}_2$ to move downward, and at same time, keep all instances $(\mathbf{x}_i, -y_i)$ are below the mirroring function. Final, the lower bound function is $f_2(\mathbf{x}) = -\bar{f}_2(\mathbf{x})$. The problem of estimating $\bar{f}_2(\mathbf{x}) = \langle \overline{\mathbf{w}}_2 \cdot \Phi(\mathbf{x}) \rangle + \bar{b}_2$ is equivalent to the following optimization problem:

$$\underset{\overline{w}_2, \bar{b}_2, \xi_{2i}}{\text{minimize}} \quad \frac{1}{2} \| \mathbf{w}_2 \|^2 + \bar{b}_2 + C_1 \sum_{i=1}^{N} \xi_{2i} \tag{20}$$

subject to $\langle \overline{\mathbf{w}}_2 \cdot \Phi(\mathbf{x}_i) \rangle + \bar{b}_2 \geq -y_i - \xi_{2i}$ and $\xi_{2i} \geq 0$ for i=1,...,N.

Similar to the above Lagrange multipliers substituting procedure, we obtain its dual problem as

$$\max \quad \frac{-1}{2} \sum_{i=1}^{N} \sum_{j=1}^{N} \alpha_{2i} \alpha_{2j} \langle \Phi(\mathbf{x}_i) \cdot \Phi(\mathbf{x}_j) \rangle - \sum_{i=1}^{N} \alpha_{2i} y_i \tag{21}$$

subject to $\quad \sum_{i=1}^{N} \alpha_{2i} = 1, \qquad \alpha_{2i} \in [0, C_1]$

After solving (21), we obtain the weight vector $\overline{\mathbf{w}}_2 = \sum_{i=1}^{N} \alpha_{2i} \Phi(\mathbf{x}_i)$. While parameter $b_2$ can be determined from the Karush-Kuhn-Tucker (KKT) conditions:

$$\alpha_{2i} \big[ \langle \overline{\mathbf{w}}_2 \cdot \Phi(\mathbf{x}_i) \rangle + \bar{b}_2 + y_i + \xi_{2i} \big] = 0, \tag{22}$$

$$(C_2 - \alpha_{2i}) \xi_{2i} = 0 \tag{23}$$

For some $\alpha_{2i} \in (0, C_2)$, we have $\xi_{2i} = 0$ and moreover the second factor in Eq. (22) has to vanish. Hence, $b_2$ can be computed as follows: $\bar{b}_2 = -y_i - \langle \overline{\mathbf{w}}_2 \cdot \Phi(\mathbf{x}_i) \rangle$ for some $\alpha_{2i} \in (0, C_2)$. Final, the lower bound function of the regressor is

$$f_2(\mathbf{x}) = -\bar{f}_2(\mathbf{x}) = -\sum_{i=1}^{N} \alpha_{2i} k(\mathbf{x}, \mathbf{x}_i) - \bar{b}_2. \tag{24}$$

After estimating the upper bound and lower bound function, the estimated regressor is constructed as follows

$$f(\mathbf{x}) = \frac{1}{2} (f_1(\mathbf{x}) + f_2(\mathbf{x}))$$

$$= \frac{1}{2} \sum_{i=1}^{N} (\alpha_{1i} - \alpha_{2i}) k(\mathbf{x}, \mathbf{x}_i) + \frac{1}{2} (b_1 - \bar{b}_2) \tag{25}$$

The training instances with $\alpha_{1i}$, $\alpha_{2i} > 0$ are termed support vectors since only those points determined the final regression function.

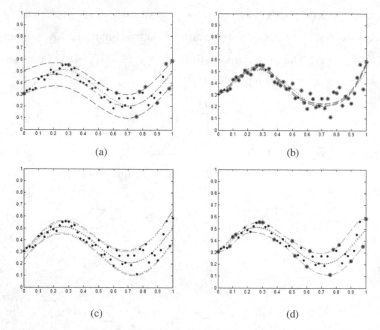

**Fig. 2.** The regression model obtained by (a) $\varepsilon$-SVR with $\varepsilon$=0.1 (b) $\varepsilon$-SVR with $\varepsilon$=0.01 (c) TSVR and (d) pair-SVR

Noted, according to the KKT conditions given by Eqs (17), (18), (22), and (23), only points outside the insensitive zone determined by the upper bound and lower bound function (or lying on the upper or lower bound function) are captured as support vectors. In general, the number of support vector is very few. Hence, pair-SVR significantly improves the sparsity than that of TSVR.

## 4     Experiments

In this section, we use a heteroscedastic dataset to verify the effectiveness of the proposed new support vector regression algorithm. The Gaussian kernel

$$k(\mathbf{x}, \mathbf{y}) = \exp\left(-\|\mathbf{x} - \mathbf{y}\|^2 / 2\sigma^2\right)$$

is used here. The optimal choice of model-parameters was tuned using a grid search mechanism. For simplicity, we set $C_1 = C_2$. The training data sets are generated by

$$y_k = 0.2\sin(2\pi x_k) + 0.2x_k^2 + 0.3 + (0.1x_k^2 + 0.05)e_k, \tag{26}$$

$$x_k = 0.02(k-1), \quad k = 1,2,....,51,$$

where $e_k$ represents a real number randomly generated in the interval $[-1; 1]$. This dataset has heteroscedastic error structure, i.e., the noise strongly depends on the input value $\mathbf{x}$. This example was also used in [7]. Figure 2 shows the regression model obtained by classical SVR ($\varepsilon$-SVR), TSVR, and the proposed pair-SVR. The support vectors are marked with circles in $\varepsilon$-SVR and pair-SVR. In $\varepsilon$-SVR and pair-SVR, the number of basis function for estimating the regression function is equal to number of support vector. However, in TSVR, the number of basis function is equal to the number of training samples. Hence, the sparsity of TSVR is worst. The $\varepsilon$-SVR algorithm relies on the assumption that the noise level is uniform throughout the domain, or at least, its functional dependency is known beforehand [8].

**Table 1.** A comparison of regression performance

|  | $\varepsilon$-SVR ($\varepsilon$=0.1) | $\varepsilon$-SVR ($\varepsilon$=0.01) | TSVR | Pair-SVR |
|---|---|---|---|---|
| RMSE | 0.068554 | 0.053653 | 0.054918 | 0.054235 |
| Training time | 0.967206 | 1.419609 | 0.390003 | 0.312002 |
| Num. of basis function | 10 | 47 | 51 | 17 |

The assumption of a uniform noise model, however, is not always satisfied. In many regression tasks, the amount of noise might depend on location. Due to the assumption that the $\varepsilon$-insensitive zone has a tube (or slab) shape, the test error (risk) in $\varepsilon$-SVR is sensitive toward the changes in $\varepsilon$ on this heteroscedastic data. As seen in Figure 2 (a) and (b), parameter $\varepsilon$ controls the trade-off between sparsity and accuracy. In the case of $\varepsilon$=0.1, we don't care about any deviation from the actual target to the regression function as long as they are less than $\varepsilon$. This leads to worse RMSE, but achieves better sparsity because the number of training samples that are outside the $\varepsilon$-tube is fewer. On the other hand, in the case of $\varepsilon$=0.01, we obtain better RMSE but loss the advantage of sparsity due to almost all training points become the support vectors (lying outside the e-insensitive tube) to derive the satisfying regression estimation. This will increase the testing time on regression, and increase the storage requirement for saving the support vectors. As shown in Figure 2 (c), the nonparallel bound functions of TSVR captures the characteristics of the data set well and obtain satisfactory regression model. However, the major drawback of TSVR is it lacks the sparsity. The prediction speed of TSVR is worst among the three approaches. Figure 2 (d) shows that the proposed pair-SVR derives the satisfying solution to estimating interval bounds and captures well the characteristics of the data set. More importantly, our approach has the advantage of TSVR, i.e., fast speed in training, and meanwhile owns the advantage of sparsity of classic $\varepsilon$-SVR, that is, fast prediction speed. In comparison to classic $\varepsilon$-SVR, our approach owns the advantage of TSVR, i.e., faster training speed than classic SVR. In comparison with TSVR, our approach preserves the advantage of sparsity of classic $\varepsilon$-SVR, i.e., faster prediction speed than TSVR. Table I reports the regression performance among those approaches. As seen from figure 2 and Table I, the pair-SVR not only has fast learning speed, but also shows good generalization performance and sparsity.

# 5    Conclusion

In this paper, the new pairing support vector regression (pair-SVR) algorithm is proposed to evaluate nonlinear regression models for crisp input and output data. In spirit of TSVR, the pair-SVR determines indirectly the regression function through a pair of nonparallel insensitive upper bound and lower bound functions solved by two smaller sized support vector machine (SVM)- type problems, which causes the pair-SVR not only have the faster learning speed than the classical SVR, but also be suitable for many cases, especially when the noise is heteroscedastic, that is, the noise strongly depends on the input value. Besides, we improve the sparsity of TSVR by introducing an insensitive zone determined by a pair of nonparallel upper bound and lower bound function. Only points outside the zone are captured as SVs, and only those SVs determine the final regression model. In general, the number of SV is very few. This makes the prediction speed of pair-SVR is obviously faster than TSVR. The experimental results indicate that the pair-SVR not only has fast learning speed, but also shows good generalization performance and sparsity.

# References

1. Cortes, C., Vapnik, V.N.: Support Vector Network. Machine Learning **20**, 1–25 (1995)
2. Vapnik, V.N.: The Nature of Statistical Learning Theory. Springer, New York (1995)
3. Khemchandani, J.R., Chandra, S.: Twin support vector machines for pattern classification. IEEE Transactions on Pattern Analysis and Machine Intelligence **29**(5), 905–910 (2007)
4. Mangasarian, O.L., Wild, E.W.: Multisurface proximal support vector classification via generalized eigenvalues. IEEE Transactions on Pattern Analysis and Machine Intellegence **28**(1), 69–74 (2006)
5. Kumar, M.A., Gopal, M.: Application of smoothing technique on twin support vector machines. Pattern Recogn. Lett. **29**(13), 1842–1848 (2008)
6. Peng, X.: TSVR: an efficient twin support vector machine for regression. Neural Networks **23**(3), 365–372 (2010)
7. Jeng, J.-T., Chuang, C.-C., Su, S.-F.: Support vector interval regression networks for interval regression analysis. Fuzzy Sets Syst. **138**, 283–300 (2003)
8. Schölkopf, B., Smola, A.J., Williamson, R., Bartlett, P.L.: New support vector algorithms. Neural Computation **12**(5), 1207–1245 (2000)

# Diversification and Entropy Improvement on the DPSO Algorithm for DTSP

Urszula Boryczka and Łukasz Strąk[✉]

Institute of Computer Science, University of Silesia, Będzińska 39,
41-205 Sosnowiec, Poland
urszula.boryczka@us.edu.pl, lukasz.strak@gmail.com

**Abstract.** This paper introduces a new Discrete Particle Swarm Optimization (DPSO) algorithm for solving the Dynamic Traveling Salesman Problem (DTSP) with entropy diversity control. An experimental environment is stochastic and dynamic. Changeability requires the algorithm to have the ability to quickly adapt. Most scientists draw attention to the correlation between the population diversity and the convergence to the optimum. Controlling population variation allows for the control of a stable convergence of the algorithm to the optimum and provides a good mechanism for avoiding stagnation. This article describes the control of this parameter by examining the pheromone matrix by using the entropy measure. The results of the research on the different variants of the measure in the context of a dynamic TSP are presented.

**Keywords:** Dynamic traveling salesman problem · Pheromone · Discrete particle swarm optimization · Entropy

## 1 Introduction

There has been a growing interest in studying evolutionary algorithms in dynamic environments in recent years due to its importance in real word applications [1]. A problem where input data are changeable depending on time is called a Dynamic Optimization Problem (DOP). The purpose of optimization for DOPs is to continuously track and adapt to changes through time and to quickly find the currently best solution [2]. Particle Swarm Optimization (PSO) is a technique based on a swarm population created by Russell Eberhart and James Kennedy in 1995 [3]. This technique is inspired by the social behavior of a bird flocking or fish schooling. The algorithm was created primarily to optimize the function of continuous space exploration. The PSO algorithm quickly became popular due to the fact that it has a small number of parameters and is easy to implement [3].

One of the reasons for premature convergence in the basic PSO is poor swarm diversity [4,5]. There are many different methods and measures to control premature convergence. Some of them were selected and tested in this article.

The aim of this article is investigate the influence of the diversity measure on algorithm convergence and to determine the best value for the pheromone

© Springer International Publishing Switzerland 2015
N.T. Nguyen et al. (Eds.): ACIIDS 2015, Part I, LNAI 9011, pp. 337–347, 2015.
DOI: 10.1007/978-3-319-15702-3_33

matrix reset procedure. Our solution uses a measure of entropy, which controls a variety of populations. We also propose values for the parameters of the most successful tests.

This paper is structured as follows: section 2 presents the basic concepts, the Dynamic Traveling Salesman Problem and different measures of diversity. Section 3 presents our DPSO algorithm proposals. The research results are shown in Section 4. Section 5 contains a summary and conclusions.

## 2    Background

The dynamic TSP is expressed through changes in both the number of vertices and a cost matrix. Every change in the input data may imply the optimum change. Formula (1) describes the distance matrix in the problem.

$$D(t) = \{d_{ij}(t)\}_{n(t) \times n(t)}, \tag{1}$$

where $t$ denote the time parameter or iteration, $i$ and $j$ are endpoints and $n$ denotes vertices count. This problem has to be understood as a series of static TSP problems. Each element contains a strong resemblance to its predecessor because just part of the data is modified. In the article, the change applies only to the distance between the vertices. The number of vertices is constant. To track the algorithm results (for testing purposes), the optimum should be estimated after the data is changed. In this article, an NEOS Server is used [6,7]. It is an online service, that takes input parameters and returns an optimal result asynchronously.

Diversity defines how the population is varied. The precise relationship between diversity and exploitation/exploration strategies in the solution space is really important. The closer the algorithm is to a local or global optimum, the more individuals are similar to each other. There are many measures of diversity in the population. We propose two classes of measures - based on the standard deviation and on entropy.

Engelbrecht and Olorunda [8] suggested the following measure of diversity: the swarm diameter and swarm radius, the average distance around the swarm center, the normalized average distance around the swarm center, the average of distance around all particles in the swarm and swarm coherence. Studies have shown that the best measure is the distance from the swarm center. Researchers Ismail and Engelbrecht developed the Cooperative Particle Swarm Optimizer for the Continuous Optimization Problem [5]. Other scientists [9] proposed a variation of different measures for the PSO based on the position of the particle, velocity and cognitive term (all references to the particle's current position are replaced with its personal best position).

The second class of diversity measure is based on the entropy definition. In information theory, entropy is a measure of unpredictability in the information content or a measure of diversification (disorder). The definition of entropy was introduced by Claude E. Shannon in 1948 (formula (2)).

$$H(X) = \sum_i P(x_i) \log_b P(\frac{1}{x_i}) = -\sum_i P(x_i) \log_b P(x_i) \qquad (2)$$

where $b$ is a base of the logarithm; entropy has different units for different values - $b = 2$ is bit, $b = e$ is nat and $b = 10$ is dit. $P(x_i)$ is a probability of $i$-th discrete random variable $X$. The two researchers Guntsch and Middendorf [10] used pheromone and entropy in the Dynamic Traveling Salesman Problem in a version with the addition and removal of vertices. They described three methods of pheromone adaptation after introducing some changes in the input data. Then, each strategy was studied in terms of the diversity that is incorporated into the population. Entropy as a diversity measure of individuals can be used in various ways, e.g. for matching parameters [11] or changes in population size [12]. Our Discrete Particle Swarm Optimization solution is based on the algorithm described in [13, 14].

## 3    DPSO with Pheromone

Edge is the form of the following three factors $(a, x, y)$, where $a$ denotes the probability of choosing the edge to the next position, $x$ and $y$ are the endpoints. Variable $a$ is constrained to a number between 0 and 1 and $(a, x, y) = (a, y, x)$. A feasible TSP solution $\{(1, 2), (2, 3), (3, 4), (4, 1)\}$ in the discrete version is equal to $\{(1,1,2), (1,2,3), (1,3,4), (1,4,1)\}$ [13, 14]. The equation of the classic PSO takes a new form given by equations (3) and (4).

$$V_i^{k+1} = c_2 rand() \cdot (gBest - X_i^k)$$
$$+ c_1 rand() \cdot (pBest_i - X_i^k) \qquad (3)$$
$$+ \omega \cdot V_i^k$$
$$X_i^{k+1} = \Delta \tau^k(V_i^{k+1}) \oplus c_3 rand() \cdot X_i^k \qquad (4)$$

where: $i$ denotes the number of particles, $k$ - iteration number and $rand()$ is a random value with uniform distribution from $[0, 1]$. Edge set $pBest$ and $gBest$ denote the particle personal best position and global best position respectively. The operations of addition and subtraction are the sum and the difference of sets and variable $w$ is called an inertia weight. Parameters $c_1$ and $c_2$ are the cognitive and social parameters for scaling $pBest$ and $gBest$ respectively. The $\oplus$ operator completes the next position with the edges from the previous position if the added edge does not make an infeasible tour. Function $\Delta \tau^k(V_i^{k+1})$ is a reinforcement probability of edge selection to the next position given by formula (5).

$$\Delta \tau^k(V_i^{k+1}) = (\tau_e - 0.5) \cdot \frac{k}{k_c}, \quad \forall e \in V_i^{k+1} \subseteq E \qquad (5)$$

where $e$ denote edge, $\tau_e$ is the edge pheromone value extracted from the pheromone matrix, $k$ denotes iteration number and $k_c$ is iterations count (assumed before algorithm run). Factor $\frac{k}{k_c}$ scales edge reinforcement depending on the progress of the algorithm. The result of equation (5) is the edge set with increased or decreased probability component $a$ for each edge. The probability of the selection is reinforced by the amount of pheromone associated with this edge. The pheromone evaporation process will be responsible for the function of penalty, if the edge does not improve the tour. To apply the pheromone, we created formula (5), which is responsible for the allocation of the pheromone for the edges. Algorithm 1 shows the technique outline.

---

**Algorithm 1.** DPSO algorithm outline

---

Create initial swarm
Create neighborhood
**for** $k = 0 \rightarrow Iterations$ **do**
    **for** $i = 0 \rightarrow SwarmSize$ **do**
        Calculate velocity $V_i^{k+1}$ (equation (3))
        ***Filtration stage***
        $X_i^{k+1} = \emptyset$
        **for all** $(a, x, y) \in V_i^{k+1}$ **do**            $\triangleright$ eq. (4)
            Increase $a$ by pheromone at edge (x,y)
            Random value $r \in [0, 1]$
            **if** $r \leq a$ **then**
                $X_i^{k+1} \cup (1, x, y)$
            **end if**
        **end for**
        **for all** $(a, x, y) \in X_i^k$ **do**            $\triangleright$ eq. (4)
            Random value $r \in [0, 1]$
            $\bar{a} = r \cdot c_3$
            Random value $r \in [0, 1]$
            **if** $r \leq \bar{a}$ **then**
                $X_i^{k+1} \cup (1, x, y)$
            **end if**
        **end for**
        ***Fulfill stage***
        Complete $X_i^{k+1}$ using neighborhood
        Set $V_i^k = V_i^{k+1}$
        Set $X_i^k = X_i^{k+1}$
        Update $pBest_i$ and $gBest$
    **end for**
    Evaporate pheromone
**end for**

---

After creating the velocity set (equation (4)), the complement to the full Hamilton cycle is begun. The operation is done by adding missing edges from the nearest neighboring heuristic. The neighborhood is created based on the nearest neighbor heuristic using $\alpha$-measure [15]. Pheromone update (evaporation) and pheromone initialization is identical to that in the MMAS algorithm [16].

## 3.1   Proposed Solution

Pseudocode 2 shows the process of the algorithm that takes into account the dynamism of DTSP.

---

**Algorithm 2.** DTSP Benchmark
___
Read static TSP problem
**while** Stop criterion is not met **do**
    Create neighborhood
    Optimize problem using DPSO                      ▷ Solve problem
    Update $gBest$
    Change input data by $n\%$
    Compare obtained value with exact value from NEOS Server
**end while**

---

The algorithms compared in the experiment section are based on the same input data (the same changes). For this reason, they can be directly compared.

The pheromone is responsible for the transfer of knowledge to the next iteration of the DTSP. In addition, the global pheromone stores information about the best solutions that are better than $gBest$ because the information is more complete.

Edge diversity in the population depends on pheromone distribution. When only some edges will accumulate greater value of pheromone, the diversity is decreasing. The aim of this article is to present a new approach to the problem of pheromone reset based on the entropy measure. Stützle and Hoos [16] proposed a pheromone reset after a fixed number of iterations. In this paper we use the measure of the entropy calculated on the basis of the pheromone matrix. The formula is as follows:

$$H = -\sum_{i=1}^{n}\sum_{j=1}^{n} \bar{\tau}_{ij} \log \bar{\tau}_{ij} \qquad (6)$$

where $H$ denotes entropy measure, $n$ indicates distance matrix size (problem size) and $\bar{\tau}_{ij}$ is the calculated normalized probability of the selection (pheromone matrix value divided by the sum of all values in the pheromone matrix).

The study is also focused on the comparison of diversity derived from the pheromone matrix and the fitness function of the swarm. For this purpose we use an algorithm called the probability distribution entropy [12].

Algorithm 1 with entropy reset works as follows: first, the swarm position is computed (DPSO one iteration). Next, the condition of an entropy sample is examined. Counting entropy is time costly. For this reason we created the sampling parameter $E_{probe}$. This factor determines how frequently the algorithm calculates the entropy value. If the entropy sample condition is satisfied, the algorithm will check the current value of entropy and the a priori minimum entropy $(E_{min})$. If the calculated value is less than the assumed $E_{min}$ value, the algorithm will start the pheromone reset procedure. All values in the matrix are assigned a value of $\tau_{min}$. After starting the reset procedure, the next restart is possible only $4 \cdot E_{probe}$ iterations afterwards.

## 4   Experimental Results

The objective of this section is to describe and report the experiments performed in this paper. Experiments were conducted on a computer with a 3.6 GHz Intel i7 processor and 16 GB of RAM memory. All tests were run on a single core. DPSO configuration is set to $c_1 = 0.5$, $c_2 = 0.5$, $c_3 = 0.5$, $\omega = 0.5$. To minimize bias, tests were repeated 30 times. The only exception is the run of the algorithm needed to draw graphs.

   The first study focuses on diversity in the current algorithm. Figure 1 presents a summary of the pheromone matrix entropy, fitness function entropy and the probability distribution entropy in the kroA100 problem.

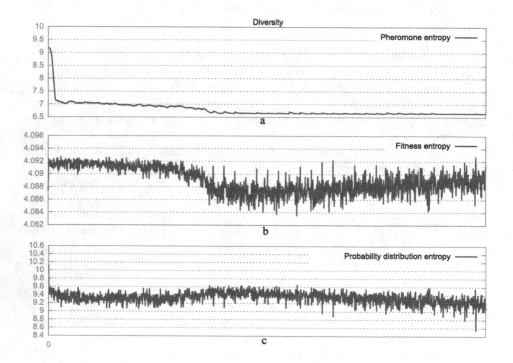

**Fig. 1.** Algorithm diversity without any reset procedure in problem kroA100. Horizontal axis denote iterations.

   Figure 1a shows the entropy pheromone without the reset procedure. Entropy decreases of the pheromone matrix may cause it to get stuck in a local optimum. Generally, there are two possibilities for the decrease of entropy: convergence to an optimum solution and getting stuck in the local optimum (stagnation). The first possibility is usually presented when the algorithm is in the final stage of the computations and it occurs in a population of the algorithm. However, in this case, this was caused by falling into a local optimum. In the figure 1b, the diversity of fitness value increases. This is due to the fact that individuals are represented by the value of the fitness function. Two different solutions that have just two different edges may have very different fitness values. The values of the fitness function may vary greatly, but their edge similarity may be high. The other case may also happen - two solutions may have the same fitness value and have completely different edges. Small differences in the tour cause the apparent increased diversity. The disadvantage of this effect improves the third measure of diversity (probability distribution entropy). However, it does not give conclusive results specifying the diversity of the population.

Figure 2 shows unique edges for iteration and pheromone entropy with the static reset procedure proposed in [16]. This solution runs a pheromone matrix reset after 50 iterations, which has not been corrected for the best solution. The reset constant factors do not take into account the context of population diversity. As the swarm begins to decline rapidly, the pheromone reset procedure is called after a fixed number of iterations. If diversity is going to significantly change in any way, it will not be taken into account in the reset procedure. Especially since the population changes very slowly after resetting the pheromone matrix. In some situations, the pheromone reset procedure is undesirable. In further iterations, the difficulties of finding better solutions increases. Frequent resets may hinder the attainment of a better solution. The first graph shows an increase in the diversity caused by restarting the pheromone values. The pheromone value also precedes the diversity of the population. Based on the value of the pheromone, the diversity of individuals can be predicted. The various values in the pheromone matrix generate new individuals. After a few iterations of the algorithm, the population begins to adapt to new pheromone values. Very large values of pheromone in the matrix unify the swarm. Frequently performing the pheromone reset procedure is related to rapid convergence of the algorithm to the optimum value (less than 1%

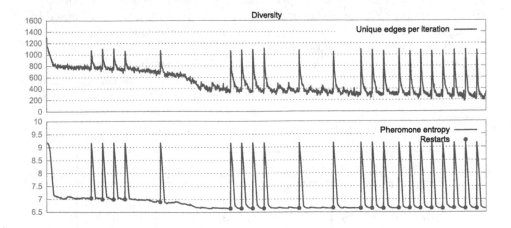

**Fig. 2.** Algorithm diversity with constant reset procedure in problem kroA100. Horizontal axis denote iterations.

optimum) and the difficulty in finding better edges and thereby a better solution. A better measure of diversity is calculated on the pheromone matrix.

Figure 3 shows two different approaches to resetting the pheromone matrix in kroA100 problem. The figure 3a shows the classic approach proposed by Stützle, Thomas and Hoos [16]. The figure 3b shows the entropy reset procedure with the parameters set to: $E_{probe} = 25$, $E_{min} = 6.7$.

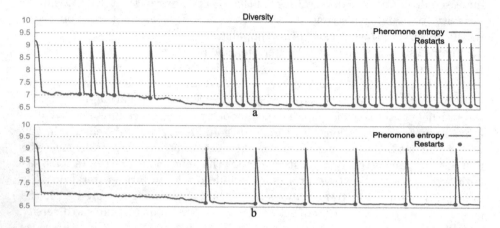

**Fig. 3.** Pheromone reset procedure with constant factor and entropy for problem kroA100

Both techniques present different numbers of the pheromone reset procedures. The algorithm with the constant factor executes the procedure more frequently. Our solution performs the procedure less frequently and only in the final iterations of the algorithm. The algorithm has more time to use the current edges to build a better solution. Regular execution of the reset procedure is caused by finding the optimum in both algorithms (first graph in the last iterations, second graph before half of the iterations).

Table 1 shows tests parameters. Common DPSO settings were set to the following values: $c_1 = 0.5$, $c_2 = 0.5$, $c_3 = 0.5$, $\omega = 0.5$ and repeated 30 times (as mentioned previously). The parameters were determined by sampling the

**Table 1.** Parameters for DPSO test optimization

| Problem | Parameters | | | | | |
|---------|-----------|--------------|--------|-------|------------|-----------|
|         | Iterations | Neighborhood | Repeat | Swarm | $E_{probe}$ | $E_{min}$ |
| kroA100 | $100 \cdot 15$ | 35 | 30 | 60 | 125 | 7.75 |
| ch130   | $130 \cdot 17$ | 35 | 30 | 60 | 50 | 6.75 |
| kroA200 | $200 \cdot 18$ | 35 | 30 | 80 | 125 | 6.75 |
| gr202   | $202 \cdot 18$ | 35 | 30 | 80 | 100 | 8 |
| gil262  | $666 \cdot 20$ | 35 | 30 | 85 | 50 | 8 |
| pcb442  | $442 \cdot 25$ | 35 | 30 | 100 | 150 | 7.5 |

different values. $E_{probe}$ is the size parameter strongly related to the convergence of the algorithm. Best values are closely related to the solved problem.

Table 2 contains the results of different versions of the algorithm in the context of the different sizes of the problems. A bold font indicates the best values. The pheromone matrix reset procedure improves the final results of the algorithm. For small problem sizes, the differences are small. For bigger problems, the differences are larger. The time required to calculate the entropy influence on running time is instance pcb442. The solution proposed by Stützle et al. [16] has not improved our algorithm. Only in gr202 did the problem result turn out better. The results achieved in the static problem (Table 2) is also reflected in the dynamic problem (Table 3). Table 3 shows the results for the dynamic problem based on the gr202 instance.

**Table 2.** DPSO result with different reset procedure variations in TSP problem instances. Bold font indicates the best values, Opt. - optimum tour, T.Dev. - time standard deviation and G.Dev. - standard deviation of distance to optimum.

| Problem | Variations | | | | | | | | Opt. |
|---|---|---|---|---|---|---|---|---|---|
| | No reset | | Stützle et al. reset | | Entropy reset | | | | |
| | Time [s] | Gap [%] | Time [s] | Gap [%] | Time [s] | T. Dev. | Gap [%] | G. Dev. | |
| kroA100 | 9.66 | 0.77 | 9.01 | 0.95 | 9.44 | 0.59 | **0.62** | 169.48 | 21282 |
| ch130 | 13.04 | 1.38 | 10.51 | 1.76 | 12.03 | 0.67 | **1.32** | 34.95 | 6110 |
| kroA200 | 56.83 | 1.99 | 59.88 | 2.02 | 58.01 | 1.46 | **1.8** | 221.28 | 29368 |
| gr202 | 62.57 | 1.79 | 60.45 | 1.58 | 48.14 | 0.68 | **1.28** | 265.25 | 40160 |
| gil262 | 127.94 | 1.07 | 122.57 | 1.22 | 114.92 | 0.96 | **0.97** | 15.37 | 2378 |
| pcb442 | 460.49 | 2.85 | 434.87 | 3.42 | 566.13 | 8.17 | **2.36** | 471.29 | 50778 |

**Table 3.** DPSO result in DTSP problem instances based on gr202 data

| Changes | Variations | | | | | |
|---|---|---|---|---|---|---|
| | Without reset | | With reset | | Entropy reset | |
| | Time [s] | Gap [%] | Time [s] | Gap [%] | Time [s] | Gap [%] |
| 0 | 57.96 | 1.55 | 55.36 | 1.62 | 58.64 | **1.49** |
| 1 | 58.93 | 1.62 | 56.12 | **1.55** | 58.86 | 1.75 |
| 2 | 55.59 | 1.74 | 52.82 | 1.78 | 55.56 | **1.68** |
| 3 | 56.13 | 1.73 | 53.67 | 1.58 | 57.13 | **1.57** |
| 4 | 56.2 | 2.57 | 53.89 | **2.36** | 56.07 | 2.41 |
| 5 | 55.41 | **1.44** | 53.83 | 1.65 | 55.53 | 1.48 |
| Avg. | 56.7 | 1.78 | 54.28 | 1.76 | 56.96 | **1.73** |

## 5   Conclusions

Entropy can be used as a measure to estimate the diversity in population algorithms. However, the measure of diversity for the traveling salesman problem based on the solution fitness value is insufficient. Better results were obtained by applying probability distribution based on entropy. The use of entropy based on the pheromone matrix improved the quality of the solutions and running time. The obtained results show that the gain in the solution quality increases with the increasing size of the problem. Even in problems where the pheromone reset procedure gave worse solutions, our modification improved the final result. Many examples have shown that entropy is suitable for controlling the pheromone matrix restart procedure. This includes the TSP problem. In the case of the dynamic version, the matrix reset gain is smaller. This is due to the fact that after the change of input data, the algorithm starts with very good edges. The disadvantage of this approach is more work involved in determining the limit of the minimum entropy.

In the future, we want to focus on a variable minimum entropy which will decrease with an increasing number of iterations.

# References

1. Branke, J.: Evolutionary approaches to dynamic environments. In: GECCO Workshop on Evolutionary Algorithms for Dynamics Optimization Problems (2001)
2. Li, W.: A parallel multi-start search algorithm for dynamic traveling salesman problem. In: Proceedings of the 10th international conference on Experimental algorithms (2011)
3. Kennedy, J., Eberhart, R.: Particle swarm optimization. In: Proceedings of the IEEE International Conference on Neural Networks, pp. 1942–1948 (1995)
4. Riget, J., Vesterstrøm, J.S.: A diversity-guided particle swarm optimizer - the arpso. Technical report (2002)
5. Ismail, A., Engelbrecht, A.P.: Measuring Diversity in the Cooperative Particle Swarm Optimizer. In: Dorigo, M., Birattari, M., Blum, C., Christensen, A.L., Engelbrecht, A.P., Groß, R., Stützle, T. (eds.) ANTS 2012. LNCS, vol. 7461, pp. 97–108. Springer, Heidelberg (2012)
6. Gropp, W., Moré, J.J.: Optimization environments and the neos server (1997)
7. Czyzyk, J., Mesnier, M.P., More, J.J.: The neos server. IEEE Computational Science Engineering 5(3), 68–75 (1998)
8. Olorunda, O., Engelbrecht, A.P.: Measuring exploration/exploitation in particle swarms using swarm diversity. In: IEEE Congress on Evolutionary Computation, 2008. CEC 2008. (IEEE World Congress on Computational Intelligence), pp. 1128–1134, June 2008
9. Shi, Y., Eberhart, R.C.: Population diversity of particle swarms. In: 2008 IEEE Congress on Evolutionary Computation, CEC 2008, pp. 1063–1067 (2008)
10. Guntsch, M., Middendorf, M.: Pheromone Modification Strategies for Ant Algorithms Applied to Dynamic TSP. In: Boers, E.J.W., Gottlieb, J., Lanzi, P.L., Smith, R.E., Cagnoni, S., Hart, E., Raidl, G.R., Tijink, H. (eds.) EvoIASP 2001, EvoWorkshops 2001, EvoFlight 2001, EvoSTIM 2001, EvoCOP 2001, and EvoLearn 2001. LNCS, vol. 2037, pp. 213–222. Springer, Heidelberg (2001)
11. Aleti, A., Moser, I.: Entropy-based adaptive range parameter control for evolutionary algorithms. In: GECCO 2013 - Proceedings of the 2013 Genetic and Evolutionary Computation Conference, pp. 1501–1508 (2013)

12. Juszczuk, P., Boryczka, U.: The Differential Evolution with the Entropy Based Population Size Adjustment for the Nash Equilibria Problem. In: Bădică, C., Nguyen, N.T., Brezovan, M. (eds.) ICCCI 2013. LNCS, vol. 8083, pp. 691–700. Springer, Heidelberg (2013)
13. Boryczka, U., Strąk, Ł.: Efficient dpso neighbourhood for dynamic traveling salesman problem, pp. 721–730 (2013)
14. Zhong, W.L., Zhang, J., Chen, W.N.: A novel set-based particle swarm optimization method for discrete optimization problems. In: Evolutionary Computation, 2007. CEC 2007, vol. 14, pp. 3283–3287. IEEE (1997)
15. Helsgaun, K.: An effective implementation of k-opt moves for the linkernighan tsp heuristic. Roskilde University, Technical report (2006)
16. Stützle, T., Hoos, H.H.: Max-min ant system. Future generation computer systems 16(8), 889–914 (2000)

# Application of Integrated Neural Network and Nature-Inspired Approach to Demand Prediction

Zhen-Yao Chen[✉]

Department of Business Administration, DE LIN Institute of Technology,
No. 1, Ln 380, Qingyun Rd., New Taipei City 23654, Taiwan
keyzyc@gmail.com

**Abstract.** This study intends to enhance the learning of radial basis function network (RBFN) for function approximation using self-organizing map network (SOMN) with artificial immune system (AIS)-based algorithm (AIA) and genetic algorithm (GA) methods (i.e., IG approach). The proposed combined of SOMN with IG approach (called: SIG) algorithm integrates the auto-clustering ability of SOMN and nature-inspired approach. The simulation results revealed that SOMN, AIA and GA methods can be integrated ingeniously and proposed a hybrid SIG algorithm which aims for obtaining a more accurate learning performance. Next, method evaluation results for two benchmark problems and demand prediction exercise showed that the SIG algorithm outperforms other algorithms and the Box-Jenkins models in accuracy.

**Keywords:** Self-organizing map network · Radial basis function network · Artificial immune system · Genetic algorithm

## 1 Introduction

In general, the widely used time series models especially auto-regressive integrated moving average (ARIMA) model [1] is applicable to linear modeling, and it hardly captures the non-linearity inherent in time series data [15]. Thus, neural network (NN) is preferred as a superior prediction model because it addresses the limitations of time series models by efficient nonlinear mapping between in- and out-of-datasets [37]. Nowadays, the mostly used soft computing approaches are evolutionary computing, fuzzy systems, NNs, swarm intelligence, etc. [9].

Afterward, the success of radial basis function (RBF) network (RBFN) could be lesser adjustable and user-specified parameters during training, tolerance to smaller sample size [33]. In addition, a self-organizing map (SOM) network (SOMN) is a nonlinear NN paradigm [20]. The SOM uses an unsupervised learning, where the network is unaware of the number of classes in which a particular backscatter dataset would be segregated [4]. Furthermore, the previous researchers have adopted the RBFN construction along with other nature-inspired approaches such as artificial immune system (AIS)-based algorithm (AIA) [10], genetic algorithm (GA) [31], and the integrated of AIS and GA (IAG) algorithm [5] to implement the learning of the RBFN.

© Springer International Publishing Switzerland 2015
N.T. Nguyen et al. (Eds.): ACIIDS 2015, Part I, LNAI 9011, pp. 348–357, 2015.
DOI: 10.1007/978-3-319-15702-3_34

Owing to hybrid models forecasting have been widely used in various applications, including price and demand forecasting, the goal is to combine different models and improve the final forecast accuracy [27]. Accordingly, this study presents a combine the auto-clustering ability of SOMN with AIA and GA methods (called: SIG algorithm) for improvement in terms of the fitting accuracy of the function approximation and further applied to a demand prediction exercise.

## 2    Literature Review

NN is a computational model that is also inspired by the structure and functional aspects of biologically neural networks [17]. It has the ability of self-learning, self-organizing and self-adapting to the datasets [15]. For prediction purpose, NN neither requires any stationary nature nor statistical information of datasets [15]. After that, RBFNs [11] have a number of advantages over other types of NNs and these include better approximation capabilities, simpler network structures and faster learning algorithms [28]. Alternatively, the SOM is constructed based on the competitor networks where inhibitory and exhibitory effects of neurons on each other provide the learning infrastructure [20]. Further, there is an increasing number of nature-inspired meta-heuristic algorithms [25] proposed, such as differential evolution (DE) algorithm [12], AIA [7], and GA [36].

Next, AIS inspired by the theory of immunology [16, 18], it is one of the recently developed evolutionary techniques. The field of AISs represents a novel computational intelligence paradigm inspired by the biological immune system [7]. Like NNs and evolutionary algorithms, AIS are extremely conceptual models of their biological counterparts applied to solve problems in different domain areas [7]. It plays a complementary role in either handling the constraints or improving the search ability [39]. For example, Diao & Passino [10] in 2002 proposed an AIA for RBFN structure and adjustment of parameters.

On the other hand, the GA was proposed by Holland in the early 1975s [14]. It follows an adaptive method simulating the evolutionary process in nature and is based on the principle of natural selection and best survival [26]. Moreover, GA method can also train RBFN to be capable of learning through the evolution procedure related with survival of the fittest [14]. For example, Sarimveis et al. [31] in 2004 proposed a GA-based approach, and the purpose is the minimization of an error function with respect to the architecture and relevant parameters of the RBFN. Further, Rajasekaran & Lavanya [30] in 2007 proposed an algorithm to solve structural optimization problems. AIS with a diversification scheme and clonal proliferation improved some drawbacks of GA [39]. Also, Chen [5] in 2013 proposed the integrated of AIS and GA (IAG) algorithm to train the RBFN for function approximation. The IAG algorithm can be utilized to make predictions in the practical Laptop sales prediction exercise [5].

Hence, this research intends to integrate SOMN with AIA and GA methods (called: SIG algorithm) for training RBFN and make suitable verification and comparison. Then, the proposed SIG algorithm can be used to learn and find out the optimal parameters in RBFN.

# 3    Methodology

Integrating the auto-clustering ability of SOMN [21] with AIA and GA methods (i.e., IG approach), this study proposed the SIG algorithm to enhance the accuracy of function approximation by RBFN. The algorithm provides the settings of the RBFN relevant parameters, such as the number of hidden neuron, width, and weight.

During the evolution of the proposed SIG algorithm, SOMN determines the number of center and its position values at first through its auto-clustering ability. Further, the results are used as the number of neuron in RBFN. The algorithm for the IG approach provides the settings of width and weight parameters in RBFN. Afterward, the parameter values of the optimal solution can be obtained and used in the SIG algorithm with RBFN to solve the problem for function approximation. On the other hand, the nonlinear Gaussian basis function is adopted in RBFN hidden layer. Then, the inverse of a root mean squared error (i.e., $RMSE^{-1}$) is used as the fitness function [8]. In the experiment, the fitness value for the SIG algorithm is computed through maximizing the $RMSE^{-1}$ value. As such, a detailed description for the IG approach was summarized and performed as follows.

*(1) Population Initialization:*
(a) Generate a population which has a random number of antibodies (Abs). Each Ab exists its own center point.
(b) With the averaged distance from itself to all the center points of other Abs, the width of each center point in the Ab can be adjusted through formula (1) as below:

$$\text{Width} = \text{Width}_{Default} \left( 1 + tan\ h \left( \frac{\text{distance}_{average} \text{ to other center points}}{\text{suppression threshold}} -1 \right) \right), \quad (1)$$

where the $Width_{Default}$ is a constant which exists dependent with the resolved problem in this initialization stage. The range of the width value was between one to two times of the $Width_{Default}$. Once the width value is set through formula (1), it could then ensure that the Abs are able to maintain a proper distance and avoid the autoimmune response in between.
(c) The value of weight for each Ab is resolved through the least-mean-square (LMS) method [38].

*(2) Fitness Evaluation:*
Similar to the response to antigens (Ags), the AIA would archive and update a pool of candidate solutions (i.e., Abs) to the problem. Thus, formula (2) is adopted to calculate the fitness value of each Ab in the population. Then, the global best solution can be gradually achieved.

$$\text{Fitness (Ab)} = \left| \text{ fitness (center point)} - \text{Mean}_{fitness} \text{ (training data)} \right| \quad (2)$$

*(3) Clone:*
A clone number $N_c$ can be generated through formula (3) ($\varphi$: clone ratio) to inherit the fitness value of each Ab in the parent population and have it as the initial local best solution.

$$N_c = \varphi. \; \lceil \text{the number of center points} \rceil \tag{3}$$

*(4) GA Method:*
The diversity of individual results in a higher chance appears to search in the direction of the global area instead of being confined in local. In the further evolution, it enhances the degree of genetic diversity and obtains a new population. Next, the GA method includes one-point crossover, one-point mutation, and Roulette wheel selection operators.

*(5) Recruitment:*
To increase the diversity in the newly generated population, a certain percentage of Abs will be randomly generated through formula (4) so then be added into the initial population:

$$N_R = \text{Min. (training set/3, 100).} \; \delta. \; \lambda \tag{4}$$

where $N_R$ is the recruitment number, Min. (training set/3, 100) is the count of the RBFN hidden neuron-centers, $\delta$ is the decaying factor, and $\lambda$ is the recruit ratio.

*(6) Update the Global Best Solution:*
Within the randomly generated population in the IG approach, the relevant evolutionary procedures would recruit more Abs to where the data character has largest RMSE, so that the global best solution can be gradually resolved.

*(7) Termination:*
The IG approach of the SIG algorithm will not stop returning to *Step (2)* unless a specific number of generations has been achieved.

Through integrating SOMN with IG approach, the proposed SIG algorithm can improve the diversity of the resolved solutions in the evolution process, and then obtained the global optimal solution. In the latter experiment, the SIG algorithm stops and the RBFN corresponding to the maximum fitness value are selected. After those critical parameter values (i.e., the number of hidden neuron, width, and weight) are set, RBFN can initiate the training and learning of approximation through two benchmark problems (i.e., B2 and Griewank continuous test functions [32]).

# 4    Experimental Simulation Results and Analysis

This research applies to two continuous test functions with many local minima that are frequently used in the literature to be the comparative benchmark of all algorithms. The experiment involves the following two benchmark problems [32], which are defined on Table 1.

Next, the Taguchi (robust design) method (Taguchi & Yokoyama, 1993) (which used in this experiment for parameter setup) is a powerful experimental design tool (Olabi, 2008) for solving the problems of optimizing the performance, quality and cost of a product or process in a simpler, more efficient and systematic manner than traditional trial-and-error processes (Lin et al., 2009). Further, the statistical software MINITAB 14 was used in the analysis of parameter design for the SIG algorithm, where the signal-to-noise (S/N) ratio (Lin et al., 2009) is used to evaluate the stability of system quality in the experiment. Subsequently, this section applies trial analysis using Taguchi experiment design method (Taguchi et al., 2005) and the relevant lite-ratures. The Taguchi trials were configured in an $L_9(3^4)$ orthogonal array for the SIG algorithm after the experiment was implemented thirty times. When S/N ratio is high-er, it indicates that the noise is less and the quality of parameters obtained is better. In this section, 10 trials of Taguchi method are applied to SIG algorithm to determine the best parameter combination. With all of these, the SIG algorithm starts with parame-ters setting shown in Table 2 to ensure consistent basis in the experiment.

**Table 1.** Two benchmark problems [32] used in this experiment

| B2 continuous test function | Griewank continuous test function |
|---|---|
| $B2(x_j, x_{j+1}) = x_j^2 + 2x_{j+1}^2 - 0.3\cos(3\pi x_j) - 0.4\cos(4\pi x_{j+1}) + 0.7$ | $GR(x_j, x_{j+1}) = \sum_{j=1}^{n} \frac{x_j^2}{4000} - \prod_{j=1}^{n} \cos(\frac{x_j}{\sqrt{j+1}}) + 1, n = 1$ |

**Table 2.** Parameter setting for the SIG algorithm in the experiment

| Description | Value | Description | Value |
|---|---|---|---|
| Population size ($S$) | 35 | The width of RBFN hidden layer ($wd_i^s$) | [1000, 40000] |
| The maximum number of genera-tions ($E$) | 1000 | Initial recruitment count ($R_i$) | 25 |
| The number of centers of SOMN ($C$) | [1, 100] | Recruit ratio ($\lambda$) | 0.09 |
| The learning rate of SOMN ($\varepsilon$) | 0.75 | Clone ratio ($\varphi$) | 0.2 |
| The radius of SOMN ($\sigma$) | 10 | Crossover rate ($P_c$) [one-point] | 0.8 |
| The max. number of generations for SOMN ($G$) | 100000 | Mutation rate ($P_m$) [one-point] | 0.1 |

In the process of train RBFN, one thousand randomly generated datasets are di-vided into three partitions: 65-% training set, 25-% testing set, and 10-% validation set [24], in which it can examines and adjusts the procedure of the parameter setting. This study used the SIG algorithm to solve the optimal RBFN parameters solution, and it randomly generates 65-% training set from one thousand generated data and input the set to network for learning. Afterward, it randomly generates 25-% testing set is used for testing process to evaluate the produced individual RBFN parameters solution in the population and individual fitness value in each epoch. After one thou-sand epochs had been progressed, the optimal RBFN parameters solution had been obtained. Finally, it randomly generates 10-% validation set to prove how the individ-ual solution approximate process and record the RMSE values to confirm the situation of RBFN. These stages were implements for 50 runs, and thus the average RMSE (i.e., $\overline{RMSE}$) values were calculated. The values of the $\overline{RMSE}$ ± SD (i.e., standard deviation) are shown in Table 3.

**Table 3.** Comparison results (i.e., $\overline{RMSE}$ ± SD) for all algorithms in the experiment

| Algorithm | B2 function experiment | | Griewank function experiment | |
|---|---|---|---|---|
| | Training set | Validation set | Training set | Validation set |
| AIA [10] | 2199.57 ± 123.76 | 2778.73 ± 438.27 | 6.92 ± 11.01 | 50.94 ± 9.98 |
| GA [31] | 24.18E-2 ± 4.19E-3 | 30.24E-2 ± 12.74E-3 | 5.20E-1 ± 117.08E-4 | 5.55E-1 ± 160.69E-4 |
| IAG [5] | 19.93E-2 ± 4.02E-3 | 23.69E-2 ± 9.84E-3 | 5.81 ± 4.39 | 15.46 ± 3.23 |
| **SIG** | **10.24E-2 ± 3.18E-3** | **12.41E-2 ± 6.29E-3** | **5.03E-1 ± 97.53E-4** | **5.38E-1 ± 86.71E-4** |

As shown in Table 3, the values of training and validation performance are consistently small, which indicates that RBFN trained through the SIG algorithm provides certain stability. Such result not only suffices for the training set and validation set, a generalization could also be made with regards to other unseen dataset. Additionally, from the numerical results comparison in Table 3, the superiority of performance results obtained from the SIG algorithm when verified under different datasets is shown.

# 5    Model Evaluation Results

RBFN has already been verified to be able to generate an accurate approximation on two benchmark problems through the proposed SIG algorithm. Furthermore, the daily sales observations of 500-cm$^3$ containers of papaya milk were offered by a corporation of convenience stores in Taiwan. The trend of papaya milk sales was assumed not affected by exogenous factors. Next, there are several parameter values within RBFN that must be set up to carry out training for the exercise of prediction analysis. In addition, this section applies trial analysis using Taguchi experiment design method (Taguchi et al., 2005) and the relevant literatures. Thus, the parameters setting for the SIG algorithm in the real exercise were shown in Table 4.

Most studies in the literatures utilize convenient ratio of splitting for in- and out-of-samples such as 70-%:30-%, 80-%:20-%, or 90-%:10-% [40]. This study adopts the ratio of 90-%:10-% here as the basis of in- and out-of-samples. Thereafter, the application example with papaya milk for historical sales is based on time series data distribution and applied to demand prediction. For confidential reasons, the data are linearly normalized between zero and one. The detailed data distribution of the real exercise is shown in Table 5.

The learning stage of RBFN will be based on 65-% training set and 25-% testing set. At this point, 90-% of the daily sales data had been adopted, which eventually generated an individual parameters solution with the most precise prediction. Also, the performance of the RBFN prediction shall be examined with a 10-% validation set. The further predicted values were generated in turn through the moving window procedure. The first 90-% of the observations were adopted for model estimation while the remaining 10-% were adopted for validation and one-step-ahead prediction.

Moreover, the research performs demand prediction based on Box-Jenkins models. The ARIMA (p, d, q) modeling procedure has three steps: (a) identifying the model order (i.e., identifying p and q); (b) estimating the model coefficients; (c) forecasting the data [2]. In the experiment, the results of model diagnosis indicate that the Q-statistic values [19] are greater than 0.05 (i.e., serial white noise) in the result of Box-Jenkins models, and it had been suitable fitted (i.e., ARMA (1, 2) model).

**Table 4.** Parameters were setting for the SIG algorithm in the real exercise

| Description | Value | Description | Value |
|---|---|---|---|
| Population size $(S)$ | 35 | The width of RBFN hidden layer $(wd_i^s)$ | [1000, 40000] |
| The maximum number of generations $(E)$ | 1000 | Initial recruitment count $(R_i)$ | 15 |
| The number of centers of SOMN $(C)$ | [1, 100] | Recruit ratio $(\lambda)$ | 0.07 |
| The learning rate of SOMN $(\varepsilon)$ | 0.75 | Clone ratio $(\varphi)$ | 0.15 |
| The radius of SOMN $(\sigma)$ | 10 | Crossover rate $(P_c)$ [One-point] | 0.7 |
| The max. number of generations for SOMN $(G)$ | 100000 | Mutation rate $(P_m)$ [One-point] | 0.2 |

**Table 5.** The observation data distribution in the real exercise

| Case study | The observation data: | |
|---|---|---|
| | Learning set (90-%) | Prediction set (10-%) |
| Papaya milk sales | 01/01/1995~12/07/1995 (341) | 12/08/1995~01/14/1996 (38) |

Alternatively, the MAE (i.e., mean absolute error) and MAPE (i.e., mean absolute percentage error) [6] are the commonly used error measures in business, and were adopted to evaluate the predict models. The prediction performances of related algorithms with the exercise data is presented in Table 6. Meanwhile, the demand prediction results of prediction set are shown in Fig. 1.

**Table 6.** The The prediction errors comparison for different algorithms in the real exercise

| Error | SIG algorithm | IAG algorithm [5] | ARMA (1, 2) model |
|---|---|---|---|
| MAE | 3.63E-2 | 7.07E-2 | 0.0556 |
| MAPE | 7.81E-1 | 11.39E-1 | 18.9163 |

Sales (normalization)

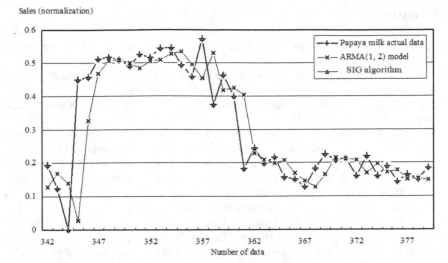

**Fig. 1.** The demand prediction results of prediction set in this exercise

As shown in Table 6 and Fig. 1, the results judged from MAE and MAPE values of the proposed SIG algorithm were the lowest ones. Thus, the SIG algorithm can substantially improve the accuracy of practical demand prediction exercise. This study further supplement an application exercise for the short-term papaya milk daily demand prediction to expound the superiority of the SIG algorithm and it makes its contribution on practical development.

## 6    Conclusions

This research for the proposed SIG algorithm through integrating the auto-clustering ability of SOMN with AIA and GA methods (i.e., IG approach), which adjusts the RBFN parameters involved. The complementation of some nature-inspired learning operations that improve the diversity of populations also enhances the accuracy of the results. Additionally, a case study and the tuning values of parameters with RBFN using the Box-Jenkins models have been given. The SIG algorithm has better parameter establishment and then enables RBFN to fulfill better training and approximation in two benchmark problems and application in the real demand prediction exercise.

In the future, the effort will continuously be made to improve clustering to get better initial kernel functions for RBFN in order to speed up the training process. Additionally, it may be promising to employ differently nature-inspired learning approaches, such as ant colony optimization (ACO), artificial bee colony (ABC), and hyper-heuristics algorithms, can be applied to improve the accuracy of function approximation problem and as well as be applied to different exercises and to further verify its prediction capability.

# References

1. Box, G.E.P., Jenkins, G.: Time Series Analysis. Forecasting and Control. Holden-Day, San Francisco (1976)
2. Babu, C.N., Reddy, B.E.: A moving-average-filter-based hybrid ARIMA-ANN model for forecasting time series data. Applied Soft Computing (2014) (in press)
3. De Castro, L.N., Timmis, J.: Artificial Immune Systems: A New Computational Approach. Springer, London UK (2002)
4. Chakraborty, B., Menezes, A., Dandapath, S., Fernandes, W.A., Karisiddaiah, S.M., Haris, K., Gokul, G.S.: Application of Hybrid Techniques (Self-Organizing Map and Fuzzy Algorithm) Using Backscatter Data for Segmentation and Fine-Scale Roughness Characterization of Seepage-Related Seafloor Along the Western Continental Margin of India. IEEE Journal of Oceanic Engineering (2014) (Accepted [JOE.2013.2294279])
5. Chen, Z.Y.: Soft Computing-based Neural Network Learning for Sales Forecasting. In: 2013 Annual Meeting of the Operations Research Society of Taiwan (ORSTW 2013), National Chiao Tung University, Hsinchu, Taiwan, October 19, 2013
6. Co, H.C., Boosarawongse, R.: Forecasting Thailand's rice export: Statistical techniques vs. artificial neural networks. Computers & Industrial Engineering 53, 610–627 (2007)
7. De Castro, L.N., Timmis, J.: Artificial Immune Systems: A Novel Paradigm to Pattern Recognition. In: Artificial Neural Networks in Pattern Recognition, pp. 67–84. University of Paisley, UK (2002)
8. DelaOssa, L., Gamez, J.A., Puetra, J.M.: Learning weighted linguistic fuzzy rules with estimation of distribution algorithms. In: IEEE Congress on Evolutionary Computation, pp. 900–907. Sheraton Vancouver Wall Centre Hotel, Vancouver, BC, Canada (2006)
9. Dey, S., Bhattacharyya, S., Maulik, U.: Quantum inspired genetic algorithm and particle swarm optimization using chaotic map model based interference for gray level image thresholding. Swarm and Evolutionary Computation 15, 38–57 (2014)
10. Diao, Y., Passino, K.M.: Immunity-based hybrid learning methods for approximator structure and parameter adjustment. Engineering Applications of Artificial Intelligence 15, 587–600 (2002)
11. Duda, R.O., Hart, P.E.: Pattern Classification and Scene Analysis. Wiley, New York (1973)
12. Elsayed, S.M., Sarker, R.A., Essam, D.L.: On an evolutionary approach for constrained optimization problem solving. Applied Soft Computing 12(10), 3208–3227 (2012)
13. Feng, H.M.: Self-generation RBFNs using evolutional PSO learning. Neurocomputing 70, 241–251 (2006)
14. Holland, J.H.: Adaptation in Natural and Artificial Systems. University of Michigan Press, Ann Arbor, MI (1975)
15. Jaipuria, S., Mahapatra, S.S.: An improved demand forecasting method to reduce bullwhip effect in supply chains. Expert Systems with Applications 41, 2395–2408 (2014)
16. Jerne, N.K.: Towards a network theory of immune system. Ann. Immunol. 125(C), 373–389 (1974)
17. Katherasan, D., Elias, J.V., Sathiya, P., Haq, A.N.: Simulation and parameter optimization of flux cored arc welding using artificial neural network and particle swarm optimization algorithm. Journal of Intelligent Manufacturing 25, 67–76 (2014)
18. Khilwani, N., Prakash, A., Shankar, R., Tiwari, M.K.: Fast clonal algorithm. Engineering Applications of Artificial Intelligence 21(1), 106–128 (2008)
19. Kmenta, J.: Elements of Econometrics, (2nd ed.). Macmillan Publishing Co, New York (1986)

20. Kohonen, T.: Self-Organizing and Associative Memory, (2nd ed.). Springer, Berlin (1987)
21. Kohonen, T.: The Self-Organizing Map. Proc. IEEE **78**(9), 1464–1480 (1990)
22. Lee, Z.J.: A novel hybrid algorithm for function approximation. Expert Systems with Applications **34**, 384–390 (2008)
23. Lin, C.F., Wu, C.C., Yang, P.H., Kuo, T.Y.: Application of Taguchi method in light-emitting diode backlight design for wide color gamut displays. Journal of Display Technology **5**(8), 323–330 (2009)
24. Looney, C.G.: Advances in feedforward neural networks: demystifying knowledge acquiring black boxes. IEEE Trans. Knowledge Data Eng. **8**(2), 211–226 (1996)
25. Mezura-Montes, E., CoelloCoello, C.A.: Constraint-handling in nature-inspired numerical optimization: Past, present and future. Swarm Evol. Comput. **1**(22), 173–194 (2011)
26. Milani, A.E., Haghifam, M.R.: An evolutionary approach for optimal time interval determination in distribution network reconfiguration under variable load. Mathematical and Computer Modelling **57**, 68–77 (2013)
27. Motamedi, A., Zareipour, H., Rosehart, W.D.: Electricity price and demand forecasting in smart grids. IEEE Trans. on Smart Grid **3**(2), 664–674 (2012)
28. Qasem, S.N., Shamsuddin, S.M., Zain, A.M.: Multi-objective hybrid evolutionary algorithms for radial basis function neural network design. Knowledge-Based Systems **27**, 475–497 (2012)
29. Olabi, A.G.: Using Taguchi method to optimize welding pool of dissimilar laser-welded components. Opt. Laser Technol. **40**, 379–388 (2008)
30. Rajasekaran, S., Lavanya, S.: Hybridization of genetic algorithms with immune system for optimization problems in structural engineering. Struct. Multidisciplinary Optimization **34**(5), 415–429 (2007)
31. Sarimveis, H., Alexandridis, A., Mazarakis, S., Bafas, G.: A new algorithm for developing dynamic radial basis function neural network models based on genetic algorithms. Computers and Chemical Engineering **28**, 209–217 (2004)
32. Shelokar, P.S., Siarry, P., Jayaraman, V.K., Kulkarni, B.D.: Particle swarm and colony algorithms hybridized for improved continuous optimization. Applied Mathematics and Computation **188**, 129–142 (2007)
33. Singh, P., Deo, M.C.: Suitability of different neural networks in daily flow forecasting. Applied Soft Computing **7**, 968–978 (2007)
34. Taguchi, G., Chowdhury, S., Wu, Y.: Taguchi's Quality Engineering Handbook. Wiley, Hoboken (2005)
35. Taguchi, G., Yokoyama, T.: Taguchi Methods: Design of Experiments. ASI Press, Dearbon, MI (1993)
36. Tessema, B., Yen, G.G.: An adaptive penalty formulation for constrained evolutionary optimization. IEEE Trans. Syst., Man, Cybern., A, Syst. Hum. **39**(3), 565–578 (2009)
37. Wei, S., Song, J., Khan, N.I.: Simulating and predicting river discharge time series using a wavelet-neural network hybrid modelling approach. Hydrological Process **26**, 281–296 (2012)
38. Widrow, B., Hoff, M.E.: Adaptive switching circuits in WESCON Convention. Institute of Radio Engineers, pp. 96–104, New York (1960)
39. Zhang, W., Yen, G.G., He, Z.: Constrained Optimization via Artificial Immune System. IEEE Trans. on Cybernetics **44**(2), 185–198 (2014)
40. Zou, H.F., Xia, G.P., Yang, F.T., Wang, H.Y.: An investigation and comparison of artificial neural network and time series models for Chinese food grain price forecasting. Neurocomputing **70**, 2913–2923 (2007)

# Detecting Entanglement in Quantum Systems with Artificial Neural Network

Joanna Wiśniewska[1](✉) and Marek Sawerwain[2]

[1] Faculty of Cybernetics, Institute of Information Systems,
Military University of Technology, Kaliskiego 2, 00-908 Warsaw, Poland
jwisniewska@wat.edu.pl
[2] Institute of Control and Computation Engineering,
University of Zielona Góra, Licealna 9, Zielona Góra 65-417, Poland
M.Sawerwain@issi.uz.zgora.pl

**Abstract.** The entanglement is an extremely important property of quantum computations. However, detecting the entanglement is a difficult problem – even if models of quantum systems are simulated with use of classical computers. It is caused by the exponential complexity of quantum systems which implies overloading the computational resources. This problem appears especially when the quantum states are expressed as the density matrices. Fortunately, regarding quantum states as patterns and using artificial neural networks to diagnose the presence of entanglement seems to be an interesting idea, because the experiment described in this chapter shows that neural networks are able to detect the entanglement with high probability of correct classification. Additionally, the process of neural network's learning needs less computational resources than complete simulation of quantum computations.

**Keywords:** Quantum states · Entanglement · Feed-forward neural network

## 1 Introduction

The phenomenon of entanglement is one of the most unusual and interesting aspects of quantum mechanics. In the quantum computing the entanglement is a feature of computational systems which allows the crushing acceleration of the calculations in comparison with classical computing. Because of the meaning role of entanglement in a description of the world in the quantum level, it is crucial to be able to recognize the entangled quantum states. There is a large set of methods [3] allowing to calculate if a quantum state is entangled or not (in this case we call it *separable*). However, it should be pointed out that quantum systems can be described by non-polynomial complexity class, so simulation of quantum effects on classic computer consumes the resources very quickly. The idea of using Artificial Neural Network (ANN) to detect the entanglement in quantum states seems then very tempting. Thanks to the well-known properties of generalization the ANN are very often used with success in recent researches

© Springer International Publishing Switzerland 2015
N.T. Nguyen et al. (Eds.): ACIIDS 2015, Part I, LNAI 9011, pp. 358–367, 2015.
DOI: 10.1007/978-3-319-15702-3_35

in many different areas, even the exotic ones like automation control [1]. We decided then to learn simple feed-forward ANN how to distinguish entangled and separable states. The results of the experiment are demonstrated in this chapter.

In Sec. 2 the basic review about quantum units of information, notations of quantum state and entanglement was presented. Sec. 3 contains description of some selected methods for entanglement detecting. The mentioned methods were used to construct the learning set for ANN training – the result of this test is presented in Sec. 5.

## 2    Definitions

Before we will consider the problem of entanglement detecting it is reasonable to recall some essential definition to introduce the reader to the subject raised in this chapter.

**Definition 1.** *A unit vector in two-dimensional Hilbert space $\mathcal{H}_2$ and also quantum circuit, with its state described by space's $\mathcal{H}_2$ vector, will be called a quantum bit (qubit).*

According to the "ket" notation, a state (vector) $x$ in Hilbert space will be represented as $|x\rangle$.

Qubits (quantum bits), just like classical bits, accept two distinct states. These two states have to be orthogonal vectors and this condition is sufficient to set the computational basis in Hilbert space. The most common basis is so-called standard basis – presented below in a form of vectors:

$$|0\rangle = \begin{bmatrix} 1 \\ 0 \end{bmatrix}, \ |1\rangle = \begin{bmatrix} 0 \\ 1 \end{bmatrix}. \tag{1}$$

In general, every quantum state can be expressed as a linear combination of vectors setting a basis, e.g. $n$-qubit state may be described as a superposition of $2^n$ components with use of standard basis:

$$|x\rangle = \alpha_0|00\ldots000\rangle + \alpha_1|00\ldots001\rangle + \alpha_2|00\ldots010\rangle + \ldots + \alpha_{(2^n-1)}|11\ldots111\rangle \tag{2}$$

on the following standardization condition:

$$\sum_{i=0}^{2^n-1} |\alpha_i|^2 = 1 \tag{3}$$

where $\alpha_i$ are complex numbers.

Furthermore, a qubit is a case of a qudit with, so-called, freedom level $d = 2$ (the quantum state is described with two orthogonal vectors). For any positive integer number $d$ greater than one, the quantum state may be expressed as

$$|x\rangle = \sum_{i=0}^{d-1} \alpha_i|i\rangle \ \text{ and } \ \sum_{i=0}^{d-1} |\alpha_i|^2 = 1 \ . \tag{4}$$

It means that for so-called qutrits (qudits with $d = 3$) standard basis vectors are:

$$|0\rangle = \begin{bmatrix} 1 \\ 0 \\ 0 \end{bmatrix}, \quad |1\rangle = \begin{bmatrix} 0 \\ 1 \\ 0 \end{bmatrix}, \quad |2\rangle = \begin{bmatrix} 0 \\ 0 \\ 1 \end{bmatrix}. \tag{5}$$

In addition to vector and superposition form, quantum state can be described as a density matrix. The density matrix $\rho$ of pure quantum state $|x\rangle$ is

$$\rho = |x\rangle\langle x| \tag{6}$$

where $\langle x|$ is the conjugate of $|x\rangle$ ($|\cdot\rangle$ represents the vertical vector and $\langle\cdot|$ horizontal vector).

The entanglement is a characteristic feature of some quantum states. This phenomenon may be shortly described as a kind of correlation between two systems: $A$ and $B$. If these systems are entangled, it means that some operations performed on $A$ may affect the system $B$. The entanglement, as one of key features of quantum mechanics, consists a base for solving problems with many quantum algorithm and protocols, e.g. quantum teleportation algorithm.

The most known examples of quantum entangled states are Bell states.

$$|\psi^{\pm}\rangle = \frac{1}{\sqrt{2}}\left(|00\rangle \pm |11\rangle\right) \text{ or } |\phi^{\pm}\rangle = \frac{1}{\sqrt{2}}\left(|01\rangle \pm |10\rangle\right) \tag{7}$$

These states can be generalized for qudits, e.g. for qutrits – higher unity of quantum information ($d = 3$) – Bell-like state can be expressed as follows:

$$|\psi_3^{\pm}\rangle = \frac{1}{\sqrt{2}}\left(|00\rangle \pm |22\rangle\right) \tag{8}$$

where 00 and 22 denote numbers in ternary number system.

We can also discuss the level of entanglement. The maximally entangled qudit state is presented by the following equation:

$$|\psi_d^{+}\rangle = \frac{1}{\sqrt{d}} \sum_{i=0}^{d-1} |i\rangle|i\rangle . \tag{9}$$

It should be added that if quantum states of two or more qubits can be directly obtained as a tensor product, e.g. two-qubit state $|00\rangle = |0\rangle \otimes |0\rangle$, are called separable states.

We have just presented very short resume of elementary information about quantum states and entanglement property. Further details can be found in numerous textbooks about Quantum Information Science e.g. [11].

# 3    The Entanglement Detection

There are many methods [3] to check, if the quantum state is entangled or separable. In this section we present some selected criteria – our choice is driven by the ease of implementation for these methods. Additionally, because of non-polynomial complexity of mathematical model of quantum systems, some experiments presented in the next section are especially constructed for bipartite systems.

In the case of a vector states, which are discussed in this work, detection of the presence of entanglement in the quantum register can be realized by the Schmidt decomposition.

Let $\mathcal{H}_A$ and $\mathcal{H}_B$ be Hilbert spaces with dimensions $N$ and $M$, respectively. Any pure quantum state associated with spaces $\mathcal{H}_A$ and $\mathcal{H}_B$, which is made by the tensor product, can be written as

$$|\psi\rangle = \sum_{ab} C_{ab}|a\rangle|b\rangle, \tag{10}$$

where vectors $|a\rangle$, $|b\rangle$ belong to the orthonormal bases of spaces $\mathcal{H}_A$ and $\mathcal{H}_B$.

The Schmidt decomposition gives complete information whether a given state $|\psi\rangle$ is entangled or not. Three cases in this context have to be distinguished:

(i) the state $|\psi\rangle$ is a separable state if and only if only one non-zero Schmidt coefficient exists and is equal to one, the remaining values of $C_{ab}$ are equal to zero,

(ii) if there are more than one non-zero values in the set of $C_{ab}$ values, then the state $|\psi\rangle$ is an entangled state,

(iii) if all values from the set $C_{ab}$ are non-zero and equal to one another, then the state $|\psi\rangle$ is a maximally entangled state.

The use of Schmidt decomposition is necessary to construct the learning set, for neural network training, containing examples of entangled and separable states.

In many cases, especially for computations in open quantum systems, the state of quantum system have to be described as a density matrix. We have tested an example of state called Horodecki's quantum state [7] described with a $2 \times 2$ density matrix:

$$\rho = \begin{pmatrix} pa^2 & 0 & 0 & pab \\ 0 & (1-p)a^2 & (1-p)ab & 0 \\ 0 & (1-p)ab & (1-p)b^2 & 0 \\ pab & 0 & 0 & pb^2 \end{pmatrix}. \tag{11}$$

It is assumed that: $a, b > 0$ and $|a|^2 + |b|^2 = 1$. It is known that this state is separable when $p = 1/2$ or $a = 0$, either $b = 0$ in other cases this state is entangled.

To verify if matrix $\rho$ represents an entangled state, the PPT criterion can be used. The PPT criterion [12] is also called a Peres and Horodecki criterion or a positive partial transposition criterion. The PPT criterion gives the final answer whether the entanglement is present or not, however only for states with lower dimensions: $2 \times 2$ and $2 \times 3$. Therefore, it is known that a state $\rho$ belonging to the space $\mathcal{H}_2 \otimes \mathcal{H}_2$ or to the space $\mathcal{H}_2 \otimes \mathcal{H}_3$ is a separable state if and only if the state $(\rho)^{T_1}$ is positive $(\rho)^{T_1} \geq 0$, where $T_1$ denotes so-called partial transposition performed only on the first subspace. From computational point of view checking whether eigenvalues of $(\rho)^{T_1}$ are negative needs finding one negative eigenvalue – it is sufficient to determine the state $\rho$ as the entangled state.

Another example of ANN usefulness to the entanglement detection refers to the Ha family states on the space $\mathcal{H}_3 \otimes \mathcal{H}_3$ described in [4]. These states are described with the following equation:

$$
\rho_\gamma = \frac{1}{N_\gamma}
\begin{pmatrix}
1 & . & . & . & 1 & . & . & . & 1 \\
. & a_\gamma & . & . & . & . & . & . & . \\
. & . & b_\gamma & . & . & . & . & . & . \\
. & . & . & b_\gamma & . & . & . & . & . \\
1 & . & . & . & 1 & . & . & . & 1 \\
. & . & . & . & . & a_\gamma & . & . & . \\
. & . & . & . & . & . & a_\gamma & . & . \\
. & . & . & . & . & . & . & b_\gamma & . \\
1 & . & . & . & 1 & . & . & . & 1
\end{pmatrix}
\tag{12}
$$

for $\gamma \in (0,1)$, and where

$$
a_\gamma = \frac{1}{3}(\gamma^2 + 2)\, b_\gamma = \frac{1}{3}(\gamma^{-2} + 2),
\tag{13}
$$

additionally, the normalization factor $N_\gamma$ is defined as

$$
N_\gamma = 7 + \gamma^2 + \gamma^{-2}.
\tag{14}
$$

It must be stressed that for parameters $\gamma_1, \gamma_2, \ldots, \gamma_i \in (0,1)$, any convex combination of states $\rho_{\gamma_k}$

$$
p_1 \rho_{\gamma_1} + p_2 \rho_{\gamma_2} + \ldots + p_i \rho_{\gamma_i}
\tag{15}
$$

also describes an entangled state which is detected by the entanglement witnesses $W_0$.

The witness operator $W_0$ also belongs to the Hilbert space $\mathcal{H}_3 \otimes \mathcal{H}_3$ and has the following matrix representation

$$
W_0 = \frac{1}{N_\gamma}
\begin{pmatrix}
1 & . & . & . & -1 & . & . & . & -1 \\
. & 1 & . & . & . & . & . & . & . \\
. & . & . & . & . & . & . & . & . \\
. & . & . & . & . & . & . & . & . \\
-1 & . & . & . & 1 & . & . & . & -1 \\
. & . & . & . & . & 1 & . & . & . \\
. & . & . & . & . & . & 1 & . & . \\
. & . & . & . & . & . & . & . & . \\
-1 & . & . & . & -1 & . & . & . & 1
\end{pmatrix}.
\tag{16}
$$

This operator will be called a witnesses of entanglement for states of the Ha family. The entanglement detection can be realised with the following function:

$$
f(\rho) = \mathrm{Tr}\,(W_0 \rho)
\tag{17}
$$

where $\mathrm{Tr}\,(\cdot)$ denotes a trace of a given matrix, and negative value of $f(\rho)$ signals the presence of entanglement in $\rho$.

The works [7], [14] show also a different criterion for detection of entanglement called entanglement witnesses. Currently, it is the most important criterion and it plays a fundamental role in the theory of quantum entanglement. The Hermitian operator $W$ is called a witnesses of entanglement for a given entangled state if $\mathrm{Tr}\,(W\rho) < 0$, and for separable states $\mathrm{Tr}\,(W\rho_{sep}) \geq 0$.

It should be noted that the source of witnesses of entanglement lays in geometry, because a convex set can be described by hyperplanes what allows to divide the learning set into two main groups of separable and entangled states. This fact allows to expect that success rate of entanglement detection should be high, what is confirmed in our numerical test described in Sec. 5.

## 4    The Type of the Neural Network Used in This Work

We have chosen the feed-forward neural network for detecting the entanglement in quantum systems. This kind of neural network is very popular because they can be used for any kind of input-to-output mapping. Besides, feed-forward neural networks are relatively simple: the signal in the network is propagated in only one direction – from input to output; the mathematical model of this neural network is relatively clear; the learning methods are easy to implement.

The feed-forward neuron networks are used for solving problems like [10]:

- function approximation,
- pattern classification,
- object recognition,
- data compression.

Detecting the entanglement in quantum systems may be considered as the two-valued classification problem (the quantum state can be recognized as entangled or separable). The mentioned features of feed-forward neuron networks and also the accessibility of ready solution (Matlab [8] function *feedforwardnet*) convinced us to use this kind of artificial neuron networks for solving the problem formulated in this chapter.

The environment of Matlab gives the possibility of using Levenberg-Marquardt method or Scaled Conjugate Gradient in the process of ANN learning. The first mentioned method is known as one of the most effective learning algorithms for feed-forward neural networks, which combines Gauss-Newton algorithm and the method of steepest descent to minimize the error in the learning process. The Scaled Conjugate Gradient is a supervised learning algorithm and is a member of the class of conjugate gradient methods – it has a worse convergence than Levenberg-Marquardt method but its implementation requires much less computationalresources.

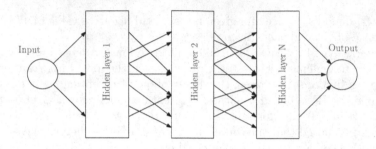

**Fig. 1.** The scheme of feed-forward ANN used to detect the presence of entanglement in quantum states. The input vectors are vector forms of quantum states or density matrices (represented as a vector containing successive matrix columns). The output of ANN is a boolean value.

## 5    Results of the Experiment

The ANN for the problem of entangled states recognition is a two-valued classifier $f$. It means that for the quantum state $q$, expressed in a form of state vector or density matrix $\rho$, we expect:

$$f(q) = \begin{cases} 1, \text{ if } q \text{ is entangled} \\ 0, \text{ if } q \text{ is separable} \end{cases} \quad \text{and} \quad f(\rho) = \begin{cases} 1, \text{ if } \rho \text{ is entangled} \\ 0, \text{ if } \rho \text{ is separable} \end{cases}. \quad (18)$$

The learning set, used in this work, contains the examples of entangled and separable quantum states. For instance the analyzed states based on qutrits are:

$$|\phi\rangle = \frac{1}{\sqrt{2}}(|00\rangle + |11\rangle + |22\rangle), \quad |\phi\rangle = \frac{1}{\sqrt{2}}(|00\rangle + |22\rangle) \quad (19)$$

The above states are maximally entangled in the context of Schmidt decomposition, but to train the ANN we have used also other randomly selected states with the lower level of entanglement. Of course the training set contains also the separable states which were generated with a following quasi-Matlab language expressions:

```
q1 = rand(3,1);
q2 = [1 ; 0 ; 0];
q = kron(q1,q2);
q = q/norm(q);
```

The first variable q1 represents a vector of random values. The second variable q2 holds the basis vector. The next two steps are to compute the tensor product and normalize it. To simplify the model, only real numbers were used as values of quantum states amplitudes.

The following code generates the set of entangled states, belonging to the Horodecki family, and creates the training set which is represented by the objects train_data and target.

```
for p=0:0.05:0.45,
   v=rand(1,2);
v=v/norm(v);
a=v(1);
b=v(2);
rho =[ p*a*a 0           0 p*a*b ; ...
0     (1-p)*a*a (1-p)*a*b 0 ; ...
0     (1-p)*a*b (1-p)*b*b 0 ; ...
p*a*b0         0         p*b*b ; ...
];
w = isHorodeckiEntangled(rho);
train_data=[train_data, reshape(rho, 4*4, 1)];
target=[target, w]; end
```

Table 1 shows the results of experiments during which feed-forward neural networks were trained to recognize the entangled quantum states. The calculations were run for networks with different number of hidden layers and neurons. According to the theorem concerning the approximation capability of multilayer feed-forward ANN [2], [6], in most cases a few-layer perceptron is able to distinguish between entangled and separable state.

It is possible to reach the efficiency of pattern recognition exceeding 90 %. Almost 100 % of efficiency was achieved for the states from Horodecki and Ha families described by density matrices. In all the experiments the Levenberg-Marquardt Method [5] was used for the process of networks learning.

It should be highlighted that the time elapsed for the process of learning with use of Levenberg-Marquardt method is relatively short – about 7 to 10 seconds for the states of Horodecki and Ha with the structure of ANN: $[20, 20, 4]$. The experiments were run on a workstation equipped with Intel i7 4790K (four computing cores and eight threads). Similar results were obtained for the network with structure $[10, 10, 10, 6, 10, 6, 10]$ which was learned to detect the entanglement for qutrit states. Naturally, the time of learning is longer for networks with more layers and more neurons, e.g. for network $[20,20,20,12,20,12,20]$ and qutrit states the time of learning process takes slightly more than 60 seconds.

We have also tried to use Graphics Processing Units (GPUs) to run the process of ANN's learning for the problem of entanglement recognition. The environment of Matlab offers scaled conjugate gradient method [9] for GPUs from Neural Network Toolbox. However, scaled conjugate gradient needs more iterations than Levenberg-Marquardt method so the time of computations with use of GPUs proved to be longer that running the process with Central Processing Unit (CPU).

In the presented experiments different test sets were created for each case of entanglement, due to the specific form (vector or matrix) of the analyzed quantum states. It should be pointed out that these sets cannot be merged, due to the fact that state vectors and density matrices are different ways of describing a quantum state. In addition, for vector states and the states represented by density matrices, separate ANNs, learning and validation sets should be prepared.

**Table 1.** Exemplary detection rates for ANNs with different structure of layers. $N_{LS}$ denotes the size of learning set – in bracket we put the number of entangled states (subscript e) and the number of separable states (subscript s). The size of testing set is denoted by $T_{LS}$. $D_E$ denotes the number of properly detected entangled states, $D_S$ is the number of properly detected separated states. $E$ stands for the final efficiency of quantum state detection (for both entangled and separable cases)

| Amount of neurons in hidden NN layers | $N_{LS}$ | $T_{LS}$ | $D_E$ | $D_S$ | E |
|---|---|---|---|---|---|
| Pure two qubit state $Q_2$ | | | | | |
| $[5, 5, 5, 3, 5, 3, 5]$ | 126 ($75_e$,$51_s$) | 2000 | 742/1000 | 1000/1000 | 95 % |
| $[7, 9, 9, 4, 5, 7]$ | ditto | ditto | 793/1000 | 1000/1000 | 90 % |
| $[8, 12, 12, 5, 6, 8]$ | ditto | ditto | 842/1000 | ditto | 92 % |
| Pure two qudit state $Q_3$ | | | | | |
| $[10, 10, 10, 6, 10, 6, 10]$ | 278 ($127_e$, $151_s$) | 4000 | 2351/2500 | 1481/1500 | 96 % |
| $[17, 19, 19, 10]$ | ditto | ditto | 2424/2500 | 1451/1500 | 96 % |
| $[20, 20, 20, 12, 20, 12, 20]$ | ditto | ditto | 2345/2500 | 1477/1500 | 95 % |
| $[5, 5, 5, 3, 5, 3, 5]$ | ditto | ditto | 1536/2500 | 902/1500 | 60 % |
| Horodecki's state | | | | | |
| $[10, 10, 5, 1]$ | 70 ($19_e$, $51_s$) | 300 | 180/200 | 100/100 | 93 % |
| $[9, 9, 9, 5, 9, 5, 9]$ | ditto | ditto | 142/200 | ditto | 80 % |
| $[20, 20, 4]$ | ditto | ditto | 185/200 | ditto | 95 % |
| Ha's state | | | | | |
| $[18, 18, 9]$ | 21 ($2_e$, $19_s$) | 100 | 9/9 | 90/91 | 99 % |
| $[9, 9, 9, 5, 9, 5, 9]$ | ditto | ditto | ditto | ditto | 99 % |
| $[5, 3, 5, 3, 5]$ | ditto | ditto | 8/9 | ditto | 98 % |

Naturally, the order of elements in the training and testing sets naturally does not matter. The training set should contain the representations of states of each subtype that occurs in the family, e.g. maximally entangled states and other entangled quantum states. The entanglement in the family of Horodecki states, as well as in the Ha family, is described for both qubit and qutrit states. This fact implies that we cannot directly compare these quantum states and also separate ANNs should be prepared for the entanglement recognition in qubit and qutrit states.

It should be emphasized that after the process of learning, ANN is able to classify the quantum state very quickly in comparison to the time needed to complete Schmidt decomposition (it is caused by the necessity of singular value decomposition) or to compute spectral distribution of density matrix when the analyzed state belongs to the Horodecki family.

# 6    Summary

The feed-forward neural networks can be used not only for detecting the entanglement but also to recognize some particular state. After performing some operations in a quantum circuit (often representing a quantum algorithm) it is needed to read out the final quantum state and to interpret the results of quantum computations – in this case using ANN could be very helpful. Moreover, when only some entries of the final state vector may be needed to read-out, using the ANN can shorten the time of operation.

Encouraged with the results of the experiment we would like to extend the Quantum Computations Simulator (QCS) [13] with additional functionalities which will allow to use ANN in some steps of quantum effects' simulation. The QCS package contains the set of methods for quantum states parallel processing with GPUs. Enriching the QCS package with a new tool for detecting the entanglement would be a valuable element in a simulator's development.

Since using feed-forward ANN results with successful process of pattern classification, we plan to check the capabilities of other types of ANNs (Hopfield Network, Radial Basis Function Network, Recurrent Neural Network, Feedback Neural Network) for entanglement's detection and particular state recognition.

# References

1. Czajkowski, A., Patan, K., Szymański, M.: Application of the state space neural network to the fault tolerant control system of the PLC-controlled laboratory stand. Engineering Applications of Artificial Intelligence **30**, 168–178 (2014)
2. Cybenko, G.: Approximations by superpositions of sigmoidal functions. Mathematics of Control, Signals, and Systems **2**(4), 303–314 (1989)
3. Gühne, O., Tóth, G.: Entanglement detection. Physics Reports **474**, 1–75 (2009)
4. Ha, K.C.: Atomic Positive Linear Maps in Matrix Algebras. Publ. RIMS, Kyoto Univ. 34, p. 591 (1998)
5. Hagan, M.T., Menhaj, M.: Training feed-forward networks with the Marquardt algorithm. IEEE Transactions on Neural Networks **5**(6), 989–993 (1994)
6. Hornik, K.: Approximation Capabilities of Multilayer Feedforward Networks. Neural Networks **4**(2), 251–257 (1991)
7. Horodecki, M., Horodecki, P., Horodecki, R.: Separability of mixed states: necessary and sufficient conditions. Phys. Lett. A **223**, 1–8 (1996)
8. Matlab by Mathworks. http://www.mathworks.com/
9. Moller, M.F.: A scaled conjugate gradient algorithm for fast supervised learning. Neural Networks **6**(4), 525–533 (1993)
10. Masters, T.: Practical Neural Network Recipies in C++. Morgan Kaufmann, San Francisco (1993)
11. Nielsen, M.A., Chuang, I.L.: Quantum computation and quantum information. Cambridge University Press, New York (2000)
12. Peres, A.: Separability Criterion for Density Matrices. Phys. Rev. Lett. **77**, 1413–1415 (1996)
13. Sawerwain, M.: GPU-Based Parallel Algorithms for Transformations of Quantum States Expressed as Vectors and Density Matrices. In: Wyrzykowski, R., Dongarra, J., Karczewski, K., Waśniewski, J. (eds.) PPAM 2011, Part I. LNCS, vol. 7203, pp. 215–224. Springer, Heidelberg (2012)
14. Terhal, B.M.: A Family of Indecomposable Positive Linear Maps based on Entangled Quantum States. Linear Algebra Appl. **326**, 61–73 (2000)

# Discovering Erasable Closed Patterns

Giang Nguyen[1], Tuong Le[2(✉)], Bay Vo[1], and Bac Le[3]

[1] Faculty of Information Technology, Ho Chi Minh City University of Technology,
Ho Chi Minh City, Vietnam
{nh.giang,vd.bay}@hutech.edu.vn
[2] Division of Data Science and Faculty of Information Technology,
Ton Duc Thang University, Ho Chi Minh City, Vietnam
lecungtuong@tdt.edu.vn
[3] Faculty of Information Technology, University of Science, VNU Ho Chi Minh City, Vietnam
lhbac@fit.hcmus.edu.vn

**Abstract.** Data mining that discovers knowledge from large datasets is more
and more popular in artificial intelligence. In recent years, the problem of mining erasable patterns (EPs) has been proposed as an interesting variant of
frequent pattern mining. There are many algorithms for solving effectively the
problem of mining EPs. However, for very big datasets, the large number of
EPs takes the large memory usage of the system, and then obstructs users' using
the system. Therefore, it is necessary to mine a condensed representation of
EPs. In this paper, we present the erasable closed patterns (ECPs) concept and
an effective algorithm for mining ECPs (MECP algorithm). The experimental
results show that the number of ECPs is much less than that of EPs. Besides,
the runtime of MECP is better than the naïve approach for mining ECPs.

**Keywords:** Data mining · Erasable closed patterns · Erasable patterns

## 1 Introduction

Data mining is the process of discovering interesting patterns in large dataset. These
patterns will be used as the knowledge in the some intelligent systems such as expert
systems, recommendation systems etc. Many problems in data mining have attracted
research attention such as association rule mining [1-2, 18-19] and classification
[6, 13-14]. Pattern mining including frequent pattern mining [7, 11, 16-17, 20], frequent closed pattern mining [15] etc. is an essential task in association rule mining.
Recently, the problem of mining erasable patterns (EPs) [3-4, 5, 8-10, 12] proposed by
Deng et al. [4] is an interesting variation of pattern mining. For details, a factory produces many products created from a number of items. Each product brings an income
to the factory. A financial resource is required to buy and store all items. However, in a
financial crisis situation, this factory has not enough money to purchase all necessary
items as usual. This problem of mining erasable patterns is defined as the following:
find the patterns which can best be erased so as to minimize the loss to factory's gain.
The managers can then utilize the knowledge of these erasable patterns to make a new
production plan. Currently, there are many algorithms for solving this problem such as

© Springer International Publishing Switzerland 2015
N.T. Nguyen et al. (Eds.): ACIIDS 2015, Part I, LNAI 9011, pp. 368–376, 2015.
DOI: 10.1007/978-3-319-15702-3_36

META [4], MERIT [3] and MEI [8] algorithms. Deng et al. (2009) proposed META, an Apriori-based algorithm, to solve the problem of mining erasable patterns. However, the execution time of META is slow because it uses a generate candidate approach which is a naïve strategy. MERIT [3] uses the concept of NC_Sets to reduce memory usage. Although the use of NC_Sets gives MERIT some advantages over META, there are still some disadvantages. First, the weight value of each node code (NC) is stored individually even though it can appear in many erasable patterns' NC_Sets, leading to a lot of duplication. Second, it uses a strategy whereby pattern s$X$'s NC_Set is assumed to be a subset of pattern $Y$'s NC_Set if $X \subset Y$. This leads to high memory consumption and high run time when patterns are combined to create new nodes. MEI [8] uses the dPidset structure to quickly determine the information of erasable patterns. Although mining time and memory usage are better than those of the above algorithms, MEI's performance for mining erasable patterns can be improved.

However, in the problem of mining EPs, the number of obtained EPs from these algorithms is so large. Therefore, intelligent systems using EPs as knowledge will be difficult situation with the large number of EPs. In reality, the set of erasable closed patterns (ECPs) can represent the set of EPs without losing information. This set does not have two or more EPs which have the same gain. They can be used to mine the non-redundant rules without pruning the redundant rules. Therefore, this paper presents ECPs concept and proposes an effective algorithm (MECP algorithm) for mining ECPs.

The rest of the paper is organized as follows. Section 2 presents basic concepts. The definition of ECPs and some theorems for fast mining ECPs are presented in Section 3. Section 4 introduces MECP algorithm for mining ECPs and an illustrated example of the MECP algorithm's process. Experimental result is presented in section 4 to show the effectiveness of MECP algorithm. The paper concluded in section 5.

## 2    Basic Concepts

### 2.1    Erasable Patterns

Deng et al. [4] proposed the problem of erasable pattern mining as follows. Let $I = \{i_1, i_2,..., i_m\}$ be a set of all items and $DB$ is a product dataset. Each product presented in the form of $\langle Items, Val \rangle$, where $Items$ are the items to conduct this product and $Val$ is the profit that the factory obtains by selling this product. Table 1 presents an example dataset ($DB_E$) which will be used throughout this article.

**Table 1.** An example dataset ($DB_E$)

| Product | Items | Val ($) |
|---------|-------|---------|
| $P_1$ | $a, b$ | 1,000 |
| $P_2$ | $a, b, c$ | 200 |
| $P_3$ | $c, e$ | 150 |
| $P_4$ | $b, d, e, f$ | 50 |
| $P_5$ | $d, e$ | 100 |
| $P_6$ | $d, e, f, h$ | 200 |

**Definition 1.** Given a threshold $\xi$ and a product dataset $DB$. A pattern $X$ is erasable if:

$$g(X) \leq T \times \xi \tag{1}$$

where:
- $g(X) = \sum_{\{P_k | X \cap P_k.Items \neq \varnothing\}} P_k.Val$ is gain of patterns $X$;
- $T = \sum_{P_k \in DB} P_k.Val$ is total gain of the factory.

Based on Definition 1, the problem of mining EPs is to find all EPs which have gain $g(X)$ less than $T \times \xi$ in dataset.

*Example 1.* We have $g(e) = P_3.Val + P_4.Val + P_5.Val + P_6.Val = 150 + 50 + 100 + 200 = 500$ dollars and $T = 1,700$ dollars. With $\xi = 30\%$, $e$ is a EPs because $g(e) = 500 \leq T \times \xi = 510$ dollars.

## 2.2    dPidset Structure

Le and Vo [8] proposed the dPidset structure for effectively mining erasable pattern as follows.

**Definition 2 (pidset).** The pidset of a pattern $X$ is denoted as follows:

$$p(X) = \bigcup_{A \in X} p(A) \tag{2}$$

where $A$ is an item in pattern $X$ and $p(A)$ is the set of product identifiers which includes $A$.

**Definition 3 (dPidset).** The dPidset of pidsets $p(XA)$ and $p(XB)$, denoted as $dP(XAB)$, is defined as follows:

$$dP(XAB) = p(XB) \setminus p(XA) \tag{3}$$

According to Definition 3, the dPidset of $p(XA)$ and $p(XB)$ is the product identifiers which only exist on $p(XB)$.

**Theorem 1.** Let $XA$ and $XB$ be two patterns and $dP(XA)$ and $dP(XB)$ be the dPidsets of $XA$ and $XB$, respectively. The dPidset of $XAB$ is computed as follows:

$$dP(XAB) = dP(XB) \setminus dP(XA) \tag{4}$$

**Theorem 2.** The gain of $XAB$ is determined based on that of $XA$ as follows:

$$g(XAB) = g(XA) + \sum_{P_k \in dP(XAB)} P_k.Val \tag{5}$$

where $g(XA)$ is the gain of $XA$ and $P_k.Val$ is the gain of the product $P_k$.

*Example 2.* We have $p(e) = \{3, 4, 5, 6\}$ and $p(c) = \{2, 3\}$, so $g(e) = 500$ and $g(c)$ = 350. We have $dP(ec) = \{2\}$ so $g(ec) = g(e) + \sum_{P_k \in dP(ec)} P_k . Val = 500 + 200 = 700$ dollars.

# 3     Erasable Closed Pattern Mining

## 3.1     Erasable Closed Patterns

Similar to the definition of frequent closed patterns [15], an erasable pattern is called an erasable closed pattern if none of its supersets has the same gain. For example, consider $DB_E$ with $\xi = 30\%$, $e$ and $edfh$ are two erasable patterns because $g(e) = g(edfh) = 500 \leq 1700 \times 30\% = 510$ dollars. However, $e$ is not an erasable closed pattern because $edfh$, one of its supersets, has the same gain as $e$.

When combining two elements $X$ and $Y$ in the same equivalence class, the algorithm will check their *dPidsets*. There are four cases as follows. If $dP(XA) = dP(XB)$ then remove $XB$ and replace $XA$ by $XA \cup XB$. If $dP(XA) \subset dP(YB)$, the algorithm will replace $XB$ by $XA \cup YB$. Conversely, no element $XA$ or $XB$ can be removed.

## 3.2     MECP Algorithm

Using *dPidset* concept and its theorems for ECPs, we propose the MECP (Mining Erasable Closed Patterns) algorithm for mining ECPs in Fig. 1.

---

**Algorithm 1.** MECP algorithm

---
**Input**: product database $DB$ and threshold $\xi$
**Output**: $E_{result}$, the set of all ECPs
 1 Scan $DB$ to determine the total profit of $DB$ ($T$), the index of gain ($G$), and erasable 1-patterns with their pidsets ($E_1$)
 2 Sort $E_1$ by the length of their pidsets in decreasing order
 3 If $E_1$ has more than one element, the algorithm will call **Expand_E**($E_1$)

 1 **Procedure Expand_E**($E_v$)
 2 **For** i ← 0 **to** $|E_v|$ **do**
 3 **Begin for**
 4     $E_{next}$ ← ∅
 5     **For** j ← i+1 **to** $|E_v|$ **do**
 6         dP(ECP) = dP($E_v$[j]) \ dP($E_v$[i])
 7         **If** ECP.val <= $\xi \times T$ **then**
 8             **If** dP($E_v$[i]) = dP($E_v$[j]) **then**
 9                 $E_v$[i] = $E_v$[i] ∪ $E_v$[j]
 10                Update $E_{next}$

---

```
11              Remove E_v[j]
12              j--
13          Else if dP(E_v[i]) ⊂ dP(E_v[j]) then
14              E_v[i] = E_v[i] ∪ E_v[j]
15              Update E_next
16          Else
17              ECP = E_v[i] ∪ E_v[j]
18              E_next ← ECP
19      End for
20      If Check_Closed_Property(E_v[i]) == true then
21          E_result ← E_v[i]
22          Add E_v[i] to Hashtable with E_v[i].val as a key
23      If |E_next| > 1 then
24          Sort E_next by the length of their dPidsets in de-
                creasing order
25          Expand_E(E_next)
26  End for

1  Function Check_Closed_Property(EI)
2  Let ECPs ← Hashtable[EI.val]
3  If ECPs is not null then
4      For each ECP in ECPs do
5          If EI ⊂ ECP then
6              Return false
7  Return true
```

**Fig. 1.** MECP algorithm

## 4    Experimental Results

All experiments presented in this section were performed on a laptop with an Intel Core i3-3110M 2.4-GHz CPU and 4 GB of RAM. MEI and MECP algorithms were coded in C# and .Net Framework Version 4.5.50709.

The experiments are conducted on Chess, Mushroom and Connect datasets[1]. To make these datasets look like product datasets, a column was added to store the profit of products. To generate values for this column, a function denoted by $N(100, 50)$, for which the mean value is 100 and the variance is 50, was created. The features of these datasets are shown in Table 2.

---

[1] Downloaded from http://fimi.cs.helsinki.fi/data/

**Table 2.** Features of datasets used in experiments

| Dataset[2] | # of Products | # of Items |
|---|---|---|
| Chess | 3,196 | 76 |
| Mushroom | 8,124 | 120 |
| Connect | 67,557 | 130 |

## 4.1    The Number of ECPs and EPs

Table 3 shows the number of ECPs and EPs on Chess, Mushroom and Connect datasets. The number of ECPs is clearly smaller than that of EPs. Therefore, the required resource in the intelligent systems is also reduced which make these systems better.

**Table 3.** The number of ECPs and EPs on experimental datasets

| Dataset | Threshold $\xi(\%)$ | Number of EPs | Number of ECPs |
|---|---|---|---|
| Chess | 10 | 665 | 523 |
| | 15 | 3,083 | 2,082 |
| | 20 | 10,913 | 6,260 |
| | 25 | 30,815 | 15,637 |
| Mushroom | 0.75 | 1,830 | 209 |
| | 1.0 | 8,368 | 549 |
| | 1.25 | 24,537 | 1,262 |
| | 1.5 | 63,033 | 2,716 |
| Connect | 0.75 | 1,677 | 1,644 |
| | 1.0 | 5,185 | 4,892 |
| | 1.25 | 13,625 | 12,220 |
| | 1.5 | 30,540 | 25,817 |

## 4.2    The Runtime

Currently, the problem of mining ECPs is unsolved. To evaluate MECP algorithm, we conduct a naïve approach which mines all EPs and then finds ECPs from the obtained EPs. This section reports the mining time of MECP algorithm and the naïve approach which show in Figs. 2-4. The experimental results show that the mining time of MECP algorithm outperforms the naïve approach.

---

[2] These datasets are available at http://sdrv.ms/14eshVm

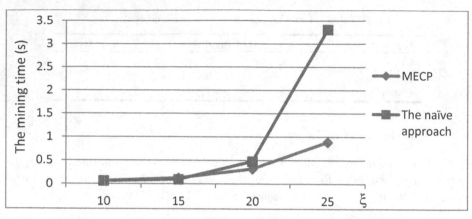

**Fig. 2.** The mining time of MECP and the naïve approach on Chess dataset

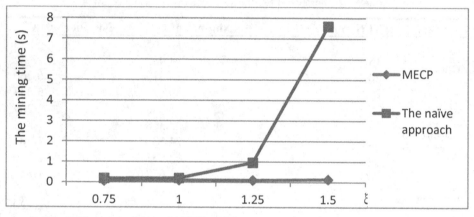

**Fig. 3.** The mining time of MECP and the naïve approach on Mushroom dataset

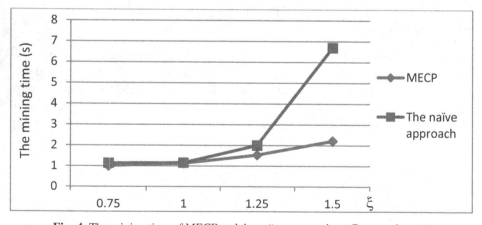

**Fig. 4.** The mining time of MECP and the naïve approach on Connect dataset

# 5    Conclusion and Future Work

Erasable pattern mining is an interesting problem is introduced in 2009. Up to now, there are many algorithms for solving effectively this problem such as META, MERIT, dMERIT+ and MEI. However, a small number of EPs is used in intelligent systems. Therefore, there is necessary to remove redundant EPs. In this paper, we present the erasable closed patterns (ECPs) concept and MECP for mining ECPs. The experiment was conducted to compare the numbers of patterns and the mining time. The results show that the number of ECPs is much smaller than the number of EPs and the mining time of ECPs is better than the naïve approach.

In future work, some issues related to EPs will be studied, such as mining EPs from huge datasets, mining top-rank-$k$ EPs, mining maximal EPs, and mining EPs from incremental datasets.

**Acknowledgments.** This research was funded by Vietnam National Foundation for Science and Technology Development (NAFOSTED) under grant number 102.01-2012.17.

# References

1. Agrawal, R., Srikant, R.: Fast algorithms for mining association rules. In: VLDB 1994, pp. 487–499 (1994)
2. Agrawal, R., Imielinski, T., Swami, A.: Mining association rules between set of items in large databases. In: SIGMOD 1993, pp. 207–216 (1993)
3. Deng, Z.H., Xu, X.R.: Fast mining erasable itemsets using NC_sets. Expert Systems with Applications **39**(4), 4453–4463 (2012)
4. Deng, Z.H., Fang, G., Wang, Z., Xu, X.: Mining erasable itemsets. In: ICMLC 2009, pp. 67–73 (2009)
5. Deng, Z., Xu, X.: An efficient algorithm for mining erasable itemsets. In: Cao, L., Feng, Y., Zhong, J. (eds.) ADMA 2010, Part I. LNCS, vol. 6440, pp. 214–225. Springer, Heidelberg (2010)
6. Do, T.N., Lenca, P., Lallich, S.: Classifying many-class high-dimensional fingerprint datasets using random forest of oblique decision trees. Vietnam Journal of Computer Science, DOI:10.1007/s40595-014-0024-7 (in press)
7. Han, J., Pei, J., Yin, Y.: Mining frequent patterns without candidate generation. In: SIGMOD 2000, pp. 1–12 (2000)
8. Le, T., Vo, B.: MEI: an efficient algorithm for mining erasable itemsets. Engineering Applications of Artificial Intelligence **27**, 155–166 (2014)
9. Le, T., Vo, B., Nguyen, G.: A survey of erasable itemset mining algorithms. WIREs Data Mining Knowl. Discov. **4**, 356–379 (2014)
10. Lee, G., Yun, U., Ryang, H.: Mining weighted erasable patterns by using underestimated constraint-based pruning technique. Journal of Intelligent and Fuzzy Systems (2014, in press)
11. Huynh, T.L.Q., Vo, B., Le, B.: An efficient and effective algorithm for mining top-rank-k frequent patterns. Expert Syst. Appl. **42**(1), 156–164 (2015)

12. Nguyen, G., Le, T., Vo, B., Le, B.: A New Approach for Mining Top-Rank-$k$ Erasable Itemsets. In: Nguyen, N.T., Attachoo, B., Trawiński, B., Somboonviwat, K. (eds.) ACIIDS 2014, Part I. LNCS, vol. 8397, pp. 73–82. Springer, Heidelberg (2014)
13. Nguyen, D., Vo, B., Le, B.: Efficient strategies for parallel mining class association rules. Expert Syst. Appl. **41**(10), 4716–4729 (2014)
14. Nguyen, L.T.T.: Mining class association rules with the difference of obidsets. In: Nguyen, N.T., Attachoo, B., Trawiński, B., Somboonviwat, K. (eds.) ACIIDS 2014, Part II. LNCS, vol. 8398, pp. 72–81. Springer, Heidelberg (2014)
15. Pasquier, N., Bastide, Y., Taouil, R., Lakhal, L.: Discovering frequent closed itemsets for association rules. In: Beeri, C., Bruneman, P. (eds.) ICDT 1999. LNCS, vol. 1540, pp. 398–416. Springer, Heidelberg (1998)
16. Song, W., Yang, B., Xu, Z.: Index-BitTableFI: An improved algorithm for mining frequent itemsets. Knowledge-Based Systems **21**, 507–513 (2008)
17. Vo, B., Coenen, F., Le, T., Hong, T.-P.: Mining frequent itemsets using the N-list and subsume concepts. International Journal of Machine Learning and Cybernetics DOI:10.1007/s13042-014-0252-2 (in press)
18. Vo, B., Hong, T.-P., Le, B.: A lattice-based approach for mining most generalization association rules. Knowledge-Based Systems **45**, 20–30 (2013)
19. Zaki, M.J.: Scalable algorithms for association mining. IEEE Transactions on Knowledge and Data Engineering **12**(3), 372–390 (2000)
20. Zaki, M.J., Gouda, K.: Fast vertical mining using diffsets. In: SIGKDD 2003, pp. 326–335 (2003)

# RBM-SMOTE: Restricted Boltzmann Machines for Synthetic Minority Oversampling Technique

Maciej Zięba$^{(\boxtimes)}$, Jakub M. Tomczak, and Adam Gonczarek

Faculty of Computer Science and Management, Wroclaw University of Technology,
Wybrzeże Wyspiańskiego 27, 50-370 Wroclaw, Poland
{maciej.zieba,adam.gonczarek}@pwr.edu.pl, jakub.tomczak@pwr.edu.pl

**Abstract.** The problem of imbalanced data, i.e., when the class labels are unequally distributed, is encountered in many real-life application, e.g., credit scoring, medical diagnostics. Various approaches aimed at dealing with the imbalanced data have been proposed. One of the most well known data pre-processing method is the Synthetic Minority Oversampling Technique (SMOTE). However, SMOTE may generate examples which are artificial in the sense that they are impossible to be drawn from the true distribution. Therefore, in this paper, we propose to apply Restricted Boltzmann Machine to learn an intermediate representation which transform the SMOTE examples to the ones approximately drawn from the true distribution. At the end of the paper we perform an experiment using credit scoring dataset.

**Keywords:** Imbalanced data · Oversampling · SMOTE · RBM

## 1 Introduction

The problem of imbalanced data became one of the key issues in the process of training a classification model [7]. A dataset is considered to be imbalanced if the class labels are strongly unequally distributed. Therefore, learning from the imbalanced data may have a negative impact on training the model which is biased toward the majority class. Recently, numerous approaches were proposed to deal with that issue. In general, they can be roughly divided into two groups: *external* and *internal* methods.

External approaches aim at sampling examples in order to balance the training set. There are several oversampling methods such as Synthetic Minority Oversampling Technique (SMOTE) [2] or its extensions, e.g., *Borderline SMOTE* [11], *LN-SMOTE* [13], *SMOTE-RSB* [15], *Safe-Level-SMOTE* [1], or recently introduced *SMOTE-IPF* [16]. Among undersampling techniques we can distinguish methods that make use of *K-NN* classifier to identify relevant instances in majority class [14], use evolutionary algorithms to balance the data [6] or make use mutual neighborhood relation called *Tomek* link [21].

In internal approaches the balancing techniques are incorporated in the training process of a classifier. Typically, ensemble classifiers are adjusted to deal with

© Springer International Publishing Switzerland 2015
N.T. Nguyen et al. (Eds.): ACIIDS 2015, Part I, LNAI 9011, pp. 377–386, 2015.
DOI: 10.1007/978-3-319-15702-3_37

the imbalanced data by either making use of oversampling techniques to diversify the base learners, such as *SMOTEBoost* [3], *SMOTEBagging* [22], *RAMOBoost* [4], or by performing undersampling before creating each of the component classifier, e.g., *UnderBagging* [20], *Roughly Balanced Bagging* [8], *RUSBoost* [18]. Beside ensemble-based approaches there are other internal balancing approaches, e.g., *active learning strategies* [5], *granular computing* [19].

In this paper, we propose an extension of SMOTE where artificial examples generated by SMOTE are projected onto the manifold of intermediate representation and then projected back to the input space. This extension results in generating new examples which are expected to be approximately drawn from the true distribution. In order to learn the manifold of the intermediate representation we propose to make use of Restricted Boltzmann Machines (RBM) [9]. RBM is usually used in feature extraction, classification [12] or collaborative filtering [17], among others. The idea of our approach is to construct artificial examples using SMOTE first, and then perform Gibbs sampling with RBM model trained using all minority examples to obtain new sample. In other words, the SMOTE-based sample is a starting point for sampling from RBM model.

The paper is organized as follows. In Section 2 RBM-SMOTE model for creating artificial samples is proposed. In Section 3 we present experimental results of the proposed approach tested on *Kaggle Give Me Some Credit*[1] dataset. This work is summarized by some conclusions and future works in Section 4.

## 2   Methodology

### 2.1   SMOTE

The SMOTE procedure is one of the most popular oversampling method for coping with the imbalanced data phenomenon. This approach generates artificial examples located on the path connecting selected minority example and one of its closest neighbor. The number of examples we want to sample is set by the parameter $P_{SMOTE}$. In SMOTE we pick randomly without replacement examples from minority class.

The procedure of generating artificial sample is described in Algorithm 1. Let us denote the dataset by $\mathbb{D}_N = \{(\mathbf{x}_n, y_n)\}_{n=1}^N$, where $\mathbf{x}_n$ is a vector of features describing $n$-th example and $y_n$ is the corresponding class label, $y_n \in \{-1, 1\}$, 1 represents positive (minority) and $-1$ negative (majority) class.[2] In the SMOTE algorithm the selected minority example $\mathbf{x}_i$ is taken as an input as well as the entire training data $\mathbb{D}_N$ to achieve a new artificial example $\tilde{\mathbf{x}}_i$. In the first step we randomly select one of the $K$ nearest neighbors of the example $\mathbf{x}_i$ (see Figure 1a). Further, the random value $r$ is generated to set the location of the new example $\tilde{\mathbf{x}}_i$ on the path connecting two points: $\mathbf{x}_i$ and $\mathbf{x}_j$ (see Figure 1b). Finally, the position of the new artificial example $\tilde{\mathbf{x}}_i$ is calculated.

---

[1] *Kaggle Give Me Some Credit* is available on-line
https://www.kaggle.com/c/GiveMeSomeCredit/

[2] In the literature it is assumed that majority class is positive and minority class is negative.

---

**Algorithm 1.** Creating artificial sample with SMOTE

---

**Input** : $\mathbb{D}_N = \{(\mathbf{x}_n, y_n)\}_{n=1}^N$: training set, $\mathbf{x}_i$: selected minority example, $K$: number of nearest neighbors

**Output**: $\tilde{\mathbf{x}}_i$: artificial example.

1 Select $k$ uniformly from $\{1, \ldots, K\}$;
2 Find $\mathbf{x}_j$, $k$-th nearest neighbor of $\mathbf{x}_i$ in $\mathbb{D}_N$;
3 Sample $r$ from $[0, 1]$;
4 $\tilde{\mathbf{x}}_i \longleftarrow \mathbf{x}_i + r \cdot (\mathbf{x}_j - \mathbf{x}_i)$;

---

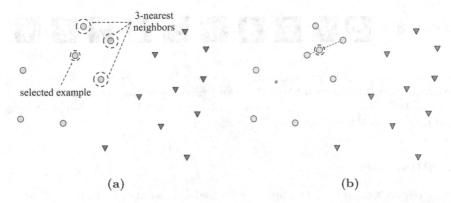

**Fig. 1.** Example of SMOTE for single example: a) selected example and 3 nearest neighbors (black circles), b) artificial example created with SMOTE

The presented version of SMOTE algorithm is dedicated to operate on real-valued features. However, it is easy to extend SMOTE to construct the generator of binary features. If we take under consideration $\mathbf{x}_i$ and $\mathbf{x}_j$ containing only binary values we may receive artificial example $\tilde{\mathbf{x}}_i$ containing real values. However, $\tilde{\mathbf{x}}_i$ may be used as a distribution of multiple Bernoullis to sample a binary vector. The detailed description of SMOTE with additional examples is presented in [2].

The main drawback of SMOTE sampling is the fact that most of the created examples are impossible to be observed in real data. The generated examples may be located far from the the true distribution. As a consequence learning a model on such generated artificial data leads to estimates biased by the noise incorporated in the newly created examples. For instance, consider the handwritten digits taken from MNIST dataset[3] presented in Figure 2. The SMOTE-based examples in most cases are far from the digits that can be written by a human.

---

[3] The MNIST is available on the Web page: http://yann.lecun.com/exdb/mnist/

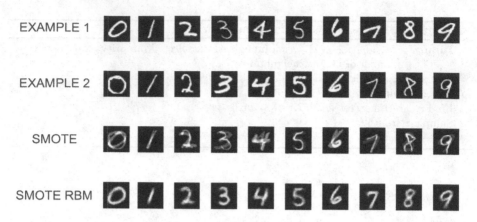

**Fig. 2.** *(Two top rows)* Pairs of examples taken from MNIST dataset that are used to generate artificial sample with SMOTE. *(Third row)* The artificial examples sampled using SMOTE and the two real examples. *(Bottom row)* Examples sampled using SMOTE and then transformed using RBM

## 2.2   RBM

In this paper, we propose to apply Restricted Boltzmann Machine (RBM) to adjust the artificial example sampled using SMOTE to the one which is approximately drawn from the true distribution. We make use of the SMOTE examples as a good starting points for Gibbs sampling from RBM model trained on minority class cases. Considering the example in Figure 2 after applying Gibbs sampling to the artificial digit generated by SMOTE we obtain objects which are easier to interpret.

Restricted Boltzmann Machine is a bipartie Markov Random Field in which visible and hidden units can be distinguished. In RBM only connections between the units in different layers are allowed, i.e., visible to hidden units. The joint distribution of binary visible and hidden units is the Gibbs distribution:

$$p(\mathbf{x}, \mathbf{h}|\boldsymbol{\theta}) = \frac{1}{Z(\boldsymbol{\theta})} \exp\big( - E(\mathbf{x}, \mathbf{h}|\boldsymbol{\theta})\big), \qquad (1)$$

with the following energy function:

$$E(\mathbf{x}, \mathbf{h}|\boldsymbol{\theta}) = -\mathbf{x}^\top \mathbf{W} \mathbf{h} - \mathbf{b}^\top \mathbf{x} - \mathbf{c}^\top \mathbf{h}, \qquad (2)$$

where $\mathbf{x} \in \{0, 1\}^D$ are the visible units, $\mathbf{h} \in \{0, 1\}^M$ are the hidden units, $Z(\boldsymbol{\theta})$ is the normalizing constant dependent on $\boldsymbol{\theta}$, and $\boldsymbol{\theta} = \{\mathbf{W}, \mathbf{b}, \mathbf{c}\}$ is the set of parameters, namely, $\mathbf{W} \in \mathbb{R}^{D \times M}$, $\mathbf{b} \in \mathbb{R}^D$, $\mathbf{c} \in \mathbb{R}^M$ are the weight matrix, visible and hidden bias vectors, respectively.

Since there are no connections among the units within the same layer, i.e., neither visible to visible, nor hidden to hidden connections, the visible units are conditionally independent given the hidden units and vice versa:

$$p(x_i = 1|\mathbf{h}, \mathbf{W}, \mathbf{b}) = \text{sigm}(\mathbf{W}_{i\cdot}\mathbf{h} + b_i), \tag{3}$$

$$p(h_j = 1|\mathbf{x}, \mathbf{W}, \mathbf{c}) = \text{sigm}((\mathbf{W}_{\cdot j})^{\top}\mathbf{x} + c_j), \tag{4}$$

where $\text{sigm}(a) = \frac{1}{1+\exp(-a)}$ is the sigmoid function, $\mathbf{W}_{i\cdot}$ is the $i$-th row of the weight matrix, and $\mathbf{W}_{\cdot j}$ is the $j$-th column of the weight matrix. Therefore, the conditional probability distributions can be presented in the following manner:

$$p(\mathbf{x}|\mathbf{h}, \mathbf{W}, \mathbf{b}) = \prod_{i=1}^{D} p(x_i|\mathbf{h}, \mathbf{W}, \mathbf{b}), \tag{5}$$

$$p(\mathbf{h}|\mathbf{x}, \mathbf{W}, \mathbf{c}) = \prod_{j=1}^{M} p(h_j|\mathbf{x}, \mathbf{W}, \mathbf{c}). \tag{6}$$

Unfortunately, in order to learn parameters $\boldsymbol{\theta}$ gradient-based optimization methods cannot be directly applied because exact gradient calculation is intractable analytically. However, we can adopt *Constrastive Divergence* algorithm which approximates exact gradient using sampling methods [10].

To train RBM we consider minimization of the negative log-likelihood in the following form:

$$\mathcal{L}(\boldsymbol{\theta}) = -\sum_{n=1}^{N} \log p(\mathbf{x}_n|\boldsymbol{\theta}). \tag{7}$$

Further, to prevent the model from overfitting, additional regularization term can be added to the learning objective:

$$\mathcal{L}_{\Omega}(\boldsymbol{\theta}) = \mathcal{L}(\boldsymbol{\theta}) + \lambda\Omega(\boldsymbol{\theta}), \tag{8}$$

where $\lambda > 0$ is the regularization coefficient, and $\Omega(\boldsymbol{\theta})$ is the regularization term. In the following, we use the *weight decay* regularization, i.e., $\Omega(\boldsymbol{\theta}) = \|\mathbf{W}\|_F$, where $\|\cdot\|_F$ is the Frobenius norm.

## 2.3    RBM-SMOTE

We have introduced SMOTE and RBM and therefore we can formulate our new oversampling scheme. The procedure of generating artificial examples is presented in Algorithm 2. First, the set of artificial examples $\mathbb{X}_{SMOTE}$ is generated using SMOTE procedure. Next, the RBM model is trained using only minority examples $\mathbb{D}_N^+$ from training data $\mathbb{D}_N$. Furthermore, for each artificial example taken from $\mathbb{X}_{SMOTE}$ we perform $K_G$ iterations of Gibbs sampling using trained RBM model. The generated example $\bar{\mathbf{x}}_n$ is included in the final set of the examples $\tilde{\mathbb{X}}_{SMOTE}$ that is returned by the procedure. One step of the procedure is also presented in Figure 3.

In the presented procedure we make use of artificial examples generated using SMOTE as a good starting points for Gibbs sampling procedure that is performed by trained RBM model. As a consequence, we achieve the examples that

---

**Algorithm 2.** Creating artificial samples with SMOTE together with application of RBM model

---

**Input** : $\mathbb{D}_N$: training set, $K_G$: number of Gibbs sampling iterations, $K$: number of nearest neighbours (SMOTE),$P_{SMOTE}$: percentage of artificial examples (SMOTE).

**Output**: $\bar{\mathbb{X}}_{SMOTE}$: set of generated artificial samples.

1  Set $\bar{\mathbb{X}}_{SMOTE} = \emptyset$;
2  Generate the set of artificial samples $\mathbb{X}_{SMOTE}$ by application of SMOTE procedure on $\mathbb{D}_N$ with parameters $K$ and $N_{SMOTE}$;
3  Estimate $\theta = \{\mathbf{W}, \mathbf{b}, \mathbf{c}\}$ by training RBM model on positive (minority) examples, i. e., $\mathbb{D}_N^+$, where $\mathbb{D}_N^+ = \{(\mathbf{x}_n, y_n) \in \mathbb{D}_N : y_n = 1\}$;
4  **foreach** $\mathbf{x}_n \in \mathbb{X}_{SMOTE}$ **do**
5  $\quad$ Set $\bar{\mathbf{x}}_n = \mathbf{x}_n$;
6  $\quad$ **for** $k = 1 \to K_G$ **do**
7  $\quad\quad$ Sample $\bar{\mathbf{h}}_n$ from $p(\mathbf{h}|\bar{\mathbf{x}}_n, \mathbf{W}, \mathbf{c})$ (see eq. (6));
8  $\quad\quad$ Sample $\bar{\mathbf{x}}_n$ from $p(\mathbf{x}|\bar{\mathbf{h}}_n, \mathbf{W}, \mathbf{b})$ (see eq. (5));
9  $\quad$ **end**
10 $\quad$ Add $\bar{\mathbf{x}}_n$ to $\bar{\mathbb{X}}_{SMOTE}$;
11 **end**

---

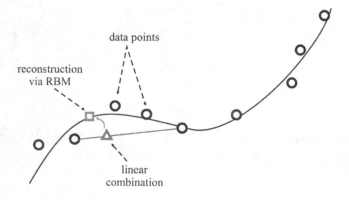

**Fig. 3.** Graphical interpretation of one step of the Algorithm 2. Circles represent observations, the triangle denote new example generated by SMOTE, and the rectangle is the SMOTE example reconstructed using RBM.

are significantly closer to the true distribution than SMOTE-based examples. The empirical studies show that even one loop of sampling procedure may result in constructing good-quality artificial examples.

## 3    Experiment

*Dataset.* The proposed solution was tested on *Kaggle Give Me Some Credit* data with the vector of the attributes transformed to the binary inputs. Each of the

150000 instances is described by 59 binary features. The considered dataset is highly influenced by the imbalanced data phenomenon with the imbalance ratio[4] equal 13.96.

*Methodology.* The goal of the experiment was to compare the performance of simple SMOTE with the same sampling method with RBM modification (further named RBM-SMOTE). As an evaluation criterion we chose *Gmean*, which is defined as a square root of the product of *True Positive Rate* (*TPR*, called *Sensitivity*)[5]:

$$TPR = \frac{TP}{TP + FN},\tag{9}$$

and the *True Negative Rate* (*TNR*, called *Specificity*):

$$TNR = \frac{TN}{TN + FP}.\tag{10}$$

This criterion is widely used to evaluate the quality of the classifiers trained on the highly imbalanced data. We also analyzed the criterion of area under ROC curve (AUC) that can be expressed as an arithmetic mean of *TPR* and *TNR*.

We compared the performance of SMOTE and RBM-SMOTE using the classifiers that are typically applied to the domain of credit risk evaluation, i.e., two decision trees (*JRip* and *CART*), Logistic Regression (*Log*) and other typically used classifiers such as *K*-nearest neighbors (*KNN*), Naïve Bayes (*NB*), Bagging (*Bag*), AdaBoost (*AdaB*), Random Forest (*RF*), LogitBoost (*LogitB*) and Multilayer Perceptron (*MLP*). For each experiment we used 90% of dataset for training and remaining 10% for testing.[6] For both methods the percentage of artificial examples was set to 1400%. The RBM model was trained using Contrastive Divergence procedure with error rate equal 0.001. The weight decay regularization was applied with regularization coefficient equal 0.001. The value of regularization coefficient was set basing on results of preliminary experiments.

*Results.* The results are presented in Table 1. It can be noticed that if no oversampling method is applied the values of *TPR* are close to 0. Comparing our approach and the SMOTE we can observe that RBM-SMOTE outperforms simple SMOTE on all classification methods (see *Gmean* and *AUC* in Table 1). The differences in results are especially visible for comprehensible models (*J48*, *CART*, *RF*). It is important to highlight that our solution operates noticeably better in detecting positive (minority) examples (see *TPR* values in Table 1) for most of the classifiers considered in the experimental studies. It is extremely important in the context of the considered credit scoring problem where the minority class stays behind the group of consumers that are unable to repay their financial liabilities.

---

[4] The ratio between negatives and positives.

[5] TP,TN,FP,FN are the elements of the confusion matrix.

[6] Due to large number of examples taken under consideration in the experiment it was unnecessary to apply other testing methodologies like *cross-validation*.

**Table 1.** The results of the experiment obtained on *Kaggle Give Me Some Credit* data

|       |        | Log | J48 | CART | KNN | NB | Bag | AdaB | RF | LogitB | MLP |
|-------|--------|-----|-----|------|-----|----|-----|------|----|--------|-----|
| TPR   | *None*   | 0.00 | 0.00 | 0.00 | 0.152 | 0.572 | 0.207 | 0.103 | 0.183 | 0.170 | 0.246 |
|       | *SMOTE*  | **0.774** | 0.320 | 0.290 | 0.400 | 0.619 | 0.385 | 0.721 | 0.251 | 0.728 | **0.933** |
|       | *RSMOTE* | 0.773 | **0.529** | **0.553** | **0.480** | **0.740** | **0.564** | **0.730** | **0.477** | **0.766** | 0.566 |
| TNR   | *None*   | 1.00 | 1.00 | 1.00 | **0.988** | **0.915** | **0.986** | **0.994** | **0.980** | **0.991** | **0.984** |
|       | *SMOTE*  | 0.788 | 0.947 | 0.949 | 0.910 | 0.897 | 0.944 | 0.798 | 0.959 | 0.803 | 0.195 |
|       | *RSMOTE* | 0.805 | 0.887 | 0.874 | 0.883 | 0.818 | 0.872 | 0.795 | 0.878 | 0.775 | 0.892 |
| Gmean | *None*   | 0.00 | 0.00 | 0.00 | 0.388 | 0.723 | 0.452 | 0.320 | 0.423 | 0.410 | 0.492 |
|       | *SMOTE*  | 0.781 | 0.551 | 0.525 | 0.603 | 0.745 | 0.603 | 0.759 | 0.491 | 0.765 | 0.427 |
|       | *RSMOTE* | **0.789** | **0.685** | **0.695** | **0.651** | **0.778** | **0.701** | **0.762** | **0.647** | **0.770** | **0.711** |
| AUC   | *None*   | 0.50 | 0.50 | 0.50 | 0.570 | 0.744 | 0.597 | 0.549 | 0.582 | 0.581 | 0.615 |
|       | *SMOTE*  | 0.781 | 0.632 | 0.620 | 0.655 | 0.758 | 0.665 | 0.760 | 0.605 | 0.766 | 0.564 |
|       | *RSMOTE* | **0.789** | **0.708** | **0.713** | **0.682** | **0.779** | **0.718** | **0.763** | **0.678** | **0.771** | **0.729** |

# 4    Conclusion and Future Work

In this paper, we present novel oversampling technique that makes use of RBM model to adjust the examples created with SMOTE to the true distribution over binary features. As a consequence, the artificial examples are expected to be approximately drawn from the true distribution. The results of the preliminary experiments performed on the selected dataset are promising in the context of more thorough analysis of the proposed solution.

For the future works we plan to evaluate the quality of the proposed solution on the large number of datasets taken from various domains. We would like also to extend the presented approach to the numerical features making the assumption that visible units are modeled with Gaussian distribution. Additionally, we consider to compare results gained by RBM-SMOTE with other SMOTE-based solutions (e.g. [16]).

**Acknowledgments.** The research conducted by the authors has been partially co-financed by the Ministry of Science and Higher Education, Republic of Poland, namely, Maciej Zięba: grant No. B40242/I32, Jakub M. Tomczak: grant No. B40020/I32, Adam Gonczarek: grant No. B40235/I32. The work conducted by Maciej Zięba is also co-financed by the European Union within the European Social Fund.

# References

1. Bunkhumpornpat, C., Sinapiromsaran, K., Lursinsap, C.: Safe-level-smote: Safe-level-synthetic minority over-sampling technique for handling the classimbalanced problem. In: Theeramunkong, T., Kijsirikul, B., Cercone, N., Ho, T.-B. (eds.) PAKDD 2009. LNCS, vol. 5476, pp. 475–482. Springer, Heidelberg (2009)
2. Chawla, N.V., Bowyer, K.W., Hall, L.O.: SMOTE: Synthetic Minority Over-sampling Technique. Journal of Artificial Intelligence Research **16**, 321–357 (2002)

3. Chawla, N.V., Lazarevic, A., Hall, L.O., Bowyer, K.W.: SMOTEBoost: improving prediction of the minority class in boosting. In: Lavrač, N., Gamberger, D., Todorovski, L., Blockeel, H. (eds.) PKDD 2003. LNCS (LNAI), vol. 2838, pp. 107–119. Springer, Heidelberg (2003)

4. Chen, S., He, H., Garcia, E.: RAMOBoost: Ranked minority oversampling in boosting. IEEE Transactions on Neural Networks 21(10), 1624–1642 (2010)

5. Ertekin, S., Huang, J., Giles, C.: Active learning for class imbalance problem. In: Proceedings of the 30th Annual International ACM SIGIR Conference on Research and Development in Information Retrieval, pp. 823–824. ACM (2007)

6. García, S., Fernández, A., Herrera, F.: Enhancing the effectiveness and interpretability of decision tree and rule induction classifiers with evolutionary training set selection over imbalanced problems. Applied Soft Computing 9(4), 1304–1314 (2009)

7. He, H., Garcia, E.A.: Learning from Imbalanced Data. IEEE Transactions on Knowledge and Data Engineering 21(9), 1263–1284 (2009)

8. Hido, S., Kashima, H., Takahashi, Y.: Roughly balanced bagging for imbalanced data. Statistical Analysis and Data Mining 2(5-6), 412–426 (2009)

9. Hinton, G.: A practical guide to training restricted boltzmann machines. Momentum 9(1), 926 (2010)

10. Hinton, G.E.: A practical guide to training restricted boltzmann machines. In: Montavon, G., Orr, G.B., Müller, K.-R. (eds.) Neural Networks: Tricks of the Trade. LNCS, vol. 7700, pp. 599–619. Springer, Heidelberg (2012)

11. Han, H., Wang, W.-Y., Mao, B.-H.: Borderline-SMOTE: a new over-sampling method in imbalanced data sets learning. In: Huang, D.-S., Zhang, X.-P., Huang, G.-B. (eds.) ICIC 2005. LNCS, vol. 3644, pp. 878–887. Springer, Heidelberg (2005)

12. Larochelle, H., Bengio, Y.: Classification using discriminative restricted boltzmann machines. In: Proceedings of the 25th International Conference on Machine Learning, pp. 536–543. ACM (2008)

13. Maciejewski, T., Stefanowski, J.: Local neighbourhood extension of smote for mining imbalanced data. In: 2011 IEEE Symposium on Computational Intelligence and Data Mining (CIDM), pp. 104–111. IEEE (2011)

14. Mani, J., Zhang, I.: KNN approach to unbalanced data distributions: a case study involving information extraction. In: Proceedings of International Conference on Machine Learning, Workshop Learning from Imbalanced Data Sets (2003)

15. Ramentol, E., Caballero, Y., Bello, R., Herrera, F.: Smote-rsb*: a hybrid preprocessing approach based on oversampling and undersampling for high imbalanced data-sets using smote and rough sets theory. Knowledge and Information Systems 33(2), 245–265 (2012)

16. Sáez, J.A., Luengo, J., Stefanowski, J., Herrera, F.: SMOTE-IPF: Addressing the noisy and borderline examples problem in imbalanced classification by a resampling method with filtering. Information Sciences (2014)

17. Salakhutdinov, R., Mnih, A., Hinton, G.: Restricted boltzmann machines for collaborative filtering. In: Proceedings of the 24th International Conference on Machine Learning, pp. 791–798. ACM (2007)

18. Seiffert, C., Khoshgoftaar, T., Van Hulse, J., Napolitano, A.: RUSBoost: A hybrid approach to alleviating class imbalance. IEEE Transactions on Systems, Man and Cybernetics, Part A: Systems and Humans 40(1), 185–197 (2010)

19. Tang, Y., Zhang, Y., Huang, Z.: Development of two-stage SVM-RFE gene selection strategy for microarray expression data analysis. IEEE/ACM Transactions on Computational Biology and Bioinformatics 4(3), 365–381 (2007)

20. Tao, D., Tang, X., Li, X., Wu, X.: Asymmetric bagging and random subspace for support vector machines-based relevance feedback in image retrieval. IEEE Transactions on Pattern Analysis and Machine Intelligence **28**(7), 1088–1099 (2006)
21. Tomek, I.: Two Modifications of CNN. IEEE Transactions on Systems, Man and Cybernetics **6**(11), 769–772 (1976)
22. Wang, S., Yao, X.: Diversity analysis on imbalanced data sets by using ensemble models. In: IEEE Symposium on Computational Intelligence and Data Mining, pp. 324–331. IEEE (2009)

# Interval Type-2 Fuzzy C-Means Clustering with Spatial Information for Land-Cover Classification

Sinh Dinh Mai and Long Thanh Ngo[(⊠)]

Department of Information Systems, Le Quy Don Technical University,
236, Hoang Quoc Viet, Hanoi, Vietnam
{maidinhsinh,ngotlong}@gmail.com

**Abstract.** The paper proposes a method to use spatial information to interval type-2 fuzzy c-Means clustering (IT2-FCM) for problems of land cover classification from multi-spectral sattelite images. The spatial information between a pixel and its neighbors on individual band is used to calculate an interval of membership grades in IT2-FCM algorithm. The proposed algorithm, called IIT2-FCM, is implemented on Landsat7 images in comparison with previous algorithms like k-Means, FCM, IT2-FCM to demonstrate the advantage of the approach in handling uncertainty or noise.

**Keywords:** Fuzzy clustering · Type-2 fuzzy sets · Land cover classification · Spatial information

## 1 Introduction

In image segmentation, the most important problem is to find a method to determine whether or not the considered pixel will belong to a certain cluster. The original algorithms like k-Means, fuzzy C-Means(FCM), interval type-2 fuzzy C-Means (IT2-FCM) exhibit the same strategy based on the Euclidean distance to compute the degree of similarity between objects and cluster centroids. Not only color based similarity of the pixels but the spatial relationship between pixels and their neighbors also certainly influence on the final clustering results.

Fuzzy clustering, especially type-2 fuzzy clustering, have exhibited the advantage in handing uncertainty of data. The fuzzy C-Means (FCM) and its variants have widely applied to various problems of image segmentation. FCMs with spatial information also proposed to enhance algorithms of image segmentation in the case of noise [9]-[12]. Z.Wang et al [11] proposed an adaptive spatial information theoretic fuzzy clustering to improve the robustness of the conventional FCM for image segmentation which handling the sensitivity to noisy data and the lack of spatial information. H. Liu et al [10] proposed a fuzzy spectral clustering with robust spatial information for image segmentation to overcome the noise sensitivity of the standard spectral clustering algorithm. K.S. Chuang et al

© Springer International Publishing Switzerland 2015
N.T. Nguyen et al. (Eds.): ACIIDS 2015, Part I, LNAI 9011, pp. 387–397, 2015.
DOI: 10.1007/978-3-319-15702-3_38

[12] introduced an algorithm of FCM with spatial information for image segmentation in which spatial function is the summation of the membership function in the neighborhood of each pixel under consideration.

Interval type-2 FCM (IT2-FCM) was proposed to handing the uncertainty [5] and have applied to various problems e.g. image segmentation, land cover classification [2,3]. L.T.Ngo et al [13] also proposed an different approach of type-2 fuzzy clustering by combining subtractive clustering and interval type-2 fuzzy sets and apply to image segmentation.

Land cover classification from multi-spectral satellite images is one of problems which have widely applied in real application and many approaches have been investigated, recently. W.Su et al [15] proposed method of object oriented information extraction for land cover classification of SPOT5 image, involves two steps: image segmentation and classification. Many other approaches based fuzzy logic also proposed in various manner such as using reformed fuzzy c-means clustering from color satellite image [14].

The paper deals with an approach to combine spatial information into IT2-FCM, called IIT2-FCM, to handle the uncertainty appearing from satellite images. The spatial information between a pixel and its neighbors on individual band is used to calculate an interval of membership grades in IT2-FCM algorithm. Experimental results on data study of Hanoi region with summarized data and validity indexes in comparison with other clustering, e.g k-Means, FCM, IT2-FCM. Especially, results is compared with survey data of the Vietnamese Center of Remote Sensing Technology (VCRST) to compare the accuracy between algorithms.

The paper is organized as follows: Section 2 is background of type-2 fuzzy sets and IT2-FCM; Section 3 introduces the IIT2-FCM algorithm; Section 4 shows experimental results of Hanoi region. Section 5 is conclusion and future works.

## 2    Background

### 2.1    Type-2 Fuzzy Sets

A type-2 fuzzy set in $X$ is denoted $\tilde{A}$, and its membership grade of $x \in X$ is $\mu_{\tilde{A}}(x, u), u \in J_x \subseteq [0, 1]$, which is a type-1 fuzzy set in $[0, 1]$. The elements of domain of $\mu_{\tilde{A}}(x, u)$ are called primary memberships of $x$ in $\tilde{A}$ and memberships of primary memberships in $\mu_{\tilde{A}}(x, u)$ are called secondary memberships of $x$ in $\tilde{A}$.

**Definition 1.** *[1] A type $-2$ fuzzy set, denoted $\tilde{A}$, is characterized by a type-2 membership function $\mu_{\tilde{A}}(x, u)$ where $x \in X$ and $u \in J_x \subseteq [0, 1]$, i. e. ,*

$$\tilde{A} = \{((x, u), \mu_{\tilde{A}}(x, u)) | \forall x \in X, \forall u \in J_x \subseteq [0, 1]\} \tag{1}$$

*or*

$$\tilde{A} = \int_{x \in X} \int_{u \in J_x} \mu_{\tilde{A}}(x, u)) / (x, u), J_x \subseteq [0, 1] \tag{2}$$

*in which $0 \leq \mu_{\tilde{A}}(x, u) \leq 1$.*

At each value of $x$, say $x = x'$, the 2-D plane whose axes are $u$ and $\mu_{\tilde{A}}(x', u)$ is called a *vertical slice* of $\mu_{\tilde{A}}(x, u)$. A *secondary membership function* is a vertical slice of $\mu_{\tilde{A}}(x, u)$. It is $\mu_{\tilde{A}}(x = x', u)$ for $x \in X$ and $\forall u \in J_{x'} \subseteq [0, 1]$, i. e.

$$\mu_{\tilde{A}}(x = x', u) \equiv \mu_{\tilde{A}}(x') = \int_{u \in J_{x'}} f_{x'}(u)/u, J_{x'} \subseteq [0,1] \tag{3}$$

in which $0 \le f_{x'}(u) \le 1$.

Type-2 fuzzy sets are called an interval type-2 fuzzy sets if the secondary membership function $f_{x'}(u) = 1 \ \forall u \in J_x$ i. e. a type-2 fuzzy set are defined as follows:

**Definition 2.** *An interval type-2 fuzzy set $\tilde{A}$ is characterized by an interval type-2 membership function $\mu_{\tilde{A}}(x, u) = 1$ where $x \in X$ and $u \in J_x \subseteq [0,1]$, i. e.*
,

$$\tilde{A} = \{((x, u), 1) | \forall x \in X, \forall u \in J_x \subseteq [0, 1]\} \tag{4}$$

Uncertainty of $\tilde{A}$, denoted FOU, is union of primary functions i. e. $FOU(\tilde{A}) = \bigcup_{x \in X} J_x$. Upper/lower bounds of membership function (UMF/LMF), denoted $\overline{\mu}_{\tilde{A}}(x)$ and $\underline{\mu}_{\tilde{A}}(x)$, of $\tilde{A}$ are two type-1 membership function and bounds of FOU which is limited by two membership functions of an type-1 fuzzy set are UMF and LMF.

## 2.2   Interval Type-2 Fuzzy C-Means Clustering

In general, fuzzy memberships in interval type-2 fuzzy C means algorithm (IT2FCM) [5] is achieved by computing the relative distance among the patterns and cluster centroids. Hence, to define the interval of primary membership for a pattern, we define the lower and upper interval memberships using two different values of $m$ . In (5), (6) and (7), $m_1$ and $m_2$ are fuzzifiers which represent different fuzzy degrees.

IT2-FCM is extension of FCM clustering by using two fuzziness parameters $m_1$, $m_2$ to make FOU, corresponding to upper and lower values of fuzzy clustering. The use of fuzzifiers gives different objective functions to be minimized as follows:

$$\begin{cases} J_{m_1}(U, v) = \sum_{k=1}^{N} \sum_{i=1}^{C} (u_{ik})^{m_1} d_{ik}^2 \\ J_{m_2}(U, v) = \sum_{k=1}^{N} \sum_{i=1}^{C} (u_{ik})^{m_2} d_{ik}^2 \end{cases} \tag{5}$$

in which $d_{ik} = \| x_k - v_i \|$ is Euclidean distance between the pattern $x_k$ and the centroid $v_i$, $C$ is the number of clusters and $N$ is the number of patterns. Upper/lower degrees of membership, $\overline{u}_{ik}$ and $\underline{u}_{ik}$ are determined as follows:

$$\overline{u}_{ik} = \begin{cases} \dfrac{1}{\sum_{j=1}^{C} \left(\dfrac{d_{ik}}{d_{jk}}\right)^{2/(m_1-1)}} & if \ \dfrac{1}{\sum_{j=1}^{C} \left(\dfrac{d_{ik}}{d_{jk}}\right)} < \dfrac{1}{C} \\ \dfrac{1}{\sum_{j=1}^{C} \left(\dfrac{d_{ik}}{d_{jk}}\right)^{2/(m_2-1)}} & otherwise \end{cases} \tag{6}$$

$$
\underline{u}_{ik} = \begin{cases} \dfrac{1}{\sum\limits_{j=1}^{C}\left(\dfrac{d_{ik}}{d_{jk}}\right)^{2/(m_1-1)}} & if\ \dfrac{1}{\sum\limits_{j=1}^{C}\left(\dfrac{d_{ik}}{d_{jk}}\right)} \geq \dfrac{1}{C} \\[2em] \dfrac{1}{\sum\limits_{j=1}^{C}\left(\dfrac{d_{ik}}{d_{jk}}\right)^{2/(m_2-1)}} & otherwise \end{cases} \tag{7}
$$

in which $i = 1, ..., C$, $k = 1, ..., N$.

Because each pattern has membership interval as the upper $\overline{u}$ and the lower $\underline{u}$, each centroid of cluster is represented by the interval between $v^L$ and $v^R$. Cluster centroids are computed in the same way of FCM as follows:

$$
v_i = (\sum_{k=1}^{N}(u_{ik})^m x_k)/(\sum_{k=1}^{N}(u_{ik})^m) \tag{8}
$$

in which $i = 1, ..., C$. After obtaining $v_i^R$, $v_i^L$, type-reduction is applied to get centroid of clusters as follows:

$$
v_i = (v_i^R + v_i^L)/2 \tag{9}
$$

For membership grades:

$$
u_i(x_k) = (u_i^R(x_k) + u_i^L(x_k))/2, j = 1, ..., C \tag{10}
$$

in which

$$
u_i^L = \sum_{l=1}^{M} u_{il}/M, u_{il} = \begin{cases} \overline{u}_i(x_k)\ if\ x_{il}\ uses\ \overline{u}_i(x_k)\ for\ v_i^L \\ \underline{u}_i(x_k)\qquad\ otherwise \end{cases} \tag{11}
$$

$$
u_i^R = \sum_{l=1}^{M} u_{il}/M, u_{il} = \begin{cases} \overline{u}_i(x_k)\ if\ x_{il}\ uses\ \overline{u}_i(x_k)\ for\ v_i^R \\ \underline{u}_i(x_k)\qquad\ otherwise \end{cases} \tag{12}
$$

Next, defuzzification for IT2FCM is made as if $u_i(x_k) > u_j(x_k)$ for $j = 1, ..., C$ and $i \neq j$ then $x_k$ is assigned to cluster $i$.

## 3 Interval Type-2 Fuzzy C-Means Clustering with Spatial Information

### 3.1 Spatial Information

In fact, the image information is stored as numeric values so the problem of image partitions is usually based on the degree of similarity among these values to decide whether an object belongs to any region in the image. Therefore the key to determine a pixel will belong to certain area is based on the similarity in these colours, which is calculated through a function of the distance in the color space $d_{ik} = \| x_k - v_i \|$ i.e Euclidean distance between the pattern $x_k$ and the centroid $v_i$.

We use a mask of size $nxn$ to slip on the image, the center pixel of mask is the considered pixel. The number of neighboring pixels is determined corresponding to the selected type of mask i.e., 8 pixels for the 8-directional mask or 4 pixels for the 4-directional mask.

To determine the degree of influence of a neighboring pixels for the center pixels, a measure spatial information $SI_{ik}$ is defined on the basic of the degree $u_{ki}$ and the attraction distance $d_{ki}$ as follows:

$$SI_{ik} = \frac{\sum_{j=1}^{N} u_{ij} d_{kj}^{-1}}{\sum_{j=1}^{N} d_{kj}^{-1}} \qquad (13)$$

in which $u_{ij}$ is the membership degree of the neighboring element $x_j$ to the cluster $i$. The distance attraction $d_{kj}$ is the squared Euclidean distance between elements $(x_k, y_k)$ and $(x_j, y_j)$. According to this formula, the value of spatial information is at greater pixel on the mask while many neighboring pixels they have similar color them and the opposite. We use the inverse distance $d_{kj}^{-1}$, because the closer neighbors $x_j$ is to the center $x_k$ the more influence it has on the result.

A new distance measure is defined as follows:

$$R_{ik} = \| x_k - v_i \|^2 (1 - \alpha e^{-SI_{ik}}) \qquad (14)$$

where $S_{ij}$ is spatial relationship information between elements $x_k$ and clusters $i$, $\alpha \in \lceil 0, 1 \rceil$ is the parameter that controls the relative factor of neighboring pixels. If $\alpha = 0$, $R_{ik}$ is the squared Euclidean distance and the algorithm becomes the original standard FCM.

The idea behinds the use of this spatial relationship information is: Consider the local $nxn$ neighborhood with the center $x_k$ has large intensity differences with the closest neighboring pixels $x_k$, which has similar intensity as the cluster centroid $v_i$. If the neighborhood attraction $SI_{ij}$, takes a large value then the expression $(1 - \alpha e^{-SI_{ik}})$ will take a small value for $\alpha \neq 0$. After each iteration of the algorithm, the central element $x_k$ will be attracted to the cluster $i$. If the neighborhood attraction $SI_{ik}$ continuously take a large value till the algorithm terminate, the central element $x_k$ will be assigned to the cluster $i$.

## 3.2 Interval Type-2 Fuzzy C-Means Clustering with Spatial Information

The main idea of the IIT2-FCM algorithm is extended from IT2-FCM by adding the spatial information to calculate distance between clusters and pixels. The steps are described as follows:

Firstly, Initialization of matrix centroid V.

Secondly, the primary memberships $\overline{u}_{ik}$ and $\underline{u}_{ik}$ for a pattern are coresponding to two fuzzifiers $m_1$ and $m_2$ which were chosen by heuristic.

Then value of spatial information of each pixels $SI_{ik}$ is computed. Because each pattern has membership interval $\overline{u}$, $\underline{u}$ with the upper bound $\overline{u}$ and the lower bound $\underline{u}$, so $SI_{ij}$ will be an interval with two bounds which are computed corresponding to the upper and lower membership grades as follows:

$$\overline{SI}_{ik} = \frac{\sum_{j=1}^{N} \overline{u}_{ij}(d_{kj})^{-1}}{\sum_{j=1}^{N}(d_{kj})^{-1}} \tag{15}$$

$$\underline{SI}_{ik} = \frac{\sum_{j=1}^{N} \underline{u}_{ij}(d_{kj})^{-1}}{\sum_{j=1}^{N}(d_{kj})^{-1}} \tag{16}$$

Then the value of spatial information is defuzzified as:

$$SI_{ik} = (\overline{SI}_{ik} + \underline{SI}_{ik})/2 \tag{17}$$

Compute the distance as the following formula:

$$R_{ik} = \| x_k - v_i \|^2 (1 - \alpha e^{-SI_{ik}}) \tag{18}$$

We define $I_k = \{i | 1 \leq i \leq C, R_{ik} = 0\}$ in which $k = \overline{1, N}$ and $\| x_k - v_i \|$ is the Euclide distance between data samples $k$ and cluster $i$ in d-dimensional space. In case of $I_k = \emptyset$, $\overline{u}_{ik}$ and $\underline{u}_{ik}$ are determined in the same way of formulas (5) and (6) by replacing the distance $d_{ij}$ by the new distance $R_{ij}$ as follows:

$$\overline{u}_{ik} = \begin{cases} \dfrac{1}{\sum\limits_{j=1}^{C}(R_{ik}/R_{jk})^{2/(m_1-1)}} & if \ \dfrac{1}{\sum\limits_{j=1}^{C}(R_{ik}/R_{jk})} < \dfrac{1}{C} \\[4ex] \dfrac{1}{\sum\limits_{j=1}^{C}(R_{ik}/R_{jk})^{2/(m_2-1)}} & if \ \dfrac{1}{\sum\limits_{j=1}^{C}(R_{ik}/R_{jk})} \geq \dfrac{1}{C} \end{cases} \tag{19}$$

$$\underline{u}_{ik} = \begin{cases} \dfrac{1}{\sum\limits_{j=1}^{C}(R_{ik}/R_{jk})^{2/(m_1-1)}} & if \ \dfrac{1}{\sum\limits_{j=1}^{C}(R_{ik}/R_{jk})} \geq \dfrac{1}{C} \\[4ex] \dfrac{1}{\sum\limits_{j=1}^{C}(R_{ik}/R_{jk})^{2/(m_2-1)}} & if \ \dfrac{1}{\sum\limits_{j=1}^{C}(R_{ik}/R_{jk})} < \dfrac{1}{C} \end{cases} \tag{20}$$

Otherwise, if $I_k \neq \emptyset$, $\overline{u}_{ik}$ and $\underline{u}_{ik}$ are determined:

$$\overline{u}_{ik} = \begin{cases} 0 & if \ i \notin I_k \\ \sum\limits_{i \in I_k} \overline{u}_{ik} = 1 & if \ i \in I_k \end{cases} \tag{21}$$

$$\underline{u}_{ik} = \begin{cases} 0 & if \ i \notin I_k \\ \sum\limits_{i \in I_k} \underline{u}_{ik} = 1 & if \ i \in I_k \end{cases} \tag{22}$$

$$\text{in which } i = \overline{1, C}, \, k = \overline{1, N}.$$

Because each pattern has membership interval $\overline{u}$ and $\underline{u}$, so each centroid of cluster is represented by the interval between $v^L$ and $v^R$.

**Algorithm 1:** Algorithm find $v^L$ and $v^R$

**Step 1**: Find $\overline{u}_{ij}, \underline{u}_{ij}$, by the equations (19)-(20).

**Step 2**: Set $m$ is a constant satisfied $m \geq 1$;

Compute $v'_j = (v'_{j1}, ..., v'_{jM})$ by the equation (8) with $u_{ij} = \frac{(\overline{u}_{ij} + \underline{u}_{ij})}{2}$.

Sort $N$ patterns on each of $M$ features in ascending order.

**Step 3**: Find index $k$ such that: $x_{kl} \leq v'_{jl} \leq x_{(k+1)l}$ with $k = 1, .., N$ and $l = 1, .., M$.

**Step 4**: Calculate $v"$ by following equation: In case $v"$ is used for finding $v^L$

$$v" = \frac{\sum\limits_{i=1}^{k} x_i \overline{\mu_A}(x_i) + \sum\limits_{i=k+1}^{N} x_i \underline{\mu_A}(x_i)}{\sum\limits_{i=1}^{k} \overline{\mu_A}(x_i) + \sum\limits_{i=k+1}^{N} \underline{\mu_A}(x_i)} \tag{23}$$

In case $v"$ is used for finding $v^R$

$$v" = \frac{\sum\limits_{i=1}^{k} x_i \underline{\mu_A}(x_i) + \sum\limits_{i=k+1}^{N} x_i \overline{\mu_A}(x_i)}{\sum\limits_{i=1}^{k} \underline{\mu_A}(x_i) + \sum\limits_{i=k+1}^{N} \overline{\mu_A}(x_i)} \tag{24}$$

**Step 5**: If $v' = v"$ go to Step 6 else set $v' = v"$ then back to Step 3.

**Step 6**: Set $v^L = v'$ or $v^R = v'$.

After obtaining $v^R$, $v^L$, to get centroid of clusters by (9).

For membership grades $u_i(x_k)$ base on the formula (10), (11) and (12).

Next, defuzzification for algorithm IIT2-FCM: if $u_i(x_k) > u_j(x_k)$ for $j = 1, ..., C$ and $i! = j$ then $x_k$ is assigned to cluster $i$.

**Algorithm 2:** IIT2-FCM algorithm

**Step 1**: Initialization

1.1 The two parameters of fuzzy $m_1$, $m_2$ $(1 < m_1, m_2)$, error $e$.

1.2 Initialization centroid $V = [v_i], v_i \in R^d$.

**Step 2**: Compute the fuzzy partition matrix $\overline{U}, \underline{U}$ and update centroid V:

2.1. Calculate the value of spatial information $SI_{ik}$ by formulas (15), (17).

2.2. Calculate matrix of membership grades $U_{ik}$ by formulas (18), (20).

2.3. Assign a pattern to a cluster.

2.4. Update the centroid of clusters $V^j = [v_1^j, v_2^j, ..., v_c^j]$ by using the algorithm of finding $v^L$ and $v^R$ and the formula (8).

**Step 3**: Verify the stop condition:

If $Max(|J^{(j+1)} - J^{(j)}|)$, go to step 4, otherwise go to step 2.

**Step 4**: Report the clustering results.

## 4    Land-Cover Classification Using IIT2-FCM

The IIT2-FCM based algorithm for land cover classification is implemented on Landsat7 images of Hanoi region, Vietnam ( 21°54'23.11"N, 105°03'06.47"E to 20°55'14.25"N, 106°02'58'.57"E) with area is 1128. $km^2$ and resolution is $30m \times 30m$. We use 6 bands of 1, 2, 3, 4, 5 and 7 as inputs of clustering algorithms and the image is classified into 6-classes corresponding to 6 types of land covers as in Figure 1.

Class1: Rivers, ponds, lakes        Class2: Rocks, bare soil

Class3: Fields, sparse tree        Class4: Planted forests, low woods

Class5: Perennial tree crops        Class6: Jungles.

**Fig. 1.** Six types of land covers

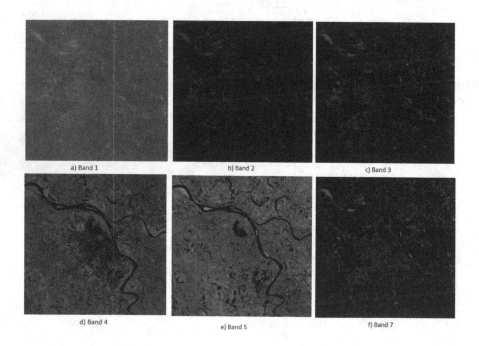

**Fig. 2.** The bands of Hanoi region

**Table 1.** Land cover classification of IIT2-FCM, IT2-FCM, FCM, k-Means algorithms

| Class | | Class 1 | Class 2 | Class 3 | Class 4 | Class 5 | Class 6 | Mean | SD |
|---|---|---|---|---|---|---|---|---|---|
| VCRST | | 94.49 | 82.65 | 205.01 | 215.54 | 282.39 | 248.78 | | |
| IIT2-FCM | area (km2) | 84.53 | 94.54 | 207.95 | 229.15 | 276.29 | 236.40 | 9.48 | 4.14 |
| | difference (km2) | 9.96 | 11.89 | 2.94 | 13.61 | 6.10 | 12.38 | | |
| IT2-FCM | area (km2) | 61.17 | 97.77 | 222.29 | 241.26 | 294.68 | 211.67 | 23.47 | 10.21 |
| | difference (km2) | 33.32 | 15.12 | 17.28 | 25.72 | 12.29 | 37.11 | | |
| FCM | area (km2) | 121.36 | 153.42 | 257.38 | 179.84 | 251.03 | 165.82 | 50.01 | 22.85 |
| | difference (km2) | 26.87 | 70.77 | 52.37 | 35.70 | 31.36 | 82.96 | | |
| k-Means | area (km2) | 167.84 | 169.52 | 239.50 | 251.34 | 206.34 | 94.30 | 76.84 | 43.86 |
| | difference (km2) | 73.35 | 86.87 | 34.49 | 35.80 | 76.05 | 154.48 | | |

Bands of Landsat7 image of Hanoi region are shown in Fig.2. The experimental results are demonstrated in Fig.3 in which (a), (b), (c), (d) images are classification result of IIT2-FCM, IT2-FCM, FCM, k-Means algorithms, respectively. Fig. 3 shows that the classification of IIT2-FCM gives the better clusters. Beside, summarized data of algorithms is compared with the survey data of VCRST which is demonstrated in Table 1 in which IIT2-FCM exhibits the best quality of clusters with the lowest standard deviation of difference, i.e. SD is 4.4.

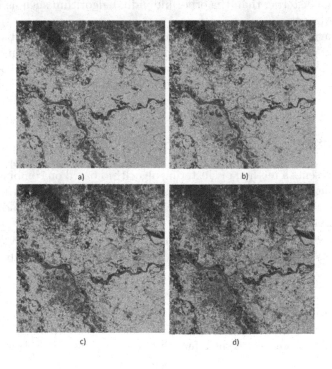

**Fig. 3.** Land cover classification. a) IIT2-FCM; b) IT2-FCM; c) FCM; d) k-Means.

To evaluate the quality of clusters, several validity indexes are computed from results of the mentioned algorithms. We considered various validity indexes such as the Bezdeks partition coefficient (PC-I) [8], the Dunns separation index (DI), the Davies-Bouldins index (DB-I), and the Separation index (S-I), Xie and Benis index (XB-I), Classification Entropy index (CE-I) [7]. The validity indexes are shown in the Table 2.

**Table 2.** The various validity indexes from classification of Hanoi region

| Validity index | k-Means | FCM | IT2-FCM | IIT2-FCM |
|----------------|---------|--------|---------|----------|
| DB-I | 4.531 | 3.4983 | 2.3981 | **1.1246** |
| XB-I | 1.761 | 1.1784 | 0.6823 | **0.1382** |
| S-I | 0.9821 | 0.6287 | 0.3834 | **0.0917** |
| CE-I | 1.323 | 0.9869 | 0.5872 | **0.1972** |
| PC-I | 0.6543 | 0.6982 | 0.7282 | **0.8628** |

Note that the validity indexes are proposed to evaluate the quality of clustering. The better algorithms exhibit smaller values of T-I, DB-I, XB-I, S-I, CE-I and the larger value of PC-I. The results in Table 2 show that the IIT2FCM have better quality clustering than the other individual algorithm such as IT2-FCM, FCM and k-Means.

In summary, classification results can be explained that, the boundary of water and soil classes are usually quite clearly, while the vegetation classes are often confused in which both grasses and trees. With the resolution of $30m \times 30m$, the difference of classification results can be acceptable to assess land cover on a large area.

## 5    Conclusion

This paper presents a method of clustering algorithm based on combining spatial information into IT2-FCM to handle uncertainty better. The experiments were carried out based on Landsat7 image of Hanoi region to assess the advantage of the proposed algorithm.

The next goal is to implement further research on hyperspectral satellite imagery for environmental classification, assessment of land surface temperature changes.

## References

1. Mendel, J.M., John, R.I.: Type-2 fuzzy sets made simple. IEEE Trans. on Fuzzy Systems **10**(2), 117–127 (2002)
2. Ngo, L.T., Nguyen, D.D.: Land cover classification using interval type-2 fuzzy clustering for multi-spectral satellite imagery. In: IEEE-SMC, pp. 2371–2376 (2012)

3. Nguyen, D.D., Ngo, L.T.: Multiple kernel interval type-2 fuzzy c-means clustering. In: IEEE Int' Conference on Fuzzy Systems, pp. 1–8 (2013)
4. Bezdek, J.C., Ehrlich, R., Full, W.: FCM: The Fuzzy C-Means clustering algorithm. Computers and Geosciences **10**(2), 191–203 (1984)
5. Hwang, C., Rhee, F.C.H.: Uncertain Fuzzy Clustering: Interval Type-2 Fuzzy Approach to C-Means. IEEE Trans. on Fuzzy Systems **15**(1), 107–120 (2007)
6. Genitha, C.H., Vani, K.: Classification of satellite images using new fuzzy cluster centroid for unsupervised classification algorithm. In: 2013 IEEE Conf' on ICT, pp. 203–207 (2013)
7. Wang, W., Zhang, Y.: On fuzzy cluster validity indices. Fuzzy Sets and Systems **158**, 2095–2117 (2007)
8. Bezdek, J., Pal, N.: Some new indexes of cluster validity. IEEE Transactions on Systems, Man and Cybernetics **28**(3), 301315 (1998)
9. Zhao, F., Fan, J., Liu, H.: Optimal-selection-based suppressed fuzzy c-means clustering algorithm with self-tuning non local spatial information for image segmentation. Expert Systems with Applications **41**, 4083–4093 (2014)
10. Liua, H., Zhao, F., Jiao, L.: Fuzzy spectral clustering with robust spatial information for image segmentation. Applied Soft Computing **12**, 3636–3647 (2012)
11. Wang, Z., Song, Q., Soh, Y.C., Sim, K.: An adaptive spatial information-theoretic fuzzy clustering algorithm for image segmentation. Computer Vision and Image Understanding **117**, 1412–1420 (2013)
12. Chuang, K.S., Tzeng, H.L., Chen, S., Wu, J., Chen, T.J.: Fuzzy c-means clustering with spatial information for image segmentation. Computerized Medical Imaging and Graphics **30**, 9–15 (2006)
13. Ngo, L.T., Pham, B.H.: Approach to image segmentation based on interval type-2 fuzzy subtractive clustering. In: Pan, J.-S., Chen, S.-M., Nguyen, N.T. (eds.) ACIIDS 2012, Part II. LNCS, vol. 7197, pp. 1–10. Springer, Heidelberg (2012)
14. Sowmya, B., Sheelarani, B.: Land cover classification using reformed fuzzy C-means. Journal of Sadhana **36**(2), 153–165 (2011). Springer
15. Su, W., Zhang, C., Zhu, X., Li, D.: A Hierarchical Object Oriented Method for Land Cover Classification of SPOT 5 Imagery. WSEAS Trans. on Information Science and Applications **6**(3), 437–446 (2009)

# Mining ICDDR, B Hospital Surveillance Data Using Locally Linear Embedding Based SMOTE Algorithm and Multilayer Perceptron

Adnan Firoze[1] and Rashedur M. Rahman[2(✉)]

[1] School of Engineering and Applied Science (SEAS),
Columbia University, New York, USA
af2728@columbia.edu
[2] Department of Electrical and Computer Engineering,
North South University, Dhaka, Bangladesh
rashedur@northsouth.edu

**Abstract.** In this paper, the authors use multilayer perceptron (MLP) on hospital surveillance data to categorize admitted patients according to their critical conditions which can be classified as - low, medium and high, to distinguish the criticality. The paper addresses the over-fitting problem in the unbalanced dataset using two distinct approaches since the frequency of instances of the class 'low' is significantly higher than other classes. Besides trimming, the unbalanced dataset is balanced by introducing the Synthetic Minority Over-sampling Technique (SMOTE) algorithm coupled with Locally Linear Embedding (LLE). We have constructed three models and applied neural classifications and compared the performances with the decision tree based models that already exist in literature. We show that one of our models outperforms prior models in classification, contingent upon performance time trade-off, giving us an efficient model that handles large scale unbalanced dataset efficiently with standard classification performance. The models developed in this research can become imperative tools to doctors during epidemics.

**Keywords:** Data mining · neural network · SMOTE · Medical surveillance · Artificial intelligence · Classification · Imbalanced data

# 1    Introduction

Machine intervention in medicine and mining large scale medical surveillance data have caught significant attention in the recent years due to epidemics and the scarcity of physicians. We have pursued this research based on a dataset that stores patients' data from January 1, 1996 to December 31, 2007 (which is hospital surveillance data of 12 years) that was collected at International Centre for Diarrhoeal Disease Research, Bangladesh (ICDDR,B) [1].

The primary objective of this article is to create an efficient classification model that serves effectively to classify the large repository of ICDDR,B hospital surveillance data into low, mid and high criticality of patients, while taking into account the intrinsic issues of an unbalanced dataset. Instead of working with the dataset

© Springer International Publishing Switzerland 2015
N.T. Nguyen et al. (Eds.): ACIIDS 2015, Part I, LNAI 9011, pp. 398–407, 2015.
DOI: 10.1007/978-3-319-15702-3_39

directly, for achieving a more meaningful system, we rejected some incomplete data and filtered data based on some assumptions suggested by domain experts and physicians, which is known as data-cleaning in the data mining paradigm, as following.

The 'outcome' field in the dataset has the following values stored: 1 = Cured, 2 = Illness continued, 3 = Died, 4 = Absconded, 5 = Others, 9 = Unknown. We have considered the records of the patients with outcome = 1 exclusively and rejected the others since most of those records were incomplete. It should be noted that there were no attribute/field called 'Criticality' in the dataset; hence, we derived it.

We supplanted the 'duration of stay' with our target variable 'Criticality'. Thus, we create a derived attribute 'Criticality' by using the following rules:

0 to ≤ 48 hours: Low,
48> to ≤96 hours: Mid,
>96 High.

In this research, we have extended and improved the approaches of Rahman and Hasan [2] that used decision tree induction algorithm on Hospital Surveillance data to classify admitted patients according to their critical conditions. We classified the data using a neural approach, specifically Multilayer Perceptron and moreover, we have introduced a novel way (in medical context) of balancing the unbalanced dataset by using Locally Linear Embedding (LLE) based Synthetic Minority Over-sampling Technique (SMOTE) algorithm to model a more accurate classifier on the dataset. Finally, we have presented a comparative analysis of our classifier models and previous models.

## 2    Multilayer Perceptron and LLE Based SMOTE

We have performed the classification on the dataset using the neural network architecture – Multilayer Perceptron. Besides the classifier, we have used LLE based SMOTE [3] algorithm to balance the unbalanced data, specifically, to synthetically increase the number of minority classes. A multilayer perceptron can classify multiple classes and has a linear activation function in all neurons, that is, a simple on-off mechanism to determine whether or not a neuron fires. More details of multilayer perceptron and learning algorithm could be found in [4].

LLE based SMOTE [3] is one of the most recent techniques to deal with unbalanced datasets and over fitting problems. It has been used widely in different areas that also include face membership authentication [5]. Our dataset is prone to the same problems since the 'Low' criticality class is a majority class whereas the 'Mid' and 'High' criticality classes have much fewer instances.

The original SMOTE algorithm was given by Chawla, Bowyer, Hall and Kegelmeyer [6] in which the minority class is oversampled by using  the k-NN graph instead of randomized sampling with replacement. The LLE based SMOTE algorithm [3] is given as follows.

**Function** LLE-BasedSMOTE(X, C_, C+, k, l, k+) Return S
**Input:**
X: the original training set $\{x_1, x_2, \ldots, x_N\}$, $X \in R^d$
C_: majority class or negative class
C_: minority class or positive class
k: the number of nearest neighbors for linear spanning in
LLE
l: the reduced dimension by LLE
k_: a threshold for choosing a negative neighbor in LLE
**Output:** S: the oversampled training set
**Main Program:**
Y•LLE (X, C_, C_, k, k+, l)
Z• generates new positive vectors by using SMOTE algorithm
over embedding set Y
S •X
**For** each vector **z** of Z
$Y_{kNN}$ (z) • Find z's k nearest neighbors in set Y
$w_z$ • Compute linear combination weights of $Y_{kNN}$(z) as given in
(Wang, Zhu, Wang and Zhang[3]).
z'• Map z from l-dimensional embedding space back to the
original input space as given in (Wang, Zhu, Wang and Zhang
[3]).
  S • S ∪ {z'}
**End**
Procedure LLE (X, k, k_, l) Return Y
Initialize weight matrix **W**
**For** each vector **x** in X
j • 0 $X_{kNN}$(**v**)  • φ Y • φ
$X^0_{kNN}$(**x**) • Find x's k number of nearest neighbors according to
Euclidean distance
**For** each v ∈ $X^0_{kNN}$(**x**) ∩ C_
If | $X^0_{kNN}$(**v**) ∩ C_ | ≥ k_
$X_{kNN}$(**v**) • $X_{kNN}$(v) ∪ { v }
**End**
**End**
$X_{kNN}$(**v**) • Add x's k-| $X_{kNN}$(**v**)| number of nearest
positive neighbors to $X_{kNN}$(**v**)
$w_x$ • compute linear spanning weights of T
W • assign the row of **W** corresponding to x with $w_x$
**End**
Y ← Compute the embedding data according to W

[Pseudocode  of the LLE-based SMOTE algorithm [3]]

# 3    Methodology

In this research, we develop an efficient classification model that classifies large repositories of hospital surveillance data into low, mid and high criticality of patients, while taking into account the intrinsic issues of unbalanced datasets. A brief overview of the sequential process is illustrated in Figure 1.

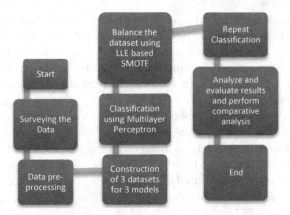

**Fig. 1.** Flowchart showing the methodology sequence

## 3.1    Surveying and Preprocessing the Data

Hospital surveillance unit of ICDDR,B keeps surveillance data and data related to patients in SPSS format. The raw dataset (from the hospital) information is as follows:

- Number of records: 26,869 patients
- Duration of data collection: January 1, 1996 to December 31, 2007
- Number of variables/features: 227

Discussing with experts of surveillance unit and a physician at ICDDR,B who have the necessary domain knowledge, we have dropped some variables and selected some variables for preprocessing, cleaning, transformation and modeling. Consequently, the variable count is decreased from 227 to 40 by eliminating incomplete entries. Non-categorical variables are removed and we keep 37 variables for accurate data modeling. Frequency distribution, histogram of the variables, and statistical information related to the variables in the selected dataset is taken to have a quick understanding about the information content and quality of the data. It must be noted that we have done all surveying and preprocessing in concordance with the research done by Rahman and Hasan [2] to facilitate a comprehensive comparative analysis. SPSS is used to find out missing values, empty values, misclassification errors, field transformation, dimensionality reduction i.e. overall data cleaning activities are carried out using SPSS. We perform this step from Rahman and Hasan [2] verbatim since we are interested to compare our models with the previous research. After preprocessing and cleaning the dataset is changed to:

- Number of records: 25,261 patients
- Number of variables/features: 22

### 3.2    Data Modeling

In this research work, we use Weka [7] for optimizing/balancing the dataset and classification. Weka is an advanced level Data Mining tool which implements various supervised learning paradigms and unsupervised clustering algorithms. In this part, we have created three distinct data models for further classification and to illustrate the classification performances over balanced and unbalanced datasets. The models are briefly described as follows.

**Model 1:** This model sports all the instances that were found once we completed the data pre-processing/surveying stage. Therefore, this is the most primitive or raw data on which we have performed classification. It must be noted that, this data model corresponds to the first model set forth by Rahman and Hasan [2] but we apply different classification algorithms and used different techniques to balance the dataset. The models' frequency distributions are shown in Figure 2.

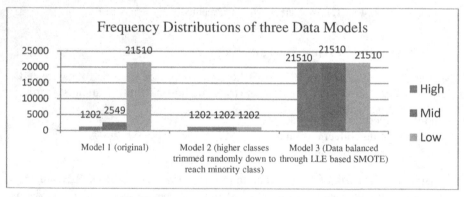

**Fig. 2.** Frequency distributions of data models

**Model 2:** This is the second model; however this is a balanced dataset i.e. all the classes have equal number of instances. We have trimmed the majority classes by choosing the number of instances from them that equals the number of instances/records of the minority class. In our Model 1, we find that class "Low" has case – 21510 records, "Mid" has 2549 records but our minority class "High" has only 1202 records. We have selected 1202 records for "Low" and "Mid" to balance the dataset for proper classification evaluation in this model. The details of this step are discussed in subsection 4.2. Notably, this model corresponds to the second model proposed by Rahman and Hasan [2] for classification comparison; however we resort to more sophisticated balancing algorithms in our Model 3, discussed in subsection 4.3.

**Model 3:** This is our novel LLE based SMOTE powered model where we balance the hospital surveillance dataset in a manner that has not been done till date and we have achieved significantly improved classification performance based on this model than any past attempts. This model is balanced but instead of trimming the majority classes down to minority classes, we use the LLE based SMOTE algorithm to generate synthetic instances for the minority classes to produce more instances to match the frequency of the majority class "Low".

Therefore, in this model, we artificially generated a balanced dataset using LLE based SMOTE algorithm to reach 21510 instances (which was originally the number of instances for the majority class – "Low") for all three classes.  See Figure 2 to find the frequency distributions of all three data models.

## 3.3   Parameters

We have used the following parameters for the techniques that were used in the research. This information is imperative when someone needs to reproduce or verify a work similar to our research.

### 3.3.1   Parameters to LLE Based SMOTE

- Nearest Neighbor Span: 5
- Percentage of Instance Generation (in 1 iteration): 100
- Random Seed: Integer valued null
- Validation Sampling: 10-fold cross validation (90% training samples and 10% testing sample for each fold and cycled through ten folds for comprehensive performance)

### 3.3.2   Parameters to Multilayer Perceptron

- Number of hidden layers: 2
- Learning rate: 0.3
- Momentum (applied to weights while updating) : 0.2
- Number of training epochs: 500
- Validation threshold: 20 misclassifications per fold
- Validation Sampling: 10-fold cross validation as in previous section 3.3.1

## 4   Evaluation and Deployment

For evaluation purposes, we use the dataset that have been pre-processed (detailed in subsection 3.2). We classify the dataset into the three classes using Multilayer Perceptron (in all three models). The first model contains the dataset with all records from the dataset but Models 2 and 3 are balanced datasets with equal number of instances in each class. However in model 3 we use LLE based SMOTE to raise the minority classes' frequencies to equal the majority class' frequency.

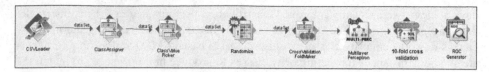

**Fig. 3.** Weka knowledge flow diagram for classification

## 4.1    Model 1 (All Instances from Preprocessed Dataset in Classification)

Since we use the originally preprocessed dataset without adding or deleting any record, we work with an unbalanced dataset in this model. We use multilayer perceptron as our second classifier using the parameters mentioned in subsection 3.3.2. For class variable 'Criticality', the Confusion Matrix for Model 1 using multilayer perceptron is given in Table 1.

**Table 1.** Confusion Matrix for Model 1 using Multilayer Perceptron

| Actual | Classified | | |
|---|---|---|---|
| | High | Mid | Low |
| High | 91 | 179 | 942 |
| Mid | 43 | 1124 | 1372 |
| Low | 56 | 194 | 21260 |

From the table we find the performance metrics that is reported in subsection 4.4.

## 4.2    Model 2 (Balanced Dataset Using Technique of Rahman and Hasan [2])

In this model, we trim our majority classes down to the size of our minority class which is "High" (Figure 2). This data modeling is similar to Rahman and Hasan's second model because we want to compare our models with the previous works. The following data processing steps are followed (Rahman and Hasan [2]):

Step 1: From frequency distribution table it is seen that instance of class "Low" has 21,510, "Mid" has 2,549 and High has 1,202 records. Percentage-wise "Low" = 85.2%, Mid = 10.1% and High = 4.8%. Discussing with domain experts, it was realized that this percentage is acceptable. Among the admitted patients this ratio Low : Mid : High = 17 : 2 : 1 represents true picture. For our research purpose, this ratio is not significantly relevant, hence, we concentrate on raising the overall accuracy of our system with uniform accuracy distribution across all classes. Thus we use random selection to select 1,202 records from "Mid" and "Low" classes, same as the instance count of class "High" to balance the surveillance dataset.

Step 2: Frequency Distribution for all variables are taken. Depth of the data is now reduced to 3606 records, with equal frequency of all classes. We observe:

(i) The variability of the values of the field "chemothy" is very high. Hence, that field is removed.

(ii) Since the number of records has been dropped down, distribution of the variable have changed drastically.

So we check for the Information Gain of the variables again. Finally, we take Cross-tabulation of "criticality" with the remaining variables. We use multilayer perceptron as our classifier using the parameters mentioned in subsection 3.3.2. The Confusion Matrix for Model 2 using multilayer perceptron  is given in Table 2.

Table 2. Confusion Matrix for Model 2 using Multilayer Perceptron

| Actual | Classified | | |
|---|---|---|---|
| | High | Mid | Low |
| High | 704 | 313 | 185 |
| Mid | 289 | 713 | 200 |
| Low | 174 | 362 | 666 |

### 4.3    Model 3 (Balanced Dataset Using LLE Based SMOTE)

We use the novel technique of LLE based SMOTE that has not been introduced in classification of hospital surveillance data till date in this model. Using this algorithm, we raise the minority classes' instances ("High" and "Mid") to the number of instances of that of the majority class i.e. "Low". We use multilayer perceptron as our classifier using the parameters mentioned in subsection 3.3.2. For class variable 'Criticality', the Confusion Matrix for Model 3 using multilayer perceptron is given in Table 3.

Table 3. Confusion Matrix for Model 3 using Multilayer Perceptron

| Actual | Classified | | |
|---|---|---|---|
| | High | Mid | Low |
| High | 19737 | 775 | 998 |
| Mid | 146 | 19862 | 1502 |
| Low | 981 | 570 | 19959 |

### 4.4    A Comparative Analysis of the Models Developed and the Previous Models of Rahman and Hasan [2]

In Table 4, we summarize the performance metrics of our own models and the models by Rahman and Hasan [2]. We note that our Model 1 and 2 can be compared with [2] since the data modeling are same; however our Model 3 is completely novel and therefore, no comparison is relevant for Model 3 with previous works. The different performance metrics across different classifiers and models are tabulated in Table 4. From this analysis we observe that our Model 3 (using LLE based SMOTE to balance the dataset) exhibits the most improved performance in classification and also in terms of convergence efficiency.

**Table 4.** A comparative analysis of the models developed and the previous models of Rahman and Hasan [2]

|  | Multilayer Perceptron | Decision Trees (Rahman and Hasan [2]) | Multilayer Perceptron | Decision Trees (Rahman and Hasan [2]) | Multilayer Perceptron |
| --- | --- | --- | --- | --- | --- |
| AC | 88.97% | 85.43% | 57.76% | 76.18% | 92.30% |
| TP(HIGH) | 7.51% | 9.82% | 58.57% | 56.57% | 91.76% |
| TP(MID) | 44.27% | 0.12% | 59.32% | 78.54% | 92.34% |
| TP(LOW) | 98.84% | 99.76% | 55.41% | 93.43% | 92.79% |
| FP(HIGH) | 92.49% | 90.18% | 41.43% | 43.43% | 8.24% |
| FP(MID) | 55.73% | 99.88% | 40.68% | 21.46% | 7.66% |
| FP(LOW) | 1.16% | 0.24% | 44.59% | 6.57% | 7.21% |
| P(HIGH) | 47.89% | 47.97% | 60.33% | 93.92% | 94.60% |
| P(MID) | 75.08% | 60.00% | 51.37% | 66.25% | 93.66% |
| P(LOW) | 90.18% | 85.80% | 63.37% | 77.08% | 88.87% |

In Table 4, AC stands for accuracy which is the proportion of the total number of predictions that were correct; the recall or true positive rate (TP) is the proportion of positive cases that were correctly identified; the false positive rate (FP) is the proportion of negatives cases that were incorrectly classified as positive; Precision (P) is the proportion of the predicted positive cases that were correct. More details of this metric could be found in [8].

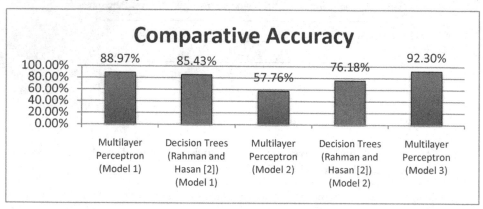

**Fig. 4.** Comparison of prediction accuracy and analysis between our models vs. older models in [2]

# 5    Conclusion and Future Work

Forecasting emerging epidemics is the primary purpose of data mining strategies on Hospital Surveillance data. Upon deployment of a system like this, precautionary measures can be taken to reduce the level of damage caused by an epidemic. Traditionally time series analysis or cluster analysis has been used in this arena to recognize any cyclic patterns in epidemic waves and to generate regression formula which can be used for forecasting. In this paper, we have used Rahman and Hasan's [2] decision tree models as foundation to introduce more modern techniques in this paradigm. Moreover, we present a novel approach of balancing unbalanced data of medical data repositories which showed better performance. Upon presenting our work, we conclude that, since classification performance has reached an optimal point, further research can be undertaken to optimize the runtime performance of the algorithms.

# References

1. International Centre for Diarrhoeal Disease Research, Bangladesh: http://www.icddrb.org/ org/orgunits.jsp?idDetails=103&searchID=103. (retrieved May 20, 2014)
2. Rahman, R.M., Hasan, F.R.M.: Using and comparing different decision tree classification techniques for mining ICDDR, B Hospital Surveillance data. Expert Syst. Appl. **38**(9), 11421–11436 (2011)
3. Wang, J., Xu, M., Wang, H., Zhang, J.: Classification of imbalanced data by using the SMOTE algorithm and locally linear embedding. In: 8th International Conference on Signal Processing, vol. 3 (2006). doi:10.1109/ICOSP.2006.345752
4. Haykin, S.: Neural Networks: A Comprehensive Foundation (2nd ed.). Prentice Hall (1998)
5. Pang, S., Kim, D., Bang, S.Y.: Face membership authentication using SVM classification tree generated by membership-based LLE data partition. IEEE Transactions on Neural Networks **16**(2), 436–446 (2005)
6. Chawla, N.V., Bowyer, K.W., Hall, L.O., Kegelmeyer, W.P.: SMOTE: Synthetic Minority Over-sampling Technique. J. Artif. Int. Res. **16**(1), 321–357 (2002)
7. Weka: The Data Mining Software in Java: http://www.cs.waikato.ac.nz/ml/weka/. (retrieved on May 20, 2014)
8. Tan, P.N., Steinbach, M., Kumar, V.: Introduction to Data Mining, 1st edn. Addison-Wesley (2006)

# Potentials of Hyper Populated Ant Colonies

Andrzej Siemiński[✉]

Faculty of Computer Science and Management,
Technical University of Wrocław, Wrocław, Poland
Andrzej.Sieminski@pwr.edu.pl

**Abstract.** The paper discusses the potentials of Hyper Populated Ant Colonies (HPAC) using the well-known Travelling Salesman Problem (TSP) as the study area. The paper starts with an examination of the simple static version of the TSP. The obtained results are later applied to its dynamic version. The carried out experiments strongly suggest that the TSP performance improves significantly with the increase of the Ant Colony size. The phenomena is especially noticeable for dynamic environments. Moreover the processing time does not necessary grow longer. The increasing size of ant colony could be compensated by the decreasing number of iterations. Both the theoretical analysis and initial experiments show that the processing time could be further reduced by the introducing parallelism. The programming technique used is the RMI - Remote Method Invocation.

## 1    Introduction

The aim of the paper is to discuss the problem of optimizing the performance of the Ant Colony Optimization (ACO) used for the Travelling Salesman Problem (TSP). Both the static and dynamic TSP environments are considered.

The operation of the ACO is controlled by a number of parameters with values that are chosen in an experimental manner. This is due to the complex interaction of individual ants which makes it extremely hard to find an analytical solution. A detailed analysis of the ACO for traditional static graphs has revealed that the most influential parameter is the size of its population. The paper studies the usefulness of the so called Hyper Populated Ant Colonies. These are colonies with ants population substantially exceeding usually used sizes. Conducted experiments clearly indicate that such colonies are especially useful for dynamic environments. As shown in the paper increasing the number of ants is not necessary result in the increasing of the processing time.

The paper is organized as follows. The next two sections briefly introduce the Travelling Salesman Problem and the Ant Colony Optimization respectively. The emphasis is laid on the stopping problem. The 4th Section introduces the concept of Hyper Populated Ant Colonies (HPAS) and reports experiments that prove their usefulness for both static and dynamic TSP environments. The Section 5 discusses the efficiency of an implementation of parallel HPAS using Java Remote Method Invocation mechanism. Plans for further work conclude the paper.

© Springer International Publishing Switzerland 2015
N.T. Nguyen et al. (Eds.): ACIIDS 2015, Part I, LNAI 9011, pp. 408–417, 2015.
DOI: 10.1007/978-3-319-15702-3_40

## 2    Travelling Salesman Problem

The TSP could be stated in a remarkably simple way: given a list of cities and the distances separating them what is the shortest possible route that visits each city exactly once and returns to the origin city? It is one of the classical problems of Artificial Intelligence.

In the original specification of the TSP the distances between nodes are symmetric and they do not change. The road structure remains relatively stable so such an assumption seems to be justified. Minimizing the distance covered by a salesman is not always the priority number one. One has often more interest in  the travelling time which is subject to the ever changing road conditions and therefore it is inherently dynamic. In what follows we will continue to use the traditionally used term distance but it may refer to travel time as well. Such an approach calls for dynamic graphs with distances changing permanently.

## 3    Ant Colony Optimization

The Ant Colony Optimization is an example of the Swarm Intelligence concept which utilizes the biological term stigmergy that is used to explain in what way insects are capable of building and of operating complex mounds.

The proposed solution is simple: the trace left the environment by an action stimulates the performance of the next action, by the same or a different ant. The ACO technique was introduced by M. Dorigo in 1992 [1] to find high quality solutions for the TSP. Since then it diversified to solve a wider class of numerical problems and is one of the most popular probabilistic techniques. An extensive account of the state of art of ACO is presented in [2].

In order to find a solution the ACO works in an iterative manner. In every iteration its ants are randomly placed on the graph and each ant tries to complete a journey through all cities taking into account the distance matrix and levels of pheromone laid by all the ants. The process of city selection is time consuming as it involves a great number of floating point operations.

### 3.1    Parameter Optimization

The operation of the ACO is controlled by 4  parameters [3]: probability of selecting exploitation over exploration mode of operation, aging factor used in the global updating rule, aging factor in the local updating rule and moderating factor for the cost measure function.

Their recommended values of the parameters are identified in an experimental way. Their recommended values have round values like 0.8 or 2.0. This gives rise to an assumption that they are not optimal. Various attempts to identify values offering better results were described e.g. in [4-6]. Parameter values were optimized by algorithms inspired by Evolutionally Programming (EP) [7], Simulated Annealing (SA) [8] and a statistical analysis of a large collection of gathered results.

The results are not quite satisfactory. It was possible to find values better than those recommended but it was not possible to correlate the them to the properties of the environment. In each particular case has to be treaded individually and the optimization of parameters values requires time consuming experiments.

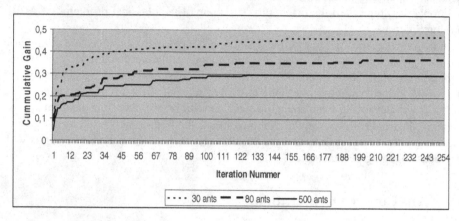

**Fig. 1.** Cumulative gain (1) achieved by Ant Colonies with sizes from 30 to 500 for the first 256 iterations, static environment, 50 nodes, 30 runs

## 3.2     Stopping Problem

No matter what the parameter values are used the processing has to be stopped at same point of time. When to stop is a question rarely raised in the discussion of the ACO. The frequency of changes of the BSF (Best So Far) route lessens as the number of iteration increases. This means that the computational effort needed to execute subsequent iterations is less and less profitable. This is clearly visible on the Fig. 1 which shows the cumulative gain for consecutive iterations for ant colonies with different sizes.

The cumulative gain is defined as follows:

$$CG(i) = (\sum\nolimits_{r=0}^{R} (len(r,0) - len(r,i)))/R \tag{1}$$

where:
R – number of runs
i – iteration number
len(r, i) – the BSF route length for the iteration i in the test run number r

The stopping problem and the lowering of the number of used iterations is especially important for the dynamic environments where the pace of changes may diminish the chances of finding a good quality solution.

# 4    Hyper Populated Ant Colonies

Many experiments have proven that the standard values of parameters provide a stable and reasonable good performance over a wide spectrum of input data. The recommended number of ants is equal to the number of nodes. For that reason the solution proposed in the paper keeps their standard values and studies the impact that the changes in the number of ants have on the performance of the ACO. To our knowledge such an approach was not examined extensively before. Increasing of the number of ants extends the processing time and this is certainly not a feature welcome in the field of computer science The Fig. 1 suggests however that with increasing the population size decreases the number of iterations that are necessary to obtain high quality results. We have therefore a trade-off between number of ants and number of iterations. Therefore we propose employing colonies with the number of ants that substantially exceeds the recommended numbers. They are called the so called HPAS - the Hyper Populated Ant Colonies.

## 4.1    Employing the HPAS for Static Graphs

The standard data set for the traditional, static version of the TSP is a symmetric matrix with random values in the range from 0.0 to 1.0. The values represent the route lengths between the nodes. Their usual number is 50 and this value is used throughout the paper for both static and dynamic graphs. The value of N may not look impressive but the number of all possible solutions is equal to approximately to 3,04141E+64 whereas the mass of an observable steady-state universe is 1,45 E+53 kg . For static graphs the performance measure is the BSF – the best so far solution found by a colony.

Studding he performance of the standard algorithm in a static graphs is a prerequisite for proposing solutions for the more complex, dynamic graph. The Table 1 shows average values of the BSF for different sizes of colony and varying number of iterations. The number of iterations is shown in square brackets. The iteration numbers in the BSF normalized column ware selected in such a way as to preserve the same the computational complexity of each raw.

**Table 1.** Observed average BSF route for varying colony populations, fixed and normalized number of iterations, number of runs 30

| Ant Colony Size | BSF fixed size | BSF normalized |
|---|---|---|
| 30 | 1.82 [1000] | 1.80 [1666] |
| 50 | 1.81 [1000] | 1.81 [1000] |
| 80 | 1.80 [1000] | 1.80 [625] |
| 120 | 1.80 [1000] | 1.85 [417] |
| 150 | 1.79 [1000] | 1.80 [333] |
| 1000 | 1.78 [1000] | 1.78 [50] |

The over populating of an ant colony does not offer a significant reduction in route length but it drastically reduces the number of iterations that are necessary to obtain a good quality result. This calls for a parallel implementation of the HPAC which is discussed in Sec. 5.

In all experiments we have used the Java implementation of ACO make available to the research community by Chirocco [3].

## 4.2     HP Ant Colonies for Dynamic Graphs

There is not a standard set of data for dynamic graphs. Therefore the comparison of different proposed algorithms requires their implementation what is a tedious task and does not guarantee reliable results.

### 4.2.1  Related Work

The schemas for graph modification that were considered in the previous studies consisted usually of a sequence of node insertions and deletions performed at certain intervals. In [5] another approach was taken. The distances change in a more complex way. The graphs are replaced by graph generators using Markov sources controlled by a number of parameters. The approach looks promising but it has one deficiency: the graphs did not guarantee keeping constant the average distance length. This in itself could happen in practice but it makes it difficult to evaluate the performance of tested algorithms and to relate it to the performance of a static environment. The graph generators used in the paper ware described in [12] and they eliminate this deficiency.

The first attempts to adopt the standard operation of ACO to the dynamic case involved global and local reset strategies [9] that is changing the accumulated pheromone levels to some default value once a graph modification has occurred. The Global Reset starts the optimization process once more again and is used mainly for reference purpose. Local reset changes only the pheromone levels for modified segments of graph. It enables the Colony to exploit at least part of data gathered so far but requires a precise data on where the change had occurred. Both of the reset strategies are less efficient or even entirely not useful when the distance changes are more subtle or occur constantly.

In an another approach the ant population diversity necessary to adopt to changing environment was achieved by the implementation of the so called immigrant schemes. There are three types of immigrants: randomly generated, elitism-based, and hybrid immigrants. The approach could use a long-term memory as in P-ACO [10] or a short-term memory as in a more recent paper [11]. The study have shown that different immigrants schemes perform well under different environmental conditions.

However all these approaches make two rather unrealistic assumptions:

1. The majority of distances do change in a single burst of graph modification activity. What is changed is the structure of the graph and not the distances.
2. It is assumed that the Ant Colony knows instantly where the changes have taken place and what is their scope. This breaches the very nature of the ACO. The colony consists of simple ants having very limited knowledge about their surroundings and there is no room for omniscient Colony Manager.

The approach proposed in the paper does not make such assumptions.

#### 4.2.2  Basic Properties of HP Ant Colonies in Dynamic Environment

In the paper the HP Colonies work with graph generators guaranteeing the stability of average distance lengths between nodes. Their detailed description and implementation could be found in [12]. Due to the feature it is now possible to compare results achieved by different colonies and relate them to the results obtained for a static graph. In the paper we work on two graph generators: CCc in which every tenth distance is subject to a random change and DCd in which most distances change but the range of changes is damped by the 0.5 factor. The changes are introduce after each iteration.

The conducted experiments consisted of 10 test runs for each generator with the number of ants ranging from 30 to 1000. Due to the dynamic nature of the graphs instead of the BSF the length of IB (Iteration Best) route is used as a performance measure. The results are shown in the Table 2.

**Table 2.** The average IB for different number of ants, number of runs =10, number of iterations 1000

| Number of ants | IB for Graph Generator CCc | IB for Graph Generator DCd |
|---|---|---|
| 30 | 2.40 | 3.83 |
| 50 | 2.34 | 3.31 |
| 80 | 2.32 | 2.98 |
| 150 | 2.29 | 2.71 |
| 1000 | 2.17 | 2.44 |

As expected, impact of increasing the colony population is more apparent for the DCd graph. In the case of truly hyper populated Colony with 1000 ants the IB route length does nor diverge much from the performance of ant colony of standard size for the much less changeable generator CCc. Even much less numerous colony with just 150 ants has performs 22% better than the standard size population in the case of DCd generator.

## 5  Parallel Implementation of HP Ant Colonies

In the case of static graphs it was possible to offset the additional computational complexity resulting from the increased number of ants by limiting the number of iterations. This is not possible for the dynamic graphs where the performance is measured by the IB route. The natural way for reducing the processing time for HP Colonies is to introduce their parallel implementation.

## 5.1    Basic Properties of Parallel ACO

We assume that the colony is implemented by a number of node computers and one host computer that coordinates their work. The operation of a parallel colony is characterized by the following factors:

- n – the number of ants
- k – the number of node computers
- m – the number of chunks of data
- ti – time required to process one chunk of data by one ant
- td – time necessary to transmit data to and from the host and a node
- tc - time for the establishing the initial connection between the host and a node.

A parallel implementation reduces the processing time only if the number of node computers satisfies the inequity:

$$k > \frac{nt_i}{nt_i - t_d - \dfrac{t_c}{m}} \tag{2}$$

and

$$nt_i - t_d - \frac{t_c}{m} > 0 \tag{3}$$

For a continues mode of operation the value of $t_c/m$ is negligible. What is really required is that the local processing time on one node is less than the time necessary to transmit the necessary data.

## 5.2    Using RMI for HP Colonies

In what follows we consider the efficiency of implementing the parallel operation mode by using the Java Remote Method Invocation (RMI) mechanism. The RMI supports direct transfer of serialized Java objects over networks.

While implementing parallelism one has to decide upon the level of granularity. The fine-grained parallelization schemes have been investigated extensively studied e.g. in [13], [14] and the experimental results showed that communication overhead could diminish the reduction of the potential processing time. Therefore in the study we have used an extreme version of coarse grained parallelization scheme where there is no communication among the sub colonies and the server just selects the best solution. Obtained results are in line with those reported by Stuetzle and of Manfrin et al. [15].

In the study a local network was used and the average measured values of factors from the Section 5.1 are shown in the Table 3. The initialization time necessary for establishing the connection between a server and a client was stable and always shorter then 1.5 sec. In the processing phase each client performed 200 iterations on a matrix of 50 cities using 50 ants. The average processing time was just over 13 sec.

and the transmitting of a serialized version of distance and pheromone matrix took only 1.5 seconds. The time necessary for sending the results to the server was negligible.

**Table 3.** Average processing time of the factors from Sec. 5.1

| Code | Short description | remote Host |
|------|------------------|-------------|
| $t_c$ | Initialization | 1.3 s. |
| $t_i$ | Data Processing | 13.1 s. |
| $t_d$ (full) | Matrix of distances | 1.6 s. |
| $t_d$ (comp.) | Results | 0.01s. |

The time for transmitting a distance matrix constitutes only 12,5% of the time necessary to process the data chunk what offers a good incentive to distribute processing. There is however a difference between having one hyper populated Colony and a number of independently working smaller colonies. Processing independently such chunks of data eliminates the interaction between ants from different colonies. The lack of interaction does not effect much the route lengths, see the data presented in the Table 4. The combined number of ants in all servers was equal to 400 in each case. Note that the execution time for the last row was almost 8 times shorter than for the first row.

**Table 4.** Average BSF for different number of servers

| Number of servers | Number of ants | Avg. BSF | Std. Dev. | Min |
|-------------------|----------------|----------|-----------|--------|
| 1 | 400 | 1.8022 | 0.0101 | 1.7588 |
| 2 | 200 | 1.7911 | 0.0113 | 1.7335 |
| 4 | 100 | 1.7884 | 0.0074 | 1.7334 |
| 8 | 50 | 1.7815 | 0.0108 | 1.7335 |

The data clearly indicates that lack if interaction does not harm substantially the results. Choosing 4 servers seems to be a reasonable choice: the average BSF value is almost minimal and it has the lowest value of the standard deviation.

# 6    Conclusions

The paper studies the relationship between the increasing of the size of an Ant Colony population and its performance. It advocates using the so called HPAS Hyper-Populated Ant Colonies. These are colonies which use the far larger number of ants than usually advocated.

In the traditional, static environment a single colony the HPAS using a remarkably low number of iterations guaranties stable and good quality results. However no significant reduction of processing time could be achieved. For more complex, dynamic environments both the theoretical analysis and a serious of conducted experiments strongly suggest that parallel, coarse grained version of HPAS is capable of reducing both the route length and the processing time in comparison to a standard Ant Colony. During the experiments the parallelism was implemented using the mechanism of Remote Method Invocation.

The next step of research work is to we plan to study whether other ACO algorithms could also benefit from the concept of over population and parallel execution.

**Acknowledgment.** This work was supported by the European Commission under the 7th Framework Programme, Coordination and Support Action, Grant Agreement Number 316097, ENGINE - European research centre of Network intelliGence for INnovation Enhancement (http://engine.pwr.wroc.pl/).

# References

1. Dorigo, M.: Optimization, Learning and Natural Algorithms, PhD thesis, Politecnico di Milano, Italie, (1992)
2. Dorigo, M., Stuetzle, T.: Ant Colony Optimization: Overview and Recent Advances, IRIDIA - Technical Report Series, Technical Report No. TR/IRIDIA/2009-013, May 2009
3. Chirico, U.: A Java Framework for Ant Colony Systems, Ants2004: Forth International Workshop on Ant Colony Optimization and Swarm Intelligence, Brussels (2004)
4. Siemiński, A.: TSP/ACO Parameter Optimization; Information Systems Architecture and Technology; System Analysis Approach to the Design, Control and Decision Support; pp. 151–161; Oficyna Wydawnicza Politechniki Wrocławskiej 2011
5. Siemiński, A.: Ant colony optimization parameter evaluation. In: Zgrzywa, A., Choroś, K., Siemiński, A. (eds.) Multimedia and Internet Systems: Theory and Practice. AISC, vol. 183, pp. 143–153. Springer, Heidelberg (2013)
6. Gaertner, D., Clark, K.L.: On Optimal Parameters for Ant Colony Optimization Algorithms. In: IC-AI pp. 83–89, June 2005
7. Eiben, A.E., Smith, J.E.: Introduction to Evolutionary Computing. Springer (2003)
8. Busetti, F.: Simulated Annealing Overview, Report (2003)
9. Guntsch, M., Middendorf, M.: Pheromone modification strategies for ant algorithms applied to dynamic TSP. In: Boers, E.J.W., Gottlieb, J., Lanzi, P.L., Smith, R.E., Cagnoni, S., Hart, E., Raidl, G.R., Tijink, H. (eds.) EvoIASP 2001, EvoWorkshops 2001, EvoFlight 2001, EvoSTIM 2001, EvoCOP 2001, and EvoLearn 2001. LNCS, vol. 2037, pp. 213–222. Springer, Heidelberg (2001)
10. Guntsch, M., Middendorf, M.: A population based approach for ACO. In: Cagnoni, S., Gottlieb, J., Hart, E., Middendorf, M., Raidl, G.R. (eds.) EvoIASP 2002, EvoWorkshops 2002, EvoSTIM 2002, EvoCOP 2002, and EvoPlan 2002. LNCS, vol. 2279, pp. 72–81. Springer, Heidelberg (2002)
11. Mavrovouniotis, M., Yang, S.: Ant colony optimization with immigrants schemes in dynamic environments. In: Schaefer, R., Cotta, C., Kołodziej, J., Rudolph, G. (eds.) PPSN XI. LNCS, vol. 6239, pp. 371–380. Springer, Heidelberg (2010)
12. Siemiński, A.: Using ACS for dynamic traveling salesman problem. In: Zgrzywa, A., Choroś, K., Siemiński, A. (eds.) New Research in Multimedia and Internet Systems. AISC, vol. 314, pp. 145–155. Springer, Heidelberg (2015)

13. Yu, B., Yang, Z.-Z., Xie, J.-X.: A parallel improved ant colony optimization for multi-depot vehicle routing problem. Journal of the Operational Research Society **62**, 183–188 (2011)
14. Doerner, K.F., Hartl, R.F., Benkner, S., Lucka, M.: Parallel cooperative saving based ant colony optimization - multiple search and decomposition approaches. Parallel Processing Letters **16**(3), 351–369 (2006)
15. Manfrin, M., Birattari, M., Stützle, T., Dorigo, M.: Parallel ant colony optimization for the traveling salesman problem. In: Dorigo, M., Gambardella, L.M., Birattari, M., Martinoli, A., Poli, R., Stützle, T. (eds.) ANTS 2006. LNCS, vol. 4150, pp. 224–234. Springer, Heidelberg (2006)

# The Selected Metaheuristics Efficacy Assessment for a Given Class of Problems

Urszula Markowska-Kaczmar[(✉)] and Adam Mróz

Wroclaw University of Technology, Wyb. Wyspianskiego 27, 50-370 Wroclaw, Poland
urszula.markowska-kaczmar@pwr.edu.pl

**Abstract.** The paper presents an efficacy comparison of selected meta-heuristics: a cuckoo algorithm, firefly algorithm, bat algorithm, invasive weed optimization and group search optimizer, genetic algorithm, bee algorithm and particle swarm optimization. All tests were conducted in homogeneous environment, specially designed and implemented for this purpose. Research was done using well known benchmark functions: De Jong, Rosenbrock and Ackley for various number of dimensions. The metaheuristics were compared with each other using Friedman test and post hoc analysis. The experimental result analysis shows that there is no statistical evidence to select the best one.

**Keywords:** Metaheuristics · Optimization · Statistical analysis

## 1 Introduction

In recent years one can observe the popularity rise of nature inspired metaheristics. As stochastic optimization methods they are an option for deterministic algorithms. They allow to find optimal solutions, although they do not guarantee to find the global ones. Searching the shortest path, job scheduling or device design with the best parameters are their popular applications. Their advantage lies in relatively quick time of searching a satisfying solution. Therefore they are frequently used in case when deterministic algorithms fail. It refers to very computationally complex problems.

Unfortunately, despite the large number of new metaheuristics developed last time, there is no objective comparison of their efficacy. Most of authors compare their proposal to older, well known methods. One of the obstacle can be the absence of a framework with implemented metaheuristics creating a homogenous research platform. Therefore, we first designed and implemented such environment and then it was used to achieve the aim of our research, i.e. comparison of efficacy of selected metaheuristcs in the same environment on the same benchmark problems. We chose to compare the following new metaheuristics: Firefly Algorithm (FA), Cuckoo Search (CS), Invasive Weed Optimisation (IWO), Group Search Optimizer (GSO) with better known ones: Genetic Algorithm (GA), Particle Swarm Optimization (PSO) and Bee Algorithm (BA).

© Springer International Publishing Switzerland 2015
N.T. Nguyen et al. (Eds.): ACIIDS 2015, Part I, LNAI 9011, pp. 418–429, 2015.
DOI: 10.1007/978-3-319-15702-3_41

The paper consists of 5 sections. In section 2 the details of the above methods are presented. Section 3 shortly presents the experiments performed in the testbed application designed and implemented for this experimental study. The last section contains conclusions and summary of the research.

## 2    The Tested Methods

In this section, we will describe the metaheuristics that were used in the comparison. Their aim is to solve optimization problems. The last decades have witnessed an increasing emphasis on the studies about nature-inspired methods. Each metaheuristics has its own representation of an individual and the way in which exploitation and exploration is made. Exploration gives the possibility of global search while exploitation performs local search. They exist in each metaheusitics but they have various form depending on an algorithm.

*Genetic algorithm.* A genetic algorithm is the oldest metaheuristics considered in our research. It mimics some processes of natural evolution [3]. A population is created by a set of individuals represented by chromosomes with encoded solutions. At the beginning the population is created randomly and each individual is evaluated. Evaluation function gives a score for each individual. It reflects how well performs a given individual. The higher the fitness, the higher the individual chance of being selected for reproduction to create one or more offsprings. Then, with the assumed probability, a crossover operation is executed. Next, the offsprings are mutated with the assumed probability. The process continues until a suitable solution has been found or a certain number of generations have passed. The method in the form of pseudocode is presented in Alg. 1.

*Particle Swarm Optimization.* Particle swarm optimization (PSO) starts its performance from creation of a random population of particles. PSO has lot common ideas with GA, like random initial population and searching optimum in subsequent generations, but the main difference is lack of genetic operators and

---

**Algorithm 1.** Pseudocode of genetic algorithm

---

Create an initial population (a set of individuals);
**repeat**
    Evaluate every individual in the current population (calculate its fitness values);
    Evaluate the stop criterion;
    **if** *the stop criterion not satisfied* **then**
        Select individuals for reproduction using assumed selection method;
        Reproduce individuals using crossover operator with probability $p_{xover}$ and mutation operator with probability $p_{mutation}$;
    **end**
**until** *stop criterion satisfied*;

---

---

**Algorithm 2.** Particle Swarm Optimization in pseudocode

Create an initial population of particles ;
**while** *the stop condition unsatisfied* **do**
> **for** *each particle* **do**
> > calculate its velocity ;
> > calculate a new position;
> > **if** *if the new position is better than the current pbest* **then**
> > > actualize *pbest*;
> >
> > **end**
> > **if** *the new position is better than current gbest* **then**
> > > actualize *gbest*;
> >
> > **end**
>
> **end**

**end**

---

particle memory [4]. A particle represents a solution and it is expressed by a vector $X_i^k[x_{i1}^k; x_{i2}^k, ..., x_{in}^k]$ in $n$ dimensional space. The vector defines a position of a particle. The evaluation of individual is made based on fitness function $f(X)$. The velocity $V_i^k[v_{i1}^k; v_{i2}^k, ..., v_{in}^k]$ is also associated with the particle. A particle knows its best position $P_{best}$ in the past and the best solution $G_{best}$ in its neighbourhood. The eq. 1 and 2 describe the way the velocity and position [6] are changed in each iteration:

$$V_i^{k+1} = wV_i^k + c_1 r_1 (P_{best} - X_i^k) + c_2 r_2 (G_{best} - X_i^k) \tag{1}$$

$$X_i^{k+1} = X_i^k + V_i^{k+1} \tag{2}$$

$V_i^k$ defines velocity of the particle $i$ in time $k$, $X_i^k$ means its position at the same time, $r_1$ i $r_2$ are random numbers from the range $\langle 0, 1 \rangle$, coefficients $w$, $c_1$, $c_2$ are defined by a user. The parametr $c_1$ defines how much the particle is attracted by ($P_{best}$). The parametr $c_2$ refers to the attraction by ($G_{best}$). The parametr $w$ describes the particle inertia. The PSO pseudocode is presented in Alg. 2.

*Bees algorithm.* It was first proposed by Pham and others in 2005 and described in the technical report [5]. Similarly to other described metaheuristics, bees algorithm starts with random population of encoded solutions that in this case are called bees. Bee $i$ in the timestep $k$ is represented by a vector $X_i^k = [x_{i1}^k, x_{i2}^k, \ldots, x_{in}^k]$ in $n$ dimensional space. The most promising $m$ bees are chosen on the basis of the fitness function $f(X)$. Next, bees are recruited to these $m$ places in the neighbourhood radius ($r$). Among $m$ solutions (bees) $e$ elite bees are searched. They recruit the most bees. Next, $m$ bees with the highest fitness value is chosen. The rest of bees are placed randomly in the solution space. The idea of the bees algorithm is presented in Alg. 3.

---

**Algorithm 3.** Bee algorithm in pseudocode

---

Initialize random bee population;
**while** *the stop condition is not satisfied* **do**
  | evaluate the population;
  | choose $m$ places (bees) to search a neighbourhood ;
  | Recruit bees to the chosen places (assign more bees to the elitesolutions $e$);
  | Choose the best solution from each of the $m$ places. Choose the best bee
  | among them. The rest of the bees will perform random search;
**end**

---

*Cuckoo Search.* This algorithm was proposed by Yang in 2009 [10]. Its pseudocode is shown in Alg. 4. The algorithm starts with random population of host nests representing a solution. Each cuckoo lays one egg (a solution) at a time, and dumps its egg in a randomly chosen nest. The new solution replaces the old one if its fitness value is greater than the old one. A fraction of the worse nets are abandoned and the new ones are created. The best nests with high quality of solutions will carry over to the next generation. A solution $k$ is represented by the vector $X_i^k = [x_{i1}^k, x_{i2}^k, \ldots, x_{in}^k]$ and evaluated by the fitness function $f(X)$. A new solution is generated according to the Lévy's distribution (eq. 3).

$$X_u^{k+1} = X_i^k + \alpha Levy(\beta) \tag{3}$$

where: $\alpha > 0$ is the scale of the step length and $Levy(\beta)$ is a vector of random numbers from the stable Lévy distribution. In the consecutive iterations a new location for eggs is searched according to the Lévy distribution. Next, eggs are overthrown with the probability $p_a$ and new nests are generated. The procedure is repeated until stop condition is met.

*Firefly Algorithm.* The firefly algorithm (FA) was inspired by the flashing behaviour of fireflies. It was proposed by Yang in 2008 [9]. Each firefly at the

---

**Algorithm 4.** Cuckoo search algorithm in psedocode [9]

---

Initialize random population of $n$ nestles;
**while** *the stop condition is not achieved* **do**
  | Choose $(i)$ cuckoo randomly according to the distribution and calculate its
  | fitness ;
  | Choose $(j)$ nest randomly and evaluate it;
  | **if** *the new solution has a higher fitness value than j one* **then**
  |   | change the nest $j$ for $i$;
  | **end**
  | Search worst nests and andreplace by new ones with probability $p_a$;
  | Save nests with the best solutions ;
  | evaluate solutions and choose the best ones ;
**end**

---

---

**Algorithm 5.** Firefly Algorithm in pseudocode

---

initialize random population of fireflies;
**while** *the stop condition is not met* **do**
  **for** *each firefly (I)* **do**
    **for** *each firefly (J)* **do**
      **if** *brightness of firefly J is greater than firefly I* **then**
      | move firefly *I* towards firefly *J*
      **end**
    **end**
    **if** *there is no brighter firefly than firefly I* **then**
    | move the firefly *I* at random direction
    **end**
  **end**
**end**

---

timestep $k$ is represented by the vector $X_i^k = \left[x_{i1}^k, x_{i2}^k \ldots x_{in}^k\right]$ that corresponds to the solution. The firefly's flash acts as a signal system to attract other fireflies. The investigations have shown that the brighter is the firefly, the more courtship success it has. More bright firefly attracts less bright ones. Algorithm 5 shows the FA algorithm in the pseudocode.

The main parameter of the metaheuristics is the light absorption ($\gamma$). It defines the firefly attractiveness. Usually, the parameter takes values from the range $\gamma = \langle 0, 1, 10 \rangle$.

Flashing corresponds to the fitness function. It decreases with the distance. The firefly attractiveness depends on its flashing. It is expressed by eq. 4.

$$\beta = \beta_0 e^{-\gamma r^2} \tag{4}$$

where: $\beta_0$ is the attractiveness in the distance $r = 0$, $\gamma$ is the light absorption, $r$ is the euclidian distance among the fireflies.

Movement of the firefly $i$ towards the firefly $j$ is defined by eq. 5.

$$X_i = X_i + \beta_0 e^{-\gamma r_{i,j}^2}(X_j - X_i) + \alpha \mathcal{E}_i \tag{5}$$

In equation 5 $\mathcal{E}_i$ is a vector of random numbers from the range $\langle -0.5, 0.5 \rangle$, $\alpha$ is a randomness coefficient from the range $\langle 0, 1 \rangle$.

*Invasive Weed Optimisation.* The algorithm was proposed by Mehrabian and Lucas in 2006 [2]. Its pseudocode is shown in Alg.6. Weed colony is represented by a solution table. In the initial phase of the method, weeds are randomly generated and then evaluated. Weeds spread the number of seeds depending on their fitness value and the smallest and the highest fitness value in the population. The relation is linear.

New seeds are spread in $d$ dimensional space randomly with normal distribution characterized by the average equal to 0 and a variable variance. It results

---

**Algorithm 6.** Weed optimization algorithm in pseudocode

---

Initialize random population of weeds;
**while** *the stop condition is not satisfied* **do**
   |   spread the seeds ;
   |   **if** *the max size of population is achieved* **then**
   |    |   eliminate the weakest individuals ;
   |   **end**
**end**

---

in spreading weeds around the parents. The variance of normal distribution is changed according to eq. 6.

$$\sigma_{iter} = \frac{(iter_{max} - iter)^n}{(iter_{max})^n}(\sigma_{initial} - \sigma_{final}) + \sigma_{final} \tag{6}$$

In eq. 6 $iter_{max}$ is the max number of iterations, $iter$ is the current iteration, $\sigma_{max}$ and $\sigma_{min}$ are respectively the max and min standard deviation, $\sigma_{iter}$ is the standard deviation in the current iteration, $n$ is a parameter being a real number. The whole weeds population is evaluated in each iteration.

*Group Search Optimizer.* In GSO a population, called a group, consists of three kinds of individuals: producers, scroungers and dispersed members who perform random walk motions. It is worth noticing that the members do not differ in their genotype. In $k$ iteration the $i$ member knows its position $X_i^k = [x_{i1}^k, x_{i2}^k, \ldots, x_{id}^k]$, angle of the head $\varphi_i^k = [\varphi_{i1}^k, \varphi_{i2}^k, \ldots, \varphi_{i(d-1)}^k]$ and search direction of the head $D_i^k(\varphi_i^k) = [d_{i1}^k, d_{i2}^k, \ldots, d_{id}^k]$, that can be calculated on the basis of the polar to Cartesian coordinate transformation. The metaheuristcs was proposed by He, Wu i Saunders in 2006 [1]. Its pseudocode is described in Alg.7. The best member of the population in the current population becomes the producer. It is assumed ([1]), that there is only one producer in a given iteration. Scroungers join the resource found by the producer. The producer and the scroungers do not differ in their relevant phenotypic characteristics. To enable producer searching resources (optima) it is equipped with visual recognition ability. If after scanning, the producer has found better position, it moves to it. In the opposite case it rotates its head. Next, the rest of the population moves. A part of members follows the producer, the others make a random walk. In the current iteration the member reaching the best solution becomes the producer. This procedure lasts until stop condition. The details of the algorithm can be found in [1].

## 3   Comparison of Considered Metaheuristics in Terms of Embedded Basic Mechanisms

On the basis of the description presented in the previous section we can distinguish their common features: they are based on a population (the size of

---

**Algorithm 7.** Group Search Optimizer algorithm in the pseudocode

Intialize random population (position and angle of head for each individual);
**while** *the stop condition is not satisfied* **do**
| Calculate the fitness function values;
| Choose the producer;
| Search for the best individual in the environment –assign the new producer
| (or rotate the head of the old one );
| Choose randomly the members that will be scroungers and move them
| towards the producer;
| Disperse other members performing random walking;
**end**

---

population is the parameter of each algorithm), an initial population is assigned randomly, an individual is represented by a vector encoding a solution, local and global random search is performed.

The main discriminants of the metaheuristics, apart from biological inspirations, are shown in Table 1. They include the way in which an individual is represented, but above all, the way of performing exploitation and exploration.

## 4    Comparison of the Metaheristics Efficacy – Experimental Study

In the first experiments we deeply investigated sensitivity of each metaheuristics to the parameter values. On this basis we assigned the most promising ones and used in this experiment. Because of the limited place the description of the parameter value adjusting experiments are skipped here. In this experiment we will focus on the efficacy comparison of metaheuristics. Efficacy is measured by the number of iterations necessary to reach a global optimum.

**Table 1.** Comparison of basic mechanisms embedded in the considered metaheristics

| Alg. | Individual representation | Exploration | Exploitation |
|------|---------------------------|-------------|--------------|
| GA | vector of genes (binary or real numbers ) | Crossover | Mutation |
| PSO | position and velocity of a particle – vectors of real numbers | movement towards $g_{best}$ | movement towards $p_{best}$ |
| BEE | vector of bee position (real numbers) | random search of a part of bees | recruitment bees to the best individuals (elite bees) |
| GSO | position and angle of a head expressed by vectors of real numbers | movement of scroungers towards the producer and random walk motion of dispersed members | neighbourhood tracking by the producer |
| IWO | vector of the weed position (real numbers) | spreading seeds (even by the weak individuals) | spreading weeds – decreasing the radius of spreading |
| FA | vector of the firefly position (real numbers) | firefly attraction by the better individuals | random movement of the best firefly |
| CS | vector of the cuckoo position | application of Lévy distribution | application of Lévy distribution |

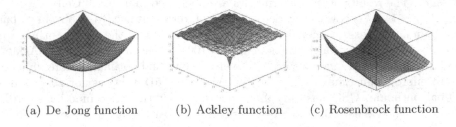

(a) De Jong function     (b) Ackley function     (c) Rosenbrock function

**Fig. 1.** Visualization of the test functions

*Course of the experiments.* The assumed values of the parameters for each meta-heuristics are shown in Tab. 3. The size of population is set to 100. The stop condition is defined as the max number of iteration=10000 or achieving global optimum with accuracy equal to $\epsilon = 0.00001$. All experiments were repeated 50 times. In each metaheuristics an individual was represented by a vector $x_i = [x_{i1}, \ldots, x_{in}]$, where $n$ is the function dimension. Initial values of this vector were assigned randomly from normal distribution.

*Test Functions.* All experiments presented in this study were made on the same functions that were used for adjusting parameters: 2, 5 and 10 dimensional De Jong, Ackley and Rosenbrock functions. They are shown in Fig. 1 and described shortly below

*De Jong function.* It is one of the simplest test functions. Formally it is expressed as follows:

$$f(X) = \sum_{i=1}^{n} x_i^2$$

where: $-5.12 \leq x_i \leq 5.12$, dla $i = 1, \ldots, n$. Global minimum $f(X) = 0$ is achieved for $x_i = 0$; $i = 1, \ldots, n$.

*Rosenbrock function.* It is described by the following equation:

$$f(X) = \sum_{i=1}^{n-1} [100(x_{i+1} - x_i^2) + (1 - x_i)^2]$$

where: $-2.048 \leq x_i \leq 2.048$, dla $i = 1, \ldots, n$. Global minimum $f(X) = 0$ is achieved for $x_i = 1$; $i = 1, \ldots, n$.

*Ackley Function.* It is expressed as follows:

$$f(X) = -a\exp(-b\sqrt{\frac{1}{n}\sum_{i=1}^{n} x_i^2}) - \exp(\frac{1}{n}\sum_{i=1}^{n} cos(cx_i)) + a + \exp(1)$$

where: $-32.768 \leq x_i \leq 32.768$. Usually $a = 20$, $b = 0.2$, $c = 2\pi$. Glomal minimum $f(X) = 0$ is achieved for $x_i = 0$; $i = 1, \ldots, n$.

*Results.* The results of the experiments are presented in Table 2. Cells $X_{ij}$ ($i = 1 \ldots b$; $j = 1 \ldots k$) show the average number of iterations needed to obtain global minimum. The cases when the algorithms did not achieve the global optima are bolded. In most cases algorithms achieved 100% of accuracy. De Jong, Ackley and 2D Rosenbrock functions are relatively easy to find optimum. More difficult is Rosenbrock function with higher dimensions. For 5D, GA was not able to find global optimum. GSO has only 92% successes. For the same function in 10D IWO, FA and CS had 100 % successes. It can be a positive sign of competitiveness of new metaheuristics in comparison to the old ones. GA and BA never once reached the global optimum. GSO had 84% successes and PSO in 90% cases got the global minimum.

*Analysis.* To compare metaheuristics efficacy Friedman test [7] was applied. Its using is justified because the following assumptions are satisfied: 1)experiments are independent (a given experiment was performed in isolation from the remaining ones), each observation can be ranked.

The test is composed of two parts. In the first one the hypothesis $H_0$ is verified. The hypothesis says that there is no statistical differences (on the assumed level of confidence) among the methods. The hypothesis ($H_1$) claims that there is at least one method that has efficacy less than at least one other method. In the experiments the confidence level $\alpha = 0,05$.

In the case the hypothesis $H_0$ is rejected, it is known that there exists at least one discriminative method but the test is not able to define difference for methods. Therefore in the second part *post hoc* analysis is made that is based on comparison of the methods by pairs. The Friedman test starts from determination of rank for each $k$ metaheuristics separately for each $b$ problem. The best method is ranked on the first place $R = 1$, the weakest – $R = k$. In the score draw, the rank is calculated as an average of ranks of both methods. The value of rank (R(X)) for each method is shown in the Table 2.

**Table 2.** Comparison of the tested metaheuristics in terms of the number of iterations needed to reach global optimum

| | GA | | PSO | | BEE | | GSO | | IWO | | FA | | CA | |
|---|---|---|---|---|---|---|---|---|---|---|---|---|---|---|
| | $X_{i1}$ | R(X) | $X_{i2}$ | R(X) | $X_{i3}$ | R(X) | $X_{i4}$ | R(X) | $X_{i5}$ | R(X) | $X_{i6}$ | R(X) | $X_{i7}$ | R(X) |
| De Jong 2 | 6,0 | 5,0 | 3,0 | 2,5 | 7,0 | 6,5 | 3,0 | 2,5 | 5,0 | 4,0 | 2,00 | 1,0 | 7,0 | 6,5 |
| De Jong 5 | 18,0 | 4,5 | 11,0 | 2,0 | 114,0 | 8,0 | 13,0 | 3,0 | 18,0 | 4,5 | 3,0 | 1,0 | 41,0 | 6,0 |
| De Jong 10 | 41,0 | 3,0 | 29,0 | 2,0 | 618,0 | 8,0 | 80,0 | 4,0 | 141,0 | 6,5 | 4,0 | 1,0 | 117,0 | 5,0 |
| Ackley 2 | 18,0 | 4,0 | 8,0 | 2,0 | 47,0 | 8,0 | 10,0 | 3,0 | 41,0 | 7,0 | 2,0 | 1,0 | 23,0 | 6,0 |
| Ackley 5 | 48,0 | 4,0 | 21,0 | 2,0 | 829,0 | 8,0 | 41,0 | 3,0 | 466,0 | 6,0 | 10,0 | 1,0 | 98,0 | 6,0 |
| Ackley 10 | 125,0 | 2,0 | 61,0 | 1,0 | 6625,0 | 8,0 | 187,0 | 3,0 | 751,0 | 6,0 | 198,0 | 4,0 | 251,0 | 5,0 |
| Rosenbrock 2 | 144,0 | 8,0 | 17,0 | 5,0 | 41,0 | 6,0 | 16,0 | 4,0 | 4,0 | 1,5 | 4,0 | 1,5 | 15,0 | 3,0 |
| Rosenbrock 5 | – **0%** | 8,0 | 1006,0 | 4,0 | 4491,0 | 5,0 | **1559,0** **92%** | 6,0 | 425,0 | 3,0 | 223,0 | 2,0 | 190,0 | 1,0 |
| Rosenbrock 10 | – **0%** | 7,0 | **7833,0** **90%** | 4,0 | – **0%** | 7,0 | **5364,0** **84%** | 5,0 | 1600,0 | 3,0 | 1135,0 | 2,0 | 591,0 | 1,0 |
| $\sum$ | | | 45,5 | | 24,5 | | 64,5 | | 33,5 | | 42,5 | | 14,5 | | 39,5 |

**Table 3.** The default parameter values for each metaheuristics used in the experiments

| Parameter name | Description | Possibilities | Default value |
|---|---|---|---|
| | GA default parameters | | |
| Selection | Selection way | Tournament /Roulette | Tournament |
| Tournament size | Number of individuals in tournament selection | $\mathbb{N}_{\leq pop.size}$ | 5 |
| Pair selection | Pair association in the selection process | Random/The best fitness/The best with the worst | Random |
| $P_{crossover}$ | Crossover probability | $\langle 0,1 \rangle$ | 0,9 |
| $P_{mutation}$ | Mutation probability | $\langle 0,1 \rangle$ | 0,01 |
| | PSO default parameters | | |
| $c_1$ | Local search level | $\mathbb{R}^+$ | 2,0 |
| $c_2$ | Global search level | $\mathbb{R}^+$ | 2,0 |
| $w$ | Inertia coefficient | $\mathbb{R}^+$ | 0,0 |
| $v_{max}$ | Particle max velocity | $\mathbb{R}^+$ | 0,5 |
| | BA default parameters | | |
| $m$ | Number of bees | $\mathbb{N}_{\leq pop.size}$ | 10 |
| $e$ | Number of elite bees | $\mathbb{N}_{\leq m}$ | 5 |
| $Recruited_e$ | Number of bees recruited to the elite bees | $\mathbb{N}$ | 10 |
| $Recruited_m$ | Number of bees recruited to the rest $(m-e)$ bees | $\mathbb{N}_{\leq recruited_e}$ | 5 |
| $r$ | Neighbourhood radius | $\mathbb{R}^+$ | 0,001 |
| | CS default parameters | | |
| $p_a$ | Probability of solution (egg) rejection | $\langle 0,1 \rangle$ | 0,25 |
| $\alpha$ | randomness coefficient of firefly movement | $\langle 0,1 \rangle$ | 1,0 |
| $\beta_0$ | attractiveness | $\mathbb{R}^+$ | 1,0 |
| | FA default parameters | | |
| $\alpha$ | Scale of the step length | $\mathbb{R}^+$ | 1,0 |
| $\beta$ | Parameter of Lévy distribution | $(0,2)$ | 1,5 |
| $\gamma$ | light absorption | $\mathbb{R}^+$ | 0,1 |
| | IWO default parameters | | |
| $N_0$ | Initial population size | $\mathbb{N}_{\leq populationsize}$ | 10 |
| $s_{min}$ | Number of seeds spread by weed with the worst value of fitness function | $\mathbb{N}_{\leq s_{max}}$ | 1 |
| $s_{max}$ | Number of seeds spread by weed with the best value of fitness function | $\mathbb{N}_{\geq s_{min}}$ | 10 |
| $\sigma_{initial}$ | Initial standard deviation value of spreading | $\mathbb{R}^+_{\geq \sigma_{final}}$ | 3,0 |
| $\sigma_{final}$ | Final standard deviation of seed spreading | $\mathbb{R}^+_{\leq \sigma_{initial}}$ | 0,001 |
| $n$ | constant used to calculate standard deviation in a given iteration | $\mathbb{R}^+$ | 3,0 |
| | GSO default parameters | | |
| Rangers | Part of the population performing random search | $\langle 0,1 \rangle$ | 0,2 |
| $\Theta_{max}$ | max angle of pursuit | $\langle 0,2\pi \rangle$ | $\frac{\Pi}{a^2}$ |
| $l_{max}$ | max distance of pursuit for producer and random search | $\mathbb{R}^+$ | $\sqrt{\sum_{i=1}^n (U_i - L_i)^2}$ |
| $\alpha_{max}$ | max angle of rotation head | $\langle 0,2\pi \rangle$ | $\frac{\Pi}{2a^2}$ |
| $a$ | max number of scannings without fitness value improvement | $\mathbb{N}$ | $round(\sqrt{n+1})$ |

Statistics $T_2$ helps us to decide to assume or to reject hypothesis $H_0$.

$$T_2 = \frac{(b-1)[B_2 - bk(k+1)^2/4]}{A_2 - B_2}$$

where: $k$ is the number of compared metaheuristics, $b$ is the number of problems, $A_2$ and $B_2$ are defined by the following:

$$A_2 = \sum_{i=1}^{b}\sum_{j=1}^{k} R(X_{ij})^2 \quad B_2 = \frac{1}{b}\sum_{j=1}^{k} R_j^2, R_j = \sum_{i=1}^{b} R(X_{ij})$$

In our case $A_2 = 1795, 5$, $B_2 = 1674, 44$ and $T_2 = 14, 304$. $T_2$ is compared with the value from $F$ distribution with the assumed confidence level $\alpha$ (in our case $\alpha = 0, 05$) with $k - 1$ and $(b - 1)(k - 1)$ level of freedom:

$$F_{1-\alpha,k-1,(b-1)(k-1)} = F_{0.095,7,56} = 2, 178$$

In our case $T_2 > F_{0.095,7,56}$ so the hypothesis $H_0$ is rejected. This means that with accuracy limited to the confidence level $\alpha = 0, 05$, there exists at least one metaheuristics with efficacy different from others. In order to define differences among the methods, *post hoc* analysis is performed. Similarly to [7], Conover's approach based on $t$ distribution was used by us. For each metaheuristics pair, the absolute value of the ranks deference is calculated. One assume that metahauristcs have different results if the following inequality is satisfied:

$$|R_i - R_j| > t_{1-\frac{\alpha}{2}}\left[\frac{2b(A_2 - B_2)}{(b-1)(k-1)}\right]^{\frac{1}{2}}$$

where $t_{1-\frac{\alpha}{2}}$ is $1 - \frac{\alpha}{2}$ quantile of $t$ distribution with $(b - 1)(k - 1)$ freedom levels. In our case $t_{0.975} = 2, 003$, and the critical value is equal to 12.494. Table 4 presents a comparison of rank ($|R_i - R_j|$). The pairs that statistically differ are underlined in this table.

If we consider rank values of the metaheuristics their list is as follows: 1) Firefly Algorithm, 2)Particle Swarm Optimization, 3)Group search Optimization, 4)Cuckoo Search, 5)Invasive Weed Optimization, 6) Genetic Algorithm, 7) Bee Algorithm.

On the first place is FA. It always has reached the global optimum. For most cases it has the highest efficacy, but Table 4 shows that there is no statistical difference between FA and the other algorithms with the confidence level $\alpha = 0, 05$. PSO has almost as good results but there was no statistical evidence that it is better than FA. PSO is one of the oldest and well known method, so we can conclude that the new methods unnecessary are challengers for the old ones.

**Table 4.** The comparison of the metaheuristics pair ranks ($|R_i - R_j|$)

|     | GA | PSO | BEE | GSO | IWO | FA | CA |
| --- | --- | --- | --- | --- | --- | --- | --- |
| GA  | -  | 21  | 19  | 12  | 3   | 31 | 6  |
| PSO | -  | -   | 40  | 9   | 18  | 10 | 15 |
| BA  | -  | -   | -   | 31  | 22  | 50 | 25 |
| GSO | -  | -   | -   | -   | 9   | 19 | 6  |
| IWO | -  | -   | -   | -   | -   | 28 | 3  |
| FA  | -  | -   | -   | -   | -   | -  | 25 |
| CA  | -  | -   | -   | -   | -   | -  | -  |

# 5 Conclusion

The motivation of our research come from growing number of new metaheuristics proposed in the recent years. In this research we have analysed seven chosen algorithms, starting from the very old to the fresh ones. We compared them experimentally using three test functions with dimensionality: 2,5 and 10.

In statistical analysis of metaheuristics there was no evidence that we can choose the best one. We went to the extra mile to the implementation of the methods in one testbed platform, as well as to the choice of the final parameter values. Our study confirms a statement from the old paper [8] about no free lunch theorems for optimization problems. It means that the best universal way of optimization does not exist. All of them achieve similar The developers of new optimization metaheuristics should strive towards recognition of sort of problems the new methods are dedicated to and where they could obtain better results than others, because the choice of metaheuristics is dependent on the solved problem [9].

**Acknowledgments.** This work was partially supported by the European Commission under the 7th Framework Programme, Coordination and Support Action, Grant Agreement Number 316097, ENGINE - European research centre of Network intelliGence for INnovation Enhancement?.

# References

1. He, S., Wu, Q., Saunders, J.: A novel group search optimizer inspired by animal behavioural ecology. In: IEEE Congress on Evolutionary Computation, CEC 2006, pp. 1272–1278. IEEE (2006)
2. Mehrabian, A.R., Lucas, C.: A novel numerical optimization algorithm inspired from weed colonization. Ecological Informatics 1(4), 355–366 (2006)
3. Michalewicz, Z.: Algorytmy genetyczne + struktury danych = programy ewolucyjne. Wydawnictwa Naukowo-Techniczne (1999)
4. Parsopoulos, K.E., Vrahatis, M.N.: Particle swarm optimization and intelligence: advances and applications. Information Science Reference Hershey (2010)
5. Pham, D., Ghanbarzadeh, A., Koc, E., Otri, S., Rahim, S., Zaidi, M.: The bees algorithm- a novel tool for complex optimisation problems. In: Proceedings of the 2nd Virtual International Conference on Intelligent Production Machines and Systems (IPROMS 2006), pp. 454–459 (2006)
6. Shi, X.H., Liang, Y.C., Lee, H.P., Lu, C., Wang, Q.: Particle swarm optimization-based algorithms for tsp and generalized tsp. Information Processing Letters 103(5), 169–176 (2007)
7. Villegas, J.G.: Using nonparametric test to compare the performance of metaheuristics (2011). http://juangvillegas.files.wordpress.com/2011/08/friedman-test-24062011.pdf
8. Wolpert, D.H., Macready, W.G.: No free lunch theorems for optimization. IEEE Transactions on Evolutionary Computation 1(1), 67–82 (1992)
9. Yang, X.S.: Nature-inspired metaheuristic algorithms. Luniver Press (2010)
10. Yang, X.S., Deb, S.: Cuckoo search via lévy flights. In: World Congress on Nature & Biologically Inspired Computing, NaBIC 2009, pp. 210–214. IEEE (2009)

# Adaptive Complex Event Processing Based on Collaborative Rule Mining Engine

O-Joun Lee[1], Eunsoon You[2], Min-Sung Hong[1], and Jason J. Jung[1(✉)]

[1] School of Computer Engineering, Chung-Ang University, Seoul, Korea
{concerto9203,minsung.holdtime,j2jung}@gmail.com
[2] Institute of Media Contents, Dankook University, Yongin-si,
Gyeonggi-do, Korea
tesniere@naver.com

**Abstract.** Complex Event Processing (CEP) detects complex events or patterns of event sequences based on a set of rules defined by a domain expert. However, it lowers the reliability of a system as the set of rules defined by an expert changes along with dynamic changes in the domain environment. A human error made by an expert is another factor that may undermine the reliability of the system. In an effort to address such problems, this study introduces Collaborative Rule Mining Engine (CRME) designed to automatically mine rules based on the history of decisions made by a domain expert by adopting a collaborative filtering approach, which is effective in mimicking and predicting human decision-making in an environment where there are sufficient data or information to do so. Furthermore, this study suggests an adaptive CEP technique, which does not hamper the reliability since it prevents potential errors caused by mistakes of domain experts and adapts to changes in the domain environment on its own as it is linked to the system proposed by Bharagavi [10]. In a bid to verify this technique, an automated stocks trading system will be established and its performance will be measured using the rate of return.

**Keywords:** Collaborative system · Human-like decision · Rule mining · Complex Event Processing

## 1 Introduction

Complex Event Processing (CEP) is a technique used to respond to requirements of time-critical applications introduced recently. CEP enables real-time event detection by performing querying, filtering and transforming tasks on events delivered from a wide variety of sources as well as validation, cleaning, enrichment and analysis, which are considered as traditional database and data mining processes. This means that the technique makes real-time event pattern monitoring possible. To be specific, it is possible to provide effective real-time services by identifying an event-response mechanism through the discovery of the correlations between events and thus by extracting meaningful and practical events.

© Springer International Publishing Switzerland 2015
N.T. Nguyen et al. (Eds.): ACIIDS 2015, Part I, LNAI 9011, pp. 430–439, 2015.
DOI: 10.1007/978-3-319-15702-3_42

The CEP infrastructure consists of three main components: event source, event processing agent (EPA) and event listener. EPAs refer to consecutive queries or pattern signatures carrying out CEP tasks, such as filtering, aggregating, event correlating and event pattern searching. EPAs are connected to event sources, event listeners or other EPAs. There are three stages in setting up the CEP application. First, register all the event sources into the system. Second, EPAs will be formed for the event sources. Third, register event listeners and link them to corresponding EPAs.

Since changing of components would not be an easy task once the CEP application is arranged and operated, it is difficult to add or alter rules for most of the existing CEP engines. In CEP, reasoning is made up of responses of the EPA/rule-based inference engine designed to handle consecutive streams of real-time events. In the beginning, the system is configured according to a set of rules established by domain experts who define complex events or patterns to be identified. These domain experts are required to define a set of primitive events to be inputted, their correlations and parameters into rules. However, defining rules is a difficult task even for experts. In addition, the fact that rules defined by domain experts have a tendency to change according to dynamic changes in the domain environment raises the risk of reliability problems of the system. Moreover, an error in rules due to mistakes made by experts is another reason that makes the system vulnerable.

Bharagavi [10] proposed the Dynamic CEP technique enabling dynamic rule updates by designing the CEP infrastructure to move dynamically. Bharagavi's system addressed the problems of existing CEP techniques mentioned above to the great extent. However, the fact that the rules can be readjusted means that it cannot be a fundamental solution to the aforementioned two vulnerabilities.

With a view to addressing such problems, this study establishes the Collaborative Rule Mining Engine (CRME), which automatically lays out rules by analyzing decision-making history of domain experts and proposes the adaptive CEP technique designed to automatically enable dynamic rule updates by combining the engine with Bhargavi's system. This technique prevents errors caused by mistakes of domain experts from entering the system and secures the reliability of a system even in the face of dynamic changes in the domain environment. CRME creates ECA (Event-Condition-Action) rules that mimic decision-making of experts by classifying events according to their influence in the decision-making process and by taking a multi-layered collaborative approach.

This study is comprised of the following sections. Section 2 looks into the trend of related works. Section 3 provides a detailed description on automatic rule mining of the Adaptive CEP technique that this study introduces while Section 4 states and analyzes experiments and verification methods and their results to verify the validity and effectiveness of this study. Finally, Section 5 summarizes what the study proposes and suggests future research direction.

## 2    Related Work

Most of the existing CEP systems are based on an event-driven architecture, which use stream processing engines and database systems to back up continuous querying, pattern recognition and real-time response generation. Esper [6] employs EsperJMX [9] for the runtime management of Esper CEP. For instance, Esper [2] supports the ad-hoc execution of EPL expression through demand query/rule facility. However, there are limitations, such as sub-queries not being able to be executed, expressions not being allowed for some clauses, and precise execution being required at all time when some of them remain unexecuted. Drools [7] uses Drools Guvnor [8] for the multiuser management. Internally, Drools uses Knowledge Agent [1], which responds to the polling mechanism, for changing rules and queries that become dynamic during the runtime. The existing CEP engines support the initial configuration of rules and the addition depending on cases. This is being done in alignment with the prescribed time interval based on the polling mechanism. According to the existing works, there are some applications [4,5] that require a dynamic rule engine, which secures the adaptability of a CEP system. Yet, the existing CEP engines do not provide a fully dynamic rule engine, which points to the fact that the system is vulnerable to dynamic changes driven by the domain environment and that there exists a possibility of critical impact on the system when rules defined by experts are accompanied with errors.

Paschke et al. [3] introduced RuleML and logic based homogenous reaction rule language, which are rule and event based middleware that combines declarative rule based programming with enterprise application technologies for CEP and SOC. They defined interval-based event algebra for event definition, event selection and event consumption. Furthermore, homogenously integrated event messaging reaction rule language, also known as Prova AA, was proposed. However, dynamic addition and update of rules are not being supported.

Bastian et al. [5] suggested CEP-based anomaly management. They discussed the need for a dynamic CEP system for some applications and suggested a CEP infrastructure framework based on incorporating anomaly identification so as to support the dynamic CEP system. But, a solution for the rule management was not proposed in the end.

Bharagavi et al. [10] proposed a dynamic CEP based on an adaptive rule engine that can dynamically update the rules of a CEP system during the runtime. To this end, they employed an event-driven approach, which improves a system's performance just by updating rules in the knowledgebase without polling or observing any resource. Their proposed dynamic infrastructure-based CEP engine executes addition or deletion of the rules without any difficulty. However, they also did not suggest a solution for errors that can be entered by a domain expert or a solution for difficulties in generating rules perse.

# 3    Adaptive Complex Event Processing Technique

This section describes a concept of Adaptive Complex Event Processing (Adaptive CEP) technique. The technique aims to automatically generate rules for a CEP system by mimicking a domain expert's decision-making. The following is how this study defines a model for an expert's decision-making. First, any agent making decisions and any target being subject to a collaborative method are defined as an object within a domain as follows:.

- Definition 1 (Object). An object refers to all the agents that are in existence and are active in the domain. The object can be a person, an entity, an organization, a corporation, or another system.

Furthermore, elements affecting the decision making of an object are classified into six categories below. An event-interaction model is defined, which holds each category as its entity.

- Definition 2 (Status). Status refers to the overall status or situation of an object included in the domain, which is an attribute of a certain object.
- Definition 3 (Environmental Event). An environmental event refers to an event that exerts global influence on most of the objects existing in the domain.
- Definition 4 (Primitive Event). A primitive event occurs when attributes of respective object collide with an environmental event. This does not affect the decision-making directly but serves as an element of a macro event.
- Definition 5 (Complex Event). A complex event is a macro event that exerts direct influence on the decision-making. This could be deemed as a certain sequence of a primitive event.
- Definition 6 (Action). An action results from decision-making based on how an object's complex event is perceived. This is shown by an object taking an actual action in the domain.
- Definition 7 (Propensity). Propensity is tendency displayed by an object while taking a certain action in a complex event. Depending on the propensity, the object selects an action for the complex event.

The aforementioned six elements change in the time-sequential manner and are event sequences that exist as an event object at each point of time. If the events that constitute each element are identified to a limited number then to be quantified, each element can be stated as vector sequence: $e = [\vec{e_1}, \vec{e_2}, \cdots, \vec{e_n}]$, $S_i = [\vec{S_{i,1}}, \vec{S_{i,2}}, \cdots, \vec{S_{i,n}}]$, $A_i = [\vec{A_{i,1}}, \vec{A_{i,2}}, \cdots, \vec{A_{i,n}}]$, $P_i = [\vec{P_{i,1}}, \vec{P_{i,2}}, \cdots, \vec{P_{i,n}}]$, $E_i = [\vec{E_{i,1}}, \vec{E_{i,2}}, \cdots, \vec{E_{i,n}}]$, $CE_{i,l} = [\vec{E_{i,n}}, \vec{E_{i,m}}, \cdots, \vec{E_{i,p}}]$.

$e$ refers to an environmental event sequence of a domain in question while it signifies a vector that refers to an environmental event at a time point of n for $\vec{e_n}$, a status sequence of an $i$th object for $S_i$, a vector that shows status of an $i$th object at a time point of n for $\vec{S_{i,n}}$, an action sequence of an $i$th object for $A_i$, a vector that shows an action of an $i$th object for $\vec{A_{i,n}}$, propensity sequence of an $i$th object for $P_i$, a vector that expresses propensity of an $i$th object at a time point of n for $\overrightarrow{P_{i,n}}$, a primitive event sequence of an $i$th object for $E_i$, a vector that shows an event of an $i$th object at a time point of n for $\overrightarrow{E_{i,n}}$ and a $l$th complex event of an $i$th object for $CE_{i,l}$. The proposed technique is mainly divided into three parts.

— Complex Event Essence Extraction: Through event sequence comparison at a time of action occurrence, a suspected event sequence cluster model is generated while essence of complex event that can represent each cluster is extracted.
— Object Propensity Estimation: Based on an action for a complex event, an object's propensity is estimated.
— Automatic Rule-generation: ECA rules are generated based on the correlations between the essence of complex events and an object's propensity as well as an action to be taken by an object.

## 3.1    Complex Event Essence Extraction

Complex Event Essence Extracting is a process to extract patterns of event sequences that are suspected of complex events. This pattern is referred to as complex event essence and takes a form of a transition probability model among primitive events. This is comprised of mainly two steps.

— Complex Event-Clustering: Each suspected event sequence similarity is estimated, which serves the basis for clustering event sequences and thus for generating complex event-clusters.
— Complex Event Essence Extracting: Essence of a complex event that can represent each complex event-cluster is extracted.

This study regards an action taken by an expert as a response to a complex event that occurred to the individual. And, the occurrence of a complex event is detected through an expert's action. Accordingly, an event sequence just before an expert takes an action is regarded as a suspected event sequence. No limit is placed on the maximum size of the suspected event sequence. The end point of the suspected event sequence is the starting point of a certain action while its starting point is the occurrence time of an action included in the same action-cluster.

### Complex Event-Clustering

Complex Event-Clustering is a process of generating a complex event cluster model by clustering the suspected event sequence. The process uses k-nearest neighbor algorithm modified for clustering. It is difficult to describe the suspected event sequences in vectors or spatial coordinates. Therefore, the similarity between them would be used to replace the distance. The process is divided into mainly three phases.

— Primitive Event and Action Clustering: Each cluster model is generated by clustering primitive events and actions.

- Similarity Estimating: Similarity between suspected event sequences is found based on an action-cluster model and an object-cluster model.
- Event Sequences Clustering Phase: This is a process of generating a complex event cluster model by clustering suspected event sequences based on their similarity.

**Complex Event Essence Extracting**

Complex Event Essence Extracting is a process of extracting complex event essence from a complex event cluster. This is done by using Markov Transition Probability Model on primitive events in the complex event cluster. The process is divided into two phases.

- Complex Event Essence Modeling: A complex event essence model on frequent events is generated based on suspected event sequences included in respective complex event cluster.
- Complex Event Essence Refining: A model is then refined by removing meaningless primitive events included in the generated complex event essence model.

### 3.2    Object Propensity Estimation

Object Propensity Estimation Part is a process of estimating propensity of each object's response to a complex event by using Collaborative Methodology. Objects with similar actions taken at similar complex events are preconditioned to have similar propensity. The similarity of object propensity is calculated based on each object's action at a complex event, which serves as the basis for the object clustering. Cosine similarity is used for estimating the similarity. The more similar the action is taken by objects regarding a suspected event sequence included in the same complex event cluster, the higher the value gets.

$$
Pw_{i,j} = \frac{\sum_{CE \in CCE}(A_{i,CE.time} - \overline{A_\iota})(A_{j,CE.time} - \overline{A_j})}{\sqrt{\sum_{CE \in CCE}(A_{i,CE.time} - \overline{A_\iota})^2}\sqrt{\sum_{CE \in CCE}(A_{j,CE.time} - \overline{A_j})^2}} \tag{1}
$$

$Pw_{i,j}$ signifies similarity weight between $O_i$ and $O_j$ while it means a set of complex events occurred both to $O_i$ and $O_j$ for $CCE$, complex event for $CE.time$, occurrence time point for $CE$, action taken at $CE.time$ by $O_i$ for $A_{i,CE.time}$, and the average of actions taken by $O_i$ regarding complex events included in $CCE$ for $\overline{A_\iota}$.

### 3.3    Automatic Rule Generation

When a sequence of a certain event is presented based on the object propensity and essence of complex events, which are extracted above, it becomes possible to predict which action would be taken by an object. This is a process of automatically generating ECA rules that can be used in the CEP system based on the model for essence of complex event and on the object propensity calculated above. ECA rules to be generated through such a process are illustrated in equation 2.

$$On\ P(CE_a|o_i) > CV_{CE_a}\quad If\ o_i \in PC_m\quad Do\ A_n \tag{2}$$

This means that, "If the likelihood of certain essence of complex event and $CE_a$ occurring to a certain object, $o_i$, becomes bigger than the critical value of $CV_{CE_a}$, a maximum value of occurrence probability of other complex events, then a recommendation shall be made for a certain action, $A_n$, to be taken or considered." This is when the object, $o_i$, is included in a certain propensity-cluster of $PC_m$. And, $A_n$ refers to a representative value of actions responding to $CE_a$ by objects included in $PC_m$, which can be calculated based on mode or average value.

## 4    System Architecture and Experimentation

This study implemented an auto stocks trading system to validate the effectiveness of the proposed technique. The system performed automatic stock trading with an aim for the highest rate of return within three months in consideration of changes in the major economic indicators, changes in the government policies, such as interest rate changes, fluctuations in the major blue stocks and stockholding by system users.

An experiment was carried out in the simulated environment using an investment program based on actual stock price data instead of in the actual stock trading market. The data for the experiment was gathered on an hourly basis from YAHOO FINANCE for NASDAQ 100 stocks in the NASDAQ market traded for the past five years (January 2007 to December 2012).

The procedural details of the experiment are as follows: (1) A simulated investments were performed using data for the past three years starting January 2007 until December 2010 accumulated by users of the stimulated stock investment program; (2) A group of 40 experts were selected based on the program use results; (3) CRME was used for rule mining by assuming changes in the economic indicators and in the benchmark interest rate to be an environmental event while respective user was assumed to be an object, each user's asset portfolio to be status, changes in each user's asset values to be primitive event and each user's stock trading to be action; (4) Through this group of experts, a common rule set was generated, which allows to run the dynamic CEP technique proposed by Bharagavi [10] and the intuitive CEP technique suggested by this Study in the same environment; and (5) each technique went through simulation based on asset portfolios of the sample user group by using data accumulated for two year between January 2011 and December 2012.

The performance evaluation of the intuitive CEP technique has been made through comparison of earning rates based on the simulation results. By looking at changes in the earing rates depending on the number of experts in each domain and on the system operation period, improvement and reliability of the system performance were evaluated.

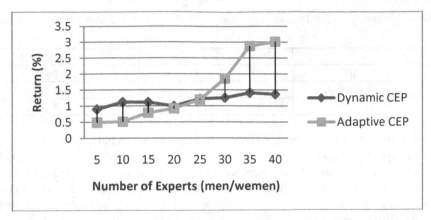

Fig. 1. Return of each Technique according to number of experts

Table 1. Average, Standard Deviation and Range of Return of each Technique

|                    | Dynamic CEP | Adaptive CEP |
|--------------------|-------------|--------------|
| Average            | 1.174       | 1.455        |
| Standard Deviation | 0.176       | 1.016        |
| Range              | 0.469       | 2.525        |

As shown in Fig. 1 and Table 1, the adaptive CEP technique generated better performance results when the number of domain experts in the training set exceeded a certain level compared to the case where domain experts entered rules directly. On average, the rate of return generated by the two techniques was 1.174% and 1.455% respectively. Compared to the dynamic CEP technique, the adaptive CEP technique exhibited performance improvements by 19.32% while the improvements were by 41.16% on average when the number of domain experts was more than 25 persons.

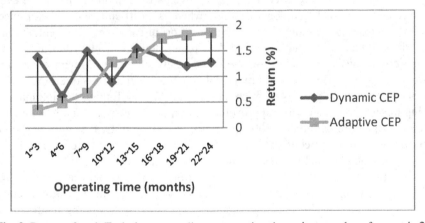

Fig. 2. Return of each Technique according to operating time when number of expert is 25

**Table 2.** Average, Standard Deviation and Range of Return of each Technique when number of expert is 25

|                    | Dynamic CEP | Adaptive CEP |
|--------------------|-------------|--------------|
| Average            | 1.223       | 1.194        |
| Standard Deviation | 0.296       | 0.571        |
| Range              | 0.657       | 1.493        |

It is proved that the adaptive CEP technique generates better performance by the domain experts than the case of rules entered by domain experts themselves provided that the experts go through sufficient training. When the number of experts was 25 persons and the system operation ran for 10 to 24 months, each technique generated 1.259% and 1.607% in rate of return. This is 21.68% higher in performance for the adaptive CEP technique compared to the dynamic CEP.

The results from the above two experiments prove that it is possible to replace the existing techniques of stating rules by experts themselves with the adaptive CEP technique. This also proves that the adaptive CEP can redress issues of performance deterioration caused by the changes in the domain environment and of errors entered due to rules defined by experts, which have been the source of troubles for the existing CEP techniques. Also, the performance improved by 41.16% when the history of a sufficient number of experts was provided while it was 21.68% improvement when sufficient learning time was given. This means that it can even improve the performance of the existing CEP techniques.

## 5    Conclusion

CEP can be deemed as a rule-based expert system optimized for data stream processing. The CEP system operates based on all the rules for summoning appropriate event listeners and for detecting complex events by combining primitive events. These rules are defined by a domain expert, which is a difficult task even for experts. The domain experts are required to offer a set of primitive events to be entered in the rules, the correlations among them and their parameters. Also, the rules are likely to change continuously depending on the dynamically changing domain environment while an error entered in the rules by mistake may exert fatal impact on the system's reliability. This study proposes Collaborative Rule Mining Engine for automatic rule mining through a collaborative approach based on the history of experts with an aim to redress the aforementioned problems. Also, the adaptive CEP system is proposed for rule-mining through continuous learning during the runtime by combining its engine with Adaptive Rule Engine suggested by Bharagavi [10].

As the future research direction, research efforts would be focused on solving issues caused by the borrowing of the collaborative filtering technique concept and enabling real-time rule learning by combining the global rule mining method of this proposed technique with the existing rule learning based on an inference engine. Also statistical analysis of experimental result will be a most urgent short-term goal.

**Acknowledgements.** This research was supported by the MSIP (Ministry of Science, ICT&Future Planning), Korea, under the ITRC(Information Technology Research Cetner) support program (NIPA-2014-H0301-14-1044) supervised by the NIPA(National ICT Industry Promotion Agency). Also, this work was supported by the National Research Foundation of Korea (NRF) grant funded by the Korea government (MSIP) (NRF-2014R1A2A2A05007154).

# References

1. Drools Expert User Guide, Version 5.4.0 CR1, JBoss Drools Team, pp. 31–33 (2012)
2. Esper Reference, Version 4.9.0, Esper Team and Esper Tech Inc, pp. 441–443 (2012)
3. Paschke, A., Kozlenkov, A., Boley, H.: A homogenous reaction rule language for complex event processing. In: Proceedings of 2nd International Workshop on Event Driven Architecture and Event Processing Systems (EDA-PS) (2007)
4. Turchin, Y., Gal, A., Wasserkrug, S.: Tuning complex event processing rules using the prediction-correction paradigm. In: Proceedings of 3rd ACM International Conference on Distributed Event-based Systems (DEBS 2009), pp. 1–12 (2009)
5. Hobbach, B., Seeger, B.: Anomaly management using complex event processing: extending database technology. In: Proceedings of the 16th ACM International Conference on Extending Database Technology (EDBT 2013), pp. 149–154 (2013)
6. http://esper.codehaus.org/
7. http://www.jboss.org/drools/
8. JBoss Drools Team, Drools Guvnor User Guide, Ver.5.4.0 CRI, pp. 1–2 (2012)
9. http://www.espertech.com/resources/sd_esperjmx.html
10. Bhargavi, R., Pathak, R., Vaidehi, V.: Dynamic complex event processing—Adaptive rule engine. In: 2013 International Conference on Information Technology (ICRTIT) Recent Trends. IEEE (2013)

# Adaptive Neuro Fuzzy Inference System for Diagnosing Dengue Hemorrhagic Fever

Nilo Legowo[1](✉), Bayu Kanigoro[1], Afan Galih Salman[1],
and Muhammad Syafii[2]

[1] School of Computer Science, Bina Nusantara University, Jakarta, Indonesia
{nlegowo,bkanigoro,asalman}@binus.edu
[2] RSUD Pasar Rebo, Jakarta, Indonesia
drsyafii@plasa.com

**Abstract.** Death of effect disease Dengue Hemorrhagic Fever (DHF) in Jakarta is high. To diagnosis patient of DHF, also render help immediately, and prevent death. Expert system can be used for the diagnosis of DHF. This research is conducted by the model system of Neuro-Fuzzy. One of them is ANFIS for conducting the diagnosis and medical procedure of DHF. Sample of research consist patient data suffering from DHF and dengue fever (DF), 32 cases of DHF and 32 cases of DF. Examination of model is conducted by some phase with data before and after verification. The examination of the model consists of amount and type of membership function. The conclusion of this research is ANFIS can be used selectively for conduct diagnosis of disease of DHF and can be used better if fulfilled minimum condition according to regulation which have been known such as haemorrhage manifestation and finding of fever. Model use the type of membership function of Gaussian yield accuracy 86,67 %, type of trapezoid yield accuracy 40 %, type of bells yield accuracy 40 % and type of triangular yield accuracy 40 %.

## 1 Introduction

In Jakarta, fatality rate of Dengue Hemorrhagic Fever (DHF) patient is still high. According to the 2004 Hospital active surveillants report of Greater Jakarta Special Capital Region there are 20.643 DHF cases where 91 persons death or 0,44% case fatality rate (CFR). Death caused by many factors one of them is diagnose delay [1]. It is excepted that earliest diagnose and immediate aid could prevent the death of Dengue Hemorrhagic Fever patient . Expert system can be used to diagnose disease such as Mycin system [2]. This system is developed in 1970?s by Edward Shortliffe Ph.D from Stanford University. Mycin include knowledge basic which storing information about the disease. A doctor can use the knowledge basic through a inference machine program [3]. In 1992, J.S.R. Jang developed Neuro-fuzzy system called adaptive neuro fuzzy inference system (ANFIS). ANFIS is adaptive system based on fuzzy inference system. In 2003 Castelano and friends used Neuro-fuzzy system named Kernel to diagnose skin diseases. This study will be done by modelled on Neuro-fuzzy system namely ANFIS to diagnose Dengue Hemorrhagic Fever disease.

© Springer International Publishing Switzerland 2015
N.T. Nguyen et al. (Eds.): ACIIDS 2015, Part I, LNAI 9011, pp. 440–447, 2015.
DOI: 10.1007/978-3-319-15702-3_43

The study's objective of developing adaptive neuro fuzzy model is making system which capable of diagnosing DHF. The expectation is that this study will useful to assist doctor in clinical diagnosing Dengue Hemorrhagic Fever particularly those who working at Rural Clinic which has very limited facility of blood cell count test laboratory that is needed to do thrombocytopenia test where the platelet count below 100,000 per $mm^3$, one of the blood count test in the evaluation of real clinical condition of the dengue hemorrhagic fever patient.

The study scope is doing acquisition of expert knowledge, making inferences by fuzzy inference system with type of Takagi-Sugeno, doing artificial neural network learning by using hybrid learning algorithm, develop Neuro-fuzzy model namely ANFIS for diagnosing Dengue Hemorrhagic Fever and undertaking model testing.

Study samples is taken from medical record of Dr. Sardjito hospital in Yogyakarta at the date of December 13th -16th 2005. Samples comprises data of Dengue Hemorrhagic Fever and Dengue Fever patient who was inpatient from January until November 2005. There are 205 cases but only 64 cases which has completed record complying with research requirement. The 64 cases consist of 32 Dengue Hemorrhagic Fever cases and 32 Dengue Fever cases. Moreover the cases is sorted into two groups that are 44 cases (70%) are used as training data and 20 cases (30%) are used as testing data.

Clinical diagnosing Dengue Hemorrhagic Fever depends on skill of the doctor. If the skill can be transferred to expert system by training then the trained expert system will have capability similar to expert?s skill. Basic concept of expert system is skill, expert, skill transfer, inference, rule, and ability to explain [4]. Skill transfer is done by conducting interview with a Dengue Hemorrhagic Fever specialist (specialist doctor). ANFIS will be used for developing model. Data is acquired from medical record of DHF patient and the data is not belong to patient who treated in hospital. The acquired data is comprised of training data and testing data. Model training is done by using training data whereas model testing using testing data. The conclusion result of model testing will be verified by diagnosis of testing data. Flow chart of the research idea framework conceptual can be seen completely in Figure 1.

## 2    Adaptive Neuro Fuzzy Inference System (ANFIS)

ANFIS is developed by J.S.R Jang in 1992. According to Jang the class of adaptive network is functionally equivalent with fuzzy inference system. ANFISis an architecture which is functionally the same as fuzzy rule base of first orde Sugeno model [5]. If it is assumed that fuzzy inference system has two inputs $x$ and $y$ as well as one output $z$, then according to first orde Sugeno model there are two rules as follow,

Rule 1 : If $x$ is $A_1$ and $y$ is $B_1$, then $f_1 = p_1 x + q_1 y + r$
Rule 2 : If $x$ is $A_2$ and $y$ is $B_2$, then $f_2 = p_2 x + q_2 y + r$

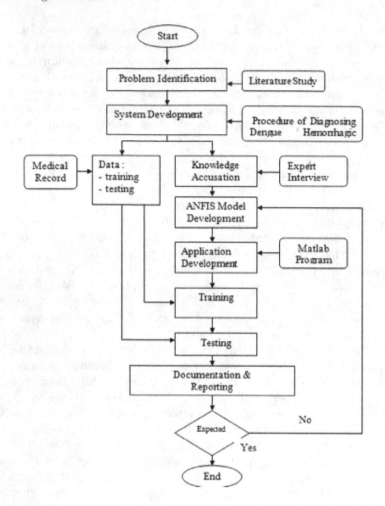

**Fig. 1.** Research Framework

The ANFIS system consist of 5 layers as shown in figure 2. At layer 1, every node in $i$ sequence of this layer is adaptive to parameter of a activation function. Output of every node in the form of membership degree that is given by input membership function. If the node output in $i$ sequence of layer 1 is symbolized as $O_{1,i}$ then the output layer 1 is :

$$O_{1,i} = \mu A_i(x), \text{for } i = 1, 2 \tag{1}$$

or

$$O_{1,i} = \mu B_{i-2}(y), \text{for } i = 3, 4 \tag{2}$$

Where $x$ or $y$ is input to node in $i$ sequence and $A_i$ or $B_{i-2}$ is fuzzy set and $O_{1,i}$ is membership degree of fuzzy set ($A_1, A_2, B_1$ or $B_2$). Example a bell function as follow:

$$\mu A(x) = \frac{1}{1 + (\frac{x - c_i}{a_i})^{2b_i}} \tag{3}$$

Where $a_i, b_i, c_i$ is parameter. If parameter value changes then the occurred bell curve will also be changed. The parameter on this layer called parameter premis.

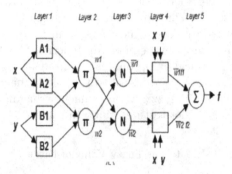

**Fig. 2.** The ANFIS Architecture [5]

Every nodes at layer 2 are in form of fixed node that their outputs are the result of all entered signal which conform below,

$$O_{2,i} = \omega_i = \mu A_i(x) \mu B_i(y), \text{for } i = 1, 2 \tag{4}$$

Every outputs of the nodes describe firing strength ($\alpha$-predicate) of a rule. AND operator is usually used.

Every nodes at layer 3 is same as layer 2 which are in form of fixed nodes. Output of the node in $i$ sequence is the result of comparison between $\alpha$-rule predicate in $i$ sequence with total of sum of $\alpha$-rule predicate,

$$O_{3,i} = \varpi_i = \frac{\omega_i}{\omega_1 + \omega_2}, \text{for } i = 1, 2 \tag{5}$$

Output from this layer is called normalised firing strength.

Every nodes at layer 4 are adaptive nodes which have equation function,

$$O_{4,i} = \varpi_i f_i = \varpi_i (p_i x + q_i y + r_i) \tag{6}$$

where $\varpi_i$ is normalized firing strength of layer 3 and $p_i, q_i, r_i$ are parameter of the node, which named consequence parameter.

Layer 5 consists of fixed single node which its output constitutes of the sum of all output,

$$O_{5,i} = \sum_i \varpi_i f_i = \frac{\sum_i \omega_i f_i}{\sum_i \omega_i} \tag{7}$$

ANFIS is trained by hybrid learning algorithm. There are two steps in hybrid namely forward step and backward step. On the foward step the premise parameter is fixed, input network will spread out forward to the fourth layer, where consequence parameter $(p, q, r)$ will be identified by using method of least-square estimator (LSE). Whereas on the backward signal, error signal between desired input and actual input will spread out backward and the premise parameters $(a, b, c)$ will be improved using method of gradient-descent.

Data structure design use fuzzy set. The criteria is clinical symptoms of Dengue Hemorrhagic Fever such as fever, pockmark, spontaneous haemorrhaging, and tourniquet test. The clinical criteria is represented as fuzzy data. Every criteria has parameter that reflected membership in fuzzy set. In fuzzy system this membership is represented in membership function (MF). The membership function value of every clinical symptoms is acquired based on interview with the expert. The fuzzy value of fever described in table 1, the fuzzy value of pockmark described in table 2, the fuzzy value of haemorrhaging described in table 3, and fuzzy value of tourniquet test described in table 4.

**Table 1.** The Fuzzy Value of Fever

| Fever | Value | Measurement |
|-------|-------|-------------|
| Low | 36.0 – 37.3 | Temperature 36.0 – 37.3°C and the duration of fever is 1 - 2 days |
| Moderate | 36.5 – 38.5 | Temperature 36.5 – 38.5°C and the duration of fever is 3 or 6 days |
| High | 38.0 – 42.0 | Temperature 38.0 – 42.0°C and the duration of 4 - 5 days |

**Table 2.** Fuzzy Value of Pockmark

| Pockrmark | Score Value | Measurement |
|-----------|-------------|-------------|
| Few | 0.00 – 0.40 | Amount of *petechiae* < 4 in circle with 2.8 cm diameter |
| Moderate | 0.25 – 0.75 | Amount of *petechiae* 4 - 9 in circle with 2.8 cm diameter |
| Many | 0.60 – 1.00 | Amount of *petechiae* ≥ 10 in circle with 2.8 cm diameter |

On the calculation of fuzzy data by ANFIS we used principle regulation of IF THEN. Regulation in made based on expert opinion. Amount of regulation in accordance with amount of the criteria and parameter. In this study there are four criteria namely fever, pockmark, spontaneous haemorrhaging and tourniquet test. Every criteria has three parameter (membership function) so that

**Table 3.** Fuzzy Value of Haemorrhaging

| Haemorrhaging | Score Value | Measurement |
| --- | --- | --- |
| Unclear | 0.00 – 0.40 | Mild haemorrhaging in the nose or at the gums |
| Clear | 0.25 – 0.75 | Severe haemorrhaging in the nose or at the gums. |
| Very Clear | 0.60 – 1.00 | haematemesis (vomiting blood) or melæna is found |

**Table 4.** Fuzzy Value of Tourniquet Test

| Tourniquet Test | Score Value | Measurement |
| --- | --- | --- |
| Negative | 0.00 – 0.40 | Amount of *petechiae* < 4 in circle with 2.8 cm diameter in *fossa cubiti* |
| Ambiguous | 0.25 – 0.75 | Amount of *petechiae* 4 - 9 in circle with 2.8 cm diameter in *fossa cubiti* |
| Positive | 0.60 – 1.00 | Amount of *petechiae* $\geq$ 10 in circle with 2.8 cm diameter in *fossa cubiti* |

the amount of regulation are $3^4 = 81$ regulations. On the ANFIS architecture, the regulation is formed adaptively according to characteristic of training data. Table 5 shows sample of regulation.

**Table 5.** Regulation Base

| | | |
| --- | --- | --- |
| Rule-1 | IF the fever is low AND pockmark is few AND spontaneous haemorrhaging is unclear AND tourniquet test is negative. | THEN It is not DHF |
| Rule-2 | IF the fever is high AND pockmark is many AND Spontaneous haemorrhaging is very clear AND Tourniquet test is positive. | THEN It is DHF |

The applied fuzzy inferential system is type of first orde Takagi-Sugeno, while the applied membership function is Gaussian. In the learning proses we use Hybrid algorithm and defuzzyfication is made by using method of Weighted Average.

## 3    Result

In fact result of model testing on various number of membership function affects conclusion value that made by the Model. From the 15 sample data that are tested, Model testing used three membership function which resulted in 86.67% accuracy with error average 0.20. Model testing used four membership function resulted in 86.67% accuracy with higher error average that is 0.83. Model testing used five membership function resulted in 86.67% accuracy with the highest error average that is 3.55.

Model testing with three membership function employs 91 rules, model testing with four membership function employs 108 rules, model testing fivemembership function employs 135 rules. Comparison of testing result can be seen in table 6. This testing shows that model that use three membership function is better than that of use $4mf$ and $5mf$.

**Table 6.** Test result of various membership function

| Sequence | Diagnosis in Medical Record | Total of Fever Membership Function | | | | | |
|---|---|---|---|---|---|---|---|
| | | $3mf$ | Error | $4mf$ | Error | $5mf$ | Error |
| 1 | 91.0000 | 89.1630 | 1.84 | 81.2544 | 9.75 | 51.5436 | 39.46 |
| 2 | 91.0000 | 90.2062 | 0.79 | 88.7245 | 2.28 | 77.8026 | 13.20 |
| 3 | 91.0000 | 90.9914 | 0.01 | 91.0192 | 0.02 | 90.9889 | 0.01 |
| 4 | 91.0000 | 91.0217 | 0.02 | 90.9742 | 0.03 | 91.0094 | 0.01 |
| 5 | 91.0000 | 91.0725 | 0.07 | 91.1028 | 0.10 | 91.3563 | 0.36 |
| 6 | 91.0000 | 90.9734 | 0.03 | 91.0016 | 0.00 | 90.9969 | 0.00 |
| 7 | 90.0000 | 90.0178 | 0.02 | 89.9913 | 0.01 | 89.9652 | 0.03 |
| 8 | 90.0000 | 89.9698 | 0.03 | 90.0006 | 0.00 | 90.0310 | 0.03 |
| 9 | 90.0000 | 90.0108 | 0.01 | 90.0150 | 0.02 | 90.0276 | 0.03 |
| 10 | 90.0000 | 90.0198 | 0.02 | 90.0321 | 0.03 | 90.0077 | 0.01 |
| 11 | 90.0000 | 89.9391 | 0.06 | 90.0838 | 0.08 | 90.0503 | 0.05 |
| 12 | 90.0000 | 89.9539 | 0.05 | 90.0217 | 0.02 | 90.0589 | 0.06 |
| 13 | 90.0000 | 89.9970 | 0.00 | 90.0182 | 0.02 | 90.0027 | 0.00 |
| 14 | 90.0000 | 89.9875 | 0.01 | 89.9874 | 0.01 | 89.9991 | 0.00 |
| 15 | 90.0000 | 89.9814 | 0.02 | 90.0427 | 0.04 | 90.0017 | 0.00 |
| Error Average | | | 0.20 | | 0.83 | | 3.55 |
| Accuration (%) | | 86.6 | | 86.6 | | 86.6 | |
| Total Rule | | 91 | | 108 | | 135 | |

Result of model testing on various type of membership function that are gaussian, trapezoid, bells and triangular has an effect on conclusion value which is made by. As a matter of fact, from the 15 tested data sample, model testing using gaussian resulted in 86.67% accuracy with error average 0.20, model testing using trapezoid resulted in 40% accuracy with error average 5.26 , model testing using bells resulted in 40% accuracy. With error average 5.01 and model testing using triangular resulted in 40% accuracy with error average 5.31. This testing show that Model that use membership function of gaussian type is better than model which use trapezoid, bells and triangular types.

# 4    Conclusion

ANFIS can be selectively employed for diagnosing Dengue Hemorrhage Fever and can be employed well if minimal condition complied in accordance with the known rule. In this research the minimal condition is the finding of one of the hemorrhaging manifestation and fever. Model that is use three membership function ($3mf$) on criteria of fever and gaussian type resulted in rate of DHF diagnosis accuracy 86.67%. Testing using $4mf$ dan 5 $mf$ resulting in the same accuracy rate that is 86.67%, but has different error average consecutively, error average of $3mf = 0.20$, error average of $4mf = 0.83$ and error average of $5mf = 3.55$.

The model needs development on measurement of clinical critera using the direct calculation result such as counting number of pockmark seen on the skin. The number of criteria should bee added including the subjective symptons such as headache, stomach ache, nausea, etc. It needs development on visualization of clinical symptoms of Dengue Hemorrhage Fever as guidance on evaluation for user.

# References

1. Sutaryo, S.: Dengue. Medika Fakultas Kedokteran Universitas Gajah Mada, Yogyakarta (2004)
2. Kusumadewi, S., Purnomo, H.: Aplikasi Logika Fuzzy Untuk Mendukung Keputusan. Graha Ilmu, Yogyakarta (2004)
3. Kristanto, A.: Jaringan Syaraf Tiruan (Konsep Dasar, Algoritma dan Aplikasinya). Gava Media, Yogyakarta (2004)
4. Turban, E.: Decision support and expert systems: management support systems. Prentice Hall PTR (1990)
5. Jang, J.S.R., Sun, C.T.: Neuro-fuzzy and Soft Computing: A Computational Approach to Learning and Machine Intelligence. Prentice-Hall Inc, Upper Saddle River (1997)

# Multiple Model Approach to Machine Learning (MMAML 2015)

# Adaptive Ant Colony Decision Forest in Automatic Categorization of Emails

Urszula Boryczka, Barbara Probierz$^{(\boxtimes)}$, and Jan Kozak

Institute of Computer Science, University of Silesia, Będzińska 39,
41-200 Sosnowiec, Poland
{urszula.boryczka,barbara.probierz,jan.kozak}@us.edu.pl

**Abstract.** In this article an approach to the automatic classification of email messages in mailboxes has been proposed. The aim of this paper is to devise methods to build decision tables from the collection of email messages on which it is possible to build Ant Colony Optimization-based ensemble classifiers, whose application allows to use the collection of emails without cleaning, at the same time improving the accuracy of the email folders classification. The proposed method has been tested by the selected algorithms on the Enron Email Dataset. The results confirm that the proposed solutions allows to improve the accuracy of classification of new emails to folders.

**Keywords:** Enron e-mail · Classification · Ensemble methods · Bagging · Boosting · Random forests · Ant Colony Decision Trees · Ant Colony Decision Forest

## 1 Introduction

Email is one of the best communication method nowadays. It is easy to use, accessible, fast, cheap and it allows to communicate with many people, over long distance, without leaving the house or work. However, for users, especially those for whom email is the basis of communication, the biggest problem is to organise emails in a proper way and to assign messages to specific folders. Especially when automatic categorization of email into folders is taken into consideration.

The aim of this paper is to invent methods to build decision tables from the available datasets that contain email messages. On the basis of decision tables it is built ensemble classifiers based on Ant Colony Optimization (ACO) it allows to improve the accuracy of the classification of email folders. The proposed method has been applied to the raw collection of email messages from selected mailboxes derived from publicity available Enron Email Dataset.

Classification of emails into folders using Ant Colony Optimization was proposed in the paper [3]. The authors have prepared for this purpose a special algorithm designed for the analysis of these emails. In the addition, the article [1] presents experiments, that were carried out by four classifiers, i.e. Maximum Entropy, Naive Bayes, Support Vector Machine and Wide-Margin Winnow.

© Springer International Publishing Switzerland 2015
N.T. Nguyen et al. (Eds.): ACIIDS 2015, Part I, LNAI 9011, pp. 451–461, 2015.
DOI: 10.1007/978-3-319-15702-3_44

However, to the best of our knowledge, until now, there was no attempt to use ensemble classifiers to classify emails.

This article is organized as follows. Section 1 comprises an introduction to the subject of this article. In section 2, classification methods are presented. Section 3 describes Enron e-mail dataset. Section 4 describes Ensemble Methods. In section 5, Ant Colony Decision Trees approach is presented. Section 6 describes Ant Colony Decision Forest algorithm. Section 7 focuses on the presented proposal of a method transforming a data set of emails into a decision table. Section 8 presents the experimental study that has been conducted to evaluate the performance of the proposed algorithm, taking into consideration Enron e-mail dataset. Finally, we conclude with general remarks on this work and a few directions for future research are pointed out.

## 2    Classification Methods

The classification is a process of assigning each document $d_i$ from a given training set $D_t = \{(d_1, c_1), \ldots, (d_m, c_m)\}$ to one of the predefined classes based on a set of values of attributes that describe a given document. Therefore, one can find a mapping that assigns one class from set $C = \{c_1, \ldots, c_m\}$ to a given document $d_i$ which is represented by the vector of features $(a_1, \ldots, a_n)$. The mapping:

$$f : R^n \ni (a_1, \ldots, a_n) \to c_m \in C \tag{1}$$

is referred to as a classifier or a classification mapping. The aim of categorizing text is to teach an algorithm to generate a classifier based on a training set.

The email foldering problem is a special case of the classification issue. It is about assigning emails to folders that are created or deleted by users over time. Such folders can be used for categorizing tasks to do, project groups or certain recipients. Email foldering is a complex problem because an automatic classification method can work for one user while for another it can lead to errors.

The first studies on methods of categorizing emails were carried out in the 1990s. D. Lewis introduced the model of concept learning for text classification systems, including systems for retrieving documents, automatic indexing and filtering electronic mail [17]. In the article [16] Kiritchenko and Matwin presented research which indicated that classification with SVM gives much better results than the naive Bayes classifier. In the article [20] M. Wang, Y. He and M. Jiang presented a categorization of email messages which was based on the information bottleneck (IB) and maximum entropy methods. The IB method was used in order to find key words, and then email subjects and address groups were used as features that were supplementary to email texts.The maximum entropy model was used to improve the classifiers accuracy.

## 3    Enron E-mail Dataset

The Enron email dataset constitutes a set of data which were collected and prepared as part of the CALO Project (a Cognitive Assistant that Learns and Organizes). It contains more than 600,000 email messages which were sent or received by 158 senior employees of the Enron Corporation. The dataset was taken over by the Federal Energy Regulatory Commission during an investigation that was carried out after the companys collapse and then it was made available to the public. A copy of the database was purchased by Leslie Kaelbling with the Massachusetts Institute of Technology (MIT), and then it turned out that there were serious problems associated with data integrity. As a result of work carried out by a team from SRI International, especially by Melinda Gervasio, the data were corrected and made available to other scientists for research purposes.

This database is considered to be one of the most valuable data sets because it consists of real email messages that are available to the public, which is usually problematic as far as other data sets are concerned due to data privacy. These emails are assigned to personal email accounts and divided into folders. There are no email attachments in the data set and certain messages were deleted because their duplicate copies could be found in other folders. The missing information was reconstructed, as far as possible, based on other information items; if it was not possible to identify the recipient the phrase no_address@enron.com was used.

The Enron dataset is commonly used for the purpose of studies that deal with social network analysis, natural language processing and machine learning. The classification of emails can have many different applications; in particular, it can be used to filter emails based on the priority criteria associated with assigning emails to folders that have been created by a user, and to identify spam.

## 4    Ensemble Methods

Ensemble methods work by running a base algorithm multiple times, and forming a vote out of the resulting hypotheses. Ensemble methods, which are popular in machine learning and pattern recognition, are learning algorithms that construct a set of many individual classifiers, called base learners, and combine them to classify new data points or samples by taking a weighted or unweighted vote of their predictions.

There are two main approaches to design ensemble learning algorithms. The first approach is to construct each hypothesis independently in such a way that the resulting set of hypotheses is accurate and diverse. The second approach designing ensembles is to construct the hypotheses on a coupled fashion so that the weighted vote of the hypothesis gives a good fit to the data. One way to fulfill this task – construct multiple hypotheses is to run the algorithm several times and provide it with somewhat different data (e.g. bootstrap samples) in each run. Ensemble methods are a collection of classifiers [7,9]. We defined the ensemble method by the following formula:

$$EM = \{d_j : X \to \{1, 2, ..., g\}\}_{j=1,2,...,J}, \tag{2}$$

where $J$ is a number of classifiers $j$ $(J \geqslant 2)$.

In ensemble method, predictions of classifiers are combined to make the overall prediction for the forest. Classification is done by a simple voting. Each classifier votes on the decision for the sample and the decision with the highest number of votes is chosen. The classifier created by a ensamble method $EM$, denoted as $dEM : X \rightarrow 1, 2, ..., g$, uses the following voting rule:

$$dEM(x) := \arg\max_k N_k(x), \tag{3}$$

where $N_k(x)$ is the number of votes for the sample $x \in X$ classification in to class $k$, such that $N_k(x) := \#\{j : d_j(x) = k\}$; $k$ a decision class, such that $k \in \{1, 2, ..., g\}$.

Boosting is an approach was firstly proposed by Schapire in 1990 [19].In particular, this approach is connected with the creation of a good learning set based on weak learning subsets. The newest boosting version, AdaBoost, was created in 1995 and presented by Freund and Schapire in publications from 1996 [14]. This algorithm is still under further development, which was shown in [18]. A weight is assigned (initially equal to $\frac{1}{n}$) for each element of the learning set. This value determines the element's probability of being chosen for the pseudo-samples. In the next step the classifier is created. Then, for each element that is wrongly classified to the decision class the weight coefficient is increased, so that during the next pseudo-sample creation this weak element (object) will more likely be chosen. In the AdaBoost approach the weights of elements belonging to the learning set are modified depending on the classification error coming from the individual classifier. This factor consists of the sum of the weak element's weights:

$$\epsilon(j) = \sum_{x_i} w_i[k_i \neq k_i^j], \tag{4}$$

where $we_i$ represents the weight of element $x_i$ and $k_i^j$ is a decision class of the analyzed object. The modification is performed when the classification error is smaller or equal to 0.5. The weights are multiplied by the coefficient 4 and then normalized:

$$\kappa(j) = \frac{1 - \epsilon(j)}{\epsilon(j)}. \tag{5}$$

Good example of such a method is Bagging – "Bootstrap Aggregating" method [13], firstly introduced by Breiman [6]. This approach works as follows. Given a set of $n$ training data (learning set), Bagging chooses in each iteration a learning set of size $n$ by sampling uniformly with replacement from the original data set. Each element of such a learning set can be chosen exactly with the same probability close to the $\frac{1}{n}$.

To aggregate the base classifiers in a consensus manner, strategy such as voting is commonly used. Assuming the result of the base classifiers are independent of each other, each of the base classifier give exactly one vote and finally the simple voting decides about the classification the samples (see formula (3)).

Some ensemble methods such as Random Forests are particularly useful for high-dimensional data sets because increased classification accuracy. It can be achieved by generating multiple prediction models each with a different subset of learning data consisted of attribute subsets [7].

Breiman provides a general framework for tree ensembles called "random forests" [7]. Each tree depends on the values of a randomly chosen attributes, independently for each node or tree and with the same distribution for all trees. Thus, a random forest is a classifier (ensemble) that consists of many decision trees. Each splitting rule is performed independently for different subset of attributes. As a result it could be chosen $m$ attributes from the $p$ descriptions of the learning samples. Assume that $m \ll p$, and according to the performed experiments, good results should be obtained when $m = \sqrt{p}$. Let we assume, that $\frac{1}{3}$ of samples cannot be chosen to the training sample (in accordance to the probability equal to $(1 - n)^n \approx e^{-n}$), so only $\frac{1}{3}$ trees in the analyzed forest are constructed without this sample. In this situation, Breiman proposed that it will be well-grounded to apply the unencumbered estimator of misclassification probability obtained by decision tree [7].

# 5    Ant Colony Decision Trees

Fortunately, the opportunity to more effectively apply (in e-mail foldering) Ant Colony Decision Trees – treated as the aggregated knowledge, insight and expertise of diverse agent-ants – has become a reality. As individuals – agent-ants become more adapt and effective sharing thoughts and experiences in virtual space (via pheromone values), firms can use these insights to address critical challenges. This form of collective intelligence can play an important role in generating new solutions, solving difficult problems, disaggregating and distributing work in new and innovative ways, and making better, more informed decisions about future classification problems.

The optimization algorithm in this paper was inspired by the previous works on Ant Systems (AS) and, in general, by the concept – stigmergy. This phenomenon was first introduced by P. P. Grasse [15]. An essential step in this direction was the development of Ant System by Dorigo et al. [12], a new type of heuristic inspired by analogies to the foraging behavior of real ant colonies, which has proven to work successfully in a series of experimental studies. Diverse modifications of AS have been applied to many different types of discrete optimization problems and have produced very satisfactory results [11]. Recently, the approach has been extended by Dorigo et al. [10] to a full discrete optimization metaheuristic, called the Ant Colony Optimization (ACO) metaheuristic.

Ant Colony Decision Trees (ACDT) algorithm [2] employs Ant Colony Optimization techniques [12] for constructing decision trees and decision forests. Ant Colony Optimization is a branch of a newly developed form of artificial intelligence called swarm intelligence. Swarm intelligence is a form of emergent collective intelligence of groups of simple individuals: ants, termites or bees in which a form of indirect communication via pheromone was observed. Pheromone values encourage the ants following the path to build good solutions of the analyzed problem and the learning process occurring in this situation is called positive feedback or auto-catalysis.

In each ACDT step an ant chooses an attribute and its value for splitting the objects in the current node of the constructed decision tree. The choice is made according to a heuristic function and pheromone values. The heuristic function is based on the Twoing criterion, which helps ants select an attribute-value pair which well divides the objects into two disjoint sets, i.e. with the intention that objects belonging to the same decision class should be put in the same subset. Pheromone values indicate the best way (connection) from the superior to the subordinate nodes – all possible combinations are taken into account.

As mentioned before, the value of the heuristic function is determined according to the splitting rule employed in CART approach, that is, in the algorithm proposed by Breiman et al. in 1984 [8]. The probability of choosing the appropriate split in the node is calculated according to a probability used in ACO:

$$p_{i,j} = \frac{\tau_{m,m_{L(i,j)}}(t) \cdot \eta_{i,j}^{\beta}}{\sum_i^a \sum_j^{b_i} \tau_{m,m_{L(i,j)}}(t) \cdot \eta_{i,j}^{\beta}}, \tag{6}$$

where $\eta_{i,j}$ is a heuristic value for the split using the attribute $i$ and value $j$; $t$ is a step of the algorithm; $\tau_{m,m_{L(i,j)}}$ is an amount of pheromone currently available at step $t$ on the connection between nodes $m$ and $m_{L(i,j)}$ (it concerns the attribute $i$ and value $j$), and $\beta$ is the relative importance of the heuristic value.

The pheromone trail is updated by increasing pheromone levels on the edges connecting each tree node with its parent node (excepting the root):

$$\tau_{m,m_L}(t+1) = (1-\gamma) \cdot \tau_{m,m_L}(t) + Q(T), \tag{7}$$

where $Q(T)$ determines the evaluation function of decision tree (see equation (8)), and $\gamma$ is a parameter representing the evaporation rate.

The evaluation function for decision trees will be calculated according to the following equation:

$$Q(T) = \phi \cdot w(T) + \psi \cdot a(T, P), \tag{8}$$

where $w(T)$ is the size (number of nodes) of the decision tree $T$; $a(T, P)$ is the accuracy of the classification object from a training set $P$ by the tree $T$; and $\phi$ and $\psi$ are constants determining the relative importance of $w(T)$ and $a(T, P)$.

## 6   Ant Colony Decision Forest

The Ant Colony Decision Forest (ACDF) algorithm is based on two approaches: Random Forest and ACDT [2]. The ACDF algorithm can be applied for difficult data set analyzes by adding randomness to the process of choosing which set of features or attributes will be distinguished during construction of the decision trees [4]. In case of the ACDF, agent-ants create a collection of hypotheses in a random manner by complying to the threshold or rule to split on. The challenge is to introduce a new random subspace method for growing collections of decision trees  this means that the agent-ants can create a collection of hypotheses from the hypothesis space by using random-proportional rules. At each node of the tree the agent-ant can choose from the random subset (random pseudo-samples) of attributes and then constrain the tree-growing hypothesis to choose its splitting rule from among this subset. Because of the re-labelled randomness proposed in our approach we have resigned from having different subsets of attributes chosen for each agent-ant or colony in favour of greater stability of the undertaken hypotheses. This is a consequence of the proposition that was firstly used in Random Forest.

In this article was used adaptive ACDF algorithm (called aACDF), described in [5], where the similarities to Random Forests are emphasised. In article [2] we have a method of generating pseudo-samples for each population of virtual ants. The on-the-go, dynamic pseudo-samples were chosen according to a previously obtained classification quality. The adaptability was focused on weak samples. The choice of objects was done by sampling with a replacement from the $n$-objects' set, and always consisted of $n$ objects.

The original probability of choice was equal to $\frac{1}{n}$. In the following population of virtual ants the value of this probability depended on the weight of this object. In case of an incorrect classification this coefficient was increased according to the formula:

$$we_i = \begin{cases} 1, & \text{if the object is well classified} \\ 1 + \lambda \cdot n, & \text{otherwise.} \end{cases} \tag{9}$$

Meanwhile, the probability of choosing the object was calculated according to the formula:

$$pp(x_i) = \frac{we_i}{\sum_{j=1}^{n} we_j}. \tag{10}$$

## 7 Proposed Method

The proposed method is based on the e-mail messages transformation into the decision table. After the previous step, decision tables are then applied to the classification by popular ensemble classifiers - including the adaptive algorithm ACDF.

The first step in the proposed method of improving the accuracy of emails classification to folders is to process the collection of emails derived from the Enron Email Dataset using decision table. Each row in the decision table contains a rule that determines the decisions that have to be made when appropriate conditions are fulfilled. The prepared decision table consists of six conditional attributes and one decision attribute, which defines the folder to which the email is assigned.

Conditional attributes were selected to define the most important information about each message. They consist of the information from the sender field, the first three words from the email subject (with the exception of basic words and copulas) where additionally, words which belong to the set of decision classes are supported. The length of the massage, and the information about the person who received the message was added to a courtesy copy (CC) (as the Boolean value) is also checked by the conditional attributes. If the person is not in a courtesy copy - the person is the recipient.

We selected seven mailboxes – the same, as in the article [1] for purposes of comparability. All data sets contain raw, not cleaned data, therefore all the emails in given dataset were taken to the research purposes, not only selected emails. Each of the seven datasets is divided into a training set (2/3 of the objects) and a test set (1/3 of the objects). On the basis of training set the decision rules has been generated. In the next step of the proposed method the effectiveness of these rules has been verified with the selected algorithms on the basis of test set.

The number of decision classes depends on the case that is being analyzed; it is provided in Tab. 1 for each data set. These data sets are very large they are composed of a large number of decision classes and have attributes with many values, mainly with continuous values. Therefore, Ant Colony Decision Forests algorithms were used to analyze this data set because they perform very well as far as such problems are concerned [2].

**Table 1.** Characteristics of analyzed data sets

| Dataset | N. of objects | N. of class (email folders) | Number of attributes from | word1 | word2 | word3 | cc | length |
|---|---|---|---|---|---|---|---|---|
| beck-s | 1971 | 101 | 390 | 527 | 670 | 549 | 2 | 1331 |
| farmer-d | 3672 | 25 | 412 | 827 | 985 | 864 | 2 | 1679 |
| kaminski-v | 4477 | 41 | 821 | 1231 | 1304 | 1058 | 2 | 2461 |
| kitchen-l | 4015 | 46 | 597 | 1170 | 1207 | 996 | 2 | 2138 |
| lokay-m | 2493 | 11 | 295 | 842 | 955 | 863 | 2 | 1654 |
| sanders-r | 1188 | 30 | 272 | 442 | 485 | 423 | 2 | 1033 |
| williams-w3 | 2769 | 18 | 196 | 523 | 597 | 540 | 2 | 1056 |

## 8 Experiments

For the research the aACDF algorithm described in section 6 has been chosen, as well as the selected algorithms from WEKA system, i.e. Waikato Environment for Knowledge Analysis [21]. The WEKA system offers a number of algorithmic methods that allow to perform data analysis. For the purposes of this article, four ensemble classifiers were selected, as listed below: AdaBoost algorithm in conjunction with CART and RandomTree; Bagging algorithm in conjunction with CART; Random Forest algorithm.

All calculations were carried out on a computer with an Intel Core i5 2.5 GHz processor, 4 GB RAM, running on theWindows operating system. The proposed algorithm aACDF was implemented in C++. We performed 30 experiments for each data set. Each experiment included 625 generations with the population size of the ant colony equal to 25. In each case the decision forest consisted of 25 trees. The parameters values employed in ACO are established in the way firstly presented in [2]: $q_0 = 0.3$, $\alpha = 3.0$, $\gamma = 0.1$, $\phi = 0.05$, $\psi = 1.0$ and $\lambda = 0.5$.

Tab. 2 presents the results of the email classification to the folders accuracy for the selected ensemble classifiers in WEKA system and aACDT algorithm. The results are the arithmetic means from all starts of different algorithms, however, for the aACDF algorithm each time obtained better results than in case of using the other ensemble classifiers presented in the Tab. 2.

Experiments confirm that the use of adaptive classifiers (aACDF) allows to classify the emails with high accuracy. Classical ensemble methods in comparison

**Table 2.** Comparision of all approaches in terms of classification accuracy

| Dataset | AdaBoost | | Bagging | Random | aACDF |
|---|---|---|---|---|---|
| | CART | RandomTree | CART | Forest | |
| beck-s | 0.384 | 0.184 | 0.416 | 0.481 | **0.517** |
| farmer-d | 0.620 | 0.658 | 0.587 | 0.670 | **0.775** |
| kaminski-v | 0.271 | 0.105 | 0.250 | 0.349 | **0.657** |
| kitchen-l | 0.292 | – | 0.321 | 0.264 | **0.583** |
| lokay-m | 0.685 | 0.638 | 0.723 | 0.461 | **0.846** |
| sanders-r | 0.480 | 0.758 | 0.457 | 0.649 | **0.759** |
| williams-w3 | 0.887 | 0.914 | 0.859 | 0.819 | **0.944** |

with proposed in this article method do not show such good results and often classify emails with much lower accuracy (up to 30-40%). This is probably caused by raw, not cleaned data. Fig. 1 and 2 present the accuracy of classification obtained on the basis of the aACDF algorithm and the best single decision tree in this set. As can be seen, despite of the weak single classifiers, aACDF algorithm obtain very good classification results.

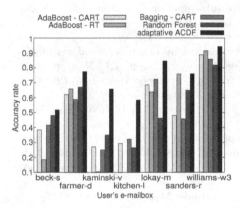 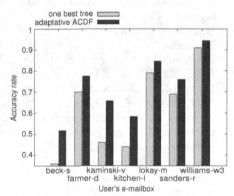

**Fig. 1.** The accuracy of classification obtained on the basis of the selected ensemble classifiers

**Fig. 2.** The best single decision tree obtained on the basis of the aACDF algorithm

## 9    Conclusions

The proposed approach led to an improvement in the classification of emails into folders. Based on the experiments that were carried out, it was confirmed that the correctness of an automatic categorization of email messages, was considerably improved when the adaptive ACDF algorithm is used.

The aim of this article has been achieved. This is particularly important because the algorithms that are described in section 6 do not required a thorough process of data cleaning. The aACDF, at the present stage, does not require that much work be carried out when preparing a data set for research purposes, and its adaptability allows one to obtain stable results even for "uncleaned", real data sets. Classification of bigger data sets may be improved on the basis of increased number of decision trees.

The above-mentioned adaptation abilities of Ant Colony Optimization algorithms additionally enhance their operational abilities while, at the same time, data sets only have to be slightly cleaned. In the future the authors of this article intend to adapt the social network mechanism for this purpose to a larger extent and to improve the process of creating decision tables. In future stages of the research, the incorporation of elements of text mining in an analysis of email message content and the direct coupling of these elements with the pheromone trail of the proposed algorithm should produce positive effects.

# References

1. Bekkerman, R., McCallum, A., Huang, G.: Automatic categorization of email into folders: Benchmark experiments on enron and sri corpora. Center for Intelligent Information Retrieval, Technical report IR (2004)
2. Boryczka, U., Kozak, J.: Ant colony decision trees – a new method for constructing decision trees based on ant colony optimization. In: Pan, J.-S., Chen, S.-M., Nguyen, N.T. (eds.) ICCCI 2010, Part I. LNCS, vol. 6421, pp. 373–382. Springer, Heidelberg (2010)
3. Boryczka, U., Probierz, B., Kozak, J.: An ant colony optimization algorithm for an automatic categorization of emails. In: Hwang, D., Jung, J.J., Nguyen, N.-T. (eds.) ICCCI 2014. LNCS, vol. 8733, pp. 583–592. Springer, Heidelberg (2014)
4. Boryczka, U., Kozak, J.: Ant colony decision forest meta-ensemble. In: Nguyen, N.-T., Hoang, K., Jędrzejowicz, P. (eds.) ICCCI 2012, Part II. LNCS, vol. 7654, pp. 473–482. Springer, Heidelberg (2012)
5. Boryczka, U., Kozak, J.: On-the-Go adaptability in the new ant colony decision forest approach. In: Nguyen, N.T., Attachoo, B., Trawiński, B., Somboonviwat, K. (eds.) ACIIDS 2014, Part II. LNCS, vol. 8398, pp. 157–166. Springer, Heidelberg (2014)
6. Breiman, L.: Bagging predictors. Machine Learning 24(2), 123–140 (1996)
7. Breiman, L.: Random forests. Machine Learning 45(1), 5–32 (2001)
8. Breiman, L., Friedman, J.H., Olshen, R.A., Stone, C.J.: Classification and Regression Trees. Chapman and Hall, New York (1984)
9. Bühlmann, P., Hothorn, T.: Boosting algorithms: Regularization, prediction and model fitting. Statistical Science 22(4), 477–505 (2007)
10. Doerner, K.F., Merkle, D., Stützle, T.: Special issue on ant colony optimization. Swarm Intelligence 3(1), 1–2 (2009)
11. Dorigo, M., Caro, G.D., Gambardella, L.: Ant algorithms for distributed discrete optimization. Artif. Life 5(2), 137–172 (1999)
12. Dorigo, M., Stützle, T.: Ant Colony Optimization. MIT Press, Cambridge (2004)
13. Efron, B.: Bootstrap methods: Another look at the jackknife. The Annals of Statistics 7(1), 1–26 (1979)
14. Freund, Y., Schapire, R.E.: Experiments with a new boosting algorithm. In: International Conference on Machine Learning, pp. 148–156 (1996)
15. Grasse, P.P.: Termitologia, vol. II. Masson, Paris (1984)
16. Kiritchenko, S., Matwin, S.: Email classification with co-training. University of Ottawa, Technical report (2002)

17. Lewis, D.D.: Representation and Learning in Information Retrieval. Ph.D. thesis, Department of Computer Science, University of Massachusetts (1992)
18. Rudin, C., Schapire, R.E.: Margin-based ranking and an equivalence between AdaBoost and RankBoost. J. Mach. Learn. Res. **10**, 2193–2232 (2009)
19. Schapire, R.E.: The strength of weak learnability. Machine Learning **5**, 197–227 (1990)
20. Wang, M., He, Y., Jiang, M.: Text categorization of enron email corpus based on information bottleneck and maximal entropy (2010)
21. Witten, I.H., Frank, E., Hall, M.A.: Data Mining: Practical Machine Learning Tools and Techniques, 3rd edn. Morgan Kaufmann Publishers Inc. (2011)

# Application of Parallel Distributed Implementation to Multiobjective Fuzzy Genetics-Based Machine Learning

Yusuke Nojima$^{(\boxtimes)}$, Yuji Takahashi, and Hisao Ishibuchi

Department of Computer Science and Intelligent Systems,
Graduate School of Engineering, Osaka Prefecture University,
1-1 Gakuen-cho, Naka-ku, Sakai, Osaka 599-8531, Japan
{nojima,hisaoi}@cs.osakafu-u.ac.jp,
yuji.takahashi@ci.cs.osakafu-u.ac.jp

**Abstract.** Fuzzy genetics-based machine learning is one of data mining techniques based on evolutionary computation. It can generate accurate classifiers with a small number of fuzzy if-then rules from numerical data. Its multiobjective version can provide a number of classifiers with a different tradeoff between accuracy and complexity. One major drawback of this method is the computation time when we use it for large data sets. In our previous study, we proposed parallel distributed implementation of single-objective fuzzy genetics-based machine learning which could drastically reduce the computation time. In this paper, we apply our idea of parallel distributed implementation to multiobjective fuzzy genetics-based machine learning. Through computational experiments on large data sets, we examine the effects of parallel distributed implementation on the search performance of our multiobjective fuzzy genetics-based machine learning and its computation time.

**Keywords:** Fuzzy genetics-based machine learning · Multiobjective genetic fuzzy systems · Parallel distributed implementation

## 1 Introduction

Genetic fuzzy systems (GFS) have been actively studied in the nearly two decades [8], [9], [16], [17]. GFS can optimize fuzzy if-then rules by using evolutionary computation (EC). There are two advantages of GFS from the viewpoint of data mining. One is that fuzzy if-then rules are intuitively understandable because numerical attributes can be converted into linguistic labels. The other is that EC can globally optimize any parameters and combinations in fuzzy if-then rules such as the shape of membership functions, feature selection, linguistic label selection, and rule selection. Besides, a fitness function in EC can be defined considering multiple objectives. The following is one of the most frequently-used fitness functions, which is defined by two conflicting objectives: accuracy and complexity.

$$fitness(S) = w_1\,Accuracy(S) - w_2\,Complexity(S), \tag{1}$$

where $S$ and $w = (w_1, w_2)$ represent a set of fuzzy if-then rules and a weight vector, respectively. The classification rate and the error rate of $S$ are often used as the

© Springer International Publishing Switzerland 2015
N.T. Nguyen et al. (Eds.): ACIIDS 2015, Part I, LNAI 9011, pp. 462–471, 2015.
DOI: 10.1007/978-3-319-15702-3_45

accuracy measures, while the number of rules and the number of conditions in $S$ are often used as the complexity measures. Well-designed interpretability measures are also used instead of the complexity measures in some studies [14], [15].

In recent years, multiobjective GFS (MoGFS) have attracted increasing attention in the research field of GFS [1]-[3], [7], [12], [23]. This is because we do not need to specify the weight vector in (1) and can obtain a number of classifiers with a different tradeoff between accuracy and complexity by only a single run of MoGFS. The typical multiobjective formulation of MoGFS is as follows:

$$\text{Maximize } Accuracy(S) \text{ and minimize } Complexity(S). \tag{2}$$

To find a number of non-dominated solutions (i.e., non-dominated classifiers), evolutionary multiobjective optimization algorithms [10] are usually used.

Although GFS in general have nice search ability, the computation time becomes a serious problem, when we apply GFS to large data sets. There are several attempts to reduce the computation time. For example, instance/feature selection, windowing, and parallel computing are representatives [4]-[6], [13], [19], [20], [24]. In our former studies [21], [25], we proposed parallel distributed implementation of single-objective GFS. The idea is very simple. When a workstation or a PC cluster with multiple CPU cores is available, we divide a population of candidate solutions and training data into subpopulations and training data subsets, respectively. Then, we assign a pair of a subpopulation and a training data subset to each CPU core as an island model of EC. In each island (i.e., each CPU core), an individual GFS is performed. The islands communicate with each other. Our parallel distributed model can drastically reduce the computation time without severe deterioration of the search ability of GFS.

In this paper, we apply our parallel distributed implementation to MoGFS and examine the effect of the number of subgroups for a population (and also training data) on the search ability of MoGFS for large data sets. This paper is organized as follows. First, we explain fuzzy classifiers and multiobjective fuzzy genetics-based machine learning in Section 2. Then, we explain our parallel distributed implementation in Section 3. In Section 4, we show some experimental results on large data sets and analyze non-dominated solutions. Finally, we conclude this paper in Section 5.

## 2    Fuzzy Classifiers and Multiobjective Optimization

Let us assume that we have $m$ training patterns. Each pattern has $n$ attribute values $x_p = (x_{p1}, ..., x_{pn})$, $p = 1, ..., m$ and a class label. For simplicity, each attribute $x_{pi}$ is normalized into a real number in [0, 1] for $i = 1, ..., n$. A classifier is composed of fuzzy if-then rules of the following type [22]:

$$\text{If } x_1 \text{ is } A_1 \text{ and ... and } x_n \text{ is } A_n \text{ then Class } C \text{ with } CF, \tag{3}$$

where $A_i$, $C$, and $CF$ represent the $i$th antecedent condition, a consequent class and a rule weight, respectively. As antecedent conditions, four fuzzy partitions with different granularities shown in Fig. 1 are used together with "*don't care*" condition to generate simple or general rules. The consequent class and rule weight of each rule are heuristically specified according to the compatibility grade of its antecedent part for the training patterns. See [22] for more detail.

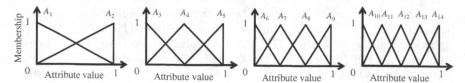

**Fig. 1.** Fuzzy membership functions for representing antecedent conditions

In this paper, we employ our multiobjective fuzzy genetics-based machine learning (MoFGBML) [23] for optimizing fuzzy classifiers. MoFGBML is basically the Pittsburgh-style approach where a classifier is coded as a string. A single iteration of the Michigan operation is also performed as local search. For solution evaluation, we use two objective functions such as the error (i.e., the misclassification rate) $f_1(S)$ and the number of rules $f_2(S)$ as:

$$\text{Minimize } f_1(S) \text{ and minimize } f_2(S). \tag{4}$$

The parent selection and the generation update are based on the non-dominated ranking and crowding distance as in NSGA-II [11]. See [23] in detail.

## 3    Parallel Distributed Implementation

In [25], we proposed the framework of parallel distributed implementation for single-objective genetic fuzzy rule selection. It was also applied to single-objective fuzzy genetics-based machine learning in [21]. Figure 2 illustrates its framework. Our approach is used for a workstation or a PC cluster with a number of CPU cores. In Fig. 2, seven CPU cores are used for GBML. In this case, a population of candidate solutions is divided into seven subpopulations. The training data is also divided into seven training data subsets. A pair of a subpopulation and a training data subset is assigned to each CPU core. Each subpopulation is optimized by GBML in each CPU core.

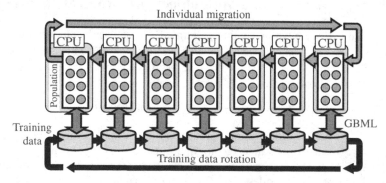

**Fig. 2.** Parallel distributed GBML

To avoid the overfitting of each subpopulation to the corresponding training data subset, training data rotation is employed periodically (e.g., every 100 generations). Besides, individual migration is also employed periodically (e.g., every 100 generations) in order to enhance the search ability like island models of EC.

After the termination of our parallel distributed model, all the solutions at the final generation are evaluated using all the training data. Then the best solution is selected with respect to the fitness value. For multiobjective formulation, non-dominated solutions are selected from the final population using all the training data.

When the population and the training data are divided into seven subgroups and seven CPU cores are used, the total computation time is expected to be decreased to 1/49. This is a quadratic speed-up comparing with the non-parallel standard model.

# 4    Computational Experiments

First, we demonstrate the effect of parallel distributed implementation on the search performance for a single-objective fuzzy GBML. We used the Satimage data (6435 patterns, 36 attributes, 7 classes). We used the following fitness function:

$$fitness(S) = w_1 f_1(S) + w_2 f_2(S) + w_3 f_3(S), \tag{5}$$

where $f_1(S)$ and $f_2(S)$ are the same as in (4). The $f_3(S)$ represents the total number of conditions in $S$. The weight vector $w$ was specified as (100, 1, 1). For the parallel distributed model, a population and training data were divided into seven subgroups. This experiment was performed by a workstation with Intel Xeon X5570 2.93GHz x 2 (i.e., eight CPU cores in total). The experimental settings are as follows:

- Population size: 210, 30 per subpopulation,
- Crossover: Uniform crossover,
- Crossover probability: 0.9 (Pittsburgh), 0.9 (Michigan),
- Mutation probability: $1/(n \times |S|)$ (Pittsburgh), $1/n$ (Michigan),
- Michigan operation probability: 0.5,
- Number of rules: 30 (Initial), 60 (Maximum),
- Rotation interval for training data: 1, 10, 50, 100, None,
- Migration interval: 1, 10, 50, 100, None,
- Termination condition: 50,000 generations,
- Number of runs: 50 (i.e., 5 x 10-fold cross validation).

We compare the non-parallel model (i.e., single-island model) with the parallel distributed model (i.e., seven-island model). Figure 3 shows the results by the parallel distributed model with different rotation interval and migration interval. The black bars in Fig. 3 (a)-(c) represent the better results than those by the non-parallel model. The black bars in Fig. 3 (d) represent that the speed-up rate is more than 49. Notice that only Fig. 3 (d) is rotated 180° around the z-axis. From Fig. 3 (a) and (b), we can observe that the classification rates for both training and test data were improved by our parallel distributed model, when we specified the rotation interval as 50 and 100. Too short rotation intervals deteriorated the search performance due to the rapid change of the training data subset. On the other hand, the obtained classifiers could not properly classify the whole training data, when we did not employ the training data rotation (i.e., rotation interval: None). As a result, the obtained classifier lost the generalization ability for the test data. We can also observe that the migration effect was low with respect to the classification rates.

From Fig. 3 (c), we can see that less frequent training data rotation and individual migration decreased the number of rules. This effect led to the improvement of the generalization ability. Besides, the computation time was more than 49 times faster than the non-parallel model, when our parallel distributed model was used with less frequent training data rotation and individual migration (Fig. 3 (d)).

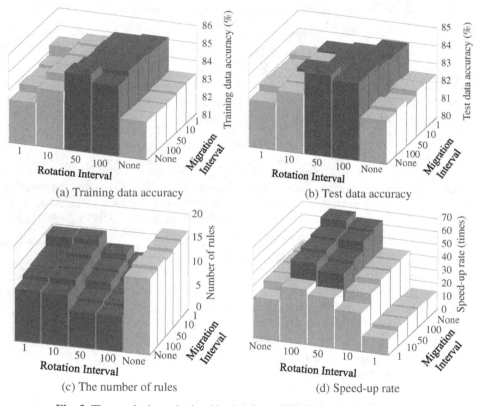

(a) Training data accuracy

(b) Test data accuracy

(c) The number of rules

(d) Speed-up rate

**Fig. 3.** The results by a single-objective fuzzy GBML for the Satimage data

Next, we show the results by the MoFGBML for the Phoneme data (5404 patterns, 5 attributes, 2 classes), the Satimage data, and the Segment data (2310 patterns, 19 attributes, 7 classes). Each solution was evaluated by (4). We examined the effects of the number of islands (i.e., subgroups). In the case of "three islands", the population and training data were divided into three subgroups. Thus, the subpopulation size is 70. In the case of "seven islands", the population and training data were divided into seven subgroups as the previous experiments. Figures 4-6 show the non-dominated solutions obtained by the MoFGBML. Due to the page limitation, we show only the average error rates on the test data. When classifiers with a particular number of rules were obtained only from 24 or less runs, the average error rate over those classifiers is not reported in this paper since such an average result is not reliable.

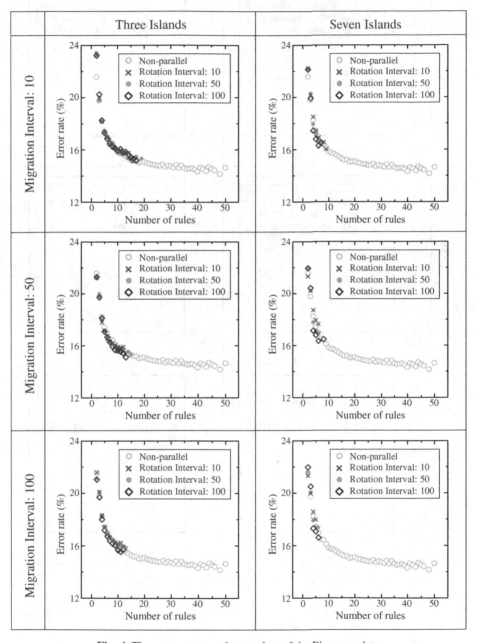

**Fig. 4.** The error rates on the test data of the Phoneme data

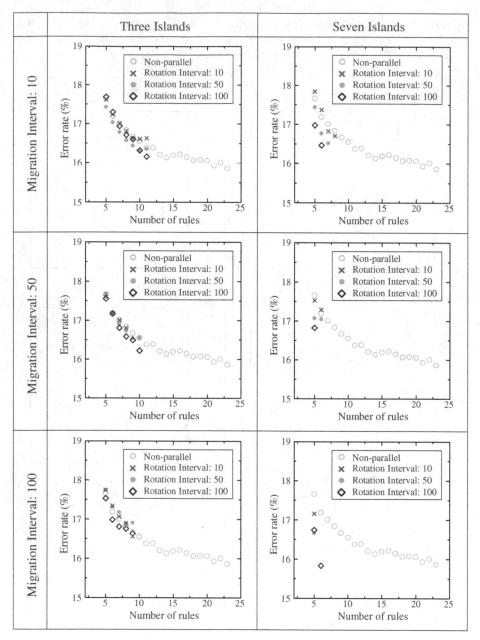

**Fig. 5.** The error rates on the test data of the Satimage data

For comparison, we include the non-dominated solutions obtained by the non-parallel model in each plot (i.e., gray circles). For all the data sets, the number of non-dominated classifiers by our parallel distributed model was smaller than that by the non-parallel model. We did not obtain classifiers with a large number of rules.

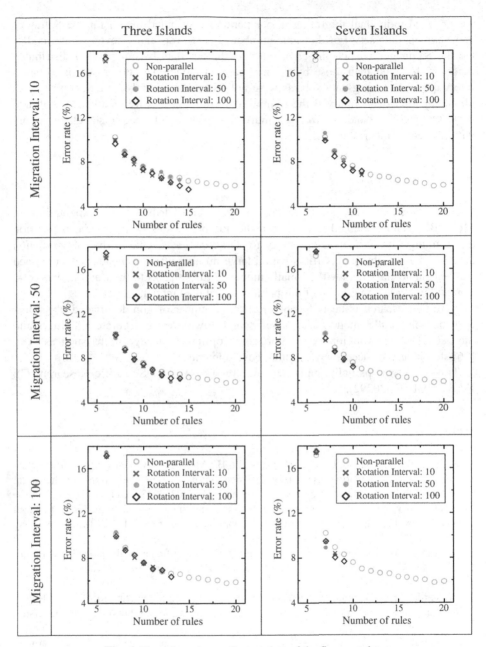

**Fig. 6.** The error rates on the test data of the Segment data

The three-island models obtained a large number of non-dominated classifiers than the seven-island models. This is probably because the subpopulation size of the three-island models is larger than that of the seven-island models. As in the case of the single-objective fuzzy GBML, we can observe the positive effects of the training data rotation and individual migration in our parallel distributed model for the Satimage and Segment data sets. For the same number of rules, especially seven-island models for the Satimage data set, the error rates on the test data by our parallel distributed model were better than those by the non-parallel model. The effects of the rotation interval of the training data subsets were not clear. But, slightly better accuracy was obtained, when we employed the training data rotation less frequently (e.g., 50 and 100). On the other hand, the frequent individual migration helped to search classifiers with a larger number of rules.

## 5    Conclusion

In this paper, we applied the idea of parallel distributed implementation to MoFGBML and examined the effect of the number of subgroups for the population and training data on the search ability for large data sets. Due to the subpopulation size, we did not obtain classifiers with a large number of rules. Instead, we obtained more accurate classifiers with a small number of rules thanks to the training data rotation and individual migration of our parallel distributed model.

A future research issue is to obtain a large number of non-dominated classifiers like the non-parallel model. The biased search toward more accurate classifiers and the use of other evolutionary multiobjective optimization algorithms such as [18], [26] should also be necessary for future developments.

This work was partially supported by Grand-in-Aid for Scientific Research (C): KAKENHI (25330292).

## References

1. Alcalá, R., Nojima, Y., Herrera, F., Ishibuchi, H.: Multiobjective genetic fuzzy rule selection of single granularity-based fuzzy classification rules and its interaction with the lateral tuning of membership functions. Soft Computing **15**, 2303–2318 (2011)
2. Antonelli, M., Ducange, P., Marcelloni, F.: Genetic training instance selection in multiobjective evolutionary fuzzy systems: A coevolutionary approach. IEEE Trans. on Fuzzy Systems **20**, 276–290 (2012)
3. Antonelli, M., Ducange, P., Marcelloni, F.: A fast and efficient multi-objective evolutionary learning scheme for fuzzy rule-based classifiers. Information Sciences **283**, 36–54 (2014)
4. Bacardit, J.: Pittsburgh Genetics-Based Machine Learning in the Data Mining era: Representations, generalization, and run-time. Doctoral disertation, Ramon Llull University, Barcelona (2004)
5. Bacardit, J., Llorà, X.: Large-scale data mining using genetics-based machine learning. WIREs Data Mining and Knowledge Discovery **3**, 37–61 (2013)
6. Cano, J.R., Herrera, F., Lozano, M.: Stratification for scaling up evolutionary prototype selection. Pattern Recognition Letters **26**, 953–963 (2005)
7. Chen, C.-H., He, J.-S., Hong, T.-P.: MOGA-based fuzzy data mining with taxonomy. Knowledge-Based Systems **54**, 53–65 (2013)

8. Cordón, O.: A historical review of evolutionary learning methods for Mamdani-type fuzzy rule-based systems: Designing interpretable genetic fuzzy systems. International Journal of Approximate Reasoning **52**, 894–913 (2011)

9. Cordón, O., Gomide, F., Herrera, F., Hoffman, F., Magdalena, L.: Ten years of genetic fuzzy systems: Current framework and new trends. Fuzzy Sets and Systems **14**, 5–31 (2004)

10. Deb, K.: Multi-objective optimization using evolutionary algorithms. Wiley, Chichester (2001)

11. Deb, K., Pratap, A., Agarwal, S., Meyarivan, T.: A fast and elitist multiobjective genetic algorithm: NSGA-II. IEEE Trans. on Evolutionary Computation **6**, 182–197 (2002)

12. Fazzolari, M., Alcalá, R., Nojima, Y., Ishibuchi, H., Herrera, F.: A review of the application of multiobjective evolutionary fuzzy systems: Current status and further directions. IEEE Trans. on Fuzzy Systems **21**, 45–65 (2013)

13. de Vega, F.F., Cantú-Paz, E (eds): Parallel and Distributed Computational Intelligence. Springer (2010)

14. Gacto, M.J., Alcalá, R., Herrera, F.: Interpretability of linguistic fuzzy rule-based systems: An overview of interpretability measures. Information Sciences **181**, 4340–4360 (2011)

15. Galende, M., Gacto, M.J., Sainz, G., Alcalá, R.: Comparison and design of interpretable linguistic vs. scatter FRBSs: GM3M generalization and new rule meaning index (RMI) for global assessment and local pseudo-linguistic representation. Information Sciences **282**, 190–213 (2014)

16. Herrera, F.: Genetic fuzzy systems: Status, critical considerations and future directions. International Journal of Computational Intelligence Research **1**, 59–67 (2005)

17. Herrera, F.: Genetic fuzzy systems: Taxonomy, current research trends and prospects. Evolutionary Intelligence **1**, 27–46 (2008)

18. Hiroyasu, T., Miki, M., Watanabe, S.: The new model of parallel genetic algorithm in multiobjective optimization problems: Divided range multi-objective genetic algorithm. In: Proceedings of 2000 IEEE Congress on Evolutionary Computation, pp. 333–340 (2000)

19. Hong, T.P., Lee, Y.C., Wu, M.T.: Using master-slave parallel architecture for GA-fuzzy data mining. In: Proceedings of 2005 IEEE International Conference on Systems, Man, and Cybernetics, pp. 3232–3237 (2005)

20. Hong, T.P., Lee, Y.C., Wu, M.T.: An effective parallel approach for genetic-fuzzy data mining. Expert Systems with Applications **41**, 655–662 (2014)

21. Ishibuchi, H., Mihara, S., Nojima, Y.: Parallel distributed hybrid fuzzy GBML models with rule set migration and training data rotation. IEEE Trans. on Fuzzy Systems **21**, 355–368 (2013)

22. Ishibuchi, H., Nakashima, T., Nii, M.: Classification and Modeling with Linguistic Information Granules: Advanced Approaches to Linguistic Data Mining. Springer, Berlin (2004)

23. Ishibuchi, H., Nojima, Y.: Analysis of interpretability-accuracy tradeoff of fuzzy systems by multiobjective fuzzy genetics-based machine learning. International Journal of Approximate Reasoning **44**, 4–31 (2007)

24. Liu, H., Motoda, H.: On issues of instance selection. Data Mining and Knowledge Discovery **6**, 115–130 (2002)

25. Nojima, Y., Ishibuchi, H., Kuwajima, I.: Parallel distributed genetic fuzzy rule selection. Soft Computing **13**, 511–519 (2009)

26. Zhang, Q., Li, H.: MOEA/D: A multiobjective evolutionary algorithm based on decomposition. IEEE Trans. on Evolutionary Computation **11**, 712–731 (2007)

# A Method for Merging Similar Zones to Improve Intelligent Models for Real Estate Appraisal

Tadeusz Lasota[1], Edward Sawiłow [1], Bogdan Trawiński[2(✉)],
Marta Roman[3], Paulina Marczuk[3], and Patryk Popowicz[3]

[1] Department of Spatial Management, Wrocław University of Environmental and Life Sciences,
ul. Norwida 25/27, 50-375 Wrocław, Poland
{tadeusz.lasota, edward.sawilow}@up.wroc.pl,
[2] Department of Information Systems, Wrocław University of Technology,
Wybrzeże Wyspiańskiego 27, 50-370 Wrocław, Poland
bogdan.trawinski@pwr.edu.pl
[3] Faculty of Computer Science and Management, Wrocław University of Technology,
Wybrzeże Wyspiańskiego 27, 50-370 Wrocław, Poland

**Abstract.** A method for property valuation based on the concept of merging different areas of the city into uniform zones reflecting the characteristics of the real estate market was worked out. The foundations of the method were verified by experimental testing the accuracy of the models devised for the prediction of real estate prices built over the merged zones. The experiments were conducted using real-world data of sales transactions of residential premises completed in a Polish urban municipality. Two machine learning techniques implemented in the WEKA environment were employed to generate property valuation models. The comparative analysis of the methods was made with the nonparametric Friedman and Wilcoxon statistical tests. The study proved the usefulness of merging of similar areas which resulted in better reliability and accuracy of predicted prices.

**Keywords:** Real estate appraisal · Predictive models · Machine learning · Linear regression · Decision trees

## 1 Introduction

The current system of real estate appraisal in Poland is not flawless and is based mainly on the analysis of real estate market which can be performed in a limited scope. Such an analysis is carried out using market transactions of similar properties located nearby the property being appraised, and the requirements as for the similarity are not unambiguous. The weakness of this system and binding standards consist in the difficulties in acquiring and assessing input data. The appraisers tend to find identical objects, however those are mostly sparse on the market or even lacking. To address this difficult problem the authors propose a method for merging different areas of an urban municipality into uniform zones reflecting the similar characteristics of the real estate market. The uniform zones may constitute the basis for elaborating more reliable and accurate models for real property valuation.

© Springer International Publishing Switzerland 2015
N.T. Nguyen et al. (Eds.): ACIIDS 2015, Part I, LNAI 9011, pp. 472–483, 2015.
DOI: 10.1007/978-3-319-15702-3_46

Real estate market modelling and predicting real property value have been an intensively developed area of research for many years. Numerous approaches to real estate appraisal ranged from statistical through operational research to computational intelligence techniques have been proposed and experimentally evaluated recently. Various methods can be found in rich literature on the topic including models built based on statistical multiple regression and neural networks [1], [2], [3], linear parametric programming [4], decision trees [5], rough set theory [6], fuzzy systems [7], and hybrid approaches [8].

For a decade we have been working out methods for generating data driven regression models to aid in real estate appraisal based on fuzzy systems and neural networks as both single models [9], [10] and multiple models built using various resampling techniques [11], [12], [13], [14], [15]. A relatively good accuracy provided evolving fuzzy models applied to cadastral data [16], [17]. We have also explored methods to predict from a data stream of real estate sales transactions based on ensembles of genetic fuzzy systems and neural networks [18], [19].

In this paper we propose a novel algorithm for property valuation based on the concept of merging different areas of an urban municipality into uniform zones reflecting the characteristics of the real estate market. The algorithm was evaluated by experimental testing the accuracy of the models for the prediction of real estate prices built over the merged zones. The experiments were conducted using real-world data of sales transactions of residential premises completed in a Polish urban municipality and linear regression and model pruned tree algorithms taken from WEKA [20].

## 2    Method for Merging Similar Zones of An Urban Municipality

The core of the proposed method is an algorithm for merging similar zones to constitute firm grounds for real estate appraisal. The area of an urban municipality was partitioned into about 250 zones based on a self-governmental land-use plan as shown in Fig. 1. The idea of merging consists in finding zones in which the prices of premises change similarly in the course of time. Such zones are then merged into bigger areas encompassing greater number of sales transactions of similar nature which allow for constructing more reliable and accurate property valuation models.

**Fig. 1.** Zones of an urban municipality based on a land-use plan considered in the paper

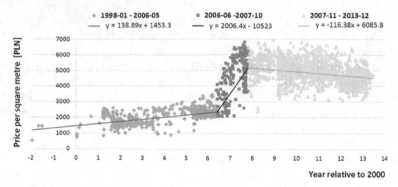

**Fig. 2.** Three periods with different trends of real estate price variability

The considered time span should be divided into a number of periods according to the dynamics of real estate prices. In Fig. 2 three periods with different linear trends of price variability are illustrated. The consecutive steps of the algorithm are as follows.

1) Partition the land into a number of uniform *zones i* (i=1,2,…,n) to consider.
2) Partition the considered time span into a number of *periods j* (j=1,2,…,m) with a uniform trend of price changes.
3) For each intersection *ij* of a zone and period set the minimum number of sales transactions to obtain a credible linear trend function.
4) Remove outliers from each intersection *ij* of a zone and period.
5) Exclude zones which do not contain the minimum number of transactions.
6) For each intersection *ij* determine the linear trend function $y=F_{ij}(t)=kt+b$, where $t$ – time, $y$ – price per square metre, $k$ – the slope, and $b$ – the intercept.
7) For each *zone i* compute the values of linear trend functions: $\boldsymbol{y_{ij}^b}$ and $\boldsymbol{y_{ij}^e}$ at the beginning $b$ and end $e$ of each *period j* as shown in Fig. 3.
8) Assign the weights $a_j$ to discriminate among individual periods in respect of their importance (age), e.g. the earlier the period the lower the weight. Then normalize the weights to sum to one according to Formula 1.

$$w_j = \frac{a_j}{\sum_{j=1}^{m} a_j} \tag{1}$$

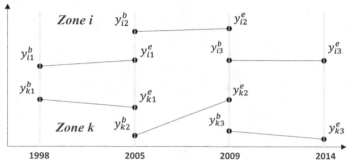

**Fig. 3.** Denotation of trend function values for two zones and three periods

9) Assign the weights $c_j$ to discriminate among individual periods in respect of their credibility to determine the trend functions, e.g. the smaller number of sales transactions within the period the lower the weight. Then normalize the weights to sum to one according to Formula 2 where $n_{ij}$ and $n_{kj}$ the number of transactions in the intersection of $j$-$th$ period and $i$-$th$ and $k$-$th$ zones respectively.

$$v_j = \frac{c_j}{\sum_{j=1}^m c_j} = \frac{min(n_{ij}, n_{kj})}{\sum_{j=1}^m min(n_{ij}, n_{kj})} \tag{2}$$

10) Determine similarity measures $D_{ik}$ between each pair of zones $i$ and $k$ applying Formulas 3, 4, 5, or 6. Each formula contains the division by 2 because two deviations are computed for each period, i.e. for the beginning and end.

$$D_{ik} = \frac{1}{2m} \sum_{j=1}^m |y_{ij}^b - y_{kj}^b| + |y_{ij}^e - y_{kj}^e| \tag{3}$$

$$D_{ik} = \frac{1}{2} \sum_{j=1}^m w_j \left( |y_{ij}^b - y_{kj}^b| + |y_{ij}^e - y_{kj}^e| \right) \tag{4}$$

$$D_{ik} = \frac{1}{2} \sum_{j=1}^m v_j \left( |y_{ij}^b - y_{kj}^b| + |y_{ij}^e - y_{kj}^e| \right) \tag{5}$$

$$D_{ik} = \frac{\sum_{j=1}^3 w_j v_j \left( |y_{ij}^b - y_{kj}^b| + |y_{ij}^e - y_{kj}^e| \right)}{2 \sum_{j=1}^3 w_j v_j} \tag{6}$$

11) Normalize the similarity measures $D_{ik}$ by dividing them by the average price per square metre in the respective zones $i$ and $k$ applying Formulas 7 and 8. In Formula 7 $N_i$ denotes the number of transactions in the *zone i* and $P_{il}$ stands for the price of individual transaction in this zone. In turn, in Formula 8 $N_k$ denotes the number of transactions in the *zone k* and $P_{kl}$ stands for the price of individual transaction in this zone.

$$d_{ik}^i = \frac{D_{ik}}{\frac{1}{N_i} \sum_{l=1}^{N_i} P_{il}} \tag{7}$$

$$d_{ik}^k = \frac{D_{ik}}{\frac{1}{N_k} \sum_{l=1}^{N_k} P_{kl}} \tag{8}$$

12) Create a similarity matrix which contains the values of the normalized similarity measures $d_{ik}^i$ and $d_{ik}^k$ between each pair of zones i and k determined by Formulas 7 and 8.

13) Set the criteria for merging zones according to the Formulas 9 and 10, where $\delta$ denotes the maximum acceptable deviation.

$$d_{ik}^i \in< 0, \delta > \tag{9}$$

$$d_{ik}^{ik} \in< 0, \delta > \tag{10}$$

14) Merge each pair of zones that satisfies the criteria (9) and (10) creating as a result bigger zones encompassing the transactions from both component zones. For each new zone remove outliers and then determine new linear trend functions and their values at the beginning and end of each period.
15) Repeat merging until no pair of zones $i$ and $k$ meets the criteria (9) and (10).

As a result the algorithm produces a new partition of the area into smaller number of zones with the greater cardinality which should allow for constructing the more credible and in consequence more accurate predictive models for real estate appraisal.

In order to create predictive models we need to update all prices for the last day of the considered timespan using trend functions. The procedure, which is very similar to our Delta method [18], is illustrated in Fig. 4. It is performed for each zone separately starting from the first period.

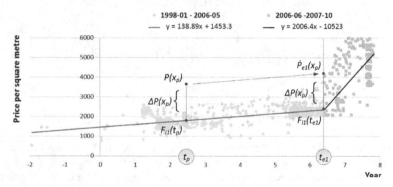

**Fig. 4.** Updating the prices of a premises for the end of *period 1* and *i-th* zone

16) At the time point $t_p$ for a given transaction $x_p$ compute the value of the trend function $F_{il}(t_p)$. Then, calculate the deviation of the price per square metre from the trend value $\Delta P(x_p)=P(x_p)- F_{il}(t_p)$. Finally work out the updated price per square metre of the premises $\dot{P}_{el}(x_p)$ by adding this deviation to the trend value in the end point of the first period $t_{el}$ using the formula $\dot{P}_{el}(x_p) =\Delta P(x_p)+ F_{il}(t_{el})$, where $F_{il}(t_{el})$ is the trend function value in the *i-th zone* at the time point $t_{el}$.
17) Update the so obtained values of $\dot{P}_{el}(x_p)$ of all premises for the end points of individual periods applying the same approach and the trend functions of successive periods. Finish the update when the last day of the considered timespan is reached.
18) Update analogously the prices of transactions from the second and next periods.

As a result all the transaction prices are updated for the last day of the considered timespan. The revised prices together with the other attributes of the premises, such as usable area, age, number of storeys in a building, and distance from the city centre, constitute the basis for generating predictive models for real estate appraisal.

# 3    Experimental Evaluation of the Proposed Method

The goal of evaluating experiments was to prove the utility of merging similar zones. The experiments aimed to show that the predictive models for real estate appraisal built over the merged zones surpass in terms of accuracy the ones created over original component zones.

The study was performed employing real-world data on sales transactions taken from a cadastral system and a public registry of real estate transactions. The dataset used in experiments was drawn from an unrefined dataset containing above 150 000 records referring to residential premises transactions accomplished in one Polish urban municipality, i.e. a city with the population of 640 000 and area of about 300 square kilometres within 16 years from 1998 to 2013. The final dataset after cleansing was confined to sales transaction data of residential premises (flats) where the land was leased on terms of perpetual usufruct and counted 31,553 samples. This dataset was then divided by the experts into four classes according to the year of a building construction as shown in Table 1. The properties encompassed by individual classes differ in construction technology and dynamics of relative market value. The differences in behaviour of property prices over time within individual classes are so big that the estimation of the values of properties should be carried out for individual classes separately.

Table 1. Classes of real estate considered in the paper

| Class | Property type | Year of construction | No. of trans. |
|---|---|---|---|
| 1 | Flats in buildings over land in perpetual usufruct | before 1945 | 13,178 |
| 2 | Flats in buildings over land in perpetual usufruct | 1946-1970 | 7,329 |
| 3 | Flats in buildings over land in perpetual usufruct | 1971-2000 | 4,017 |
| 4 | Flats in buildings over land in perpetual usufruct | 2001-2013 | 7,029 |

The algorithm for merging similar zones was run for individual classes of properties presented in Table 1 with the following parameters:

– three periods of accomplished transactions: 1998-2004; 2005-2008; 2009-2013,
– importance weights of individual periods: $a_1=0.6$, $a_2=0.7$, and $a_3=0.9$,
– threshold value of the deviation to merge zones $\delta=0.2$.

Next, all transactional prices were updated for the last day of the considered time-span, i.e. for June 30, 2013. The update was performed within both merged zones and component ones using the algorithm described in the previous section.

For further comparative analysis four datasets comprising merged zones were employed. They are shown in Table 2 where the numbers of datasets 1, 2, 3, and 4 correspond to the denotation of classes in Table 1. Due to removing outliers after merging the number of transactions in merged zones is usually less than the sum of transactions over component zones.

**Table 2.** Datasets used in evaluation experiments

| Dataset | Merged zones | Component zones. | Trans. in merged zones | Transactions in component zones |
|---|---|---|---|---|
| Set-1 | 2 | 8 | 2582 | 1517, 122, 329, 117, 730, 167, 46, 142 |
| | | 12 | 6036 | 364, 165, 375, 65, 1216, 224, 401, 3998, 58, 37, 194, 366 |
| Set-2 | 1 | 10 | 4709 | 2000, 178, 276, 53, 104, 57, 42, 2700, 318, 41 |
| Set-3 | 2 | 6 | 880 | 121, 94, 269, 480, 127, 90 |
| | | 2 | 931 | 537, 539 |
| Set-4 | 4 | 9 | 940 | 19, 74, 202, 96, 373, 96, 36, 278, 67 |
| | | 2 | 289 | 343, 33 |
| | | 6 | 1033 | 160, 359, 152, 394, 159, 79 |
| | | 5 | 386 | 30, 117, 268, 22, 83 |

In order to perform experiments the holdout method was employed. The transactions within each considered zone were randomly split into two parts in such way that 70% of transactions constituted the training set and remaining 30 % formed the test set. Two following types of predictive models were built in each zone using WEKA procedures over training sets [20]:

*LRM – Linear Regression Model.* The algorithm is a standard statistical approach to build a linear model which uses the least mean square method in order to adjust the parameters of the linear function.

*M5P – Pruned Model Tree.* The algorithm implements routines for generating M5 model trees. It is based on decision trees, however, instead of having values at tree's nodes, it contains a multivariate linear regression model at each node. The input space is divided into cells using training data and their outcomes, then a regression model is built in each cell as a leaf of the tree.

Five following attributes of premises proposed by professional appraisers were applied to the *LRM* and *M5P* machine learning algorithms. As four input features: usable area of a flat (*Area*), age of a building construction (*Age*), number of storeys in a building (*Storeys*), the distance of the building from the city centre (*Centre*) were taken, in turn, price per square metre (*Price*) was the output variable.

For comparing the accuracy of the models we used subsets of *Set-1, Set-2, Set-3,* and *Set-4,* namely the test sets of all merged zones. From each test subset 100 transactions were randomly drawn and the performance measure *Mean Absolute Percentage Error (MAPE)* was computed according to Formula 11, where $y_i^a$ and $y_i^p$ denote the actual and predicted values respectively and $n$ stands for the number of transactions in the test set.

$$MAPE = \frac{1}{n} * \sum_{i=1}^{n} \left| \frac{y_i^p - y_i^a}{y_i^a} \right| * 100\% \tag{11}$$

The drawing procedure was repeated 50 times and the final score was computed as the arithmetic mean of the results obtained in individual iterations.

## 4    Analysis of Experimental Results

The performance of *LRM* and *M5P* models built over training datasets for the zones before and after merging is depicted in Figures 5 and 6. It is clearly seen that the models constructed over merged zones surpass the ones over component zones in terms of accuracy expressed in *MAPE*. Moreover, the *M5P* models reveal better accuracy than *LRM* ones over transaction datasets both before and after merging . The individual *MAPE* values ranging from 4 to 11 per cent were regarded by professional appraisers as a good result especially when you take into account the confined number of features which could be drawn from cadastral systems.

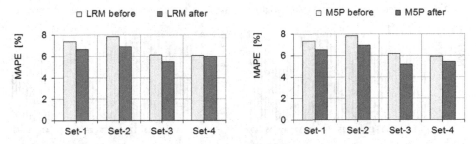

**Fig. 5.** Average *MAPE* values for *LRM* and *M5P* before and after merging over 50 iterations

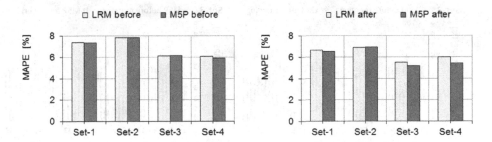

**Fig. 6.** Comparison of *MAPE* values for *LRM* and *M5P* before and after merging

The performance of *LRM* and *M5P* models for individual iterations for the zones before and after merging are illustrated in Figure 7. It is also visible that the majority of models constructed over merged zones outperform the ones built over component zones. However, in order to prove the significance of differences one should refer to statistical tests. Based on the output of individual iterations statistical tests of significance were made. We employed nonparametric tests, namely the Friedman test followed by the paired Wilcoxon test [21].

Average rank positions of models determined during Friedman test for the comparison of *LRM* before and after merging, *M5P* before and after merging, and *LRM* and *M5P* after merging are presented in Tables 3, 4, and 5, respectively. The ranks produced by the Friedman test mean the lower rank value the better model. In turn, the results of the paired Wilcoxon test indicate whether the differences are statistically significant.

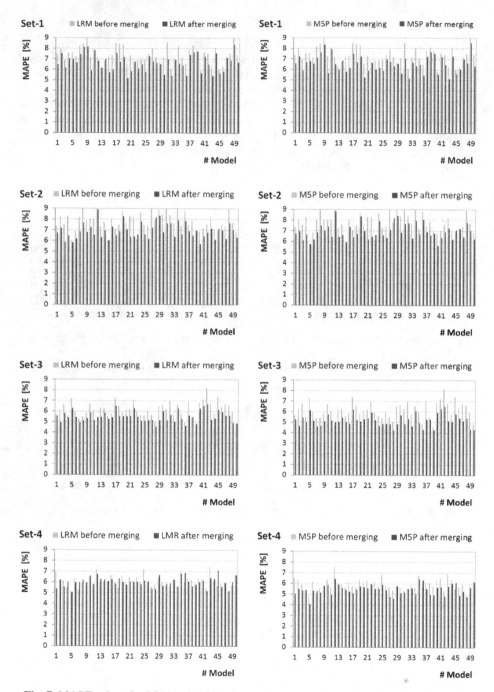

**Fig. 7.** MAPE values for *LRM* and *M5P* before and after merging for individual 50 iterations

The zero hypothesis stated there were not significant differences in accuracy between a given pair of models. In Tables 3, 4, and 5 + denotes that the model in the left column performed significantly better than, – significantly worse than, and ~ statistically equivalent to the one in the right column, respectively. The significance level considered for the null hypothesis rejection was 5%. The main outcome is as follows: *LRM* after merging showed significantly better performance than *LRM* before merging over all datasets but one; *M5P* after merging surpassed significantly *M5P* before merging over all datasets. In turn, *M5P* after merging revealed significantly lower *MAPE* than *LRM* after merging over all datasets but one.

**Table 3.** Results of statistical tests for comparison *LRM* before and after merging zones

| Dataset | Friedman | | | Wilcoxon | Result |
|---------|-----------|----------|---------|----------|--------|
| | LRM after avg. rank | LRM before avg. rank | p-value | p-value | |
| Set-1 | 1.06 | 1.94 | 0.0000 | 0.0000 | + |
| Set-2 | 1.08 | 1.92 | 0.0000 | 0.0000 | + |
| Set-3 | 1.08 | 1.92 | 0.0000 | 0.0000 | + |
| Set-4 | 1.40 | 1.60 | 0.1573 | 0.1397 | ~ |

**Table 4.** Results of statistical tests for comparison *M5P* before and after merging zones

| Dataset | Friedman | | | Wilcoxon | Result |
|---------|-----------|----------|---------|----------|--------|
| | M5P after avg. rank | M5P before avg. rank | p-value | p-value | |
| Set-1 | 1.02 | 1.98 | 0.0000 | 0.0000 | + |
| Set-2 | 1.07 | 1.93 | 0.0000 | 0.0000 | + |
| Set-3 | 1.02 | 1.98 | 0.0000 | 0.0000 | + |
| Set-4 | 1.08 | 1.92 | 0.0000 | 0.0000 | + |

**Table 5.** Results of statistical tests for comparison *LRM* and *M5P* after merging zones

| Dataset | Friedman | | | Wilcoxon | Result |
|---------|-----------|----------|---------|----------|--------|
| | LRM after avg. rank | M5P after avg. rank | p-value | p-value | |
| Set-1 | 1.68 | 1.32 | 0.0109 | 0.0001 | – |
| Set-2 | 1.38 | 1.62 | 0.0897 | 0.1423 | ~ |
| Set-3 | 1.98 | 1.02 | 0.0000 | 0.0000 | – |
| Set-4 | 2.00 | 1.00 | 0.0000 | 0.0000 | – |

## 5    Conclusions and Future Work

The problem of spatial object classification for the purposes of real estate appraisal was tackled in the paper. A novel algorithm for merging similar zones of the area of the city to obtain bigger homogeneous regions was devised. The similarity among zones was considered in respect of analogous changes of real property prices over time. The measure of zone similarity based on the values of the linear trend functions within the selected time periods was also proposed.

The proposed method was evaluated experimentally using real-world data of sales transactions of residential premises derived from a cadastral system and a public registry of real estate transactions. Machine learning procedures were acquired from the WEKA data mining system to build linear regression models (*LRM*) and pruned model trees (*M5P*) over merged and corresponding component zones. Mean absolute percentage error (*MAPE*) was employed as the performance measure and the statistical nonparametric Friedman and Wilcoxon tests were applied to examine the significance of the outcome. Experimental results proved the utility of the proposed algorithm. The models built over zones after merging revealed significantly better accuracy than the ones constructed on the basis of transactional data taken from the zones before merging.

The results of our research constitute a step towards the development an intelligent system to support appraisers' work providing them with the tool for accomplishing more reliable and accurate property valuations. Moreover, they may be utilized for preparing the maps of real property values for the purposes of mass taxation.

It is planned to conduct further experiments to tune the parameters of the proposed algorithm using the other real-world datasets. Moreover, the other machine learning techniques will be used to construct real estate models such as neural networks, support vector regression, and genetic fuzzy systems. The comparison with the valuation methods routinely employed by professional appraisers will be also accomplished.

# References

1. Kontrimas, V., Verikas, A.: The mass appraisal of the real estate by computational intelligence. Applied Soft Computing **11**(1), 443–448 (2011)
2. Zurada, J., Levitan, A.S., Guan, J.: A Comparison of Regression and Artificial Intelligence Methods in a Mass Appraisal Context. Journal of Real Estate Res. **33**(3), 349–388 (2011)
3. Peterson, S., Flangan, A.B.: Neural Network Hedonic Pricing Models in Mass Real Estate Appraisal. Journal of Real Estate Research **31**(2), 147–164 (2009)
4. Narula, S.C., Wellington, J.F., Lewis, S.A.: Valuating residential real estate using parametric programming. European Journal of Operational Research **217**, 120–128 (2012)
5. Antipov, E.A., Pokryshevskaya, E.B.: Mass appraisal of residential apartments: An application of Random forest for valuation and a CART-based approach for model diagnostics. Expert Systems with Applications **39**, 1772–1778 (2012)
6. D'Amato, M.: Comparing Rough Set Theory with Multiple Regression Analysis as Automated Valuation Methodologies. Int. Real Estate Review **10**(2), 42–65 (2007)
7. Kusan, H., Aytekin, O., Özdemir, I.: The use of fuzzy logic in predicting house selling price. Expert Systems with Applications **37**(3), 1808–1813 (2010)
8. Musa, A.G., Daramola, O., Owoloko, A., Olugbara, O.: A Neural-CBR System for Real Property Valuation. Journal of Emerging Trends in Computing and Information Sciences **4**(8), 611–622 (2013)

9.  Król, D., Lasota, T., Nalepa, W., Trawiński, B.: Fuzzy system model to assist with real estate appraisals. In: Okuno, H.G., Ali, M. (eds.) IEA/AIE 2007. LNCS (LNAI), vol. 4570, pp. 260–269. Springer, Heidelberg (2007)
10. Król, D., Lasota, T., Trawiński, B., Trawiński, K.: Comparison of Mamdani and TSK fuzzy models for real estate appraisal. In: Apolloni, B., Howlett, R.J., Jain, L. (eds.) KES 2007, Part III. LNCS (LNAI), vol. 4694, pp. 1008–1015. Springer, Heidelberg (2007)
11. Lasota, T., Telec, Z., Trawiński, B., Trawiński, K.: Exploration of bagging ensembles comprising genetic fuzzy models to assist with real estate appraisals. In: Corchado, E., Yin, H. (eds.) IDEAL 2009. LNCS, vol. 5788, pp. 554–561. Springer, Heidelberg (2009)
12. Lasota, T., Telec, Z., Trawiński, B., Trawiński, K.: A multi-agent system to assist with real estate appraisals using bagging ensembles. In: Nguyen, N.T., Kowalczyk, R., Chen, S.-M. (eds.) ICCCI 2009. LNCS, vol. 5796, pp. 813–824. Springer, Heidelberg (2009)
13. Krzystanek, M., Lasota, T., Telec, Z., Trawiński, B.: Analysis of bagging ensembles of fuzzy models for premises valuation. In: Nguyen, N.T., Le, M.T., Świątek, J. (eds.) Intelligent Information and Database Systems. LNCS, vol. 5991, pp. 330–339. Springer, Heidelberg (2010)
14. Kempa, O., Lasota, T., Telec, Z., Trawiński, B.: Investigation of bagging ensembles of genetic neural networks and fuzzy systems for real estate appraisal. In: Nguyen, N.T., Kim, C.-G., Janiak, A. (eds.) ACIIDS 2011, Part II. LNCS, vol. 6592, pp. 323–332. Springer, Heidelberg (2011)
15. Lasota, T., Telec, Z., Trawiński, G., Trawiński, B.: Empirical comparison of resampling methods using genetic fuzzy systems for a regression problem. In: Yin, H., Wang, W., Rayward-Smith, V. (eds.) IDEAL 2011. LNCS, vol. 6936, pp. 17–24. Springer, Heidelberg (2011)
16. Lasota, T., Telec, Z., Trawiński, B., Trawiński, K.: Investigation of the eTS Evolving Fuzzy Systems Applied to Real Estate Appraisal. Journal of Multiple-Valued Logic and Soft Computing 17(2–3), 229–253 (2011)
17. Lughofer, E., Trawiński, B., Trawiński, K., Kempa, O., Lasota, T.: On Employing Fuzzy Modeling Algorithms for the Valuation of Residential Premises. Information Sciences 181, 5123–5142 (2011)
18. Trawiński, B.: Evolutionary Fuzzy System Ensemble Approach to Model Real Estate Market based on Data Stream Exploration. Journal of Universal Computer Science 19(4), 539–562 (2013)
19. Telec, Z., Trawiński, B., Lasota, T., Trawiński, G.: Evaluation of neural network ensemble approach to predict from a data stream. In: Hwang, D., Jung, J.J., Nguyen, N.-T. (eds.) ICCCI 2014. LNCS, vol. 8733, pp. 472–482. Springer, Heidelberg (2014)
20. Witten, I.H., Frank, E., Hall, M.A.: Data Mining: Practical Machine Learning Tools and Techniques, 3rd edn. Morgan Kaufmann, San Francisco (2011)
21. Trawiński, B., Smętek, M., Telec, Z., Lasota, T.: Nonparametric Statistical Analysis for Multiple Comparison of Machine Learning Regression Algorithms. International Journal of Applied Mathematics and Computer Science 22(4), 867–881 (2012)

# Pruning Ensembles of One-Class Classifiers
# with X-means Clustering

Bartosz Krawczyk and Michał Woźniak[✉]

Department of Systems and Computer Networks, Wrocław University of Technology,
Wybrzeże Wyspiańskiego 27, 50-370 Wrocław, Poland
{bartosz.krawczyk,michal.wozniak}@pwr.edu.pl

**Abstract.** In this paper, we present a novel approach for pruning ensembles of one-class classifiers. One-class classification is among the most challenging topics in the contemporary machine learning. Creating multiple classifier systems for this task is one of the most effective ways of improving the quality and robustness in case of lack of counterexamples. However, very often we are faced with the problem of redundant or weak classifiers in the pool, as one-class ensembles tend to overproduce the base learners. To tackle this problem a dedicated pruning scheme must be employed, which will allow to discard classifiers that do not contribute to the formed ensemble. We propose to approach this problem as a clustering task. We discover groups of classifiers according to their support function values for the target class. For each group, we select the most representative classifier and discard the remaining ones. We apply an efficient x-means clustering algorithm, that automatically establishes the optimal number of clusters with the use of the Bayesian Information Criterion. Experimental results carried out on a set of benchmarks prove, that our proposed method is able to provide an efficient pruning mechanism for one-class problems.

**Keywords:** Machine learning · One-class classification · Classifier ensemble · Ensemble pruning · Clustering · X-means

## 1 Introduction

One-class classification (OCC) is considered as one of the most challenging topics in the contemporary machine learning. It is based on the assumption that for some problems positive samples are abundant, while it is hard, costly, time-consuming, or simply impossible to gather meaningful counterexamples. In OCC problems, we have only samples coming from a single class available during the training phase. This class is known as the target class or target concept, and denoted as $\omega_T$. A classifier must be trained using only samples from the positive class. At the same time, one must acknowledge that during the exploitation phase they may appear both new samples from $\omega_T$, as well as some unknown, negative and malicious objects. We have no *apriori* knowledge about the nature of these objects, but they must be rejected during the classification step. We denote

© Springer International Publishing Switzerland 2015
N.T. Nguyen et al. (Eds.): ACIIDS 2015, Part I, LNAI 9011, pp. 484–493, 2015.
DOI: 10.1007/978-3-319-15702-3_47

this unknown class as outliers $\omega_O$. OCC has gained a significant attention from the research community in last years, due to its complexity and many potential real-life applications [7].

There is a plethora of different methods dedicated for constructing one-class classifiers. However, due to the lack of counterexample at the training step proper model and parameter selection is not a trivial problem [12]. As one have no information about the possible nature of $\omega_O$, it is hard to decide which model will be able to deal with the problem at hand. This lead to a successful application of ensemble learning in OCC [9]. Multiple Classifier Systems [13] allowed for utilizing a pool of classifiers in order to exploit their strengths and reduce weaknesses. This is especially useful for OCC, as one does not need to search for a single-best model, and protects from selecting the weakest model from the pool. Most popular methods for creating one-class ensembles are based on Bagging, Random Subspace and atomic subsets of data [3].

However, these methods usually use the "overproduce" scheme, which outputs many classifiers with risk that some of them are weak or similar to each other [5]. Therefore, applying a classifier selection or ensemble pruning procedures is necessary [2,8,9]. Methods used so far need some kind of measure for evaluating which classifier should be selected and which discarded. Due to the lack of counterexamples, these measures only approximate classifier's performance [2]. Additionally, a search engine must be applied in order to check all of the possible combinations of classifiers - most popular ones are full-search [2] and evolutionary algorithms [8]. However, they are time consuming or require a number of parameters to be set.

In this paper, we propose an alternative approach for pruning one-class classification ensembles based on clustering. As a search engine, we apply an clustering method that allows to detect a number of groups in data. This groups correspond to the number of classifiers with different competencies. If classifiers are within a single group, one may deem them as similar. This approach allows to prune the ensemble with very low computational cost. We use an information-based criterion for an automatic selection of a number of clusters. Furthermore, we propose to cluster the support functions outputted by individual classifiers from the pool. With this, we search for some classification patterns in their decisions. This allows us to abstain from using any direct measure of one-class classifier's performance. We show, that the proposed scheme is superior to state-of-the art one-class pruning methods.

## 2    Clustering-Based One-Class Ensemble Pruning

In this paper, we propose to approach the ensemble pruning problem as the clustering task. We aim at finding the atomic groups of classifiers formed within the pool of available models. Such groups should have lowest inter-group and highest intra-group variances. With this assumptions, classifiers within a single group should have similar competencies to each other and low diversity. Therefore, using all of them would not contribute to the formed ensemble but only increase its computational complexity. On the other hand, different clusters of classifiers should be formed in complementary competence spaces - thus covering the entire decision space. Selecting a single most representative classifier from each of the established clusters will be similar to choosing classifiers with best individual quality and high diversity. Remaining classifiers in each cluster can be discarded, thus achieving the pruning effect.

The clustering-based approach for pruning has few advantages over the standard methods used in multi-class and one-class problems:

- It requires just a single run of the clustering algorithm, while evolutionary-based schemes need a high number of iterations, and schemes based on ranking/performance evaluation require a full-search over the possible combinations of classifiers. This property is especially beneficial for cases, in which the pool of available models is very large.
- It uses a relatively low number of parameters to be tuned, and in special cases these parameters can be automatically selected by the proper clustering algorithm - as in the proposed pruning scheme, where the number of clusters is established automatically.
- Any kind of metric can be used for the clustering phase. One may use the overall accuracy, support function values, diversity among classifiers or any other measure of performance. This allows to obtain a diverse clusters of models, suitable for the considered problem.
- Evolutionary and ranking methods need a specific metric to be used. This is problematic in one-class classification due to the lack of counterexamples at the training phase. Several proposed measures, such as consistency or one-class AUC are only approximations of the performance. Diversity measures for one-class classifiers tend to work well, but work under some assumptions about the structure of outlier class. Therefore, methods that are independent from such metrics / assumptions are of a high value to this problem. We show, that one do not need such metrics in clustering-based scheme.

In the following subsection, we will present the underlying mechanisms behind the clustering-based ensemble pruning for one-class classification ensembles.

## 2.1   Preliminaries for One-Class Cluster-Based Pruning

To apply the pruning step, we need to have separate training set $TS$ and validation set $VS$. The former is used for constructing the pool of classifiers, while the latter is utilized to evaluate the classifiers and conduct the pruning step.

For the clustering algorithm, we need to form a data matrix on which the group discovery will be performed. It is straightforward to apply clustering on a datasets, as we look for dependencies between objects and can use their features to measure the distance between samples. In this work, we aim at clustering classifiers. Therefore, we need some kind of classifier's features description in order to form a matrix. One can use several different measures specific for one-class classification - such as consistency, one-class AUC or diversity. However, as mentioned before, they are just approximations of the potential performance on unseen data - and may be misleading. In this paper, we propose to work on support function values outputted by one-class classifiers. We are interested in support of a given classifier $\Psi$ that analyses object $x$ belongs to the target concept $\omega_T$. It can be denoted as $F(x, \omega_T)$. Please note, that we only need this value as the support of a given classifier $\Psi$ for object $x$ belonging to the outlier class $\omega_O$ is $F(x, \omega_O) = 1 - F(x, \omega_T)$.

Working on direct supports outputted by one-class classifiers offers several different benefits:

- We alleviate the problem of selection of a proper metric for one-class classifier evaluation.
- It can work efficiently for both homogeneous and heterogeneous ensembles.
- Classifiers can form support templates for given objects. Therefore, classifiers outputting similar supports for training samples can be deemed as similar and all but one can be easily discarded. This can be seen as indirect assurance of diversity in the ensemble pool.

According to these assumptions, we propose to form a new data matrix consisting of support functions outputted by available classifiers for objects from $\mathcal{VS}$. We assume, that we have $L$ classifiers in the pool and $\mathcal{VS}$ consists of $N$ objects. The formed classifier support matrix $\mathcal{CSM}$ can be defined as follows:

$$
\mathcal{CSM} = \begin{pmatrix} F_{\Psi_1}(x_1, \omega_T) & F_{\Psi_1}(x_2, \omega_T) & \dots & F_{\Psi_1}(x_N, \omega_T) \\ F_{\Psi_2}(x_1, \omega_T) & F_{\Psi_2}(x_2, \omega_T) & \dots & F_{\Psi_2}(x_N, \omega_T) \\ \vdots & \vdots & \ddots & \vdots \\ F_{\Psi_L}(x_1, \omega_T) & F_{\Psi_L}(x_2, \omega_T) & \dots & F_{\Psi_L}(x_N, \omega_T) \end{pmatrix}, \tag{1}
$$

where $\Psi_l(x_n, \omega_T)$ stands for the support given by $l$-th classifier for $n$-th object belonging to the target concept $\omega_T$.

Our method uses the $\mathcal{CSM}$ for conducting the clustering step.

## 2.2 X-means Clustering Scheme

From the preliminaries, one may easily see that the number of clusters will have a crucial impact on the quality of the pruning scheme. The detected number of groups will be directly translated to the number of selected classifiers. Therefore, we need to use an efficient clustering algorithm that will be able to select the optimal number of competence areas.

From a plethora of clustering algorithms, we decided to apply an efficient $x$-means algorithm [10]. It is a powerful modification of the very popular $k$-means algorithm and can be used for our pruning task in a very straightforward way. Let us present below the main advantages of $x$-mean method:

- Selection of the parameter responsible for the number of clusters for many algorithms is very difficult and time consuming. X-means offer an efficient and fully automatic procedure for eastablishing the optimal number of clusters, based on computation of the Bayesian Information Criterion (BIC).

– $X$-means offers efficient speed-up for large datasets, by improving the model selection scheme. It calculates the membership to a given centroid not for a single point, but for a subset of points in the constructed $kd$-tree. This is done by considering the geometry of the bounding box for each centroid and their current location, in order to eliminate some of the centroids from the list of potential candidates by proving that they cannot own any point from the current node of the $kd$-tree.

– By noticing that some centroids do not change the location during evaluating different clustering models and no new centroid move into their positions, $x$-means reduces the time spend on traversing the $kd - tree$ in subsequent iterations.

Let us now describe the most important part (from our point of view) of the $x$-means algorithm, responsible for automatic selection of the number of clusters.

We assume that we have at our disposal data in form of $\mathcal{CSM}$ and a family of alternative clustering models $M_j$. In our case, different models from this family correspond to the different number of clusters $K$.

To select the best parameter setting, $x$-means uses the *a posteriori* probabilities $P[M_j|\mathcal{CSM}]$ to score the models. We assume that models are spherical Gaussians. One may approximate the posteriors with the use of following formula, commonly known as Schwarz criterion:

$$BIC(M_j) = \hat{l}_j(\mathcal{CSM}) - \frac{p_j}{2} \cdot \log R, \tag{2}$$

where $\hat{l}_j(\mathcal{CSM})$ is the log-likelihood of the data according to the $j$-th model and taken at the maximum likelihood point, $p_j$ is the number of parameters in $M_j$, and $R = |\mathcal{CSM}|$.

The maximum likelihood estimate (MLE) for the variance, under spherical Gaussians assumption, can be expressed as follows:

$$\hat{\sigma}^2 = \frac{1}{R - K} \sum_i (x_i - \mu_{(i)})^2, \tag{3}$$

where $K$ is the considered number of clusters, and $\mu_i$ is the $i$-th centroid. The point probabilities are:

$$\hat{P}(x_i) = \frac{R_{(i)}}{R} \cdot \frac{1}{\sqrt{2\pi}\hat{\sigma}^M} \exp\left(-\frac{1}{2\hat{\sigma}^2} \| x_i - \mu_{(i)} \|^2\right), \tag{4}$$

where $R_i = |\mathcal{CSM}_i|$, and $\mathcal{CSM}_i$ is a number of points with $\mu_i$ as their closest centroid.

The log-likelihood of the data is:

$$l(\mathcal{CSM}) = \log \prod_i P(x_i) = \sum_i \left(\log \frac{1}{\sqrt{2\pi}\hat{\sigma}^M} - \frac{1}{2\hat{\sigma}^2} \| x_i - \mu_{(i)} \|^2 + \log \frac{R_{(i)}}{R}\right). \tag{5}$$

Assume that $1 \leq k \leq K$. When we consider a set of $\mathcal{CSM}_k$, which belong to the centroid $k$ and plunging in the maximum likelihood estimate gives us:

$$\hat{l}(\mathcal{CSM}_k) = -\frac{R_k}{2} \log(2\pi) - \frac{R_k \cdot M}{2} \log(\hat{\sigma}^2) - \frac{R_k - K}{2} + R_k \log R_k - R_k \log R. \quad (6)$$

One should notice, that the number of free parameters $p_j$ is simply the sum of $K - 1$ class probabilities, $M \cdot K$ centroid coordinates, and one variance estimate. This can be extended to more than one centroid by using the fact, that log-likelihood of the points belonging to all of the analyzed centroids is the sum of the log-likelihoods of the individual centroids. With this, we can replace $R$ from above equations with the number of points that belong to the analyzed centroid.

This allows to select the best model from the considered family of models $M_j$.

## 2.3   Selecting Representative Classifiers

Having established the number of clusters with the $x$-means algorithm, one need to conduct the pruning procedure. There are three main ways, in which one can select the representative classifier for each cluster:

- Select the classifier that is the closest one to the corresponding centroid.
- Select the classifier that is the furthest one from all of the remaining centroids.
- Train a new classifier for each cluster.

In this work, we use the first methodology. We select a classifier that lies closest to the respective centroid in hope, that it will be the best representation of the established support template:

$$\Psi^k = \arg\min_{k=1}^{K} dist(\Psi_l, C_k), \quad (7)$$

where $\Psi^k$ is the classifier selected for $k$-th cluster, $C_k$ is the location of the centroid of $k$-th cluster and $dist$ denotes the Euclidean metric.

## 3   Experimental Study

The aims of the experiment was to check the quality of the proposed method on several benchmark datasets and compare the proposed FA-based pruning algorithm with other existing classifier selection methods, dedicated to OCC problems. We also checked, if applying a weighted fusion procedure for combining outputs of individual one-class classifiers leads to a significantly better performance of the entire ensemble.

**Table 1.** Details of datasets used in the experimental investigation. Numbers in parentheses indicates the number of objects in the minor class in case of binary problems.

| No. | Name | Objects | Features | Classes |
|---|---|---|---|---|
| 1 | Breast-cancer | 286 (85) | 9 | 2 |
| 2 | Breast-Wisconsin | 699 (241) | 9 | 2 |
| 3 | Colic | 368 (191) | 22 | 2 |
| 4 | Diabetes | 768 (268) | 8 | 2 |
| 5 | Heart-statlog | 270 (120) | 13 | 2 |
| 6 | Hepatitis | 155 (32) | 19 | 2 |
| 7 | Ionosphere | 351(124) | 34 | 2 |
| 8 | Sonar | 208 (97) | 60 | 2 |
| 9 | Voting records | 435 (168) | 16 | 2 |
| 10 | CYP2C19 isoform | 837 (181) | 242 | 2 |

### 3.1   Datasets

We have chosen 10 binary datasets in total. Details of the chosen datasets are given in Tab. 1.

Due to the lack of one-class benchmarks we use the canonical multi-class ones. We used a method for testing one-class classifiers, presented in [9]. The objects from the majority class were used as the target concept, while objects from the minority class as outliers. The training set was composed from the part of objects from the target class (according to cross-validation rules), while the testing set consisted of the remaining objects from the target class and outliers (to check both the false acceptance and false rejection rates).

### 3.2   Set-up

For the experiment a Support Vector Data Description [11] with a polynomial kernel is used as a base classifier. The pool of classifiers were homogeneous, i.e. consisted of classifiers of the same type.

The pool of classifiers was created in a fixed way to allow a proper exploitation of properties of different classifier selection criteria. It consisted in total of 60 models for each of the classes build on the basis of a Random Subspace approach with each subspace consisting of 50 % of original features.

To put the obtained results into context, we compared our proposed method with the following reference approaches and and the following state-of-the-art methods for selecting one-class classifiers: an unpruned pool of classifiers (to check, if classifier selection will lead to any improvement over using all available classifiers); a genetic algorithm pruning procedure, that uses diversity measure as fitness function with settings from [8]; a consistency-based one-class ensemble pruning [2], and an AUC-based ensemble pruning [2].

Each pruning method worked on the same pool of classifiers.

In order to present a detailed comparison among a group of machine learning algorithms, one must use statistical tests to prove, that the reported differences among classifiers are significant [6]. We use both pairwise and multiple comparison tests. For a pairwise comparison, we use a 5x2 combined CV F-test [1]. It repeats five-time two fold cross-validation so that in each of the folds the size of the training and testing sets is equal. From each training fold, we separate 20% of objects to be used as the validation set $VS$. For assessing the ranks of classifiers over all examined benchmarks, we use a Friedman ranking test [4]. Shaffer post-hoc test [6] is used to find out which of the tested methods are distinctive among an $n$ x $n$ comparison.

We fix the significance level $\alpha = 0.05$ for all comparisons.

## 3.3   Results and Discussion

The results are presented in Table 2. $UNPR$ stands for the unpruned pool of classifiers, $GA$ for a genetic algorithm pruning, $CONS$ for a consistency-based pruning, $AUC$ for an AUC-based pruning, and $CBP$ for the proposed approach.

**Table 2.** Results of the experimental results with the respect to the accuracy [%] and statistical significance. Small numbers under each method stands for the indexes of models from which the considered one is statistically better. The last row presents ranks according to the Friedman test.

| Dataset | UNPR[1] | GA[2] | CONS[3] | AUC[4] | CBP[6] |
|---|---|---|---|---|---|
| Breast-cancer | 55.23 − | 62.41 1,3,4 | 60.05 1 | 61.20 1,3 | **64.11** ALL |
| Breast-Wisconsin | 81.36 − | 89.82 1,3 | 85.38 1 | 89.27 1,3 | **90.02** ALL |
| Colic | 64.92 − | 75.63 1,3,4,5 | 70.56 1 | 73.62 1,3 | **76.14** ALL |
| Diabetes | 55.45 − | 58.65 1 | 58.21 1 | 59.65 1,2,3 | **60.89** ALL |
| Heart-statlog | 83.01 − | **86.54** 1,3,5 | 84.26 1 | 86.20 1,3 | 86.02 1,3 |
| Hepatitis | 56.84 − | **64.87** ALL | 60.89 1 | 61.43 1,3 | 62.89 1,3 |
| Ionosphere | 73.42 − | 79.17 1,4 | 78.64 1,4 | 76.18 1 | **82.03** ALL |
| Sonar | 85.18 − | 90.35 1,4 | 86.72 1,4 | 88.84 1 | **92.08** ALL |
| Voting records | 83.45 − | **90.74** ALL | 86.19 1 | 89.52 1,3 | 88.83 1,3 |
| CYP2C19 isoform | 72.84 − | 80.52 1,3,4,5 | 77.61 1 | 78.30 1,3 | **82.14** ALL |
| Rank | 5.00 | 2.35 | 3.45 | 2.90 | 1.30 |

Results of the Shaffer post-hoc test between the proposed FA-based one-class ensemble pruning and reference methods are depicted in Table 3.

**Table 3.** Shaffer test for comparison between the proposed clustering-based one-class ensemble pruning and reference methods. Symbol '=' stands for classifiers without significant differences, '+' for situation in which the method on the left is superior and '-' vice versa.

| hypothesis | $p$-value |
|---|---|
| CBP vs UNPR | + (0.0024) |
| CBP vs GA | + (0.0217) |
| CBP vs CONS | + (0.0116) |
| CBP vs AUC | + (0.0159) |

Experimental analysis has assessed a high quality of the proposed pruning method. For 7 out of 10 considered benchmarks the proposed clustering-based pruning was significantly better than all of the reference methods, which was confirmed by the statistical pairwise test. Additionally using two statistical tests for multiple comparison, we were able to prove that our method is statistically superior to all other pruning approaches when considering its performance over a set of benchmarks.

The high quality of our method can be explained by an efficient combination of automatic clustering algorithm ($x$-means) and using support functions outputted by individual classifiers as input for pruning step. Bayesian Information Criterion allows to quickly and automatically select an optimal number of clusters for each considered model, which is of crucial importance to the pruning step (as the number of clusters will results in the final size of the pruned pool of classifiers). This allows to pre-select the ensemble size beforehand, while other methods use trail-and-error schemes to choose the number of classifiers. By applying clustering over the support functions, we are able to detect templates in decisions of classifiers. This allows us to chose similar groups of learners and quickly discard irrelevant / similar models. Other methods rely directly on some metrics, and as they require some approximations of the outlier class they may mislead the pruning algorithm.

## 4    Conclusions

In this paper a novel scheme for pruning one-class classifier ensembles based on clustering was presented. It combined a powerful $x$-means clustering algorithm with a search over templates formed by values of support functions outputted by individual one-class classifiers. $X$-means method allowed to find an optimal clustering model in a reduced computational time, due to the use of information criterion and accelerated search over the possible clustering trees. Using support values as input for clustering alleviated the need for using a dedicated measure for evaluating the performance of classifiers. Experimental analysis, backed-up with statistical testing, had confirmed that the proposed clustering-based pruning is able to deliver highly satisfactory results in one-class problems.

Future works will concentrate on checking different criteria for measuring the similarity between models and selecting the representatives for each cluster.

**Acknowledgments.** This work was supported by funds for young researchers from Faculty of Electronics, Wroclaw University of Technology.

# References

1. Alpaydin, E.: Combined 5 x 2 cv f test for comparing supervised classification learning algorithms. Neural Computation **11**(8), 1885–1892 (1999)
2. Cheplygina, V., Tax, D.M.J.: Pruned random subspace method for one-class classifiers. In: Sansone, C., Kittler, J., Roli, F. (eds.) MCS 2011. LNCS, vol. 6713, pp. 96–105. Springer, Heidelberg (2011)
3. Cyganek, B.: One-class support vector ensembles for image segmentation and classification. Journal of Mathematical Imaging and Vision **42**(2–3), 103–117 (2012)
4. Demsar, J.: Statistical comparisons of classifiers over multiple data sets. Journal of Machine Learning Research **7**, 1–30 (2006)
5. Galar, M., Fernández, A., Barrenechea Tartas, E., Bustince Sola, H., Herrera, F.: Dynamic classifier selection for one-vs-one strategy: Avoiding non-competent classifiers. Pattern Recognition **46**(12), 3412–3424 (2013)
6. García, S., Fernández, A., Luengo, J., Herrera, F.: Advanced nonparametric tests for multiple comparisons in the design of experiments in computational intelligence and data mining: Experimental analysis of power. Inf. Sci. **180**(10), 2044–2064 (2010)
7. Kang, P., Kim, D., Cho, S.: Evaluating the reliability level of virtual metrology results for flexible process control: a novelty detection-based approach. Pattern Analysis and Applications **17**(4), 863–881 (2014)
8. Krawczyk, B., Woźniak, M.: Combining diverse one-class classifiers. In: Corchado, E., Snášel, V., Abraham, A., Woźniak, M., Graña, M., Cho, S.-B. (eds.) HAIS 2012, Part II. LNCS, vol. 7209, pp. 590–601. Springer, Heidelberg (2012)
9. Krawczyk, B., Woźniak, M.: Diversity measures for one-class classifier ensembles. Neurocomputing **126**, 36–44 (2014)
10. Pelleg, D., Moore, A.W.: X-means: extending k-means with efficient estimation of the number of clusters. In: Proceedings of the Seventeenth International Conference on Machine Learning (ICML 2000), June 29 - July 2, 2000, pp. 727–734. Stanford University, Stanford (2000)
11. Tax, D.M.J., Duin, R.P.W.: Support vector data description. Machine Learning **54**(1), 45–66 (2004)
12. Tax, D.M.J., Muller, K.: A consistency-based model selection for one-class classification. In: Proceedings - International Conference on Pattern Recognition, vol. 3, pp. 363–366 (2004). Cited By (since 1996):12
13. Woźniak, M., Grana, M., Corchado, E.: A survey of multiple classifier systems as hybrid systems. Information Fusion **16**(1), 3–17 (2014)

# Static Classifier Selection with Interval Weights of Base Classifiers

Robert Burduk[✉] and Krzysztof Walkowiak

Department of Systems and Computer Networks, Wroclaw University of Technology,
Wybrzeze Wyspianskiego 27, 50-370 Wroclaw, Poland
{robert.burduk,krzysztof.walkowiak}@pwr.wroc.pl

**Abstract.** The selection of classifiers is one of the important problems in the creation of an ensemble of classifiers. The paper presents the static selection in which a new method of calculating the weights of individual classifiers is used. The obtained weights can be interpreted in the context of the interval logic. It means that the particular weights will not be provided precisely but their lower and upper values will be used. A number of experiments have been carried out on several data sets from the UCI repository.

**Keywords:** Classifier fusion · Interval logic · Static classifiers selection · Multiple classifier system

## 1 Introduction

The pattern recognition task is one of the trends in research on machine learning [1]. In the case of the supervised classification we have a set of data in which a class label is assigned for each observation. In this issue we can consider a lot of research trends that are associated with problems such as: feature selection, extraction of features, selection of the training set, classifier selection and more. The classification task can be accomplished by a single classifier or by a team of classifiers. In the literature, the use of multiple classifiers for a decision problem is known as multiple classifier systems (MCS) or an ensemble of classifiers EoC [3], [8], [25]. These methods are popular for their ability to fuse together multiple classification outputs for better accuracy of classification.

The output of an individual classifier can be divided into three types [16].

- The abstract level – the classifier $\psi$ assigns the unique label $j$ to a given input $x$.
- The rank level – in this case for each input $x$, each classifier produces an integer rank array. Each element within this array corresponds to one of the defined class labels. The array is usually sorted and the label at the top being the first choice.
- The measurement level – the output of a classifier is represented by a measurement value that addresses the degree of assigning the class label to the given output $x$. An example of such a representation of the output is a posteriori probability returned by Bayes classifier.

© Springer International Publishing Switzerland 2015
N.T. Nguyen et al. (Eds.): ACIIDS 2015, Part I, LNAI 9011, pp. 494–502, 2015.
DOI: 10.1007/978-3-319-15702-3_48

According to these three types of outputs of the base classifier, various problems of combination function of classifiers outputs are considered. The problems studied in [17], [23] belong to the abstract level. The combining outputs for the rank level are presented in [11] and problems studied in [14], [15] belong to the last level.

The selection of classifiers is one of the important problems in the creation of EoC [12], [22]. This task is related to the choice of a set of classifiers from all the available pool of classifiers. Here you can distinguish between the static or dynamic selection [19], [24]. In the static classifier selection one set of classifiers is selected to create an EoC. This EoC is used in the classification of all the objects from the testing set. The main problem in this case is to find a pertinent objective function for selecting the classifiers. One of the best objective functions for the abstract level of classifier outputs is the simple majority voting error [21]. In the dynamic classifier selection for each unknown sample a specific subset of classifiers is selected [2]. It means that we are selecting different EoCs for different objects from the testing set. In this type of the classifier selection, the classifier is chosen and assigned to the sample based on different features [26] or different decision regions [4], [13].

In this work we will consider the static approach to build the EoC. In detail we propose the new method to select classifiers from the available pool. This method is based on the incorrect prediction of base classifiers and can be interpreted in the contents of the interval logic. The presented results are compared with the oracle concept [5] and base classifiers. The oracle classifier is used as the possible upper limit of classification accuracy of the EoC.

The text is organized as follows: in Section II the ensemble of classifiers and combination functions of classifiers outputs are presented. Section III contains the new method for assigning weights of individual base classifiers and the proposed static selection of classifiers. Section IV includes the description of research experiments comparing the suggested algorithms with base classifiers. Finally, conclusions from the experiments are presented.

## 2   Ensemble of Classifiers

Let us assume that we possess $K$ of different classifiers $\Psi_1, \Psi_2, \ldots, \Psi_K$. Such a set of classifiers, which is constructed on the basis of the same learning sample is called an ensemble of classifiers or a combining classifier. However, any of $\Psi_i$ classifiers is described as a component or base classifier. As a rule $K$ is assumed to be an odd number and each of $\Psi_i$ classifiers makes an independent decision. As a result, of all the classifiers' action, their $K$ responses are obtained. Having at the disposal a set of base classifiers one should determine the procedure of making the ultimate decision regarding the allocation of the object to the given class. It implies that the output information from all $K$ component classifiers is applied to make the ultimate decision.

In this work we consider the situation when each base classifier returns the estimation of a posteriori probability. This means that the output of all the base classifiers is at the measurement level. Let us denote a posteriori probability estimation by $\hat{p}_k(i|x)$, $k = 1, 2, \ldots, K$, $i = 1, 2, \ldots, M$, where $M$ is the number of the class labels. One of the possible methods for such outputs is the linear combination method. This method makes use of the linear function like Sum, Prod or Mean for the combination of the outputs. In the sum method the score of the group of classifiers is based on the application of the following sums:

$$s_i(x) = \sum_{k=1}^{K} \hat{p}_k(i|x), \qquad i = 1, 2, \ldots, M. \tag{1}$$

The final decision of the group of classifiers is made following the maximum rule and is presented accordingly, depending on the sum method (1):

$$\Psi_S(x) = \arg \max_i s_i(x). \tag{2}$$

In the presented method (2) the discrimination functions obtained from the individual classifiers take an equal part in building the combined classifier. Also, the weighted versions of these methods can be created. In this approach each of the classifiers has an allocated weight, which is taken into account by reaching the final decision of the group. Weights depend largely on the quality of their base classifiers. In the case when each classifier has one weight for all the possible classes an adequate group classification formula for the sum method is presented as follows:

$$sw_i(x) = \sum_{k=1}^{K} w_k * \hat{p}_k(i|x), \qquad i = 1, 2, \ldots, M, \tag{3}$$

where $w_k = 1 - Pe_{\Psi_k}$, and $Pe_{\Psi_k}$ is the empirical error of $\Psi_k$ classifier estimated on the testing set. In the case when the error is estimated on the learning set, we can talk about the estimation error based on the resubstitution method. Then $w_k$ weight of each component classifier is calculated depending on the:

$$w_k = \frac{\sum_{n=1}^{N} I(\Psi_k(x_n) = i, j_n = i)}{N}. \tag{4}$$

The $N$ value refers to the number of the learning set observations, which is used for estimating classifiers' weights, and $j_n$ is the class number of the object with $n$ index.

The obtained weights are normalised according to the formula:

$$\sum_{k=1}^{K} w_k = 1, \tag{5}$$

which means that the sum of weights of all classifiers from the ensemble is equal to unity. In this case the final decision of the ensemble of classifiers is the following:

$$\Psi_{wS}(x) = \arg \max_i sw_i(x). \tag{6}$$

Within the basic version (1) we have $w_k = 1$ for all $k = 1, \ldots, K$.

Another approach to obtain weights, is the calculation in each class separately. Then the corresponding weight is calculated from the equation:

$$w_{ki} = \frac{\sum_{n=1}^{N} I(\Psi_k(x_n) = i, j_n = i)}{\sum_{n=1}^{N} I(j_n = i)}. \tag{7}$$

Decision rules are clear, they are multiplied by the weights (7), and the sum method assumes designation of $\Psi_{wcS}(x)$.

## 3   The Proposal of Calculation Classifier Weights

We will suggest now the method for determining weights for the individual base classifiers. The values of these weights are the basis for the selection of classifiers. These weights can be seen in the context of the interval logic. It means that the particular weights will not be provided precisely but their lower and upper values will be used. Therefore, each $w_k$ weight of $K$-component classifier will be represented by the upper $\overline{w}_k^{ss}$ and lower $\underline{w}_k^{ss}$ value.

The method for calculating weights for individual classifiers is based on incorrect prediction of base classifiers. The upper value $\overline{w}_k^{ss}$ of $k$-classifier weight refers to the situation in which $k$-classifier was correct, while the other committee classifiers proved the incorrect prediction, in other words, while the other committee classifiers pointed to the incorrect decision. The lower value $\underline{w}_k^{ss}$ describes the situation in which $k$-classifier made errors, while the other committee classifiers made errors as well. The upper value is obtained from the dependence:

$$\overline{w}_k^{ss} = \frac{\sum_{n=1}^{N} UC_k^{ss_n}}{\arg \max_{l \in \mathcal{K}} \sum_{n=1}^{N} UC_l^{ss_n}}, \tag{8}$$

where

$$UC_k^{ss_n} = \sum_{n=1}^{N} I(\Psi_k(x_n) = i, j_n = i) * \frac{1}{K-1} \sum_{l=1, l \neq k}^{K} I(\Psi_l(x_n) \neq i, j_n = i). \tag{9}$$

However, the lower value is obtained from the dependence:

$$\underline{w}_k^{ss} = \frac{\sum_{n=1}^{N} LC_k^{ss_n}}{\arg \max_{l \in \mathcal{K}} \sum_{n=1}^{N} LC_l^{ss_n}}, \tag{10}$$

where

$$LC_k^{ss_n} = \sum_{n=1}^{N} I(\Psi_k(x_n) \neq i, j_n = i) * \frac{1}{K-1} \sum_{l=1, l \neq k}^{K} I(\Psi_l(x_n) \neq i, j_n = i). \quad (11)$$

Similarly, as in equation (7), we can calculate the weighted in class appropriate lower and upper values.

## 3.1   Classifier Selection

Given $K$ classifiers from the initial pool of classifiers now we select $L$, $L \leq K$ classifiers to the ensemble. The final decision is made on the basis of $L$ classifiers. It means that we select $L$ best classifiers. The advantage of this method is that it is very cheap computationally [21].

In the selection process, we set the value of $L$ and we select the upper $\overline{w}^{ss}$ or lower $\underline{w}^{ss}$ limit of the weights. Then the available pool $L$ classifiers with the largest coefficients $\overline{w}^{ss}$ or $\underline{w}^{ss}$ are chosen. For example, if we use upper value $\overline{w}^{ss}$, then we define the set $\overline{w}^{uSS} = \{\overline{w}_k^{uSS} : \overline{w}_k^{uSS} \geq \overline{w}_k^{ss-L}\}$. Then we create coefficient $\alpha_k^{uSS}$ according to the formula:

$$\alpha_k^{uSS} = \begin{cases} \overline{w}_k^{uSS} & \text{if } \overline{w}_k^{uSS} \in \overline{w}^{uSS} \\ 0 & \text{otherwise} \end{cases}. \quad (12)$$

If we use the sum method for the final combination of classifier outputs, then the score of the selected group of classifiers is the following:

$$s_i^{uSS}(x) = \sum_{k=1}^{K} \alpha_k^{uSS} * \hat{p}_k(i|x), \qquad i = 1, 2, \ldots, M. \quad (13)$$

The final decision of the selected group of classifiers is made according to the formula:

$$\Psi_{wS}^{uSS}(x) = \arg\max_i s_i^{uSS}(x). \quad (14)$$

Before making the final decision the coefficients $\alpha_1^{uSS}, \ldots, \alpha_K^{uSS}$ are normalised to unity.

## 4   Experimental Studies

In the experiential research 12 data sets were tested. Ten data sets come from the UCI repository [7]. Two of them have random observations in accordance with a certain assumed distribution. One of them has objects generated according to the procedure [6], this is the so called banana distribution. The second one, instead, has random objects drawn in accordance with the procedure [10] – Higleyman distribution. In both cases the a priori probability distribution of classes amounted to 0.5, and for each class 200 elements were drawn. The numbers of attributes, classes and available examples of the investigated data sets are introduced in Tab. 1. In the study the feature selection [9], [20] was not performed. The aim of the experiments was to compare the quality of classifications of the proposed static selection method algorithms with the sum method and base classifiers.

**Table 1.** Description of data sets selected for the experiments

| Data set | example | attribute | class |
|---|---|---|---|
| Banana | 400 | 2 | 2 |
| Breast Tissue | 106 | 10 | 6 |
| Dermatology | 366 | 33 | 6 |
| Glass Identification | 214 | 10 | 6 |
| Haberman's Survival | 306 | 3 | 2 |
| Highleyman | 400 | 2 | 2 |
| Ionosphere | 351 | 34 | 2 |
| Irys | 150 | 4 | 3 |
| Pima Indians Diabetes | 768 | 8 | 2 |
| Sonar (Mines vs. Rocks) | 208 | 60 | 2 |
| Vertebral Column | 310 | 6 | 3 |
| Wine | 178 | 13 | 3 |

The research assumes that the group of classifiers is composed of 7 elementary classifiers. Three of them work according to the $k - NN$ rule where the $k$ parameter is from the set $k \in 3, 5, 7$. For the four remaining base classifiers the decision trees are used, with the number of branches denoted as 2 and the depth of the precision tree having at most 6 levels. In the decision-making nodes the Gini index or entropy are used. The results are obtained via 10-fold-cross-validation method.

Tab. 2 show the results of the classification for the initial ensemble of classifiers $L = 7$ and the results after the classifier selection process $L = 5$, $L = 3$. Additionally this table shows the result for all the base classifiers.

The obtained results confirm the validity of the application for the proposed method of the classifiers selection. Additionally, it can be confirmed that the method proposed in the work can improve the quality of the classification in comparison with the other methods used in the work. The experiments did not confirm the existence of a statistical difference in the obtained results. The critical difference for the post-hoc Nemenyi test at $p = 0.05$ is equal $CD = 10.5$. In the obtained results we do not have difference greater that $CD$.

**Table 2.** Average rank positions for the base classifiers ($\Psi_1, ..., \Psi_7$) and EoC classifiers ($\Psi_{wS}^{SS}, ..., \Psi_S$) before ($L = 7$) and after ($L = 5$ or $L = 3$) selection produced by Friedman test

| Classifier | L=7 | L=5 | L=3 |
|---|---|---|---|
| $\Psi_{wS}^{SS}$ | 11.01 | 12.22 | 16.56 |
| $\Psi_{wcS}^{SS}$ | 12.32 | 11.21 | 13.08 |
| $\Psi_{wS}^{uSS}$ | 10.69 | 9.21 | 14.44 |
| $\Psi_{wcS}^{uSS}$ | 9.23 | $9.01^{(1)}$ | 12.12 |
| $\Psi_{wS}^{lSS}$ | 12.32 | 10.03 | 10.38 |
| $\Psi_{wcS}^{lSS}$ | 13.32 | 13.30 | 14.62 |
| $\Psi_S$ | 12.73 | - | - |
| $\Psi_1$ | 12.96 | - | - |
| $\Psi_2$ | 16.22 | - | - |
| $\Psi_3$ | 15.29 | - | - |
| $\Psi_4$ | 18.54 | - | - |
| $\Psi_5$ | 15.05 | - | - |
| $\Psi_6$ | 14.11 | - | - |
| $\Psi_7$ | 19.38 | - | - |

# 5    Conclusion

This paper discusses static classifier selection from a pool of base classifiers. The stage of the selection process uses a learning phase of the classifiers ensemble, in which information is obtained about the correction classification of each base classifier. This information provides the basis to obtain the weights for each base classifier.

Experimental studies were carried out on the data sets available from the UCI repository. They show that using the proposed in the work classifier selection is a good way. For the selected method for calculating weights, we obtained improvement of the classification quality measured by average values from Friedman test.

The advantage of EoC presented in the paper is the possibility to work in parallel and distributed environment. The distributed computing approaches enable efficient and parallel processing of the complicated data analysis task, also in the context of classification systems with multiple classifiers. In more detail, if a particular task can be separated into independent subtasks, the processing of each individual task can be executed on a separate machine. In a local environment such as a single data center, the separate machines are located quite close to each other in distances measured in kilometers. In the case of a more distributed approach, the computing systems can be located in geographically spread data centers located in distances measured in thousands of kilometers. First of all, the parallel processing provides a possibility to speed up the completion task what results in the decreased response time of the classification [18], what in some applications could be a key factor.

**Acknowledgments.** This work was supported by the Polish National Science Center under the grant no. DEC-2013/09/B/ST6/02264 and by the statutory funds of the Department of Systems and Computer Networks, Wroclaw University of Technology.

# References

1. Bishop, C.M.: Pattern Recognition and Machine Learning (Information Science and Statistics). Springer-Verlag New York Inc., Secaucus (2006)
2. Cavalin, P.R., Sabourin, R., Suen, C.Y.: Dynamic selection approaches for multiple classifier systems. Neural Computing and Applications **22**(3–4), 673–688 (2013)
3. Cyganek, B.: One-class support vector ensembles for image segmentation and classification. Journal of Mathematical Imaging and Vision **42**(2–3), 103–117 (2012)
4. Didaci, L., Giacinto, G., Roli, F., Marcialis, G.L.: A study on the performances of dynamic classifier selection based on local accuracy estimation. Pattern Recognition **38**, 2188–2191 (2005)
5. dos Santos, E.M., Sabourin, R.: Classifier ensembles optimization guided by population oracle. In: IEEE Congress on Evolutionary Computation, pp. 693–698 (2011)
6. Duin, R., Juszczak, P., Paclik, P., Pekalska, E., de Ridder, D., Tax, D., Verzakov, S.: PR-Tools4.1, A Matlab Toolbox for Pattern Recognition. Delft University of Technology (2007)
7. Frank, A., Asuncion, A.: UCI machine learning repository (2010)
8. Giacinto, G., Roli, F.: An approach to the automatic design of multiple classifier systems. Pattern Recognition Letters **22**, 25–33 (2001)
9. Guyon, I., Elisseeff, A.: An introduction to variable and feature selection. The Journal of Machine Learning Research **3**, 1157–1182 (2003)
10. Highleyman, W.H.: The design and analysis of pattern recognition experiments. Bell System Technical Journal **41**, 723–744 (1962)
11. Ho, T.K., Hull, J.J., Srihari, S.N.: Decision combination in multiple classifier systems. IEEE Trans. Pattern Anal. Mach. Intell. **16**(1), 66–75 (1994)
12. Jackowski, K., Krawczyk, B., Woźniak, M.: Improved adaptive splitting and selection: The hybrid training method of a classifier based on a feature space partitioning. International Journal of Neural Systems **24**(03) (2014)
13. Jackowski, K., Woźniak, M.: Method of classifier selection using the genetic approach. Expert Systems **27**(2), 114–128 (2010)
14. Kittler, J., Alkoot, F.M.: Sum versus vote fusion in multiple classifier systems. IEEE Trans. Pattern Anal. Mach. Intell. **25**(1), 110–115 (2003)
15. Kuncheva, L.I.: A theoretical study on six classifier fusion strategies. IEEE Trans. Pattern Anal. Mach. Intell. **24**(2), 281–286 (2002)
16. Kuncheva, L.I.: Combining Pattern Classifiers: Methods and Algorithms. John Wiley and Sons Inc. (2004)
17. Lam, L., Suen, C.Y.: Application of majority voting to pattern recognition: an analysis of its behavior and performance. IEEE Transactions on Systems, Man, and Cybernetics, Part A **27**(5), 553–568 (1997)
18. Przewoźniczek, M., Walkowiak, K., Woźniak, M.: Optimizing distributed computing systems for k-nearest neighbours classifiersevolutionary approach. Logic Journal of IGPL **19**(2), 357–372 (2010)
19. Ranawana, R., Palade, V.: Multi-classifier systems: Review and a roadmap for developers. International Journal of Hybrid Intelligent Systems **3**(1), 35–61 (2006)

20. Rejer, I.: Genetic algorithms in EEG feature selection for the classification of movements of the left and right hand. In: Burduk, R., Jackowski, K., Kurzynski, M., Wozniak, M., Zolnierek, A. (eds.) CORES 2013. AISC, vol. 226, pp. 579–589. Springer, Heidelberg (2013)
21. Ruta, D., Gabrys, B.: Classifier selection for majority voting. Information Fusion 6(1), 63–81 (2005)
22. Smętek, M., Trawiński, B.: Selection of heterogeneous fuzzy model ensembles using self-adaptive genetic algorithms. New Generation Comput. 29(3), 309–327 (2011)
23. Suen, C.Y., Legault, R., Nadal, C.P., Cheriet, M., Lam, L.: Building a new generation of handwriting recognition systems. Pattern Recognition Letters 14(4), 303–315 (1993)
24. Trawiński, K., Cordon, O., Quirin, A.: A study on the use of multiobjective genetic algorithms for classifier selection in furia-based fuzzy multiclassifiers. International Journal of Computational Intelligence Systems 5(2), 231–253 (2012)
25. Ulas, A., Semerci, M., Yildiz, O.T., Alpaydin, E.: Incremental construction of classifier and discriminant ensembles. Information Science 179(9), 1298–1318 (2009)
26. Woloszyński, T., Kurzyński, M.: A probabilistic model of classifier competence for dynamic ensemble selection. Pattern Recognition 44(10–11), 2656–2668 (2011)

# Pruning Ensembles with Cost Constraints

Bartosz Krawczyk and Michał Woźniak[✉]

Department of Systems and Computer Networks, Wrocław University of Technology,
Wybrzeże Wyspiańskiego 27, 50-370 Wrocław, Poland
{bartosz.krawczyk,michal.wozniak}@pwr.edu.pl

**Abstract.** The paper presents a cost-sensitive classifier ensemble pruning method, which employs a genetic algorithm to choose the most promising ensemble. In this study the pruning algorithm considers constraints put on the cost of selected features, which is the one of the key-problems in the real-life decision support systems, especially dedicated medical support systems. The proposed method takes into consideration both the overall classification accuracy and the cost constraints, returning balanced solution for the problem at hand. Additionally, also to boost the value of the exploitation cost, we propose to use cost-sensitive decision trees as the base classifiers. The pruning algorithm was evaluated on the basis of the comprehensive computer experiments run on cost-sensitive medical benchmark datasets.

**Keywords:** Pattern classification · Machine learning · Classifier ensemble · Classifier ensemble pruning · Cost-sensitive classification

## 1 Introduction

The classification systems are widely used to solve practical problems coming from different areas of human activities. Basically, for most of them we are able to propose an appropriate classifier to make a high-quality decision. However, its choice is usually limited by several constrains related to the classification cost, because the question is what does mean "the high-quality". The criteria used to asses the classifier are strongly related to the misclassification cost. Bayes decision theory bases on so-called loss function [3], which returns the cost (loss) connected with the wrong assigning object from class $j$ to class $i$. Let's assume that the set of possible labels is $\mathcal{M} = \{1, ..., M\}$, then

$$L : \mathcal{M} \times \mathcal{M} \to \mathcal{X} \tag{1}$$

where $\mathcal{X}$ is the feature space (we assume that $\mathcal{X} \subseteq \mathrm{R}^d$). For such formulated loss function the optimal Bayes classifier $\Psi^*$ should minimizes the average (overall) classification risk

$$\min_{\Psi} Risk(\Psi) = Risk(\Psi^*) \tag{2}$$

© Springer International Publishing Switzerland 2015
N.T. Nguyen et al. (Eds.): ACIIDS 2015, Part I, LNAI 9011, pp. 503–512, 2015.
DOI: 10.1007/978-3-319-15702-3_49

where

$$Risk(\Psi) = E[L(i,j)] = \int_{\mathcal{X}} \sum_{j=1}^{M} L(\Psi(x),j)p_j f_j(x)dx \qquad (3)$$

and $x \in \mathcal{X}$ stands for example, $p_j$ is the posterior probability of the class $j$, and $f_j(x)$ is the conditional probability density function.

In this work we will use such the specific but very popular case of the 0-1 loss function

$$L(i,j) = \begin{cases} 0 & if \ i = j \\ 1 & if \ i \neq j \end{cases} \qquad (4)$$

because usually we have a problem with assessing the loss values, e.g., for medical diagnosis, we are not able to answer the question, whether the loss is higher as patient suffering from flu is assigned strep throat or patient suffering from laryngitis is assigned to flu suffers. It leads to the following decision rule of the optimal (Bayes) recognition algorithm $\Psi^*$, minimizing the probability of misclassification

$$\Psi^*(x) = i \ if \ p_i(x) = \max_{k \in \mathcal{M}} p_k(x), \qquad (5)$$

We can find several propositions on how to deal with it e.g., Peng et al. proposed how to create cost-sensitive ensemble for the medical decision support system [16]. But we should also mention that cost could be considered during training or decision making process. In this work we will focus on the cost associated with the acquisition cost of the attributes necessary to make a decision. Such an approach is usually called cost-sensitive classification [18].The cost may be measured in a given currency as a price of medical tests or time, in case of measurements, which repetitions require significant time. Such a meaning of cost-sensitive classification arises frequently in many fields of human activities as an industrial production process [19], robotics [17], technological diagnosis [11] to enumerate only a few. But this problem is most clearly visible for the medical diagnosis [14]. As we mentioned above for the most of decision problems we have a necessary tools at our disposal to make a high quality decision, but from the practical point of view, we have to notice the assumption that a decision-maker has unlimited budget is unreal. Therefore in real cases, e.g., physicians have to balance the costs of various medical tests with the expected benefits. Often doctors have to make the diagnosis fast on the basis of (low cost) features acquired form measurements that do not require much time to conduct - because therapeutic action has to be taken without any delay.

We should also mention that we realize of other costs appeared during classifier training or testing. Usually, they are associated with the time necessary to build the structure used by the classifier in the future (as the decision tree) or time necessary for decision making, which could also be very high (e.g., for lazy classifiers). In this work we will not handle this problem, but we will be focusing on the problem of how to prune a classifier ensemble taking into consideration on the one hand the cost of necessary attributes using by individuals and on the other hand the overall classifier system accuracy. Additionally, we will take into consideration a fixed cost limit of attribute acquisition. To solve this problem the genetic algorithm is employed.

In this article, as the base classifier the cost-sensitive decision tree trained with limited budget of exploitation [15] is used, therefore let's present shortly its idea. Its authors proposed to modify the function of decision tree induction algorithm, which is responsible for the evaluation the information gain of chosen attribute (using information gain, information ratio, or any other criterion), what was presented in Alg. 1.

---

**Algorithm 1..** Modification of the information gain evaluation

---
1: **if** set of attributes is empty or cost of previous chosen attributes and cost of any remaining one exceed the COST_LIMIT **then**
2:     return the single node tree Root with label = most common value of label in examples and return
3: **end if**
4: Choose the best attribute A from the set of attributes for which summarized cost of attributes does not exceed the cost limit

---

where COST-LIMIT means the maximum price which could be paid for the information about attributes' values (as a general initial COST-LIMIT is fixed by the expert), This work is the continuation of the previous works of authors on combined classifiers [8,10]. We have already presented the genetic approach to form an ensemble with minimal classification error within a fixed cost bounds [10], and [9], where we proposed the extension of the previously developed AdaSS (Adaptive Splitting and Selection) algorithm which uses the criterion based on EG2 [13].

The contributions of this work is as follows:

- The proposition of the new pruning criterion which takes into consideration on the one hand the classification accuracy and on the other hand the cost of the individual classifiers' runs.
- The pruning algorithm based on genetic approach.
- The comprehensive computer experiments on quality of the developed method.

The content of the work is as follows. The next section presents the chosen decision rule using by the classifier ensemble and the method of the classifier ensemble pruning using genetic approach. Then the experiments on the pruning quality are presented. The last section concludes the work and proposes the future research directions.

## 2    Algorithm

Let's us shortly describe the main components of the proposed method.

## 2.1  Combination Rule

As the combination rule we choose the majority voting, which uses the following formulae

$$\Psi(x) = \arg\max_{i \in \mathcal{M}} \sum_{k=1}^{n} [\Psi_n(x) = i] \tag{6}$$

where $[\ ]$ denotes Inverson's bracket, $n$ is the number of individuals, and $\Psi_k$ stands for the $k$th individual classifier.

Of course we realize that more sophisticated combination rules could be used, especially based on the weighted voting or aggregating, but this work focuses on the cost-sensitive pruning efficiency and we do not want to complicate the representation of the optimization task. In the next works, we are going to use more advanced combiners.

## 2.2  Training Criterions

Two main objectives are defined for the training algorithm:

1. maximization of the classification accuracy,
2. minimization of the data acquisition cost required for classification.

The first one is based on the frequency of correct classifications:

$$Q(\Psi) = \frac{1}{K} \sum_{n=1}^{K} (\delta(\Psi(x_n), j_n)), \tag{7}$$

where $\delta$ denotes Kronecker's delta. This is computed over a learning set:

$$\mathcal{LS} = \{(x_1, j_1), (x_2, j_2), ..., (x_K, j_K)\}, \tag{8}$$

where $x_i$ denotes observations described in the $i$-th object and $j_i$ denotes its correct class label [1].
Because we would like to assure the pool diversity by the individual classifier training on the basis of different subsets of available features, therefore let us propose the following representation of the classifier $\Psi(x)$:

$$\mathcal{A} = \begin{pmatrix} a^{(1)}(x^{(1)}) & \cdots & a^{(1)}(x^{(d)}) \\ \vdots & \ddots & \vdots \\ a^{(n)}(x^{(1)}) & \cdots & a^{(n)}(x^{(d)}) \end{pmatrix}, \tag{9}$$

where $a^{(p)}(x^{(q)}) = 1$ if the $q$-th feature is used by the $p$-th individual classifier used by $\Psi(x)$, otherwise $a^{(p)}(x^{(q)}) = 0$ means that mentioned above attribute

---

[1] Let us note that instead of misclassification error estimator we can use the cost of misclassification what requires a different lost function definition only.

is not used by it. It allows us to formulate classification cost of the combined classifier:

$$cost(\Psi) = \sum_{l=1}^{d} \left( sgn \left( \sum_{k=1}^{n} a^{(k)}(x^{(l)}) \right) cost(x^{(l)}) \right), \qquad (10)$$

where $cost(x^{(l)})$ denotes the acquisition cost of $l$-th feature.

Both objectives are included in the proposed criterion, which must be subject to a maximization procedure:

$$\hat{Q}(\Psi) = \frac{Q(\Psi)}{(1 + cost(\Psi))^{\omega}}, \qquad (11)$$

where $\omega$ is a parameter which controls the weight of the cost criterion. It may take values in range [0,1]. The mentioned above proposition is similar to the split criterion used by Núñez in the cost-sensitive decision tree induction algorithm EG2 [13].

## 2.3   Pruning Algorithm

As the pruning method we chose so-called optimization-based approach presented in Alg. 2, and the methods consider ensemble pruning as an optimization problem and most of them use heuristic techniques [1], evolutionary algorithms [5], or competitive techniques based on cross-validation [2] to enumerate only a few. For classification tasks, the cost of acquiring feature values (which could be interpreted as the price for examination or time required to collect the data for decision making) plays a key role, and this must be taken into consideration during the ensemble pruning.

---

**Algorithm 2..** Ensemble pruning framework

---

**Require:** pool of elementary classifiers $\Pi_{init}$
**Ensure:** pool of selected elementary classifier $\Pi$, where $|\Pi| \leq |\Pi_{init}|$
 1: $\Pi = \varnothing$
 2: **for** each possible combination of $\Pi_{init}$ **do**
 3:    **if** a given combination of $\Pi_{init}$ *is better* than $\Pi$ **then**
 4:       $\Pi$ = a given combination of $\Pi_{init}$
 5:    **end if**
 6: **end for**

---

The process of searching for maximum value of criterion (11Alpaydin:2010) was treated as a compound optimization problem solved by the genetic algorithm (GA) [12]. It relays on an iterative processing of a population of individuals which represents possible solutions in the form of a chromosome consisting selected features used by individual classifiers:

$$Chromosome = [\mathcal{A}] \qquad (12)$$

For the purpose of GA implementation $\mathcal{A}$ matrix is transformed into vector. Training procedure consists of following steps:

– Initialization - random generation of the initial population and setting of parameters which control the algorithm: $N_c$ - upper limit of algorithm cycles, $N_p$ - size of population, $\beta$ - mutation probability, $\gamma$ - crossover probability, $\Delta_m$ - mutation range factor.
– Selection and reproduction process of drawing the best individuals from the population according to the roulette selection with elite.

$$P_a(t) = 2\beta \frac{t}{N_c} \qquad (13)$$

where $t$ is the iteration index of the algorithm and $P_a(t)$ is the probability of mutation.
– Crossover procedure for exchanging data between the two parent individuals according to the two-point rule.
– Protecting against overfitting procedure which cancels training process when the accuracy of classification (controlled at each generation over a validation data set) deteriorates.

# 3   Experimental Investigations

The main objective of the experiment was to examine the behavior of the proposed versions of the cost-sensitive ensemble pruning algorithm (CSEP1 uses as the pruning criterion accuracy only, while CSEP2 uses the combination of the cost and accuracy in the form of eq.11) in comparison with single-model of the cost sensitive classifier (CSTree) described shortly in the first section.

## 3.1   Set-up

In this study medical benchmark datasets from [4] with features described by cost values are used:

1. Heart disease (303 examples, 13 features, 2 classes),
2. Hepatitis (155 examples, 19 features, 2 classes),
3. Liver disorders (345 examples, 5 features, 2 classes),
4. Pima Indian diabetes (768 examples, 8 features, 2 classes).

As we mention above, as a base classifier, we used C4.5 model which considers the classification (attribute acquisition) cost limit. Each classifier was trained on the basis of Random Subspace method [7], using 20% of original feature space size. We generated 50 base classifiers for each experiment.

Genetic algorithm, used for the cost-sensitive pruning procedure utilized the following parameters: upper limit of algorithm cycles $N_c = 100$, population size $N_p = 50$, mutation probability $\beta = \{0.7; 0.3\}$, crossover probability $\gamma = \{0.3; 0.7\}$, mutation range factor $\Delta_m = 0.2$ and upper limit of iterations without improvement $V = 15$. The $\omega$ parameter, responsible for weight of cost criterion, was set as $\omega = 0.7$.

For results of the experiments were evaluated using 10CV and the Shaffer post-hoc test [6] to find out which of the tested methods are distinctive among an $n \times n$ comparison. The post-hoc procedure is based on a specific value of the significance level $\alpha$. Additionally, the obtained $p$-values should be examined in order to check how different given two algorithms are. We set $\alpha = 0.05$.

## 3.2   Results and Discussion

The results of experiments are presented in the Figure 1. The plots show the correlation between the maximum cost threshold and overall ensemble error, with respect to the examined classifier.

In Table 1, we present the average costs required by each classifier, while the results for Shaffer test are given in Table 2.

**Table 1.** Average cost required by each of the examined classifiers.

| Dataset | CSTree | CSEP1 | CSEP2 |
|---------|--------|-------|-------|
| Heart disease | 292.38 | 290.52 | **242.18** |
| Hepatitis | 24.05 | 24.18 | **19.87** |
| Liver disorders | 49.88 | 49.56 | **40.68** |
| Pima Indians diabetes | 48.20 | 48.72 | **42.04** |

**Table 2.** Shaffer test for comparison between the methods. Symbol '=' stands for classifiers without significant differences, '+' for situation in which the method on the left is superior and '-' vice versa.

| hypothesis | $p$-value |
|---------|--------|
| CSEP1 vs CSTree | + (0.0149) |
| CSEP2 vs CSTree | + (0.0164) |
| CSEP1 vs CSEP2 | = (0.7389) |

From the results, one may clearly see that the proposed pruning methods for forming cost-sensitive ensembles return models with superior accuracy in comparison to single-model classifiers. Additionally, they still follow the imposed cost constraints. We are able to select classifiers with highest individual accuracies without increasing the cost required for the final diagnosis. It is important to note, that due to using a binary sum method when calculating the overall cost displayed by the ensemble, we avoid the situation in which one feature would be counted several times in the cost summarization procedure (as we "pay" the cost of the feature during its acquisition process - once extracted, it can be used by many classifiers).

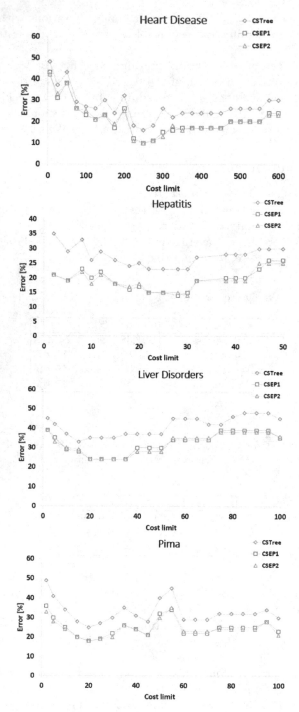

**Fig. 1.** Classification errors with the respect to the maximum classification cost for the tested datasets

Shaffer test proved, that both versions of our ensemble pruning algorithm are statistically superior to the reference classifier. However, differences between them are very small and deemed as statistically insignificant. But the main advantage of the CSEP2 algorithm lies somewhere else - in reduced cost requirements. When examining Table 1, one can see that CSEP2 average utilized cost is much lowet than ones used by CSTree and CSEP1. At the same time, CSEP2 is able to maintain accuracy on similar level as CSEP1. This shows, that embedding cost as a part of the pruning criterion is worthwhile research direction.

## 4   Final remarks

The paper concentrated on the cost-sensitive classifier ensemble pruning. In this work the cost was associated with acquisition cost of the features selected for each of classifiers in the pool. The cost-sensitive criterion was embedded in the pruning criterion which took into consideration both the overall classification accuracy and the cost constraints, returning a balanced solution for the problem at hand.

The experimental results are quite promising, because the proposed pruning methods usually returned combined classifiers which outperformed the individual models. We believe that the proposed method may be very useful, especially for medical decision support systems in which physicians require at the same time a high accuracy and a lowest possible diagnosis cost. Therefore, our future studies will focus on:

- increasing the classifier ensemble diversity by employing a diversity measure in he pruning criterion,
- using more sophisticated combination rule based on weighted voting or weighted aggregation, but it will require the deep rebuilding of the pruning algorithm by adding the weight calculating procedure,
- carrying out the experiments on real datasets not only from the medical domain.

**Acknowledgment.** This work was supported by the Polish National Science Center under the grant no. DEC-2013/09/B/ST6/02264.

## References

1. Banfield, R.E., Hall, L.O., Bowyer, K.W., Kegelmeyer, W.P.: Ensemble diversity measures and their application to thinning. Information Fusion **6**(1), 49–62 (2005)
2. Dai, Q.: A competitive ensemble pruning approach based on cross-validation technique. Knowl.-Based Syst. **37**, 394–414 (2013)
3. Duda, R.O., Hart, P.E., Stork, D.G.: Pattern Classification, 2nd edn. Wiley, New York (2001)
4. Frank, A., Asuncion, A.: UCI machine learning repository (2010). http://archive.ics.uci.edu/ml

5. Gabrys, B., Ruta, D.: Genetic algorithms in classifier fusion. Appl. Soft Comput. **6**(4), 337–347 (2006)
6. García, S., Fernández, A., Luengo, J., Herrera, F.: Advanced nonparametric tests for multiple comparisons in the design of experiments in computational intelligence and data mining: Experimental analysis of power. Inf. Sci. **180**(10), 2044–2064 (2010)
7. Ho, T.K.: The random subspace method for constructing decision forests. IEEE Trans. Pattern Anal. Mach. Intell. **20**, 832–844 (1998)
8. Jackowski, K., Krawczyk, B., Wozniak, M.: Improved adaptive splitting and selection: The hybrid training method of a classifier based on a feature space partitioning. International Journal of Neural Systems **24**(03), 1430007 (2014)
9. Jackowski, K., Krawczyk, B., Woźniak, M.: Cost-sensitive splitting and selection method for medical decision support system. In: Yin, H., Costa, J.A.F., Barreto, G. (eds.) IDEAL 2012. LNCS, vol. 7435, pp. 850–857. Springer, Heidelberg (2012)
10. Krawczyk, B., Woźniak, M.: Designing cost-sensitive ensemble – genetic approach. In: Choraś, R.S. (ed.) Image Processing and Communications Challenges 3. AISC, vol. 102, pp. 227–234. Springer, Heidelberg (2011)
11. Lirov, Y., Yue, O.-C.: Automated network troubleshooting knowledge acquisition. Applied Intelligence **1**, 121–132 (1991)
12. Michalewicz, Z.: Genetic algorithms + data structures = evolution programs, 3rd edn. Springer, London (1996)
13. Núñez, M.: The use of background knowledge in decision tree induction. Mach. Learn. **6**(3), 231–250 (1991)
14. Núñez, M.: Economic induction: A case study. In: EWSL, pp. 139–145 (1988)
15. Penar, W., Wozniak, M.: Cost-sensitive methods of constructing hierarchical classifiers. Expert Systems **27**(3), 146–155 (2010)
16. Peng, Y., Huang, Q., Jiang, P., Jiang, J.: Cost-sensitive ensemble of support vector machines for effective detection of microcalcification in breast cancer diagnosis. In: Wang, L., Jin, Y. (eds.) FSKD 2005. LNCS (LNAI), vol. 3614, pp. 483–493. Springer, Heidelberg (2005)
17. Tan, M., Schlimmer, J.C.: Cost-sensitive concept learning of sensor use in approach and recognition. In: Proceedings of the Sixth International Workshop on Machine Learning, pp. 392–395. Morgan Kaufmann Publishers Inc., San Francisco (1989)
18. Turney, P.D.: Cost-sensitive classification: empirical evaluation of a hybrid genetic decision tree induction algorithm. J. Artif. Int. Res. **2**(1), 369–409 (1995)
19. Verdenius, F.: A method for inductive cost optimization. In: Proceedings of the European Working Session on Learning on Machine Learning, EWSL 1991, pp. 179–191. Springer-Verlag New York Inc., New York (1991)

# Kernel-Based Regularized Learning for Time-Invariant Detection of Paddy Growth Stages from MODIS Data

Sidik Mulyono[1], Harisno[2(✉)], Mahfudz Amri[2], M. Ivan Fanany[1],
and T. Basaruddin[1]

[1] Faculty of Computer Science, Universitas Indonesia, Depok, Indonesia
sidik.mulyono@bppt.go.id, {ivan,chan}@cs.ui.ac.id
[2] Graduate Program in Information System Management,
Bina Nusantara University, Jakarta 1530, Indonesia
harisno@binus.edu, mahfudz.amri@gmail.com

**Abstract.** Most current studies have been applying high temporal resolution satellite data for determining paddy crop phenology, that derive into a certain vegetation indices, by using some filtering and smoothing techniques combined with threshold methods. In this paper, we introduce a time invariant detection of paddy growth stages using single temporal resolution satellite data instead of high temporal resolution with complex cropping pattern. Our system is a kernel-based regularized learner that predicts paddy growth stages from six-bands spectral of Moderate Resolution Image Spectroradiometer (MODIS) satellite data. It evaluates three Kernel-based Regularized (KR) classification methods, i.e. Principal Component Regression (KR-PCR), Extreme Learning Machine (KR-ELM), and Support Vector Machine with radial basis function (RBF-SVM). All data samples are divided into training (25%) and testing (75%) sampling, and all models are trained and tested through 10-rounds random bootstrap re-sampling method to obtain more variety on hypothesis models during learning. The best model for each classifier method is defined as the one which has the highest kappa coefficient during testing. The experimental results show that the classification accuracy of each classifiers on testing are high competitive, i.e. 84.08%, 84.04%, and 84.95% respectively.

**Keywords:** Remote sensing · MODIS · Phenology · Paddy growth stages · Kernel based learner

## 1 Introduction

As one of the major staple foods in the world, rice plays an important role in supporting more than one billion people ([3]). It has a unique phenology profile during its growing, characterized by a cycle that begins from planting stage, followed by heading stage, and ending at harvesting stage. This paddy phenology profile is vary depend on cultivar and planting area, and heavily influenced by surrounding environment conditions, that represents information for irrigation schedule or fertilizer management ([4]), and information for pest and disease attacks, in order to minimize yield looses.

© Springer International Publishing Switzerland 2015
N.T. Nguyen et al. (Eds.): ACIIDS 2015, Part I, LNAI 9011, pp. 513–525, 2015.
DOI: 10.1007/978-3-319-15702-3_50

In tropical region, paddy may be planted 2 or 3 times a year for a certain cultivar in irrigated sites, and once a year in rain-fed site that can be shown by a number of cycle waves of the phenology profile around a year. In Indonesia, paddy plantation is not constrained only by some physical factors (such as air temperature and climate), but also constrained on a schedule of irrigation and arbitrary decisions (local wisdom) managed by farmers. In case of rain-fed paddy fields, depend on climate condition and customer demands, sometimes they switch paddy crop with other valuable crops such as peanuts, cassava, and soy beans. As a consequence, such condition will cause complex cropping patterns, which means spatial crop rotation may change arbitrary in every planting season. Meanwhile, widely used statistical estimation of rice production, which is largely based on direct human observation, often causes some irregularities since the results tend to be excessive or over-estimated. More reliable and rapid harvest yield estimation for paddy fields is a critical issue to support the National Food Security Program that have been promoted and coordinated by Indonesian government. In addition, an accurate and timely rice conditions, monitoring, and rapid rice harvest area estimation are certainly needed ([1], [2]).Studies on paddy phenology prediction using remote sensing data have been proposed by many researchers. The time-series of three vegetation indices derived from MODIS images, including Normalized Difference Vegetation Index (NDVI), Normalized Difference Soil Index (NDSI), Enhanced Vegetation Index (EVI), and Land Surface Water Index (LSWI) can be effectively applied to build an algorithm for large scale mapping of flooding and transplanting paddy rice in summer season ([5]). The algorithm maps paddy rice fields in 13 countries of South and Southeast Asia ([6]). [7] extended and adapted the MODIS based approach for paddy rice detection and mapping, as used in [6], with more variables, mixed and local rice cropping systems, including single, early, and late rice. [8] improved this algorithm by developing the variable threshold model for the threshold (T) in conjunction with MODIS EVI utilizing Synthetic Aperture Radar (SAR) -based irrigated paddy field maps.

[9] developed three procedures for remotely determining paddy phenological growth stages: (i) prescription of multi-temporal MODIS data; (ii) filtering time-series EVI profile by time-frequency analysis; and (iii) specifying the phenological stages by detecting the maximum point, minimum point, and inflection point from the smoothed time-based EVI profile. Based on this procedures, [9] also defined paddy growth stages into three dates, (a) heading date, characterized by the maximum point of EVI time profile, (b) flooding and planting date, indicated by the minimum point of EVI time profile and the first derivative changes from negative to positive, and (c) harvesting date, presented by the minimum point of first derivative. Meanwhile [5] used LSWI and EVI combination to provide sensitivity to standing water during flooding and transplanting date. However, the relationship between EVI and LSWI may become complicated due to the growth of aquatic plants during flooding and transplanting date [7]. Therefore, in order to simplify the detection of this date, [5] used a global threshold value of 0.05 using $LSWI + 0.05 \geq EVI$ approach to detect flooding and transplanting date.

Based on these three dates proposed by [9], [10] developed a simple algorithm by defining the threshold values of EVI and LSWI derived from MODIS 500-m for each stage for detecting paddy growth stages with complex cropping pattern. [20] proposed the threshold value of EVI and NDWI to estimate rice planting time with complex cropping pattern. [11] presented a comparative study on satellite and model based crop phenology in West Africa using 4 dates of EVI phenology crops, i.e. green-up or start of season (SOS); maturity or start of maximum (SMAX), senescence or end of maximum (EMAX), and dormancy or end of season (EOS).

Time-based phenology profile constructed from remote sensing data have noises caused by cloud cover and atmospheric condition. Some filter and smoothing process should be applied before further analysis. Wavelet and Fourier transform were applied to remove noise in time-based EVI profile by dividing the noise components and reconstructing time-based EVI profile ([9], [12]). In another study, a progressive iteration approximation (PIA) by providing a Bézier curve ([13]) and Savitzky-Golay filter ([21]) can also be used for signal smoothing and reducing noise representation in time-based NDVI profile. [4] performed a cubic interpolation method on discrete NDVI points to obtain a smoothed NDVI curve and to calculate its slope.

From this point of view, most researchers have been successfully utilizing various time-based vegetation indices derived from various remote sensing data to determine paddy crop phenological stages. To have a better smoothed paddy phenology profile they have to apply high temporal resolution (several time-series) of remote sensing images, that will require a huge capacity of RAM to store remote sensing image's files for predicting the existing paddy growth condition in certain wide coverage area. This paper addresses the following main issues on: (i) how to determine the existing paddy growth stages by using some limited series of temporal resolution of remote sensing data, and (ii) how to realize a general prediction model derived from this limited series that is applicable in time invariant manner for detecting paddy growth stages at any given time with high accuracy. Instead of relying on time-series data, we pursue a spectral-based time-invariant prediction model.

To address these two issues, this paper focuses on utilizing the spectral properties of remote sensing images instead of relying on time series or temporal images. Initially we construct the paddy crop time-based phenology profiles from a limited six-years time series of EVI and LSWI derived from MODIS. We then construct an algorithm for indicating three-labeled paddy growth stages as defined by [9], i.e., flooding & transplanting date; heading date; and harvesting & post harvest, which contain the LSWI threshold to detect water existence, and EVI threshold to detect greenness level. The results of our algorithm have been validated with 32 monitoring sites of field survey held during June-September 2012 laid around Karawang and Indramayu district of West Java Indonesia. The results showed that paddy growth stages detected by using this algorithm were almost matched with field survey data. Therefore they can be considered as ground reference data for further analysis. Henceforward, regardless of prediction time, paddy growth stages prediction models are constructed from six-band spectral of MODIS using three Kernel-based Regularized (KR) classification methods, i.e. Principal Component Regression (KR-PCR), Extreme Learning Machine (KR-ELM), and Support Vector Machine with radial basis function

(RBF-SVM). All data samples are divided into training (25%) and testing (75%) sampling, and all models are trained and tested through 10-round random bootstrap resampling methods to obtain more variety on hypothesis models. The best model for each classifier method is defined as the one which has the highest kappa coefficient during testing. The best models for each classifier is defined as the one which has the highest kappa coefficient during testing.

The time-invariant models built from this study are expected to be applied directly to the latest MODIS imagery with a single temporal resolution to predict the existing paddy growth stages, the harvesting time, and harvested area spatially.

## 2    Data Samples

Field campaign for paddy growth stages observation had been conducted during June - October 2012 around Karawang and Indramayu District, lying in Northern part of West Java-Indonesia. Figure 1 presents the location of study area, which comprises 2 locations, where the first one is located in Karawang district that has 16 sites with 3-times observation; and the second one is located in Indramayu district that has 16 sites with 2-times observation.

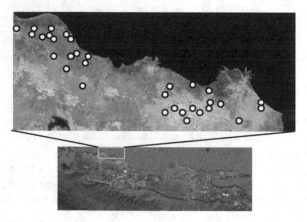

**Fig. 1.** Area of study laid on Karawang and Indramayu district of West Java

A series of 8-day composite MODIS images used in this study are taken during 2008 - 2014. We processed MODIS surface reflectance product computed from the MODIS bands 1, 2, 3, 4, 5, 6, and 7 (centered at 648 nm, 858 nm, 470 nm, 555 nm, 1240 nm, 1640 nm, and 2130 nm, respectively) with 500 meter spatial resolution, and all images were also preprocessed in two stages for geometric and atmospheric corrections. In this study, we neglected band 5 of MODIS due to an error caused by its sensor, and used the rest of six bands for analysis.

From these data, we derive time series of EVI and LSWI using equation (1) and (2),

$$EVI = 2.5 \frac{R_{NIR} - R_{Red}}{R_{NIR} + R_{Red} - 7.5 R_{Blue} + 1} \tag{1}$$

$$LSWI = \frac{R_{NIR} - R_{SWIR}}{R_{NIR} + R_{SWIR}} \tag{2}$$

where $R_{Blue}$, $R_{Red}$, $R_{NIR}$, $R_{SWIR}$ are surface reflectance value from blue, red, near infrared, and short wave infrared band of MODIS respectively.

## 3    Methodology

### 3.1    The Algorithm for Detecting Paddy Growth Stages

In this paper, we divided paddy growth stages into three dates, i.e. flooding & transplanting date (1); heading date (2); and harvesting & post-harvest date (3); that will be detected by our proposed algorithm, where hereinafter these dates are indicated by index 1, 2, and 3 respectively. We use a global threshold value of 0.05 using $LSWI + 0.05 \geq EVI$ approach to detect flooding & transplanting date, and $LSWI + 0.05 \leq EVI$ to detect heading date and harvesting & post harvest date, while they can be separated easily by using EVI threshold=0.28. We propose a heuristic algorithm for detecting three dates of paddy growth stages as shown in Fig. 2.

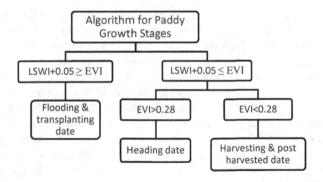

**Fig. 2.** Heuristic algorithm for detecting three dates of paddy growth stages

Instead of using commonly used paddy growing dates, we used the terminology of dates that commonly used by most farmers during field campaign, i.e. plowing, first vegetative, second vegetative, generative, harvested and post harvest (hereinafter indicated by A, B, C, D, and E respectively), which have slice different with dates used in this algorithm. The adjustment of date terminology between field campaign and algorithm is shown in table 1. In the result session, we validate the dates detected by this algorithm with field campaign dates using this adjustment.

**Table 1.** The adjustment dates between field campaign and algorithm

| Dates used in Algorithm | Dates used in field campaign |
|---|---|
| 1. Flooding and transplanting date | A. Plowing date |
| | B. First vegetative date |
| 2. Heading date | C. Second vegetative date |
| | D. Generative date |
| 3. Harvesting and post harvest date | E. Harvested and bare land date |

For learning the classifiers, all paddy growth stages detected by this algorithm using MODIS data are divided into train and test samples as shown in table 2. Moreover, these train samples are divided again into 25% for learning and 75% for validation during learning through 10-rounds random bootstrap re-sampling method to obtain more variety on hypothesis models. The best model for each classifier method is defined as the one which has the highest kappa coefficient during testing.

**Table 2.** Data samples

| No | Class | Number of samples | | |
|---|---|---|---|---|
| | | Train | Test | Total |
| 1 | Flooding & transplanting | 500 | 8,611 | 9,111 |
| 2 | Heading | 500 | 501 | 1,001 |
| 3 | Harvesting & post harvest | 500 | 7,610 | 8,110 |

## 3.2    Kernel-Based Classification Method

The support vector machine (SVM) is a popular technique and useful for data classification. Basically, SVM classifies binary data by determining the separating hyperplane or decision surface, which maximizes the margin between the two classes in the training data [15]. One major disadvantage, however, is that SVM classification is binary. This problem can be resolved in a simple and robust procedure by training several SVMs simultaneously in any multiclass approaches. Kernel function provides SVMs with powerful ability of efficiently determining the nonlinear decision surfaces after transforming the input space into kernel space. In addition, SVM is able to transform a non-convex problem into a convex problem with a single global minima thanks to its quadratic optimization scheme. Given a training set of instance-label pairs $(x_i, y_i), i = 1, ..., l$, where $x_i \in R^n$ and $y \in \{1, -1\}^l$, the SVM requires the solution of the following optimization problem:

$$\min_{w,b,\xi} \quad \frac{1}{2} w^T w + C \sum_{i=l}^{l} \xi_i$$

$$\text{Subject to} \quad y_i(w^T \phi(x_i) + b) \geq 1 - \xi_i$$
$$\xi_i \geq 0$$

$$(1)$$

Here training vectors $x_i$ are mapped into a higher dimensional space by the function $\phi$. SVM finds a linear separating hyperplane with the maximal margin this higher dimensional space, whereas $C > 0$ is the penalty parameter of the error term.

The development of SVMs classification for paddy growth stages is done by using a library for support vector machines (LIBSVM) based on sequence minimal optimization (SMO) [16]. This library was integrated into Interactive Data Language (IDL) program for linking to the data samples. The kernel trick used in this paper is radial basis function (RBF).

$$K(\mathbf{x}_i, \mathbf{x}_j) = \exp\left(-\gamma \|\mathbf{x}_i - \mathbf{x}_j\|^2\right), \gamma > 0 \qquad (2)$$

Compared to other kernel tricks (linear, polynomial, and sigmoid), the $\gamma$ parameter of RBF kernel provides greater flexibility in controlling the desired classification accuracy. Such accuracy control mechanism is highly important when we want to combine the SVM as weak classifiers in boosting learning to overcome the over-fitting problem.

Principal component regression (PCR) is a regression analysis that uses the principal component analysis (PCA) to estimate regression coefficients and proposed to overcome problems which arise when the independent variables are closed to being collinear. In PCR, the role of the independent variables, which are directly used in common regression on the dependent variable, is replaced by the principal components of the independent variables. The basic idea in PCR is that after choosing a set of suitable PCs that is indexed $g$, the input features of $X$ have been retained by score matrix $V_g$[17] and then perform a multiple linear regression (MLR) with $V_g$ instead of $X$ for $n \times m$ calibration data matrix $Y$

$$y = V_g \beta + \epsilon \qquad (3)$$

where in this paper, $y$ is related to (i) the output vector in term of regression, and (ii) the output matrix of labeled classification in binary code [-1,1]. The least squares method then gives coefficient of regression

$$\hat{\beta} = V_g^T (V_g^T V_g)^{-1} y \qquad (4)$$

So in the development of prediction models using PCR, the most decisive factors are score matrix $V_g$ and coefficient of regression $\hat{\beta}$ only. Finally the parameters of the model are computed for the selected PC. Consider a new sample $z$ and predicted value by $\hat{y}$ (both uncentered), and let $\bar{x}$ and $\bar{y}$ be the mean value of calibration samples. By improving equation (3) then the prediction for new sample $h$ takes the form.

$$\hat{y} = \bar{y} + (h - \bar{x})V_g \hat{\beta} \qquad (5)$$

Meanwhile, ELM originally developed by [18], is designed as a generalization of the single-hidden-layer feed forward neural networks (SLFNs), where all the parameter of networks do not need to be updated iteratively. The ELM randomly chooses instead a number of hidden nodes and analytically determines the output weights of SLFNs, so that it can learn extremely faster than traditional SLFNs. The decision function of ELM with $n$ hidden nodes can be expressed as

$$f(x) = \sum_{i=1}^{n} \beta_i g_i(x), \qquad x \in R^d, \beta_i \in R \tag{6}$$

where $g_i(x)$ is the ouput hidden layer corresponding to the input samples x, and $\beta_i$ is the output weight vector between output hidden layer and output. In this paper, we apply Radial Basis Function (RBF) as a hidden nodes function, which defined as

$$g_i(x) = g(b_i \| x - a_i \|), a_i \in R^d, b_i \in R^+ \tag{7}$$

where $a_i$ and $b_i$ are the center and impact factor of the $i$th hidden nodes respectively. Equation (1) can be written in matrix form as

$$F = G\beta \tag{8}$$

If the number of hidden layer is equal to the number of training samples, equation (8) can be easily resolved by using conventional least square method. However, in most cases the number of hidden nodes much less than the number of training samples, then this problem can be accomplished by

$$\beta = G^T (G^T G)^{-1} F \tag{9}$$

Compared to equation (4), the equation (9) has the same form of pseudo inverse matrix, which called as the moore-penrose generalized inverse of matrix.

In the same manner with SVM, in order to have better generalization performance of PCR and ELM, the regularization is used based on ridge regression [19], which adding a positive value $I/C$ as a penalty into regression coefficient in equation (4) and (9). By minimizing errors of both equations, the solution to the ridge regression problem is given. Furthermore, by defining a RBF kernel matrix (2), therefore the decision function for PCR and ELM can be written in equation (10) and (11) respectively.

$$f(x) = K(h, v) \left( \frac{I}{C} + K(v_i, v_j) \right)^{-1} y \tag{10}$$

$$F(x) = K(g(h), g) \left( \frac{I}{C} + K(g_i, g_j) \right)^{-1} y \tag{11}$$

Here parameter $C$ and $\gamma$ of each base learner for classification must be defined in advanced by using grid-search technique, and the best parameters are shown in table 3.

**Table 3.** The best parameters of learner

| Base learner | Parameter $C$ | Parameter $\gamma$ |
|---|---|---|
| RK-PCR | 128 | 2 |
| RK-ELM | 128 | 4 |
| RBF-SVM | 32 | 32 |

# 4     Results and Discussion

Paddy growth stages in case of segment 321782105 are located in Karawang district. Paddy growth stages detection result from time series EVI profile using our algorithm is shown in figure 3, where blue line and green line indicate LSWI and EVI profile respectively, and paddy growth stages indicated by index 1, 2, and 3 on green line. To compare visually, dates in this figure were overlaid with dates collected from field campaign indicated by three blue dot lines. The comparison of all samples in 32 segments were evaluated using adjusted confusion matrix shown in table 4, and total accuracy = 94.74%. This result shows that our proposed algorithm has high confidence for detecting paddy growth stages, and all samples can be considered as ground reference data for building the prediction models correspond to six-bands spectral of MODIS data in time invariant.

After learning through 10-rounds random bootstrap re-sampling method to obtain more variety on hypothesis models, the best model for each classifier method is defined as the one which has the highest kappa coefficient during testing. Table 5 shows the best learning results during training, validation, and testing for each base learner, and all base learners (RK-PCR, RK-ELM, and RBF-SVM) have competitive accuracy in testing. In addition, learning in testing can give better accuracy than learning in training, even though number of training data is much less than testing. It means that our constructed models using kernel-based learner have general prediction model for single temporal resolution of MODIS data.

Henceforward, the models built in this study are applied into the latest MODIS imagery with a single temporal resolution (acquired date: 12 July 2014) around paddy field area in Karawang district, West of Java. Figure 4 shows distribution map of paddy growth stages prediction, where blue, green, and red colors indicate plowing & vegetative stage, heading date, and harvesting & post-harvested date respectively. Distribution map of a paddy growth stages detected from a heuristic algorithm (figure 4.a) proposed in this study is used as a reference to compare with three learners of models, and the comparison results are shown in table 6. Although all of the models have competitive in accuracy, however, it is contrast in time consuming during prediction. KR-PCR and KR-ELM almost have the same computational cost, while RBF-SVM has a highest computational cost, since for multi-class approach, we have to consider SVM learning through in each two-class classifier.

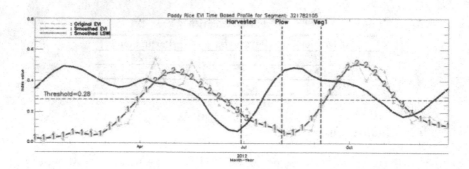

**Fig. 3.** Paddy growth stages detection using heuristic algorithm (time variant) in case of segment 321782105

**Table 4.** Confusion matrix for all samples

|  |  | Field campaign | | |
|---|---|---|---|---|
|  |  | **A+B** | **C+D** | **E** |
| Our Algorithm | **1** | 31 | 2 | 1 |
|  | **2** | 1 | 26 | 0 |
|  | **3** | 0 | 1 | 15 |

**Table 5.** Learning results

| No | Base Learner | Kappa (%) | | |
|---|---|---|---|---|
|  |  | **Training** | **Validation** | **Testing** |
| 1 | RK-PCR | 81.20 | 84.72 | 84.08 |
| 2 | RK-ELM | 85.55 | 84.91 | 84.04 |
| 3 | RBF-SVM | 79.17 | 84.72 | 84.95 |

**Table 6.** Implementation models on MODIS Image (date: 12 July 2014)

| No | Base Learner | Kappa (%) | Computational cost (sec) |
|---|---|---|---|
| 1 | RK-PCR | 87.46 | 42.55 |
| 2 | RK-ELM | 87.76 | 51.79 |
| 3 | RBF-SVM | 87.34 | 748.95 |

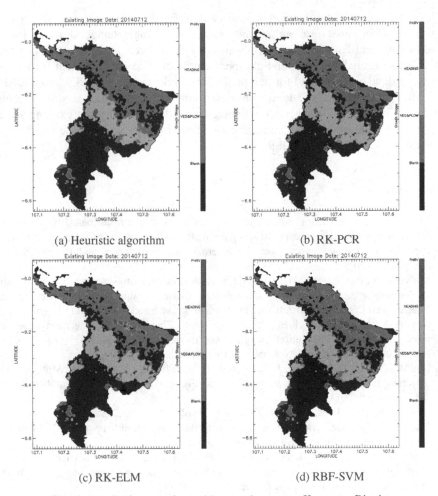

(a) Heuristic algorithm          (b) RK-PCR

(c) RK-ELM          (d) RBF-SVM

**Fig. 4.** Distribution map for paddy growth stages at Karawang District

# 5    Conclusion

In this study, we proposed a new algorithm for detecting paddy growth stages from time series LSWI and EVI profile derived from MODIS data. Even though initially we used multi temporal time series MODIS data, we pursue a general model which can be applied for any given time. Instead of relying one time series data, we pursue the extraction of spectral data. After validating with ground referenced data, we use all the samples to build prediction models in time invariant manner by introducing three kernel-based learners (RK-PCR, RK-ELM, and RBF-SVM). The results show that all kernel-based learner have good accuracy (more than 84%) in testing and have better accuracy than training, even though number of training data is much less than testing. We successfully build prediction models for detecting three dates of paddy growth stage in time invariant manner using only single temporal resolution of MODIS data, and advocating the application of the general prediction model which can give more reliable, faster, and cheaper computation cost requirement (in term of processing load and memory) for real prediction model using MODIS remote sensing data. Since RK-PCR has the least time consuming in prediction, it is strong recommended to be used as a general prediction model for detecting the existing paddy growth stages, harvesting area, and harvesting time by implementing the latest single temporal resolution of MODIS image.

# References

1. Mulyono, S., Ivan Fanany, M., Basaruddin, T.: Genetic algorithm based new sequence principal component regression (ns-pcr) for feature selection and yield prediction using hyperspectral remote sensing data. In: International Geosciences and Remote Sensing Symposium (2012)
2. Mulyono, S., Ivan Fanany, M., Basaruddin, T.: A paddy growth stages classification using modis remote sensing images with balanced branches support vector machines. In: International Conference on Advanced Computer Science and Information Systems (2012)
3. Sun, H.-S., Huang, J.-F., Huete, A.R., Peng, D.-l., Zhang, F.: Mapping paddy rice with multi-date moderate-resolution imaging spectroradiometer (MODIS) data in China Journal of Zhejiang University Science A 10(10), 1509–1522 (2009)
4. Lin, W., Fu-cun, Z., Yuan-shu, J., Xiao-dong, J., Shen-bin, Y., Xiao-mei, H.: Multi-temporal detection of rice phonological stages using canopy spectrum. ScienceDirect, Rice Science 21(2), 108–115 (2014)
5. Xiao, X., Stephen Boles, T., Liu, J., Zhuang, D., Frolking, S., Li, C., Salas, W., Moore, B.: III Mapping paddy rice agriculture in southern China using multi-temporal MODIS images. Remote Sensing of Environment 95 480–492 (2005)
6. Xiao, X., Boles, S., Frolking, S., Li, C., Babu, J.Y., Salas, W.: Berrien Moore III,: Mapping paddy rice agriculture in South and Southern Asia using multi-temporal MODIS images. Remote Sensing of Environment 100, 96–113 (2006)
7. Peng, D., Huete, A.R., Huang, J., Wang, F., Sun, H.: Detection and estimation of mixed paddy rice cropping patterns with MODIS data. International Journal of Applied Earth Observation and Geoinformation 13, 13–23 (2011)
8. Jeong, S., Kang, S., Jang, K., Lee, H., Hong, S., Ko, D.: Development of Variable Threshold Models for detection of irrigated paddy rice fields and irrigation timing in heterogeneous land cover. Agricultural Water Management 115, 83–91 (2012)
9. Sakamoto, T., Yokozawa, M., Toritani, H., Shibayama, M., Ishitsuka, N., Ohno, H.: A crop phenology detection method using time-series MODIS data. Remote sensing of environment 96, 366–374 (2005)
10. Sari, D.K., Ismullah, I.H., Sulasdi, W.N., Harto, A.B.: Detecting rice phenology in paddy fields with complex cropping pattern using time series MODIS data - A case study of northern part of West Java Indonesia. ITB Journal Science 42A(2), 91–106 (2010)
11. Vintrou, E., Bégué, A., Baron, C., Saad, A., LoSeen, D., Traoré, S.B.: A Comparative Study on Satellite and Model-Based Crop Phenology in West Africa. Remote Sensing Journal 6, 1367–1389 (2014). doi:10.3390/rs6021367

12. Galford, G.L., Mustard, J.F., Melillo, J., Gendrin, A., Cerri, C.C., Cerri, C.E.P.: Wavelet analysis of MODIS time series to detect expansion and intensification of row-crop agriculture in Brazil. Remote Sensing of Environment **112**, 576–587 (2008)
13. Khobkhun, B., Prayote, A., Rakwatin, P., Dejdumrong, N.: Rice phenology monitoring using PIA time series MODIS imagery. In: 10th International conference computer graphics, Imaging and visualization (2013)
14. Meng, J., Wu, B., Li, Q., Du, X., Jia, K.: Monitoring crop phenology with MERIS data - A case study of winter wheat in North China plain, Progress In electromagnetics research symposium, Beijing, China, March 23–27, 2009
15. Archibald, R., Fann, G.: Feature Selection and Classification of Hyperspectral Images With Support Vector Machines. IEEE Geoscience and Remote Sensing Letter **4**(4) (October 2007)
16. Hsu, C.-W., Chang, C.-C., Lin, C.-J.: A Practical Guide to Support Vector Classification, Software available at. http://www.csie.ntu.edu.tw/~cjlin, last updated, April 15, 2010
17. Camps-Valls, G., Bruzzone, L.: Kernel Based Method for Hyperspectral Image Classification. IEEE Trans. on Geosci. and RS **43**(6) (2005)
18. Huang, G.-b., Zhu, Qin-Yu., Siew, C.-K.: Extreme learning machine: Theory and applications. Elsevier Neurocomputing **70**, 489–501 (2006)
19. Hoerl, A.E., Kennard, R.W.: Ridge reggression: biased estimation for nonorthogonal problem. Technometrics **12**(1), 55–67
20. Uchida, S.: Monitoring of Planting Paddy Rice with Complex Cropping Pattern in the Tropical Humid Climate Region Using LANDSAT and MODIS data - A Case of West Jave, Indonesia, International Archives of the Photogrammetry, Remote Sensing and Spatial Information Science, Volume **XXXVIII**, Part 8, Kyoto Japan (2010)

# Data Classification with Ensembles of One-Class Support Vector Machines and Sparse Nonnegative Matrix Factorization

Bogusław Cyganek[1(✉)] and Bartosz Krawczyk[2]

[1] AGH University of Science and Technology, Al. Mickiewicza 30, 30-059 Kraków, Poland
cyganek@agh.edu.pl
[2] Wrocław University of Technology, Wybrzeże Wyspiańskiego 27, 50-370 Wrocław, Poland
bartosz.krawczyk@pwr.edu.pl

**Abstract.** The paper presents a method for data classification with ensemble of one-class classifiers based on data segmentation. Each data class is partitioned with the nonnegative matrix factorization (NMF) algorithm with sparse constraints. It allows splitting of the input data into compact and consistent data clusters with automatic determination of a number of clusters. Data partitions are fed to an ensemble composed of a number of one-class support vector machine (SVM) classifiers. The proposed method shows high accuracy and fast classification.

**Keywords:** Ensemble of classifiers · One-class support vector machine · Sparse nonnegative matrix factorization

## 1 Introduction

Ensembles of classifiers gain much attention due to superior results [17][19]. However, there are a number of design issues when considering construction of an ensemble for given classification tasks. This concerns choice of the base classifiers, their training method, as well as the method of response fusing, just to name a few [13][12]. It was demonstrated that the best results are obtained if an ensemble has high degree of diversity [17][24].

In the paper a method for construction of ensemble of classifiers for nonnegative datasets is proposed. Diversity is achieved thanks to partitioning of the training data with the nonnegative matrix factorization (NMF) method, originally proposed by Lee and Seung [18]. The presented method is a variant of our previous systems in which the k-means clusterization was used [8][15]. As was reported in literature NMF with the sparsity constraint allows computation of more compact and consistent data clusters [14]. However, compared with the k-means, the NMF allows determination of a number of clusters from data to achieve cluster consistency. In the second stage, the ensemble of one-class SVM classifiers is constructed with each member classifier trained with a separate data cluster. The method is suitable for one-class or imbalanced datasets, as well as for multi-class problems, as outlined in our previous

© Springer International Publishing Switzerland 2015
N.T. Nguyen et al. (Eds.): ACIIDS 2015, Part I, LNAI 9011, pp. 526–535, 2015.
DOI: 10.1007/978-3-319-15702-3_51

work [15]. Also, the proposed method leads to a parallel algorithm which results in much faster response than a serial realization.

The rest of the paper is organized as follows. In section 2 the overall view on the architecture of the proposed method is presented. Section 3 presents details of the nonnegative matrix factorization for data clustering. Base classifiers are presented in section 4. Experimental results and conclusions are presented in sections 5 and 6, respectively.

## 2    Architecture of the Ensemble of Classifiers

Fig. 1 shows the architecture of the proposed clustering based ensemble of classifiers. The most important features of the system are

1. For a single class, data is clustered which leads to separate data partitions, each used to train the ensemble of one-class classifier (OCC);
2. In the case of multi-class problems, each class is clusterized and trained independently;

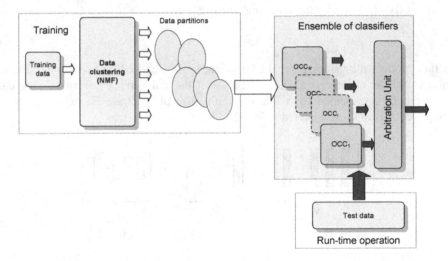

**Fig. 1.** Architecture of clustering based ensemble of one-class classifiers (OCC). Clustering is performed by the NMF algorithm with the sparsity constraint.

There are two modes of operation: training and the response. For the ensembles with the same type of classifiers, to achieve higher degree of diversity, the pool of input data is frequently split into partitions. This can be achieved by bagging or clustering. In our previous works, in which the one-class support vector machines were used (OC-SVM), the latter approach for data splitting showed better efficiency [8][15].

In the run-time operation, the arbitration unit takes into consideration support values returned by each of the member classifiers. This fusion method is described in [15]. However, other fusion methods can be also used [23].

# 3     Nonnegative Matrix Factorization for Data Clustering

The Nonnegative Matrix Factorization was first proposed by Lee and Seung as a method for computation of the linear part-based representation of nonnegative data [18]. Given a dataset of nonnegative observations, NMF tries to represent that dataset with an assumed accuracy as a product of lower rank nonnegative matrices. For some datasets such representation can lead to discovery of latent variables and in many cases it allows easier interpretation of the results due to only additive components.

Let us now assume a series of $N$ measurements, or data, each represented by an $L$-dimensional column vector $d_i$, all arranged in a matrix $D$. Now the question is whether we can find an equivalent representation of $D$ which will exhibit some hidden factors, or base information, as well as a number of coefficients. Such representation can be written as follows

$$D \approx BC, \tag{1}$$

where $B$ is a matrix of hidden factors, also called a base matrix, and $C$ is a matrix of components (coefficients). Let us now observe that each data point $d_i$ can be equivalently expressed by the following product

$$d_i \approx Bc_i. \tag{2}$$

In this product all values from the base matrix $B$ are used and only one selected column $c_i$ from the matrix $C$, i.e. the one with the same index $i$. That is, each data point can be represented by a linear combination of the base $B$, weighted by the components from $C$, as depicted in Fig. 2.

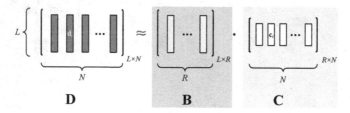

**Fig. 2.** Visualization of the factorization of the data matrix $D$ into base matrix $B$ and matrix of sparse coefficients $C$ (after [10]). NMF additionally assumes nonnegative components in $B$ and $C$.

However, in NMF approach, factorization (1) is endowed with the constraints imposing nonnegative components of the two matrices, which is denoted as follows

$$B \geq 0, \quad C \geq 0. \tag{3}$$

The two most frequently used in computation of NMF are the squared Frobenius norm, as well as the generalized Kullback-Leibler divergence [10][7]. In this work we assume only the former one. Since (1) is only an approximation then it will be denoted as follows

$$\tilde{\mathbf{D}} = \mathbf{BC}. \tag{4}$$

We are interested in minimizing the difference in the light of the norm $D_F$ between the original matrix $\mathbf{D}$ and its approximation (4). This can be expressed as follows

$$\min_{\tilde{\mathbf{D}}} D_F^2 \left( \mathbf{D} - \tilde{\mathbf{D}} \right) = \min_{\mathbf{B},\mathbf{C}} D_F^2 \left( \mathbf{D} - \mathbf{BC} \right) = \min_{\mathbf{B},\mathbf{C}} \left\| \mathbf{D} - \mathbf{BC} \right\|_F^2, \tag{5}$$

subject to the nonnegative condition (3).

Interestingly enough, Kim and Park showed that then NMF as formulated in the above can be used for data clustering [14]. They showed that (5) can be interpreted as an objective function for the k-means clustering algorithm in which matrix $\mathbf{B}$ denotes cluster centers and $\mathbf{C}$ is a binary matrix containing cluster assignment weights. In this case, $C_{ij}=1$ if the $i$-th data point is assigned to the $j$-th cluster, and $C_{ij}==0$ otherwise. However, the NMF formulation of the clustering problem differs from the k-means in certain aspects. First, the base vectors in $\mathbf{B}$ need not be the centroids of the data clusters. Second, the bases $\mathbf{B}$ are also not orthonormal, although they can be used for additional data classification, as will be discussed. Thirdly, the NMF approach gives much better and more consistent results for various datasets and in effect it allows easier determination of the unknown number of clusters in data. However, Kim and Park showed that properly used, NMF outperforms k-means and approximates well the Affinity Propagation algorithm [10]. This is one of the reasons of investigating the NMF for ensemble classification. In the rest of this section we will focus on the computation of NMF for data clustering.

The solution to (5) was found by Lee and Seung [18]. This can be solved based on the Karush-Kuhn-Tucker first-order optimality conditions, which lead to the following multiplicative update rules

$$b_{nr} \leftarrow b_{nr} \frac{\left[ \mathbf{DC}^T \right]_{nr}}{\left[ \mathbf{BCC}^T \right]_{nr} + \mu}, \tag{6}$$

$$c_{rm} \leftarrow c_{rm} \frac{\left[ \mathbf{B}^T \mathbf{D} \right]_{rm}}{\left[ \mathbf{B}^T \mathbf{BC} \right]_{rm} + \mu}, \tag{7}$$

where $\mu$ denotes a small constant value to assure the numerical stability. From the above we see that the optimization process is a twofold process. That is, elements of the matrices $\mathbf{B}$ and $\mathbf{C}$ are obtained alternatively. However, this strategy does not guarantee a global minimum and sometimes (6) and (7) converge to a local minimum, similarly to the alternating least squares method (ALS) [10]. Also, the method relies on proper initialization of the matrices $\mathbf{B}$ and $\mathbf{C}$, as will be discussed.

However, as alluded to previously, clustering with the NMF assumes *sparse* representation of the coefficient matrix $\mathbf{C}$. This is to assure proper membership assignment of a data to its cluster which is achieved if each column of $\mathbf{C}$ contains only one large value with all other being 0. In practice, however, we will expect one large value and few lower. Such a situation is rather similar to the fuzzy c-means method [1]. Nevertheless, the sparsity constraint necessary for clusterization cannot be directly obtained from (6) and (7). In this respect Cichocki *et al.* propose a modification to incorporate over-relaxation and sparsity control (ISRA-NMF) to the standard multiplicative method given in (6) and (7). Their method modifies the optimization problem (5) and requires additional three scalars. In this case the optimization problem with additional control parameters is as follows

$$\min D_{Fr}\left(\mathbf{D}-\tilde{\mathbf{D}}\right) = \min_{\mathbf{B,C}}\left\{\frac{1}{2}\left\|\mathbf{D}-\mathbf{BC}\right\|_F^2 + s_\mathbf{B}J_\mathbf{B}\left(\mathbf{B}\right) + s_\mathbf{C}J_\mathbf{C}\left(\mathbf{C}\right)\right\}, \tag{8}$$

subject to the nonnegative condition (3). In the above, $J_\mathbf{B}(\mathbf{B})$ and $J_\mathbf{C}(\mathbf{C})$ are penalty functions that enforce additional requirements on the matrices $\mathbf{B}$ and $\mathbf{C}$, while $s_\mathbf{B}$ and $s_\mathbf{C}$ are constant values that control their influence on the solution. The higher these constants, the degree of the corresponding constraints is also higher. Using the standard gradient descent method the following multiplicative update rules for the NMF problem can be obtained [10]

$$b_{nr} \leftarrow b_{nr}\left[\frac{\left[\mathbf{DC}^T\right]_{nr} - s_\mathbf{B}J'_\mathbf{B}\left(\mathbf{B}\right)}{\left[\mathbf{BCC}^T\right]_{nr} + \mu}\right]_+^q, \tag{9}$$

$$c_{rm} \leftarrow c_{rm}\left[\frac{\left[\mathbf{B}^T\mathbf{D}\right]_{rm} - s_\mathbf{C}J'_\mathbf{C}\left(\mathbf{C}\right)}{\left[\mathbf{B}^T\mathbf{BC}\right]_{rm} + \mu}\right]_+^q, \tag{10}$$

where $q$ is a positive parameter (usually in the range 0.5-2) which role is to speed up the convergence, and $[.]_+$ denotes an operation of assure nonnegativity. It is defined as follows

$$\left[x\right]_+ = \max\left(x, \mu\right), \tag{11}$$

for sufficiently low positive threshold $\mu$ (usually in the range $10^{-9}$-$10^{-12}$). This is necessary due to the subtraction in the numerator. In the above we have used the element-wise derivatives of the control functions

$$J'_\mathbf{B}\left(\mathbf{B}\right) = \frac{\partial J_\mathbf{B}\left(\mathbf{B}\right)}{\partial b_{nr}} \text{ and } J'_\mathbf{C}\left(\mathbf{C}\right) = \frac{\partial J_\mathbf{C}\left(\mathbf{C}\right)}{\partial c_{rm}}. \tag{12}$$

As alluded to previously, the functions $J_\mathbf{B}(\mathbf{B})$ and $J_\mathbf{C}(\mathbf{C})$ are chosen to add additional properties on the output matrices. For the interesting sparsity and orthogonality constraints, they and their derivatives are as follows [10]

Sparsity:
$$J\left(\mathbf{M}\right) = \left\|\mathbf{M}\right\|_1 = \sum_{r,s} \left| m_{rs} \right|, \qquad J'\left(\mathbf{M}\right) = \mathbf{1}_{R \times S}. \qquad (13)$$

Orthogonality:
$$J\left(\mathbf{M}\right) = \sum_{i=1}^{S-1} \sum_{j=i+1}^{S} \mathbf{m}_i^T \mathbf{m}_j, \qquad J'\left(\mathbf{M}\right) = \mathbf{M}\left(\mathbf{1}_{S \times S} - \mathbf{I}\right). \qquad (14)$$

In the above, a matrix $\mathbf{M}$ is assumed to be of size $R \times S$, then $\mathbf{1}_{S \times S}$ is a matrix of dimensions $S \times S$ with all ones, and $\mathbf{I}$ is the identity matrix. As it is easily observed, the sparsity constraint (13) involves the $L_1$ norm.

In the proposed ensemble of classifiers, the clustering is performed offline in accordance with the formulas (9)-(10) before training the base classifiers. Always the sparsity constraint (13) is turned on, choosing the positive coefficient $s_C$ in (10) in the range 0.2-0.5. The orthogonality constraint (14) was also tested. However, its inclusion did not add to the clustering results. Therefore investigation of its influence is left for the further research.

As already mentioned, NMF allows semi-automatic determination of the number of proper clusters for datasets. This is due to the consistency properties of the NMF, as described in [14]. This means that for a broad range of tested number of clusters $R$, the points stay relatively well assigned to the same clusters. To measure property of the clusterization process for a chosen number of clusters $R$ in the assumed range, the average connectivity matrix was used, as suggested in [4]. The connectivity matrix $\mathbf{Q}$ has dimensions $N \times N$, where $N$ denotes a number of data. For each pair of data $\mathbf{x}_i$ and $\mathbf{x}_j$, and cluster index $R$, elements of $\mathbf{Q}$ take two values: $\mathbf{Q}_R(i,j)=1$ if $\mathbf{x}_i$ and $\mathbf{x}_j$ are assigned to the same cluster $k$, and 0 otherwise. For a given $k$ and a number of trials $t$ (in our experiments this $t=100$), an average $\overline{\mathbf{Q}}$ is computed. In the next step, the dispersion coefficient is computed as follows [4]

$$\delta_R = \frac{1}{N^2} \sum_{i=1}^{N} \sum_{j=1}^{N} 4\left(\overline{\mathbf{Q}}_R(i,j) - 0.5\right)^2. \qquad (15)$$

Then, an optimal number of clusters corresponds to the highest dispersion (1 denotes a perfect consistency), as follows

$$R^* = \arg\max_R \left(\delta_R\right). \qquad (16)$$

For implementation the *DeRecLib* C++ software was used, which is available at the webpage of the book [10]. The detailed algorithm is also described in [10]. However, in the framework of streaming data such an approach needs to be modified to account for the changes of the incoming data. To overcome a problem of a conventional NMF algorithm in case of online processing of large datasets, Bucak and Gunsel proposed a method for incremental subspace learning via non-negative matrix factorization [5]. This can be considered in the streaming data classification frameworks.

## 4    Structure of the Base One-Class Classifier

In the proposed ensemble, the base classifier are the OC-SVM since they operate well with the clustered datasets, as was shown in our previous works [8][15]. They exhibit many useful properties from which the most important are as follows:

- Only one-class data are necessary for training.
- After training, the whole class is represented exclusively by the few support vectors (data can be discarded).
- Kernel mapping allows better separation of the class of interest.
- Very fast response time (real-time operation).
- For multi-class classification, each separate class is classified by its dedicated ensemble of OC-SVMs [15].

An OC-SVM can be modeled as a hypersphere enclosing the training class data [21]. It can be characterized by its centre $\mathbf{a}$ and a radius $r$. The volume of the hypersphere is proportional to $r^n$ and should be kept as minimal as possible to tightly encompass the data. These requirements lead to the minimization with respect to $r^2$. Thus, the minimization functional $\Theta$ is as follows

$$\Theta(\mathbf{a}, r) = r^2 \tag{17}$$

with the constraint

$$\forall_i : \ \|\mathbf{x}_i - \mathbf{a}\| \le r^2 , \tag{18}$$

where $\mathbf{x}_i$ is a data point. Due to outliers present in real datasets, the slack variables $\xi_i$ are introduced to allow existence of points distant further than $r$ (the outliers). This modifies (17) as follows

$$\Theta(\mathbf{a}, r) = r^2 + C \sum_i \xi_i \tag{19}$$

with the new constraints

$$\forall_i : \ \|\mathbf{x}_i - \mathbf{a}\| \le r^2 + \xi_i , \qquad \xi_i \ge 0 . \tag{20}$$

The parameter $C$ denotes controls the optimization process - the larger $C$, the less outliers are possible, at the larger volume of the hypersphere. Given a set of training points $\{\mathbf{x}_i\}$, a solution to the equations (19) and (20) can be computed with the Lagrange multipliers. From this a distance $d$ from the centre $\mathbf{a}$ of the hypersphere to a test point $\mathbf{x}_x$ can be obtained.

During classification an unknown test point $\mathbf{x}_x$ is classified as an inlier to the class if it is inside the hypersphere, that is

$$d^2(\mathbf{x}_x, \mathbf{a}) \le r^2, \tag{21}$$

After solving (19) and (20), the above classification rule can be expressed as

$$\sum_{i \in Idx(SV)} \alpha_i K(\mathbf{x}_x, \mathbf{x}_i) \ge \sum_{i \in Idx(SV)} \alpha_i K(\mathbf{x}_s, \mathbf{x}_i) = \delta . \tag{22}$$

where $Idx(SV)$ is a set of the support vectors and $\alpha_i$ are scalar coefficients; $K$ denotes a kernel function, which in our system is the RBF, as follows

$$K_{RBF}(\mathbf{x}_i, \mathbf{x}_j) = e^{-\gamma \|\mathbf{x}_i - \mathbf{x}_j\|^2} , \tag{23}$$

where $\gamma$ denotes a spread parameter. The right side of (22) is a constant in the recognition stage. Thus it can be precomputed to a value $\delta$ which denotes a cumulative kernel-distance of a SV to all other SVs. Equation (22) is used to test a pattern $\mathbf{x}_x$ if it belongs to a class represented a given OC-SVM. In our system, (22) was implemented in CUDA environment to speed up response on large datasets. Additionally, all OC-SVMs in the ensemble operate in parallel. Details with code fragments are provided in [9].

# 5    Experimental Results

The presented method was implemented in C++ using the *DeRecLib* software libraries [10]. Experiments were run on the PC with 8 GB RAM and Pentium Q 820 microprocessor. The method was tested on five datasets *Iris*, *Wisconsing Breast Cancer (WBC)*, and *Wine* from the UCI repository, as well as *ColorSamples* [8]. Results are presented in Tab. 1.

**Table 1.** Obtained results for test datasets

| Dataset | Data | Features | Classes | Acc [%] |
|---|---|---|---|---|
| Iris | 150 | 4 | 3 | 93.78 |
| ColorSamples | 923 | 3 | 1 | 86.3 |
| Wisconsing Breast Cancer (WBC) | 699 | 9 | 2 | 90.43 |
| Wine | 178 | 13 | 3 | 94.34 |

The obtained results show that the method has accuracy comparable with the best results presented in the work [15]. Only the ensembles with the ECOC fusion and kernel clusterization show better results. However, the NMF clusterization method allows determination of the number of clusters in the datasets. It also has potential for additional improvements due to subspace decomposition obtained as a by-product of the NMF decomposition. However, as alluded to previously, the obtained databases are not orthogonal. Thus additional orthogonalization process needs to be employed. This is left for further research.

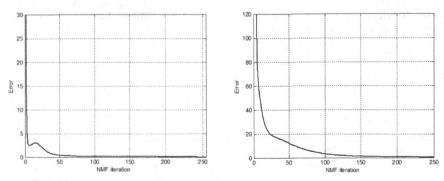

**Fig. 3.** Approximation error for the NMF with sparsity constraint and $s_C$=0.5 for the *Iris* (left) and *ColorSamples* (right) datasets

Performance of the NMF clusterization with the sparsity constraint was analyzed in terms of convergence of the alternating algorithm. Fig. 3 depicts approximation error for the NMF with sparsity constraint and $s_C=0.5$ for the *Iris* (a) and *ColorSamples* (b) datasets, respectively. The error is defined as $D_F^2\left(\mathbf{D}-\tilde{\mathbf{D}}\right)$ from (5). However, it is computed as a maximum of the differences of the consecutive approximations of the matrices $\mathbf{B}$ and $\mathbf{C}$, computed in consecutive iteration steps. Details are in [10]. From Fig. 3 we notice that the convergence process can be quite specific to a given dataset. Nevertheless, for each of the tested datasets sufficiently low error was obtained after a hundred of iterations.

# 6     Conclusions

In this paper the system for construction of the ensemble of classifiers is presented. The ensemble consists of OC-SVM classifiers trained on the clustered partitions of the input data. These are obtained with help of the nonnegative matrix factorization with sparsity constraint. NMF allows determination of a number of clusters and produces more compact clusters due to its high consistency. Thanks to this property, it is easier to determine a proper number of clusters based on the data consistency matrix criteria. The method is suitable for one-class or imbalanced datasets, as well as for multi-class problems. Also, it allows parallel response which results in fast response. Summarizing, the advantages of the proposed method are as follows: Usually more compact clusters compared to k-means. The number of clusters can be found thanks to the cluster consistency conditions, which stays more stable than in the case of k-means. Also, data partitions are amenable for easier interpretation due to only the additive rule. The drawbacks: NMF operates on nonnegative data requiring a significant number of iterations and it is not guaranteed to converge to a global minimum. In further research we plan to address the issues of automatic choice of the clustering method and its parameters, automatic choice of the best features for clusterization and a set of best features for ensemble building.

**Acknowledgment.** This work was supported by the Polish National Science Center under the grant no. DEC-2013/09/B/ST6/02264.

# References

1. Bezdek, J.: Pattern Recognition with Fuzzy Objective Function Algorithms. Plenum Press, New York (1981)
2. Bicego, M., Figueiredo, M.A.T.: Soft clustering using weighted one-class support vector machines. Pattern Recognition **42**, 27–32 (2009)
3. Blake, C., Keogh, E., Merz, C.: UCI repository of machine learning databases. University of California, Irvine, Department of Information and Computer Science (http://www.ics.uci.edu/~mlearn/MLRepository.html) (1998)

4. Brunet, J., Tamayo, P., Golub, T., Mesirov, J.: Metagenes and molecular pattern discovery using matrix factorization. Proceedings of the National Academy of Sciences **101**(12), 4164–4169 (2004)
5. Bucak, S.S., Gunsel, B.: Incremental subspace learning via non-negative matrix factorization. Pattern Recognition **42**, 788–797 (2009)
6. Cichocki, A., Zdunek, R., Amari, S.: Nonnegative Matrix and Tensor Factorization. IEEE Signal Processing Magazine **25**(1), 142–145 (2008)
7. Cichocki, A., Zdunek, R., Phan, A.H., Amari, S-I.: Nonnegative Matrix and Tensor Factorizations. Applications to Exploratory Multi-way Data Analysis and Blind Source Separation. Wiley (2009)
8. Cyganek, B.: Image segmentation with a hybrid ensemble of one-class support vector machines. In: Graña Romay, M., Corchado, E., Garcia Sebastian, M. (eds.) HAIS 2010, Part I. LNCS, vol. 6076, pp. 254–261. Springer, Heidelberg (2010)
9. Cyganek, B., Wiatr, K.: Design of parallel architectures of classifiers suitable for intensive data processing. In: The 9th International Conference on Networking and Services, LMPCNA 2013, Lisbon, Portugal, pp. 14–19 (2013)
10. Cyganek, B.: Object Detection and Recognition in Digital Images, Theory and Practice. Wiley (2013)
11. Frey, B.J., Dueck, D.: Clustering by passing messages between data points. Science **315**, 972–976 (2007)
12. Horzyk, A.: Self-optimizing neural network 3. In: Franco, L., Elizondo, D.A., Jerez, J.M. (eds.) Constructive Neural Networks. SCI, vol. 258, pp. 83–101. Springer, Heidelberg (2009)
13. Jackowski, K., Woźniak, M.: Algorithm of designing compound recognition system on the basis of combining classifiers with simultaneous splitting feature space into competence areas. Pattern Analysis and Applications **12**, 415–425 (2009)
14. Kim, J., Park, H.: Sparse Nonnegative Matrix Factorization for Clustering. Georgia Institute of Technology Technical Report GT-CSE-08-01 (2008)
15. Krawczyk, B., Woźniak, M., Cyganek, B.: Clustering-based ensembles for one-class classification. Information Sciences **264**, 182–195 (2014). Elsevier
16. Kruse, R., Döring, C., Lesot, M-J.: Fundamentals of fuzzy clustering. In: de Oliveira, J.V., Pedrycz, W. (eds.) Advances in Fuzzy Clustering and its Applications, pp. 3–30. Wiley (2007)
17. Kuncheva, L.I.: Combining Pattern Classifiers. Wiley (2004)
18. Lee, D.D., Seung, H.S.: Learning the parts of objects by non-negative matrix factorization. Nature **401**, 788–791 (1999)
19. Polikar, R.: Ensemble Based Systems in Decision Making. IEEE Circuits and Systems Magazine, 21–45 (2006)
20. Tax, D., Duin, R.: Support vector domain description. Pattern Recognition Letters **20**, 1191–1199 (1999)
21. Tax, D., Duin, R.: Support Vector Data Description. Machine Learning **54**, 45–66 (2004)
22. University of California, Database (2011). (*ftp://ftp.ics.uci.edu/pub/machine-learning-databases/*)
23. Wilk, T., Woźniak, M.: Soft computing methods applied to combination of one-class classifiers. Neurocomputing **75**, 185–193 (2012)
24. Woźniak, M., Graña, M., Corchado, E.: A survey of multiple classifier systems as hybrid systems. Information Fusion **16**, 3–17 (2014). Elsevier

# Truck Loading Schedule Optimization Using Genetic Algorithm for Yard Management

Tadeusz Cekała, Zbigniew Telec, and Bogdan Trawiński[✉]

Department of Information Systems, Wroclaw University of Technology,
Wybrzeże Wyspiańskiego 27, 50-370 Wrocław, Poland
{zbigniew.telec,bogdan.trawinski}@pwr.edu.pl

**Abstract.** A new information system for order and yard management was implemented and deployed in a timber products company. The system was equipped with an innovative mechanism which automatically updates loading appointment schedule on the basis of current data of truck arrivals and departures. During a day the current schedule often becomes outdated due to various unexpected difficulties in loadings and unpredicted delays in trucks arrival. In the paper a genetic algorithm which is the core of the updating mechanism was presented. Penalty functions were employed in order to protect its solution against violating constraints. The algorithm was enhanced by additional processing just before computing the value of the fitness function. The improved genetic algorithm was experimentally evaluated both in terms of correctness and speed of producing the loading appointment schedule for a test problem. Moreover the simulation of its planned exploitation was performed using real-world data. The proposed genetic algorithm revealed better performance than the competitive particle swarm optimisation method as well as rescheduling made by the dispatchers manually.

**Keywords:** Loading schedule · Genetic algorithm · Particle swarm optimization · Yard management · Manufacturing company

## 1    Introduction

The problem of dock assignment and truck scheduling problem has been drawing the attention of many researchers recently. Most of works are devoted to the problem of cross docking where shipments are transferred directly from incoming to outgoing trucks without storage in between. In a cross docking model customers are known before the goods get to the warehouse and hence the two most expensive warehousing operations , i.e. storing and retrieving are eliminated. The reviews of earlier literature on o mathematical models cross-dock planning provided Agustina et al. [1], and and quantitative approaches for dock door assignment in cross-docking elaborated Shuib et al. [2]. The authors  develop integer linear programming as well as non-linear dynamic programming models which typical objective is to minimize the total penalty of earliness and tardiness in incoming and outgoing trucks. Since dock assignment and truck scheduling problems are NP-hard various meta heuristic procedures have been proposed in literature to solve these models.

© Springer International Publishing Switzerland 2015
N.T. Nguyen et al. (Eds.): ACIIDS 2015, Part I, LNAI 9011, pp. 536–548, 2015.
DOI: 10.1007/978-3-319-15702-3_52

Vahdani and Zandieh [3] utilized five meta-heuristic algorithms: genetic algorithm), tabu search, simulated annealing, electromagnetism-like algorithm, and variable neighbourhood search to schedule the trucks in cross-dock terminals to minimize total operation time. In turn, Boloori Arabani et al. [4] applied  five meta-heuristics for scheduling of trucks in a cross-docking, namely genetic algorithm, tabu search, particle swarm optimization, ant colony optimization, and differential evolution. Liao et al. [5] proposed and experimentally evaluated three hybrid ant colony optimization and two hybrid simulated annealing algorithms, Madani-Isfahani [6] devised simulated annealing and firefly algorithms, Kuo [7] worked out  variable neighborhood search heuristics and compared it with four simulated annealing algorithms. Naderi et al. [8] addressed a biobjective problem of truck scheduling in a cross-docking system and solved it using a multiobjective iterated greedy algorithm. Next they compared their algorithm with the subpopulation particle swarm optimization and strength Pareto evolutionary algorithm.

Particle swarm optimization (PSO) approach was also applied to solve scheduling problems. PSO is an alternative to genetic algorithm population-based evolutionary computation technique which simulates the social behaviour of bird flocks or fish schools. A number of researchers utilize PSO to solve the discrete problem of timetable scheduling [9, 10, 11, 12]. In this paper we used the PSO approach for building an alternative truck loading schedule to conduct comparative analysis.

The authors have long experience in using genetic algorithms for machine learning methods, e.g. to obtain optimal predictive ensemble models. We applied the self-adapting genetic algorithms to compose heterogeneous bagging ensembles [13].We have also explored the methods to predict from a data stream of real estate sales transactions based on ensembles of genetic fuzzy systems [14].

One of the main functions of the newly implemented and deployed information system in a timber products company is the management of a company yard. It includes noting the times when trucks arrive, enter loading bays, and leave the company yard. The loading appointment schedule is prepared in advance. The loading time is determined when an order is taken and then the transport operator is notified about the scheduled appointment. However during a working day it often happens that the current schedule becomes outdated. For example the time of loading a truck can be estimated imprecisely or unexpected difficulties can occur during loading process. Moreover, some trucks may be delayed and the others can arrive before the scheduled time. In such situations the dispatcher is obliged to improve the plan manually. This additional activity takes time and may cause unnecessary extra delays.

The main purpose of the method presented in the paper is to update a daily loading plan automatically. In the paper a genetic algorithm, which is the core of the updating mechanism, was worked out and tested. Penalty functions were employed to protect the solution against violating constraints. An improvement introduced into the genetic algorithm significantly accelerates the minimization of evaluation function and reduces time needed to find a solution, was also proposed and evaluated. It is planned that proposed GA is executed systematically before each predetermined unit of time and produces an improved schedule for the rest of a working day. Such routine should result in shortening an average wait time of trucks at the company yard.

## 2    Genetic Algorithm for Loading Schedule Optimization

The main properties of the proposed genetic algorithm for loading schedule optimization and the particle swarm optimization approach used for comparative evaluation are as follows.

*Coding chromosomes.* Integer-coded chromosome representing the whole solution, i.e. the complete version of a daily loading appointment schedule was employed. The chromosome embraces all subsequent appointments (see Fig. 1). Each appointment is represented by two integer genes where the first one is the number of a loading bay (*LB*) and the second one indicates time (*T*) at which outbound truck enters the bay. The latter is expressed by the number of a half hour. Such representation of individuals protects the algorithm against generating unfeasible solutions during crossover operation. Crossover will produce neither doubling nor vanishing appointments because they are determined by the position in the chromosome and the length of the individual does not change during this operation.

| Appointment 1 | | Appointment 2 | | Appointment 3 | | ... | Appointment i | | ... | Appointment n | |
|---|---|---|---|---|---|---|---|---|---|---|---|
| LB1 | T1 | LB2 | T2 | LB3 | T3 | ... | LBi | Ti | ... | LBn | Tn |
| 1 | 14 | 2 | 14 | 3 | 14 | ... | 10 | 20 | ... | 15 | 40 |

**Fig. 1.** Structure of a chromosome

*Genetic operators.* Classic crossover operator is used and performed at the boundary points between appointments. In consequence crossover results in swapping a number of appointments between the individuals. In turn, mutation consists in replacing a given appointment by a new one comprising randomly drawn loading bay number and truck arrival time.

*Constraint handling.* Two types of constraints are distinguished: strong and weak ones. The strong constraint says that only one truck can be assigned to a given loading bay at a given point of time, what means that the appointments cannot overlap. In turn, the weaker constraints concern the order the trucks should enter loading bays as well as the discrepancy between a new schedule and the original one. Penalty functions are used to handle constraints. Their values are proportional to the number of constraint violations by the schedule encoded in the chromosome and penalty coefficients depend on the type of a constraint. The stronger constraint the bigger value of this coefficient.

*Fitness function.* The ultimate goal of the algorithm is to minimize the overall wait time during a day. Therefore, the fitness function is expressed by equation (1).

$$F(x) = \sum_{x_i \in x} g(x_i) + P(x) \tag{1}$$

where $x$ denotes a given individual in the population, $F(x)$ stands for the fitness function, $x_i$ – $i$-th appointment, $g(x_i)$ – rating of $i$-th appointment, and $P(x)$ – penalty function. In turn, $g(x_i)$ is expressed by equation (2).

$$g(x_i) = \begin{cases} t_{ns}(x_i) - t_a(x_i) & \text{if the truck has already arrived} \\ |t_{ns}(x_i) - t_s(x_i)| & \text{if the truck has not arrived yet} \end{cases} \quad (2)$$

where $x_i$ denotes an *i-th* appointment, $t_a(x_i)$ – time of actual arrival of a truck, $t_s(x_i)$ – time of originally scheduled entry of a truck, $t_{ns}(x_i)$ – time of newly scheduled entry of a truck. The penalty function $P(x)$ is expressed by equation (3) as the sum of all penalties imposed on the individual.

$$P(x) = \sum_{j=1}^{5} p_j v_j(x) \quad (3)$$

where $x$ denotes a given individual in the population, $p_j$ stands for a penalty coefficient for the j-th constraint, and $v_j(x)$ – the number of violations of the j-th constraint by the schedule encoded in the individual.

Five different penalties are taken into account. Penalty 1 is the penalty for overlapping appointments, $v_1(x)$ is the number of time units, i.e. half hours when the appointments overlap. Penalty 2 is the penalty for scheduling an appointment for a truck which has not arrived yet before that truck which stays already at a parking lot; $v_2(x)$ is the number of such pairs of appointments. Penalty 3 is the penalty for arranging an appointment for a delayed truck before that truck which has arrived on time; $v_3(x)$ is the number of such pairs of appointments. Penalty 4 is the penalty for making an appointment for a truck which has already arrived but its appointment time has not passed yet before that truck which entry time for a loading bay has been postponed; $v_4(x)$ is the number of such pairs of appointments. Penalty 5 is the penalty for moving an appointment for a truck which has not arrived yet to the time point which is earlier than originally scheduled one; $v_5(x)$ is the number of time units of such shift.

*Improvement of GA.* An enhancement of the genetic algorithm applied to the preparation of the loading appointment schedule was devised. In each iteration each individual is improved just before the calculation of the fitness function. The scheduled appointments are moved to earlier time points provided:
-    a loading bay is vacant in time units before a given appointment,
-    if a truck has already arrived and its appointment is booked later than a current time point,
-    if a truck has not arrived yet and its appointment is booked for a later time point than previously scheduled.

The appointments are moved to the earliest points of time when above conditions are still satisfied. This improvement prevents from the situation when a truck stays at a parking lot and at the same point there is an unoccupied loading bay.

*Particle Swarm Optimization.* The competitive particle swarm optimisation method was elaborated to produce a loading appointment schedule. The particle is represented in the same way as in the GA chromosome. Constraints and fitness function are formulated analogously. The schema of PSO used in the experiments reported in the paper differed from classic one. The concept of self-mutation proposed by Chu et al. [10] was implemented. Moreover, the velocity vector to change particle positions was not used either. The exchange of genes, which represent appointments, among individuals was employed instead. The algorithm was composed of the following steps.

1. Initialize randomly the population.
2. Evaluate the fitness function for each particle.
3. Memorize by the particles their previous best positions and the best position in the neighborhood.
4. Change randomly selected appointment in each particle.
5. For each particle copy one randomly selected appointment from its previous best position
6. For each particle copy one randomly selected appointment from the previous best position in its neighborhood
7. Evaluate the fitness function for each particle.
8. If a termination criterion is not met go to Step 3.
9. Return the best particle as the solution

The random change of an appointment in a particle corresponds to the movement of a particle by a velocity vector in the classical PSO algorithm. In turn, copying an appointment from the particle's best solution and from the best solution in the particle's neighborhood are the equivalent of moving the particle toward its best solution and the best solution in its neighborhood respectively.

## 3     Experimental Evaluation of Proposed Algorithm

Two series of experiments were conducted to evaluate the proposed algorithm. First, single runs of the algorithm for a test problem in a fixed time of a day were done. The test data were created based on the experience of an operating timber products company and reflected main events which led to deteriorating the appointment schedule. Second, the algorithm was run for real-world data to simulate its daily routine. In this case the algorithm was executed each half an hour to produce an optimized loading plan for the rest of a working day.

### 3.1     Test Problem and Results

The outdated schedule used in the experiments is illustrated in Fig. 2 in the form of a screenshot taken from the dispatcher module of the information system for yard management. The time point for running the algorithms was set to 12:00. At this hour all possible events which can distort the current loading schedule are considered, namely:

- three trucks, which have arrived in time, cannot enter any loading bay according to the appointment because of delays in loading process,
- two trucks are late, but one of them has already arrived,
- one truck, which has arrived in time, will not be loaded on time because its appointed loading bay will be still occupied,
- two trucks have arrived before their scheduled time.

The task of tested algorithms is to reschedule all remaining appointments including the outdated appointments. The goal of subsequent series of runs was to tune GA parameters such as crossover rate, mutation rate, population size, crossover count, elite count, tournament size, and selection method. Finally, GA and PSO as well as GA and improved GA algorithms with selected parameters were contrasted. Within each series the tested algorithms were executed ten times and the mean values of fitness, rating, and penalties were compared and their convergence was analysed.

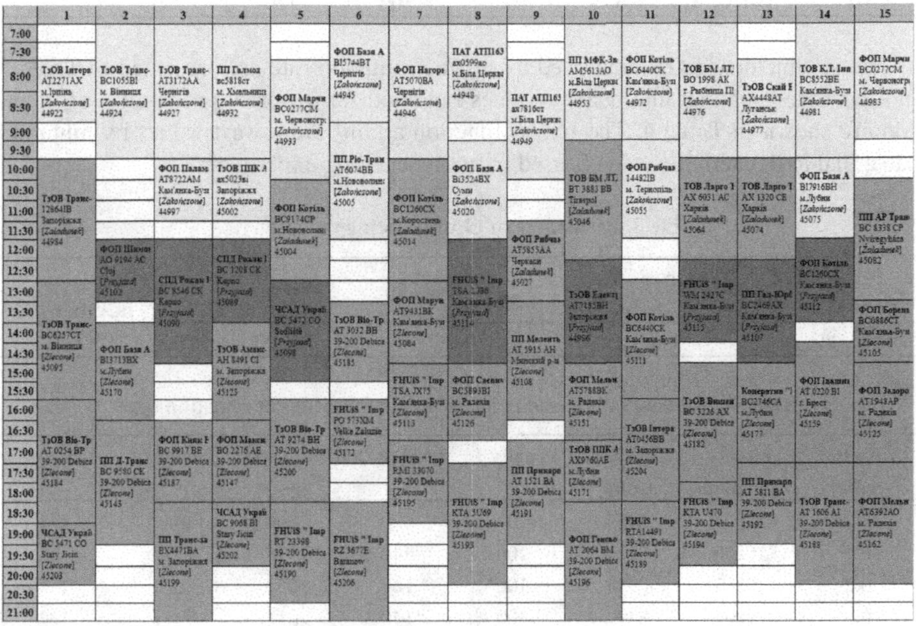

**Fig. 2.** Illustration of outdated appointment schedule used in a test problem

The parameters of GA used by testing crossover rate are shown in Table 1 and the fitness, rating, and penalty values obtained for different probability of crossover are placed in Table 2. The fitness function attains the lowest values by crossover rate equal to 0.8 and 0.9. Due to faster convergence of GA by crossover rate 0.9, this value is used in the next experiments.

**Table 1.** Parameters of GA by testing crossover rate

| Population size | Crossover type | Mutation rate | Selection method | Iteration number |
|---|---|---|---|---|
| 80 | One-point | 0.1 | Roulette wheel | 1000 |

**Table 2.** Fitness, rating, and penalty values for different crossover rate

| Crossover rate | Fitness function | Rating (sum) | Penalty 1 | Penalty 2 | Penalty 3 | Penalty 4 | Penalty 5 |
|---|---|---|---|---|---|---|---|
| 0.5 | 10789 | 9066 | 500 | 260 | 246 | 270 | 447 |
| 0.7 | 11065 | 9567 | 100 | 480 | 174 | 282 | 462 |
| 0.8 | 10435 | 8829 | 400 | 360 | 228 | 264 | 354 |
| 0.9 | 10638 | 8871 | 300 | 600 | 162 | 336 | 369 |
| 0.95 | 10795 | 8967 | 500 | 500 | 120 | 330 | 378 |

The parameters of GA employed by testing mutation rate are presented in Table 3 and the fitness, rating, and penalty values obtained for different probability of mutation are shown in Table 4. The fitness function attains the lowest values by mutation rate equal to 0.3 and this value is used in the succeeding experiments.

**Table 3.** Parameters of GA by testing mutation rate

| Population size | Crossover type | Crossover rate | Selection method | Iteration number |
|---|---|---|---|---|
| 80 | One-point | 0.9 | Roulette wheel | 1000 |

**Table 4.** Fitness, rating, and penalty values for different mutation rate

| Mutation rate | Fitness function | Rating (sum) | Penalty 1 | Penalty 2 | Penalty 3 | Penalty 4 | Penalty 5 |
|---|---|---|---|---|---|---|---|
| 0.05 | 11553 | 9495 | 700 | 500 | 360 | 354 | 144 |
| 0.1 | 10236 | 8826 | 200 | 340 | 318 | 372 | 180 |
| 0.2 | 9883 | 8745 | 100 | 180 | 282 | 432 | 144 |
| 0.3 | 9800 | 8499 | 300 | 140 | 318 | 411 | 132 |
| 0.4 | 10704 | 9096 | 300 | 360 | 294 | 516 | 138 |

The parameters of GA applied by testing population size are presented in Table 5 and the fitness, rating, and penalty values obtained for different sizes are shown in Table 6. The fitness function gains the lowest values by population size equal to 80 chromosomes and this number is used in the subsequent experiments.

**Table 5.** Parameters of GA by testing population size

| Crossover type | Crossover rate | Mutation rate | Selection method | Iteration number |
|---|---|---|---|---|
| One-point | 0.9 | 0.3 | Roulette wheel | 1000 |

**Table 6.** Fitness, rating, and penalty values for different population size

| Population size | Fitness function | Rating (sum) | Penalty 1 | Penalty 2 | Penalty 3 | Penalty 4 | Penalty 5 |
|---|---|---|---|---|---|---|---|
| 20 | 13467 | 9624 | 1900 | 980 | 174 | 240 | 549 |
| 40 | 11182 | 8865 | 1200 | 280 | 150 | 300 | 387 |
| 60 | 9962 | 8436 | 500 | 120 | 132 | 336 | 438 |
| 80 | 9800 | 8499 | 300 | 140 | 132 | 318 | 411 |
| 100 | 9954 | 8649 | 200 | 220 | 150 | 366 | 369 |

The parameters of GA applied by testing the elite count are presented in Table 7 and the fitness, rating, and penalty values obtained for different elite count are shown in Table 8. The application of elitism resulted in substantial improvement of the fitness value. The best output was obtained by the elite count equal to 8, i.e. 10 percent of the population size. Moreover, the elitism eliminated the Penalty 1, i.e. the penalty for overlapping appointments; other penalties were also decreased.

**Table 7.** Parameters of GA by testing the elite count

| Population size | Crossover type | Crossover rate | Mutation rate | Selection method | Iteration number |
|---|---|---|---|---|---|
| 80 | One-point | 0.9 | 0.3 | Roulette wheel | 1000 |

**Table 8.** Fitness, rating, and penalty values for different elite count

| Elite count | Fitness function | Rating (sum) | Penalty 1 | Penalty 2 | Penalty 3 | Penalty 4 | Penalty 5 |
|---|---|---|---|---|---|---|---|
| 0 | 9931 | 8640 | 200 | 200 | 156 | 420 | 315 |
| 1 | 5343 | 4860 | 0 | 60 | 78 | 192 | 153 |
| 2 | 4816 | 4440 | 0 | 40 | 48 | 138 | 150 |
| 4 | 4610 | 4224 | 0 | 20 | 24 | 174 | 168 |
| 8 | 4180 | 3798 | 0 | 40 | 66 | 174 | 102 |
| 16 | 4248 | 3780 | 0 | 0 | 90 | 216 | 162 |

The parameters of GA applied by testing the number of crossover points are presented in Table 9 and the fitness, rating, and penalty values obtained for different sizes are shown in Table 10. The differences were insignificant therefore the two-point crossover was is used in the next experiments.

**Table 9.** Parameters of GA by testing the number of crossover points

| Population size | Crossover rate | Mutation rate | Selection method | Iteration number |
|---|---|---|---|---|
| 80 | 0.9 | 0.3 | Roulette wheel | 1000 |

**Table 10.** Fitness, rating, and penalty values for different number of crossover points

| Crossover points | Fitness function | Rating (sum) | Penalty 1 | Penalty 2 | Penalty 3 | Penalty 4 | Penalty 5 |
|---|---|---|---|---|---|---|---|
| 1 | 9800 | 8499 | 300 | 140 | 132 | 318 | 411 |
| 2 | 9931 | 8640 | 200 | 200 | 156 | 420 | 315 |
| 4 | 9828 | 8550 | 200 | 220 | 180 | 330 | 348 |

The parameters of GA applied by comparing roulette and tournament selection methods are presented in Table 11 and the fitness, rating, and penalty values obtained for different sizes are shown in Table 12. The optimal value of tournament size equal to 5 was determined in an additional test. The faster and more efficient minimization of the fitness function occurred by tournament selection method.

**Table 11.** Parameters of GA by comparing roulette and tournament selection methods

| Population size | Crossover type | Crossover rate | Mutation rate | Tourn. size | Elite | Iteration number |
|---|---|---|---|---|---|---|
| 80 | Two-point | 0.9 | 0.3 | 5 | 8 | 1000 |

**Table 12.** Fitness, rating, and penalty values for compared selection methods

| Selection | Fitness function | Rating (sum) | Penalty 1 | Penalty 2 | Penalty 3 | Penalty 4 | Penalty 5 |
|---|---|---|---|---|---|---|---|
| Roulette | 3984 | 3657 | 0 | 0 | 129 | 102 | 85 |
| Tournament | 3599 | 3262 | 0 | 40 | 75 | 192 | 73 |

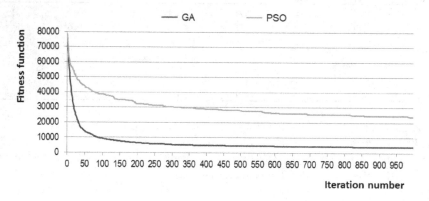

**Fig. 3.** Performance comparison of GA and PSO algorithms

The performance of GA in comparison with the competitive PSO algorithm is depicted in Fig. 3. It is clearly seen that GA outperforms PSO significantly in terms of both fitness and convergence rate. The GA and PSO algorithms were run with previously tuned parameters.

The performance of primary GA compared with the improved version of GA for the number of iterations up to 1000 is illustrated Fig. 4. Fitness, rating, and penalty values obtained for 1000 iterations are given in Table 13. It is clearly seen that improved GA surpasses substantially primary GA in respect of both fitness function and convergence rate.

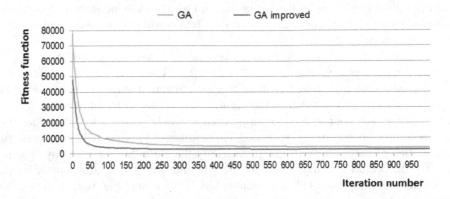

**Fig. 4.** Performance comparison of primary and improved GAs

**Table 13.** Fitness, rating, and penalty values for compared primary and improved GAs

| Selection | Fitness function | Rating (sum) | Penalty 1 | Penalty 2 | Penalty 3 | Penalty 4 | Penalty 5 |
|-----------|------------------|--------------|-----------|-----------|-----------|-----------|-----------|
| Primary | 3599 | 3262 | 0 | 40 | 75 | 192 | 73 |
| Improved | 2857 | 2433 | 0 | 40 | 96 | 270 | 18 |

The primary and improved GAs were also tested with other stopping conditions, namely 100 iterations without fitness improvement. The results for fitness function, number of iterations, and execution time are depicted in Figures 5 a), b), and c) respectively. They confirm the superiority of the improved GA over the primary one.

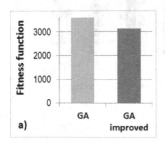 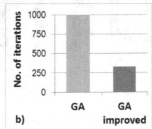 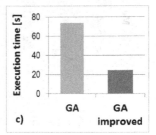

**Fig. 5.** Performance comparison of GA and GA improved algorithms

## 3.2    Simulation for an Operating Company

The ultimate goal of the proposed genetic algorithm is the regular correction of a loading appointment schedule during a working day. The algorithm is intended for running before a given time unit, e.g. each half an hour, and producing an optimized schedule for the rest of a day. The main evaluation criterion in this case is the mean wait time of all trucks during the whole day. In order to examine if the proposed improved GA ensures the shorter mean wait time real-world data were derived from an operating company.

The data of a primary schedule and actual loading times were taken for the period of ten working days from the 5th to 16th May, 2014. The mean wait time was calculated for individual days. Based on these data the utilization of the proposed algorithm in real conditions was simulated. The algorithm was run 10 minutes before and 20 minutes after each hour to produce an updated schedule. The trucks entered loading bays according to the newly generated appointments. This procedure allowed for computation of the mean wait time provided by the algorithm for each day. The results for the operating company, where the schedule was corrected manually by the dispatchers, and improved GA are compared in Table 14 and Fig. 6. It is shown that thanks to GA the mean wait time decreased by 5 minutes i.e. almost 10 per cent. For a company the improvement of a given operating factor of 10% can be very important.

**Table 14.** Comparison of mean wait time in the operating company and provided by GA

| Individual days | 1 | 2 | 3 | 4 | 5 | 6 | 7 | 8 | 9 | 10 | **Mean** |
|---|---|---|---|---|---|---|---|---|---|---|---|
| Operating company | 75 | 51 | 42 | 77 | 49 | 37 | 60 | 64 | 60 | 72 | **58.7** |
| Improved GA | 66 | 47 | 36 | 71 | 42 | 38 | 54 | 60 | 56 | 67 | **53.7** |

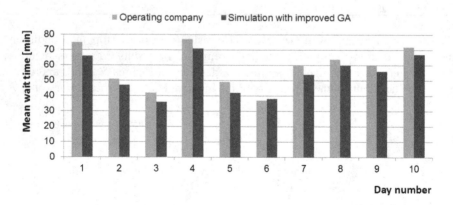

**Fig. 6.**   Comparison of mean wait time in the operating company and provided by GA

The t-test was performed in respect of mean wait time determined for the operating company and provided by the improved GA algorithm. The *p-value* was equal to 0.407, i.e. greater than the alpha level 0.05, thus the null hypothesis could not be rejected. It means that the differences between both results were not statistically significant. We could apply the t-test because we conducted earlier the Shapiro-Wilk and Levene tests. The former did not reject the null hypothesis that the samples came from a normally distributed population and the latter did not reject the null hypothesis that the population variances were equal. Hence, the assumptions underlying a t-test were satisfied.

# 4    Conclusions and Future Work

A method for optimizing truck loading appointment schedule using a genetic algorithm was devised. The chromosomes represent the whole solution, i.e. loading appointment schedules. Each appointment is encoded in two genes comprising the number of a loading bay and time at which outbound truck should enter the bay respectively. Five different penalties were employed to protect the solution against violating constraints. The primary version of GA was enhanced by additional processing just before computing the value of the fitness function. The scheduled appointments are moved to earlier time points whenever it may result in a better solution.

The proposed algorithm was evaluated experimentally using a test problem based on the experience of an operating timber products company. Moreover the simulation of its planned exploitation was performed using real-world data of loading trucks within two weeks. The results proved the usefulness of the algorithm to optimize the loading appointment schedule which becomes outdated during a working day. The proposed GA method outperformed an alternative approach to streamline loading trucks based on particle swarm optimization. During simulation of everyday exploitation at an operating company it turned out that the improved GA provided better appointment schedules than the dispatchers did manually.

It is planned to compare our GA based method with other nature inspired techniques such as ant colony optimization, artificial bee colony, and firefly algorithms and as well as with traditional simulated annealing and tabu search approaches. Moreover, the algorithm will be incorporated into the information system for yard management at the timber products company what allows for examining it during a longer period of time.

# References

1. Agustina, D., Lee, C.K.M., Piplani, R.: A Review: Mathematical Models for Cross Docking Planning. International Journal of Engineering Business Management 2(2), 47–54 (2010)
2. Shuib, A., Ashikin, W.N., Fatthi, W.A.: A Review on Quantitative Approaches for Dock Door Assignment in Cross-Docking. International Journal on Advanced Science, Engineering, Information Technology 2(5), 30–34 (2012)
3. Vahdani, B., Zandieh, M.: Scheduling trucks in cross-docking systems: Robust meta-heuristics. Computers & Industrial Engineering 58, 12–24 (2010)
4. Boloori Arabani, A.R., Fatemi Ghomi, S.M.T., Zandiehb, M.: Meta-heuristics implementation for scheduling of trucks in a cross-docking system with temporary storage. Expert Systems with Applications 38(3), 1964–1979 (2011)

5. Liao, T.W., Chang, P.C., Kuo, R.J., Liao, C.-J.: A comparison of five hybrid metaheuristic algorithms for unrelated parallel-machine scheduling and inbound trucks sequencing in multi-door cross docking systems. Applied Soft Computing **21**, 180–193 (2014)
6. Madani-Isfahani, M., Tavakkoli-Moghaddam, R., Naderi, B.: Multiple cross-docks scheduling using two meta-heuristic algorithms. Computers & Industrial Engineering **74**, 129–138 (2014)
7. Kuo, Y.: Optimizing truck sequencing and truck dock assignment in a cross docking system. Expert Systems with Applications **40**, 5532–5541 (2013)
8. Naderi, B., Rahmani, S., Rahman, S.: A Multiobjective Iterated Greedy Algorithm for Truck Scheduling in Cross-Dock Problems. Journal of Industrial Engineering, Article ID 128542, 12 (2014)
9. Chen, R.-M., Shih, H.-F.: Solving University Course Timetabling Problems Using Constriction Particle Swarm Optimization with Local Search. Algorithms **6**(2), 227–244 (2013)
10. Chu, S.-C., Chen, Y.-T., Ho, J.-H.: Timetable scheduling using particle swarm optimization. In: Proceedings of the First International Conference on Innovative Computing, Information and Control, ICICIC 2006, Vol. 3, pp. 324-327 (2006)
11. Kumar, A., Singh, K., Sharma, N.: Automated Timetable Generator Using Particle Swarm Optimization. International Journal on Recent and Innovation Trends in Computing and Communication **1**(9), 686–692 (2013)
12. Tassopoulos, I.X., Beligiannis, G.N.: A hybrid particle swarm optimization based algorithm for high school timetabling problems. Applied Soft Computing **12**(11), 3472–3489 (2012)
13. Smętek, M., Trawiński, B.: Selection of Heterogeneous Fuzzy Model Ensembles Using Self-adaptive Genetic Algorithms. New Generation Computing **29**(3), 309–327 (2011)
14. Trawiński, B.: Evolutionary Fuzzy System Ensemble Approach to Model Real Estate Market based on Data Stream Exploration. Journal of Universal Computer Science **19**(4), 539–562 (2013)

# Innovations in Intelligent Systems and Applications

# A New Fuzzy Interpolative Reasoning Method
# Based on the Ratio of Fuzziness of Rough-Fuzzy Sets

Shyi-Ming Chen[1(✉)], Shou-Hsiung Cheng[2,3], and Ze-Jin Chen[1]

· [1] Department of Computer Science and Information Engineering,
National Taiwan University of Science and Technology, Taipei, Taiwan
smchen@mail.ntust.edu.tw
[2] Department of Information Management,
Chienkuo Technology University, Changhua, Taiwan
[3] Department of Kinesiology Health Leisure Studies,
Chienkuo Technology University, Changhua, Taiwan

**Abstract.** In this paper, we propose a new fuzzy interpolative reasoning method for sparse fuzzy rule-based systems based on the ratio of fuzziness of polygonal rough-fuzzy sets, where the values of the antecedent variables and the consequence variables in the fuzzy rules are represented by polygonal rough-fuzzy sets. The experimental results show that the proposed fuzzy interpolative reasoning method outperforms the existing method for fuzzy interpolative reasoning in sparse fuzzy rule-based systems.

**Keywords:** Degrees of fuzziness · Fuzzy interpolative reasoning · Rough-fuzzy sets · Ratio of fuzziness

## 1 Introduction

Fuzzy interpolative reasoning is a very important research topic in sparse fuzzy rule-based systems. In recent years, some fuzzy interpolative reasoning methods [1]-[12] based on fuzzy sets [13] have been presented for sparse fuzzy rule-based systems. In [3], Chen and Shen presented a fuzzy interpolative reasoning method based on triangular rough-fuzzy sets. However, Chen and Shen [3] pointed out that their fuzzy interpolated reasoning method only can be used in the situations of triangular rough-fuzzy sets. Moreover, the fuzzy interpolative results of the examples shown in [3] are not reasonable enough. Therefore, we need to develop a new fuzzy interpolative reasoning method based on polygonal rough-fuzzy sets, to overcome the drawbacks of Chen and Shen's method [3].

In this paper, we propose a new fuzzy interpolative reasoning method for sparse fuzzy rule-based systems based on the ratio of fuzziness of polygonal rough-fuzzy sets, where the values of the antecedent variables and the consequence variables in the fuzzy rules are represented by polygonal rough-fuzzy sets. The experimental results show that the proposed fuzzy interpolative reasoning method outperforms Chen and Shen's method [3] for fuzzy interpolative reasoning in sparse fuzzy rule-based systems.

© Springer International Publishing Switzerland 2015
N.T. Nguyen et al. (Eds.): ACIIDS 2015, Part I, LNAI 9011, pp. 551–561, 2015.
DOI: 10.1007/978-3-319-15702-3_53

## 2    Preliminaries

In this section, we propose the definition of polygonal rough-fuzzy sets, the definition of the representative value of a polygonal rough-fuzzy set, and the definition of the degree of fuzziness of a polygonal rough-fuzzy set.

A polygonal rough-fuzzy set $\tilde{A}$ in the universe of discourse $X$ can be characterized by $n$ characteristic points $\overline{a_0}$, $\overline{a_1}$, ..., $\overline{a_{n-2}}$ and $\overline{a_{n-1}}$ of the upper approximation fuzzy set $\overline{\tilde{A}}$ and $n$ characteristic points $\underline{a_0}$, $\underline{a_1}$, ..., $\underline{a_{n-2}}$ and $\underline{a_1}$ of the lower approximation fuzzy set $\underline{\tilde{A}}$, as shown in Fig. 1, where $\tilde{A} = <(\underline{a_0}, \underline{a_1}, ..., \underline{a_{n-1}}; \underline{\mu_0}, \underline{\mu_1}, ..., \underline{\mu_{n-1}}), (\overline{a_0}, \overline{a_1}, ..., \overline{a_{n-1}}; \overline{\mu_0}, \overline{\mu_1}, ..., \overline{\mu_{n-1}})>$, the degrees of membership of the characteristic points $\overline{a_0}$, $\overline{a_1}$, ..., $\overline{a_{n-2}}$ and $\overline{a_{n-1}}$ are $\overline{\mu_0}$, $\overline{\mu_1}$, ..., $\overline{\mu_{n-2}}$ and $\overline{\mu_{n-1}}$, respectively, the degrees of membership of the characteristic points $\underline{a_0}$, $\underline{a_1}$, ..., $\underline{a_{n-2}}$ and $\underline{a_{n-1}}$ are $\underline{\mu_0}$, $\underline{\mu_1}$, ..., $\underline{\mu_{n-2}}$ and $\underline{\mu_{n-1}}$, respectively, and $n \geq 1$. In Fig. 1, $\overline{a_0}$ and $\overline{a_{n-1}}$ are called the "left extreme point" and the "right extreme point" of the upper approximation fuzzy set $\overline{\tilde{A}}$, respectively, $\underline{a_0}$ and $\underline{a_{n-1}}$ are called the "left extreme point" and the "right extreme point" of the lower approximation fuzzy set $\underline{\tilde{A}}$, respectively, $\overline{a_l}$ and $\overline{a_r}$ are called the "left normal point" and the "right normal point" of the upper approximation fuzzy set $\overline{\tilde{A}}$, respectively, $\underline{a_l}$ and $\underline{a_r}$ are called the "left normal point" and the "right normal point" of the lower approximation fuzzy set $\underline{\tilde{A}}$, respectively, where $n \geq 1$, $l = \left\lfloor \frac{n-1}{2} \right\rfloor$, and $r = \left\lceil \frac{n-1}{2} \right\rceil$.

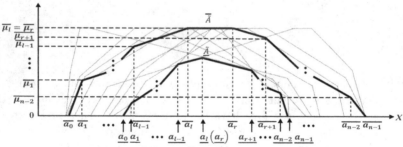

**Fig. 1.** A polygonal rough-fuzzy set $\tilde{A} = < \underline{\tilde{A}}, \overline{\tilde{A}} >$

In the following, we propose the definition of the representative value $Rep(\tilde{A})$ of the polygonal rough-fuzzy set $\tilde{A} = < \underline{\tilde{A}}, \overline{\tilde{A}} >$ shown in Fig. 1. The representative value $Rep(\tilde{A})$ of the polygonal rough-fuzzy set $\tilde{A} = < \underline{\tilde{A}}, \overline{\tilde{A}} >$ shown in Fig. 1 is defined as follows:

$$Rep(\tilde{A}) = \begin{cases} \dfrac{Rep\left(\overline{\tilde{A}}\right) \times Area\left(\overline{\tilde{A}}\right) - Rep(\underline{\tilde{A}}) \times Area(\underline{\tilde{A}})}{Area\left(\overline{\tilde{A}}\right) - Area(\underline{\tilde{A}})}, & if \left| Area\left(\overline{\tilde{A}}\right) - Area(\underline{\tilde{A}}) \right| > 0 \\ Rep(\tilde{A}), & if \left| Area\left(\overline{\tilde{A}}\right) - Area(\underline{\tilde{A}}) \right| = 0 \end{cases} \quad (1)$$

where $Rep(\underline{\tilde{A}}) = \frac{a_0 + a_1 + a_2 + \cdots + a_{n-1}}{n}$, $Rep\left(\overline{\tilde{A}}\right) = \frac{\overline{a_0} + \overline{a_1} + \overline{a_2} + \cdots + \overline{a_{n-1}}}{n}$, $\underline{a_0}$, $\underline{a_1}$, ..., $\underline{a_{n-2}}$ and $\underline{a_{n-1}}$ are the characteristic points of the lower approximation fuzzy set $\underline{\tilde{A}}$ of the polygonal rough-fuzzy set $\tilde{A}$, respectively, $\overline{a_0}$, $\overline{a_1}$, ..., $\overline{a_{n-2}}$ and $\overline{a_{n-1}}$ are the characteristic points of the upper approximation fuzzy set $\overline{\tilde{A}}$ of the polygonal rough-fuzzy set $\tilde{A}$, respectively, $Area(\underline{\tilde{A}})$ denotes the area of the lower approximation fuzzy set $\underline{\tilde{A}}$ of the polygonal rough-fuzzy set $\tilde{A}$, $Area\left(\overline{\tilde{A}}\right)$ denotes the area of the upper approximation fuzzy set $\overline{\tilde{A}}$ of the polygonal rough-fuzzy set $\tilde{A}$, $Area(\underline{\tilde{A}}) \geq 0$, and $Area\left(\overline{\tilde{A}}\right) \geq 0$.

Let $\tilde{A}_1$, ..., $\tilde{A}_{z-1}$, $\tilde{A}_z$, $\tilde{A}_{z+1}$, ..., and $\tilde{A}_n$ be rough-fuzzy sets, where the representative values of $\tilde{A}_1$, ..., $\tilde{A}_{z-1}$, $\tilde{A}_z$, $\tilde{A}_{z+1}$, ..., and $\tilde{A}_n$ are $Rep(\tilde{A}_1)$, ..., $Rep(\tilde{A}_{z-1})$, $Rep(\tilde{A}_z)$, $Rep(\tilde{A}_{z+1})$, ..., and $Rep(\tilde{A}_n)$, respectively, and $Rep(\tilde{A}_1) \leq \ldots \leq Rep(\tilde{A}_{z-1}) \leq Rep(\tilde{A}_z) \leq Rep(\tilde{A}_{z+1}) \leq \ldots \leq Rep(\tilde{A}_n)$. Because $Rep(\tilde{A}_{z-1}) \leq Rep(\tilde{A}_z) \leq Rep(\tilde{A}_{z+1})$, the rough-fuzzy set $\tilde{A}_{z-1}$ and the rough-fuzzy set $\tilde{A}_{z+1}$ are called the left closest rough-fuzzy set and the right closest rough-fuzzy set of the rough-fuzzy set $\tilde{A}_z$, respectively.

The degree of fuzziness $Fuzziness(\tilde{A})$ of the polygonal rough-fuzzy set $\tilde{A} = \langle \underline{\tilde{A}}, \overline{\tilde{A}} \rangle$ shown in Fig. 1 is defined as follows:

$$Fuzziness(\tilde{A}) = \frac{Area(\underline{\tilde{A}}) + Area(\overline{\tilde{A}})}{2}, \tag{2}$$

where $Area(\underline{\tilde{A}})$ denotes the area of the lower approximation fuzzy set $\underline{\tilde{A}}$ of the polygonal rough-fuzzy set $\tilde{A}$, $Area\left(\overline{\tilde{A}}\right)$ denotes the area of the upper approximation fuzzy set $\overline{\tilde{A}}$ of the polygonal rough-fuzzy set $\tilde{A}$, $Area(\underline{\tilde{A}}) \leq Fuzziness(\tilde{A}) \leq Area\left(\overline{\tilde{A}}\right)$, $Area(\underline{\tilde{A}}) \geq 0$, and $Area\left(\overline{\tilde{A}}\right) \geq 0$.

## 3    A New Fuzzy Interpolative Reasoning Method Based on the Ratio of Fuzziness of Polygonal Rough-Fuzzy Sets

In this section, we propose a new fuzzy interpolative reasoning method based on the ratio of fuzziness of polygonal rough-fuzzy sets. Let us consider the following multiple fuzzy rules fuzzy interpolation scheme based on polygonal rough-fuzzy sets:

**Rule 1**: If $X_1$ is $\tilde{A}_{11}$ and $X_2$ is $\tilde{A}_{12}$ and $\cdots$ and $X_m$ is $\tilde{A}_{1m}$ **Then** $Y$ is $\tilde{B}_1$
**Rule 2**: If $X_1$ is $\tilde{A}_{21}$ and $X_2$ is $\tilde{A}_{22}$ and $\cdots$ and $X_m$ is $\tilde{A}_{2m}$ **Then** $Y$ is $\tilde{B}_2$
**Rule q**: If $X_1$ is $\tilde{A}_{q1}$ and $X_2$ is $\tilde{A}_{q2}$ and $\cdots$ and $X_m$ is $\tilde{A}_{qm}$ **Then** $Y$ is $\tilde{B}_q$
**Observation**: $X_1$ is $\tilde{A}_1^*$ and $X_2$ is $\tilde{A}_2^*$ and $\cdots$ and $X_m$ is $\tilde{A}_m^*$
**Conclusion**: $Y$ is $\tilde{B}^*$

where $X_j$ denotes the $j$th antecedent variable; $Y$ denotes the consequence variable; the antecedent polygonal rough-fuzzy set $\tilde{A}_{ij}$ of *Rule i* is represented by $\tilde{A}_{ij} = \langle(\underline{a_{ij,0}}, \underline{a_{ij,1}}, ..., \underline{a_{ij,n-1}}; \underline{\mu_{ij,0}}, \underline{\mu_{ij,1}}, ..., \underline{\mu_{ij,n-1}}), (\overline{a_{ij,0}}, \overline{a_{ij,1}}, ..., \overline{a_{ij,n-1}}; \overline{\mu_{ij,0}}, \overline{\mu_{ij,1}}, ...,$

$\overline{\mu_{ij,n-1}}$)>, where the upper approximation fuzzy set $\tilde{A}_{ij}$ of the antecedent polygonal rough-fuzzy set $\tilde{A}_{ij}$ can be characterized by $n$ characteristic points $\overline{a_{ij,0}}$, $\overline{a_{ij,1}}$, ..., $\overline{a_{ij,n-2}}$ and $\overline{a_{ij,n-1}}$, and $\overline{\mu_{ij,0}}$, $\overline{\mu_{ij,1}}$, ..., $\overline{\mu_{ij,n-2}}$ and $\overline{\mu_{ij,n-1}}$ are the degrees of membership of the characteristic points $\overline{a_{ij,0}}$, $\overline{a_{ij,1}}$, ..., $\overline{a_{ij,n-2}}$ and $\overline{a_{ij,n-1}}$, respectively. The lower approximation fuzzy set $\underline{\tilde{A}}_{ij}$ of the antecedent polygonal rough-fuzzy set $\tilde{A}_{ij}$ can be characterized by $n$ characteristic points $\underline{a_{ij,0}}$, $\underline{a_{ij,1}}$, ..., $\underline{a_{ij,n-2}}$ and $\underline{a_{ij,n-1}}$, and $\underline{\mu_{ij,0}}$, $\underline{\mu_{ij,1}}$, ..., $\underline{\mu_{ij,n-2}}$ and $\underline{\mu_{ij,n-1}}$ are the degrees of membership of the characteristic points $\underline{a_{ij,0}}$, $\underline{a_{ij,1}}$, ..., $\underline{a_{ij,n-2}}$ and $\underline{a_{ij,n-1}}$, respectively, $1 \le i \le q$, $1 \le j \le m$, and $n \ge 1$. The observation polygonal rough-fuzzy set $\tilde{A}_j^*$ is represented by $\tilde{A}_j^* =$ <($\underline{a_{j,0}^*}$, $\underline{a_{j,1}^*}$, ..., $\underline{a_{j,n-1}^*}$; $\underline{\mu_{j,0}^*}$, $\underline{\mu_{j,1}^*}$, ..., $\underline{\mu_{j,n-1}^*}$), ($\overline{a_{j,0}^*}$, $\overline{a_{j,1}^*}$, ..., $\overline{a_{j,n-1}^*}$; $\overline{\mu_{j,0}^*}$, $\overline{\mu_{j,1}^*}$, ..., $\overline{\mu_{j,n-1}^*}$)>, where the upper approximation fuzzy set $\tilde{A}_j^*$ of the observation polygonal rough-fuzzy set $\tilde{A}_j^*$ can be characterized by $n$ characteristic points $\overline{a_{j,0}^*}$, $\overline{a_{j,1}^*}$, ..., $\overline{a_{j,n-2}^*}$ and $\overline{a_{j,n-1}^*}$, and $\overline{\mu_{j,0}^*}$, $\overline{\mu_{j,1}^*}$, ..., $\overline{\mu_{j,n-2}^*}$ and $\overline{\mu_{j,n-1}^*}$ are the degrees of membership of the characteristic points $\overline{a_{j,0}^*}$, $\overline{a_{j,1}^*}$, ..., $\overline{a_{j,n-2}^*}$ and $\overline{a_{j,n-1}^*}$, respectively. The lower approximation fuzzy set $\underline{\tilde{A}}_j^*$ of the observation polygonal rough-fuzzy set $\tilde{A}_j^*$ can be characterized by $n$ characteristic points $\underline{a_{j,0}^*}$, $\underline{a_{j,1}^*}$, ..., $\underline{a_{j,n-2}^*}$ and $\underline{a_{j,n-1}^*}$, and $\underline{\mu_{j,0}^*}$, $\underline{\mu_{j,1}^*}$, ..., $\underline{\mu_{j,n-2}^*}$ and $\underline{\mu_{j,n-1}^*}$ are the degrees of membership of the characteristic points $\underline{a_{j,0}^*}$, $\underline{a_{j,1}^*}$, ..., $\underline{a_{j,n-2}^*}$ and $\underline{a_{j,n-1}^*}$, respectively, $1 \le j \le m$, and $n \ge 1$; the consequence polygonal rough-fuzzy set $\tilde{B}_i$ of each fuzzy rule *Rule i* is represented by $\tilde{B}_i =$ <($\underline{b_{i,0}}$, $\underline{b_{i,1}}$, ..., $\underline{b_{i,n-1}}$; $\underline{\mu_{i,0}}$, $\underline{\mu_{i,1}}$, ..., $\underline{\mu_{i,n-1}}$), ($\overline{b_{i,0}}$, $\overline{b_{i,1}}$, ..., $\overline{b_{i,n-1}}$; $\overline{\mu_{i,0}}$, $\overline{\mu_{i,1}}$, ..., $\overline{\mu_{i,n-1}}$)>, where the upper approximation fuzzy set $\tilde{B}_i$ of the consequence polygonal rough-fuzzy set $\tilde{B}_i$ can be characterized by $n$ characteristic points $\overline{b_{i,0}}$, $\overline{b_{i,1}}$, ..., $\overline{b_{i,n-2}}$ and $\overline{b_{i,n-1}}$, and $\overline{\mu_{i,0}}$, $\overline{\mu_{i,1}}$, ..., $\overline{\mu_{i,n-2}}$ and $\overline{\mu_{i,n-1}}$ are the degrees of membership of the characteristic points $\overline{b_{i,0}}$, $\overline{b_{i,1}}$, ..., $\overline{b_{i,n-2}}$ and $\overline{b_{i,n-1}}$, respectively. The lower approximation fuzzy set $\underline{\tilde{B}}_i$ of the consequence polygonal rough-fuzzy set $\tilde{B}_i$ can be characterized by $n$ characteristic points $\underline{b_{i,0}}$, $\underline{b_{i,1}}$, ..., $\underline{b_{i,n-2}}$ and $\underline{b_{i,n-1}}$, and $\underline{\mu_{i,0}}$, $\underline{\mu_{i,1}}$, ..., $\underline{\mu_{i,n-2}}$ and $\underline{\mu_{i,n-1}}$ are the degrees of membership of the characteristic points $\underline{b_{i,0}}$, $\underline{b_{i,1}}$, ..., $\underline{b_{i,n-2}}$ and $\underline{b_{i,n-1}}$, respectively, $1 \le i \le q$, and $n \ge 1$; the consequence polygonal rough-fuzzy set $\tilde{B}^*$ is represented by $\tilde{B}^* =$ <($\underline{b_0^*}$, $\underline{b_1^*}$, ..., $\underline{b_{n-1}^*}$; $\underline{\mu_0^*}$, $\underline{\mu_1^*}$, ..., $\underline{\mu_{n-1}^*}$), ($\overline{b_0^*}$, $\overline{b_1^*}$, ..., $\overline{b_{n-1}^*}$; $\overline{\mu_0^*}$, $\overline{\mu_1^*}$, ..., $\overline{\mu_{n-1}^*}$)>, where the upper approximation fuzzy set $\tilde{B}^*$ of the consequence polygonal rough-fuzzy set $\tilde{B}^*$ can be characterized by $n$ characteristic points $\overline{b_0^*}$, $\overline{b_1^*}$, ..., $\overline{b_{n-2}^*}$ and $\overline{b_{n-1}^*}$, and $\overline{\mu_0^*}$, $\overline{\mu_1^*}$, ..., $\overline{\mu_{n-2}^*}$ and $\overline{\mu_{n-1}^*}$ are the degrees of membership of the characteristic points $\overline{b_0^*}$, $\overline{b_1^*}$, ..., $\overline{b_{n-2}^*}$ and $\overline{b_{n-1}^*}$, respectively. The lower approximation fuzzy set $\underline{\tilde{B}}^*$ of the consequence polygonal rough-fuzzy set $\tilde{B}^*$ can be characterized by $n$ characteristic points $\underline{b_0^*}$, $\underline{b_1^*}$, ..., $\underline{b_{n-2}^*}$ and $\underline{b_{n-1}^*}$, and $\underline{\mu_0^*}$, $\underline{\mu_1^*}$, ..., $\underline{\mu_{n-2}^*}$ and $\underline{\mu_{n-1}^*}$ are the degrees of membership of the characteristic points $\underline{b_0^*}$, $\underline{b_1^*}$, ..., $\underline{b_{n-2}^*}$ and $\underline{b_{n-1}^*}$, respectively, and $n \geq 1$; $m$ denotes the number of antecedent variables appearing in the fuzzy rules; $q$ denotes the number of fuzzy rules.

The proposed fuzzy interpolative reasoning method based on the ratio of fuzziness of polygonal rough-fuzzy sets is now presented as follows:

**Step 1:** Calculate the left placement factor $\gamma_{lj}$ and the right placement factor $\gamma_{rj}$ of each observation polygonal rough-fuzzy set $\tilde{A}_j^*$, respectively, where $1 \leq j \leq m$, shown as follows:

$$\gamma_{lj} = \begin{cases} 1 - \frac{Rep(\tilde{A}_j^*) - Rep(\tilde{A}_{lj})}{Rep(\tilde{A}_{rj}) - Rep(\tilde{A}_{lj})}, & \text{if } Rep(\tilde{A}_{lj}) \leq Rep(\tilde{A}_j^*) \leq Rep(\tilde{A}_{rj}) \\ 1, & \text{if } Rep(\tilde{A}_{ij}) < Rep(\tilde{A}_j^*), \text{where } 1 \leq i \leq q \end{cases}, \quad (3)$$

$$\gamma_{rj} = \begin{cases} 1 - \frac{Rep(\tilde{A}_{rj}) - Rep(\tilde{A}_j^*)}{Rep(\tilde{A}_{rj}) - Rep(\tilde{A}_{lj})}, & \text{if } Rep(\tilde{A}_{lj}) \leq Rep(\tilde{A}_j^*) \leq Rep(\tilde{A}_{rj}) \\ 1, & \text{if } Rep(\tilde{A}_j^*) < Rep(\tilde{A}_{ij}), \text{where } 1 \leq i \leq q \end{cases}, \quad (4)$$

where $Rep(\tilde{A}_{ij})$ denotes the representative value of the antecedent polygonal rough-fuzzy set $\tilde{A}_{ij}$, $Rep(\tilde{A}_j^*)$ denotes the representative value of the observation polygonal rough-fuzzy set $\tilde{A}_j^*$, $\tilde{A}_{lj}$ is the left closest polygonal rough-fuzzy set of the observation polygonal rough-fuzzy set $\tilde{A}_j^*$, $\tilde{A}_{rj}$ is the right closest polygonal rough-fuzzy set of the observation polygonal rough-fuzzy set $\tilde{A}_j^*$, where $1 \leq i \leq q$ and $1 \leq j \leq m$.

**Step 2:** Based on [8], calculate the weight $W_i$ of each fuzzy rule *Rule i*, where $1 \leq i \leq q$, shown as follows:

$$W_i = \frac{min_{j=1,2,...,m} \gamma_{ij}}{\sum_{i=1}^h min_{j=1,2,...,m} \gamma_{ij}}, \quad (5)$$

where $1 \leq i \leq q, 1 \leq j \leq m, 0 \leq \gamma_{ij} \leq 1, 0 \leq min_{j=1,2,...,m} \gamma_{ij} \leq 1, 0 \leq W_i \leq 1$, and $\sum_{i=1}^q W_i = 1$.

**Step 3:** Get the intermediate polygonal rough-fuzzy set $\tilde{B}'$. Firstly, calculate the characteristic points $\underline{b_0'}$, $\underline{b_1'}$, ..., $\underline{b_{n-2}'}$ and $\underline{b_{n-1}'}$ of the lower approximation fuzzy set $\underline{\tilde{B}'}$ of the intermediate polygonal rough-fuzzy set $\tilde{B}'$ and calculate the degrees of membership $\underline{\mu_0'}$, $\underline{\mu_1'}$, ..., $\underline{\mu_{n-2}'}$ and $\underline{\mu_{n-1}'}$ of the characteristic points $\underline{b_0'}$, $\underline{b_1'}$, ..., $\underline{b_{n-2}'}$ and $\underline{b_{n-1}'}$, respectively, where $n \geq 1$, shown as follows:

$$\underline{b_y'} = \sum_{i=1}^q W_i \times \underline{b_{i,y}}, \quad (6)$$

$$\underline{\mu_y'} = \sum_{i=1}^q W_i \times \underline{\mu_{i,y}}, \quad (7)$$

where $1 \leq i \leq q, 0 \leq y \leq n-1, 0 \leq W_i \leq 1$, and $\sum_{i=1}^q W_i = 1$. Then, calculate the characteristic points $\overline{b_0'}$, $\overline{b_1'}$, ..., $\overline{b_{n-2}'}$ and $\overline{b_{n-1}'}$ of the upper approximation fuzzy set $\overline{\tilde{B}'}$ of the intermediate polygonal rough-fuzzy set $\tilde{B}'$ and calculate the degrees of membership $\overline{\mu_0'}$, $\overline{\mu_1'}$, ..., $\overline{\mu_{n-2}'}$ and $\overline{\mu_{n-1}'}$ of the characteristic points $\overline{b_0'}$, $\overline{b_1'}$, ..., $\overline{b_{n-2}'}$ and $\overline{b_{n-1}'}$, respectively, where $n \geq 1$, shown as follows:

$$\overline{b_y'} = \sum_{i=1}^q W_i \times \overline{b_{i,y}}, \tag{8}$$

$$\overline{\mu_y'} = \sum_{i=1}^q W_i \times \overline{\mu_{i,y}}, \tag{9}$$

where $1 \leq i \leq q, 0 \leq y \leq n-1, 0 \leq W_i \leq 1$, and $\sum_{i=1}^q W_i = 1$. Then, we can get the intermediate polygonal rough-fuzzy set $\tilde{B}' = <(\underline{b_0'}, \underline{b_1'}, ..., \underline{b_{n-1}'}; \underline{\mu_0'}, \underline{\mu_1'}, ..., \underline{\mu_{n-1}'})$, $(\overline{b_0'}, \overline{b_1'}, ..., \overline{b_{n-1}'}; \overline{\mu_0'}, \overline{\mu_1'}, ..., \overline{\mu_{n-1}'})>$.

**Step 4:** Calculate the average degree of fuzziness $Average\_Fuzziness(\tilde{A}_1^*, \tilde{A}_2^*, ..., \tilde{A}_m^*)$ of the observation polygonal rough-fuzzy sets $\tilde{A}_1^*, \tilde{A}_2^*, ..., \tilde{A}_{m-1}^*$ and $\tilde{A}_m^*$ and calculate the average degree of fuzziness $Average\_Fuzziness(\tilde{A}_{i1}, \tilde{A}_{i2}, ..., \tilde{A}_{im})$ of the antecedent polygonal rough-fuzzy sets $\tilde{A}_{i1}, \tilde{A}_{i2}, ..., \tilde{A}_{i(m-1)}$ and $\tilde{A}_{im}$ of the fuzzy rule *Rule i*, respectively, where $1 \leq i \leq q$ and $1 \leq j \leq m$, shown as follows:

$$Average\_Fuzziness(\tilde{A}_1^*, \tilde{A}_2^*, ..., \tilde{A}_m^*) = \sum_{j=1}^m \frac{Fuzziness(\tilde{A}_j^*)}{m}, \tag{10}$$

$$Average\_Fuzziness(\tilde{A}_{i1}, \tilde{A}_{i2}, ..., \tilde{A}_{im}) = \sum_{j=1}^m \frac{Fuzziness(\tilde{A}_{ij})}{m}, \tag{11}$$

where $1 \leq i \leq q$ and $1 \leq j \leq m$.

**Step 5:** Calculate the ratio $R$ of the average degree of fuzziness of the observation polygonal rough-fuzzy sets to the average degree of fuzziness of the antecedent polygonal rough-fuzzy sets of the $q$ fuzzy rules, shown as follows:

$$R = \frac{Average\_Fuzziness(\tilde{A}_1^*, \tilde{A}_2^*, ..., \tilde{A}_m^*)}{\sum_{i=1}^q W_i \times Average\_Fuzziness(\tilde{A}_{i1}, \tilde{A}_{i2}, ..., \tilde{A}_{im})}, \tag{12}$$

where $R \geq 0$, $W_i$ is the weight of each fuzzy rule *Rule i*, $1 \leq i \leq q, 0 \leq W_i \leq 1$, and $\sum_{i=1}^q W_i = 1$.

**Step 6:** Get the consequence polygonal rough-fuzzy set $\tilde{B}^*$ from the following calculation. Firstly, calculate the characteristic points $\underline{b_0^*}$, $\underline{b_1^*}$, ..., $\underline{b_{n-2}^*}$ and $\underline{b_{n-1}^*}$ of the lower approximation fuzzy set $\underline{\tilde{B}^*}$ of the consequence polygonal rough-fuzzy set $\tilde{B}^*$, respectively, where $n \geq 1$, shown as follows:

$$\underline{b_c'} = (\underline{b_l'} + \underline{b_r'})/2, \tag{13}$$

$$\underline{b_l^*} = \underline{b_c'} - R \times \frac{\underline{b_r'} - \underline{b_l'}}{2}, \tag{14}$$

$$\underline{b_r^*} = \underline{b_c'} + R \times \frac{\underline{b_r'} - \underline{b_l'}}{2}, \tag{15}$$

$$\underline{b_s^*} = \begin{cases} \underline{b_l^*} - \sum_{q=s}^{l-1} R \times (\underline{b_{q+1}'} - \underline{b_q'}), & if\ 0 \leq s \leq l-1 \\ \underline{b_r^*} + \sum_{q=r}^{s-1} R \times (\underline{b_{q+1}'} - \underline{b_q'}), & if\ r+1 \leq s \leq n-1 \end{cases}, \tag{16}$$

where $0 \leq s \leq n-1$. Then calculate the characteristic points $\overline{b_0^*}$, $\overline{b_1^*}$, ..., $\overline{b_{n-2}^*}$ and $\overline{b_{n-1}^*}$ of the upper approximation fuzzy set $\tilde{\overline{B}}^*$ of the consequence polygonal rough-fuzzy set $\tilde{B}^*$, respectively, where $n \geq 1$, shown as follows:

$$\overline{b_c'} = (\overline{b_l'} + \overline{b_r'})/2, \tag{17}$$

$$\overline{b_l^*} = \overline{b_c'} - R \times \frac{\overline{b_r'} - \overline{b_l'}}{2}, \tag{18}$$

$$\overline{b_r^*} = \overline{b_c'} + R \times \frac{\overline{b_r'} - \overline{b_l'}}{2}, \tag{19}$$

$$\overline{b_s^*} = \begin{cases} \overline{b_l^*} - \sum_{q=s}^{l-1} R \times \left( \overline{b_{q+1}'} - \overline{b_q'} \right), if\ 0 \le s \le l-1 \\ \overline{b_r^*} + \sum_{q=r}^{s-1} R \times \left( \overline{b_{q+1}'} - \overline{b_q'} \right), if\ r+1 \le s \le n-1 \end{cases}, \tag{20}$$

where $0 \le s \le n-1$. Finally, calculate the degrees of membership $\underline{\mu_0^*}$, $\underline{\mu_1^*}$, ..., $\underline{\mu_{n-2}^*}$ and $\underline{\mu_{n-1}^*}$ of the characteristic points $\underline{b_0^*}$, $\underline{b_1^*}$, ..., $\underline{b_{n-2}^*}$ and $\underline{b_{n-1}^*}$, and calculate the degrees of membership $\overline{\mu_0^*}$, $\overline{\mu_1^*}$, ..., $\overline{\mu_{n-2}^*}$ and $\overline{\mu_{n-1}^*}$ of the characteristic points $\overline{b_0^*}$, $\overline{b_1^*}$, ..., $\overline{b_{n-2}^*}$ and $\overline{b_{n-1}^*}$, respectively, where $n \ge 1$, shown as follows:

$$\underline{\mu_y^*} = \sum_{i=1}^{q} W_i \times \underline{\mu_{i,y}}, \tag{21}$$

$$\overline{\mu_y^*} = \sum_{i=1}^{q} W_i \times \overline{\mu_{i,y}}, \tag{22}$$

where $1 \le i \le q$, $0 \le y \le n-1$, $0 \le W_i \le 1$, and $\sum_{i=1}^{q} W_i = 1$. Thus, we can get the consequence polygonal rough-fuzzy set $\tilde{B}^* = <(\underline{b_0^*}, \underline{b_1^*}, ..., \underline{b_{n-1}^*}; \underline{\mu_0^*}, \underline{\mu_1^*}, ..., \underline{\mu_{n-1}^*}), (\overline{b_0^*}, \overline{b_1^*}, ..., \overline{b_{n-1}^*}; \overline{\mu_0^*}, \overline{\mu_1^*}, ..., \overline{\mu_{n-1}^*})>$.

## 4    Experimental Results

In this section, we use some example to compare the experimental results of the proposed fuzzy interpolative reasoning method with the ones of Chen and Shen's method [3]. The fuzzy interpolative reasoning schemes of the following examples based on triangular rough-fuzzy sets are shown as follows:

> ***Rule 1*: If $X_I$ is $\tilde{A}_1$ Then $Y$ is $\tilde{B}_1$**
> ***Rule 2*: If $X_I$ is $\tilde{A}_2$ Then $Y$ is $\tilde{B}_2$**
> ***Observation*: $X_1$ is $\tilde{A}^*$**
> ***Conclusion*: $Y$ is $\tilde{B}^*$**

***Example 4.1*** [3]: Let us consider the situation that the antecedents and the observation of the given fuzzy rules are triangular rough-fuzzy sets, as shown in    Fig. 2, where $\tilde{A}_1 = <(1, 3.5, 4; 0, 0.7, 0), (0, 4, 5; 0, 1, 0)>$, $\tilde{A}_2 = <(12, 13, 13.5; 0, 0.7, 0), (11, 13, 14; 0, 1, 0)>$, $\tilde{A}^* = <(6.5, 8, 9.5; 0, 0.6, 0), (6, 8, 10; 0, 1, 0)>$, $\tilde{B}_1 = <(1.5, 2, 3; 0, 0.5, 0), (0, 2, 4; 0, 1, 0)>$, $\tilde{B}_2 = <(11, 11.5, 12; 0, 0.5, 0), (10, 11, 13; 0, 1, 0)>$. From Fig. 2, we can see that the fuzzy interpolative result $\tilde{B}^*$ of Chen and Shen's method [3] is $\tilde{B}^* = <(6.28, 6.70, 7.95; 0, 0.43, 0), (5.31, 6.28, 8.83; 0, 1, 0)>$ and the fuzzy interpolative result $\tilde{B}^*$ of the proposed method is $\tilde{B}^* = <(6.37, 7.02, 7.67; 0, 0.5, 0), (5.07, 6.90, 8.73; 0, 1, 0)>$. Therefore, the fuzzy interpolative result of the proposed method is more reasonable than Chen and Shen's method [3] in terms of the shapes of the observations.

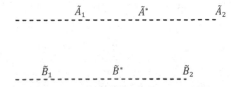

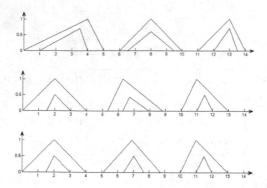

**Fig. 2.** A comparison of fuzzy interpolative reasoning results of *Example 4.1* for different methods based on triangular rough-fuzzy sets

***Example 4.2*** **[3]:** Let us consider the situation that the antecedents and the observation of the given fuzzy rules are triangular rough-fuzzy sets, as shown in    Fig. 3, where $\tilde{A}_1$ = <(1, 3.5, 4; 0, 0.7, 0), (0, 4, 5; 0, 1, 0)>, $\tilde{A}_2$ = <(12, 13, 13.5; 0, 0.7, 0), (11, 13, 14; 0, 1, 0)>, $\tilde{A}^*$ = <(7, 8, 9; 0, 0.6, 0), (6, 8, 10; 0, 1, 0)>, $\tilde{B}_1$ = <(1.5, 2, 3; 0, 0.5, 0), (0, 2, 4; 0, 1, 0)>, $\tilde{B}_2$ = <(11, 11.5, 12; 0, 0.5, 0), (10, 11, 13; 0, 1, 0)>. From Fig. 3, we can see that the fuzzy interpolative result $\tilde{B}^*$ of Chen and Shen's method [3] is $\tilde{B}^*$ = <(6.52, 6.81, 7.64; 0, 0.43, 0), (5.32, 6.31, 8.83; 0, 1, 0)> and the fuzzy interpolative result $\tilde{B}^*$ of the proposed method is $\tilde{B}^*$ = <(6.44, 7.02, 7.61; 0, 0.5, 0), (5.26, 6.90, 8.54; 0, 1, 0)>. Therefore, the fuzzy interpolative result of the proposed method is more reasonable than Chen and Shen's method [3] in terms of the shapes of the observations.

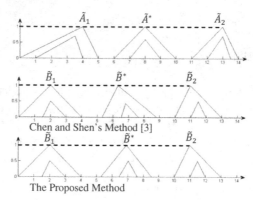

**Fig. 3.** A comparison of fuzzy interpolative reasoning results of *Example 4.2* for different methods based on triangular rough-fuzzy sets

***Example 4.3*** **[3]:** Let us consider the situation that the antecedents and the observation of the given fuzzy rules are triangular rough-fuzzy sets, as shown in    Fig. 4, where $\tilde{A}_1$ = <(3, 3, 3; 0, 1, 0), (3, 3, 3; 0, 1, 0)>, $\tilde{A}_2$ = <(12, 13, 13.5; 0, 0.6, 0), (11,

13, 14; 0, 1, 0)>, $\tilde{A}^* = <(6, 7, 8; 0, 0.6, 0), (5, 7, 9; 0, 1, 0)>$, $\tilde{B}_1 = <(4, 4, 4; 0, 1, 0),$ (4, 4, 4; 0, 1, 0)>, $\tilde{B}_2 = <(10.5, 11.5, 12; 0, 0.5, 0), (10, 11.5, 13; 0, 1, 0)>$. From Fig. 4, we can see that the fuzzy interpolative result $\tilde{B}^*$ of Chen and Shen's method [3] is $\tilde{B}^* = <(5.98, 7.04, 7.98; 0, 0.57, 0), (5.27, 6.66, 9.27; 0, 1, 0)>$ and the fuzzy inter- polative result $\tilde{B}^*$ of the proposed method is $\tilde{B}^* = <(6.02, 7.02, 8.02; 0, 0.79, 0),$ (5.13, 7.13, 9.13; 0, 1, 0)>. Therefore, the fuzzy interpolative result of the proposed method is more reasonable than Chen and Shen's method [3] in terms of the shapes of the observations.

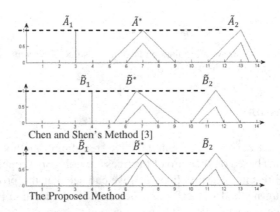

**Fig. 4.** A comparison of fuzzy interpolative reasoning results of *Example 4.3* for different me- thods based on triangular rough-fuzzy sets

*Example 4.4* [3]: Let us consider the situation that the antecedents and the observa- tion of the given fuzzy rules are triangular rough-fuzzy sets, as shown in     Fig. 5, where $\tilde{A}_1 = <(0, 5, 6; 0, 1, 0), (0, 5, 6; 0, 1, 0)>$, $\tilde{A}_2 = <(11, 13, 14; 0, 1, 0), (11, 13, 14; 0, 1, 0)>$, $\tilde{A}^* = <(7, 8, 9; 0, 1, 0), (7, 8, 9; 0, 1, 0)>$, $\tilde{B}_1 = <(0, 2, 4; 0, 1, 0), (0, 2, 4; 0, 1, 0)>$, $\tilde{B}_2 = <(10, 11, 13; 0, 1, 0), (10, 11, 13; 0, 1, 0)>$. From Fig. 5, we can see that the fuzzy interpolative result $\tilde{B}^*$ of Chen and Shen's method [3] is $\tilde{B}^* = <(5.83, 6.26, 7.38; 0, 1, 0), (5.83, 6.26, 7.38; 0, 1, 0)>$ and the fuzzy interpolative result $\tilde{B}^*$ of the proposed method is $\tilde{B}^* = <(5.80, 6.57, 7.35; 0, 1, 0), (5.80, 6.57, 7.35; 0, 1, 0)>$. Therefore, the fuzzy interpolative result of the proposed method is more reasonable than Chen and Shen's method [3] in terms of the shapes of the ob- servations.

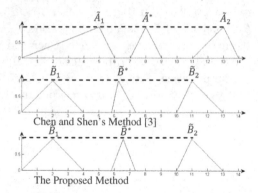

**Fig. 5.** A comparison of fuzzy interpolative reasoning results of *Example 4.4* for different methods based on triangular rough-fuzzy sets

# 5    Conclusions

We have proposed a new fuzzy interpolative reasoning method for sparse fuzzy rule-based systems based on the ratio of fuzziness of polygonal rough-fuzzy sets. From the experimental results shown in Figs. 2-5, we can see that the proposed fuzzy interpolative reasoning method outperforms Chen and Shen's method [3] for fuzzy interpolative reasoning in sparse fuzzy rule-based systems.

**Acknowledgements.** This work was supported in part by the Ministry of Science and Technology, Republic of China, under Grant MOST 103-2221-E-011-108-MY2.

# References

1. Chang, Y.C., Chen, S.M., Liau, C.J.: Fuzzy interpolative reasoning for sparse fuzzy-rule-based systems based on the areas of fuzzy sets. IEEE Transactions on Fuzzy Systems **16**(5), 1285–1301 (2008)
2. Chang, Y.C., Chen, S.M.: A new method for multiple fuzzy rules interpolation with weighted antecedent variables. In: Proceedings of 2008 IEEE International Conference on Systems, Man, and Cybernetics, Singapore, pp. 76–81 (2008)
3. Chen, C., Shen, Q.: A new method for rule interpolation inspired by rough-fuzzy sets. In: Proceedings of 2012 IEEE International Conference on Fuzzy Systems, Brisbane, Australia (2012)
4. Chen, S.M., Chang, Y.C.: A new method for weighted fuzzy interpolative reasoning based on weights-learning techniques. In: Proceedings of 2010 IEEE International Conference on Fuzzy Systems, Barcelona, Spain (2010)
5. Chen, S.M., Chang, Y.C.: Weighted fuzzy rule interpolation based on GA-based weight-learning techniques. IEEE Transactions on Fuzzy Systems **19**(4), 729–744 (2011)
6. Chen, S.M., Chang, Y.C.: Weighted fuzzy interpolative reasoning for sparse fuzzy rule-based systems. Expert Systems with Applications **38**(8), 9564–9572 (2011)

7.  Chen, S.M., Chang, Y.C.: Fuzzy rule interpolation based on principle membership functions and uncertainty grade functions of interval type-2 fuzzy sets. Expert System with Applications **38**(9), 11573–11580 (2011)
8.  Chen, S.M., Chang, Y.C.: Fuzzy rule interpolation based on the ratio of fuzziness of interval type-2 fuzzy sets. Expert Systems with Applications **38**(10), 12202–12213 (2011)
9.  Chen, S.M., Chang, Y.C., Chen, Z.J., Chen, C.L.: Multiple fuzzy rules interpolation with weighted antecedent variables in sparse fuzzy rule-based systems. International Journal of Pattern Recognition and Artificial Intelligence **27**(5), 1359002-1–1359002-15 (2013)
10. Chen, S.M., Chang, Y.C., Pan, J.S.: Fuzzy rules interpolation for sparse fuzzy rule-based systems based on interval type-2 Gaussian fuzzy sets and genetic algorithms. IEEE Transactions on Fuzzy Systems **21**(3), 412–425 (2013)
11. Chen, S.M., Hsin, W.C., Yang, S.W., Chang, Y.C.: Fuzzy interpolative reasoning for sparse fuzzy rule-based systems based on the slopes of fuzzy sets. Expert Systems with Applications **39**(15), 11961–11969 (2012)
12. Chen, S.M., Lee, L.W.: Fuzzy interpolative reasoning for sparse fuzzy rule-based systems based on interval type-2 fuzzy sets. Expert Systems with Applications **38**(8), 9947–9957 (2011)
13. Zadeh, L.A.: Fuzzy sets. Information and Control **8**(3), 338–353 (1965)

# The Cooperation Mechanism of Multi-agent Systems with Respect to Big Data from Customer Relationship Management Aspect

Luyao Xu[1] and Hai-Cheng Chu[2(✉)]

[1] Department of Bilingual Teaching and Research at Business School,
Sias International University, Xinzheng 451150, Henan, China
xluyao@yahoo.com
[2] Department of International Business,
National Taichung University of Education, Taichung 40306, Taiwan
hcchu@mail.ntcu.edu.tw

**Abstract.** Unarguably, with the unparalleled emergence of metamorphic utilization of mobile computing gadgets combining with the social networks. Hefty and massive amount of data are unprecedentedly generated within a second. Search engines host diversified streams of information have created unprecedented scattered data. Hence, effective management and the capability to process large-scale data pose an interesting but critical challenge for contemporary business organizations. Substantively, customers are expanding their online footprints extensively, which makes it hard to extract data value through data collection and data mining. Due to the distributed databases embedded based on heterogeneous platforms, business organizations are facing problematic challenges. It becomes urgent research issues to efficiently and effectively conducting data mining mechanisms with respect to massive amount of data to meet the organizational strategic objectives. Evidently, Big Data era has witnessed the rigorous challenges concerning data transferring, integration, and data-processing technologies. Proverbially, the commonly known Intelligent Agents (IAs), as the autonomous entities to direct its actions towards diverse goals in order to satisfy the implicit requirements for high-speed data integration as well as cooperation mechanisms among different heterogeneous databases. Literally, a Multi-Agent System (MAS) can deal with the flexible communication and cooperation among distributed intelligent agents as an information processor. This paper will introduce multi-agent systems and their applications from data mining aspect, followed by the value of data mining from Customer Relationship Management (CRM) aspect. At last, we propose a three-step data-mining model, which can help business organizations to dig out potential value to manage CRM optimally including using K-means to cluster massive data. In addition, we generalize data to focus on relevant attributes via using information gained and information entropy calculation method to make decision trees for extracting potential valuable knowledge purpose.

**Keywords:** Big data · Multi-agent system · Intelligent agent · Data mining · Customer relationship management

© Springer International Publishing Switzerland 2015
N.T. Nguyen et al. (Eds.): ACIIDS 2015, Part I, LNAI 9011, pp. 562–572, 2015.
DOI: 10.1007/978-3-319-15702-3_54

# 1    Introduction

With the increasing use of mobile devices and development of Internet, expansive stream of information has created unprecedented diverse data. Instead of traditional data-computing devices, massive information requires a superior data-processing technology to achieve high-speed data transferring and integration. Such a wave of information that has overcapacity on data mining and integration compared with conventional database management system is big data.

Big data is also called "massive data". Big data processing has higher requirement on speed. Under big data era, everyone is a big contributor of data. The expansive resources of data include but not just limit on the following: social media such as Weibo, WeChat, email, video, audio, web scanning, GPS, traffic monitoring systems, and other media. Users are expanding their digital footprint and creating more data whenever they use these media. Personal information, consuming habit, preferences, and even relevant social network will be identified under big data era. Undoubtedly, such a data collection and mining are beneficial for business through providing valuable clues for future strategic management.

Faced with mushrooming diverse data, how to continue data mining accurately has challenged enterprises to probe for new technologies with stronger processing power to find the value that hidden inside data. The MAS makes it easier to retrieve the most suitable results for users by processing massive data into explicit reasoning engine.

Enterprises can have a better understanding of customers through integrating and analyzing various data such as historical transaction records, geography, frequency of web scanning and other data. Retail can use big data processing and analysis to achieve demand forecasting, price and merchandizing optimization; Manufacturing can use big data analysis to work towards product customization, new product development, and supply chain management; Medical agencies can use data processing to implement disease management and preliminary diagnosis; Government can use data analysis to achieve crime prevention, fraud detection, and revenue optimization; Business can customize service or product with the highest return on investment through targeting on potential customers accurately. All of these situations stated above can be attributed to benefits brought by effective customer relationship management through big data mining. It is no wonder that effective customer relationship management (CRM) is exceedingly important for any business based on its potential economic benefits.

Some literatures suggest that CRM is understood as a customer-oriented management approach in which information systems provide information to support operational, analytical, and collaborative CRM processes and thus contribute to customer profitability and retention [2]. With the mushrooming various data resources, collecting customer information and analyzing the information using data mining techniques are the primary processes of customer relationship management. This paper will focus on the discussion about data mining model to extract data value effectively. This data mining approach has three steps: Using K-means to cluster massive data; Generalize data to focus on relevant attributes; Using information gain and information entropy calculation method to make decision trees for extracting potential valuable knowledge purpose.

## 2    Literature Review

### 2.1    The Cooperation Mechanism of Multi-agent Systems

A M.A.S is a computerized system composed of multiple interacting intelligent agents within an environment. The intelligent agent is a part of Distributed Artificial Intelligence (DAI), which carries out a goal-directed behavior through sensors-based observation and actuators-orientated action as the information processor. By means of contemporary interrelated networking systems, an IA is able to communicate, collaborate, or even to negotiate with other IAs with semantic reasoning capability [1, 3, 4, 6]. MAS has step further on data transferring and integration compared with IA. MAS can accelerate the cooperation and communication among different intelligent agents. For MAS, how to ensure the autonomous characteristic of every single IA as a coherent behavior has become a challenge. Based on the related researches, social autonomy is believed to be one of the most important behaviors concerning the interactions among the agents in MAS [7, 14]. In other words, adoption of goals is a reflection of social autonomy. On the contrary, it is hard for individual IA to integrate massive data in a timely manner to provide users the most useful information accurately based on its limited capability. Fortunately, the MAS can be the intelligent director as the information processor due to its capability of directing the cooperation or negotiation between different IAs.

In other words, the MAS can call for different IAs to work together for problem solving in an efficient way. One IA can delegate its task or responsibility to the other IAs based on cooperation purpose or specialized function. Also, IA can choose its intelligent partner based on interactive platforms and databases support. Of course, these IAs act like a real problem-solving team. There maybe some conflicts between IAs during this cooperation process. It is time-consuming to deal with negotiation between different IA platforms. One intermediary system that acted as the communicator and decision maker among IAs is exceedingly important and meaningful under big data era. Luckily, the MAS acts as such a role to direct IAs to cooperate and negotiate with each other to provide users the most useful information in a timely manner. Under big data era, high accuracy and high efficiency are main points for data mining technology.

### 2.2    Big Data and Data Mining in CRM

With the increasing use of mobile devices, mushrooming massive data requires a superior data-processing technology to achieve high-speed data transferring and integration. Such a wave of information that has overcapacity on data mining and integration compared with conventional database management system is big data. Big data era advocates higher requirement on data integration and analysis, which push the development of more advanced data analysis technologies. The process of extracting valuable information through probing for regular patterns of data based on effective data analysis technologies is data mining.

Data mining is a computational process that aims to processing uncertainty through discovering regular patterns from amounts of data based on different technologies such as artificial intelligence, statistics and other methods. In addition, from the CRM aspect, based on customer's changing need and habit behaviors, how to extract useful data to identify target customers has become the main point of business. For business, data mining can be regarded as the technology that can be used for categorizing customers based on different attributes such as purchasing behavior, attitudes to serve for future effective business strategy and customer relationship management strategy [9].

As far as the CRM in business, how to set up customer profile through finding patterns in a customer database is the key point. Business can achieve targeted marketing strategies through targeting on specific promotions to existing and potential customers. Also, business can continue market-based analysis accurately. With data mining, retailers can determine which products to stock in and how to display them within a store. In addition, business can hold customer retention by determining characteristics of a customer who is likely to leave for a competitor. Furthermore, some financial institutions such as bank, insurance company can achieve fraud detection through identifying potentially fraudulent transactions…the achievement of the above purposes cannot live without effective data mining functions. The commonly used data mining methods include but not just limited on classification, clustering, regression, association rules, the abnormal detection and other methods or models.

# 3    Research Analyses

## 3.1    Data Mining Model from Customer Relationship Management Respect

Customer relationship management relies on segmenting customers into groups based on profiles developed through a firm's data mining activities. Collecting customer information and analyzing the information using data mining techniques are the primary processes of customer relationship management. CRM optimizes profitability, revenue, and customer satisfaction by organizing around customer segments, fostering customer-satisfying behaviors, and implementing customer-centric business models [13]. CRM is understood as a customer-oriented management approach in which information systems provide information to support operational, analytical, and collaborative CRM processes and thus contribute to customer profitability and retention [2].

In view of the importance of customer relationship management in business, this paper will propose a model to elaborate its contributions for business. This model is a three-step process: Using K-means to cluster massive data; Generalize data to focus on relevant attributes; Using information gain and information entropy calculation method [10] to make decision trees for extracting potential valuable knowledge purpose. Here is a list of assumed data based on research on a chain coffee company. This company is planning for launching a new product, however, to target extensive customers, this company is hesitating on its marketing strategies for instant coffee and brewed coffee. Consequently, massive data about the popularity between instant coffee and freshly brewed coffee has been collected as follows (See Table 1). This paper will use this assumed database as an example to illustrate the application of proposed model.

*First Step: Clustering—K-means Calculation*
Clustering is the process that dividing various data into different groups, satisfying highly similarities within a group, but highly difference among different groups. The similarities within a group can be expressed using the equation (1) shown below. The goal of K-means is grouping data into K clusters based on parameter K and target data. After picking up several data randomly, we can image that each data stands for one average number of one cluster. Then, grouping other data based on their distance to these identified average number. Again, repeat this process to get new average number for each cluster, until the criterion function is converged. Here, we propose to use function (2) shown below to evaluate the effectiveness of K-means clustering:

$$d\left(x_a, x_b\right) = \sqrt{\sum_{m=1}^{d}\left(x_{am} - x_{bm}\right)^2} \tag{1}$$

$$E = \sum_{i=1}^{k}\sum_{a \subset A}\left\|a - q_i\right\|^2 \tag{2}$$

Here, $q_i$ is the average number of the A cluster. a is one target data in a cluster. After repeating this computing process, massive data can be categorized into different clusters.

*Second Step: Attribute Selection for Each Cluster—Generalized Data Processing*
Integrating and generalized processing data based on different clusters to extract the most influential informative attributes makes it easier to extract the data value. Use attribute value of "age" shown in Table 1 as an example, we can attribute these massive data to four different groups: Youth is the group whose members' age is <=30、middle-aged group is the group of members whose age is between 31 and 60、old-aged group is a group of members whose age is above 60. The same, we can generalize other attributes such as "City Size" and "Annual Revenue" (See Table 2). Such a generalized data processing makes it easier to cluster massive data to analyze the regular pattern.

*Step 3: Explaining attributes—Decision Trees*
Several approaches have been proposed in the literature including association rules [11, 12], and self-organizing feature maps [9], however, to achieve the visualization of analysis, this paper mainly propose the use of decision trees to explain the value hidden in big data. This paper recommends the use of the information gain-ID3 calculation method. The specifics are shown as following:

From the Chart 1, it is easy to know that there are two different values for the type attribute: instant or brewed (m=2). Here we can identify C1 is instant, and C2 is brewed. So 18 samples are included in C1, and 8 samples are included in C2. Our goal is probing for the most important test attribute of decision trees as the root node. Based on infor-mation-gain based calculation, this root node can be obtained through calculating and comparing the information gain among different attributes. We propose the following function about information entropy's calculation. The specifics are as follows:

**Table 1.** Attributes value of customers' sample

|  | Age | City Size | Sex | Annual Revenue | Type (Instant & Brewed) |
|---|---|---|---|---|---|
| 1 | 18 | Beijing | Female | 24000 | Instant |
| 2 | 25 | Shanghai | Female | 12890 | Brewed |
| 3 | 30 | Guangzhou | Male | 33210 | Instant |
| 4 | 27 | Shenzhen | Male | 11530 | Brewed |
| 5 | 22 | Luoyang | Female | 92000 | Brewed |
| 6 | 19 | Jilin | Female | 21540 | Brewed |
| 7 | 25 | Wenzhou | Male | 18980 | Brewed |
| 8 | 13 | Shenyang | Female | 235700 | Brewed |
| 9 | 29 | Wuhan | Female | 40080 | Brewed |
| 10 | 24 | Dalian | Male | 36900 | Brewed |
| 11 | 21 | Shenyang | Female | 25000 | Brewed |
| 12 | 35 | Ningbo | Female | 40000 | Brewed |
| 13 | 38 | Wuhan | Male | 39780 | Brewed |
| 14 | 42 | Zhengzhou | Female | 23710 | Instant |
| 15 | 46 | Shanghai | Female | 85600 | Brewed |
| 16 | 50 | Beijing | Female | 32150 | Brewed |
| 17 | 37 | Guangzhou | Female | 24000 | Brewed |
| 18 | 42 | Guangzhou | Male | 150000 | Instant |
| 19 | 47 | Beijing | Male | 26000 | Instant |
| 20 | 49 | Zhuhai | Female | 120000 | Brewed |
| 21 | 39 | Wenzhou | Female | 28900 | Brewed |
| 22 | 50 | Nanjing | Male | 160000 | Instant |
| 23 | 36 | Wuhan | Female | 460000 | Brewed |
| 24 | 64 | Shenyang | Female | 500000 | Instant |
| 25 | 62 | Nanjing | Male | 85000 | Instant |

1.  Based on the information entropy calculation function:

$$H(x_1, x_2 \ldots x_n) = -\sum_i^n P_i \log_2(P_i); \quad (i=1, 2, \ldots n) \tag{2}$$

The information entropy can be got as follows:

$$H(x_1, x_2) = H(17, 8) = -\frac{17}{25} \log_2 \frac{17}{25} - \frac{8}{25} \log_2 \frac{8}{25} = \frac{17}{25} \log_2 \frac{25}{17} + \frac{8}{25} \log_2 \frac{25}{8} = 0.9044$$

2. Calculating the entropy of every attribute, including customers' age, city size, sex and revenue (Calculating the entropy of age as a detailed example shown as follows):

When age<=30 : $x_{11}$=9 , $x_{21}$=2,   $H(x_{11}, x_{21}) = -\frac{9}{11} \log_2 \frac{9}{11} - \frac{2}{11} \log_2 \frac{2}{11} = 0.6840$

When age is 31-60: $x_{12}$=8,  $x_{22}$=4,   $H(x_{12}, x_{22}) = -\frac{8}{12} \log_2 \frac{8}{12} - \frac{4}{12} \log_2 \frac{4}{12} = 0.9183$

When age>=60 : $x_{13}$=0,  $x_{23}$=2,   $H(x_{13}, x_{23}) = -\frac{2}{3} \log_2 \frac{2}{3} = 0.3899$

**Table 2.** Generalized Attributes value of sampled customers group

| Attribute | Age | City Size | Sex | Revenue | Type |
|---|---|---|---|---|---|
| 1 | <=30 | First-tier City | Female | Low | Instant |
| 2 | <=30 | First-tier City | Female | High | Brewed |
| 3 | <=30 | First-tier City | Male | Medium | Instant |
| 4 | <=30 | First-tier City | Male | High | Brewed |
| 5 | <=30 | Third-tier City | Female | High | Brewed |
| 6 | <=30 | Third-tier City | Female | Low | Brewed |
| 7 | <=30 | Third-tier City | Male | Low | Brewed |
| 8 | <=30 | Second-tier City | Female | High | Brewed |
| 9 | <=30 | Second-tier City | Female | Medium | Brewed |
| 10 | <=30 | Second-tier City | Male | Medium | Brewed |
| 11 | <=30 | Second-tier City | Female | Low | Brewed |
| 12 | 31-60 | Second-tier City | Female | Medium | Brewed |
| 13 | 31-60 | Second-tier City | Male | Medium | Brewed |
| 14 | 31-60 | Second-tier City | Female | Low | Instant |
| 15 | 31-60 | First-tier City | Female | High | Brewed |
| 16 | 31-60 | First-tier City | Female | Medium | Brewed |
| 17 | 31-60 | First-tier City | Female | Low | Brewed |
| 18 | 31-60 | First-tier City | Male | High | Instant |
| 19 | 31-60 | First-tier City | Male | Low | Instant |
| 20 | 31-60 | Third-tier City | Female | High | Brewed |
| 21 | 31-60 | Third-tier City | Female | Low | Brewed |
| 22 | 31-60 | Second-tier City | Male | High | Instant |
| 23 | 31-60 | Second-tier City | Female | High | Brewed |
| 24 | >60 | Second-tier City | Female | High | Instant |
| 25 | >60 | Second-tier City | Male | High | Instant |

So the information entropy of age sample is:

$$E(age) = \frac{11}{25} H(x_{11}, x_{21}) + \frac{12}{25} H(x_{12}, x_{22}) + \frac{2}{25} H(x_{13}, x_{23}) = 0.7929$$

With the same method, it is easier to know the information entropy of other attributes:

$$E(sex) = \frac{16}{25}\left(-\frac{13}{16}\log_2\frac{13}{16} - \frac{3}{16}\log_2\frac{3}{16}\right) + \frac{9}{25}\left(-\frac{4}{9}\log_2\frac{4}{9} - \frac{5}{9}\log_2\frac{5}{9}\right) = 0.8024$$

$$E(citysize) = \frac{9}{25}\left(-\frac{5}{9}\log_2\frac{5}{9} - \frac{4}{9}\log_2\frac{4}{9}\right) + \frac{11}{25}\left(-\frac{7}{11}\log_2\frac{7}{11} - \frac{4}{11}\log_2\frac{4}{11}\right) + \frac{5}{25}\left(-\frac{5}{5}\log_2\frac{5}{5} - \frac{0}{5}\log_2\frac{0}{5}\right) = 0.7729$$

$$E(revenue) = \frac{11}{25}\left(-\frac{7}{11}\log_2\frac{7}{11} - \frac{4}{11}\log_2\frac{4}{11}\right) + \frac{6}{25}\left(-\frac{5}{6}\log_2\frac{5}{6} - \frac{1}{6}\log_2\frac{1}{6}\right) + \frac{8}{25}\left(-\frac{5}{8}\log_2\frac{5}{8} - \frac{3}{8}\log_2\frac{3}{8}\right) = 0.8775$$

So the information gain can be calculated as follows :

Gain (sex) = H $(x_1,x_2)$ -E (sex) =0.9044-0.8024=0.102, Gain (age) = H$(s_1,s_2)$ − E(age) =0.1115

Gain (city size) = H $(x_1,x_2)$- E(city size) =0.1315, Gain(revenue) = $I(s_1,s_2)$-E(revenue) =0.0269

Based on the above calculation result, "city size" has the highest information gain to be the root node of decision tree. (See Figure 1)

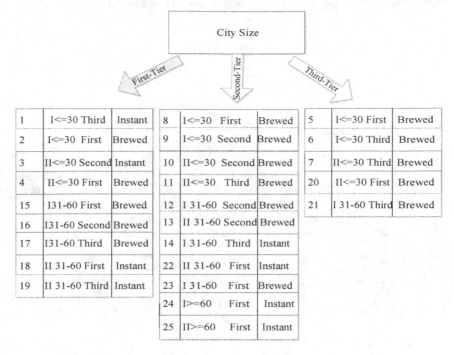

| 1 | I<=30 Third | Instant | | 8 | I<=30 First | Brewed | | 5 | I<=30 First | Brewed |
|---|---|---|---|---|---|---|---|---|---|---|
| 2 | I<=30 First | Brewed | | 9 | I<=30 Second | Brewed | | 6 | I<=30 Third | Brewed |
| 3 | II<=30 Second | Instant | | 10 | II<=30 Second | Brewed | | 7 | II<=30 Third | Brewed |
| 4 | II<=30 First | Brewed | | 11 | II<=30 Third | Brewed | | 20 | II<=30 First | Brewed |
| 15 | I31-60 First | Brewed | | 12 | I 31-60 Second | Brewed | | 21 | I 31-60 Third | Brewed |
| 16 | I31-60 Second | Brewed | | 13 | II 31-60 Second | Brewed | | | | |
| 17 | I31-60 Third | Brewed | | 14 | I 31-60 Third | Instant | | | | |
| 18 | II 31-60 First | Instant | | 22 | II 31-60 First | Instant | | | | |
| 19 | II 31-60 Third | Instant | | 23 | I 31-60 First | Brewed | | | | |
| | | | | 24 | I>=60 First | Instant | | | | |
| | | | | 25 | II>=60 First | Instant | | | | |

**Fig. 1.** Decision tree model

Here, we can get three new subsets:

P1={1,2,3,4,15,16,17,18,19}, P2={8,9,10,11,12,13,14,22,23,24,25} ; P3={5,6,7, 20,21}

It is easier to calculate the information gain of other attributes in these subsets using the same

calculation method that is stated above: For P1, $Gain(age)=H(x_1,x_2)-E(age)=0.007$

$Gain(revenue)=H(x_1,x_2)-E(revenue)=0.093$

$Gain(sex)=H(x_1,x_2)-E(sex)=0.23$

Here, attribute "sex" has the highest information gain, so this attribute will be regarded as another new node for further establishment of decision trees. After dealing with P2 using the same method, the final decision tree is formed as follows:

Now, it is easier for business to find regular pattern through analyzing the potential valuable information. From the above decision tree, we can find the following patterns:

1. The brewed coffee is more popular for high-revenue young customers
2. Instant coffee is more popular among middle-revenue young customers and low-revenue young customers.
3. Among different cities, brewed coffee is more popular for young customers, but instant coffee is more popular for middle-aged customers and old-aged customers.

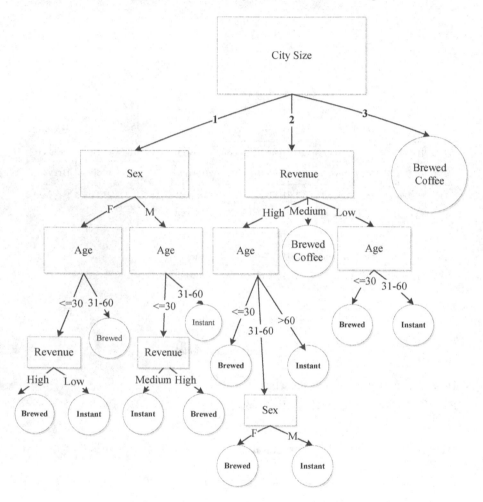

**Fig. 2.** Decision tree-based clustering model

Based on the above analysis, business can sense that the target group of brewed coffee is young customers with high-revenue, and the target groups of instant coffee should be middle-aged customers and old-aged customers. Consequently, it is the right time to accelerate advertisement about newly brewed coffee among young people, especially among young people with high revenue and promotes newly instant coffee among middle-aged customers and old-aged customers.

Based on such a data-mining model, business can extract valuable information to achieve accurate marketing strategies. Business can promote personalized marketing strategies through finding inner regular patterns on the basis of data clustering and integration. It is no doubt that business can lower the operation cost, and optimize the customers' retention rate for better customer relationship management.

# 4    Conclusions

Based on the above discussion about this proposed three-step data mining model including using K-means to cluster massive data, generalizing data to focus on relevant attributes, and making decision trees to abstract and visualize data patterns, we have sensed the value of effective data analysis. This proposed data-mining model makes it more persuasive to make decision based on its descriptiveness, predictability, and intuitive illustration. More importantly, such a data-mining model simplifies the data analysis process, and provides logical basis for analysis. Unarguably, data mining model can bring amazing value for CRM, no matter for business, government, or for public institutions. CRM will work effectively with the speedy data integration and data mining. This model also sets up a good foundation for future data research under "big data era". With the confidence about more newly data research methods and highly expectation about modified data mining model, I firmly believe that data mining will make our life easier and more interesting.

**Acknowledgement.** The authors would like to acknowledge the funding support of National Taichung University of Education of Taiwan concerning the research grant of year 2013 (NTCU-F102104).

# References

1. Bittencourt, I., Costa, E., Silva, M., Soares, E.: A Computational Model for Developing Semantic Web-based Educational Systems. Knowledge Based Systems **22**, 302–315 (2009)
2. Bueren, A., Schierholz R., Kolbe L., Brenner, W.: Customer Knowledge Management: Improving Performance of Customer Relationship Management with Knowledge Management. In: Proceedings of the 37th IEEE Hawaii International Conference on System Sciences. IEEE Computer Society Press, Big Island, HI
3. Chen, D., Vachharajani, N., Hundt, R., Li, X., Eranian, S., Chen, W., Zheng, W.: Taming Hardware Event Samples for Precise and Versatile Feedback Directed Optimizations. IEEE Transactions on Computers **62**(2), 376–389 (2013)
4. Chen, X., Zheng, Z., Liu, X., Huang, Z., Sun, H.: Personalized QoS-Aware Web Service Recommendation and Visualization. IEEE Transactions on Services Computing **6**(1), 35–47 (2013)
5. Fung, G., Mangasarian, L.O.: 'Proximal Support Vector Machine Classifiers' Knowledge Discovery and Data Mining, pp. 77–86, New York, NY, USA (2001)

6. Hsu, C.H., Hsu, C.G., Chen, S.C., Chen, T.L.: Message Transmission Techniques for Low Traffic P2P Services. International Journal of Communication Systems **22**(9), 1105–1122 (2009)
7. Hsu, C., Chen, Y., Kang, H.: Performance-Effective and Low-Complexity Redundant Reader Detection in Wireless RFID Networks. EURASIP Journal on Wireless Communications and Networking 1–9 (2008)
8. Kuoa, R.J., Ana, Y.L., Wanga, H.S., Chungbi, W.J.: Integration of Self-Organizing Feature Maps Neural Network and Genetic K-means Algorithm for Market Segmentation. Expert System 313–324 (2006)
9. Romdhane, L.B., Nadia, F., Ayeb, B.: Building Customer Models From Business Data: An Automatic Approach Based on Fuzzy Clustering and Machine Learning. International Journal of computational intelligence and application. **8**(4), 445–465 (2009)
10. Guo, J., Xu, M.: The Implementation of Enterprise CRM Based on Big Data Mining Technologies (Chinese). http://www.chinadmd.com/file/uei3uaosocwevsetuziuocxr_1.html
11. Giudici, P., Passerone, G.: Data Mining of Association Structures to Model Consumer Behavior. Computer Statistics Data Analysis 533–541 (2002)
12. Mitra, S., Pal, S.K., Mitra, P.: Data Mining in Soft Computing Framework: A Survey. IEEE Trans. Neural Networks 3–14 (2002)
13. Soukakos, P.I., Georgopoulos, N.B., Pekka Economou, V.: Interrelated Frame-Works Proposed for Mapping and Performance Measurement of Customer Relationship Management Strategies. International Journal of Knowledge and Learning 299–315 (2007)
14. Thomas, A.M., Shah, H., Moore, P., Rayson, P.: E-Education 3.0: Challenges and Opportunities for the Future of iCampuses. In: International Conference on Digital Object Identifier, pp. 953–958 (2012)

# An Intelligence Maximum Power Point Tracking Controller for Human Power System

Meng-Hui Wang[1](✉), Wei-Jhe Jiang[1], and Mei-Ling Huang[2]

[1] Department of Electrical Engineering, National Chin-Yi University of Technology,
Taichung, Taiwan
wangmh@ncut.edu.tw, f127860@gmail.com
[2] Industrial Engineering and Management, National Chin-Yi University of Technology,
Taichung, Taiwan
huangml@ncut.edu.tw

**Abstract.** This study is to design the intelligence maximum power point track-ing controller (IMPPTC) based on the Boost power converter for Human Power Generation System. Through the new power electronics technology, IMPPTC overcomes the instability of human defects, achieves a smooth power output by using conversion technologies, and effectively improve the power conversion efficiency. Manpower generators were not made through the power converter for maximum power point tracking control in the past. Therefore, this paper proposes the intelligence maximum power point tracking (IMPPT) controller based on the characteristics of human power generation system. The most commonly used control methods include Perturb and observe algorithm (P&O), Sliding mode control (SMC) and Incremental algorithm. With the fast response and high efficiency, SMC outperforms P&O on the maximum power point tracking. This article first uses Sliding mode control (SMC) for maximum pow-er point tracking. In addition, one of the drawbacks of SMC is the inability of suppressing noises. By PSIM software that SMC at the maximum power point tracking than P&O fast response and high efficiency. In addition, one of the drawbacks of SMC is the inability of suppressing noises. This study propose Extension theory suppress noise for SMC. The simulation results proved that the proposed method can provide better response on the maximum power point tracking performance and possess higher conversion efficiency.

**Keywords:** Human power generation · Extension sliding mode control (ESMC) · Perturb and observe algorithm (P&O) · Incremental algorithm

## 1    Introduction

In 21st century, green living represents a better friendly lifestyle for human and environmental. How to reduce the environmental impact and resource consumption is the new trend of international. The Causes of global warming is carbon dioxide emis-sions from Large. Therefore use does not emissions of carbon dioxide, will not be depleted, and there is no danger of human power. Has developed renewable energy

© Springer International Publishing Switzerland 2015
N.T. Nguyen et al. (Eds.): ACIIDS 2015, Part I, LNAI 9011, pp. 573–582, 2015.
DOI: 10.1007/978-3-319-15702-3_55

technologies which includes wind power, solar power, fuel cell and human power. Many countries research projects focus on human power. Human power generation system consists of Brushless DC motors (BLDCM), three-phase bridge rectifier and power converters. Human power generation system is kinetic energy converted into electrical energy by the BLDC generator. The human power system is using Boost converter. Since human stampeded at different speeds, affecting power generation system power output changes. Different characteristic curves has an optimal operating point, the operating point is maximum power point (MPP). In the past Human power generation are used constant voltage regulator output have not yet to do Maximum power point tracking. This paper presents the human power system to do maximum power point tracking control system. Traditional MPP tracking are used P&O, extremmum seeking control (ESC), etc. These methods are disadvantageous, because the maximum power point in the steady state is proportional to the shock. To overcome the above drawbacks is proposed another method of SMC. SMC is based on the characteristic curve define $dP_m/dV_m=0$ of generator mathematical models. If Operating point fall in the MPP of the left or right, operating point will moves in the opposite direction toward the MPP. But for SMC's own perturbation problem, so many scholars have proposed artificial means reduce the degree of disturbance. So this paper extension sliding mode control (ESMC) to reduce the degree of disturbance,added power output stability and speed of response. Finally, use PSIM simulation and compared to other methods to verify its practicality.

## 2    BLDCM Dynamic Modeling

Human power generation system uses a three-phase brushless DC generator. Generator equivalent circuit was given in figure 1. Rs is resistance of the coil; L is self – inductance of the coil; M is mutual inductor of the coil; $e_a$, $e_b$ and $e_c$ represent armature Reaction electromotive; $i_a$, $i_b$, and $i_c$ are the phase current; $V_a$, $V_b$ and $V_c$ are terminal voltage. BLDCM equation of state represents as follows:[2]

$$\begin{bmatrix} V_a \\ V_b \\ V_c \end{bmatrix} = \begin{bmatrix} e_a \\ e_b \\ e_c \end{bmatrix} - \begin{bmatrix} R_s & 0 & 0 \\ 0 & R_s & 0 \\ 0 & 0 & R_s \end{bmatrix} \begin{bmatrix} i_a \\ i_b \\ i_c \end{bmatrix} - \begin{bmatrix} L-M & 0 & 0 \\ 0 & L-M & 0 \\ 0 & 0 & L-M \end{bmatrix} \frac{d}{dt} \begin{bmatrix} i_a \\ i_b \\ i_c \end{bmatrix} \tag{1}$$

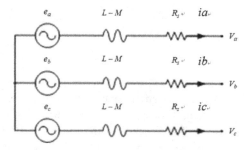

**Fig. 1.** Generator equivalent circuit

Generator internal electromotive and current of the converter control relationship, can define a three-phase commutation function $S_a(t), S_b(t)$ and $S_c(t)$ in eq.(2).[3]

$$S_a(t) = \sum_{n=0}^{\infty}\left[u(\omega_e t - 2n\pi) - u\left(\omega_e t - \frac{2\pi}{3} - 2n\pi\right) - u(\omega_e t - \pi - 2n\pi) + u\left(\omega_e t - \frac{5\pi}{3} - 2n\pi\right)\right], S_b(t) = S_a\left(t - \frac{2\pi}{3\omega_e}\right), S_c(t) = S_a\left(t - \frac{4\pi}{3\omega_e}\right) \qquad (2)$$

Definition eq. (2), u(t) is expressed as unit step function, and n is an integer. Observe the distribution of three-phase current, eq.(3) is to define the equivalent generator DC armature current $i_{eq}(t)$ in (3) as expressed.

$$i_{eq}(t) = \frac{1}{2}[S_a(t)S_b(t)S_c(t)][i_a(t)i_b(t)i_c(t)]^T \qquad (3)$$

According to the commutation function, armature voltage $e_{eq}(t)$ and voltage across $V_{eq}(t)$ of BLDC generator were expressed in eqs. (4) and (5), respectively.

$$e_{eq}(t) = \frac{1}{2}[S_a(t)S_b(t)S_c(t)][e_a(t)e_b(t)e_c(t)]^T \qquad (4)$$

$$V_{eq}(t) = \frac{1}{2}[S_a(t)S_b(t)S_c(t)][V_a(t)V_b(t)V_c(t)]^T \qquad (5)$$

The dynamic modeling of BLDC equivalent DC generators in eq. (6) was achieved when substituting eqs. (4) and (5) in eq. (1).

$$V_{eq} = e_{eq} - R_s i_{eq} - (L - M)\frac{di_{eq}}{dt} \qquad (6)$$

According to dynamic modeling provided above, PSIM simulates human power generation and observes the P-V curves. Figure 2 shows the P-V characteristic curve at different speeds. In Figure 2, human power generation can be found at different speeds produce different MPP.

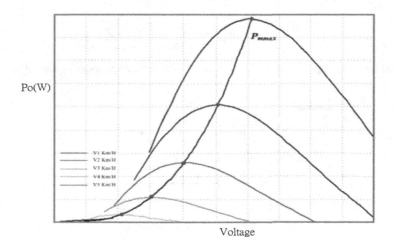

**Fig. 2.** Human power generator P-V characteristic curve

# 3     Boost Converter Design

Human power generation system is using Boost converter. Boost converter architecture as shown in Figure 3. [4]

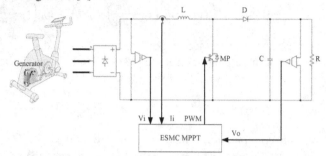

**Fig. 3.** Human generation system architecture diagram

The Boost converter possesses the step-up effect when pulse width modulation (PWM) controls the switches of converter. As illustrated in Figure 4, the output is related to the input by Eq. (7). It is assumed that, at steady state, the current remains constant and there is no voltage drop across transistors and diodes. Thus when the transistor is turned on, $V_s=0$; otherwise $V_s=V_o$. The output voltage $V_m$ can be expressed as in Equation (7), where $t_{on}$ ($t_{off}$) represents the time interval during which the transistor is on (off), respectively, and $T = t_{on} + t_{off}$.

$$\begin{cases} 0 = \int_{on} V_m dt + \int_{off} (V_m - V_o)\, dt \\ V_m t_{on} + (V_m - V_o) t_{off} = 0 \end{cases} \tag{7}$$

Rearrangement of eq. (7), can get the output voltage and input voltage duty cycle such as equation (8).

$$\frac{V_o}{V_m} = \frac{t_{on} + t_{off}}{t_{off}} = \frac{D}{(1-D)} \tag{8}$$

There are two steps status of the step-up converter with respect to the MP states:

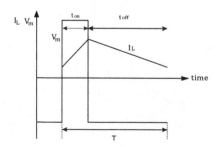

**Fig. 4.** Voltage and duty ratio

Step Status 1:

During the time interval, the switch is on, and the output voltage and current are expressed as:

$$\dot{I}_{L(MP\_on)} = \frac{V_m}{L}$$ (9)

$$\dot{V}_{o(MP\_on)} = -\frac{V_o}{CR_L}$$ (10)

Step Status 2:

When the time interval the switch is off, and the output voltage and current are expressed as:

$$\dot{I}_{L(MP\_off)} = \frac{V_m}{L} - \frac{V_o}{L}$$ (11)

$$\dot{V}_{o(MP\_on)} = -\frac{V_o}{CR_L}$$ (12)

Setting the duty cycle of the switch by PWM leads to eq. (13), representing a dynamic system. Using the state space averaging method and substituting eqs. (9) to (12) to eq. (13), we have eqs. (14) and (15).

$$\dot{X} = U\dot{X}_1 + (1 - U)\dot{X}_2$$ (13)

Where $\dot{X}_1 = [\dot{I}_{L(MP\_on)} \quad \dot{V}_{o(MP\_on)}]$, $\dot{X}_2 = [\dot{I}_{L(MP\_off)} \quad \dot{V}_{o(MP\_off)}]$, and duty ratio $U \in [0 \quad 1]$. Thus the resultant state equations are as follow:

$$\dot{I}_L = (U - 1)\frac{V_o}{L} + \frac{V_m}{L}$$ (14)

$$\dot{V}_o = (1 - U)\frac{I_L}{C} + \frac{V_o}{CR}$$ (15)

## 4 Controller Design

### 4.1 Sliding Mode Control Mathematical Derivation

SMC was first discovered by a Russian scholar, and commonly practices today in a variety of control design. The purpose of the control design is first defining the sliding surface, and then designing the control laws [5]. The system state can be attracted to the sliding surface, and not detach it. Combined with a chosen sliding function satisfying $\frac{\partial P_g}{\partial V_m} = 0$, SMC is expressed as eq.(16):

$$\frac{\partial P_g}{\partial V_m} = \frac{\partial (I_m V_m)}{\partial V_m} = V_m \cdot dI_m + I_m dV_m, (I_m = i_{eq} \, , V_m = V_{eq})$$ (16)

According to eq.(16), define the sliding surface as eq.(17).

$$S = V_m \cdot dI_m + I_m dV_m$$ (17)

Based on the characteristic curve, the sliding surface should be $S = 0$, and is expressed as eq.(17).

$$S = V_m. dI_m + I_m dV_m = 0 \tag{18}$$

As illustrated in Figure 5, the duty cycle and operation voltage at the tracking maximum power U is related to S, where U represents the controller. In order to obtain equivalent control, a derivative of S is gained as.[6]

$$\dot{S} = \left[\frac{\partial S}{\partial x}\right]^T \dot{X} = \left[\frac{\partial S}{\partial x}\right]^T (f(X) + g(X)U_{eq}) = 0, U_{eq} = -\frac{\left[\frac{\partial S}{\partial x}\right]^T \cdot f(X)}{\left[\frac{\partial S}{\partial x}\right]^T \cdot g(X)} = 1 - \frac{V_m}{V_o} \tag{19}$$

With $U(0 \leq U \leq 1)$, the control is represented as eq.(19).

$$U = \begin{cases} 1 & , & U_{eq} + \sigma S \geq 1 \\ U_{eq} + \sigma S & , for \ 0 \leq U_{eq} + \sigma S < 1 \\ 0 & , & U_{eq} + \sigma S \leq 0 \end{cases} \tag{20}$$

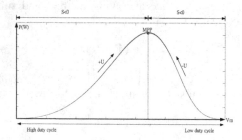

**Fig. 5.** Duty cycle versus operation region

Setting $\dot{S} = 0$ and the input of $U_{eq}$, within the context of MPP and $\sigma S$, U will track the MPP. Finally along sliding mode controller design by Lyapunov theorem for validation, therefore eq.(21) is required.

$$\dot{V} = S\dot{S} < 0 \tag{21}$$

Which must satisfy $\sigma > 0$, the control curve will meet the maximum power on sliding surfaces. However SMC dithers in the vicinity of MPP. This paper proposes the use of extension theory to significantly reduce the steady of jitter and deermine $\sigma$ values.

## 4.2   Extension Theory

Extension theory is a new kind of knowledge system based on the concepts of matter-elements and extension sets [7]. It was first proposed by Cai to solve contradictions and incompatibility problems in 1983. The hard core of extension theory is two theoretical pillars that include matter-element theory and the theory of extension set. Assessment methods are divided into 5 steps.

Step 1:

A matter-element model involves three elements: a name, denoted by N, a feature, denoted by C, and the value of a feature, denoted by V. The matter can be expressed as follows:

$$R = (N \cdot C \cdot V) \tag{22}$$

The name, feature, and value are the three elements of a matter. If we assume that $R = (N, C, V)$ is a multi-dimensional matter-element, $C = [c_1 \quad c_2 \quad \cdots \quad c_n]^T$ a feature vector and $V = [v_1 \quad v_2 \quad \cdots \quad v_n]^T$ a value vector of $C$, then a multi-dimensional matter-element is defined as

$$R = \begin{bmatrix} N, & c_1, & u_1 \\ & c_2, & u_2 \\ & \vdots & \vdots \\ & c_n, & u_n \end{bmatrix} \tag{23}$$

To determine the matter-element to be measured, divide an unknown matter-element into various grade set characteristics, the number of grade sets must be equal to the number of sets of classical domain and joint domain as follows:

$$R_k = \begin{bmatrix} N & C_1 & <a_1, b_1> \\ & C_2 & <a_2, b_2> \\ & \vdots & \vdots \\ & C_n & <a_n, b_n> \end{bmatrix} \tag{24}$$

Where N represents the matter-element name of the divided, all the characteristics of the matter-element name are expressed as Ci. The characteristic value range is represented by Vi. This characteristic value range is the distribution range of the i-th characteristic, and the characteristic value is denoted by ai ,indicating the maximum value of characteristics of this matter-element grade set. On the contrary, bi denotes the minimum value of characteristics of this matter-element grade set.

Step 2:

This step is to select $\sigma$ value, so as to reduce the system chattering control law is as eq. (25).

$$U = \begin{cases} \sigma S & , S \in E \\ U_{eq} + \sigma S & , S \notin E \end{cases} \tag{25}$$

Where E is the extension region in extension theory, appropriate control laws are designed in the rational region and extension region respectively, and then the system stability must be insured. The extension correlation function and the corresponding weights are calculated according to the defined characteristic value S.

Step 3:

Calculate the degree of association between the data to be measured and various sets, and establish the evaluation degree. The degree of association is established as follows:

$$k(x_i) = \begin{cases} \frac{-\rho(x_i, X_i)}{|X_i|} & , x_i \in X_i \\ \frac{\rho(x_i, X_i)}{\rho(x_i, X_{pi}) - \rho(x_i, X_i)} & , x_i \notin X_i \end{cases}$$

(26)

Where

$$\rho(x_i, X_i) = \left| x_i - \frac{a_i + b_i}{2} \right| - \frac{b_i - a_i}{2}, \rho(x_i, X_{pi}) = \left| x_i - \frac{c_i + d_i}{2} \right| - \frac{c_i - d_i}{2}$$

(27)

Step 4:

By calculating the relative values of degree of association of various grade sets, this paper uses the normalization of Eq. (26) to keep the degree of association of each grade set within <1, -1>. This step is for identifying power tracking control as follows:

$$k(q)^* = \frac{2k(q) - k(q)_{max} - k(q)_{min}}{k(q)_{max} - k(q)_{min}}$$

(28)

Step 5:

This step is to determine which grade set the matter-element is in. If K(q)* equals to 1, the extension identifies this degree of correlation and the evaluation result is type i. The probability of other set types is determined according to the degree of association. This step is used to identify the $\sigma$ value of controller, so as to track the maximum power point.

## 5    Simulation Result

PSIM simulates output power and the generator voltage characteristic curves of different speed on juman power generation, Fig. 6. Figure 7 (a) (b) (c) illustrate the

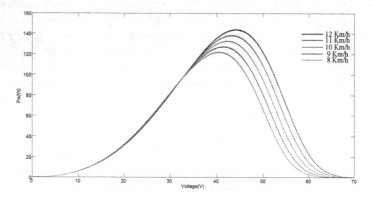

**Fig. 6.** Human power generation characteristic under different speed

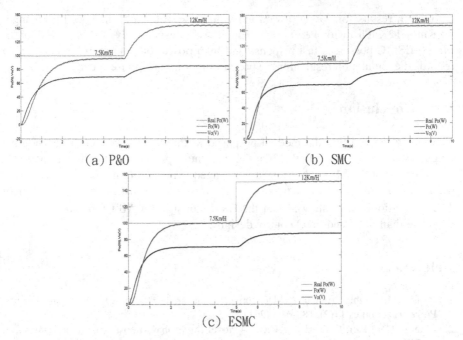

(a) P&O                              (b) SMC

(c) ESMC

**Fig. 7.** Simulation results of the three control methods

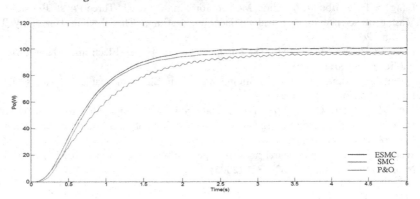

**Fig. 8.** Comparison of three control power output results

**Table 1.** Simlation results for three control methods

| Item          Control method | P&O | SMC | ESMC |
|---|---|---|---|
| $\eta$ (%) | 95.83% | 96.78% | 99.53% |
| Response Time(s) | 3.55s | 3.27s | 2.77s |
| Po(W) | 95.83W | 96.78W | 99.53W |
| Pi(W) | 100W | 100W | 100W |

simulation results for output power and voltage waveform at the speed of change ar 7.5Km/h-12Km/h in for P&O, SMC and ESMC, respectively. At speed 7.5Km/h in Fig. 8, ESMC possesses fast response and high power output among the all. Table 1 records the simulation results for the power output and response speed.

# 6    Conclusions

The results proved that the method proposed in this study can control the nonlinear characteristic behavior of fuel cell at the maximum power, and the human power can be stabilized at the MPP when speed is changed. The sliding mode was used to track the MPP, and the proper key parameter $\sigma$ was selected by using extension theory. The simulation results showed that the ESMC design for MPPT has better transient response than SMC and P&O control designs.

# References

1. Ryu, H.M.: Highly efficient AC-DC converter for small wind power generators. Journal of Power Electronics **11**(2), 188–193 (2011)
2. Chang, T.Y., Pan, C.T., Hsiao, C.C.: A novel three-phase boost type ac/dc converter for BLDC wind power generation systems. In: The 2010 TWEC, pp. 144–148 (2010)
3. Lung, S.T.: Harmonic Injection Method for Single-Switch Three-Phase Boost Rectifier, Master of science in electrical engineering. National Cheng Kung University Tainan, Taiwan (2003)
4. Erickson, R.W., Maksimovic, D.: Fundamentals of Power Electronics. Kuwer Academic Publicshers (2001)
5. Utkin, V., Guldner, J., Shi, J.: Sliding Modes in Electromechanical Systems. Taylor & Francis, London, U.K (1999)
6. Yau, H.T., Chen, C.L.: Fuzzy sliding mode controller design for maximum power point tracking control of a solar energy system. Transactions of the Institute of Measurement and Control (2012)
7. Wang, M.H.: Extension neural network-type 2 and its applications. IEEE Transactions on Neural Networks **16**(6), 1352–1361 (2005)

# Rainfall Estimation in Weather Radar Using Support Vector Machine

Bo-Jhen Huang[1], Teng-Hui Tseng[2], and Chun-Ming Tsai[1(✉)]

[1] Department of Computer Science, University of Taipei,
No. 1, Ai-Kuo W. Road, Taipei 100, Taiwan, ROC
{joe.huang74,cmtsai2009}@gmail.com
[2] Department of Communication Engineering, Oriental Institute of Technology,
New Taipei City 220, Taiwan, ROC
alex@mail.oit.edu.tw

**Abstract.** Estimation of rainfall is a very important issue for weather and flood forecasting. However, the traditional rainfall estimation is not precise enough. The traditional rainfall estimation method used the Z-R relation to estimate the rainfall rate. However, when applying the Z-R relation in the real rainfall estimation, there are many limitations. Thus, this paper proposes a method to estimate the rainfall in weather radar and to solve above-mentioned problems. The proposed method first extracts the radar reflectivity and radial velocity in a region which based on the Taipei weather station as the features. And then, these features are trained by support vector machine (SVM) to obtain the rainfall estimation model. Last, this model is used to estimate the rainfall in the weather radar. Experimental results show that the proposed method can estimate the rainfall and achieving approximately 70% rainfall estimation rates.

**Keywords:** Rainfall estimation · Weather radar · Radar reflectivity · Radial velocity · Support vector machine

## 1 Introduction

Accurate rainfall estimation is one of the most important issues in meteorological and hydrologic analyses [1]. For example, in Taiwan, rainfall estimation is used to let the government to make a decision when to declare heavy rain holidays. And if the decision is wrong, the government will be criticized. But estimation methods are inadequate. In the traditional method, rain gauge-based estimates are usually inadequate because of their limited sampling distribution [1]. Furthermore, the system of rain gauges is not widespread enough to cover the wide areas which the meteorologists and hydrologists are interested in.

Weather radar has been used to estimate rainfall for a long time. Quantitative estimation of rainfall from radar observations is a complex process. However, Weather radar can offer an unprecedented opportunity to improve our ability of observing extreme storms and quantifying their associated precipitation [2].

© Springer International Publishing Switzerland 2015
N.T. Nguyen et al. (Eds.): ACIIDS 2015, Part I, LNAI 9011, pp. 583–592, 2015.
DOI: 10.1007/978-3-319-15702-3_56

However, while an accurate estimation of rainfall from weather radar may greatly improve not only numerical weather prediction but also flash flood prediction models, it remains a challenging problem due to the inherent complexity of precipitation process.

Many authors [2]-[7] have discussed radar-based observations of rainfall. From their definitions, radar measures the power of electromagnetic waves backscattered by rain-drops which is directly related to a physical quantity called reflectivity, $Z$. Traditional rainfall estimation method used the $Z$-$R$ relation [8]-[9] to estimate the rainfall rate ($R$) as a function of reflectivity ($Z$) shown as follows:

$$Z = a \times R^b, \tag{1}$$

where $a$ and $b$ are empirical coefficients that vary from different locations and different seasons [9]. However, this $Z$-$R$ relation is very highly dependent on raindrop size distribution [9]-[11]. Furthermore, this raindrop size distribution cannot be derived from the reflectivity. Moreover, the fact that different precipitation types have different raindrop size distribution is a leading cause for more than one $Z$-$R$ relation applying within the same climatic zone [12].

Accurate estimation of ground rainfall from radar measurements is an important topic of current interest [13]. Traditionally, researchers made experiments to compare rainfall estimates from radar and rain gauge [14]-[15]. Radar observation at one elevation angle yields nearly an instantaneous snap shot of the horizontal reflectivity structure. However, different ranges from the radar have different altitude height and different reflectivity. By contrast, the rain gauge has to accumulate or count sufficient number of raindrops to obtain an accurate estimate of rainfall accumulation [13].

Figure 1 shows an example of radar volume scanning at elevation $\phi$ angle. A typical scan of radar rotates 360 degrees horizontally at a given elevation $\phi$ angle. The elevation angle usually starting from 0.5°, 1.5°, 2.4°, 3.4°, 4.3°, and etc.; the radar takes about six minutes to finish a volume scan for all defined elevations before it starts over for a new cycle. The flat radar image is obtained by projecting the conic plane from above the radar station to the ground. Geometrically, a pixel in the radar image with a range $d$ from the radar station to its corresponding altitude height, $H$, depends on the elevation $\phi$ angle.

Figure 1 shows three ranges $d1$, $d2$, and $d3$ and their corresponding altitude heights are $H1$, $H2$, and $H3$, respectively. When the range is $d3$, an instantaneous snap shot of the horizontal reflectivity is obtained at the altitude height $H3$. However, that measure indicates the estimated rainfall rate corresponding to the observed reflectivity, above the ground (at altitude height $H3$) instead of the actual rain falling to the ground, which is, of course the real concern for estimation. In this condition, it is difficult to verify the radar results with actual ground observations of rain.

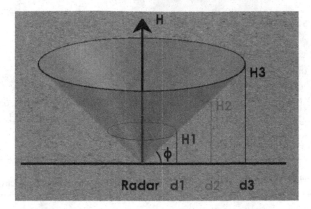

**Fig. 1.** Example of radar volume scanning at elevation $\phi$ angle

In addition, radar's reflectivity values are affected by various types of particles. Non-hydrometeors can severely contaminate the estimated precipitation. Furthermore, different types of hydrometeors have different Z-R relation. But all the radar observation gives a single value of reflectivity.

Rainfall on the ground is generally dependent on the four-dimensional structure of precipitation aloft (i.e. three-dimensional and time). In principle, one can obtain a functional relation between rain rate on the ground and the four-dimensional radar observations aloft. However, it is difficult to express this functional relationship in a useful form [16].

Neural networks provide a mechanism to solve the above-mentioned complex functional approximation problem [13] [16]. There are two important aspects of using neural networks for radar rain rate estimation: training and applying the network. In the training stage, radar data are input to the network and the corresponding rain gauge data are used as the desired output. The network is modified based on the backward error propagation according to a learning algorithm, and this process is repeated until the error between the network output and the desired output meets the prescribed accuracy requirement. When this training process is complete, the network is ready for application [16].

In this paper, a different approach, namely, a support vector machine (SVM) based technique [17], is introduced to discuss the rainfall estimation problems in weather radar. In this proposed approach, the Z-R relation is not used. Instead, this approach maps the three dimensional radar measurements to the ground rain gauge measurements directly, using rain gauge measurements as the expert experiences to label the target set of observations on the ground. The features at the rain gauge place corresponding to the weather radar data points are extracted to be the training data set. When the SVM is trained appropriately, it generalizes the relations to generate a model. This model is used to predict rainfall, using information of other new testing data sets.

## 2    Data Set and Learning Method

A convenient way to input training data to SVM is the radar measurement. These radar training data were conducted in Taipei region from March of 2014 to May of 2014 and were collected by Taiwan's Wu Fen Shan radar station. The rain gauge data were collected by the Taipei weather station in Taiwan which located at (25.0394N, 121.5067E). Figure 2 shows the flow diagram of the proposed method. The detail will be described as follows.

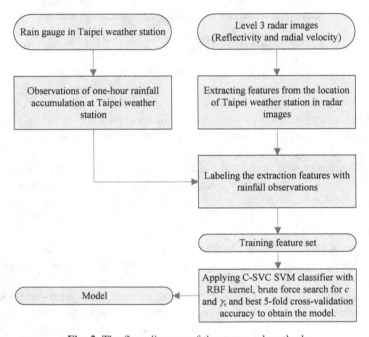

**Fig. 2.** The flow diagram of the proposed method

### 2.1    Data Set Preparation

The interested location of this study is Taipei weather station [18], which is located at (25.0394N, 121.5067E). Its one-hour rainfall accumulation observations are used to label the extracted feature vectors of level-3 radar products. The Level-3 radar images including reflectivity and radial velocity of several elevation angles in Taiwan's Wu Fen Shan radar station [19] are used as the sources of feature vectors.

The radar takes about 6 minutes to finish volume scanning for all defined elevation angles. This means that we have a new set of radar images for every 6 minutes. That is, there are about 10 frames per one hour. In this study, only radar images approximately 30 minutes before rainfall observation are selected. Thus, from one hour of radar data, only one or two images are used to as the source of the training set.

Figure 3 shows an example of the labeling method over Taipei weather station. In this figure, Taipei weather station is located at (25.0394N, 121.5067E) and is marked with a small red circle. The reflectivity (level-3 radar image) is obtained from the Wu Fen Shan (RCWF) radar station at 23:38Z March 07, 2014. The one-hour precipitation accumulation is collected at 00:00Z March 08, 2014 and is used to label the extracted feature vectors.

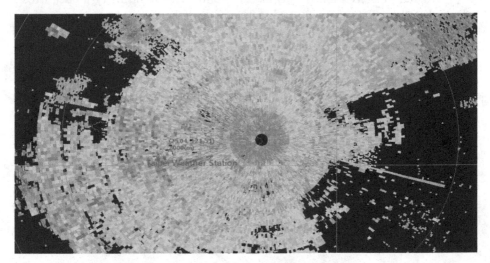

**Fig. 3.** Example of the labeling method over Taipei weather station

The one-hour rainfall accumulations of March, April, and May in the year of 2014 come from Taipei weather station. The level-3 radar images on rainy days are selected. A total number of 377 feature vectors are extracted to be the training set.

## 2.2    Learning Model

In this study, the class labels used simply rain measurements. For example, the rainfall value of 1.5(mm) will go directly to the class label field of feature vectors as one of the classes. Thus is, to classify the feature vectors in the level-3 image into the proper rainfall class is a multi-classification problem. Herein, a support vector machine is used to solve this problem. The type C-SVC of the support vector machine is defaulted with RBF kernel function and is used to solve the multi-class classification problem. The feature vectors are scaled to the range of [0, 1]. To ensure generality of the results, a 5-fold cross-validation is applied to measure accuracy. To get the best parameters $c$ and $gamma$ for RBF kernel, a brute force procedure is applied to search for various $c$- $gamma$ combinations. The $c$ parameter is increased from -5 to 15 with an interval of 2 and the $gamma$ parameter is decreased from 3 to -15 also with an interval of 2. The learning experiments include four models which used different feature vectors as the training set. These four models are described as follows.

(1) Reflectivity Model: In this model, the reflectivity is obtained from the radar with elevation angle of 0.5 degrees and is used as the source of feature vectors. The feature vectors are extracted from the location of the Taipei weather station on the level-3 radar image. The size is $5 \times 5$ and uses the location of the Taipei weather station as the center. Thus, there are 25-dimension feature vectors that are extracted from a level-3 radar image.

(2) Radial Velocity Model: This model is similar to Reflectivity Model, except that, the radial velocity is used as the source of the feature vectors.

(3) Combining Model: In this model, both reflectivity and radial velocity are used as the source of the feature vectors. The 25-dimension feature vectors in the Reflectivity Model are further used to calculate 4 statistical values: mean, standard deviation, maximum, and minimum reflectivity. In the same way, the 25-dimension feature vectors in Radial Velocity Model are used to compute the same 4 statistical values. Combing above-mentioned 8 statistical values, 8-dimension feature vectors are extracted.

(4) Different Elevations Model: In this model, both reflectivity and radial velocity are used as the source of the feature vectors. Furthermore, the radar scanned with different elevation angles. The different scanning elevation angles for the radar are varied from the elevation angle of 0.5 degrees to another 4 elevation angles, 1.5, 2.4, 3.4, and 4.3 degrees. For each elevation angle, 4 statistical values for reflectivity and radial velocity are computed, respectively. Thus, 40-dimension feature vectors are extracted.

# 3   Experimental Results

The proposed training method was performed using LIBSVM and Java 7 on an Intel(R) Atom(TM) CPU 330 @ 1.60 GHz Asus Eee PC.

To demonstrate the performance of the proposed method, the rain observations from Taipei weather station and the level-3 radar images from Wu Fen Shan radar station are used as the experiment label and data, respectively. The experimental results for the four learning model are described as follows.

(1) Reflectivity Model: In this model, the 25-dimension feature vectors of reflectivity ($Z$) are extracted from around the location of Taipei weather station as shown in the level-3 radar image and are then used as the input training set. After applying the brute force procedure, the ($c$, $gamma$) parameters are (2, 0.5). Then, after the best 5-fold cross validation is applied, the accuracy is 65.252%. Figure 4 shows the training result of the Reflectivity Model.

(2) Radial Velocity Model: In this model, the 25-dimension feature vectors of radial velocity ($V$) are extracted from around the location of Taipei weather station in the level-3 radar image and are used to as the input training set. After applying the brute force procedure, the ($c$, $gamma$) parameters are (8, 2). Then, after the best 5-fold cross validation is applied, the accuracy is 62.8647%. Figure 5 shows the training result of the Radial Velocity Model.

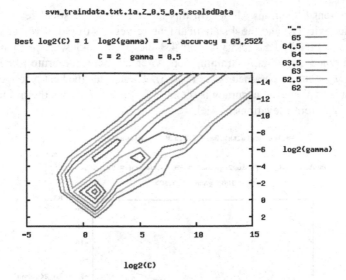

**Fig. 4.** Training result of the Reflectivity Model

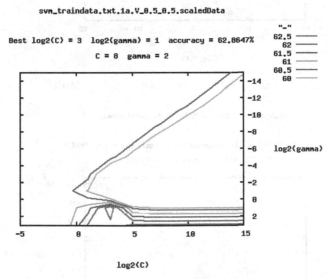

**Fig. 5.** Training result of the Radial Velocity Model

(3) Combining Model: In this model, the 4 statistical feature vectors of reflectivity (*Z*) and the 4 statistical feature vectors of radial velocity (*V*) are extracted at 0.5-degree elevation angle and are combined as the input training set. After applying the brute force procedure, the (*c*, *gamma*) parameters are (128, 0.125). Then, after the best 5-fold cross validation is applied, the accuracy is 67.9045%. Figure 6 shows the training result of the Combining Model.

(4) Different Elevations Model: In this model, the 4 statistical feature vectors of reflectivity ($Z$) and the 4 statistical feature vectors of radial velocity ($V$) are extracted at 0.5, 1.5, 2.4, 3.4, and 4.3 elevation angles, respectively, and are combined as the input training set. After applying the brute force procedure, the ($c$, *gamma*) parameters are (2, 0.5). Then, after the best 5-fold cross validation is applied, the accuracy is 69.2308%. Figure 7 shows the training result of the Different Elevations Model.

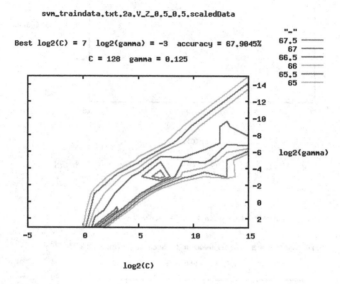

**Fig. 6.** Training result of the Combining Model

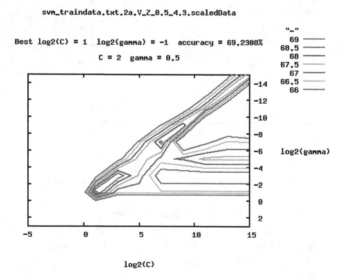

**Fig. 7.** Training result of the Different Elevations Model

Table 1 shows the summary of the four above-mentioned learning modes. From this table, the accuracy of the reflectivity mode is seen to be larger than the accuracy of the radial velocity mode. That is, reflectivity is probably a better indicator of rainfall than radial velocity. The accuracy of the different elevation models is larger than the accuracy of the combining mode.

**Table 1.** Summary of the proposed four learning modes

| Modes \ Results | Reflectivity | Radial Velocity | Combining | Different Elevations |
|---|---|---|---|---|
| c | 2 | 8 | 128 | 2 |
| *gamma* | 0.5 | 2 | 0.125 | 0.5 |
| accuracy | 65.252% | 62.8647% | 67.9045% | 69.2308% |

These experiments show that the proposed method can estimate the rainfall in the weather radar without use of the Z-R formulation. Overall, the most accurate learning mode is the different elevations mode. It achieves approximately 70% rainfall estimation rates.

## 4    Conclusions

This study proposes a learning based method to estimate the rainfall shown in the weather radar. In this study, four learning modes are proposed. The experimental results show that reflectivity is a better feature to use for rainfall estimation than radial velocity. Furthermore, the mode combining reflectivity and radial velocity in different elevation angles can achieve approximately 70% rainfall estimation rates.

In the future, more effective features will be used to enhance the accuracy of the rainfall estimation. Furthermore, the observation time of rain and radar will be also improved to be the labelled and training data, respectively. Lastly, dual-polarization radar products will be used to as the source of training features to improve the accuracy of the rainfall estimation.

**Acknowledgements.** The authors would like to express his gratitude to Walter Slocombe and Dr. Jeffrey Lee, who assisted editing the English language for this article. The authors also thank for Taiwan Central Weather Bureau to support the experimental data: Wu-Fen-Shan level-3 radar data and one-hour rainfall accumulation.

## References

1. Morin, E., Gabella, M.: Radar-based quantitative precipitation estimation over Mediterranean and dry climate regimes. J. Geophys. Res. **112**, D20108 (2007). doi:10.1029/2006JD008206
2. Krajewski, W.F., Smith, J.A.: Radar hydrology: rainfall estimation. Advances in Water Resources **25**, 1387–1394 (2002)

3. Battan, L.J.: Radar observation of the atmosphere. The University of Chicago Press (1973)
4. Doviak, R.J., Zrnic, D.S.: Doppler radar and weather observations. Academic Press Inc., San Diego, CA (1993)
5. Reinhart, R.: Radar for meteorologists. Reinhart Publications (1997)
6. Sauvageot, H.: Radar Meteorology. Artech House, Inc. (1991)
7. Austin, P.M.: Relation between measured radar reflectivity and surface rainfall. Monthly Weather Review 115, 1053–1070 (1987)
8. Marshall, J.S., Langille, R.C., Palmer, W.M.: Measurement of rainfall by radar. Journal of Meteorology 4, 186–192 (1947)
9. Uijlenhoet, R.: Raindrop size distributions and radar reflectivity-rain rate relationships for radar hydrology. Hydrology and Earth System Sciences 5(4), 615–627 (2001)
10. Spilhaus, A.F.: Drop size, intensity, and radar echo of rain. Journal of Meteorology 5, 161–164 (1948)
11. Marshall, J.S., Palmer, W.M.: The distribution of raindrops with size. Journal of Meteorology 5, 165–166 (1948)
12. Reddy, K.K., Kozu, T.: Measurements of raindrop size sidtribution over Gadanki during south-west and north-east monsoon. Indian Journal of Radio & Space Physics 32, 286–295 (2003)
13. Xiao, R.R., Chandrasekar, V.: Development of a neural network based algorithm for rainfall estimation from radar observations. IEEE Transaction on Geoscience and Remote Sensing 35(1), 160–171 (1997)
14. Seo, D.J.: Real-time estimation of rainfall fields using radar rainfall and rain gauge. Journal of Hydrology 208(1–2), 37–52 (1998)
15. Aydin, K., Lure, Y.M., Seliga, T.A.: Polarimetric radar measurements of rainfall compared with ground-based rain gauges during MAYPOLE'84. IEEE Transaction on Geoscience and Remote Sensing 28, 443–449 (1990)
16. Bringi, V.N., Chandrasekar, V.: Polarimetric Doppler weather radar: principles and applications, pp. 560–569. Cambridge University Press (2004)
17. Chang, C.C., Lin, C.J.: LIBSVM: a library for support vector machines. ACM Transactions on Intelligent Systems and Technology 2(27), 1–27 (2011). Software. http://www.csie.ntu.edu.tw/~cjlin/libsvm
18. Taipei weather station. http://www.cwb.gov.tw/V7/eservice/docs/overview/organ/stations/46692/
19. Wu-Fen-Shan weather radar station. http://www.cwb.gov.tw/V7/eservice/docs/overview/organ/stations/46685/

# Explicitly Epistemic Contraction by Predicate Abstraction in Automated Theorem Finding: A Case Study in NBG Set Theory

Hongbiao Gao, Yuichi Goto, and Jingde Cheng$^{(\boxtimes)}$

Department of Information and Computer Sciences, Saitama University, Saitama
338-8570, Japan
{gaohongbiao,gotoh,cheng}@aise.ics.saitama-u.ac.jp

**Abstract.** In automated theorem finding by forward reasoning, there are many redundant theorems as intermediate results. This paper proposes an approach of explicitly epistemic contraction by predicate abstraction for automated theorem finding by forward reasoning in order to remove redundant theorems in a set of obtained theorems, and shows the effectiveness of the explicitly epistemic contraction by a case study of automated theorem finding in NBG set theory.

**Keywords:** Automated theorem finding · Predicate abstraction · Forward reasoning · Strong relevant logic · NBG set theory

## 1 Introduction

The problem of automated theorem finding (ATF for short) is one of 33 basic research problems in automated reasoning which was originally proposed by Wos in 1988 [7,8], and it is still an open problem until now. The most important and difficult requirement of the problem is that, in contrast to proving conjectured theorems supplied by the user, it asks for criteria that an automated reasoning program can use to find some theorems in a field that must be evaluated by theorists of the field as new and interesting theorems. The significance of solving the problem is obvious because an automated reasoning program satisfying the requirement can provide great assistance for scientists in various fields [1].

Theorem finding/proving processes in mathematics must include some concept/notion abstraction processes. In any mathematical field, definitions and axioms mean simple concepts. Mathematicians continue to define more complex concepts by using previously given definitions and axioms, and already defined concepts. Then, mathematicians think, assume, prove propositions by using the defined complex concepts. After that, they obtain new theorems. For example, predicate "∈" is the most basic predicate in the set theory. Mathematicians define predicate "⊆" which is a higher level predicate than "∈", and abstract from "∈" by the definition of "⊆": $\forall x \forall y (\forall u((u \in x) \Rightarrow (u \in y)) \Leftrightarrow (x \subseteq y))$. In addition, they define predicate "=" which is a higher level predicate than "⊆",

© Springer International Publishing Switzerland 2015
N.T. Nguyen et al. (Eds.): ACIIDS 2015, Part I, LNAI 9011, pp. 593–602, 2015.
DOI: 10.1007/978-3-319-15702-3_57

and abstracts from "$\subseteq$" by the axiom: $\forall x \forall y(((x \subseteq y) \wedge (y \subseteq x)) \Leftrightarrow (x = y))$. Moreover, mathematicians think, assume, and prove own beliefs by using simple representation rather than complex representation if both representation show same meaning. For example, after mathematicians defined "$=$", they will think, assume, and prove propositions by using $\forall x \forall y(x = y)$ rather than $\forall x \forall y((x \subseteq y) \wedge (y \subseteq x))$. In theorem finding/proving process, a complex representation is replaced with a simple representation, but not preserved, after mathematicians define the simple representation as the complex representation.

On the other hand, a concept/notion abstraction process is also a part of an epistemic process for scientific discovery. Any scientific discovery must include an epistemic process to gain knowledge of or to ascertain the existence of some empirical and/or logical entailments previously unknown or unrecognized [1]. Review the human beings' history of science, not only in Mathematics but also in Physics and Chemistry, when scientists explore and discover in a new field by brain, they always make definition from low level to high level so that they can represent the complex things by simple definitions.

We have proposed a systematic methodology of ATF by forward reasoning based on the strong relevant logic [4]. To show the effectiveness of the methodology, we also presented two case studies of ATF in von Neumann-Bernays-Godel (NBG) set theory and Peano's arithmetic by using the methodology. Under the above first observation about the abstraction processes, we introduced abstraction processes into the methodology. As results, we obtained theorems that consists of from low-level predicate to high-level predicate step by step. By contrast with the first observation, the second observation about the replacement is not introduced into the methodology. There are complex representations that consist of low-level predicate in a set of obtained theorems while there are simple representations as same as the complex representations in the set of obtained theorems. Under the second observation, the complex representations are redundant. Moreover, they cause the increase of execution time and used memory space of automated forward reasoning in ATF according to the methodology, and cost for excavation of new and interesting theorems from obtained theorems.

This paper proposes an approach of explicitly epistemic contraction by predicate abstraction for automated theorem finding by forward reasoning. By the explicitly epistemic contraction, it is possible to remove the complex representations in a set of obtained theorems. The paper also shows the effectiveness of the explicitly epistemic contraction by a case study of ATF in NBG set theory.

The rest of the paper is organized as follows: Section 2 explains the terminology used in the paper. Section 3 reviews the proposed methodology. Section 4 shows a procedure of explicitly epistemic contraction by predicate abstraction for the set of obtained theorems. Section 5 presents a case study of ATF in NBG set theory by the proposed method. Finally, some concluding remarks are given in Section 6.

## 2     Basic Notions and Notations

A formal logic system $L$ is an ordered pair $(F(L), \vdash_L)$ where $F(L)$ is the set of well formed formulas of $L$, and $\vdash_L$ is the consequence relation of $L$ such that for a set $P$ of formulas and a formula $C$, $P \vdash_L C$ means that within the framework of $L$ taking $P$ as premises we can obtain $C$ as a valid conclusion. $Th(L)$ is the set of logical theorems of $L$ such that $\phi \vdash_L T$ holds for any $T \in Th(L)$. According to the representation of the consequence relation of a logic, the logic can be represented as a Hilbert style system, a Gentzen sequent calculus system, a Gentzen natural deduction system, and so on [1].

Let $(F(L), \vdash_L)$ be a formal logic system and $P \subseteq F(L)$ be a non-empty set of sentences. A formal theory with premises $P$ based on $L$, called an $L$-theory with premises $P$ and denoted by $T_L(P)$, is defined as $T_L(P) =_{df} Th(L) \cup Th_L^e(P)$ where $Th_L^e(P) =_{df} \{A | P \vdash_L A$ and $A \notin Th(L)\}$, $Th(L)$ and $Th_L^e(P)$ are called the logical part and the empirical part of the formal theory, respectively, and any element of $Th_L^e(P)$ is called an empirical theorem of the formal theory [1].

Based on the definition above, the problem of ATF can be said as "for any given premises $P$, how to construct a meaningful formal theory $T_L(P)$ and then find new and interesting theorems in $Th_L^e(P)$ automatically" [1].

The notion of the degree [3] of a connective is defined as follows: Let $\theta$ be an arbitrary $n$-ary $(1 \leq n)$ connective of logic $L$ and $A$ be a formula of $L$, the degree of $\theta$ in $A$, denoted by $D_\theta(A)$, is defined as follows: (1) $D_\theta(A) = 0$ if and only if there is no occurrence of $\theta$ in $A$, (2) if $A$ is in the form $\theta(a_1, a_2, ..., a_n)$ where $a_1, a_2, ..., a_n$ are formulas, then $D_\theta(A) = max\{D_\theta(a_1), D_\theta(a_2), ..., D_\theta(a_n)\} + 1$, (3) if $A$ is in the form $\sigma(a_1, a_2, ..., a_n)$ where $\sigma$ is a connective different from $\theta$ and $a_1, a_2, ..., a_n$ are formulas, then $D_\theta(A) = max\{D_\theta(a_1), D_\theta(a_2), ..., D_\theta(a_n)\}$, and (4) if $A$ is in the form $QB$ where $B$ is a formula and $Q$ is the quantifier prefix of $B$, then $D_\theta(A) = D_\theta(B)$.

The notion of a fragment of a logic [3] is defined as follows: Let $\theta_1, \theta_2, ..., \theta_n$ be connectives of logic $L$ and $k_1, k_2, ..., k_n$ be natural numbers, the fragment of $L$ about $\theta_1, \theta_2, ..., \theta_n$ and their degrees $k_1, k_2, ..., k_n$, which are denoted by $Th^{(\theta_1, k_1, \theta_2, k_2, ..., \theta_n, k_n)}(L)$, is a set of logical theorems of $L$ which is inductively defined as follows (in the terms of Hilbert style axiomatic system): (1) if $A$ is an axiom of $L$ and $D_{\theta_1}(A) \leq k_1, D_{\theta_2}(A) \leq k_2, ..., D_{\theta_n}(A) \leq k_n$, then $A \in Th^{(\theta_1, k_1, \theta_2, k_2, ..., \theta_n, k_n)}(L)$, (2) if $A$ is the result of applying an inference rule of $L$ to some members of $Th^{(\theta_1, k_1, \theta_2, k_2, ..., \theta_n, k_n)}(L)$ and $D_{\theta_1}(A) \leq k_1, D_{\theta_2}(A) \leq k_2, ..., D_{\theta_n}(A) \leq k_n$, then $A \in Th^{(\theta_1, k_1, \theta_2, k_2, ..., \theta_n, k_n)}(L)$, (3) Nothing else is in $Th^{(\theta_1, k_1, \theta_2, k_2, ..., \theta_n, k_n)}(L)$. Similarly, the notion of degree of formal theory about conditional can also be generally extended to other logic connectives, and a fragment of a formal theory with premises $P$ based on the logic fragment $Th^{(\theta_1, k_1, \theta_2, k_2, ..., \theta_n, k_n)}(L)$ denoted by $T_{Th^{(\theta_1, k_1, ..., \theta_n, k_n)}(L)}^{(\eta_1, j_1 ... \eta_s, j_s)}(P)$ is also similarly defined as the notion of a fragment of a logic.

The notion of predicate abstract level is defined as follows: (1) Let $pal(X) = k$ denote that an abstract level of a predicate $X$ is $k$ where $k$ is a natural number, (2) $pal(X) = 1$ if $X$ is the most primitive predicate in a target field,

(3) $pal(X) = max(pal(Y_1), pal(Y_2), ..., pal(Y_n)) + 1$ if a predicate $X$ is defined by other predicates $Y_1, Y_2, ..., Y_n$ in the target field where $n$ is a natural number. A predicate $X$ is called $k$-level predicate, if $pal(X) = k$. If $pal(X) < pal(Y)$, then the abstract level of predicate $X$ is lower than $Y$, and $Y$ is higher than $X$.

The notion of abstract level of a formula is defined as follows: (1) $lfal(A) = k$ denote that an abstract level of a formula $A$ is $k$, where $k$ is a natural number, (2) $k = max(pal(Q_1), pal(Q_2), ..., pal(Q_n))$ where $Q_i$ is a predicate and occurs in $A$ $(1 \leq i \leq n)$. A formula $A$ is $k$-level formula, if $lfal(A) = k$.

$k$-level fragment of premises $P$, denoted by $P_k$, is a set of all formulas in $P$ that consists of only $m$-level formulas where $m$ and $k$ are natural numbers $(1 \leq m \leq k)$.

If two theorems are same theorems when we use the most primitive predicate of the target field to represent them, then we defined them as "equivalence theorems".

FreeEnCal [3] is a forward reasoning engine for general purpose, that provides an easy way to customize reasoning task by providing different axioms, inference rules and facts. Users can set the degree of logical connectives to make FreeEnCal to reason out in principle all logical theorem schemata of the fragment $Th^{(\theta_1, k_1, \theta_2, k_2, ..., \theta_n, k_n)}(L)$. FreeEnCal can also reason out in principle all empirical theorems of $T_{Th^{(\theta_1, k_1, ..., \theta_n, k_n)}(L)}^{(\eta_1, j_1 ... \eta_s, j_s)}(P)$ from $Th^{(\theta_1, k_1, ..., \theta_n, k_n)}(L)$ and $P$ with inference rules of $L$.

# 3    A Systematic Methodology for ATF with Forward Reasoning

A systematic methodology [4] for ATF consists of five phases as shown in Fig. 1. Phase 1 is to prepare logical fragments of strong relevant logic for various empirical theories. The prepared logic fragments are independent from any target field, therefore they can be reused for ATF in different fields. Phase 2 is to prepare empirical premises of the target theory and draw up a plan to use collected empirical theorems. In detail, we prepare $k$-level $(k \geq 1)$ fragment of collected empirical premises in the target field to define a semi-lattice. A set of the prepared fragments and inclusion relation on the set is a partial order set, and is a finite semi-lattice. Moreover, a set of formal theories with the fragments and inclusion relation on the set is also a partial order set, and is also a finite semi-lattice. Partial order of the set of the prepared fragments can be used for a plan to reason out fragments of formal theories with collected empirical premises. According to the partial order, we can systematically do ATF from simple theorems to complex theorems.

Phase 3 to phase 5 are performed repeatedly until deducing all fragments of formal theory that have been planed in phase 2. In this methodology, *loop* means doing phase 3 to phase 5 at once. We use one of $k$-level $(k \geq 1)$ fragment of collected empirical premises to perform phase 3 to phase 5 in one loop. In detail, we firstly use 1-level fragment of premises to reason out empirical theorems. Then, we enter into phase 4 to abstract deduced empirical theorems, and then

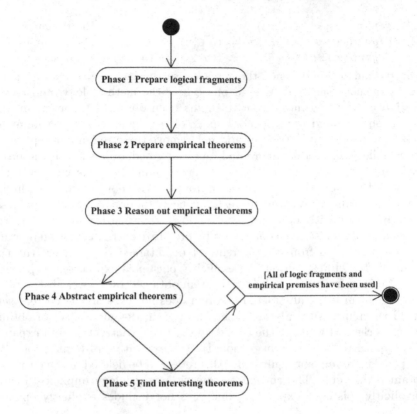

**Fig. 1.** The activity diagram of the systematic method for ATF

enter into phase 5 to find new and interesting theorems from empirical theorems. After that, we go back to the phase 3, and use 2-level fragment of premises and theorems obtained in last loop as premises of this loop. Then, we enter into phase 4 and phase 5 to abstract theorems and find interesting theorems again. We repeat the loops until all of the $k$-level ($k \geq 1$) fragments of collected empirical premises have been used.

## 4   The Explicitly Epistemic Contraction by Predicate Abstraction

In ATF by using the systematic methodology we mentioned in section 3, there are many redundant theorems as intermediate results. Equivalence theorems we mentioned in section 2 cause redundant theorems. We need one theorem represented by higher level predicates. Equivalence theorems of the target theorem are redundant from the viewpoint of meaning of the theorem. For example, the following three equivalence theorems can be unified to only one theorem $\forall x \forall y ((x = y) \Rightarrow (x \subseteq y))$.

(1) $\forall x \forall y ((x = y) \Rightarrow (x \subseteq y))$      [Represent by $=$, $\subseteq$]

(2) $\forall x \forall y \forall u ((x = y) \Rightarrow ((u \in x) \Rightarrow (u \in y)))$      [Represent by $=$, $\in$]

(3) $\forall x \forall y \forall u (((x \subseteq y) \wedge (y \subseteq x)) \Rightarrow ((u \in x) \Rightarrow (u \in y)))$ [Represent by $\subseteq$, $\in$]

The redundant equivalence theorems cause the increase of execution time and used memory space of ATF in phase 3 of the methodology, and cost for excavation of new and interesting theorems from obtained theorems in phase 5. Execution time and used memory space of reasoning depend on number of premises given as input data, if we did not remove those redundant equivalence theorems, they will produce more and more intermediate results in the process of reasoning. On the other hand, if we have found one theorem, equivalence theorems of the found theorem are waste from the viewpoint of theorem finding.

To solve the above problem and realize concept/notion abstraction processes, we introduced explicitly epistemic contraction/expansion by predicate abstraction. An explicitly epistemic contraction by predicate abstraction is an operation to remove a theorem from the current set of obtained theorems, if the theorem can be $k$-level abstraction. An explicitly epistemic expansion by predicate abstraction is an operation to add a theorem obtained by $k$-level abstraction to a source theorem if the abstracted theorem does not exist in the current set of obtained theorems. After phase 4, each theorem in the current set of obtained theorems is checked whether the explicitly epistemic contraction and expansion can be applied to the theorem or not. By the explicitly epistemic contraction and expansion, redundant equivalence theorems can be deleted in the current set of obtained theorems. The current systematic methodology implicitly includes the explicitly epistemic expansion, but does not includes explicitly epistemic contraction.

The procedure of $k$-level abstraction is as follows. An $i$-level abstraction rule $r_i$ is defined as $r_i = < p_i, w_j >$ where $p_i$ is $i$-level predicate, $w_j$ is a $j$-level formula, $i$, $j$, and $k$ are natural numbers, and $(1 \leq j < i \leq k)$.

1. $S$ is a set of all $i$-level abstraction rules.
2. $r_i \in S$ if $S \neq \emptyset$.
3. if a theorem $A$ includes $w_j$ of $r_i$, then replace all $w_j$ in $A$ with $p_i$ of $r_i$.
4. $S \leftarrow S - \{r_i\}$
5. Go back to 2.

Abstraction rules are created from definitions of predicates in a certain mathematical fileds. For example, if an $i$-level predicate $p$ is defined as $\forall x_1 \dots \forall x_n (p(x_1, \dots, x_n) \Leftrightarrow B)$ where $B$ is $j$-level formula, $i, j$, and $n$ are natural numbers, and $(1 \leq j < i)$, then an $i$-level abstraction rule is $< p(x_1, \dots, x_n), B >$.

By introducing explicitly epistemic contraction by predicate abstraction to the methodology, we can expect two positive effects, but it may be a side effect. The first positive effect is that we can reduce execution time and amount of used memory space of automated forward reasoning for ATF, because by unifying those equivalence theorems, lots of intermediate results in the process of reasoning of ATF have been reduced. The second positive effect is that we can save the cost of excavation of new and interesting theorems from obtained theorems, because lots of redundant equivalence theorems have been removed, and the set of obtained theorems have been contracted. On the other hand, the explicitly epistemic contraction by predicate abstraction may also lead to a side effect, that is some new and interesting theorems may not be deduced due to the removed equivalence theorems. Some new and interesting theorems may be deduced from the removed equivalence theorems or the theorems deduced from the removed equivalence theorems.

## 5    Case Study in NBG Set Theory

Purpose of this case study is to confirm the effectiveness of explicitly epistemic contraction by predicate abstraction.

### Method

We tried to investigate two things. First is how amount of redundant results the explicitly epistemic contraction by predicate abstraction can reduce. Second is whether the explicitly epistemic contraction has a side effect or not. To investigate the two things, we did two case studies. One is a case study of ATF according to the systematic methodology in NBG set theory, "case study 1" for short. Other is a case study of ATF according to the systematic methodology with the epistemic contraction by predicate abstraction in NBG set theory, "case study 2" for short. In both case studies, we used same logical fragment and empirical premises. The logic fragment we used is $Th^{(\Rightarrow,2,\neg,1,\wedge1)}(EcQ)$, which has been prepared [4]. The empirical premises are all axioms and definitions of NBG set theory in Quaife's book [5]. Quaife recorded the axioms, definitions, and more than 400 known theorems of NBG set theory in his book.

As operations of phase 2 of the methodology we mentioned in section 3, we summarized all of the abstract levels of collected definitions and axioms of NBG set theory. Table 1 shows all of the abstract level of predicates in collected definitions and axioms of NBG set theory. Then, we defined the $k$-level fragments of the axioms and definitions according to Table 2.

In the both of case studies, we did 6 loops from phase 3 to phase 5 because the number of highest level of predicate is 6. In phase 3, when we were doing $i$-th loop, we used the $i$-level fragments of the axioms and definitions, and results obtained in last loop as empirical premises, and obtained empirical theorems of NBG set theory by using FreeEnCal with setting the degree $\Rightarrow$ to 2, $\neg$ to 1, and $\wedge$ to 1. Note that output of FreeEnCal includes not only empirical theorems of NBG set theory, but also intermediate results that are used for reasoning the empirical theorems. In phase 4, when we were doing $i$-th loop, we applied $i$-level abstraction to each of obtained empirical theorems and intermediate results. In case study 1, both a theorem/intermediate result and a result of $i$-level abstraction of the theorem/intermediate result are preserved. However, in case study 2, only a result of $i$-level abstraction of the theorem/intermediate result is preserved because of the explicitly epistemic contraction by predicate abstraction.

In both case studies, we counted the number of obtained results (including intermediate results) after 6th loop for checking how amount of redundant results the explicitly epistemic contraction by predicate abstraction can

**Table 1.** Predicate abstract level in NBG set theory

| Predicate | Abstract from | Level |
|:---:|:---:|:---:|
| ∈ | none | 1 |
| ⊆ | ∈ | 2 |
| = | ⊆ | 3 |
| INDUCTIVE | ∈, ⊆ | 3 |
| SINGVAL | ⊆ | 3 |
| FUNCTION | ⊆, SINGVAL | 4 |
| ONEONE | FUNCTION | 5 |
| OPERATION | FUNCTION, =, ⊆ | 5 |
| COMPATIBLE | FUNCTION, =, ⊆ | 5 |
| HOM | OPERATION, COMPATIBLE, =, ∈ | 6 |

**Table 2.** The abstract level of axioms and definitions in NBG set theory

| Abstract level | Axiom and definition |
|:---:|:---:|
| 1 | Axiom B1, B2, B3, B5, C2, C3 |
| 2 | Axiom A2, B6, B7, B8, Definition of ⊆, ∘ |
| 3 | Axiom A3, A4, B4, C1, D, Definition of { }, ∪, +, <, >, succ, restrict, inverse, U, diag, R, ", INDUCTIVE, SINGVAL, P, ', cantor |
| 4 | Axiom C4, Axiom E, Definition of FUNCTION |
| 5 | Definition of ONEONE, OPERATION, COMPATIBLE |
| 6 | Definition of HOM |

reduce. To investigate whether the explicitly epistemic contraction has a side effect or not, we counted the number of $i$-level theorems obtained after $i$-th loop ($4 \leq i \leq 6$). Moreover, we counted the number of each level theorems obtained after each loop in case study 1.

## Result

We showed the results of the two case studies. First, we counted the number of theorems included intermediate results of the two case stuies in Table 3. To investigate the side effect, we counted the number of $i$-th level ($4 \leq i \leq 6$) theorems in $i$-th loop ($4 \leq i \leq 6$) as shown in Table 4, and we also counted the theorems in $i$-th loop ($1 \leq i \leq 6$) of case study 1 as shown in Table 5.

**Table 3.** The number of theorems included intermediate results

|  | Case study 1 | Case study 2 |
|:---:|:---:|:---:|
| 6th loop | 1,855 | 1,516 |

**Table 4.** The number of theorems of $i$ level theorems $(4 \leq i \leq 6)$

|  | Case study 1 | Case study 2 |
|---|---|---|
| 6th loop, 6 level theorems | 20 | 20 |
| 5th loop, 5 level theorems | 43 | 37 |
| 4th loop, 4 level theorems | 21 | 16 |

**Table 5.** The number of theorems in case study 1

|  | 1 level | 2 level | 3 level | 4 level | 5 level | 6 level |
|---|---|---|---|---|---|---|
| 1st loop | 111 |  |  |  |  |  |
| 2nd loop | 220 | 18 |  |  |  |  |
| 3rd loop | 228 | 19 | 62 |  |  |  |
| 4th loop | 228 | 57 | 98 | 21 |  |  |
| 5th loop | 228 | 57 | 98 | 41 | 43 |  |
| 6th loop | 228 | 57 | 98 | 41 | 43 | 20 |

We analyzed the results from the two aspects. First one is for the two expected positive effects. The current results show that the number of obtained theorems and the intermediate results have been indeed reduced by the proposed approach as shown Table 3, that means the approach can reduce a lot of redundant results and reduce the cost for excavation of new and interesting theorems. Second investigation is for the side effect. The higher abstract level theorems are more possible to become new and interesting theorems. Based on the opinion, we counted the highest abstract level from 4th loop to 6th loop as shown in Table 4. We found almost all of the highest abstract level theorem in each loops were not removed by our approach, such as the highest abstract level theorems in 6th loop. Although some theorems cannot be obtained, some of these theorems are removed as equivalence theorems, so it is not side effect. For example, the six highest level theorems cannot be obtained in 5th loop in the case study 2, but four theorems are removed as equivalence theorems. Some theorems cannot be deduced due to the removed equivalence theorems, however, we consider that the possibility of those theorems found as new and interesting theorems is low, because the removed theorems cannot successively impact on the deduction of the higher abstract level theorems as shown in Table 5. Therefore, the experiment results show our approach is effective.

## 6    Concluding Remarks

We have proposed an approach of explicitly epistemic contraction by predicate abstraction for ATF, and showed effectiveness of the explicitly epistemic contraction by a case study of ATF in NBG set theory. We can conclude that the approach is useful for ATF by forward reasoning in many mathematical fields. On the other hand, from the viewpoint of a strong relevant logic model of epistemic processes in scientific discovery [1], the approach of the explicitly epistemic contraction of predicate abstraction is an automation approach of an epistemic processes. The proposed approach is also the first study about how to operate abstraction processes based on the strong relevant logic model.

There are many interesting and challenging research problems in our future works. First, we will do more case studies in different fields to verify the generality of the proposed approach, such as number theory, graph theory, and combinatorics. Second, we will extend the proposed approach to function abstraction. Third, we will use EPLAS [6] which is an epistemic programming language for all scientists to perform the proposed approach in future. Finally, the proposed approach only focus on the known predicates defined by mathematicians, in future we will make the system to automatically or semi-automatically define new predicates and use the explicitly epistemic contraction by predicate abstraction for new predicates.

# References

1. Cheng, J.: A Strong Relevant Logic Model of Epistemic Processes in Scientific Discovery. In: Information Modelling and Knowledge Bases XI, Frontiers in Artificial Intelligence and Applications, vol. 61, pp. 136–159 (2000)
2. Cheng, J.: A Semilattice Model for the Theory Grid. In: Proc. 3rd International Conference on Semantics, Knowledge and Grid, IEEE Computer Society, pp. 152–157 (2007)
3. Cheng, J., Nara, S., Goto, Y.: FreeEnCal: A Forward Reasoning Engine with General-Purpose. In: Apolloni, B., Howlett, R.J., Jain, L. (eds.) KES 2007, Part II. LNCS (LNAI), vol. 4693, pp. 444–452. Springer, Heidelberg (2007)
4. Gao, H., Goto, Y., Cheng, J.: A Systematic Methodology for Automated Theorem Finding. Theoretical Computer Science **554**, 2–21 (2014)
5. Quaife, A.: Automated Development of Fundamental Mathematical Theories. Kluwer Academic (1992)
6. Takahashi, I., Nara, S., Goto, Y., Cheng, J.: EPLAS: An Epistemic Programming Language for All Scientists. In: Shi, Y., van Albada, G.D., Dongarra, J., Sloot, P.M.A. (eds.) ICCS 2007, Part I. LNCS, vol. 4487, pp. 406–413. Springer, Heidelberg (2007)
7. Wos, L.: Automated Reasoning: 33 Basic Research Problem. Prentic-Hall (1988)
8. Wos, L.: The Problem of Automated Theorem Finding. Journal of Automated Reasoning **10**(1), 137–138 (1993)

# Mining Sequential Patterns with Pattern Constraint

Show-Jane Yen$^{(\boxtimes)}$, Yue-Shi Lee, Bai-En Shie, and Yeuan-Kuen Lee

Department of Computer Science and Information Engineering, Ming Chuan University
5 De Ming Rd., Gui Shan District, Taoyuan County 333, Taiwan
{sjyen,leeys}@mail.mcu.edu.tw

**Abstract.** Mining *sequential patterns* is to find the sequential purchasing behaviors for most of the customers. There were many algorithms proposed for discovering all the sequential patterns. However, users may be only interested in certain items or behaviors. The items or patterns specified by users are called "*pattern constraints.*" If we first find all the sequential patterns and then filter out the patterns which the users are not interested in, then it will take much more time to find interesting sequential patterns. Therefore, the challenge for mining interesting sequential patterns is how to avoid searching for uninteresting sequential patterns, such that the mining time can be reduced. In this paper, we propose a query expression to represent the pattern constraints and an efficient mining algorithm to find sequential patterns which satisfy user specified pattern constraints. In our experiments, we compare our algorithm with well-known SPIRIT(R) algorithm on a real dataset. The experimental results show that our algorithm is more efficient than SPIRIT(R) algorithm.

**Keywords:** Data mining · Sequential pattern · Pattern constraint

## 1    Introduction

The definitions about mining sequential patterns [1, 4] are as follows:   Let $I = \{i_1, i_2, \ldots, i_r\}$ be a set of all of the items, and $i_j$ $(1 \leq j \leq r)$ is an *item*.   An *itemset* is a set of one or more items, which is denoted as $(a_1, a_2, \ldots, a_m)$, where $a_k \in I$ and $1 \leq k \leq m$.    A *sequence* is an ordered list of one or more itemsets, which is denoted as $<s_1, s_2, \ldots, s_n>$, where $s_p$ $(1 \leq p \leq n)$ is an itemset, and the items in $s_p$ were bought in one transaction.    The *length* of a sequence is the number of items in the sequence. A sequence with length $l$ is called an *l-sequence*.    For two sequences $X = <r_1, r_2, \ldots, r_n>$ and $Y = <s_1, s_2, \ldots, s_m>$, if there exists integers $1 \leq j_1 < j_2 < \ldots < j_n \leq m$, such that $r_1 \subseteq s_{j1}$, $r_2 \subseteq s_{j2}, \ldots, r_n \subseteq s_{jn}$, then we say that $Y$ *contains* $X$.

   Before performing sequential pattern mining process, a transaction database needs to be transformed into a *customer sequence database*. A customer sequence database includes *customer identifier* and *customer sequence*. A customer sequence is an ordered list of the transactions purchased by a customer, which are ordered by increasing transaction date.    For example, the transaction database in Table 1 can be transformed into a customer sequence database in Table 2.    The *support* for a sequence is

© Springer International Publishing Switzerland 2015
N.T. Nguyen et al. (Eds.): ACIIDS 2015, Part I, LNAI 9011, pp. 603–613, 2015.
DOI: 10.1007/978-3-319-15702-3_58

defined as the ratio of the number of the customer sequences which contain the sequence to the number of total customer sequences in the database. The *support count* for a sequence is the number of the customer sequences which contain the sequence. If the support of a sequence is no less than a user-specified *minimum support threshold*, then this sequence is a *sequential pattern*. A sequential pattern with length $l$ is called an *l-sequential pattern*.

**Table 1.** A transaction database

| SID | TID | Date | Items |
|-----|-----|------|-------|
| 10 | 06 | 2005/08/23 | a d |
| 10 | 07 | 2005/09/01 | b c |
| 10 | 09 | 2005/09/10 | b c d |
| 20 | 02 | 2005/07/07 | a b d |
| 20 | 04 | 20050804 | b f |
| 30 | 08 | 20050906 | b d g |
| 30 | 10 | 20050921 | e |
| 40 | 05 | 20050807 | b c d |

**Table 2.** A customer sequence database

| SID | Customer sequence |
|-----|-------------------|
| 10 | <(ad)(bc)(bcd)> |
| 20 | <(abd)(bf)> |
| 30 | <(bdg)e> |
| 40 | <(bcd)> |

There are many researches [1, 4] for mining sequential patterns from a transaction database. However, users may be only interested in some sequential patterns related to certain items or behaviors. In this paper, we adopt the regular expression to be the basic form of the pattern constraint. The regular expression [2, 3] has strongly expressive power. For example, it can express the logical operator "or", such as the regular expression R = <b [ a | c ] d>. The sequences in the pattern constraint are the *base sequences*. For the regular expression R, the sequences beginning with item b followed by items a or c, and ending with item d are the base sequences, that is, the two sequences <bad> and <bcd> are base sequences. The sequences which involve the other items in any itemset of a base sequence are *derived sequences*. For example, the two sequences <(bf)(ag)d> and <b(ce)(dg)> satisfy the pattern constraint R, but the sequence <b f a d> does not satisfy the pattern constraint, because the items after item b is item a or c, but not item f in the regular expression R.

Therefore, we define the problem of *mining sequential patterns with pattern constraints* as follows: "Given a transaction sequence database, a user-specified minimum support threshold and a pattern constraint, mining sequential patterns with pattern constraint is to find the complete set of sequential patterns in the database, which satisfies the user-specified minimum support and the pattern constraint." For mining sequential patterns with regular expression, SPIRIT(R) first retrieves all the base sequences from the regular expression and scans the original database to compute the supports for these base sequences. After finding all the frequent base sequences, SPIRIT(R) recursively generates all the derived sequences from these frequent base sequences.

For each frequent base sequence with length $k$ $(k \geq 1)$, all the frequent items are appended to the frequent base sequence to generate candidate (k+1)-sequences, that is, derived sequences. The frequent (k+1)-derived sequences can be generated after scanning the original database to count the support for each derived sequences. These frequent (k+1)-derived sequences are regarded as the frequent (k+1)-base sequences, and SPIRIT(R) uses the same way to generate (k+2)-derived sequences and find the frequent (k+2)-sequences. Therefore, SPIRIT(R) needs to generate a large number of candidates and scan the original database many times to count the supports for the generated sequences.

## 2 Our Approach

For the pattern constraint, we improve the regular expression [2, 3] and propose a flexible representation by adding some operators in the regular expression. The operators used in the pattern constraint and their meanings are shown in Table 3. For example, if user-specified pattern constraint is "$*$ $*$ computer$^\wedge$ scanner iPod\$", then the sequential patterns the user wants to find is that "most of the customers bought computers, and then they could buy any items, after that they bought scanners, and then they could buy any items, and finally they bought iPods."

Table 3. The operators in pattern constraints

| Op | Comment |
|---|---|
| x* | Itemset x can appear zero, one or more times continuously |
| x$^+$ | Itemset x can continuously appear at least once |
| x \| y | Itemset x or itemset y is permitted |
| _ | Any itemset is permitted |
| x? | Itemset x can appear zero or once |
| s$^\wedge$ | Sequence s must appear in the beginning of the pattern |
| s\$ | Sequence s must appear in the last of the pattern |
| $*$ | Any sequence is permitted |
| $*$ $*$ | This operator must appear in front of a pattern constraint. Any sequence can appear in any place of the pattern. |

Before processing a pattern constraint, we first transform the pattern constraint into a finite state automaton. Our algorithm sequentially takes each itemset in a pattern constraint as a state transition condition from a state to the next state for a finite state automaton. Our algorithm sequentially compares the itemset on the transition with the sequences generated by the sequential pattern mining process. If a sequence reaches the final state of the finite state automaton, then the sequence satisfies the pattern constraint, and it is one of the sequences which the user wants to find. For example, the pattern constraint P = (ad) [ b | c ] means that the itemset (ad) must be followed by item b or c, which can be transformed into the finite state automaton as shown in Fig. 1, in which the first transition condition is itemset (ad). If the first itemset of a sequence matches this condition (ad), that is, contains (ad), it will be moved to the next state (i.e., state 2), and then the current transition condition is "b or c". If the second itemset of this sequence contains item b or c, and no more itemset exists in the sequence, then the sequence will reach the final state (i.e., state 3) of the finite state automaton.

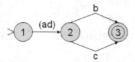

**Fig. 1.** Finite state automaton for P = (ad) [ b | c ]

In the following, we explain how to generate the finite state automaton if the pattern constraint includes the operators in Table 3. If a pattern constraint with operator "*", then the state transition will point to the original state, that is, the next state is itself. For example, pattern constraint P = a*[ (cd) | e ] can be transformed into the finite state automaton as shown in the Fig. 2, in which the label "ε" on the self-transition means that the transition is an arbitrary itemset or sequence. If a sequence matches the condition "a", it will be moved to state 2, and it will reach the final state 3 if the last itemset of this sequence contains (cd) or e.

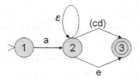

**Fig. 2.** Finite state automaton for P = a*[ (cd) | e ]

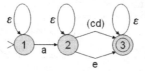

**Fig. 3.** Finite state automaton for P = **a [(cd)| e ]

If a pattern constraint beginning with the operator "**", then all the sequences which contain the sequence in the pattern constraint satisfy the pattern constraint. In this case, there will be self-transition on all the state in the corresponding finite state automaton. For example, the pattern constraint P = ＊＊a [(cd)|e] can be transformed into the finite state automaton as shown in the Fig. 3. In this example, if the user wants to find the sequences beginning with a certain sequence, then the pattern constraint is ＊＊a^ [(cd)|e], which the corresponding finite state automaton is shown in Fig. 4. A pattern constraint ending with the operator "$" means that the user wants to find the sequence ending with the last itemset of the pattern constraint. For example, the finite state automaton for P = ＊＊[(cd)| e ] f g$ is shown in Fig. 5.

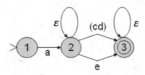

**Fig. 4.** Finite state automaton for P = ＊＊a^ [(cd)|e]

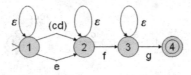

**Fig. 5.** Finite state automaton for P = ＊＊[(cd)| e ] f g$

An itemset x with superscript "*" in the pattern constraint means that itemset x could continuously appear zero, once or more times in the sequences which the user wants to find. For example, the corresponding finite state automaton for P = a* [(cd)|e] is shown in Fig. 6. An itemset x with superscript "+" in the pattern constraint means that itemset x could continuously appear at least once in the sequences which the user wants to find. For example, the corresponding finite state automaton for P = a+ [(cd)| e] is shown in Fig. 7.

**Fig. 6.** Finite state automaton for P = a* [(cd)| e ]

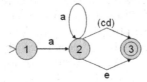

**Fig. 7.** Finite state automaton for P = a⁺ [(cd)| e ]

The operator "?" behind an itemset x in a pattern constraint means that x could appear zero times or once in the sequences which the user wants to find. For example, the finite state automaton for the pattern constraint P = a? [(cd)| e ] is shown in Fig. 8, in which the label "ψ" on the transition from state 1 to state 2 means that the state could pass to the next state without matching any itemset. The operator "_" in a pattern constraint means arbitrary itemset. The finite state automaton for the pattern constraint P = a_[(cd)|e] is shown in Fig. 9, in which the label "σ" on the transition from state 2 to state 3 means that the transition is an arbitrary itemset.

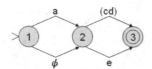

**Fig. 8.** Finite state automaton for P = a ? [(cd)| e ]

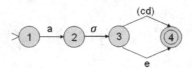

**Fig. 9.** Finite state automaton for P = a _[(cd)| e ]

## 2.1    Mining Sequential Patterns with Pattern Constraint

In this subsection, we describe how to find the sequential patterns which satisfy the user-specified pattern constraint from a customer sequence database. For a pattern constraint, our algorithm first removes the itemsets with the operator "|" (i.e., "or") on them and all the other operators, and retrieves the remaining sequence S from the pattern constraint without any operator. After retrieving the sequence S from a pattern constraint, we collect the customer sequences containing the sequence S from the customer sequence database to generate a small database DBs including all the customer sequences containing the sequence S, because these customer sequences may contain the sequential patterns satisfying the pattern constraint. For example, we can retrieve the sequence S = <afgh> from the pattern constraint P = a [ ( b c d ) | e ] f* g h$, and then scan the customer sequence database to collect the customer sequences which contain sequence <afgh>.

Our algorithm is based on PrefixSpan [4], which generates the *projected databases* recursively. For a customer sequence *CS* and a sequence *Pre* as a *Prefix*, the *projected sequence PCS* for *Pre* is the remaining sequence after removing sequence *Pre* and all the items before *Pre* from *CS*. If the last item x of sequence *Pre* is in an itemset (e.g., {xy}) in *CS*, then the itemset will be {_y} in *PCS*. A *projected database Pre_DBs* for a Prefix *Pre* contains the projected sequences of all the customer sequences in *DBs* for *Pre*. For a projected database *Pre_DBs*, our algorithm appends the frequent items in Pre_DBs to the sequence *Pre* with length k (k ≥1) to generate the sequences with length k+1, and then constructs the projected databases for the extended sequences which satisfy the minimum support.

Therefore, our algorithm first scans all the customer sequences in DBs to count the support for each item and then find the frequent items in DBs, that is, all the 1-sequential patterns in DBs can be generated. For each 1-sequential pattern <i>, there are three cases to match the sequential pattern <i> with the finite state automaton for a pattern constraint: Case 1: if only an item on the transition from start state to the next state and this item is i, then the 1-sequential pattern <i> can be moved to the next state; Case 2: if the length of the itemset Y on the transition is greater than one and item i is the same as or its lexicographic order is smaller than the first item of itemset Y, then the 1-sequential pattern <i> still remains at the start state and waits for combining with the other items.   Case 3: if the lexicographic order of i is greater than the first item of itemset Y on the transition, then the 1-sequential pattern <i> does not need to be extended, since the extended sequential pattern must cannot satisfy the pattern constraint.

If a 1-sequential pattern satisfies Case 2, then a parameter *match_count* is attached to record how many items in the current transition have been matched. If the *match_count* of a sequential pattern X is equal to the length of the itemset on the current transition, then the sequential pattern X satisfies the transition condition and it can be moved to the next state in the finite state automaton. For example, if the itemset on the current transition is (bdf), then the *match_count* of the sequential pattern <(bcd)> is 2 because the itemset of the sequence matches the first two items (bd) of the itemset (bdf) on the transition.

Our algorithm constructs the projected database for each sequential pattern satisfying Case 1 and Case 2. If a sequential pattern satisfies Case 2, then there must be some items in the itemset on the current transition have not been matched by the sequential pattern. Therefore, if the first itemset of a projected sequence for the sequential pattern does not contain these items, then the projected sequence does not need to be put in the projected database for the sequential pattern, because the projected sequence does not contain the sequence satisfying the pattern constraint. When the projected sequences are added into the projected database, our algorithm counts the support for each item from all the projected sequences, and counts the number of the projected sequences in the projected database. After generating the projected database, the frequent items in the projected database also can be obtained. If the number of the projected sequences in the projected database for a sequential pattern is less than the minimum support, then this projected database can be removed, because the sequential pattern does not need to be extended.

For a $k$-sequential pattern SP = $<i_1, i_2, ...i_k>$ ($k \geq 1$) reaching state $t$ of a finite state automaton, our algorithm continues to find frequent items from the projected database for SP. After appending each frequent item $j$ to the $k$-sequential pattern SP, the $(k+1)$-sequential pattern $<i_1, i_2, ...i_k, j >$ can be generated. Similar to 1-sequential patterns, there are three cases to match the $(k+1)$-sequential pattern $<i_1, i_2, ...i_k, j >$ with the finite state automaton for a pattern constraint: Case 1: if only an item on the transition from state $t$ to the next state and this item is $j$, then the $(k+1)$-sequential pattern can be moved to the next state; Case 2: if the length of the itemset Y on the transition from state t to the next state is greater than one and item j is the same as or its lexicographic order is smaller than the first item of the remaining itemset Z which has not been matched by SP, then the $(k+1)$-sequential pattern still remains at the state t and waits for combining with the other items. Case 3: if the lexicographic order of j is greater than the first item of itemset Z on the transition, then the $(k+1)$-sequential pattern does not need to be extended, since the extended sequential pattern must cannot satisfy the pattern constraint.

For example, the user wants to find the sequential patterns with pattern constraint P = (ad) [ b | c ] from the customer sequence database Table 2, and the minimum support is set to be 50%. The corresponding finite state automaton for the pattern constraint is shown in Fig. 1. We first retrieve the sequence S = <(ad)> from the pattern constraint, and collect the customer sequences containing <(ad)> to generate the database DBs as shown in Table 4, in which the frequent items are a, b and d, that is, the 1-sequential patterns are <a>, <b> and <d>.

After matching the three 1-sequential patterns with the finite state automaton, 1-sequential pattern <a> satisfies Case 2 and the other two 1-sequential patterns satisfy Case 3. Therefore, <a> still stays in the start state and the match_count of <a> is 1. Only the projected database for <a> needs to be created, which is shown in Table 5. The frequent items in the projected database is b and (_d), that is, the 2-sequential patterns <ab> and <(ad)> can be generated. Because the remaining item which has not been matched is d on the current transition, <(ad)> satisfies the transition condition and is moved to state 2. The 2-sequential pattern <ab> is removed, because it cannot satisfy the transition condition.

**Table 4.** A Database DBs

| SID | Customer sequence |
|-----|-------------------|
| 10  | <(ad)(bc)(bcd)>   |
| 20  | <(abd)(bf)>       |

**Table 5.** A projected database for <A>

| SID | Projected sequence |
|-----|--------------------|
| 10  | <(_d)(bc)(bcd)>    |
| 20  | <(_bd)(bf)>        |

After generating the projected database for <(ad)> which is shown in Table 6, the frequent item b in the projected database can be obtained which satisfies Case 1 after matching the 3-sequential pattern <(ad)b> with the finite state automaton. Therefore, the 3-sequential pattern <(ad)b> satisfies the pattern constraint, because it reaches the final state 3.

**Table 6.** Projected database for <A>

| SID | Projected sequence |
|-----|--------------------|
| 10  | < (bc)(bcd)>       |
| 20  | < (bf)>            |

## 3    Experimental Results

In this section, we evaluate the performance of our algorithm by comparing it with the well-known algorithm SPIRIT(R) [2, 3] in terms of execution time and memory usage. The real dataset Gazelle.com which we used in the experiments comes from KDD-Cup 2000. This real dataset includes 6650 web pages and 87493 customer traversal sequences. The average length of the customer traversal sequences is 7.39.

Because SPIRIT(R) only can process the regular expression with "or" operator, it cannot process the other operators in Table 3. Therefore, in the experiment, we only use the pattern constraints with "or" operators to compare our method with SPIRIT(R). For a pattern constraint, the number of the itemsets in the "or" operation will affect the efficiency for the two algorithm. The more the number of the itemsets in the "or" operation, the more the number of the candidates generated by SPIRIT(R) and the less the number of the sequences filtered out by our method. Therefore, the number of the itemsets in the "or" operation is set to be 1, 2 and 3 for the pattern constraints in the experiments. Fig. 10 and Fig. 11 show the execution time and memory usages for our algorithm and SPIRIT(R), in which Number of Constraints = k means that the number of the itemset in the "or" operation is k.

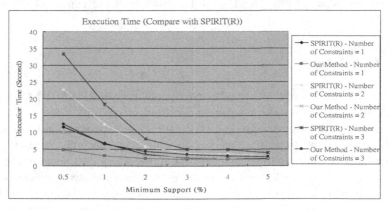

**Fig. 10.** Execution times for the two algorithms

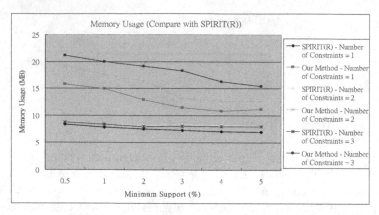

**Fig. 11.** Memory usages for the two algorithms

From Fig. 10, we can see that our method outperforms SPIRIT(R), since SIRIT(R) needs to scan the original database many times, but our method only needs to scan the original database once. The performance gap increases as the minimum support decreases, since the number of the candidates increases and the search space increases for SPIRIT(R). Besides, the execution time increases as the number of the itemsets in the "or" operation increases, since the number of the generated sequences increases for the two algorithms.

For memory usages, our method needs to recursively generate projected database and store these projected databases in main memory, but only the candidate sequences need to be stored for SPIRIT(R). Therefore, our method takes more memory space than SPIRIT(R). From Fig. 11, we can see that the memory usages increase as the number of the itemsets in the "or" operation increases for the two algorithms in the same minimum support threshold, since the number of the generated sequences increases.

## 4    Conclusions

In this paper, we improve the regular expression used in SPIRIT(R) to represent a pattern constraint and transform a pattern constraint into a finite state automaton. We propose an algorithm for mining sequential patterns with pattern constraint according to the corresponding finite state automaton. Our algorithm only needs to scan the original database once and then removes the unnecessary customer sequences from the original database, such that the size of the further generated projected databases can be reduced. Our algorithm recursively generates projected databases and finds the frequent items from the projected database to generate longer sequential patterns without generating any candidates. The experimental results show that our algorithm is more efficient than SPIRIT(R).

# References

1.  Agrawal, R., Srikant, R.: Mining Sequential Patterns. In: Proceedings of the 11th International Conference on Data Engineering, pp. 3–14, March 1995
2.  Garofalakis, M.N., Rastogi, R., Shim, K.: SPIRIT: Sequential Pattern Mining with Regular Expression Constraints. In: Proceedings of the 25th International Conference on Very Large Data Bases, pp. 223–234, September 1999
3.  Garofalakis, M.N., Rastogi, R., Shim, K.: Mining Sequential Patterns with Regular Expression Constraints. In: IEEE Transactions on Knowledge and Data Engineering, vol. 14, No. 3, May/June 2002. (TKDE 2002)
4.  Pei, J., Han, J., Mortazavi-Asl, B., Pinto, H., Chen, Q., Dayal, U., Hsu, M.C.: Mining Sequential Patterns by Pattern-Growth: The PrefixSpan Approach. In: IEEE Transactions on Knowledge and Data Engineering, vol.16, No. 10 (2004)
5.  Pei, J., Han, J., Wang, W.: Mining Sequential Patterns with Constraints in Large Databases. In: Proceedings of CIKM, 4–9 November 2002

# An Intelligent Saliva Recognition System
# for Women's Ovulation Detection

Hui-Ching Wu[1], Ching-Yi Lin[2], Shing-Han Huang[3], and Ming-Hseng Tseng[3]($\boxtimes$)

[1] School of Medical Sociology and Social Work, Chung Shan Medical University,
Taichung, Taiwan, R.O.C.
[2] Department of Obstetrics and Gynecology, Chung Shan Medical University Hospital,
Taichung, Taiwan, R.O.C.
[3] School of Medical Informatics, Chung Shan Medical University,
Taichung, Taiwan, R.O.C.
mht@csmu.edu.tw

**Abstract.** This study presents using image processing and data mining approaches to develop a saliva image automatic recognition system for woman's ovulation prediction. Detect woman's ovulation can sterility treatment, diagnose certain diseases and avoid undesired pregnancies. Saliva ferning test is a technique that monitors woman's saliva and looks for patterns related to ovulation. We use a digital camera with a 100-time microscope to take saliva images. In the proposed method, six important features in dried saliva images are automatically extracted by employing some image processing techniques at first, and an intelligent system is developed by using the decision tree J48 algorithm for the detection of ovulation. In this study, the result of the best classification accuracy is 84% in 100 saliva samples. The proposed method has very important aspects of human, medical and economical. In addition, the proposed system can detect woman's ovulation more safe, natural, convenient and efficient.

**Keywords:** Image processing · Data mining · Ovulation · Saliva ferning

## 1   Introduction

A reliable method of predicting ovulation can effectively help a woman ascertain fertility period for pregnancy as well as for right of autonomy for birth control. An automatically diagnostic system for monitoring various personal physiological conditions has a urgent demand for the prediction of ovulation for women.

Generally, a woman's menstrual cycle lasts from 27 to 30 days, while menstruation lasts from 3 to 7 days and fertile period lasts about six days. It would be too late to test based upon detection of ovulation on the ending day of ovulation. Also, it is not useful in determining the fertility time for planning. In recent years, most people used advanced predictions is a urine test, which focus on the concentration of luteinizing hormone (LH) and can detect ovulation before 1-2 days. The defect of urine test is not sufficient to detect the entire fertile period of three to six day. There are many other methods for predicting a woman's ovulation, for example, measuring a woman's body

© Springer International Publishing Switzerland 2015
N.T. Nguyen et al. (Eds.): ACIIDS 2015, Part I, LNAI 9011, pp. 614–623, 2015.
DOI: 10.1007/978-3-319-15702-3_59

temperature (Basal Body Temperature, BBT) which increases with estrogen's rise to detect fertile times. It has been demonstrated that shortly after menstruation begins the body temperature decreases until ovulation starts, and after that the temperature increases. During the menstruation period, the vaginal secretions also becomes increasingly viscous and to peak at the fertile period. Many methods are based on certain properties of cervical mucus which contains a lot of information related to the female's fertility and not only the ovulation. These body temperature and cervical mucus measurements methods are not reliable in determining fertile periods for personal operating and misjudging. Other ovulation prediction methods include a blood test and a urine test for detecting a surge on estrogen-related hormone. These tests can detect whether the woman is at ovulation instead of providing signal of impending ovulation. [1]

Saliva is a complicated body fluid containing several different electrolytes including potassium chloride, salts of sodium and non-electrolyte parts including proteins, enzymes, and immunoglobulins. In recent, some investigates [2,3] found that saliva ferning is a newer symptom that has high correlation with ovulation. They suggested that saliva ferning test is a newer technique that can monitor female saliva and look for patterns related to ovulation and fertility.

To reduce the aged problem, several international and national organizations support research on fertility detection and periodic abstinence methods. An interesting research is on finding new devices, chemical kits, computer programs and methods that would be more accurate and convenient to use while asserting the ovulation and fertility conditions. The ferning test is a microscope-based method. It is relatively inexpensive and is used, especially in less developed countries, as the most reliable clinical ovulation test [4].

To our knowledge there are no reports on image processing and data mining techniques to be applied to the enhancement and recognition of full-color saliva ferning patterns. The present study is then a first attempt to develop an intelligent system to automatic detection and prediction of woman's ovulation based on the digital saliva ferning test.

## 2 Background Research

### 2.1 Ovulation Detection

During each menstrual cycle, the structure and possibly the function of genital organs and ovaries are modulated by cyclic hormonal changes of the demonstrated hypothalamic–pituitary–ovarian axis [5]. Some variations can be evaluated by basal body temperature (BBT), ferning examination, ultrasound, particularly trans-vaginal sonography alone or combined with the three-dimensional (3D) ultrasound and Doppler technique [6]. The day of ovulation is designated either as the day of maximum follicular growth, or as the day of follicle rupture [7]. The numerous methods of hormonal tests to predict ovulation [11] are all retrospective. Hormones used to measure in blood have been now estimated in saliva, though the quantities are comparatively less [8]. Hence, saliva is considered as the best non-invasive source for chemical and biochemical study [9].

Research indicates that saliva is a very good source for both hormones and enzymes and their levels changed in accordance with the phases of menstrual cycle [10].

Perkowski et al. [4] reported initial experiments with a micro-computer system that totally automates the image processing and decision making processes of the ferning test on gray-level cervical mucus image. Kuo [1] presented an algorithm in the microprocessor that analyzes the gray-level saliva image and calculates pixel density of dark image of salt contents to translate into a density index value for establishing a trend curve for predicting ovulation.

## 2.2    Saliva Ferning Test

Galati et al. [2] performed the saliva ferning test on 328 women using the Saliva Tester and confirmed ovulation by using a trans-vaginal ultrasonography. Their result showed that the correlation rate is 98% between ultrasound images and saliva ferning images. Xu et al. [3] confirmed that the cervical mucus of the menstrual cycle appears fern crystals, while the saliva fern crystals also occur synchronously. Their result also indicated that the correlation rate is 100% between the cervical mucus ferning test and the saliva ferning test.

The saliva ferning test is a newer method for determining the ovulation that based by visual examination of a woman's dried saliva. The term "ferning" is used because of the similarity of the patterns of ferns. The method is based on observations of crystallized salt pattern in dried saliva, which is referred as ferning patterns. Some research results relate the crystallization pattern to increases in the chloride content, changes in ionic strength and/or the content of sodium or potassium in the saliva. Corona [12] indicated that saliva crystallization appears when the blood folliculin level has reached a certain height that coincides with the third or fourth day before ovulation. The crystallization pattern is visible under 100-fold magnification of a saliva sample on a slide. The crystallization lasts until 3 or 4 days after ovulation, when the presence of lutein inhibits the crystallization. At fertile times, microscopic viewing of dried saliva reveals a structure of salt distribution pattern that starts to form chains. This method of examination of saliva offers a reliable way to determine fertility. [1]

Saliva ferning displays crystal patterns, looks like ferns, and can be classified into three different types, named "No Ferning", "Partial Ferning" and "Full Ferning". During infertile stage of menstrual cycle, the saliva should not exhibit any crystallization or ferning patterns. As fertility increases, the ferning patterns should begin to develop and become more significant. Based on 100 times microscope full-color images, an example of saliva ferning patterns is shown in Fig.1 to Fig. 3. During infertile stage of the cycle, "No Ferning" patterns with only dots are present as shown in Fig. 1. During the transitional stage of the cycle, Figure 2 shows "Partial Ferning" patterns that small fern-like crystals start to appear in isolated areas. Immediately before and during ovulation, more intense ferning will be apparent as displayed in Fig. 3.

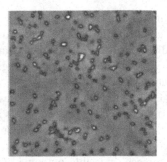 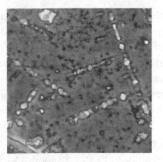

**Fig. 1.** 100x image of no ferning patterns  **Fig. 2.** 100x image of partial ferning patterns

**Fig. 3.** 100x image of full ferning patterns

## 2.3    Data Mining and Decision Tree

In medicine and the delivery of health services, large volumes of data have been collected routinely in the activity of day-to-day management. An important research area has been developed called data mining for mining helpful information and useful knowledge from these large data sets. [13]

Data mining is defined as the process of extracting unknown predictive information from large amounts of raw data. It can also be the procedure of extracting unseen knowledge from large databases. Data mining is more intuitive and increases insight beyond data warehousing. Most hospitals collect, refine and deduce massive medical data. An application of data mining in a hospital will serve as a guide to discovery intrinsic tendencies and trends in patient's historical records. It will also take into account statistical analyses, classifications and predictions of data.

Data mining methods can be employed rapidly on existing hardware and software platforms to improve the usefulness of existing data resources, and can be combined with new products and systems, as they become part of the system. Some usefulness results can be discovered by applying data mining tools to provide answers to many different kinds of predictive demands. There are many jobs associated with data mining such as regression, clustering, classification and association rule mining. [13]

Decision trees are popular and powerful techniques for prediction and classification jobs. In medical applications, both the ability to clarify the cause of a decision and the accuracy of a classification are equally necessary. The advantage of decision trees is due to the fact that, in contradiction of support vector machines and neural networks, decision trees can generate clear prediction rules. Rules can promptly be expressed so that humans can straightforwardly realize them and they can be employed directly in applications. Decision trees also can deliver a clear suggestion of which subjects are most important for classification or prediction.

The purpose of decision tree learning is to generate a model that classifies the value of a decision variable based on some input variables. The decision tree is a type of classifier that has a tree structure form, where each node is either a leaf node or a decision node. Each decision node specifies some test to be completed on a single attribute-value, with one branch and sub-tree for each possible outcome of the test. Each leaf denotes a value of the decision variable given the values of the input variables expressed by the path from the root to the leaf. A decision tree can be applied to categorize a new case by starting at the root of the tree and moving through it until a leaf node, which delivers the classified class of the new case. [13]

In this paper, the J48 algorithm developed in Weka software was applied to build up an intelligent system for automatic classification on saliva ferning images.

## 3    Material and Methods

### 3.1    Data Collection

We prospectively enrolled patients who had undergone physicians diagnose in the obstetrics and gynecology clinic of a 1,312-bed academic urban tertiary-care referral center because of clinical symptoms.

A trained worker interviewed these subjects aged between 20-45 years old for eligibility and invited them to participate in the examination day. Subjects were included if (1) unable to answer questions from investigator; (2) with consciously gynecological symptoms or irregular menstruation; (3) during pregnancy or breastfeeding need; (4) suffering from ovarian disease or oral disease; (5) using drugs that may affect the menstrual cycle.

After being approved by Institutional Review Board (IRB), the study was completely de-identified to all subjects and all subjects signed written informed consent before participation. This study collected 100 dried saliva samples from 5 different women volunteers during various reproductive phases. Of the 100 samples, 32 were no ferning, 27 partial ferning, and 41 full ferning cases, respectively.

### 3.2    Methods

To develop a saliva image automatic recognition system for woman's ovulation prediction, three algorithms were proposed in this study based on the different complex level of image processing.

The process of analyzing the full-color saliva image and the prediction of ovulation includes six steps in the proposed Algorithm 1 as demonstrated in Fig. 4, eight steps in the proposed Algorithm 2 as showed in Fig. 5, and nine steps in the proposed Algorithm 3 as exhibited in Fig. 6, respectively. All proposed algorithms in this paper were developed in MATLAB software.

## Algorithm 1

Figure 4 shows the proposed Algorithm 1 in this study, it is a flow chart of image processing for calculating dark pixel density and for the detection of woman's ovulation. Algorithm 1 comprises six steps: (1) Capturing a digital image of a dried saliva sample. (2) Transforming a full-color saliva image into an image with 256 gray-scale values. (3) Using median filter to cleans-up the spot noise in the image, without much destruction of the edges in the image. (4) Applying high-boost filter to emphasize high frequency components representing the image details as sharpening features while still keeping the low frequency components representing the basic form of the signal. In this case, the low spatial frequency components (global, large black background and bight areas) are suppressed while the high spatial frequency components (the texture of the fur and the whiskers) are enhanced. (5) Binarizing the image. This produces a black and white (0 or 255 pixel value) image. A threshold value is requested from the user, or is selected automatically. This threshold is used to set the new pixel value, if below or equal to the threshold the pixel will get a value of 0 black), if greater, the pixel will get a value of 255 (white). The threshold value changes from image to image and is decided based on the amount of edge information and noise that results from the procedures [4]. (6) Calculating density of dark pixel in the saliva image. The density of dark pixels is the ratio of the number of dark pixels to the number of total pixels in the image area [1]. The number of all pixels in the image area is the sums of the number of dark pixels and the number of white pixels.

## Algorithm 2

Figure 5 demonstrates the proposed Algorithm 2 in this paper; it is a flow chart of image processing for calculating dark pixel density and for the detection of woman's ovulation. We increase two steps in Algorithm 2 that comprises eight steps. Two additional steps are (1) Thinning the image to get one pixel wide lines. This procedure will take a binary image (0 or 255 only pixel values) and reduce the width of a line (or area) to a 1 pixel wide line. (2) Using the Hough transformation to find lines in the saliva image.

## Algorithm 3

Figure 6 displays the proposed Algorithm 3 in this article; it is a flow chart of image processing for extraction line features and decision tree for the detection of woman's ovulation. Two new steps are employed in Algorithm 3 that comprises nine steps. Two new steps are (1) Extracting six line features of the saliva image. Six features include total lines, long lines, short lines, parallel lines, percentage of long line, and percentage of parallel lines [4]. (2) Building a decision tree classifier based on J48 of Weka software.

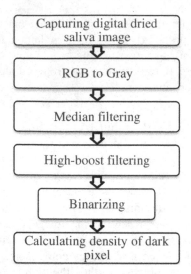

**Fig. 4.** Algorithm 1 for the detection of woman's ovulation

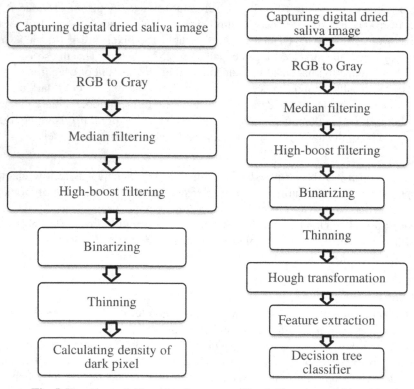

**Fig. 5.** Flowchart of Algorithm 2          **Fig. 6.** Flowchart of Algorithm 3

# 4    Results and Discussion

After capturing digital full-color saliva images, every image first has to be transferred into an image with 256 gray-level values. A series of image processing then were performed to distinguish between possible line points and a background, which applied median filter, high-boost filter, and binarizing procedure. The results of applying binarizing procedure were shown in Figures 7 through 9 for no ferning, partial ferning and full ferning patterns, respectively.

After applying binarizing procedure on the selected image, thinning procedure was performed in the proposed Algorithm 2 and Algorithm 3. Figures 10 to 12 demonstrated that corresponding results for no ferning, partial ferning and full ferning patterns, respectively. By using the proposed image processing, one can found that the line-like features have been successful recognition, which represents the full ferning, partial ferning and no ferning characteristics.

Table 1 listed the overall accuracy of three proposed algorithms; one can see that the proposed Algorithm 3 has the best classification rate 84%. The corresponding accuracy for the proposed Algorithm 2 and Algorithm 1 is 65% and 60%, respectively. Although calculating the density of dark pixels is a simpler method for monitoring and predicting ovulation, but its performance was not enough well for the need of clinical application. By integrating the data mining approach - decision tree J48 algorithm with appropriate image processing subroutines can great improve the performance of the proposed system for woman's ovulation prediction in this paper.

**Table 1.** Overall accuracy of three proposed algorithms

| Approach | Description | Accuracy |
|---|---|---|
| Algorithm 1 | Binarizing + dark pixel density | 60% |
| Algorithm 2 | Binarizing + thinning + dark pixel density | 65% |
| Algorithm 3 | Binarizing + thinning + Hough transform + decision tree | 84% |

  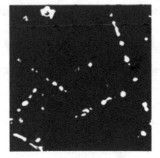

**Fig. 7.** Result of applying binarizing on Fig. 1    **Fig. 8.** Result of applying binarizing on Fig. 2

**Fig. 9.** Result of applying binarizing on Fig. 3    **Fig. 10.** Result of applying thinning on Fig. 7

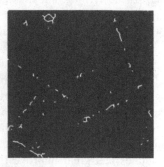

**Fig. 11.** Result of applying thinning on Fig. 8    **Fig. 12.** Result of applying thinning on Fig. 9

## 5    Conclusions and Future Work

Based on image processing techniques and decision tree analysis, an intelligent saliva recognition system for women's ovulation detection has been developed in this paper. The proposed system can provide an easy, economical, and efficient way to detect ovulation automatically, which can helpful determine a woman's fertility period for birth control as well as for pregnancy.

Future directions for our work include the following two points. First, collection larger sample size is important to further enhance the performance of our proposed algorithms. Second, development of a home diagnosis system on smartphones for the predicting of ovulation for women is needed in future study.

**Acknowledgments.** This paper was supported by the Ministry of Science and Technology, Taiwan, R.O.C., under grants MOST 103-2629-E-040-001.

## References

1. Kuo, Y.: Ovulation-prediction devices with image processing system. Google Patents US20080255472 (2007)
2. Galati G, Trapani, E., Yacoub, M., Toccaceli, M.R., Toccaceli, M.G., Galati, C.F.: A New Test for Human Female Ovulation Diagnosis. International Review od Medical Sciences **6**(1) (1994)

3. Xu, X., Shi, H.-M., Zao, H., Zao, M.-M.: Monitoring saliva crystallization observed 32 cases of infertile women ovulate. Journal of Ningxia Medical (in Chinese) **22**(5), 300 (2000)
4. Perkowski, M., Wang, S., Spiller, W.K., Legate, A., Pierzchata, E.: Ovulo-computer: application of image processing and recognition to mucus ferning patterns. In: Proceedings of Third Annual IEEE Symposium on Computer-Based Medical Systems 1990, pp 52–59. IEEE (1990)
5. Gonzalez, R.C., Woods, R.E., Eddins, S.L.: Digital image processing using MATLAB, vol 2. Gatesmark Publishing Knoxville (2009)
6. Bakos, O., Lundkvist, Ö., Wide, L., Bergh, T.: Ultrasonographical and hormonal description of the normal ovulatory menstrual cycle. Acta obstetricia et gynecologica Scandinavica **73**(10), 790–796 (1994)
7. Pearlstone, A.C., Surrey, E.S.: The temporal relation between the urine LH surge and sonographic evidence of ovulation: determinants and clinical significance. Obstetrics and Gynecology **83**(2), 184–188 (1994)
8. Braat, D.D., Smeenk, J.M., Manger, A.P., Thomas, C.M., Veersema, S., Merkus, J.M.: Saliva test as ovulation predictor. The Lancet **352**(9136), 1283–1284 (1998)
9. Freundl, G., Bremme, M., Frank-Herrmann, P., Baur, S., Godehardt, E., Sottong, U.: The CUE fertility monitor compared to ultrasound and LH peak measurements for fertile time ovulation detection. Advances in contraception **12**(2), 111–121 (1996)
10. Flynn, A.M., Lynch, S.: Cervical mucus and identification of the fertile phase of the menstrual cycle. BJOG: An International Journal of Obstetrics & Gynaecology **83**(8), 656–659 (1976)
11. Strott, C., Yoshimi, T., Ross, G., Lipsett, M.: Ovarian Physiology: Relationship Between Plasma LH and Steroidogenesis by the Follicle and Corpus Luteum; Effect of HCG 1. The Journal of Clinical Endocrinology & Metabolism **29**(9), 1157–1167 (1969)
12. Corona, L.F.O.: Optical apparatus with a slide lighting system for detecting a woman's fertile period during her menstrual cycle. Google Patents US4815835 (1989)
13. Wu, H.-C., Chang, C.-J., Lin, C.-C., Tsai, M.-C., Chang, C.-C., Tseng, M.-H.: Developing Screening Services for Colorectal Cancer on Android Smartphones. Telemedicine and e-Health **20**(8), 687–695 (2014)

# A Hybrid Predicting Stock Return Model Based on Logistic Stepwise Regression and CART Algorithm

Shou-Hsiung Cheng[⊠]

Department of Information Management, Chienkuo Technology University,
Changhua 500, Taiwan
shcheng@ctu.edu.tw

**Abstract.** This study presents a hybrid model to predict stock returns. The following are four main steps in this study: First, logistic stepwise regression theory is used to find out the core financial indicators by computing the importance of financial indicators affecting the ups and downs of a stock price. Second, based on the core of the financial indicators coupled with the technology of classification and regression tree, a hybrid classificatory model is established. Third, the predictable rules that affect the ups and downs of a stock price are obtained by employing the proposed hybrid classificatory model. Fourth, we use the established rules to sift out the sound investing targets to invest and calculate the rates of investment. These results of simulated investment reveal that the average rates of reward are far larger than the mass investment rates.

**Keywords:** Stock returns · Financial indicators · Logistic stepwise regression · Classification and regression tree

## 1 Introduction

The problem of predicting stock returns has been an important issue for many years. Advancement in computer technology has allowed many recent studies to utilize machine learning techniques or data mining approaches to predict stock returns. Generally, there are two instruments to aid investors for doing prediction activities objectively and scientifically, which are technical analysis and fundamental analysis. Technical analysis considers past financial market data, represented by indicators such as Relative Strength Indicator (RSI), Moving Average Convergence-Divergence (MACD) and field-specific charts, to be useful in forecasting price trends and market investment decisions. In particular, technical analysis evaluates the performance of securities by analyzing statistics generated from various marketing activities such as past prices and trading volumes[1]. The fundamental analysis can be used to compare a firm's performance and financial situation over a period of time by carefully analyzing the financial statements and assessing the health of a business. Using ratio analysis, trends and indications of good and bad business practices can be easily identified. To this end, fundamental analysis is performed on both historical and present data in order to perform a company stock valuation and hence, predict its probable price evolution. Financial ratios including profitability, liquidity, coverage, and leverage can be

© Springer International Publishing Switzerland 2015
N.T. Nguyen et al. (Eds.): ACIIDS 2015, Part I, LNAI 9011, pp. 624–633, 2015.
DOI: 10.1007/978-3-319-15702-3_60

calculated from the financial statements [2]. Therefore, selective a good stock is the first and the most important step for intermediate- or even long-term investment planning. In order to reduce risk, in Taiwan, the public stock market observation of permit period will disclose regularly and irregularly the financial statements of all listed companies. Based on fundamental analysis, this study proposes a hybrid classification models to extract out the meaningful decision rules to sift out the sound investing targets. Employing the proposed meaningful decision rules, a useful predicting stock return system for intermediate- or long-term investors is proposed in this study.

In general, some related work considers a feature selection step to examine the usefulness of their chosen variables for effective stock prediction. This is because not all of features are informative or can provide high discrimination power. This can be called as the curse of dimensionality problem [3]. As a result, feature selection can be used to filter out redundant and/or irrelevant features from a chosen dataset resulting in more representative features for better prediction performances [4]. The idea of combining multiple feature selection methods is derived from classifier ensembles [5]. The aim of classifier ensembles is to obtain highly accurate ones. They are intended to improve the classification performance of a single classifier. The idea of combining multiple feature selection methods is derived from multiple classifiers [6]. The aim of hybrid classifier is to obtain highly accurate classifiers by combining less accurate ones. They are intended to improve the classification performance of a single classifier. That is, the combination is able to complement the errors made by the individual classifiers on different parts of the input space. Therefore, the performance of hybrid classifier is better than one of the best single classifiers. Therefore, a hybrid classification models is proposed in this study to extract out the meaningful decision rules to sift out the sound investing targets.

The rest of the paper is organized as follows: In Section 2 an overview of the related works is introduced, while Section 3 presents the proposed procedure and briefly discusses its architecture. Section 4 describes analytically the experimental results. Finally, Section 5 shows conclusions of this paper.

## 2    Related Works

This study proposes a new stock selective system applying the logistic stepwise regression and classification and regression tree algorithm to verify that whether it can be helpful on prediction of the shares rose or fell for investors. Thus, this section mainly reviews related studies of the association rules, cluster analysis and classification and regression tree.

### 2.1    Logistic Stepwise Regression

Multiple regression analysis is strain relationship between variables and forecast numbers. When the results of strain number that we discuss are discrete data, the most commonly used method is Roger's logistic regression [7]. Because 14 kinds of financial ratios is a very important contribution to the regression model, we can extract the largest financial indicators, which in turn can simplify the complexity of classification and prediction issues.

The difference between Roger's logistic regression analysis and multiple regression analysis lies in the number of data types on different strain, which makes both differ from the assumptions in the parameter estimates. When performing multiple regression analysis, regression analysis type typically needs to meet the scale of assumed normality; but the assumption of Roger's logistic regression is to observe the probability of the numbers on the sample, and then to distribute them into a S-shaped distribution, which is known as Roger's allocation. Furthermore, in terms of parameter estimation, multiple regression analysis is usually minimal squares through classical methods, and let residuals minimize to get the best estimate of the value of the variable parameters. However, Roger's logistic regression analysis is to estimate (Maximum Likelihood Estimation; MLE) through the maximum probability, so the chances of the strain maximizes, and then to get the best estimate of the value of the forecast variable parameters.

Consider a Roger's logistic regression analysis with k predictor variables, which can be expressed as:

$$P(S \mid X_1, X_2, \cdots, X_k) = \frac{1}{1 + e^{-(\beta_0 + \beta_1 X_1 + \cdots + \beta_k X_k)}} \tag{1}$$

Model that seems very strong likelihood to function estimator is as

$$L = L(b_0, b_1, \cdots, b_{p-1}) \tag{2}$$

Among them, $b_0, b_1, \ldots, b_{p-1}$ are the largest estimator $\beta_0, \beta_1, \ldots, \beta_{p-1}$ respectively. Besides, if $(X_1, Y_1), (X_2, Y_2), \ldots, (X_n, Y_n)$ is a set of random samples, using models to test the overall amount of deviation fit of the model, then deviation of the model is defined as

$$DEV = -2 \ln L(b_0, b_1, \cdots, b_{p-1}) = -2 \sum_{i=1}^{n} \left[ Y_i \ln \overline{P_i} + (1 - Y_i) \ln(1 - \overline{P_i}) \right] \tag{3}$$

In here $\overline{P_i} = \{1 + \exp[-(b_0 + b_1 X_{i1} + \cdots + b_{p-1} X_{i,p-1})]\}$,    $i = 1, 2, \cdots, n$.

$p$ is the number of model parameters. Deviation model $-2 \ln L$ is similar to the $\chi^2_{n-p}$.

This study is to use stepwise selection method to extract 14 kinds of financial ratios and use forward selection procedure to gradually increase the predictive variables. This program is mainly based on score test and condition parameters estimation to gradually select the most significant predictor variation that can exact from. When performing Roger's logistic regression, all the variables are not included in the regression model in the initial stage. However, if score carries out a significant parameter estimates for each parameter, and if the coefficient values of all variables are below the forecast level of significance, it means that all predictive variables corresponding to the variables can't explain and predict the effect, and then we should stop the logistic regression analysis. If there is at least one prediction variable coefficients reaching a significant level, it would further select the predictive variables into the regression model, and carry out the parameter estimates of Roger's logistic regression, select or weed out individually predictive variables until the significant values that enter the regression model predicting are noticeable.

The predictive variables in Roger's logistic regression analysis strike a strong resemblance between the variables associated with the strain numbers and multiple regression analysis values. However, logistic regression analysis shows the strength of association mainly in the regression model to predict the intensity of variables and the number of strain, but it can't explain the variation in the amount of strain of the number of the predicted variables.

## 2.2    Classification and Regression Tree

The most commonly used decision tree algorithms include chi-square automatic interaction detection (CHAID), Classification and Regression Trees (CART) and C5.0 and so on. Chi-squared automatic interaction detection of variables is limited to categories of variables, and if you must use a continuous variable segment classification approach, then you need convert it into class variables. Another difference is the way to prune decision tree. Classification and Regression Trees, is taken after the pruning, but the chi-square automatic interaction detection is taken beforehand pruning.

This study adopts classification and regression tree algorithm [8]. At the beginning of construction of a decision tree, there must be a good training group pre-classified data. Classification and Regression Trees separate data at each node by a single input variable function, and construct a dichotomous tree. Each division will be divided into two subsets of data, and then repeat the finding by each subset in the next test attribute through constant data into two subsets of the way to construct the tree until it can't do split up. In this study, Gini index method is adopted. And based on the minimum time required for classification and regression tree impurities, it changes the volume segmentation model.

Suppose a collection that contains $D_n$ samples, in which the value of a property's value for T, Gini index method will find a dividing point in the range T of the value of the property, such as t, then the sample is divided into less than and greater than the two subsets, so that it is $D_1$ and $D_2$ , respectively, containing $N_1$ and $N_2$ samples. If the sample contains a collection of $D_n$ type sample, the Gini index method will define sample set D of Gini index value as

$$Gini(D) = 1 - \sum_{j=1}^{n} P_j^2 \qquad (4)$$

$P_j$ belongs to class j, and is an occurrence that is rather frequent in D.

After the collection D branching point is cut into $D_1$ and $D_2$, Gini index value is defined as:

$$Gini(D) = \frac{N_1}{N} Gini(D_1) + \frac{N_2}{N} Gini(D_2) \qquad (5)$$

Steps of calculus can be summarized as follows:

1. Establish a tree structure: In the beginning, put all the sample data on the root node, then sort into different subsets of the data according to the selected test conditions.

2. If the sample within a certain subset of all belongs to the same category tags, so they produce a leaf node to represent the group of sample classification mark until all the samples belonging to the same class can be divided into subsets after the completion of the tree structure.
3. Tree pruning: After the tree structure is established, then trim the trees that comes along, that is, cut the trees that are symbolic of noises or some special branch of the tree to prevent regression tree that comes along from excessive accommodating specific sample data.
4. After completion of the installation tree structure, classification and regression trees will produce a series of classification rules, and the rules will correctly divide the new sample data into appropriate groups.

# 3    Methodology

In this study, using the rate of operating expenses, cash flow ratio, current ratio, quick ratio, operating costs, operating income, accounts receivable turnover ratio (times), payable accounts payable cash (days), the net rate of return (after tax)net turnover ratio, earnings per share, operating margin, net growth rate, rate of return of total assets and 14 financial ratios, the use of data mining technology on the sample of the study association rules, cluster analysis, classification and regression tree analysis taxonomy induction a simple and easy-to-understand investment rules significantly simplify the complexity of investment rules. The goal of this paper is proposes a straightforward and efficient stock selective system to reduce the complexity of investment rules.

## 3.1    Flowchart of Research Procedure

The study proposes a new procedure for a stock selective system. Figure 1 illustrates the flowchart of research procedure in this study.

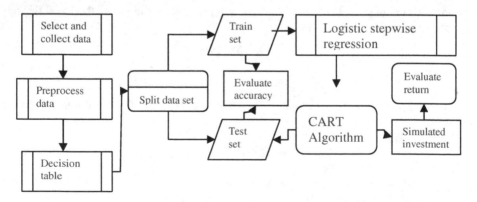

**Fig. 1.** Flowchart of research procedure

## 3.2    Algorithms

This subsection further explains the algorithms of the proposed hybrid model to predict stock returns. The proposed algorithms of the proposed hybrid model to predict stock returns can be divided into six steps in detail, and its computing process is introduced systematically as follows:

Step 1: Select and collect the data.
>    Firstly, this study selects the target data that is collected from Taiwan stock trading system.

Step 2: Preprocess data.
>    To preprocess the dataset to make knowledge discovery easier is needed. Thus, firstly delete the records that include missing values or inaccurate values, eliminate the clearly irrelative attributes that will be more easily and effectively pro-cesed for extracting decision rules to select stock. The main jobs of this step includes data integration, data cleaning and data transformation.

Step 3: Build decision table.
>    The attribute sets of decision table can be divided into a condition attribute set and a decision attribute set. Use financial indicators as a condition attribute set, and whether the stock prices up or down as a decision attribute set.

Step 4: Logistic stepwise regression theory.
>    Logistic stepwise regression theory is used to find out the core financial indicators by computing the importance of financial indicators affecting the ups and downs of a stock price.

Step 5: Extract decision rules.
>    Based on the core attributes obtained in step 4 and a decision attribute (i.e. the stock prices up or down), generate decision rules by classification and regression tree algorithm.

Step 6: Evaluate and analyze the results and simulated investment.
>    For verification, the dataset is split into two sub-datasets: The 67% dataset is used as a training set, and the other 33% is used as a testing set. Furthermore, evaluate return of the simulated investment,

# 4    Empirical Analysis

## 4.1    The Computing Procedure

A practically collected dataset [9] is used in this empirical case study to demonstrate the proposed procedure: The dataset for 993 general industrial firms listed in Taiwan stock trading system from 2008/01 to 2013/12 quarterly. The dataset contains 11916 instances which are characterized by the following 14 condition attributes: (i) operating expense ratio(A1), (ii) cash flow ratio(A2), (iii) current ratio(A3), (iv) quick ratio(A4), (v) operating costs(A5), (vi) operating profit(A6), (vii) accounts receivable turnover ratio(A7), (viii) the number of days to pay accounts(A8), (ix) return on equity (after tax) (A9), (x) net turnover(A10), (xi) earnings per share(A11), (xii) operating margin(A12), (xiii) net growth rate(A13), and (xiv) return on total assets growth rate(A14); all attributes are continuous data in this dataset.

**Table 1.** The Definition of various Attribute

| No. | Attribute name | Notation | Definition |
|---|---|---|---|
| 1 | Operating expense ratio | A1 | Operating expenses / net sales |
| 2 | Cash flow ratio | A2 | Net cash flow from operating activities / current liabilities |
| 3 | current ratio | A3 | current assets / current liabilities, to measure short-term solvency. Current assets (cash + marketable securities + funds + inventory + should be subject to prepayment), current liabilities within one year, short-term liabilities which must be spent on current assets to pay.    The higher the current ratio is, the better short-term liquidity is, which is often higher than 100%. |
| 4 | Quick ratio | A4 | liquid assets / current liabilities = (cash + marketable securities + should be subject to payment) divided by current liabilities = (Current assets - Inventories - Prepaid expenses) divided by current liabilities, to measure very short-term liabilities as of capacity. The higher it is, the better short-term solvency is. |
| 5 | Operating costs | A5 | because of regular business activities and sales of goods or services, enterprises should pay the costs in a period of operating time, which mainly include: cost of goods sold, labor costs. Cost of goods sold can be divided into two major categories: product cost of self-made goods and purchased products. For manufacturing sector, the former usually accounts for the most majority, while other industries the latter. By the definition of accounting costs, operating costs are cost that arises throughout the manufacturing process, also known as product costs or manufacturing costs.    Manufacturing costs are composed of direct materials + direct labor + manufacturing costs (including indirect materials, indirect labor and factory operations and related product manufacturing or other indirect costs). |
| 6 | Operating profit | A6 | Operating profit / paid-up capital |
| 7 | Accounts receivable turnover ratio | A7 | net credit (or net sales) / average receivables. Measure whether the speed of the current collection of accounts receivable and credit policy is too tight or too loose. The higher receivables turnover ratio is, the better the efficiency of collection representatives is. |
| 8 | the number of days to pay accounts | A8 | Average accounts payable / operating costs * day |
| 9 | Return on equity (after tax) | A9 | Shareholders 'equity is shareholders' equity for the growth rate of the year.    The net income refers to the dividend earnings deducted the special stock, while equity refers to the total common equity. From the equity growth rate, we can see whether the company's business class objectives are consistent with shareholder objectives, based on shareholders' equity as the main consideration. Return on equity is acquired due to companies retain their earnings, and hence show a business can also promote the ability to grow their business even not to rely on external borrowing.    It is calculated as: ROE = (Net income - dividend number) / Equity |
| 10 | net turnover | A10 | net operating income / average net worth |
| 11 | Earnings per share | A11 | (Net income - Preferred stock dividends) / numbers of public ordinary shares |
| 12 | Operating margin | A12 | operating margin / revenue, often used to    compare the competitive strength and weakness of the same industrial, showing the company's products, pricing power, the ability to control manufacturing costs and market share can also be used to compare different industries industry trends change. |
| 13 | Net growth rate | A13 | the net price will fluctuate with the market increase or decrease the asset, and its upper and lower rate of increase or decrease, then is known as the net growth rate. |
| 14 | Return on total assets growth rate | A14 | which represent in a certain period of time (usually one year), companies use total assets to create profits for shareholders over the previous period the growth rate. |
| 15 | Decision attribute | D1 | the stock prices up or down |

The computing process of the stock selective system can be expressed in detail as follows:

Step 1: Select and collect the data.

This study selects the target data that is collected from Taiwan stock trading system. Due to the different definitions of industry characteristics and accounting subjects, the general industrial stocks listed companies are considered as objects of the study. The experiment dataset contains 11916 instances which are characterized by 14 condition attributes and one decision attribute.

Step 2: Preprocess data.

Delete the 31 records (instances) that include missing values, and eliminate the 10 irrelative attributes. Accordingly, in total the data of 637 electronic firms that consist of 15 attributes and 6793 instances are included in the dataset.

Step 3: Build decision table.

The attribute sets of decision table can be divided into a condition attribute set and a decision attribute set. Use financial indicators as a condition attribute set, and whether the stock prices up or down as a decision attribute set.

Step 4: Extract core attributes by logistic stepwise regression

The use of logistic stepwise regression attributes reduction, the core financial attributes can be obtained. The core financial attributes are: (1) return on total assets, (2) the net rate of return after tax, (3) earnings per share, (4) net growth rate, (5) current ratio, and (6) cash flow ratio.

Step 5: Extract decision rules.

Based on core attributes extracted in step 4 and a decision attribute (i.e. the stock prices up or down), generate decision rules by classification and regression tree algorithm. Classification and regression tree algorithm, is to build a classification and regression tree recursive relationship between the interpretation of the field with the output field data divided into a subset of the and export classification and regression tree rules, try a different part in the interpretation of the data with output field or the relationship of the results. Financial ratios and price change decision rules are shown in Table 2, as follows:

**Table 2.** The Decision rule set

| No. | Decision rule set |
|-----|-------------------|
| 1 | If the return on total assets growth rate <= 0.075 and the return on total assets growth rate <= -0.425 then the stock price down. |
| 2 | If the return on total assets growth rate <= 0.075 and the return on total assets growth rate> -0.425 and earnings per share <= 0.905 then the stock price up. |
| 3 | If the return on total assets growth rate <= 0.075 and the return on total assets growth rate> -0.425 and earnings per share of > 0.905 and the cash flow ratio of<= 36.255 then the stock price down. |
| 4 | If the return on total assets growth rate <= 0.075 and the return on total assets growth rate> -0.425 and earnings per share of > 0.905 and the cash flow ratio of> 36.255 and the return on total assets growth rate<= -0.235 then the stock price up. |
| 5 | If the return on total assets growth rate <= 0.075 and the return on total assets growth rate<=2.960 and earnings per share of <= -0.785 and the cash flow ratio of > 36.255 and the return on total assets growth rate> -0.235 then the stock price up. |
| 6 | If the return on total assets growth rate > 0.075 and the return on total assets growth rate> -0.425 and earnings per share of > 0.905 and the cash flow ratio of <=13.825 and the return on total assets growth rate> -0.235 then the stock price down. |
| 7 | If the return on total assets growth rate > 0.075 and the return on total assets growth rate> -0.425 and earnings per share of > 0.905 and the cash flow ratio of <=13.825 and the return on total assets growth rate> -0.235 then the stock price up. |
| 8 | If the return on total assets growth rate> 0.075 and the cash flow ratio of <= 13.825 and return on total assets growth rate > 2.960 then the stock price up. |
| 9 | If the return on total assets growth rate> 0.075 and the cash flow ratio of >13.825 then the stock price up. |

## 4.2    Simulated Investment

The rules generating from Table 2 get down on stock selection from the listed companies rise in the rules in year 2011~2013. There are 24 companies in the first quarter in line with the rise in the rules, the average quarter rate of return of 11.51 %; 5 companies in second quarter 2, the average quarter rate of return of 4.64%; 2 companies in the third quarter, the average quarter rate of return of 2.12%; 11 companies in the fourth quarter, the average quarter rate of return of 4.92% in year 2011. There are 43 in the first quarter in line with the rise in the rules, the average quarter rate of return of 35.10 %; 3 companies in second quarter 2, the average quarter rate of return of 1.15%; 34 companies in the third quarter, the average quarter rate of return of 14.42%; 123 companies in the fourth quarter, the average quarter rate of return of 24.42% in year 2012. There are 50 in the first quarter in line with the rise in the rules, the average quarter rate of return of 12.20 %; 22 companies in second quarter 2, the average quarter rate of return of 9.45%; 120 companies in the third quarter, the average quarter rate of return of 10.72%; 131 companies in the fourth quarter, the average quarter rate of return of 15.52% in year 2013. The comparsion of the average quarter rate of return and the broader market quarter rate of return is shown as Table 3. These evidences reveal that the average rates of reward are far larger than the mass investment rates.

**Table 3.** The average quarter rate of return

|      |                | The average rates of reward | The mass investment rates |
|------|----------------|-----------------------------|---------------------------|
| 2011 | First quarter  | 11.51 %                     | -5.32%                    |
|      | Second quarter | 4.64%                       | -0.55%                    |
|      | Third quarter  | 2.12%                       | -14.62%                   |
|      | Fourth quarter | 4.92%                       | -7.11%                    |
| 2012 | First quarter  | 35.10 %                     | 10.32%                    |
|      | Second quarter | 1.15%                       | -8.25%                    |
|      | Third quarter  | 14.42%                      | -1.62%                    |
|      | Fourth quarter | 24.42%                      | 7.51%                     |
| 2013 | First quarter  | 12.20 %                     | 3.32%                     |
|      | Second quarter | 9.45%                       | 2.25%                     |
|      | Third quarter  | 10.72%                      | 0.32%                     |
|      | Fourth quarter | 15.52%                      | 1.51%                     |

## 5    Conclusion

This paper presents a hybrid predicting stock return model based on logistic stepwise regression and classification and regression tree algorithm. From the results of empirical analysis obtained in this study, some conclusions can be summarized as follows:

(1.) By the dependence of each company's financial indicators and by the ups and downs of the stock, the use of logistic stepwise regression and classification and regression tree, we can gain a simple set of classification and prediction rules.

(2.) The average rate of return derives from the empirical results show that return on investment on stock price in the research is obvious higher than general market average.

# References

1. Murphy, J.J.: Technical Analysis of the Financial Markets. Institute of Finance, New York (1999)
2. Bernstein, L., Wild, J.: Analysis of Financial Statements. McGraw-Hill (2000)
3. Abraham, A., Baikunth, N., Mahanti, P.K.: Hybrid Intelligent Systems for Stock Market Analysis. In: Alexandrov, V.N., Dongarra, J.J., Juliano, B.A., Renner, R.S., Kenneth Tan, C.J. (eds.) Computational Science - ICCS 2001. Lecture Notes in Computer Science, vol. 2074, pp. 337–345. Springer, Heidelberg (2001)
4. Huang, C.L., Tsai, C.Y.: A hybrid SOFM-SVR with a filter-based feature selection for stock market forecasting. Expert System with Applications 36(2), 1529–1539 (2009)
5. Chang, P.C., Liu, C.H.: A TSK type fuzzy rule based system for stock price prediction. Expert Systems with Application 34(1), 135–144 (2008)
6. Yu, L., Wang, S., Lai, K.K.: Mining stock market tendency using GA-based support vector machines. Lecture Notes in Computer Science 3828, 336–345 (2005)
7. Kim, K.J.: Financial time series forecasting using support vector machines. Neurocomputing 55, 307–319 (2003)
8. Vapnik, V.: Statistical Learning Theooy. John Wiley & Sons Inc, New York (1998)
9. Breiman, L., Friedman, J., Stone, C.J., Olshen, R.A.: Classification and Regression Trees. Taylor & Francis (1984)

# A Bidirectional Transformation Supporting Tool for Formalization with Logical Formulas

Shunsuke Nanaumi, Kazunori Wagatsuma, Hongbiao Gao,
Yuichi Goto, and Jingde Cheng[✉]

Department of Information and Computer Sciences,
Saitama University, Saitama 338-8570, Japan
{shunsuke,wagatsuma,gaohongbiao,gotoh,cheng}@aise.ics.saitama-u.ac.jp

**Abstract.** In many applications in computer science and artificial intelligence, logical formulas are used as a formal representation to represent and/or specify various objects and relationships among them. However, transforming the informal propositional statements, e.g., declarative sentences and mathematical formulas, into logical formulas is not an easy task for most people. Moreover, when people obtain new logical formulas as results of deduction/reasoning based on logic, investigating the obtained formulas is also not an easy task for them. Although a tool to support transformation from the informal propositional statements of a target domain into logical formulas, and vice versa, is demanded, there is no such tool until now. This paper presents a bidirectional transformation method for formalization with logical formulas, and its supporting tool we are developing.

**Keywords:** Bidirectional transformation · Logical formula · Formalization · Knowledge representation

## 1 Introduction

In many applications in computer science and artificial intelligence, logical formulas are used as a formal representation to represent and/or specify various objects and relationships among them. Transforming declarative sentences into logical formulas is not an easy task for most people. Schubert revealed a characteristic for transforming declarative sentences into logical formulas [9]. The characteristic is that a declarative sentence must have elements transformed into terms, functions, predicates, conjunctions, and quantifiers in logical formula. If a declarative sentence have those elements, the sentence can be transformed into a logical formula. However, most of declarative sentences do not satisfy that characteristic. Thus, we have to change declarative sentences to satisfy the Schubert's characteristic before the transformation. On the other hand, when people obtain new logical formulas as results of deduction/reasoning based on logic, investigating the obtained logical formulas is also not an easy task for them. For example, in automated theorem finding [4], a lot of theorems as logical formulas

© Springer International Publishing Switzerland 2015
N.T. Nguyen et al. (Eds.): ACIIDS 2015, Part I, LNAI 9011, pp. 634–643, 2015.
DOI: 10.1007/978-3-319-15702-3_61

are deduced from axioms and definitions in a mathematical field by reasoning. People have to judge which logical formulas are interesting and new theorems in the deduced logical formulas.

Although a tool to support transformation from the informal propositional statements (e.g., declarative sentences and mathematical formulas) into logical formulas, and vice versa, is demanded, there is no such tool proposed and developed until now. Fuchs proposed a method to transform English declarative sentences into logical formulas through an intermediate language [3], but it is not convenient to use Fuchs's method because of difficulty of transformation into an intermediate language. We have developed a supporting tool for transformation between declarative sentences and logical formulas [7]. However, this supporting tool is corresponded to transformation between only English/Japanese declarative sentences and logical formulas. Some application domains may deal with not only declarative sentences, but also mathematical formulas, e.g., the automated theorem finding, and also deal with other natural languages. Thus, supporting tool of transformation between informal propositional statements and logical formulas is needed, but no one has tackled the tool.

This paper presents a bidirectional transformation method for formalization with logical formulas and its supporting tool we are developing. At first, we analyzed transformation process between English declarative sentences/Japanese declarative sentences/mathematical formulas and logical formulas to find general transformation activities. Then, we summarized the found general transformation activities. After that, we defined requirement of the supporting tool, proposed its design, presented its current implementation and a use case.

The rest of this paper is organized as follows. Section 2 presents analysis of transformation process, section 3 shows bidirectional transforming method, section 4 explains the bidirectional transformation supporting tool, and concluding remarks is given in section 5.

## 2  Analysis of Transforming Process

### 2.1  English Declarative Sentences and Logical Formulas

A demonstrative pronoun and a personal pronoun, a synonym, a comparative expression, a participle construction, a passive, and an indefinite pronouns in English declarative sentences are expressions to prevent the sentences from satisfying the Schubert's characteristic [7]. A demonstrative pronoun is one of the words such as "this", "that", "these", and "those". A personal pronoun is one of the pronoun such as "I", "you", "she", and "they". In transforming English declarative sentences into logical formulas, from view point of meaning of words, assignment of the indicative expressions and words/phrases they indicate in a sentence to symbols in a logical formula may be unsuitable. An indicative expression and an indicated word are represented in different expressions. As the result, ones may assign the indicated expression and the indicated word to the different symbols in logical formulas. However, the meaning of the indicative expression and the indicated word is same. Thus, they should be assigned to a same symbol in logical formulas. A synonym is a word or expression which means the same as another word or expression. Similar to demonstrative, from view point of meaning of words, assignment of the synonyms in a sentence to symbols in a logical formula may be unsuitable. A comparative expression is the expression that the form of an adjective or adverb that shows an increase in size, degree, and so on, when something is considered in relation to something else is used. A comparative expression is interpreted differently by people because verbs or clauses in a sentence can be omitted. In transforming English declarative sentences into logical formulas, ones cannot assign symbols to omitted verbs or clauses. A participle construction is the expression that one of the forms of a verb that are used to make tenses is used. In transforming English declarative sentences into logical formulas, ones cannot assign symbols to omitted clauses. A passive is the expression that the passive form of a verb is used. In transforming English declarative sentences into logical formulas, ones cannot assign a symbol to an omitted subject or verb. An indefinite pronoun is a word such as "some", "any", or "either" that is used instead of a noun, but an indefinite pronoun does not say exactly which person or thing is meant. Moreover, words or clauses that follow pronouns is omitted. In transforming English declarative sentences into logical formulas, ones cannot assign symbols to omitted words or clauses.

Before transforming English declarative sentences into logical formulas, we have to remove and replace the six expressions. The process of transforming English declarative sentences into logical formulas is as follows.

1. Replace demonstrative pronouns and personal pronouns with words/phrases they indicate.
2. Unify synonyms.
3. Complement verbs in comparative expressions.
4. Complement words or clauses in participle constructions.
5. Complement subjects in passives.
6. Complement clauses in indefinite pronouns.
7. Assign the words/phrases of English declarative sentences to the symbols of logical formulas.
8. Generate the logical formulas from the English declarative sentences.

On the other hand, the process to transform logical formulas into English declarative sentences is as follows.

1. Replace the symbols of logical formulas with the words/phrases of English declarative sentences according to the assignment of symbols to words/phrases used in transforming English declarative sentences into logical formulas.
2. Change word order of the replaced formula according to a grammar of the English declarative sentences.

## 2.2    Japanese Declarative Sentences and Logical Formulas

A demonstrative pronoun and a personal pronoun, a synonym, a passive, an omission of the subject, and non plural noun in Japanese declarative sentences

are expressions to prevent the sentences from satisfying the Schubert's characteristic [7]. Demonstrative pronouns, personal pronouns, synonyms, and passives
are as same as English one. An omission of the subject is the expression to omit a
subject in a sentence. In transforming Japanese declarative sentences into logical
formulas, ones cannot assign a symbol to an omitted subject. A non plural noun
is a word that can be used to refer to a person, place, thing, quality, or idea,
and so on. In transforming Japanese declarative sentences into logical formulas,
ones should complement quantitative expressions to non plural nouns. However,
there are no quantitative expression generally.

Similarly to English declarative sentences, we have to remove and replace the
five expressions before transforming Japanese declarative sentences into logical
formulas. The process of transforming Japanese declarative sentences into logical
formulas is as follows.

1. Replace demonstrative pronouns and personal pronouns with they indicate.
2. Unify synonyms.
3. Complement subjects.
4. Complement the quantitative expression to non plural noun.
5. Assign the words/phrases of Japanese declarative sentences to the symbols
   of logical formulas.
6. Generate the logical formulas from the Japanese declarative sentences.

On the other hand, the process of transform logical formulas into Japanese
declarative sentences is as follows.

1. Replace the symbols of logical formulas with the words/phrases of Japanese
   declarative sentences according to the assignment of symbols to
   words/phrases used in transforming Japanese declarative sentences into logical formulas.
2. Change word order of the replaced formula according to a grammar of the
   Japanese declarative sentences.

### 2.3   Mathematical Formulas and Logical Formulas

We can regard mathematical formulas as a kind of informal propositional statement as same as English/Japanese declarative sentences, and mathematical formulas. A quantifier and a scope of variables prevent the sentences from transforming into logical formulas. Following formulas are for explanation.

$$x + y = 0, x = 1, y + 1 = 0 \tag{1}$$
$$x = 0 \tag{2}$$

A quantifier is a word or phrase that is used with a noun to show quantity.
In mathematical formula (1), quantifiers are omitted. Formula (1) is as same as
$\exists x \exists y (x+y = 0$ and $x = 0$ and $x+1 = 0)$. In transforming mathematical formulas
into logical formulas, ones should complement quantifiers to the mathematical
formulas. A scope of variables is a scope to which a variable is applied. We

**Table 1.** Type of expression that should be changed

| | Dem. | Syn. | Pasv. | Comprt. | Prtcpl const. | Indef. pron. | Omis. sub. | Non plur. noun | Quant. | Scp. of var. |
|---|---|---|---|---|---|---|---|---|---|---|
| Unif. | ✓ | ✓ | | | | | | | | |
| Compl. | | | ✓ | ✓ | ✓ | ✓ | ✓ | ✓ | ✓ | ✓ |

sometimes use the same variable to represent different things. In mathematical formulas (1) and formula (2), we use the same symbol $x$, but each formula mean $x = 1$ in formula (1), but $x = 0$ in formula (2). In transforming mathematical formulas into logical formulas, ones should define and complement a scope of variables.

Before transforming mathematical formulas into logical formulas, we have to remove and replace the two expressions. The process of transforming mathematical formulas into logical formulas is as follows.

1. Complement quantifiers.
2. Complement a scope of variables.
3. Assign the symbols of mathematical formulas to the symbols of logical formulas.
4. Generate the logical formulas from the mathematical formulas.

On the other hand, the process of transform logical formulas into mathematical formulas is as follows.

1. Replace the symbols of logical formulas with the symbols of mathematical formulas according the assignment of symbols to words/phrases used in transforming mathematical formulas into logical formulas.
2. Change word order of the replaced formula according to a rule of the mathematical formulas.

## 3   Bidirectional Transforming Method

Through the analysis from section 2, we can conclude that transforming informal propositional statements into logical formulas consists of three stages: satisfying the Schubert's characteristic, assigning words/phrases in informal propositional statements to symbols in logical formulas, and generating logical formulas from the informal propositional statements.

Expressions to prevent the statements from satisfying the Schubert's characteristic can be classified into two types: unifying and complementing. Unifying is the process unifying the expression of the words/phrases which have same meaning. Complementing is the process complementing the expression of the words/phrases omitted in the sentence. Table 1 shows the assignment of the types to the expressions that satisfies the Schubert's characteristic.

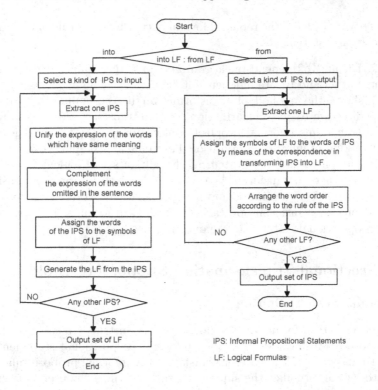

**Fig. 1.** Method of transformation

On the other hand, we can conclude that transforming logical formulas into informal propositional statements consists of two stages: replacing the symbols with words/phrases, and changing the word order. We can conclude that transformation process between informal propositional statements and logical formulas. Transforming informal propositional statements into logical formulas is as follows:

a1. Unify expressions which have same meaning.
a2. Complement omitted expressions.
a3. Assign words/phrases of the informal propositional statements to symbols of logical formulas.
a4. Generate the logical formulas from the informal propositional statements.

Transforming logical formulas into informal propositional statements is as follows:

b1. Replace symbols of logical formulas with words/phrases of a target informal propositional statements according to the assignment of symbols to words/phrases used in transforming the target informal propositional statements into logical formulas.

b2. Change word order of the replaced formula according to a rule of the informal propositional statements.

   a1, a2, a3, and b2 depend on a grammar of target informal propositional statement, but a4 and b1 are independent from the grammar.

   Fig. 1 shows the method of transformation in the supporting tool. One selects the direction to do transformation, and do transformation tasks in order. When one transforms informal propositional statements into logical formulas, one selects a kind of informal propositional statements to input. After that, one do transformation tasks every statement. Finally, a set of logical formulas are outputted. When one transforms logical formulas into informal propositional statements, one selects a kind of informal propositional statements to output. After that, one do transformation tasks every logical formula. Finally, a set of informal propositional statements are outputted.

# 4    Bidirectional Transformation Supporting Tool

## 4.1    Requirement Definition

A bidirectional transformation supporting tool instructs a user and gives supports to do transformation between informal propositional statements and logical formulas easily. When a user transforms the informal propositional statements into logical formulas, the supporting tool instructs the user to transform inputted informal propositional statements into expressions that can be transformed into logical formulas. After that, the supporting tool suggests assignment of words/phrases of the informal propositional statements to symbols of logical formulas. Then, the supporting tool generates the logical formulas from the informal propositional statements according to user's decision. When the user transforms logical formulas into the informal propositional statements, the supporting tool instructs the user to do transformation. Then, the supporting tool replaces symbols of logical formulas with words/phrases of a target informal propositional statements according to the assignment of symbols to logical formulas and words/phrases used informal propositional statements in transforming informal propositional statements into logical formulas.

   Moreover, the supporting tool changes word order of the replaced formula according to the grammar of the informal propositional statements. Furthermore, it is necessary to share those tasks because the amount of tasks become large in a great deal of statements.

   We enumerated seven requirements for this supporting tool.

**R1:** The supporting tool should instruct a user to unify expressions which have same meaning.
**R2:** The supporting tool should instruct the user to complement omitted expressions.
**R3:** The supporting tool should give the user assignment of words/phrases of the informal propositional statements to symbols of logical formulas.

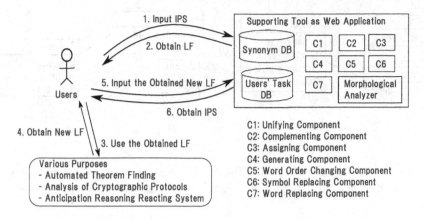

**Fig. 2.** Overview of a supporting tool

**R4:** The supporting tool should generate the logical formulas.
**R5:** The supporting tool should change word order according to the rule of the informal propositional statements.
**R6:** The supporting tool should replace symbols of logical formulas with words/phrases, and vice versa, according to the assignment of symbols to words/phrases.
**R7:** The supporting tool should be used by users every time.

### 4.2   Design

As shown in Fig. 2, the supporting tool is a Web application to satisfy R7. The supporting tool consists of ten components: unifying component to satisfy R1, complementing component to satisfy R2, assigning component to satisfy R3, generating component to satisfy R4, word order changing component to satisfy R5, symbol replacing component to satisfy R6, word replacing component to satisfy R6, a morphological analyzer to satisfy R1 and R2, and two databases. A morphological analyzer is an external tool to do morphological analysis in declarative sentences. A synonym DB is a database that synonyms in declarative sentences are saved and to find synonyms in sentences. A user's task DB is a database that transforming data are stored.

The supporting tool is independent with different kinds of informal propositional statements from the view point of data structure. It is not necessary to change the data structure if a function to transform a new kind of informal propositional statements is add into the supporting tool because each data table in the supporting tool is independent with different kinds of informal propositional statements.

Fig. 3 shows data and the relation among them in the supporting tool. There are six types of data to be stored. Users table manages data of users of the tool. Projects table manages data of projects. Roles table manages data of roles in projects. In this tool, a project is a unit of transformation tasks. Statements

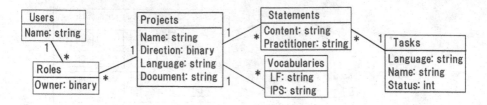

**Fig. 3.** Data model

table manages data of statements. Tasks table manages data of transformation tasks. Vocabularies table manages assignment of symbols to words/phrases.

### 4.3 A Use Case

In case of formal analysis of cryptographic protocols with reasoning [2], analysts need to transform a specification of cryptographic protocol into logical formulas. The result of transformation is used as premises for reasoning, and new logical formulas are deduced. After that, analysts need to transform the deduced logical formulas into a form of specification for analyzing the deduced result easily.

The supporting tool is useful for analysts to perform formal analysis of cryptographic protocol with reasoning easily. In formal analysis of cryptographic protocols with reasoning, it is time-consuming task for analysts to transform a specification of cryptographic protocol into logical formulas. By using the supporting tool, analysts can easily transform a specification of cryptographic protocol into logical formulas. Furthermore, it is time-consuming task for analysts to analyze deduced logical formulas because it is difficult to interpret logical formulas. By using the supporting tool, analysts can obtain a form of specification from deduced logical formulas by reasoning. Therefore, analysts can easily interpret and analyze the deduced result.

### 4.4 Current Implementation

We implemented the function for bidirectional transformation for formalization between English/Japanese declarative sentences and mathematical formulas and logical formulas. However, we did not implemented the function for bidirectional transformation for formalization between the other informal propositional statements and logical formulas. We implemented it by using Java. Moreover, it is needed to find synonyms, to parse and to conduct morphological analysis for transformation. Then, we used WordNet (Japanese [1] and English [6]) as a large lexical database and Enju [8] and Cabocha [5] as a morphological analyzer.

## 5   Concluding Remarks

We analyzed bidirectional transformation process between informal propositional statements (English/Japanese declarative sentences and mathematical formulas) and logical formulas, proposed a bidirectional transformation method for formalization with logical formulas, and presented a supporting tool we are developing. Our method is general in the sense that a bidirectional transformation between declarative sentences of any language as same as any mathematical formulas and logical formulas must consist of six activities that are clarified and defined in our method. Our tool is easy to improve and extend in the sense that data structure of our tool is independent with different kinds of informal propositional statements. We can improve and extend the tool for a new kind of informal propositional statements by adding only components to transform for the kinds of informal propositional statements without changing data structure.

As the future works, we will implement the whole of a bidirectional transformation supporting tool and evaluate the supporting tool in order to prove its effectiveness.

# References

1. Bond, F., Isahara, H., Fujita, S., Uchimoto, K., Kuribayashi, T., Kanzaki, K.: Enhancing the Japanese WordNet. In: Proceedings of the 7th Workshop on Asian Language Resources, pp. 1–8 (2009)
2. Cheng, J., Miura, J.: Deontic Relevant Logic as the Logical Basis for Specifying, Verifying, and Reasoning about Information Security and Information Assurance. In: Proceedings of the 1st International Conference on Availability, Reliability and Security, pp. 601–608. Vienna (2006)
3. Fuchs, N.E., Kaljurand, K., Kuhn, T.: Attempto Controlled English for Knowledge Representation. In: Baroglio, C., Bonatti, P.A., Małuszyński, J., Marchiori, M., Polleres, A., Schaffert, S. (eds.) Reasoning Web. LNCS, vol. 5224, pp. 104–124. Springer, Heidelberg (2008)
4. Gao, H., Goto, Y., Cheng, J.: A Systematic Methodology for Automated Theorem Finding. Theoretical Computer Science **554**, 2–21 (2014)
5. Kudo, T., Matsumoto, Y.: Japanese Dependency Analysis Using Cascaded Chunking. In: Proceedings of the 6th Workshop on Computational Language Learning, pp. 63–69. Taipei (2002)
6. Miller, G.A.: WordNet: A Lexical Database for English. Communications of the ACM **38**(11), 39–41 (1995)
7. Nanaumi, S., Wagatsuma, K., Goto, Y., Cheng, J.: Development of a Supporting Tool for Translation between Declarative Sentences and Logical Formulas. In: Proceedings of the 12th International Conference on Machine Learning and Cybernetics, pp. 1179–1184. Tianjin (2013)
8. Ninomiya, T., Matsuzaki, T., Miyao, T., Tsuji, J.: A Log-linear Model with an N-gram Reference Distribution for Accurate HPSG Parsing. In: Proceedings of the 10th Conference on Parsing Technologies 2007, pp. 60–68. Prague (2007)
9. Schubert, L.K., Pelletier, F.J.: From English to Logic Context-free Computation of Conventional Logical Translation. Journal Computational Linguistics **8**(1), 26–44 (1982)
10. Woodcock, J., Larsen, P.G., Bicarregui, J., Fitzgerald, J.: Formal Methods Practice and Experience. ACM Computing Surveys **41**(4), 1–36 (2009)

# Author Index

Printed in the United States
By Bookmasters